Saddlepoint Approximations with Applications

Modern statistical methods use models that require the computation of probabilities from complicated distributions, which can lead to intractable computations. Saddlepoint approximations can be the answer. Written from the user's point of view, this book explains in clear, simple language how such approximate probability computations are made, taking readers from the very beginnings to current applications.

The book aims to make the subject accessible to the widest possible audience by using graduated levels of difficulty in which the core material is presented in chapters 1–6 at an elementary mathematical level. Readers are guided in applying the methods in various computations that will build their skills and deepen their understanding when later complemented with discussion of theoretical aspects. Chapters 7–9 address the p^* and r^* formulas of higher order asymptotic inference, developed through the Danish approach to the subject by Barndorff-Nielsen and others. These provide a readable summary of the literature and an overview of the subject beginning with the original work of Fisher. Later chapters address special topics where saddlepoint methods have had substantial impact through particular applications. These include applications in multivariate testing, applications to stochastic systems and applied probability, bootstrap implementation in the transform domain, and Bayesian computation and inference.

No previous background in the area is required as the book introduces the subject from the very beginning. Many data examples from real applications show the methods at work and demonstrate their practical value. Ideal for graduate students and researchers in statistics, biostatistics, electrical engineering, econometrics, applied mathematics, and other fields where statistical and probabilistic modeling are used, this is both an entry-level text and a valuable reference.

RONALD W. BUTLER is the C. F. Frensley Professor of Mathematical Sciences and Professor of Statistical Science at Southern Methodist University, Dallas, Texas.

CAMBRIDGE SERIES IN STATISTICAL AND
PROBABILISTIC MATHEMATICS

Editorial Board
R. Gill (Department of Mathematics, Leiden University)
B. D. Ripley (Department of Statistics, University of Oxford)
S. Ross (Department of Industrial & Systems Engineering, University of Southern California)
B. W. Silverman (St. Peter's College, Oxford)
M. Stein (Department of Statistics, University of Chicago)

This series of high-quality upper-division textbooks and expository monographs covers all aspects of stochastic applicable mathematics. The topics range from pure and applied statistics to probability theory, operations research, optimization, and mathematical programming. The books contain clear presentations of new developments in the field and also of the state of the art in classical methods. While emphasizing rigorous treatment of theoretical methods, the books also contain applications and discussions of new techniques made possible by advances in computational practice.

Already published
1. *Bootstrap Methods and Their Application*, by A. C. Davison and D. V. Hinkley
2. *Markov Chains*, by J. Norris
3. *Asymptotic Statistics*, by A. W. van der Vaart
4. *Wavelet Methods for Time Series Analysis*, by Donald B. Percival and Andrew T. Walden
5. *Bayesian Methods*, by Thomas Leonard and John S. J. Hsu
6. *Empirical Processes in M-Estimation*, by Sara van de Geer
7. *Numerical Methods of Statistics*, by John F. Monahan
8. *A User's Guide to Measure Theoretic Probability*, by David Pollard
9. *The Estimation and Tracking of Frequency*, by B. G. Quinn and E. J. Hannan
10. *Data Analysis and Graphics using R*, by John Maindonald and John Braun
11. *Statistical Models*, by A. C. Davison
12. *Semiparametric Regression*, by D. Ruppert, M. P. Wand, R. J. Carroll
13. *Exercises in Probability*, by Loic Chaumont and Marc Yor
14. *Statistical Analysis of Stochastic Processes in Time*, by J. K. Lindsey
15. *Measure Theory and Filtering*, by Lakhdar Aggoun and Robert Elliott
16. *Essentials of Statistical Inference*, by G. A. Young, R. L. Smith
17. *Elements of Distribution Theory*, by Thomas A. Severini
18. *Statistical Mechanics of Disordered Systems*, by Anton Bovier
19. *Random Graph Dynamics*, by Rick Durrett

Saddlepoint Approximations with Applications

Ronald W. Butler

Department of Statistical Science,
Southern Methodist University

CAMBRIDGE UNIVERSITY PRESS
Cambridge, New York, Melbourne, Madrid, Cape Town, Singapore, São Paulo

Cambridge University Press
The Edinburgh Building, Cambridge CB2 8RU, UK

Published in the United States of America by Cambridge University Press, New York

www.cambridge.org
Information on this title: www.cambridge.org/9780521872508

© Ronald W. Butler 2007

This publication is in copyright. Subject to statutory exception
and to the provisions of relevant collective licensing agreements,
no reproduction of any part may take place without
the written permission of Cambridge University Press.

First published 2007

Printed in the United Kingdom at the University Press, Cambridge

A catalog record for this publication is available from the British Library

ISBN 978-0-521-87250-8 hardback

Cambridge University Press has no responsibility for the persistence or accuracy of URLs for external or third-party internet websites referred to in this publication, and does not guarantee that any content on such websites is, or will remain, accurate or appropriate.

Contents

Preface		*page* ix
1	**Fundamental approximations**	1
1.1	Univariate densities and mass functions	1
1.2	Univariate cumulative distribution functions	12
1.3	Failure (hazard) rate approximation	28
1.4	Final remarks	30
1.5	Computational notes	30
1.6	Exercises	31
2	**Properties and derivations**	38
2.1	Simple properties of the approximations	38
2.2	Saddlepoint density	41
2.3	Saddlepoint CDF approximation	49
2.4	Further topics	54
2.5	Appendix	66
2.6	Exercises	70
3	**Multivariate densities**	75
3.1	Saddlepoint density and mass functions	75
3.2	Development of the saddlepoint density	83
3.3	Properties of multivariate saddlepoint densities	91
3.4	Further examples	93
3.5	Multivariate CDFs	101
3.6	Exercises	102
4	**Conditional densities and distribution functions**	107
4.1	Conditional saddlepoint density and mass functions	107
4.2	Conditional cumulative distribution functions	113
4.3	Further examples: Linear combinations of independent variables	123
4.4	Further topics	126
4.5	Appendix	132
4.6	Exercises	136

5	**Exponential families and tilted distributions**	145
5.1	Regular exponential families	145
5.2	Edgeworth expansions	151
5.3	Tilted exponential families and saddlepoint approximations	156
5.4	Saddlepoint approximation in regular exponential families	158
5.5	Exercises	179
6	**Further exponential family examples and theory**	183
6.1	Logistic regression and LD50 estimation	183
6.2	Common odds ratio in 2×2 tables	193
6.3	Times series analysis of truncated count data	208
6.4	Exponential families of Markov processes	209
6.5	Truncation	212
6.6	Exercises	213
7	**Probability computation with p^***	219
7.1	The p^* density in regular exponential families	219
7.2	Conditional inference and p^* in group transformation models	225
7.3	Approximate conditional inference and p^* in curved exponential families	230
7.4	Appendix	250
7.5	Exercises	254
8	**Probabilities with r^*-type approximations**	259
8.1	Notation, models, and sample space derivatives	259
8.2	Scalar parameter approximations	260
8.3	Examples	261
8.4	Derivation of (8.1)	265
8.5	Other versions of $\hat{u}$	266
8.6	Numerical examples	271
8.7	Properties	278
8.8	Appendix	279
8.9	Exercises	282
9	**Nuisance parameters**	285
9.1	Approximation with nuisance parameters	285
9.2	Examples	286
9.3	Derivation of (9.3) and (9.4)	291
9.4	Exact and approximate sample space derivatives	292
9.5	Numerical examples	294
9.6	Variation independence, conditional likelihood, and marginal likelihood	297
9.7	Examples	304
9.8	Properties of (9.3)	314
9.9	Exercises	315
10	**Sequential saddlepoint applications**	323
10.1	Sequential saddlepoint approximation	323

10.2	Comparison to the double-saddlepoint approach	324
10.3	Examples	325
10.4	P-values for the Bartlett–Nanda–Pillai trace statistic	330
10.5	Exercises	334

11 Applications to multivariate testing — 341

11.1	P-values in MANOVA	342
11.2	P-values in tests of covariance	348
11.3	Power functions for multivariate tests	355
11.4	Some multivariate saddlepoint densities	363
11.5	Appendix	365
11.6	Exercises	366

12 Ratios and roots of estimating equations — 374

12.1	Ratios	375
12.2	Univariate roots of estimating equations	384
12.3	Distributions for vector ratios	392
12.4	Multivariate roots of estimating equations	401
12.5	The conditional CDF of R_m given $R_1, \ldots, R_{m-1}$	411
12.6	Appendix	420
12.7	Exercises	422

13 First passage and time to event distributions — 430

13.1	Semi-Markov and Markov processes with finite state space	430
13.2	Passage times with a single destination state	435
13.3	Passage times with many possible destination states	452
13.4	Birth and death processes and modular systems	454
13.5	Markov processes	461
13.6	A redundant and repairable system	462
13.7	Appendix	466
13.8	Exercises	469

14 Bootstrapping in the transform domain — 474

14.1	Saddlepoint approximation to the single bootstrap distribution	475
14.2	Saddlepoint approximations for double bootstrap confidence bands	482
14.3	Semiparametric bootstrap	487
14.4	Indirect saddlepoint approximation	494
14.5	Empirical saddlepoint approximations	500
14.6	Appendix	500
14.7	Exercises	502

15 Bayesian applications — 506

15.1	Bayesian prediction with intractable predictand distribution	507
15.2	Passage time examples for Markov processes	510
15.3	Passage time examples for semi-Markov processes	517

15.4	Conclusions, applicability, and alternative methods	522
15.5	Computational algorithms	524
15.6	Exercises	525

16 Nonnormal bases — 528
16.1	Nonnormal-based saddlepoint expressions	528
16.2	Choice of base distribution	532
16.3	Conditional distribution approximation	538
16.4	Examples	539
16.5	Exercises	545

References — 548
Index — 560

Preface

Among the various tools that have been developed for use in statistics and probability over the years, perhaps the least understood and most remarkable tool is the saddlepoint approximation. It is remarkable because it usually provides probability approximations whose accuracy is much greater than the current supporting theory would suggest. It is least understood because of the difficulty of the subject itself and the difficulty of the research papers and books that have been written about it. Indeed this lack of accessibility has placed its understanding out of the reach of many researchers in both statistics and its related subjects.

The primary aim of this book is to provide an accessible account of the theory and application of saddlepoint methods that can be understood by the widest possible audience. To do this, the book has been written at graduated levels of difficulty with the first six chapters forming the easiest part and the core of the subject. These chapters use little mathematics beyond the difficulty of advanced calculus (no complex variables) and should provide relatively easy reading to first year graduate students in statistics, engineering, and other quantitative fields. These chapters would also be accessible to senior-level undergraduate mathematics and advanced engineering majors. With the accessibility issue in mind, the first six chapters have been purposefully written to address the issue and should assure that the widest audience is able to read and learn the subject.

The presentation throughout the book takes the point of view of users of saddlepoint approximations; theoretical aspects of the methods are also covered but are given less emphasis than they are in journal articles. This is why, for example, on page 3 of chapter 1 the basic saddlepoint density approximation is introduced without a lot of fuss and used in many computations well before the reader understands what the formulas actually mean. In this way, readers gain practical experience that deepens their understanding when later complemented with the discussion of theoretical aspects. With users in mind, a wide range of practical applications has been included.

Chapters 7–9 address the p^* and r^* formulas of higher order asymptotic inference that have been developed out of the Danish approach to the subject by Barndorff-Nielsen and others. It is unavoidable that the difficulty level must increase somewhat by the very nature of the topics covered here. However the account given for p^* and r^* is considerably simpler and more transparent than much of the literature and these chapters should still be quite readable to the better first year graduate students in statistics. The chapter shows the evolution of the ideas starting from Fisher's original formulation of the p^* formula, through the work of Fraser, and Efron and Hinkley and ending with r^* and related approximations due to Skovgaard, and Fraser and Reid.

Chapters 10–16 address important special topics where saddlepoint methods have had substantial impact in particular applications or subjects. For example, chapter 11 provides a comprehensive survey of the use of saddlepoint approximations in multivariate testing. The majority of commonly used multivariate tests may now be implemented by using the more accurate saddlepoint methods instead of the usual statistical package procedures. All of these results are organized in one place with a common notation.

Applications to stochastic systems and applied probability are presented in chapter 13. The emphasis here is on approximation to first passage times in stochastic systems because such computations underlie the subject of system reliability. The subject is also basic to transfer function computation and inversion, which are encountered with many electrical and control systems in the engineering sciences. This chapter should appeal directly to electrical engineers, who increasingly are embracing saddlepoint methods out of the need to invert Laplace transforms that often arise in the mathematical models of the engineering sciences.

Saddlepoint methods are also useful in avoiding much of the simulation requisite when implementing another modern statistical tool, the bootstrap. Chapter 14 shows how the bootstrap may be implemented in the transform domain thus forgoing the usual resampling. The emphasis is on multistate survival models that are used to represent degenerative medical conditions in biostatistics. Chapter 15 shows how Bayesian computation and inference may also benefit from using saddlepoint approximations particularly in settings for which the likelihood function may not be tractable when using the standard methods.

The audience for the book includes graduate students and faculty in statistics, biostatistics, probability, engineering sciences, applied mathematics, and other subjects wherein more sophisticated methods of statistical and probabilistic modeling are used. The book is an entry level text from which readers may learn the subject for the first time. The book does not attempt to cover the most advanced aspects of the subject as one would find in Jensen (1995), Field and Ronchetti (1990), and Kolassa (1994) but rather provides much of the background needed to begin to understand these more advanced presentations. These more advanced monographs also require a relatively strong background in complex analysis, something that the majority of statisticians and biostatisticians lack. While complex analysis cannot be avoided in an advanced treatment of the subject, the use of such methods assures a rather narrow audience and this is contrary to the aims of this book.

Acknowledgements

During the years needed to write this book, my own understanding of saddlepoint methods and their potential uses has deepened and broadened. With this I have gained both an appreciation of the efforts by fellow researchers and a deep respect and admiration for the original work by the late Henry Daniels who led the way for us all. My hope is that readers from very diverse backgrounds will be able to acquire similar appreciation while putting forth far less effort.

I would like to thank all those who have helped in some manner to make this book possible. In particular, James Booth, Douglas Bronson, Robert Paige, Marc Paolella, and Andrew Wood provided invaluable help for which I am very grateful. Chapters 7–10 were written on a sabbatical leave at the University of Sydney and I thank John Robinson for an enjoyable and productive visit. Financial aid for this work was provided by the National Science Foundation.

1

Fundamental approximations

This chapter introduces the basic saddlepoint approximations without the burden of undue formality or rigor. The chapter is designed to present the most fundamental and straightforward of the techniques with some associated factual information so the reader may begin using the methods in practical settings with only minimal theoretical understanding. For this reason, the presentation emphasizes implementation and computation and keeps theoretical discussion to a minimum.

1.1 Univariate densities and mass functions

1.1.1 Preliminaries

The most fundamental saddlepoint approximation was introduced by Daniels (1954) and is essentially a formula for approximating a density or mass function from its associated moment generating function or cumulant generating function. We present this expression after first defining the relevant quantities involved.

Suppose continuous random variable X has density $f(x)$ defined for all real values of x. The *moment generating function* (*MGF*) of density f (also for X) is defined as the expectation of $\exp(sX)$ or

$$M(s) = E(e^{sX}) = \int_{-\infty}^{\infty} e^{sx} f(x) \, dx$$

over values of s for which the integral converges. With real values of s, the convergence is always assured at $s = 0$. In addition, we shall presume that M converges over an open neighborhood of zero designated as (a, b), and that, furthermore, (a, b) is the largest such neighborhood of convergence. This presumption is often taken as a requirement for the existence of the MGF in many textbooks. For simplicity we shall conform to this convention but then relax the assumption when later applications demand so. The *cumulant generating function* (*CGF*) of f (also X) is defined as

$$K(s) = \ln M(s) \qquad s \in (a, b). \tag{1.1}$$

The terminology arises from the fact that the Taylor series coefficients of M and K (the collection of higher order derivatives evaluated at $s = 0$) give rise to the moments and cumulants respectively for random variable X. A simple exercise shows that the kth derivative of M evaluated at $s = 0$ is $M^{(k)}(0) = E(X^k)$, the kth moment of X. Also, using these results, one can easily show that the first two cumulants of X are given as $K'(0) = E(X)$ and $K''(0) = \text{var}(X)$ with higher order derivatives leading to the more complicated higher order cumulants.

When random variable X is discrete and integer valued rather than continuous, the same functions can be defined and we mention only those aspects that differ. Assume X has mass function $p(k)$ for integer k and define its MGF as

$$M(s) = E\left(e^{sX}\right) = \sum_{k=-\infty}^{\infty} e^{sk} p(k) \qquad s \in (a, b),$$

with (a, b) as the maximal neighborhood of convergence about zero. The CGF is defined as in (1.1) and the moments and cumulants are again the Taylor coefficients of M and K.

One of the more important facts discussed in probability courses and proven in mathematical analysis is the one-to-one (1-1) correspondence that exists between the collection of probability density and mass functions and their associated MGFs (assuming the latter exist). Indeed, the distribution of a random variable is often determined using this correspondence by recognizing the MGF as that associated with a particular density or mass function. Unfortunately, settings where such recognition may be used are more often the exception than the rule. Quite often MGFs or CGFs of random variables can be determined out of context but their density/mass functions cannot. In such instances, highly accurate approximations to these density/mass functions can be computed by using the saddlepoint methods introduced in Daniels (1954) which are based entirely on the known CGFs.

An example of this is the determination of the distribution for a sum of independent random variables. Suppose $X_1, \ldots, X_n$ is a sequence of independent variables for which X_i has MGF M_i and CGF K_i defined over (a_i, b_i). The CGFs of $X = \sum_{i=1}^{n} X_i$ and $\bar{X} = X/n$ are

$$K_X(s) = \sum_{i=1}^{n} K_i(s) \qquad K_{\bar{X}}(s) = \sum_{i=1}^{n} K_i(s/n) \qquad s \in (\max_i a_i, \min_i b_i).$$

Recognition of the form for the CGFs specifies their associated distributions. For example, if X_i has the Binomial (m_i, θ) mass function

$$p_i(k) = \binom{m_i}{k} \theta^k (1-\theta)^{m_i-k} \qquad k = 0, \ldots, m_i \qquad (1.2)$$

with CGF

$$K_i(s) = m_i \ln\{\theta(e^s - 1) + 1\} \qquad s \in (-\infty, \infty), \qquad (1.3)$$

then $K_X(s) = \sum_{i=1}^{n} K_i(s)$ is recognized from (1.3) as the CGF of a Binomial $(m., \theta)$ mass function with $m. = \sum_i m_i$. Uniqueness of the CGF specifies that X must have this distribution and this is indicated by writing $X \sim \text{Binomial}(m., \theta)$.

If, however, these distributions are changed so they do not have common parameter θ and $X_i \sim \text{Binomial}(m_i, \theta_i)$, then the mass function of X becomes rather intractable. This particular computation arises in reliability analysis when determining the reliability of a k-out-of-m. heterogeneous system (Høyland and Rausand, 1994, p. 130). Suppose the system consists of m. independent components among which m_i have common reliability θ_i for $i = 1, \ldots, n$. Variable X represents the number of components working in the system. Suppose the system functions if and only if at least k of the components work. Then, $\Pr(X \geq k)$ is the reliability of the structure. This computation is illustrated as Example 4 in sections 1.1.6 and 1.2.4.

Four suggestions for computing the mass function of X in this reliability context are:

(1) enumerate the exact probabilities,
(2) use a normal density approximation,
(3) use brute force simulation, and
(4) use a saddlepoint approximation.

Option (1) may lead to intractable computations apart from small values of n, and (2) may not result in the desired accuracy particularly when $\{m_i\}$ are small and $\{\theta_i\}$ are not near $1/2$. Option (3) can be time consuming, even with the speed of modern computers. This same option in a continuous setting, when used to approximate a density, also requires kernel density smoothing techniques which can be inaccurate even when applied with relatively large simulations. The saddlepoint option (4) is examined below and is shown to result in highly accurate approximation without the need for placing constraints or guidelines on the values of $\{m_i\}$ and $\{\theta_i\}$. Another advantage of saddlepoint methods is that the required computational times are essentially negligible as compares with simulation.

1.1.2 Saddlepoint density functions

For continuous random variable X with CGF K and unknown density f, the saddlepoint density approximation to $f(x)$ is given as

$$\hat{f}(x) = \frac{1}{\sqrt{2\pi K''(\hat{s})}} \exp\{K(\hat{s}) - \hat{s}x\}. \tag{1.4}$$

Symbol $\hat{s} = \hat{s}(x)$ denotes the unique solution to the equation

$$K'(\hat{s}) = x \tag{1.5}$$

over the range $\hat{s} \in (a, b)$, and is an implicitly defined function of x. Expression (1.5) is referred to as the *saddlepoint equation* and $\hat{s}$ the *saddlepoint* associated with value x. The approximation is meaningful for values of x that are interior points of $\{x : f(x) > 0\} = \mathcal{X}$, the *support* of density f (or of random variable X). We adopt the convention of referring to $\hat{f}$ as the *saddlepoint density* even though it isn't really a density since generally

$$c = \int_{\mathcal{X}} \hat{f}(x)\,dx \neq 1.$$

The *normalized saddlepoint density*

$$\bar{f}(x) = c^{-1}\hat{f}(x) \quad x \in \mathcal{X}$$

is however a proper density on $\mathcal{X}$.

1.1.3 Examples

(1) Normal (0, 1) density

A standard normal distribution has CGF $K(s) = s^2/2$ defined for $s \in (-\infty, \infty)$. The saddlepoint equation is explicit in this case as $\hat{s} = x$. Simple computation shows that

$$\hat{f}(x) = \phi(x) = \frac{1}{\sqrt{2\pi}} \exp\left(-\tfrac{1}{2}x^2\right) \qquad x \in \mathcal{X} = (-\infty, \infty) \tag{1.6}$$

and the saddlepoint approximation exactly reproduces the standard normal density ϕ.

(2) Gamma (α, 1) density

The density in this instance is

$$f(x) = \frac{1}{\Gamma(\alpha)} x^{\alpha-1} e^{-x} \qquad x > 0$$

with CGF

$$K(s) = -\alpha \ln(1-s) \qquad s \in (-\infty, 1).$$

This leads to the explicit saddlepoint expression $\hat{s} = 1 - \alpha/x$ for $x > 0$. The term

$$K''(\hat{s}) = \alpha(1-\hat{s})^{-2} = x^2/\alpha$$

so that for $x > 0$,

$$\begin{aligned}
\hat{f}(x) &= \frac{1}{\sqrt{2\pi x^2/\alpha}} \exp\{-\alpha \ln(1-\hat{s}) - \hat{s}x\} \\
&= \frac{1}{\sqrt{2\pi x^2/\alpha}} (x/\alpha)^{\alpha} \exp(-x+\alpha) \\
&= \left(\sqrt{2\pi}\alpha^{\alpha-1/2} e^{-\alpha}\right)^{-1} x^{\alpha-1} e^{-x}.
\end{aligned} \tag{1.7}$$

The shape of $\hat{f}$ in (1.7) is the same as that of f but differs from f in the normalization constant. Using Stirling's approximation for $\Gamma(\alpha)$,

$$\hat{\Gamma}(\alpha) = \sqrt{2\pi}\alpha^{\alpha-1/2} e^{-\alpha} \simeq \Gamma(\alpha), \tag{1.8}$$

then

$$\hat{f}(x) = \frac{\Gamma(\alpha)}{\hat{\Gamma}(\alpha)} f(x) \qquad x > 0 \tag{1.9}$$

and differs by a constant relative error determined as the relative error of $\hat{\Gamma}(\alpha)$ in approximating $\Gamma(\alpha)$. The normalized saddlepoint density is exact in this setting.

(3) Normal–Laplace convolution

Consider plotting the density of $X = X_1 + X_2$ where $X_1 \sim$ Normal (0, 1) independently of $X_2 \sim$ Laplace (0, 1) with density

$$f(x) = \tfrac{1}{2} e^{-|x|} \qquad x \in (-\infty, \infty). \tag{1.10}$$

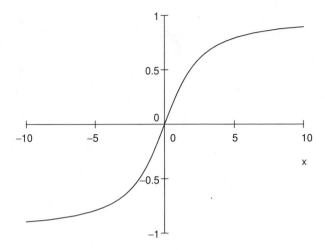

Figure 1.1. $\hat{s}(x)$ vs. x.

The CGF of X takes the particularly simple form

$$K(s) = \tfrac{1}{2}s^2 - \ln(1 - s^2) \qquad s \in (-1, 1)$$

and the saddlepoint equation is

$$K'(\hat{s}) = \hat{s}\left(1 + \frac{2}{1 - \hat{s}^2}\right) = x \qquad x \in (-\infty, \infty). \tag{1.11}$$

The saddlepoint solution is a root of a cubic polynomial that admits one real root and a complex conjugate pair of roots. The unique real root $\hat{s} = \hat{s}(x)$ has been determined numerically and a plot of $\hat{s}$ vs. x is shown in figure 1.1.

The plot shows that the saddlepoint $\hat{s}(x)$ is an odd function in x, a fact confirmed by noting that K' in (1.11) is also odd. Inspection of the plot, as well as consideration of the saddlepoint equation in (1.11) reveal that $\hat{s}(x)$ approaches asymptote $\hat{s} = -1$ as $x \to -\infty$ and asymptote $\hat{s} = 1$ when $x \to \infty$; thus for $x \in (-\infty, \infty)$, the saddlepoint equation can always be solved to find a saddlepoint within $(-1, 1)$. Figure 1.2 shows a comparative plot of the "true" density f (solid line) with the unnormalized saddlepoint density $\hat{f}$ (dotted line), and the normalized saddlepoint density $\bar{f}$ (dashed line).

The "true" density was computed using numerical convolution of the densities involved. A complicated exact expression in terms of special functions has been given in Johnson and Kotz (1970, chap. 1, Eq. (28)), however, numerical computation suggests that it is incorrect since it differs substantially from the numerical convolution. The normalization constant for $\bar{f}$ is most easily computed numerically by making the substitution $dx = K''(\hat{s})d\hat{s}$ so that

$$\int_{-\infty}^{\infty} \hat{f}(x)dx = \int_{-1}^{1} \sqrt{K''(\hat{s})/(2\pi)} \exp\{K(\hat{s}) - \hat{s}K'(\hat{s})\}d\hat{s}$$

$$\simeq 0.8903.$$

The graphical difference between the normalized saddlepoint approximation $\bar{f}$ and f is slight since $\hat{f}$ mostly captures the proper shape of f but not the correct scaling.

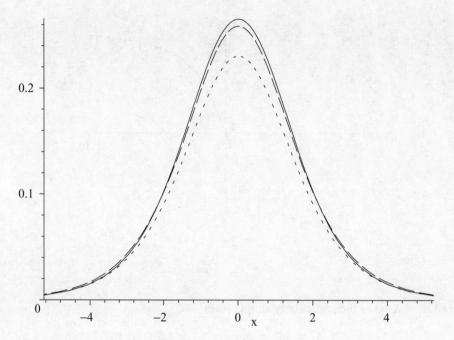

Figure 1.2. $f(x)$ (*solid*), $\bar{f}(x)$ (*dashed*), and $\hat{f}(x)$ (*dotted*) vs. x for the normal-Laplace convolution.

(4) Gumbel (0, 1) *density*

This distribution is also called the extreme value distribution and has CDF

$$F(x) = \exp(-e^{-x}) \quad x \in (-\infty, \infty),$$

with CGF

$$K(s) = \ln \Gamma(1-s) \quad s \in (-\infty, 1).$$

Saddlepoint computation involves first and second derivatives of the $\ln \Gamma$ function which are the di- and tri-gamma functions respectively. Both functions are in the Maple V library and were computed using these routines. Figure 1.3 compares f, $\hat{f}$, and $\bar{f}$. The degree of accuracy of $\hat{f}$ is striking and $\hat{f}$ integrates to about 0.9793. The plot of $\bar{f}$ is virtually indistinguishable from f.

1.1.4 Remarks

A cursory understanding of the approximation in (1.4) and (1.5) requires clarification of a number of presupposed facts that can be supported with the examples above:

(1) Function K is always a strictly convex function when evaluated over (a, b) so $K''(\hat{s}) > 0$ and the square root is well-defined.

(2) Consider the solvability of (1.5) for the saddlepoint. An appropriate choice of x guarantees that there is a unique solution to (1.5) as now explained. If $\mathcal{X}$ is the support of random variable X, defined as $\mathcal{X} = \{x : f(x) > 0\}$ for continuous density f, let $\mathcal{I}_\mathcal{X}$ be the

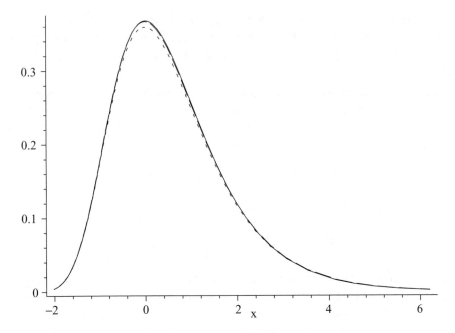

Figure 1.3. $f(x)$ (*solid*), $\bar{f}(x)$ (*dashed*), and $\hat{f}(x)$ (*dotted*) vs. x for the Gumbel density.

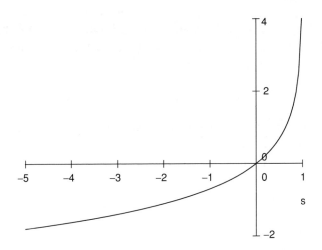

Figure 1.4. The CGF for an Exponential (1).

interior of the span of this support. For example, if X has a uniform density over support $\mathcal{X} = [0, 1] \bigcup (2, 3)$, then the span of the support is $[0, 3)$ and $\mathcal{I}_\mathcal{X} = (0, 3)$. The mapping $K' : (a, b) \to \mathcal{I}_\mathcal{X}$ is 1-1 and onto (a bijection), and K' is strictly increasing, as mentioned in remark (1), so that a unique solution exists. Other solutions to (1.5) may exist outside of (a, b), but such roots are not allowed since solutions to the saddlepoint equation are restricted to the neighborhood (a, b) about zero. If a value of $x \notin \mathcal{I}_\mathcal{X}$ is chosen, then a solution to (1.5) cannot exist in (a, b) but may exist outside of this range.

Figure 1.4 plots $K(s)$ vs. s for an Exponential (1) or Gamma (1,1) density. The slopes of the graph range from 0 as $s \downarrow -\infty$ to ∞ as $s \uparrow 1$ and span the range of support $(0, \infty)$. The corresponding s-values on the horizontal axis are the associated saddlepoints spanning

$(-\infty, 1)$, the convergence neighborhood of K. The choice of $x < 0$ outside of $\mathcal{X}$, the density's support, does not yield a solution to the saddlepoint equation since K' cannot be negative over $(-\infty, 1)$; nevertheless, a meaningless solution can be found for some $s \in (1, \infty)$.

A sum of two independent exponential random variables provides an example that illustrates the possibility of erroneous solutions to the saddlepoint equation. Suppose $X = X_1 + X_2$ where X_1 and X_2 are independent exponentials with means of 1 and 2. The MGF of X is

$$M_X(s) = (1-s)^{-1}(1-s/2)^{-1}$$

and the convergence strip is $(-\infty, 1)$. Solution to the saddlepoint equation

$$K'_X(\hat{s}) = \frac{1}{1-\hat{s}} + \frac{1}{2-\hat{s}} = x \in (0, \infty) \tag{1.12}$$

when restricted to $\hat{s} \in (-\infty, 1)$ is unique and is the smaller root of the quadratic equation determined in (1.12). The larger root is within the range (1.5, 2) and is an erroneous solution. The mapping $K'_X : (-\infty, 1) \to (0, \infty)$ is a bijection, however, the mapping $K'_X : (1.5, 2) \to (0, \infty)$ is also a bijection, yielding erroneous saddlepoints when root finding is not restricted.

(3) Since $K'(0) = E(X)$, the saddlepoint associated with $x = E(X)$ must be $\hat{s} = 0$ by uniqueness. Using the monotonicity of K' then $\hat{s}(x)$ must have the same sign as $x - E(X)$ or

$$\text{sgn}\{\hat{s}(x)\} = \text{sgn}\{x - E(X)\}.$$

Furthermore, saddlepoints for x-values in the right (left) tail of f are necessarily close but smaller than b (greater than a).

Note that for the Exponential (1) example with $x > 0$, $\hat{s} = 1 - 1/x$ and $\text{sgn}(\hat{s}) = \text{sgn}(x-1)$.

1.1.5 Saddlepoint mass functions

For discrete integer-valued random variable X, the saddlepoint approximation for its mass function $p(k)$, based on CGF K, is the same expression as (1.4) and (1.5) when evaluated over the integer values of k. It is written as

$$\hat{p}(k) = \frac{1}{\sqrt{2\pi K''(\hat{s})}} \exp\{K(\hat{s}) - \hat{s}k\} \tag{1.13}$$

where

$$K'(\hat{s}) = k \tag{1.14}$$

and $k \in \mathcal{I}_\mathcal{X}$, the interior of the span of the support of X. Saddlepoint expression (1.13) is computable for any value in $\mathcal{I}_\mathcal{X}$ whether real or integer-valued, but the plot of $\hat{p}(k)$ is meaningful as an approximation to $p(k)$ only for integer-valued arguments.

1.1.6 Examples

(1) Poisson (λ)

For fixed and known λ, the CGF is

$$K(s) = \lambda(e^s - 1) \qquad s \in (-\infty, \infty).$$

The saddlepoint equation

$$K'(\hat{s}) = \lambda e^{\hat{s}} = k \qquad k = 1, 2, \ldots$$

has an explicit saddlepoint solution $\hat{s} = \ln(k/\lambda)$. The saddlepoint equation cannot be solved at $k = 0$ but note that $0 \notin \mathcal{I}_\mathcal{X} = (0, \infty)$ and lies on the boundary of support. The saddlepoint density is

$$\begin{aligned}
\hat{p}(k) &= \frac{1}{\sqrt{2\pi k}} \exp\{k - \lambda - k\ln(k/\lambda)\} \\
&= \left(\sqrt{2\pi k} k^k e^{-k}\right)^{-1} \lambda^k e^{-\lambda} \\
&= \lambda^k e^{-\lambda} / \widehat{k!}
\end{aligned} \qquad (1.15)$$

where

$$\widehat{k!} = \sqrt{2\pi} k^{k+1/2} e^{-k} \simeq k! \qquad (1.16)$$

is Stirling's approximation to the factorial function. This factorial approximation is not exactly the equivalent of the gamma function approximation $\hat{\Gamma}(k+1)$ in (1.8) but they are related by

$$\widehat{k!} = k\hat{\Gamma}(k) \neq \hat{\Gamma}(k+1).$$

(This distinction has been a common source of confusion in the saddlepoint area.) Thus, $\hat{p}$ is related to the true mass function p by

$$\hat{p}(k) = \frac{k!}{\widehat{k!}} p(k) \qquad k = 1, 2, \ldots \qquad (1.17)$$

Normalization requires $p(0)$ to be known so that

$$\bar{p}(k) = \begin{cases} p(0) & k = 0 \\ \{1 - p(0)\}\hat{p}(k) / \sum_{j \geq 1} \hat{p}(j) & k \geq 1 \end{cases}$$

is a normalized saddlepoint approximation. In practical applications, the calculation of boundary probabilities like $p(0)$ is usually possible. The saddlepoint approximations are not exact in this setting and the relative error is that of Stirling's approximation to $k!$. Table 1.1 compares Poisson (1) probabilities for $k = 1, 3, \ldots, 19$ with the values given by $\hat{p}$ and $\bar{p}$. The values of the normalized density $\bar{p}$ are the same as the exact values to the four significant digit accuracy displayed and are listed along with the exact values. Unnormalized values of $\hat{p}$ are given in the third column with their associated relative percentage errors in the fourth column. The diminishing relative error with larger k might have been anticipated since it reflects the relative error of Stirling's approximation $\widehat{k!}$ in (1.16) which is known to decrease with k.

Table 1.1. *Poisson (1) probabilities p with saddlepoint approximations $\hat{p}$ (unnormalized) and $\bar{p}$ (normalized)*

k	$p(k) \simeq \bar{p}(k)$	$\hat{p}(k)$	% relative error
1	.3679	.3989	8.426
3	.06131	.06303	2.805
5	$.0^2 3066$	$.0^2 3117$	1.663
7	$.0^4 7299$	$.0^4 7387$	1.206
9	$.0^5 1014$	$.0^5 1023$	0.8875
11	$.0^8 9216$	$.0^8 9286$	0.7595
13	$.0^{10} 5908$	$.0^{10} 5946$	0.6432
15	$.0^{12} 2813$	$.0^{12} 2829$	0.5688
17	$.0^{14} 1034$	$.0^{14} 1039$	0.4836
19	$.0^{17} 3024$	$.0^{17} 3037$	0.4299

(2) Binomial (n, θ)

The binomial mass function in (1.2) with CGF (1.3) admits an explicit saddlepoint solution

$$\hat{s} = \ln \left\{ \frac{k(1-\theta)}{(n-k)\theta} \right\} \quad k = 1, \ldots, n-1. \tag{1.18}$$

The term $K''(\hat{s}) = k(n-k)/n$ and the saddlepoint mass function works out to be

$$\hat{p}(k) = \widehat{\binom{n}{k}} \theta^k (1-\theta)^{n-k} \quad k = 1, \ldots, n-1 \tag{1.19}$$

where the notation

$$\widehat{\binom{n}{k}} = \frac{\hat{n}!}{\hat{k}! \widehat{(n-k)}!} \simeq \binom{n}{k}$$

is used based on Stirling's approximation in (1.16). The relationship of $\hat{p}(k)$ to $p(k)$ in (1.19) exhibits a structure like that of the Poisson example in (1.17), and which arises consistently in applications of saddlepoint approximations: In the range of examples where combinatorics, factorials or gamma functions are normalization constants, the saddlepoint approximations $\hat{f}$ and $\hat{p}$ are often equal to f or p times a factor consisting of ratios of various Stirling approximations. In this example, the relationship is

$$\hat{p}(k) = \widehat{\binom{n}{k}} \binom{n}{k}^{-1} p(k) \quad k = 1, \ldots, n-1.$$

(3) Negative Binomial (n, θ)

Suppose X is the number of failures occurring in a sequence of Bernoulli trials before the nth success where θ is the probability of a single success. Exercise 9 specifies the details

Table 1.2. *Mass function p for the sum of binomials with saddlepoint approximations $\hat{p}$ and $\bar{p}$.*

k	$\hat{s}(k)$	$p(k)$	$\hat{p}(k)$	$\bar{p}(k)$	Poisson(μ)	$\phi(k;\mu,\sigma^2)$
1	−0.4445	.3552	.3845	.3643	.3384	.3296
3	0.9295	.1241	.1273	.1206	.1213	.1381
5	1.7587	$.0^2 7261$	$.0^2 7392$	$.0^2 7004$	.01305	$.0^2 2221$
7	2.488	$.0^3 1032$	$.0^3 1052$	$.0^4 9963$	$.0^3 6682$	$.0^5 1369$
9	3.300	$.0^6 3633$	$.0^6 3738$	$.0^6 3541$	$.0^4 1996$	$.0^{10} 3238$
11	4.643	$.0^9 2279$	$.0^9 2472$	$.0^9 2342$	$.0^6 3904$	$.0^{16} 2938$

for this example and asks the reader to show that

$$\hat{p}(k) = \frac{\hat{\Gamma}(n+k)}{\widehat{k!}\hat{\Gamma}(n)} \binom{n+k-1}{k}^{-1} p(k) \quad k = 1, \ldots$$

Unlike the binomial example, the combinatoric is approximated using Stirling's approximations for both the gamma and factorial functions.

(4) Sum of binomials

Consider the previously mentioned example for the sum of random variables. Let X_1, X_2, and X_3 be independent variables such that $X_i \sim$ Binomial (m_i, θ_i) with $m_i = 8 - 2i$ and $\theta_i = 2^i/30$. The CGF of $X = \sum_i X_i$ is

$$K(s) = 2 \sum_{i=1}^{3} (4-i) \ln \left\{ \frac{2^i}{30} (e^s - 1) + 1 \right\} \quad s \in (-\infty, \infty)$$

which leads to the saddlepoint equation

$$K'(\hat{s}) = \frac{6\hat{t}}{\hat{t} + 14} + \frac{8\hat{t}}{2\hat{t} + 13} + \frac{8\hat{t}}{4\hat{t} + 11} = k, \quad (1.20)$$

where $\hat{t} = \exp(\hat{s})$. Saddlepoint solutions are the implicitly defined roots of the cubic equations implied in (1.20). Table 1.2 compares the saddlepoint mass function $\hat{p}(k)$ and the normalized approximation $\bar{p}(k)$ with their exact counterpart $p(k)$ over the odd integers. Since $\mathcal{I}_\mathcal{X} = (0, 12)$ the saddlepoint equation cannot be solved at the endpoints of the support, 0 and 12; consequently a normalized approximation is computed as

$$\bar{p}(k) = \begin{cases} p(0) & \text{if } k = 0 \\ \{1 - p(0) - p(12)\} \hat{p}(k) / \sum_{1 \leq j \leq 11} \hat{p}(j) & \text{if } 1 \leq k \leq 11 \\ p(12) & \text{if } k = 12. \end{cases}$$

Over the range $k \in (0, 12)$ the saddlepoint spans $(-\infty, \infty)$, the range of convergence for the CGF. Saddlepoints are listed as $\hat{s}(k)$ in the table. The theoretical limits of solvability for saddlepoint equation (1.20) are

$$\lim_{s \to -\infty} K'(s) = 0 \quad \lim_{s \to \infty} K'(s) = 6 + \frac{8}{2} + \frac{8}{4} = 12$$

which confirms that solutions are indeed within $\mathcal{I}_\mathcal{X} = (0, 12)$.

The Poisson (μ) column in the table is the fit of a Poisson variable with mean $\mu = K'(0) = 22/15$. The last column is a fitted normal density approximation with mean μ and variance $\sigma^2 = K''(0) = 92/75$ of the form

$$\phi(k; \mu, \sigma^2) = \frac{1}{\sigma} \phi\left(\frac{k-\mu}{\sigma}\right) \quad \phi(k) = \frac{1}{\sqrt{2\pi}} \exp(-\tfrac{1}{2} k^2).$$

The numerical results indicate that the Poisson and normal approximations are considerably less accurate than either of the saddlepoint approximations. Normalization, which usually improves $\hat{p}$, has mixed benefits here with some values of $\bar{p}$ closer and other values of $\hat{p}$ closer.

1.2 Univariate cumulative distribution functions

Saddlepoint approximations to CDFs, whether continuous or discrete, require only minor additional computational effort beyond that involved in their density/mass function counterparts. Such approximation affords a simple, fast, and highly accurate method for approximating $\Pr(c < X \leq d)$ as $\hat{F}(d) - \hat{F}(c)$, where $\hat{F}$ denotes the saddlepoint approximation to the CDF F. We shall see below that two saddlepoint equations need to be solved in order to perform this computation. By comparison, this approximation, when based on the saddlepoint density/mass function, requires two numerical integration summations to achieve comparably high accuracy: one over the interval (c, d) and the other for normalization.

1.2.1 Continuous distributions

Suppose continuous random variable X has CDF F and CGF K with mean $\mu = E(X)$. The saddlepoint approximation for $F(x)$, as introduced in Lugannani and Rice (1980), is

$$\hat{F}(x) = \begin{cases} \Phi(\hat{w}) + \phi(\hat{w})(1/\hat{w} - 1/\hat{u}) & \text{if } x \neq \mu \\ \frac{1}{2} + \frac{K'''(0)}{6\sqrt{2\pi} K''(0)^{3/2}} & \text{if } x = \mu \end{cases} \quad (1.21)$$

where

$$\hat{w} = \operatorname{sgn}(\hat{s})\sqrt{2\{\hat{s}x - K(\hat{s})\}} \qquad \hat{u} = \hat{s}\sqrt{K''(\hat{s})} \quad (1.22)$$

are functions of x and saddlepoint $\hat{s}$, where $\hat{s}$ is the implicitly defined function of x given as the unique solution to $K'(\hat{s}) = x$ as in (1.5). Symbols ϕ and Φ denote the standard normal density and CDF respectively and $\operatorname{sgn}(\hat{s})$ captures the sign $\pm$ for $\hat{s}$.

The bottom expression in (1.21) defines the approximation at the mean of X or when $\hat{s} = 0$. In this case, $\hat{w} = 0 = \hat{u}$ and the last factor in the top expression of (1.21) is undefined. As $x \to \mu$, the limiting value of the top expression is the bottom expression; see section 2.5.2 for a proof. Thus, the entire expression is continuous and, more generally, continuously differentiable or "smooth." Apart from the theoretical smoothness, any practical computation that uses software is vulnerable to numerical instability when making the computation of $\hat{F}(x)$ for x quite close to μ.

1.2 Univariate cumulative distribution functions

Table 1.3. *Left (and right) tail probabilities F with saddlepoint approximation $\hat{F}$ for a* Gamma $(1/2, 1)$.

x	$F(x)$	$\hat{F}(x)$	% relative error
1/64	.1403	.1517	8.112
1/16	.2763	.2902	5.037
1/8	.3829	.3959	3.401
x	$1 - F(x)$	$1 - \hat{F}(x)$	% relative error
1/4	.4795	.4695	-2.080
1/2	.3173	.3119	-1.6905
1	.1573	.1560	-0.8257
3	.01431	.01454	1.628
5	$.0^2 1565$	$.0^2 1614$	3.132

1.2.2 Examples

(1) Normal $(0, 1)$

From (1.6), the saddlepoint is $\hat{s} = x$ so $\hat{w} = x$, $\hat{u} = x$, and $\hat{F}(x) = \Phi(x)$. The saddlepoint approximation is exact.

(2) Gamma $(\alpha, 1)$

Using the saddlepoint $\hat{s} = 1 - \alpha/x$, then

$$\hat{w} = \text{sgn}(x - \alpha) \sqrt{2 \left\{ x - \alpha \left(1 + \ln \frac{x}{\alpha}\right)\right\}} \qquad \hat{u} = \frac{x - \alpha}{\sqrt{\alpha}} \qquad (1.23)$$

for $x > 0$. With these values, expression (1.21) is explicit in x and yields a very simple approximation for the gamma CDF which is also the incomplete gamma function $P(\alpha, x)$ as defined in Abramowitz and Stegun (1970, 6.5.1). Table 1.3 compares $\hat{F}(x)$ vs. $F(x)$ {or $1 - \hat{F}(x)$ vs. $1 - F(x)$, whichever is smaller} for the very skewed Gamma $(1/2, 1)$ distribution with $\alpha = 1/2$. Relative error remains very small for the computations in the right tail and accuracy appears quite good but not quite as accurate in the left tail.

For larger values of α, the gamma approaches a normal shape as a central limit. Since the approximation is exact in the normal setting which is the limit for large α, one might expect the saddlepoint approximation to gain in accuracy with increasing α. This can be seen numerically by reproducing Table 1.3 with $\alpha = 1$, an Exponential (1) distribution, as suggested in Exercise 3.

The rate of convergence for relative error is $O(\alpha^{-3/2})$ as $\alpha \to \infty$ for a fixed value of the standardized variable $z = (x - \alpha)/\sqrt{\alpha}$ (Daniels, 1987). What this $O(\alpha^{-3/2})$ relative error statement means precisely is that

$$\alpha^{3/2} \left\{ \frac{\hat{F}(\sqrt{\alpha}z + \alpha)}{F(\sqrt{\alpha}z + \alpha)} - 1 \right\}$$

remains bounded as $\alpha \to \infty$ for fixed z.

Table 1.4. *Saddlepoint accuracy for the normal–Laplace convolution.*

x	$1 - F(x)$	$1 - \hat{F}(x)$	% relative error
.25	.4350	.4332	−0.3978
.5	.3720	.3689	−0.8478
1.0	.2593	.2550	−1.674
2.0	.1084	.1065	−1.745
3.0	.04093	.04086	−0.1907
4.0	.01510	.01525	1.028
5.0	$.0^2 5554$	$.0^2 5645$	1.636
6.0	$.0^2 2043$	$.0^2 2083$	1.957
7.0	$.0^3 7517$	$.0^3 7680$	2.161
8.0	$.0^3 2765$	$.0^3 2829$	2.314
9.0	$.0^3 1017$	$.0^3 1042$	2.440

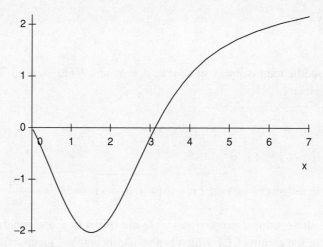

Figure 1.5. Percentage relative error $R(x)$ vs. x for the normal–Laplace convolution.

(3) Normal–Laplace convolution

This distribution is symmetric about 0 and, as we now argue, the saddlepoint CDF approximation in (1.21) must also exhibit the same symmetry. The argument consists of noting that saddlepoint $\hat{s}(x)$ is an odd function in x, so $\hat{w}(x)$ and $\hat{u}(x)$ are also odd which implies that $\hat{F}(x)$ must agree with $1 - \hat{F}(-x)$. Because of this symmetry, only values for which $x > 0$ are considered in table 1.4.

Figure 1.5 plots $R(x)$, the percentage relative error in approximation, or

$$R(x) = 100 \left\{ \frac{1 - \hat{F}(x)}{1 - F(x)} - 1 \right\}$$

as a function of x, for $x > 0$. It can be shown analytically that the asymptote of the plot is $\lim_{x \to \infty} R(x) = 100\{1/\hat{\Gamma}(1) - 1\} = 8.444\%$.

1.2 Univariate cumulative distribution functions

Table 1.5. *Saddlepoint accuracy for the Gumbel (0, 1) distribution.*

x	$F(x)$	$\hat{F}(x)$	% relative error
−3	$.0^81892$	$.0^81900$	1.110
−2	$.0^36180$	$.0^36228$	0.7760
−1	.06599	.06678	1.195
0	.3679	.3723	1.200

x	$1 - F(x)$	$1 - \hat{F}(x)$	% relative error
1	.3078	.3029	−1.602
2	.1266	.1242	−1.867
3	.04857	.04787	−1.447
4	.01815	.01801	−0.7630
5	$.0^26715$	$.0^26709$	−0.08890
6	$.0^22476$	$.0^22488$	0.4812
7	$.0^39115$	$.0^39200$	0.9399
8	$.0^33354$	$.0^33398$	1.307
9	$.0^31234$	$.0^31254$	1.604
10	$.0^44540$	$.0^44624$	1.850
11	$.0^41670$	$.0^41705$	2.058

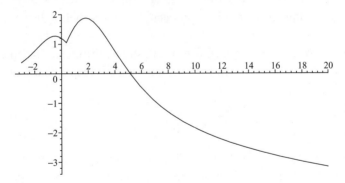

Figure 1.6. Percentage relative error $R(x)$ vs. x for the Gumbel.

(4) Gumbel (0, 1)

The saddlepoint density approximations for this example were extremely accurate and table 1.5 shows that the saddlepoint CDF approximation shares the same accuracy. The probabilities below agree with those found in tables of Daniels (1992) in which the CDF of $-X$ has been tabulated.

A plot of the percentage relative error

$$R(x) = 100 \left[\frac{\hat{F}(x) - F(x)}{\min\{F(x), 1 - F(x)\}} \right] \qquad (1.24)$$

is shown in figure 1.6. It can be shown analytically that the asymptote of the plot is $\lim_{x \to \infty} R(x) = 1 - 1/\hat{\Gamma}(1) \simeq -8.444\%$.

Table 1.6. *Exact and saddlepoint computation of the CDF of a Beta $(\alpha, 0.8)$ at $x = 0.4$.*

α	$F_{\alpha, 0.8}(0.4)$	$\hat{F}_{\alpha, 0.8}(0.4)$
.05	.9417	.9586
.1	.8878	.8919
.15	.8376	.8370
.2	.7910	.7884
.3	.7066	.7037
.4	.6327	.6307
.5	.5675	.5667
.6	.5098	.5100
.7	.4585	.4596
.8	.4128	.4146
.9	.3720	.3743
1.0	.3355	.3381

(5) Beta (α, β)

A direct computation of the MGF of this distribution leads to the confluent hypergeometric function. Such special functions make this particular approach to saddlepoint approximation more difficult so a simpler alternative approach is pursued below that leads to explicit saddlepoints.

The distribution of $X \sim$ Beta (α, β) can be constructed from two independent gamma variates as $X = Y_\alpha/(Y_\alpha + Y_\beta)$ where $Y_\alpha \sim$ Gamma $(\alpha, 1)$ independently of $Y_\beta \sim$ Gamma $(\beta, 1)$. The CDF of X is

$$F_{\alpha, \beta}(x) = \Pr\left(\frac{Y_\alpha}{Y_\alpha + Y_\beta} \leq x\right) = \Pr(Z_x \leq 0) \quad (1.25)$$

where $Z_x = (1-x)Y_\alpha - xY_\beta$ is determined from a simple rearrangement of terms. For a fixed value of x, the MGF of Z_x is easily computed and may be used to determine a CDF approximation for this variable at value 0 as indicated in (1.25). Exercise 7 asks the reader to find the explicit saddlepoint for Z_x at 0 as $\hat{s} = \{x(1-x)\}^{-1}\{x - \alpha/(\alpha+\beta)\}$. The inputs for the CDF approximation are therefore

$$\begin{aligned}\hat{w} &= \text{sgn}(\hat{s})\sqrt{2\left[\alpha \ln\left\{\frac{\alpha}{(\alpha+\beta)x}\right\} + \beta \ln\left\{\frac{\beta}{(\alpha+\beta)(1-x)}\right\}\right]} \\ \hat{u} &= \left(x - \frac{\alpha}{\alpha+\beta}\right)\sqrt{(\alpha+\beta)^3/(\alpha\beta)}.\end{aligned} \quad (1.26)$$

Table 1.6 shows exact computation of

$$F_{\alpha, 0.8}(0.4) = \Pr\{\text{Beta }(\alpha, 0.8) < 0.4\}$$

for the various values of α listed in the first column along with the saddlepoint approximation $\hat{F}_{\alpha, 0.8}(0.4)$.

The very small values of α with $\beta = 0.8$ present a daunting challenge for the approximations to maintain their accuracy. For $\alpha \geq 0.1$ the approximations are extremely accurate.

1.2 Univariate cumulative distribution functions

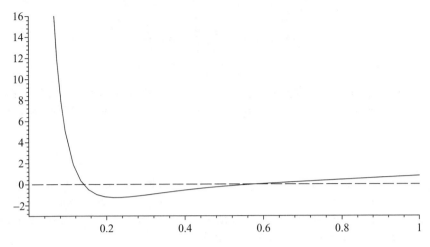

Figure 1.7. Percentage relative error vs. α for $\hat{F}_{\alpha,0.8}(0.4)$ as an approximation to $F_{\alpha,0.8}(0.4)$.

Figure 1.7 plots the relative error as in (1.24). The relative error for $\alpha = 0.01$ is 436. 2% and further computations indicate that this error grows to ∞ as $\alpha \to 0$.

1.2.3 Discrete distributions

Discrete CDF approximation requires modification to the formula for the continuous CDF approximation in order to achieve the greatest accuracy. Daniels (1987) introduced two such continuity-corrected modifications which are presented below. Such adjustment was not needed, as discussed in section 1.1.5, when dealing with saddlepoint mass functions and they share a common analytical expression with the continuous setting.

Suppose X has CDF $F(k)$ with support on the integers and mean μ. Rather than considering the CDF value $F(k)$, the right tail probability $\Pr(X \geq k)$, or the left-continuous survival function, is approximated instead which helps to avoid what would otherwise be some difficult notational problems.

The two continuity corrections of Daniels (1987) along with a third approximation that he suggested are given below. Both the first and second continuity corrections will ultimately be seen as procedures that are consistently accurate. Which of the two is most accurate depends upon the specific application. In chapter 2, it will be shown that the second approximation possesses the more advantageous theoretical properties that the others lack.

First continuity correction

Suppose $k \in \mathcal{I}_X$ so that the saddlepoint equation can be solved at value k. The first approximation is

$$\widehat{\Pr}_1 (X \geq k) = \begin{cases} 1 - \Phi(\hat{w}) - \phi(\hat{w})(1/\hat{w} - 1/\tilde{u}_1) & \text{if} \quad k \neq \mu \\ \frac{1}{2} - \frac{1}{\sqrt{2\pi}} \left\{ \frac{K'''(0)}{6K''(0)^{3/2}} - \frac{1}{2\sqrt{K''(0)}} \right\} & \text{if} \quad k = \mu \end{cases} \qquad (1.27)$$

where

$$\hat{w} = \text{sgn}(\hat{s})\sqrt{2\{\hat{s}k - K(\hat{s})\}}$$
$$\tilde{u}_1 = \{1 - \exp(-\hat{s})\}\sqrt{K''(\hat{s})} \tag{1.28}$$

and $\hat{s}$ solves $K'(\hat{s}) = k$. The expression for $\hat{w}$ agrees with its counterpart (1.22) in the continuous setting whereas $\tilde{u}_1 \neq \hat{u}$. According to this approximation, the CDF is approximated as

$$\hat{F}_1(k-1) = 1 - \widehat{\Pr}_1(X \geq k). \tag{1.29}$$

This produces the rather odd situation in which the saddlepoint $\hat{s}$ associated with value k is used to approximate the CDF at value $k-1$. Such are the notational difficulties and confusion that can occur for the discrete setting but these problems are mostly avoided by working with survival expression (1.27) instead.

The bottom expression in (1.27), when μ is integer valued, is defined by continuity to be the limit of the top expression as shown in section 2.5.2.

To understand the difference associated with using $\tilde{u}_1$ instead of $\hat{u}$, consider a Taylor series expansion of the leading factor in $\tilde{u}_1$ as $1 - \exp(-\hat{s}) \simeq \hat{s} - \hat{s}^2/2$. To first order, $\tilde{u}_1 \simeq \hat{u}$ from the continuous setting and the approximations are the same. However, to second order, $\tilde{u}_1 < \hat{u}$, which implies that $\widehat{\Pr}_1(X \geq k) > 1 - \hat{F}(k)$, where the latter term refers to the continuous version in (1.21) and (1.22) evaluated at k. Thus, the use of the smaller $\tilde{u}_1$ in place of $\hat{u}$ adjusts the tail probability in a direction that is consistent with a continuity correction.

Second continuity correction

Define $k^- = k - 1/2 \in \mathcal{I}_\mathcal{X}$ as the continuity-corrected or offset value of k. The second approximation solves the offset saddlepoint equation

$$K'(\tilde{s}) = k^- \tag{1.30}$$

for continuity-corrected saddlepoint $\tilde{s}$. Saddlepoint $\tilde{s}$ and k^- are used to alter the inputs into the CDF approximation according to

$$\tilde{w}_2 = \text{sgn}(\tilde{s})\sqrt{2\{\tilde{s}k^- - K(\tilde{s})\}}$$
$$\tilde{u}_2 = 2\sinh(\tilde{s}/2)\sqrt{K''(\tilde{s})}. \tag{1.31}$$

This leads to the second continuity-corrected approximation

$$\widehat{\Pr}_2(X \geq k) = \begin{cases} 1 - \Phi(\tilde{w}_2) - \phi(\tilde{w}_2)(1/\tilde{w}_2 - 1/\tilde{u}_2) & \text{if } k^- \neq \mu \\ \frac{1}{2} - \frac{K'''(0)}{6\sqrt{2\pi}K''(0)^{3/2}} & \text{if } k^- = \mu. \end{cases} \tag{1.32}$$

Third approximation

This approximation is denoted as $\widehat{\Pr}_3(X \geq k)$ and uses expression (1.32) with $\tilde{w}_2$ as in (1.31) and $\tilde{u}_2$ replaced with

$$\tilde{u}_3 = \tilde{s}\sqrt{K''(\tilde{s})}.$$

1.2 Univariate cumulative distribution functions

Table 1.7. *The various continuity-corrected CDF approximations for the right tail of a Poisson (1).*

k	$\Pr(X \geq k)$	$\widehat{\Pr}_1(X \geq k)$	$\widehat{\Pr}_2(X \geq k)$	$\widehat{\Pr}_3(X \geq k)$
1	.6321	.6330*	.6435	.6298
3	.08030	.08045	.08102	.08529
5	.0²3660	.0²3667	.0²3683	.0²4103
7	.0⁴8324	.0⁴8339	.0⁴8363	.0⁴9771
9	.0⁵1125	.0⁵1127	.0⁵1129	.0⁵1374
11	.0⁷1005	.0⁷1006	.0⁷1008	.0⁷1270
13	.0¹⁰6360	.0¹⁰6369	.0¹⁰6375	.0¹⁰8291
15	.0¹²3000	.0¹²3004	.0¹²3006	.0¹²4021
17	.0¹⁴1095	.0¹⁴1096	.0¹⁴1097	.0¹⁴1505
19	.0¹⁷3183	.0¹⁷3187	.0¹⁷3188	.0¹⁷4480

Note: * Requires a computation at the mean.

This approximation may be motivated as simply the application at k^- of the continuous version of the Lugannani and Rice approximation. On intuitive grounds, one might suggest this correction because it is the same one used for normal approximation of the binomial. The second and third approximations are related according to the first order of the Taylor approximation $2\sinh(\tilde{s}/2) \simeq \tilde{s} + \tilde{s}^3/24$. Since the expansion lacks an $\tilde{s}^2$ term, $\tilde{u}_2 \simeq \tilde{u}_3$ to second order so the two approximations agree to second order. While these two approximations are similar in accuracy near the mean where $\tilde{s} \approx 0$, they begin to deviate in accuracy as $|\tilde{s}|$ increases and the Taylor approximation deteriorates. This is seen in the examples below.

1.2.4 Examples

(1) Poisson (λ)

Table 1.7 compares values of $\Pr(X \geq k)$ and the three suggested values of $\widehat{\Pr}_i(X \geq k)$ for the Poisson (1) example used previously. The general expressions involved in the approximations are

Approximation 1	Approximations 2 & 3
$\hat{s} = \ln(k/\lambda)$	$\tilde{s} = \ln(k^-/\lambda)$
$\hat{w}(k) = \sqrt{2\{k\ln(k/\lambda) - (k-\lambda)\}}$	$\tilde{w}_2 = \hat{w}(k^-)$
$\tilde{u}_1 = (1 - k/\lambda)\sqrt{k}$	$\tilde{u}_2 = (1 - \lambda/k^-)k^-/\sqrt{\lambda}$

The first approximation is slightly better than the second one which is clearly better than the third. Deterioration of the third approximation is most pronounced in the tails as anticipated from the theoretical comparison of methods two and three above. The starred entry for

Table 1.8. *Continuity-corrected CDF approximations for the right tail of a Binomial* (100, 0.03).

k	$\Pr(X \geq k)$	$\widehat{\Pr}_1(X \geq k)$	$\widehat{\Pr}_2(X \geq k)$	$\widehat{\Pr}_3(X \geq k)$
1	.9525	.9523	.9547	.9471
2	.8054	.8053	.8075	.8015
3	.5802	.5803	.5819	.5800
5	.1821	.1822	.1827	.1851
7	.03123	.03125	.03131	.03240
9	.0²3216	.0²3219	.0²3224	.0²3413
11	.0³2149	.0³2151	.0³2154	.0³2333
13	.0⁵9875	.0⁵9885	.0⁵9894	.0⁴1096
15	.0⁶3258	.0⁶3261	.0⁶3264	.0⁶3694
17	.0⁸7980	.0⁸7989	.0⁸7993	.0⁸9241
19	.0⁹1490	.0⁹1491	.0⁹1492	.0⁹1761
21	.0¹¹2164	.0¹¹2167	.0¹¹2167	.0¹¹2610

$\widehat{\Pr}_1(X \geq k)$ identifies a computation for which $k = \mu$ so that $\tilde{w}_1 = 0 = \tilde{u}_1$ and the lower expression for $\widehat{\Pr}_1(X \geq k)$ was used.

(2) Binomial (n, θ)

For this example the saddlepoint is $\hat{s} = \ln[k(1 - \theta)/\{(n - k)\theta\}]$, as also given in (1.18). This leads to

$$\hat{w} = \text{sgn}(\hat{s})\sqrt{2\left\{k \ln \frac{k}{n\theta} + (n - k) \ln \frac{n - k}{n(1 - \theta)}\right\}}$$

$$\tilde{u}_1 = \frac{k - n\theta}{k(1 - \theta)}\sqrt{\frac{k(n - k)}{n}} \tag{1.33}$$

and

$$\tilde{u}_2 = \left(\frac{k^-}{\theta} - \frac{n - k^-}{1 - \theta}\right)\sqrt{\frac{\theta(1 - \theta)}{n}} \tag{1.34}$$

$$\tilde{u}_3 = \ln\left\{\frac{k^-(1 - \theta)}{(n - k^-)\theta}\right\}\sqrt{\frac{k^-(n - k^-)}{n}}.$$

Suppose $\theta = 0.03$ and $n = 100$ so the distribution is quite skewed. Table 1.8 compares the three saddlepoint approximations with the exact probabilities. In agreement with the Poisson (1) example, $\widehat{\Pr}_1$ is slightly more accurate than $\widehat{\Pr}_2$ which is distinctly better than $\widehat{\Pr}_3$.

The Binomial (100, 0.03) is quite skewed so one might expect a poor central limit approximation and a good Poisson approximation based on the law of rare events. Table 1.9 compares the worst of the saddlepoint approximations $\widehat{\Pr}_3$ with a Poisson (3) approximation

Table 1.9. *Comparison with Poisson and normal approximation.*

k	$\Pr(X \geq k)$	$\widehat{\Pr}_3(X \geq k)$	Poisson (3)	Normal $(3\frac{1}{2}, \sigma^2)$
3	.5802	.5800	.5768	.6153
5	.1821	.1851	.1847	.1896
7	.03123	.03240	.03351	.02010
9	.0²3216	.0²3413	.0²3803	.0³6317
11	.0³2149	.0³2333	.0³2923	.0⁵5499
13	.0⁵9875	.0⁴1096	.0⁴1615	.0⁷1281
15	.0⁶3258	.0⁶3694	.0⁶6704	.0¹¹7842
17	.0⁸7980	.0⁸9241	.0⁷2165	.0¹⁴1248
19	.0⁹1490	.0⁹1761	.0⁹5588	.0¹⁹5125
21	.0¹¹2164	.0¹¹2610	.0¹⁰1179	.0²⁴5409

and a normal approximation with continuity correction given by

$$\Pr(X \geq k) \simeq 1 - \Phi\{(k^- - 3)/\sigma\},$$

where $\sigma = \sqrt{100(0.03)(0.97)}$.

Saddlepoint approximation $\widehat{\Pr}_3$ is quite a bit better than the Poisson (3) approximation in a situation for which the latter might have been expected to perform well. For $k \geq 9$, the normal approximation has excessive error which increases with larger k.

(3) Negative Binomial (n, θ)

Details of the CDF approximations in this setting are given in Exercise 9. Relative errors for the approximations of the form

$$100 \frac{\widehat{\Pr}_i(X \geq k) - \Pr(X \geq k)}{\min\{\Pr(X \geq k), 1 - \Pr(X \geq k)\}} \qquad i = 1, 2, 3 \qquad (1.35)$$

are plotted in figure 1.8 for a Negative Binomial $(2, 0.2)$ distribution using $k = 1, \ldots, 30$. In the right tail above $\mu = 8$, the plot suggests a preferential ordering of the approximations: second, third and first. This is in contrast with the previous examples in which the first approximation was the best. For this particular example, the relative errors continue to increase very slowly to

$$\lim_{k \to \infty} 100 \left(\frac{\widehat{\Pr}_i(X \geq k)}{\Pr(X \geq k)} - 1 \right) = 100 \left(\frac{\Gamma(2)}{\hat{\Gamma}(2)} - 1 \right) \approx 4.22\% \qquad i = 1, 2.$$

To judge this slow convergence, compare this limit with the relative error for $\widehat{\Pr}_1$ at $k = 500$ which is 2.42%.

In the left tail, the first approximation shows the smallest absolute relative error with the second and third approximations comparable but of opposite sign.

22   *Fundamental approximations*

Table 1.10.   *Continuity-corrected approximations for the sum of binomials.*

k	$\Pr(X \geq k)$	$\widehat{\Pr}_1(X \geq k)$	$\widehat{\Pr}_2(X \geq k)$	$\widehat{\Pr}_3(X \geq k)$	Poisson (μ)
1	.7994	.8007	.8064	.7892	.7693
2	.5558	.5548	.5531	.5527	.4310
3	.1687	.1693	.1695	.1752	.1828
4	.04464	.04485	.04483	.04818	.06153
5	$.0^2 8397$	$.0^2 8452$	$.0^2 8428$	$.0^2 9482$	.01705
6	$.0^2 1137$	$.0^2 1147$	$.0^2 1140$	$.0^2 1351$	$.0^2 4003$
7	$.0^3 1109$	$.0^3 1123$	$.0^3 1112$	$.0^3 1399$	$.0^3 8141$
8	$.0^5 7732$	$.0^5 7863$	$.0^5 7742$	$.0^4 1044$	$.0^3 1458$
9	$.0^6 3753$	$.0^6 3846$	$.0^6 3751$	$.0^6 5508$	$.0^4 2334$
10	$.0^7 1205$	$.0^7 1252$	$.0^7 1199$	$.0^7 1968$	$.0^5 3372$
11	$.0^9 2299$	$.0^9 2488$	$.0^9 2266$	$.0^9 4392$	$.0^6 4441$
12	$.0^{11} 1973$	undefined	$.0^{11} 1861$	$.0^{11} 5186$	$.0^7 5373$

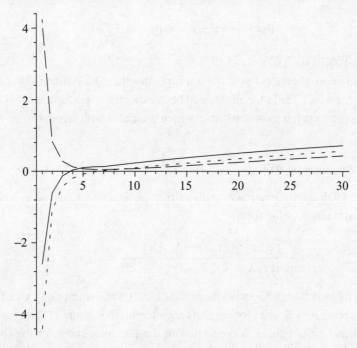

Figure 1.8.   Percentage relative error of the right tail approximations $\widehat{\Pr}_1$ (*solid*), $\widehat{\Pr}_2$ (*dashed*), and $\widehat{\Pr}_3$ (*dotted*) for a Negative Binomial (2, 0.2) distribution.

(4) Sum of binomials

Table 1.10 compares the accuracy of the three tail approximations for this example using the parameters considered in section 1.1.6. The last column is a Poisson approximation with $\mu = K'(0)$.

Approximation one is undefined at $k = 12$ which is on the boundary of the support. Figure 1.9 plots the percentage relative errors of the three approximations as in (1.35).

1.2 Univariate cumulative distribution functions

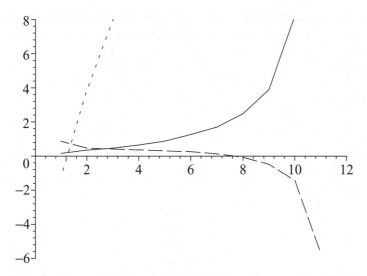

Figure 1.9. Percentage relative errors for $\widehat{\Pr}_1$ (*solid*), $\widehat{\Pr}_2$ (*dashed*), and $\widehat{\Pr}_3$ (*dotted*).

Approximation two is slightly better than one and both are much better than three whose graph for $k \geq 4$ has been truncated.

(5) Uniform $\{0, 1, \ldots, 9\}$

Accuracies for the two recommended continuity corrections are compared for a Uniform $\{0, \ldots, 9\}$ distribution. In this example, the first correction method is more accurate in the right tail but the second method is more consistently accurate when both tails are considered.

The CGF is

$$K(s) = \ln \frac{1 - e^{10s}}{10(1 - e^s)} \qquad s \neq 0$$

and is defined by continuity as 0 at $s = 0$. Table 1.11 compares the two continuity corrections $\widehat{\Pr}_1$ and $\widehat{\Pr}_2$ as well as a third CDF approximation $\hat{F}_{1a}$ that is introduced next in section 1.2.5.

1.2.5 Alternative left-tail CDF approximations for discrete distributions

The probability approximations of section 1.2.3 were introduced by Daniels (1987) for the right tails of the distributions, and were meant to be applied with values of k for which $k > \mu$. They were not necessarily intended for survival computation when the value of k is in the left tail or $k < \mu$. Of course the left tail of X is the right tail of $-X$ so that the theoretical results of Daniels may be easily applied to $-X$ and then re-expressed to apply to the left tail of X. When this is done for the second continuity-corrected approximation and the third approximation, the resulting approximations agree with those suggested in section 1.2.3. More specifically for the $i = 2, 3$ methods, if $\hat{F}_i(k)$ is the CDF approximation when $k + 1/2 < \mu$, then

$$\hat{F}_i(k) = 1 - \widehat{\Pr}_i(X \geq k + 1), \qquad (1.36)$$

24 Fundamental approximations

Table 1.11. *Survival and CDF function approximations for the uniform mass function.*

k	$\Pr(X \leq k)$	$\widehat{\Pr}_1(X \leq k)$	$\hat{F}_{1a}(k)$	$\widehat{\Pr}_2(X \leq k)$
0	.1	.0861	undefined	.09095
1	.2	.1679	.2009	.1769
2	.3	.2640	.2898	.2734
3	.4	.3730	.3960	.3830
4	.5	.4889	.5111	.5
k	$\Pr(X \geq k)$	$\widehat{\Pr}_1(X \geq k)$		$\widehat{\Pr}_2(X \geq k)$
5	.5	.5111		.5
6	.4	.3960		.3830
7	.3	.2898		.2734
8	.2	.2009		.1769
9	.1	undefined		.09095

where $\widehat{\Pr}_i$ is the corresponding survival approximation from section 1.2.3. Thus, for these two approximations, CDF approximations and survival approximations are completely compatible which makes the right-tail formulas of section 1.2.3 sufficient for characterizing all aspects of the approximations. Interestingly enough, the same cannot be said when applied to the first continuity-corrected approximation. A left-tailed CDF approximation that results from $-X$ is different from the right-tail approximation given in section 1.2.3.

This subsection explores the nature of these equivalences and differences in terms of numerical accuracy with particular emphasis on the first continuity correction. After all has been said about these differences, the recommended versions for continuity correction are the first two that were introduced in section 1.2.3. Thus, the material of this subsection is somewhat unnecessary at first reading and is presented mostly for the sake of curiosity.

Left-tail probabilities for the CDF of X can be determined as right-tail probabilities for $-X$ and the reader is asked to carry this out in Exercise 8 and derive the left-tail versions of the three continuity corrections that are given below and denoted as $\hat{F}_{1a}$, $\hat{F}_2$, and $\hat{F}_3$.

Alternative left tail for the first continuity correction

If $k \in \mathcal{I}_X$, then an alternative approximation to the CDF $\Pr(X \leq k)$ is

$$\hat{F}_{1a}(k) = \begin{cases} \Phi(\hat{w}) + \phi(\hat{w})(1/\hat{w} - 1/\tilde{u}_{1a}) & \text{if } k < \mu \\ \frac{1}{2} + \frac{1}{\sqrt{2\pi}} \left\{ \frac{K'''(0)}{6K''(0)^{3/2}} + \frac{1}{2\sqrt{K''(0)}} \right\} & \text{if } k = \mu \end{cases} \quad (1.37)$$

where

$$\hat{w} = -\sqrt{2\{\hat{s}k - K(\hat{s})\}}$$

$$\tilde{u}_{1a} = \{\exp(\hat{s}) - 1\}\sqrt{K''(\hat{s})}$$

and $\hat{s}$ solves $K'(\hat{s}) = k$. The left-tail expression $\hat{F}_{1a}(k)$ allows approximation for $F(k)$-values over $k \in (-\infty, \mu]$. An expansion of the leading factor in $\tilde{u}_{1a}$ is $\exp(\hat{s}) - 1 \simeq \hat{s} +$

$\hat{s}^2/2$ and to first order $\tilde{u}_{1a} \simeq \hat{u}$ from the continuous setting. To second order, $\tilde{u}_{1a} > \hat{u}$ which implies that $\widehat{\Pr}_{1a}(X \leq k) > \hat{F}(k)$, where the latter term is the continuous formula in (1.21) and (1.22). Thus the replacement of $\hat{s}$ with $\exp(\hat{s}) - 1$ in the $\hat{u}$-term inflates the probability in a manner consistent with a continuity adjustment for the left tail.

This approximation is clearly not the same as the left-tail CDF approximation specified in (1.29) so

$$\hat{F}_{1a}(k) \neq \hat{F}_1(k) = 1 - \widehat{\Pr}_1(X \geq k+1). \tag{1.38}$$

The right side uses the saddlepoint at $k+1$ whereas the left side uses the saddlepoint at k.

Left tail for the second continuity correction

Suppose $k^+ = k + 1/2 \in \mathcal{I}_\mathcal{X}$ is the left offset value of k. The second approximation is

$$\hat{F}_2(k) = \begin{cases} \Phi(\tilde{w}_2^+) + \phi(\tilde{w}_2^+)\left(1/\tilde{w}_2^+ - 1/\tilde{u}_2^+\right) & \text{if } k^+ < \mu \\ \frac{1}{2} + \frac{K'''(0)}{6\sqrt{2\pi}K''(0)^{3/2}} & \text{if } k^+ = \mu, \end{cases} \tag{1.39}$$

where

$$\begin{aligned} \tilde{w}_2^+ &= -\sqrt{2\{\tilde{s}^+ k^+ - K(\tilde{s}^+)\}} \\ \tilde{u}_2^+ &= 2\sinh(\tilde{s}^+/2)\sqrt{K''(\tilde{s}^+)} \end{aligned} \tag{1.40}$$

and $\tilde{s}^+$ is the left continuity-corrected saddlepoint defined as the unique solution to

$$K'(\tilde{s}^+) = k^+. \tag{1.41}$$

It is a simple exercise to show that

$$\hat{F}_2(k) = 1 - \widehat{\Pr}_2(X \geq k+1) \tag{1.42}$$

whenever $k^+ \leq \mu$. In the case that $k^+ = \mu$, the equality in (1.42) is also preserved, since the bottom expression in (1.39) is compatible with the bottom expression in (1.32) when evaluated at $(k+1)^- = k^+$.

Left tail for the third approximation

The third approximation $\hat{F}_3(k)$ uses the same expression as $\hat{F}_2(k)$ but replaces $\tilde{u}_2^+$ with

$$\tilde{u}_3^+ = \tilde{s}^+ \sqrt{K''(\tilde{s}^+)}.$$

Whenever $k^+ \leq \mu$, it is easy to show that

$$\hat{F}_3(k) = 1 - \widehat{\Pr}_3(X \geq k+1).$$

Examples

(1) Poisson (5)   Table 1.12 compares the various continuity-corrected CDF approximations in the left tail of this distribution. The expression $\hat{F}_1$ in (1.29) of section 1.2.3 is most accurate with $\hat{F}_2$ a close second.

26 Fundamental approximations

Table 1.12. *CDF approximations for the Poisson (5) in its left tail.*

k	$F(k)$	$\hat{F}_{1a}(k)$	$\hat{F}_1(k)$	$\hat{F}_2(k)$	$\hat{F}_3(k)$
0	$.0^2 6738$	undefined	$.0^2 6812$	$.0^2 6376$	$.0^2 8020$
1	.04043	.04335	.04053	.03984	.04272
2	.1247	.1283	.1248	.1239	.1272
3	.2650	.2691	.2651	.2642	.2666
4	.4405	.4443	.4405*	.4398	.4406

Note: * Requires a computation at the mean.

Table 1.13. *CDF approximations for the Binomial (23, 0.2) in its left tail.*

k	$F(k)$	$\hat{F}_{1a}(k)$	$\hat{F}_1(k)$	$\hat{F}_2(k)$	$\hat{F}_3(k)$
0	$.0^2 5903$	undefined	$.0^2 5910$	$.0^2 5591$	$.0^2 7200$
1	.03984	.04272	.03975	.03930	.04253
2	.1332	.1371	.1329	.1324	.1364
3	.2965	.3009	.2961	.2958	.2984
4	.5007	.5047	.5002	.5001	.5003

Table 1.14. *CDF approximations for the Negative Binomial (2, 0.2) in its left tail.*

k	$F(k)$	$\hat{F}_{1a}(k)$	$\hat{F}_1(k)$	$\hat{F}_2(k)$	$\hat{F}_3(k)$
0	.04000	undefined	.04103	.03831	.04176
1	.1040	.1114	.1047	.1031	.1051
2	.1808	.1863	.1810	.1803	.1816
3	.2627	.2674	.2626	.2624	.2632
4	.3446	.3486	.3443	.3444	.3449
5	.4233	.4266	.4228	.4231	.4234
6	.4967	.4995	.4960	.4965	.4966
7	.5638	.5662	.5631*	.5636	.5636

Note: *Requires a computation at the mean.

(2) Binomial (23, 0.2)   Table 1.13 compares the left-tail approximations for this distribution. Expression $\hat{F}_1$ is most accurate with $\hat{F}_2$ a close second.

(3) Negative Binomial (2, 0.2)   Probabilities for the left tail of this distribution with mean 8 are given in Table 1.14. Expression $\hat{F}_1$ is most accurate with $\hat{F}_2$ a close second.

1.2.6 Performance summary for continuity corrections

For the most part, the continuity-corrected approximations are more accurate in the right tails than the left tails. However, for all the examples above, the right tails are longer than the left tails since the distributions are skewed to the right. This suggests an intuitive reason as to why the approximations are more accurate in the right tail than the left tail. Simply put, the continuity corrections are going to exhibit more error in places where the CDF is more steeply ascending, as in the left tail, than in the more gradual right tail.

The examples seem to suggest that both $\widehat{\Pr}_1$ and $\widehat{\Pr}_2$ are consistently accurate with $\widehat{\Pr}_1$ often slightly more accurate. Later examples to be presented, however, suggest the opposite, that $\widehat{\Pr}_2$ is more consistently accurate than $\widehat{\Pr}_1$. When all past and future examples are considered together, there does not appear to be a clear cut answer as to which approximation, $\widehat{\Pr}_1$ or $\widehat{\Pr}_2$, is better: this depends upon the application. For the most part, $\widehat{\Pr}_3$ and $\hat{F}_{1a}$ are not competitors. We therefore suggest computation of both $\widehat{\Pr}_1$ and $\widehat{\Pr}_2$. If the two approximations are close, then this reinforces both of their accuracies; when they differ, then a choice must be made and this may depend more on the particular application.

Discrete nonlattice distributions

The discussion has excluded nonlattice mass functions and distributions since most formal theory has concentrated on these two settings. There is, however, considerable empirical evidence that saddlepoint methods are useful and maintain most of their accuracy when used for nonlattice mass functions perhaps mixed with a continuous density. For example, when considering a bootstrap mass or distribution function, the continuous versions of the saddlepoint approximations fit quite well (see Davison and Hinkley, 1988 and 1997, §9.5). The bootstrap and other nonlattice examples are considered further in sections 3.4.1 and 4.3.1.

1.2.7 Relationships of the various CDF approximations

The two different CDF versions $\hat{F}_1$ and $\hat{F}_{1a}$ in the left tail are related according to a certain modification that is presented in the first subsection.

In the second subsection, the continuous CDF approximation $\hat{F}$ for a gamma is shown to be the same as the continuity-corrected survival approximation $\widehat{\Pr}_1$ for a Poisson distribution. This mimics the true relationship for these distributions that arises under the assumptions of a Poisson process.

Interconnections of $\hat{F}_1$ and $\hat{F}_{1a}$

The left-tail approximation $\hat{F}_1(k-1)$ for the Poisson distribution may be determined from $\hat{F}_{1a}(k)$ by removing the the saddlepoint mass at the boundary point k, or

$$\hat{F}_1(k-1) = \hat{F}_{1a}(k) - \widehat{\Pr}(X = k). \tag{1.43}$$

The last term is the saddlepoint mass function at k which, from (1.13), is conveniently expressed in terms of $\hat{w}$ as

$$\widehat{\Pr}(X = k) = \phi(\hat{w})/\sqrt{K''(\hat{s})}. \tag{1.44}$$

To prove this identity, use expression (1.37) for $\hat{F}_{1a}(k)$ and (1.44) so that the right side of (1.43) is

$$\Phi(\hat{w}) + \phi(\hat{w})\left(\frac{1}{\hat{w}} - \frac{1}{\tilde{u}_{1a}} - \frac{1}{\sqrt{K''(\hat{s})}}\right)$$

$$= \Phi(\hat{w}) + \phi(\hat{w})\left(\frac{1}{\hat{w}} - \frac{1}{(1-e^{-\hat{s}})\sqrt{K''(\hat{s})}}\right)$$

$$= 1 - \widehat{\Pr}_1(X \geq k) = \hat{F}_1(k-1). \tag{1.45}$$

This proves the analytical agreement expressed in (1.43).

Discrete and continuous CDF interconnections

(1) Gamma (k, λ) and Poisson (λt) The CDFs of these distributions are directly related in terms of characteristics of a Poisson process with rate parameter $\lambda > 0$. Suppose N_t is the number of arrivals up to time t, so $N_t \sim$ Poisson (λt). Let Y_k be the waiting time for the kth arrival so $Y_k \sim$ Gamma (k, λ). Then the events {the kth arrival time is before time t} and {at least k arrivals in time t} are identical, so that

$$\Pr(Y_k \leq t) = \Pr(N_t \geq k). \tag{1.46}$$

These CDFs are interrelated by way of opposite tails which raises the question as to whether any of the continuity-corrected approximations for the Poisson tail probability agree with the gamma tail approximation in the manner suggested in (1.46). The answer is yes, exact agreement is achieved for the first continuity correction. The reader is asked to show in Exercise 12 that

$$\widehat{\Pr}(Y_k \leq t) = \widehat{\Pr}_1(N_t \geq k) \tag{1.47}$$

where $\widehat{\Pr}(Y_k \leq t)$ is $\hat{F}(t)$ in (1.21) for the gamma distribution and $\widehat{\Pr}_1(N_t \geq k)$ is given in (1.27).

(2) Beta $(k, n - k + 1)$ and Binomial (n, t) A probability relationship similar to (1.46) holds which relates the CDFs of $Y_k \sim$ Beta $(k, n - k + 1)$ and $N_t \sim$ Binomial (n, t). Exercise 13 asks the reader to show that the two corresponding saddlepoint CDF approximations do not exhibit a similar relationship for any of the approximations.

1.3 Failure (hazard) rate approximation

Suppose the failure time of a system is a continuous random variable X with density f, CDF F, and CGF K. If the system has not failed at time x, the probability that it should fail in the next instant $(x, x + dx)$ is

$$\Pr\{X \in (x, x + dx) \mid X > x\} = \Pr\{X \in (x, x + dx)\} / \Pr(X > x)$$

$$= f(x)\,dx / \{1 - F(x)\}$$

$$= z(x)\,dx$$

1.3 Failure (hazard) rate approximation

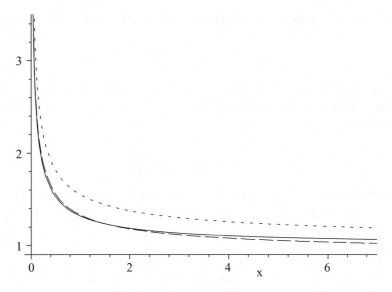

Figure 1.10. Plot of $z(x)$ (*solid*), $\bar{z}(x)$ (*dashed*), and $\hat{z}(x)$ (*dotted*) vs. x for a Gamma(1/2, 1).

where $z(x)$ is defined as the failure or hazard rate at time x. Since the failure rate depends on both the density and CDF, it may be estimated by combining either $\bar{f}$ or $\hat{f}$ with $\hat{F}$ to get

$$\bar{z}(x) = \frac{\bar{f}(x)}{1 - \hat{F}(x)}, \quad \hat{z}(x) = \frac{\hat{f}(x)}{1 - \hat{F}(x)} \quad x \in \mathcal{I}_X$$

which are the normalized and unnormalized approximations respectively.

As an example, suppose X has a Gamma $(1/2, 1)$ distribution. Figure 1.10 plots $z(x)$ (solid), $\bar{z}(x)$ (dashed), and $\hat{z}(x)$ (dotted) vs. x.

The relative error is assessed by noting that $\bar{f}(x) \equiv f(x)$ for the Gamma $(\alpha, 1)$, so

$$\frac{\bar{z}(x)}{z(x)} = \frac{1 - F(x)}{1 - \hat{F}(x)}$$

in this specific setting. The relative error of $\bar{z}(x)$ as $x \to \infty$ is the same as that of approximation $\hat{F}(x)$ which is

$$100\{\hat{\Gamma}(\alpha)/\Gamma(\alpha) - 1\} \tag{1.48}$$

or -14.22% with $\alpha = 1/2$. The figure demonstrates the superior accuracy achieved when the normalized density is used with smaller values of x up to 7. However, for very large x, the use of $\hat{f}$ makes $\hat{z}(x)$ more accurate asymptotically. For general α, both $\hat{z}(x)$ and $z(x) \to 0$ at the same rate so

$$\lim_{x \to \infty} \frac{\hat{z}(x)}{z(x)} = 1 \tag{1.49}$$

and the limiting relative error is 0%. The asymptotic bias of (1.48) for $\bar{z}$ is apparent from the figure by the fact that the dashed line continues to stay about 14% below the solid line. The asymptotic consistency of $\hat{z}$ in (1.49) is not apparent since this equivalence is achieved for quite large x.

With discrete waiting times on the integer lattice, the failure rate is defined as

$$z(k) = \frac{p(k)}{\sum_{j=k}^{\infty} p(j)}. \tag{1.50}$$

The same issues arise here as in the continuous setting with the added caveat that the denominator of (1.50) may be approximated by using the two different continuity-corrected survival estimates.

1.4 Final remarks

(1) Saddlepoint approximations, for both density/mass functions and CDFs, are usually extremely accurate over a wide range of x-values and maintain this accuracy far into the tails of the distributions. Often an accuracy of two or three significant digits in relative error is achieved.

(2) They also achieve unmatched accuracy in situations where other approximations, such as the normal (central limit) and Poisson, might have been used. When saddlepoints are explicit, such as in the binomial setting, they entail no more computational effort than a normal approximation; when not explicit, the additional burden of univariate root finding is practically negligible.

(3) The 1-1 correspondence between the density or CDF of a random variable and its CGF may be considered as a theoretical bijective mapping from the collection of CGFs $\{K(\cdot)\}$ onto the collection of densities $\{f(\cdot)\}$ or CDFs $\{F(\cdot)\}$ that admit convergent CGFs. In this context, saddlepoint methods may be viewed as highly accurate approximate mappings defined on the domain of CGFs. The nature of this approximate mapping is specified in the saddlepoint formulas themselves.

(4) Plots for univariate saddlepoint densities and CDFs *do not require root-finding* if the parametric plotting capability of Maple V is used. The graph of the density is the locus of $y = \hat{f}(x)$ vs. $x \in \mathcal{I}_X$. Upon substituting $x = K'(\hat{s})$, the graph takes on the parametric form used by Maple V when it is written as $y(\hat{s}) = \hat{f}\{K'(\hat{s})\}$ versus $x(\hat{s}) = K'(\hat{s})$ for $\hat{s} \in (a, b)$. Root-finding is not necessary since the software itself chooses a grid of saddlepoint values $\hat{s}$ that best constructs the plot. The plotting routine used by Maple V is made more efficient by selecting a higher concentration of grid points at locations where the plot has larger curvature. In the saddlepoint density context, this leads to a concentration of points near the density mode and a sparseness of points in the tails. The usage of an equally spaced grid of saddlepoints over (a, b) tends to place almost all of the points in the tails of the distribution and leaves only a few points to capture the shape in the higher portion of the density.

1.5 Computational notes

Most of the computations in this book were carried out using Maple V software and Fortran programming. The accuracy of saddlepoint computations in general is a result of the following three factors:

(1) the intrinsic accuracy of the computational routines in the software,

Table 1.15. *Extreme tail probability computations retaining 10, 20, and 30 digits in the intermediate computations.*

k	10 digits	20 digits	30 digits
7	$.0^41029$	$.0^41029$	$.0^41029$
9	$.0^61117$	$.0^61118$	$.0^61118$
11	$.0^98262$	$.0^98338$	$.0^98338$
13	$-.0^{11}4042$	$.0^{11}4530$	$.0^{11}4530$
15	$-.0^{13}1815$	$.0^{13}1871$	$.0^{13}1871$
17	$-.0^{16}6336$	$.0^{16}6073$	$.0^{16}6074$
19	$-.0^{18}1767$	$.0^{18}1533$	$.0^{18}1590$

(2) the manner in which computations were performed including such factors as the stability of various analytical expressions used, and

(3) the number of digits carried in the computations to counteract round-off error.

Issue (1) has been addressed by performing computations that might be unstable in integer arithmetic rather than floating point computation. For example, in the sum of binomials example, "exact" extreme tail probabilities were negative in the right tail when computed in floating point. For this reason they were instead computed symbolically by expanding the probability generating function and using its rational coefficients to determine exact probabilities.

Issue (3) has been addressed by increasing the number of digits of computation until there is stabilization. Table 1.15 illustrates this need in the context of computing values of $\widehat{\Pr}_2(X \geq k)$ for the Poisson (1) example. The saddlepoint is explicit in this setting but the issue is no different when implicit root finding is necessary. When carrying 10 digits in computation, the last four probabilities are negative and turn positive with 20 digits. Computation with 30 digits demonstrates that the computations have stabilized out to $k = 17$.

Generally speaking, round-off error and numerical stability can be increasingly troublesome as probabilities are computed further out in the tail. Root finders may also exhibit increasing difficulty in finding saddlepoints further into the tails. The root finders in Maple V were capable of finding all but the most extreme saddlepoints in the example convoluting a Normal (0, 1) and Laplace (0, 1). For these most extreme cases, Newton's method was successfully iterated to find the roots.

1.6 Exercises

1. If $X \sim$ Normal (μ, σ^2), show that the saddlepoint density and CDF of X are exact.
2. The Laplace (0, 1) density in (1.10) has MGF $M(s) = 1/(1 - s^2)$ for $s \in (-1, 1)$.
   (a) Show that $\hat{s}$ has the explicit form
   $$\hat{s}(x) = \frac{x}{\sqrt{y} + 1}$$
   where $y = x^2 + 1$ and the saddlepoint density is
   $$\hat{f}(x) = \frac{1}{2\sqrt{2\pi}} \sqrt{1 + 1/\sqrt{y}} \exp\left(1 - \sqrt{y}\right).$$

(b) Show that the Lugannani and Rice approximation has input values

$$\hat{w} = \mathrm{sgn}(x)\sqrt{2\left(\sqrt{y} - 1 + \ln\frac{2}{\sqrt{y}+1}\right)} \qquad \hat{u} = \frac{x}{\sqrt{1+1/\sqrt{y}}}.$$

Tabulate this approximation for $x = -3, -2, \ldots, 3$ and compare it with the exact CDF.

3. Consider table 1.3 for an Exponential (1) distribution with $\alpha = 1$.
   (a) Reproduce the table and compare the change in relative errors with the setting $\alpha = 1/2$.
   (b) Compare with a normal as a central limit approximation.
   (c) If $X \sim \mathrm{Gamma}(\alpha, 1)$, then X is an estimator of parameter α, if the parameter is assumed to be unknown. As $\alpha \to \infty$, $Z = (X - \alpha)/\sqrt{\alpha}$ converges in distribution to a Normal $(0, 1)$ so X has an approximate Normal (α, α) distribution for large α. Determine the variance stabilizing transformation and compare the resulting normal approximation to the approximations in part (a) for $\alpha = 1$.

4. The noncentral chi-square distribution is defined as the distribution of

$$X = \sum_{i=1}^{n}(X_i + a_i)^2$$

where $X_1, \ldots, X_n$ are independent and identically distributed (i.i.d.) as Normal $(0, 1)$. Designate this distribution as $X \sim \chi^2(n, \lambda)$ where $\lambda = \sum_i a_i^2$.

(a) Show that it has MGF

$$M(s) = (1 - 2s)^{-n/2} \exp\frac{\lambda s}{1 - 2s} \qquad s \in \left(-\infty, \tfrac{1}{2}\right)$$

and that the saddlepoint is explicit as the root of a quadratic and is

$$\hat{s}(x) = -\frac{1}{4x}\left\{n - 2x + \sqrt{(n^2 + 4x\lambda)}\right\} \qquad x > 0.$$

(Hougaard, 1988). Why is the saddlepoint function not specified by the negative root?

(b) Choose values for n and λ and make a comparative plot of $\hat{f}, \bar{f}$, and f using a truncated version of the true density expressed as

$$f(x) = \exp\left(-\tfrac{1}{2}\lambda\right)\left\{\sum_{i=0}^{\infty}\frac{1}{i!}\left(\frac{\lambda}{2}\right)^i \frac{x^{n/2+i-1}}{2^{n/2+i}\Gamma(n/2+i)}\right\}\exp\left(-\tfrac{1}{2}x\right), \qquad (1.51)$$

a mixture of central χ^2 densities with Poisson weights.

(c) Integrate (1.51) term by term to determine a series expression for the noncentral χ^2 CDF. Verify the following table of values for the exact and approximate computation when $n = 2$ and $\lambda = 8$:

x	$F(x)$	$\hat{F}(x)$
1	.01630	.01601
4	.1480	.1471
7	.3557	.3547

x	$1 - F(x)$	$1 - \hat{F}(x)$
10	.4351	.4360
20	.06648	.06666
30	$.0^2 5821$	$.0^2 5839$
40	$.0^3 3623$	$.0^3 3635$

5. Suppose $X = X_1 X_2$ where X_1 and X_2 are i.i.d. Normal $(0, 1)$.
   (a) Derive the MGF of X as
   $$M(s) = \frac{1}{\sqrt{1-s^2}} \qquad s \in (-1, 1).$$
   (Hint: Derive M by first conditioning on the variable X_1.)
   (b) Compute an explicit form for $\hat{s}$, $\hat{f}$, and $\hat{F}$.
   (c) Express f as a one-dimensional integral and compare $\hat{f}$, $\bar{f}$, and f using a numerical approximation for f.
   (d) Prove that $\lim_{x \to 0} f(x) = \infty$. Why is it not possible for the saddlepoint density $\hat{f}$ to reproduce the unboundedness of the true density at 0?
   (e) Express F as an integral and compare values of $\hat{F}$ and F at $x = 0.5, 1, 1.5, \ldots, 3$ using a numerical approximation to F.

6. Suppose X has a uniform density over $(-1, 0) \cup (1, 2)$ with the density
$$f(x) = \begin{cases} 1/2 & \text{if } x \in (-1, 0) \cup (1, 2) \\ 0 & \text{if } x \notin (-1, 0) \cup (1, 2). \end{cases}$$

   (a) Show the MGF of X is
   $$M(s) = \begin{cases} s^{-1}(e^s - 1) \cosh s & \text{if } s \neq 0 \\ 1 & \text{if } s = 0. \end{cases}$$

   (b) Prove that
   $$\lim_{s \to -\infty} K'(s) = -1 \qquad \lim_{s \to \infty} K'(s) = 2.$$

   (c) Show that
   $$K''(s) = \left\{ \frac{1}{s^2} - \frac{e^s}{(e^s - 1)^2} \right\} + (1 - \tanh^2 s) \qquad s \neq 0$$
   and $\lim_{s \to 0} K''(s) = 3/4$. Verify graphically that $K''(s) > 0$ for all s. Note that together, parts (b) and (c) show that K' is a bijection from $(-\infty, \infty)$ onto $(-1, 2)$.

7. Derive the saddlepoint approximation for the CDF of the Beta (α, β) distribution using the approach suggested in (1.25). Verify the explicit saddlepoint and the values for $\hat{w}$ and $\hat{u}$ in (1.26).

8. The continuity corrections in Daniels (1987) were presented only for right-tail probabilities and his survival approximations are given in (1.27) and (1.32). Derive the left-tail CDF approximations in section 1.2.5 that result from these expressions.
   (a) Suppose $k > \mu$. Daniels' (1987) first approximation is
   $$\Pr(X \geq k) \simeq \widehat{\Pr}_1(X \geq k)$$
   as given in (1.27) and (1.28). For $k < \mu$, use this expression for the right tail of $-X$ to derive the left tail approximation $\hat{F}_{1a}(k)$ for $X \sim F$ in (1.37).
   (b) Suppose $k^- = k - 1/2 > \mu$. Daniels' (1987) second approximation is
   $$\Pr(X \geq k) \simeq \widehat{\Pr}_2(X \geq k)$$
   as given in (1.32) and (1.31). For $k^+ = k + 1/2 < \mu$, use this expression for the right tail of $-X$ to derive the left tail approximation $\hat{F}_2(k)$ for $X \sim F$ in (1.39).

9. The mass function for the Negative Binomial (n, θ) is
$$p(k) = \frac{\Gamma(n+k)}{k! \Gamma(n)} \theta^n (1-\theta)^k \qquad k = 0, 1, \ldots$$

(a) Derive the MGF as
$$M(s) = \left\{\frac{\theta}{1-(1-\theta)e^s}\right\}^n \qquad s < -\ln(1-\theta).$$

(b) Show the saddlepoint is
$$\hat{s} = \ln\left\{\frac{k}{(1-\theta)(n+k)}\right\} \qquad k = 1, 2, \ldots$$
and that
$$\hat{p}(k) = \frac{\hat{\Gamma}(n+k)}{\hat{k}!\hat{\Gamma}(n)}\theta^n(1-\theta)^k$$
where $\hat{k}!$ and $\hat{\Gamma}(n)$ are defined in (1.16) and (1.8) respectively.

(c) Show that for $k > (1-\theta)(n+k)$,
$$\hat{w} = \sqrt{2\left\{k\ln\frac{k}{(1-\theta)(n+k)} + n\ln\frac{n}{\theta(n+k)}\right\}}$$
$$\tilde{u}_1 = \left\{1 - \frac{(1-\theta)(n+k)}{k}\right\}\sqrt{\frac{k(n+k)}{n}}$$
for approximation $\widehat{\Pr}_1$. For $\widehat{\Pr}_2$, $\tilde{w}_2$ is the same as $\hat{w}$ but based on $k^- = k - 1/2$ instead of k, and
$$\tilde{u}_2 = \left(\sqrt{\tilde{t}} - \sqrt{\frac{1}{\tilde{t}}}\right)\sqrt{\frac{k^-(n+k^-)}{n}} \qquad \tilde{t} = \frac{k^-}{(1-\theta)(n+k^-)}.$$

(d) Tabulate the true values $\Pr(X \geq k)$ and approximations $\widehat{\Pr}_i(X \geq k)$ for $i = 1, 2, 3$ and $k = 1, 4, 7, \ldots, 28$ using $\theta = 0.2$ and $n = 2$. Compare with the entries of table 1.14. In such a setting, the normal approximation should not be very accurate. Verify this using a continuity correction on the central limit approximation for large n. Compare the accuracy of the saddlepoint approximations with the central limit approximation.

(e) Repeat part (d) with $\theta = 1/2$ and $n = 10$. Verify that the normal approximation has improved and compare it to the saddlepoint approximations.

10. Suppose $X \sim \text{Binomial}(n, \theta)$ where $\theta < 1/2$. According to Exercise 8, a left-tail approximation to the CDF of X is based on the right-tail approximation of variable $-X$. It would seem more natural in this instance to work instead with the right tail of $n - X \sim \text{Binomial}(n, 1-\theta)$. Show, if one does so, that the continuity corrections for the left tail of X remain the same.

11. The Logarithmic Series (θ) distribution with $0 < \theta < 1$ has mass function
$$p(k) = \frac{-\theta^k}{k\ln(1-\theta)} \qquad k = 1, 2, \ldots \qquad (1.52)$$

(a) Show that (1.52) is indeed a mass function for $0 < \theta < 1$.
(b) Show the MGF is
$$M(s) = \frac{\ln(1-\theta e^s)}{\ln(1-\theta)} \qquad s \in (-\infty, -\ln\theta).$$

(c) Choose a value of θ and tabulate $p(k)$, $\bar{p}(k)$, and $\hat{p}(k)$ for sufficiently many values of k that the first four digits of $\bar{p}(k)$ stabilize. Plot the relative errors of $\bar{p}$ and $\hat{p}$.

(d) For the value of θ in part (c), compare the three continuity-corrected approximations to the exact survival function computed by summing (1.52). Take care in the summing since round-off error may be a problem.

(e) Prove that
$$\lim_{s \to -\infty} K'(s) = 1 \qquad \lim_{s \to -\ln\theta} K'(s) = \infty.$$

(f) Use Maple to show that
$$K''(s) = -\theta e^s \frac{\ln(1-\theta e^s) + \theta e^s}{(1-\theta e^s)^2 \ln^2(1-\theta e^s)}.$$

Show that $K''(s) > 0$ for any $s \in (-\infty, -\ln\theta)$. (Hint: Use Taylor expansions to show that
$$e^x < \frac{1}{1-x} \qquad x \in (0,1)$$

and hence that $\ln(1-x) + x < 0$.) This, combined with part (e), proves that K' is a bijection from $(-\infty, -\ln\theta) \leftrightarrow (1, \infty)$.

12. Show that the equality in (1.47) holds when comparing approximate tail probabilities for the Gamma (k, λ) with the first continuity correction $\widehat{\Pr}_1$ applied to the Poisson (λt). A helpful fact in showing this is that $Y_k \sim$ Gamma $(k, \lambda) \Rightarrow \lambda Y_k \sim$ Gamma $(k, 1)$.

13. Consider a Poisson process with rate parameter 1 and let Y_k and Y_{n+1} be the waiting times for the kth and $(n+1)$st occurrences for $k \leq n$. Show the following results:
    (a) The conditional distribution of Y_k given $Y_{n+1} = 1$ has a Beta $(k, n-k+1)$ distribution.
    (b) If N_t is the number of occurrences at time $t < 1$, then $N_t | Y_{n+1} = 1 \sim$ Binomial (n, t). Events $Y_k \leq t$ and $N_t \geq k$ are the same so the two conditional distributions are related according to
    $$\Pr(Y_k \leq \theta | Y_{n+1} = 1) = \Pr(N_\theta \geq k | Y_{n+1} = 1)$$
    or
    $$B(k, n-k+1)^{-1} \int_0^\theta y^{k-1}(1-y)^{n-k} \, dy = \sum_{j=k}^n \binom{n}{j} \theta^j (1-\theta)^{n-j},$$
    where $B(k, n-k+1)$ is the beta function.
    (c) None of the continuity-corrected approximations for the binomial tail probability are related analytically to the saddlepoint CDF of the beta in (1.26).

14. A Neyman type A (λ, ϕ) distribution (Johnson and Kotz, 1969) is a contagious distribution constructed as follows. Suppose X given $Y = j$ has a Poisson $(j\phi)$ distribution and $Y \sim$ Poisson (λ). The marginal distribution of X is a compound distribution and is Neyman type A.
    (a) Show that the marginal CGF of X is
    $$K(s) = -\lambda[1 - \exp\{-\phi(1-e^s)\}] \qquad s \in (-\infty, \infty). \tag{1.53}$$
    (b) Use the recurrence relation for the mass function
    $$p(k) = \frac{\lambda\phi}{ke^\phi} \sum_{j=0}^{k-1} \frac{\phi^j}{j!} p(k-j-1)$$
    given in Johnson and Kotz (1969, chap. 9, Eq. 5.1), along with
    $$p(0) = \exp\{-\lambda(1-e^{-\phi})\}$$
    to determine the mass function for some chosen values of λ and ϕ.
    (c) The probability generating function of X is defined as $P(t) = E(t^X)$ and takes the value
    $$P(t) = \exp(-\lambda[1 - \exp\{-\phi(1-t)\}]) \qquad t > 0$$

in this setting. By expanding $P(t)$ in a Taylor series about $t = 0$, show in general that
$$p(k) = \frac{P^{(k)}(0)}{k!} \qquad k = 0, 1, \ldots$$
Compute these values in Maple for the chosen values of λ and ϕ and compare them to the recursive computations of part (b).

(d) Compare $\hat{p}$ and $\bar{p}$ to the values obtained in parts (b) and (c).

(e) Use Maple to compute the CDF in integer arithmetic from part (c). Compare these exact results with the three continuity-corrected CDF approximations and plot their relative errors.

15. Solutions to the Poisson saddlepoint equation for noninteger values of k are not meaningful in approximating the Poisson mass function. However, consider approximating a continuous density that is proportional to the Poisson mass function at integer values. Such a density is described in Johnson and Kotz (1970, p. 289) as
$$f(x) = \frac{c_\lambda}{\Gamma(x+1)} \lambda^x e^{-\lambda} \qquad x \in (0, \infty), \tag{1.54}$$
where c_λ is the appropriate normalization constant for $\lambda > 0$.

(a) In determining a saddlepoint approximation to (1.54), is it sufficient to replace k with x in (1.15), and determine the saddlepoint approximation as
$$\hat{f}(x) = \hat{p}(x) \qquad x \in (0, \infty)?$$
Why or why not? If not, what is the saddlepoint approximation?

(b) Compare the normalized suggestion(s) in (a) to the true density.

16. Suppose that X has a Negative Binomial $(2, .1)$ distribution. Plot the true failure rate $z(k)$ and the various estimated failure rates $\hat{z}(k)$ and $\bar{z}(k)$ using both of the continuity-corrected survival functions.

17. Consider a single server queue in which each customer is serviced in fixed time T. Let customers arrive according to a Poisson process with rate $\lambda > 0$, where $\lambda T < 1$. If the queue initially starts with n customers, then the dynamics of the queue determine a unique distribution for the number of customers X serviced before the queue reaches length 0. This distribution is the Borel–Tanner (n, α) mass function with $\alpha = \lambda T$ as described in Johnson and Kotz (1969, §10.6) with mass function
$$p(k) = \frac{n}{(k-n)!} k^{k-n-1} \alpha^{k-n} e^{-\alpha k} \qquad k \geq n.$$
Haight and Breuer (1960) have determined that an implicit relationship holds for the MGF of X. If M is the MGF and K is the CGF, then they specify K as solving
$$K(s) = n\alpha \sqrt[n]{M(s)} + n(s - \alpha).$$

(a) Determine the convergence region for M.

(b) Derive expressions for the various saddlepoint ingredients used to compute the saddlepoint density.

(i) Show that
$$K(\hat{s}) = n \ln\left(\frac{k-n}{k\alpha}\right) \qquad k \geq n+1$$
so that
$$\hat{s} = (\alpha - \ln \alpha) - \{1 - n/k - \ln(1 - n/k)\}.$$

(ii) Show that
$$K''(\hat{s}) = k^2 (k-n)/n^2.$$

(c) Derive an expression for the saddlepoint mass function and show it satisfies the property that
$$\hat{p}(k) = \frac{(k-n)!}{\widehat{(k-n)!}} p(k) \qquad k \geq n+1.$$

(d) Derive simple expressions for the two continuity-corrected survival approximations. These expressions should be explicit in k and k^-.

(e) Specify values for n and α and plot the various survival approximations along with their true counterparts.

(f) If $\alpha > 1$, then there is a nonzero probability that X is unbounded. In this case,
$$\Pr(X = \infty) = 1 - (\beta/\alpha)^n,$$
where $\beta < 1$ is the root of
$$\beta e^\beta = \alpha e^\alpha$$
that is less than 1. Repeat the analysis in (a)–(d) but, in order that M is convergent in a neighborhood of $s = 0$, perform the analysis conditional on the event that $X < \infty$.

2

Properties and derivations

Chapter 1 introduced expressions which define the various saddlepoint approximations along with enough supplementary information to allow the reader to begin making computations. This chapter develops some elementary properties of the approximations which leads to further understanding of the methods. Heuristic derivations for many of the approximations are presented.

2.1 Simple properties of the approximations

Some important properties possessed by saddlepoint density/mass functions and CDFs are developed below. Unless noted otherwise, the distributions involved throughout are assumed to have MGFs that are convergent on open neighborhoods of 0.

The first few properties concern a linear transformation of the random variable X to $Y = \sigma X + \mu$ with $\sigma \neq 0$. When X is discrete with integer support, then Y has support on a subset of the σ-lattice $\{\mu, \mu \pm \sigma, \mu \pm 2\sigma, \ldots\}$. The resulting variable Y has a saddlepoint mass and CDF approximation that has not been defined and there are a couple of ways in which to proceed. The more intriguing approach would be based on the inversion theory of the probability masses, however, the difficulty of this approach places it beyond the scope of this text. A more expedient and simpler alternative approach is taken here which adopts the following convention and which leads to the same approximations.

Lattice convention. The saddlepoint mass function and CDF approximation for lattice variable Y, with support in $\{\mu, \mu \pm \sigma, \mu \pm 2\sigma, \ldots\}$ for $\sigma > 0$, are specified in terms of their equivalents based on $X = (Y - \mu)/\sigma$ with support on the integer lattice.

Note that the ordering of masses has not been reversed by taking $\sigma < 0$. This is important in order to maintain uniqueness of the probability computation as discussed below under invariance. The lattice convention is relaxed in section 2.4.4 as it pertains to continuity-corrected CDF approximation, where the various approximations on a δ-lattice are catalogued.

2.1.1 Equivariance

Proposition 2.1.1 *Saddlepoint density functions are equivariant under linear transformation. The terminology refers to the property in which the two saddlepoint densities are related according to the usual method of Jacobian transformation.*

Proof. Suppose X has CGF K, so that $Y = \sigma X + \mu$ has CGF $K_{\mu,\sigma}(s) = \mu s + K(\sigma s)$. The saddlepoint density of Y at y, or $\hat{f}(y)$, has saddlepoint $\hat{s}_y$ solving the equation

$$x = (y - \mu)/\sigma = K'(\sigma \hat{s}_y). \tag{2.1}$$

This is the same equation as $x = K'(\hat{s}_x)$ which defines the saddlepoint value $\hat{s}_x$ when determining the saddlepoint density of X at x, or $\hat{f}(x)$. Uniqueness of the roots in both instances implies that $\hat{s}_x$ and $\hat{s}_y$ are related by $\hat{s}_y = \hat{s}_x/\sigma$. It is now an exercise (Exercise 1) to show that $\hat{f}(y)$ is related to the saddlepoint density of X at x according to

$$\hat{f}(y) = \hat{f}(x)/\sigma. \tag{2.2}$$

□

This property would regrettably also hold for lattice variables if they were not constrained by the adoption of the lattice convention above. For example, if $X \sim$ Poisson (λ) and $Y := X/4$, then direct application of the mass function expression (1.13) to approximate the mass function of Y at value $k/4$ yields

$$\widehat{\Pr}(Y = k/4) = 4\hat{p}(k) \tag{2.3}$$

where $\hat{p}(k)$ is the saddlepoint mass of X at k. Since $\sum_k \hat{p}(k) \simeq 1$, expression (2.3) must sum to about 4. Indeed $Y_m = X/m$ would have a saddlepoint mass function determined from (1.13) which sums to about m, if the lattice convention were not adopted. Thus the indiscriminate use of (1.13), without regard for the defined convention, leads to an inaccurate computation; the equivariance property would hold regardless of whether the variable is discrete or continuous.

Examples

(1) Suppose that $Y = \sigma X + \mu \sim$ Normal (μ, σ^2). The saddlepoint density for X is $\phi(x)$ so, by the equivariance in (2.2),

$$\hat{f}(y) = \frac{1}{\sigma} \phi\left(\frac{y - \mu}{\sigma}\right)$$

and is exact.

(2) If $Y = X/\beta \sim$ Gamma (α, β) with mean α/β and variance α/β^2, then,

$$\hat{f}(y) = \frac{\Gamma(\alpha)}{\hat{\Gamma}(\alpha)} f(y),$$

where $f(y)$ is the true density.

2.1.2 Symmetry

Proposition 2.1.2 *If random variable X has a symmetric density/mass function about μ, then the saddlepoint density/mass function exhibits the same symmetry.*

Proof. Suppose that $\mu = 0$. The distribution has a CGF K that is an even function which converges on a open interval symmetric about 0. Derivative K' must be an odd function which leads to $\hat{s}(-x) = -\hat{s}(x)$. The terms in the saddlepoint density, $K(\hat{s})$, $\hat{s}x$, and $K''(\hat{s})$, are all even so the density must also be even.

When $\mu \neq 0$ then $X - \mu$ has a saddlepoint density symmetric about 0. According to the equivariance property or the lattice convention, $X = \mu + (X - \mu)$ has this same symmetric saddlepoint density/mass function translated so it is centered about μ. In this instance, the saddlepoint for X exhibits the inflective relationship $\hat{s}(x) = -\hat{s}(2\mu - x)$, because x is the same distance $x - \mu$ above the center μ as $2\mu - x$ is below. □

Proposition 2.1.3 *The following saddlepoint CDFs exhibit the same symmetry as their true underlying symmetric CDFs:*

(i) the continuous CDF approximation in (1.21),
(ii) continuity-corrected survival approximations $\widehat{\Pr}_2$ and $\widehat{\Pr}_3$ in (1.32) and (1.39), and
(iii) CDF approximation

$$\begin{array}{ll} \hat{F}_{1a}(k) & \text{if } k \leq \mu \\ 1 - \widehat{\Pr}_1(X \geq k+1) & \text{if } k+1 \geq \mu, \end{array} \quad (2.4)$$

where $\hat{F}_{1a}$ is given in (1.37) and $\widehat{\Pr}_1$ is given in (1.27).

Unlike (2.4), the first continuity correction $\widehat{\Pr}_1$, which uses the same formula in both tails, does not have this symmetric property.

Proof. See Exercise 2. □

The symmetric property exhibited by $\widehat{\Pr}_2$ and the lack of such for $\widehat{\Pr}_1$ supplies a theoretical reason for preferring $\widehat{\Pr}_2$.

Examples

(1) The Normal (μ, σ^2) example above and the normal–Laplace convolution in chapter 1 provide examples. For the latter, the symmetry is about $\mu = 0$ and the saddlepoint is an odd function as shown in figure 1.1.

(2) Suppose $X \sim$ Binomial $(10, 1/2)$ and let $\hat{s}(x)$ denote the solution to $K'(\hat{s}) = x$ where K is the CGF of X. Then continuity-corrected $\tilde{s}(k)$ is determined from (1.41) for $k = 1, \ldots, 4$ as

$$\tilde{s}^+(k) = \hat{s}(k^+) = -\hat{s}\{(10-k)^-\} = -\tilde{s}(10-k).$$

Consequently,

$$\widehat{\Pr}_i(X \leq k) = \widehat{\Pr}_i(X \geq 10 - k) \quad i = 2, 3.$$

The same identity holds for the CDF in (2.4) but does not hold for $i = 1$ since, for example,

$$\widehat{\Pr}_1(X \geq 6) = 0.3805 = \hat{F}_{1a}(4) \qquad \widehat{\Pr}_1(X \leq 4) = 0.3738.$$

2.1.3 Invariance

Proposition 2.1.4 *Probabilities for continuous X, computed by integrating the saddlepoint density or using the saddlepoint CDF approximation, are invariant under linear transformation.*

Proposition 2.1.5 *Probabilities for lattice variable X, determined by summing its mass function, are invariant under linear transformation. Those from any of the three continuity-corrected CDFs, are invariant under increasing linear transformation.*

Proof of propositions 2.1.4 and 2.1.5. Invariance in the case of lattice variables is true as a result of the specified convention. When $Y = \sigma X + \mu$ transforms continuous variables, the invariance as a result of integrating densities is self evident from the equivariance property. For continuous CDFs, one needs to compare the approximation to $\Pr(Y \leq y)$, using saddlepoint inputs $\hat{w}_y$ and $\hat{u}_y$, with the approximation to $\Pr(X \leq x)$, with inputs $\hat{w}_x$ and $\hat{u}_x$. In Exercise 1, the reader is asked to show that $\hat{w}_y = \hat{w}_x$ and $\hat{u}_y = \hat{u}_x$ so their saddlepoint approximations are the same. □

Example Consider figure 1.7 in which relative errors in approximating $F_{\alpha, 0.8}(0.4) = \Pr\{\text{Beta}(\alpha, 0.8) \leq 0.4\}$ are plotted against $\alpha \in (0, 1)$. Since $1 - \text{Beta}(\alpha, 0.8)$ has distribution Beta $(0.8, \alpha)$, the relative error plot obtained from varying the other parameter would be exactly the same.

Proposition 2.1.6 *Consider invariance of the continuity-corrected CDF approximations under decreasing linear transformations. Approximations $\widehat{\Pr}_2$ and $\widehat{\Pr}_3$ are invariant whereas $\widehat{\Pr}_1$ is not.*

Proof. See Exercise 2. □

Example Consider $X \sim \text{Binomial }(23, 0.2)$ and the transformation $Y = 23 - X \sim \text{Binomial }(23, 0.8)$. The exact computation has

$$\Pr(X \leq 3) = 0.2965 = \Pr(Y \geq 20).$$

The invariant continuity-corrected approximations that use the CGFs for X and Y yield

$$\widehat{\Pr}_2 (X \leq 3) = 0.2958 = \widehat{\Pr}_2 (Y \geq 20)$$
$$\widehat{\Pr}_3 (X \leq 3) = 0.2984 = \widehat{\Pr}_3 (Y \geq 20)$$

when computed from left-tail and right-tail approximations respectively. The lack of invariance when using the first continuity correction may be seen from

$$\widehat{\Pr}_1 (X \leq 3) = 0.2961 \neq 0.3009 = \widehat{\Pr}_1 (Y \geq 20).$$

2.2 Saddlepoint density

The continuous saddlepoint density in (1.4) is derived below. The arguments make use of Laplace's approximation, which is introduced first.

2.2.1 Laplace's approximation

Suppose that $g(x)$ is a smooth function on (c, d) with a global minimum at interior point $\hat{x} \in (c, d)$. Then

$$\int_c^d e^{-g(x)} dx \simeq \frac{\sqrt{2\pi} e^{-g(\hat{x})}}{\sqrt{g''(\hat{x})}}. \tag{2.5}$$

The idea here is that the integral value is largely determined by the local properties of g at critical value $\hat{x}$ as expressed through its value $g(\hat{x})$ and curvature $g''(\hat{x})$ at that point. As a general guideline, although there may be exceptions, the sharper the peak at $\hat{x}$, as exhibited in the plot of $\exp\{-g(x)\}$ vs. $x \in (c, d)$, the greater the accuracy that can be anticipated with the approximation. Some examples are considered before giving a proof.

Examples

Normal integral The integral

$$I = \int_{-\infty}^{\infty} e^{-\frac{1}{2}x^2} dx$$

has the Laplace approximation $\sqrt{2\pi}$, which is exact, when using $g(x) = x^2/2$.

Gamma function Define

$$\Gamma(\alpha) = \int_0^{\infty} y^{\alpha-1} e^{-y} dy$$

$$= (\alpha - 1)^{\alpha} \int_0^{\infty} \exp\{-(\alpha - 1)(x - \ln x)\} dx$$

upon transforming $x = y/(\alpha - 1)$. Taking $g(x) = (\alpha - 1)(x - \ln x)$, then $\hat{x} = 1$ locates the minimum of g when $\alpha > 1$ and locates the maximum when $\alpha < 1$. The approximation is

$$\Gamma(\alpha) \simeq (\alpha - 1)^{\alpha} \frac{\sqrt{2\pi} e^{-(\alpha-1)}}{\sqrt{\alpha - 1}} = (\alpha - 1)\hat{\Gamma}(\alpha - 1). \tag{2.6}$$

This expression is meaningful and justified as a Laplace approximation only when $\alpha > 1$. Since $\Gamma(\alpha) = (\alpha - 1)\Gamma(\alpha - 1)$, the approximation achieves the accuracy of Stirling's approximation at $(\alpha - 1)$.

A change of variable in the integration before applying the approximation leads to a more accurate approximation. Substitute $dx = y^{-1} dy$ as satisfied by the change of variables $x = \ln(y/\alpha)$. Then

$$\Gamma(\alpha) = \alpha^{\alpha} \int_{-\infty}^{\infty} \exp\{\alpha(x - e^x)\} dx \tag{2.7}$$

and $g(x) = \alpha(e^x - x)$. The minimum is at $\hat{x} = 0$ for all $\alpha > 0$ and Laplace's approximation is

$$\Gamma(\alpha) \simeq \hat{\Gamma}(\alpha) \tag{2.8}$$

and justified for all $\alpha > 0$. With this change of variable, the Laplace and Stirling approximations are the same for argument α. The change of variable has also increased the accuracy, since Stirling's approximation at α is more accurate than at $\alpha - 1$.

Beta (α, β) function This function is defined as

$$B(\alpha, \beta) = \frac{\Gamma(\alpha)\Gamma(\beta)}{\Gamma(\alpha+\beta)} = \int_0^1 y^{\alpha-1}(1-y)^{\beta-1}\,dy. \tag{2.9}$$

The integrand achieves a maximum at $\hat{y} = (\alpha-1)/(\alpha+\beta-2)$ when $\alpha > 1$ and $\beta > 1$. Direct Laplace approximation gives

$$B(\alpha, \beta) \simeq \frac{(\alpha-1)\hat{\Gamma}(\alpha-1)(\beta-1)\hat{\Gamma}(\beta-1)}{(\alpha+\beta-2)^2\hat{\Gamma}(\alpha+\beta-2)}. \tag{2.10}$$

The approximation may be improved with a change of variable. Take $dx = y^{-1}(1-y)^{-1}\,dy$ so that

$$x = \int_{1/2}^y t^{-1}(1-t)^{-1}\,dt$$

$$= \ln\{y/(1-y)\}.$$

The integral becomes

$$B(\alpha, \beta) = \int_{-\infty}^{\infty} \exp\{\alpha x - (\alpha+\beta)\ln(1+e^x)\}\,dx \tag{2.11}$$

and the integrand achieves a maximum at $\hat{x} = \ln(\alpha/\beta)$ for any $\alpha > 0$ and $\beta > 0$. With this change of variable, the approximation is

$$B(\alpha, \beta) \simeq \frac{\hat{\Gamma}(\alpha)\hat{\Gamma}(\beta)}{\hat{\Gamma}(\alpha+\beta)} = \hat{B}(\alpha, \beta) \tag{2.12}$$

Approximation $\hat{B}(\alpha, \beta)$ is more accurate than the expression in (2.10).

Proof of (2.5). Smooth function g has Taylor expansion about $\hat{x}$ given by

$$g(x) = g(\hat{x}) + g'(\hat{x})(x-\hat{x}) + \tfrac{1}{2}g''(\hat{x})(x-\hat{x})^2 + R$$

where R is the remainder. Since $\hat{x}$ locates a local minimum, $g'(\hat{x}) = 0$ and $g''(\hat{x}) > 0$. Ignoring R for the moment, the integral is

$$e^{-g(\hat{x})}\int_c^d \exp\left\{-\tfrac{1}{2}g''(\hat{x})(x-\hat{x})^2\right\}dx \simeq e^{-g(\hat{x})}\sqrt{2\pi/g''(\hat{x})}, \tag{2.13}$$

since the left side of (2.13) is approximately a normal integral. A more rigorous development that accounts for R is outlined in section 2.5.1. □

Remarks on asymptotics

(1) Often the approximation is stated as an asymptotic expansion as a parameter $n \to \infty$. Typically $g(x)$ is replaced with $ng(x)$ in the exponent. Larger values of n concentrate the

peak of $\exp\{-ng(x)\}$ about a stationary maximum $\hat{x}$. In this setting, the reader is asked to show in Exercise 3 that

$$\int_c^d e^{-ng(x)}dx = \frac{\sqrt{2\pi}e^{-ng(\hat{x})}}{\sqrt{ng''(\hat{x})}}\{1 + O(n^{-1})\}. \tag{2.14}$$

(2) A more careful handling of the remainder term R leads to an expansion for figuring the $O(n^{-1})$ term in (2.14). This term is

$$O(n^{-1}) = \frac{1}{n}\left(\frac{5}{24}\hat{k}_3^2 - \frac{1}{8}\hat{k}_4\right) + O(n^{-2}), \tag{2.15}$$

where

$$\hat{k}_3 = \frac{g^{(3)}(\hat{x})}{\{g''(\hat{x})\}^{3/2}} \qquad \hat{k}_4 = \frac{g^{(4)}(\hat{x})}{\{g''(\hat{x})\}^2}. \tag{2.16}$$

The argument for this has been given in Tierney and Kadane (1986) and is also developed in section 2.5.1.

Examples

Gamma function The $O(\alpha^{-1})$ term in the expansion for $\Gamma(\alpha)$ in (2.7) is

$$\alpha^{-1}\left(\tfrac{5}{24} - \tfrac{1}{8}\right) + O(\alpha^{-2}) \tag{2.17}$$

so that

$$\Gamma(\alpha) = \hat{\Gamma}(\alpha)\{1 + \tfrac{1}{12}\alpha^{-1} + O(\alpha^{-2})\},$$

which is the Stirling expansion in Abramowitz and Stegun (1972, 6.1.37).

Beta function The higher-order terms for the $B(\alpha, \beta)$ approximation determined from (2.11) are

$$O(\alpha^{-1} + \beta^{-1}) = \tfrac{1}{12}\{\alpha^{-1} + \beta^{-1} - (\alpha + \beta)^{-1}\} + O(\alpha^{-2} + \beta^{-2}), \tag{2.18}$$

as the reader is asked to show in Exercise 4.

(3) Laplace's approximation is sometimes stated with an extra positive factor $h(x) > 0$ in the integrand. In this case,

$$\int_c^d e^{-g(x)}h(x)\,dx \simeq \frac{\sqrt{2\pi}e^{-g(\hat{x})}}{\sqrt{g''(\hat{x})}}h(\hat{x}), \tag{2.19}$$

where $\hat{x}$ locates the minimum of $g(x)$ over (c, d). If h is relatively diffuse and flat over the neighborhood of $\hat{x}$ in which $\exp\{-g(x)\}$ is peaked, then (2.19) should generally achieve comparable accuracy with (2.5). When $g(x)$ in (2.19) is replaced with $ng(x)$, so that g'' is also replaced by ng'', then the approximation achieves relative accuracy $O(n^{-1})$ and the $O(n^{-1})$ term in this expansion is

$$n^{-1}\left\{\left(\frac{5\hat{k}_3^2}{24} - \frac{\hat{k}_4}{8}\right) + \frac{h''(\hat{x})}{2h(\hat{x})g''(\hat{x})} - \frac{\hat{k}_3 h'(\hat{x})}{2h(\hat{x})\sqrt{g''(\hat{x})}}\right\} + O(n^{-2}). \tag{2.20}$$

2.2 Saddlepoint density

The terms $\hat{\kappa}_3$ and $\hat{\kappa}_4$ are computed from g as in (2.16). This expansion term along with its approximation can be derived in two ways. The more routine argument makes use of a Taylor expansion of $h(x)$ about $h(\hat{x})$ and is left as Exercise 3. The more interesting route is the change of variables argument based on $dy = h(x) dx$. Upon integrating dy, the simpler approximation in (2.14) and (2.15) may be used. This simpler expansion in $\hat{y}$ turns out to be the same as expansion (2.19), a fact noted in the proposition below.

Examples The $\Gamma(\alpha)$ function may be approximated by using (2.19) with

$$h(x) = x^{-1} \qquad e^{-g(x)} = x^\alpha e^{-x}. \tag{2.21}$$

Also $B(\alpha, \beta)$ may be approximated with

$$h(x) = x^{-1}(1-x)^{-1} \qquad e^{-g(x)} = x^\alpha (1-x)^\beta. \tag{2.22}$$

These two approaches result in the more accurate approximations of (2.8) and (2.12) based upon Stirling approximations. In addition, the expansion terms determined from (2.20) are the same as those given in (2.17) and (2.18) as the reader is asked to show in Exercise 4.

Proof of (2.19). Change variables $dy = h(x) dx$ so that $y = H(x)$ is the antiderivative of h. The integral becomes

$$\int_{H(c)}^{H(d)} \exp[-g\{H^{-1}(y)\}] dy \tag{2.23}$$

to which (2.5) may be applied. The critical value is the root of

$$\frac{dg\{H^{-1}(y)\}}{dy} = \frac{dg/dx}{dy/dx} = \frac{g'(x)}{h(x)}.$$

If $\hat{x}$ is the root of g' and the value used in (2.19), then $\hat{y} = H(\hat{x})$ must be the critical value in variable y. Further differentiation gives

$$\left. \frac{d^2 g\{H^{-1}(y)\}}{dy^2} \right|_{x=\hat{x}} = \frac{g''(\hat{x})}{h(\hat{x})^2}.$$

Now approximation (2.5) applied in variable y results in (2.19) when expressed in terms of $\hat{x}$. An outline for the derivation of the expansion term (2.20), as the term (2.15) for the integral in (2.23), is given in section 2.5.1. □

Proposition 2.2.1 *The Laplace expansion of the integral (2.23), applied in variable y as in (2.5), is identical to its more complicated Laplace expansion in (2.19) and (2.20) applied in variable x.*

Proof. Suppose these two expansions, with $ng(x)$ replacing $g(x)$, are respectively

$$I = \hat{I}_y\{1 + n^{-1} b_1(\hat{y}) + n^{-2} b_2(\hat{y}) + \cdots\}$$
$$I = \hat{I}_x\{1 + n^{-1} a_1(\hat{x}) + n^{-2} a_2(\hat{x}) + \cdots\}.$$

The argument above has shown that $\hat{I}_y = \hat{I}_x$. To show equality of the next pair of terms, write

$$b_1(\hat{y}) = \lim_{n \to \infty} n(I - \hat{I}_y) = \lim_{n \to \infty} n(I - \hat{I}_x) = a_1(\hat{x}),$$

so that the $O(n^{-1})$ terms agree. This same result is also shown in section 2.5.1 using direct expansions. Induction on this argument now gives that $b_i(\hat{y}) = a_i(\hat{x})$ for each $i = 1, 2, \ldots$ so the expansions are identical. □

The discussion indicates that expressions (2.19) and (2.20) are actually unnecessary since they are (2.5) and (2.15) using the change of variables $dy = h(x) dx$. However, the former expressions are still presented because they are often convenient when performing the computations.

The examples and discussion have shown that the accuracy of Laplace's approximation in (2.5) varies according to the particular variable of integration. The best choice of this variable, in each example, has to do with finding a variable of integration in which the integral more closely approximates a normal integral in some sense. For the gamma function, in its role as the normalization constant for the gamma density, the transformation $y = \ln x$ is variance stabilizing with respect to a scaling parameter (but not α). The resulting log-gamma density is known to have a more "normal" shape than that of the original gamma; therefore integration in variable y might be expected to result in less error than with variable x. The same argument applies to the beta integral. If X is a Beta (α, β) random variable constructed from independent gammas Y_α and Y_β as

$$X = \frac{Y_\alpha}{Y_\alpha + Y_\beta},$$

then

$$W = \ln \frac{X}{1 - X} = \ln Y_\alpha - \ln Y_\beta$$

is a convolution of the two independent variables $\ln Y_\alpha$ and $\ln Y_\beta$ whose distributions are more normally shaped. The linear transformation $Z = \{W - \ln(\alpha/\beta)\}/2$ has distribution $\frac{1}{2} \ln F_{2\alpha, 2\beta}$ and is referred to as the z-distribution of Fisher (1990, *Statistical Methods and Scientific Inference*, §47). It is a well-known fact that Z is much closer to normality than X; see Johnson and Kotz (1970, p. 80) and Aroian (1939).

The discussion indicates that the best variable of integration when using (2.5) is equivalent to some optimal apportionment of the integrand into factors $h(x)$ and $\exp\{-g(x)\}$ for use in (2.19). In the examples, these choices have been determined on a case by case basis. It remains to be seen whether a more general theory can be devised to successfully guide these choices. However, variance stabilizing transformations are known to improve the performance of normal approximations and have proved to be successful in our examples and elsewhere; see the additional discussion in Butler and Wood (2002).

In practical settings, Laplace approximations are often used as an alternative to the exact computation in instances where the true integral value is difficult to determine. As a result, there is a need to determine a variable of integration for Laplace that leads to an accurate approximation without also performing the exact computation. In addition, a method for assessing the accuracy of the Laplace approximation without knowledge of the exact result would be useful. Some guidance in these matters is provided by computing the second-order correction terms in the expansions. Consider the gamma function for example. The correction term in variable $y = \ln x$ is reported as $\alpha^{-1}/12$ in (2.17), whereas in variable x it is $(\alpha - 1)^{-1}/12$ for $\alpha > 1$. Indeed, the correction term in y is smaller than its counterpart in x uniformly in $\alpha > 1$. As a second example, consider the beta function. The two second-order corrections in variables $y = \ln\{x/(1-x)\}$ and x are

$$\tfrac{1}{12}\{\alpha^{-1} + \beta^{-1} - (\alpha + \beta)^{-1}\}$$
$$\tfrac{1}{12}\{(\alpha - 1)^{-1} + (\beta - 1)^{-1} - 13(\alpha + \beta - 2)^{-1}\}$$

respectively with the latter requiring $\alpha > 1 < \beta$. These values are 0.0734 and -0.472 with $\alpha = 1.5$ and $\beta = 2$. The magnitude of the second-order correction may often reflect the accuracy of the first-order approximation although the extent to which this correspondence holds is not well-known.

2.2.2 Derivation of the saddlepoint density

The saddlepoint density approximation is often stated as it applies to the density of $\bar{X} = n^{-1}\sum_{i=1}^{n} X_i$ where $X_1, \ldots, X_n$ are independent and identically distributed (i.i.d.) random variables with common CGF K. In the statement of approximation, the saddlepoint density is the leading term of an asymptotic expansion as $n \to \infty$ of the form

$$f(\bar{x}) = \hat{f}(\bar{x})\{1 + O(n^{-1})\}, \tag{2.24}$$

where the term $O(n^{-1})$ is the relative error of the asymptotic order indicated. The saddlepoint density in (2.24) is

$$\hat{f}(\bar{x}) = \sqrt{\frac{n}{2\pi K''(\hat{s})}} \exp\{nK(\hat{s}) - n\hat{s}\bar{x}\}, \tag{2.25}$$

where $\hat{s}$ solves the saddlepoint equation

$$K'(\hat{s}) = \bar{x}.$$

This expression is easily derived from (1.4), when applied to random variable $\bar{X}$, as suggested in Exercise 9.

Note that while (2.25) was derived from (1.4), expression (1.4) is the special case of (2.25) which takes $n = 1$. Thus, this new expression is equivalent to (1.4) and really offers nothing extra except perhaps the possibility of forgetting where to correctly put the factors of n in the expression.

The MGF of random variable $n\bar{X}$ is $\exp\{nK(s)\}$, which may be expressed as an integral involving $f(\bar{x})$, the density of $\bar{X}$. Approximation to this integral using the method of Laplace provides an interrelationship between the CGF and density. This is the essential idea in the derivation below which first appeared informally in the discussion of Barndorff-Nielsen (1988).

Proof of (2.25). The CGF of $\sum_{i=1}^{n} X_i = n\bar{X}$ is $nK(s)$ and defined as

$$e^{nK(s)} = \int_{-\infty}^{\infty} e^{sn\bar{x} + \ln f(\bar{x})} d\bar{x} = \int_{-\infty}^{\infty} e^{-g(s,\bar{x})} d\bar{x}$$

for $g(s, \bar{x}) = -sn\bar{x} - \ln f(\bar{x})$. With s fixed, Laplace's approximation for the integral is

$$e^{nK(s)} \simeq \sqrt{\frac{2\pi}{g''(s, \bar{x}_s)}} e^{sn\bar{x}_s} f(\bar{x}_s) \qquad (2.26)$$

where $g''(s, \bar{x}) = \partial^2 g(s, \bar{x})/\partial \bar{x}^2 = -\partial^2 \ln f(\bar{x})/\partial \bar{x}^2$, and $\bar{x}_s$ minimizes $g(s, \bar{x})$ over $\bar{x}$ for fixed s. As a critical value, $\bar{x}_s$ solves

$$0 = -g'(s, \bar{x}_s) = ns + \frac{\partial \ln f(\bar{x}_s)}{\partial \bar{x}_s}. \qquad (2.27)$$

The presumption made when using the approximation is that $g''(s, \bar{x}_s) > 0$ which assures that $-\partial^2 \ln f(\bar{x}_s)/\partial \bar{x}_s^2 > 0$. We now examine the implicit relationship between s and $\bar{x}_s$ determined from (2.27). Partial differentiation of (2.27) with respect to $\bar{x}_s$ gives

$$\frac{\partial s}{\partial \bar{x}_s} = -\frac{1}{n} \frac{\partial^2 \ln f(\bar{x}_s)}{\partial \bar{x}_s^2} > 0 \qquad (2.28)$$

and shows a monotone increasing relationship between s and $\bar{x}_s$. The main task in deciphering (2.26) is in further determining the relationship between s and the value $\bar{x}_s$ through the solution of (2.27). An approximation to $\partial \ln f(\bar{x}_s)/\partial \bar{x}_s$ in this expression may be determined by first solving for $\ln f(\bar{x}_s)$ in (2.26) as

$$\ln f(\bar{x}_s) \simeq n\{K(s) - s\bar{x}_s\} - \tfrac{1}{2} \ln \frac{2\pi}{-\partial^2 \ln f(\bar{x}_s)/\partial \bar{x}_s^2}. \qquad (2.29)$$

This expression relates $\bar{x}_s$ to s by way of $K(s)$. If the last term is assumed to be approximately constant in $\bar{x}_s$ and therefore negligible upon differentiation with respect to $\bar{x}_s$, then

$$\frac{\partial}{\partial \bar{x}_s} \ln f(\bar{x}_s) \simeq n\{K'(s) - \bar{x}_s\} \frac{\partial s}{\partial \bar{x}_s} - ns \qquad (2.30)$$

follows from the product rule. To the degree of approximation made in obtaining (2.30),

$$0 = ns + \frac{\partial \ln f(\bar{x}_s)}{\partial \bar{x}_s} \iff 0 = n\{K'(s) - \bar{x}_s\} \frac{\partial s}{\partial \bar{x}_s}.$$

Accordingly, s and $\bar{x}_s$ must be related through the saddlepoint equation $K'(s) = \bar{x}_s$ to the degree of approximation considered. The final term in determining (2.26) is

$$g''(s, \bar{x}_s) = -\frac{\partial^2 \ln f(\bar{x}_s)}{\partial \bar{x}_s^2} = n \frac{\partial s}{\partial \bar{x}_s} = n \left(\frac{\partial \bar{x}_s}{\partial s}\right)^{-1} = n\{K''(s)\}^{-1} \qquad (2.31)$$

as determined from (2.28) and differentiation of the saddlepoint equation. From this, expression (2.26) becomes

$$e^{nK(s)} \simeq \sqrt{\frac{2\pi K''(s)}{n}} e^{ns\bar{x}_s} f(\bar{x}_s)$$

or

$$f(\bar{x}_s) \simeq \sqrt{\frac{n}{2\pi K''(s)}} \exp\{nK(s) - ns\bar{x}_s\}$$

where $K'(s) = \bar{x}_s$. This is the saddlepoint density in (2.25) that has been modified to have a fixed saddlepoint s which determines its associated density argument $\bar{x}_s = K'(s)$

and density $f(\bar{x}_s)$. The proof provides a simple demonstration of the analytical relationship that exists between the density and the local properties of its CGF at the associated saddlepoint. □

Remarks on asymptotics and higher-order terms

(1) The next term in the saddlepoint expansion of $f(\bar{x})$ of order $O(n^{-1})$ is given as

$$f(\bar{x}) = \hat{f}(\bar{x}) \left\{ 1 + \frac{1}{n} \left(\frac{1}{8} \hat{\kappa}_4 - \frac{5}{24} \hat{\kappa}_3^2 \right) + O(n^{-2}) \right\} \tag{2.32}$$

where

$$\hat{\kappa}_i = \frac{K^{(i)}(\hat{s})}{\{K''(\hat{s})\}^{i/2}} \qquad i = 3, 4 \tag{2.33}$$

and $\hat{f}(\bar{x})$ is given in (2.25). This term is not easily verified from the arguments above and will be derived using Edgeworth expansion methods in section 5.3.1. The difficulty with the present arguments is that the approach contributes two separate terms of this order: the $O(n^{-1})$ term from Laplace's approximation and a complicated term of the same order that results from the term that has been ignored in (2.29). The order of this latter term is considered in Exercise 9 wherein the consequence of ignoring the last term in (2.29) leads to a discrepancy between $K'(s)$ and $\bar{x}_s$ of order $O(n^{-1})$.

(2) The second-order saddlepoint density/mass function approximation for arbitrary random variable X having CGF K is given by expression (2.32) using $n = 1$. This second-order approximation preserves all the properties of the first-order approximation presented in section 2.1 including equivariance under linear transformation and symmetry.

2.3 Saddlepoint CDF approximation

The continuous CDF approximation is derived below based on an approximation due to Temme (1982). The approach involves first expressing the CDF approximation as the finite cumulative integral of the saddlepoint density. Then, using the Temme approximation, this integral is further approximated by using integration by parts. Thus, there are two layers of approximation needed in deriving (1.21). Our approach to this derivation applies to the continuous setting and could also be extended to the lattice setting by using the idea of summation by parts, however this is not developed. Formal derivations of these continuity corrections have already been presented in Daniels (1987) by using an inversion theory approach.

The proofs that the various CDF approximations have removable discontinuities at the mean and the derivations for their limiting values, as given in (1.21), (1.27), and (1.32), are deferred to the appendix in section 2.5.2.

2.3.1 Temme approximation

Suppose Z_{w_0} has a Normal $(0, 1)$ distribution truncated to attain values no larger than w_0. The Temme approximation gives an approximation for the numerator in the computation of $E\{h(Z_{w_0})\}$ of the form

$$\int_{-\infty}^{w_0} h(w)\phi(w)\,dw \simeq h(0)\Phi(w_0) + \phi(w_0)\left\{\frac{h(0) - h(w_0)}{w_0}\right\}. \tag{2.34}$$

Most often, some sort of transformation is needed to bring a cumulative integral of interest into the form of (2.34). Practically, the approximation depends on only the three values w_0, $h(w_0)$, and $h(0)$.

Examples

Incomplete gamma function This function is the CDF of a Gamma $(\alpha, 1)$ variable and is

$$\Gamma(y; \alpha) = \frac{1}{\Gamma(\alpha)} \int_0^y x^{\alpha-1} e^{-x}\,dx. \tag{2.35}$$

The transformation $w = \hat{w}(x)$, for $\hat{w}$ given in (1.23), puts the integral into the form of (2.34). Exercise 10 provides the steps in the Temme approximation that shows

$$\Gamma(y; \alpha) \simeq \frac{\hat{\Gamma}(\alpha)}{\Gamma(\alpha)} \hat{F}(y) \tag{2.36}$$

where $\hat{F}(y)$ is (1.21) with $\hat{w}(y)$ and $\hat{u}(y)$ as its inputs.

Incomplete beta function This function is the CDF of a Beta (α, β) variable and is

$$B(y; \alpha, \beta) = \frac{1}{B(\alpha, \beta)} \int_0^y x^{\alpha-1}(1-x)^{\beta-1}\,dx. \tag{2.37}$$

Upon transforming $w = \hat{w}(x)$ for $\hat{w}$ as given in (1.26), the integral in (2.37) may be expressed as

$$B(y; \alpha, \beta) = \int_{-\infty}^{\hat{w}(y)} \frac{\hat{B}(\alpha, \beta)}{B(\alpha, \beta)} \frac{w}{\hat{u}(x)} \phi(w)\,dw, \tag{2.38}$$

where $\hat{B}(\alpha, \beta)$ is the beta function approximation in (2.12). The term $\hat{u}(x)$ is given in (1.26) and should be thought of as an implicit function of $w = \hat{w}(x)$. The Temme approximation to integral (2.38) is

$$B(y; \alpha, \beta) \simeq \frac{\hat{B}(\alpha, \beta)}{B(\alpha, \beta)} \hat{F}(y) \tag{2.39}$$

where $\hat{F}(y)$ is the beta CDF approximation using (1.21); see Exercise 11.

Proof of (2.34).

$$\int_{-\infty}^{w_0} h(w)\phi(w)\,dw = h(0)\Phi(w_0) + \int_{-\infty}^{w_0} \frac{h(w) - h(0)}{w} w\phi(w)\,dw$$

$$= h(0)\Phi(w_0) - \int_{-\infty}^{w_0} \frac{h(w) - h(0)}{w}\,d\phi(w)$$

upon using the fact that $w\phi(w)\,dw = -d\phi(w)$. Integrating by parts and ignoring the resulting integral portion leads to

$$\int_{-\infty}^{w_0} h(w)\phi(w)\,dw \simeq h(0)\Phi(w_0) - \left\{\frac{h(w)-h(0)}{w}\phi(w)\right\}_{w=-\infty}^{w=w_0}$$

which is the right side of (2.34). □

2.3.2 Derivation of saddlepoint CDF

The proof is a generalization of the methods used in the two previous examples.

Proof of (1.21). An approximation to $F(y)$ is

$$\int_{-\infty}^{y} \hat{f}(x)\,dx = \int_{-\infty}^{y} \frac{1}{\sqrt{2\pi K''(\hat{s})}} \exp\{K(\hat{s}) - \hat{s}x\}\,dx$$
$$= \int_{-\infty}^{y} \frac{1}{\sqrt{K''(\hat{s})}} \phi(\hat{w})\,dx \tag{2.40}$$

where both $\hat{s}$ and $\hat{w}$ are functions of x whose dependence has been suppressed. A change of variable in (2.40) from dx to $d\hat{w}$ puts the integral in a form for which the Temme approximation can be applied. First, however, the differential of the mapping $x \leftrightarrow \hat{w}$ needs to be computed.

Lemma 2.3.1 *The mapping $x \leftrightarrow \hat{w}$ is a smooth monotonic increasing transformation with derivative*

$$\frac{dx}{d\hat{w}} = \begin{cases} \hat{w}/\hat{s} & \text{if } \hat{s} \neq 0 \\ \sqrt{K''(0)} & \text{if } \hat{s} = 0. \end{cases} \tag{2.41}$$

Proof. The mapping $x \leftrightarrow \hat{s}$ is determined from the saddlepoint equation as smooth and monotonic increasing with differential $dx = K''(\hat{s})d\hat{s}$. Also the mapping $\hat{s} \leftrightarrow \hat{w}$ is monotonic increasing with differential $d\hat{s} = \hat{w}/\{\hat{s}K''(\hat{s})\}d\hat{w}$ as we now indicate. The relationship between $\hat{w}^2$ and $\hat{s}$ is

$$\hat{w}^2/2 = \hat{s}x - K(\hat{s}) = \hat{s}K'(\hat{s}) - K(\hat{s}).$$

This is differentiated implicitly to give

$$\hat{w}d\hat{w} = \hat{s}K''(\hat{s})d\hat{s}. \tag{2.42}$$

Since $\text{sgn}(\hat{w}) = \text{sgn}(\hat{s})$, then $d\hat{w}/d\hat{s} > 0$ for $\hat{s} \neq 0$. Exercise 12 asks the reader to show that $\lim_{\hat{s}\to 0} \hat{w}/\hat{s}$ exists and is $\sqrt{K''(0)}$. Once it is known that the limit exists, there is a very simple way to determine its value. From l'Hôpital's rule,

$$\left(\lim_{\hat{s}\to 0}\frac{\hat{w}}{\hat{s}}\right)^2 = \lim_{\hat{s}\to 0}\frac{\hat{w}}{\hat{s}} \times \lim_{\hat{s}\to 0}\frac{d\hat{w}}{d\hat{s}} = \lim_{\hat{s}\to 0}\frac{\hat{w}}{\hat{s}}\frac{d\hat{w}}{d\hat{s}} = K''(0)$$

by (2.42). □

Implementing this change of variables in (2.40) yields

$$\int_{-\infty}^{y} \frac{1}{\sqrt{K''(\hat{s})}} \phi(\hat{w})\,dx = \int_{-\infty}^{\hat{w}_y} \frac{\hat{w}}{\hat{s}\sqrt{K''(\hat{s})}} \phi(\hat{w})\,d\hat{w} \tag{2.43}$$

where $\hat{w}_y$ is the value of $\hat{w}$ determined by solving the saddlepoint equation at y. The Temme approximation is applied to (2.43) with $h(\hat{w}) = \hat{w}/\{\hat{s}\sqrt{K''(\hat{s})}\}$. The value $h(0)$ is determined as in (2.41) by continuity to be

$$h(0) = \lim_{\hat{s} \to 0} \frac{\hat{w}}{\hat{s}\sqrt{K''(\hat{s})}} = 1$$

so (2.43) is

$$\hat{F}(y) = \Phi(\hat{w}_y) + \phi(\hat{w}_y)\frac{1 - h(\hat{w}_y)}{\hat{w}_y}$$

which is (1.21). □

Remarks on asymptotics and higher-order terms

(1) The Temme approximation in (2.34) is often expressed as the leading term in an asymptotic expansion as a parameter $n \to \infty$. Suppose the standard normal kernel $\phi(w)$ in (2.34) is replaced with

$$\phi_n(w) = \sqrt{n}\phi(\sqrt{n}w),$$

the density of a Normal $(0, n^{-1})$ variable. This new kernel heavily weights the integrand near $w = 0$ and helps to explain the presence of $h(0)$ in the asymptotic expansion (2.34) which becomes

$$\int_{-\infty}^{w_0} h(w)\phi_n(w)\,dw = h(0)\Phi(\sqrt{n}w_0) \qquad (2.44)$$
$$+ n^{-1/2}\phi(\sqrt{n}w_0)\left\{\frac{h(0) - h(w_0)}{w_0}\right\} + R.$$

The reader is asked to verify this relationship in Exercise 13 and show that the exact error incurred is

$$R = n^{-1}\int_{-\infty}^{\sqrt{n}w_0} h_2\left(v/\sqrt{n}\right)\phi(v)\,dv \qquad (2.45)$$

where

$$h_2(w) = \frac{1}{w}\left\{h'(w) - \frac{h(w) - h(0)}{w}\right\}.$$

In applications where $h(w)$ does not depend on n, the remainder R is $O(n^{-1})$.

(2) Some understanding of the Temme approximation is gained by noting the form of h_2 as a mixed second difference expression. The essential feature of the argument has been to extract the linear aspects of h near the concentration point $w = 0$ for incorporation into the computation and sweep the curved aspects (second and higher order derivatives) into the error term R of order $O(n^{-1})$. Alone, such techniques cannot altogether reduce the error to the extraordinarily small values seen in the previous chapter. However, the method is often complemented by the nature of the $h(w)$ functions involved, which, in the applications, have been rather flat or approximately linear near the point of concentration $w = 0$. Exercises

2.3 Saddlepoint CDF approximation

10 and 11 ask the reader to investigate the local linearity of h at $w = 0$ in the incomplete gamma and beta examples.

(3) The CDF approximation is often stated as it applies to the CDF of $\bar{X} = n^{-1} \sum_{i=1}^{n} X_i$ where $X_1, \ldots, X_n$ are i.i.d. continuous variables with common CGF K. In the statement of approximation, the saddlepoint CDF is the leading term in the asymptotic expansion

$$\Pr(\bar{X} \le \bar{x}) = \Phi(\hat{w}_n) + \phi(\hat{w}_n) \left(\frac{1}{\hat{w}_n} - \frac{1}{\hat{u}_n} \right) + O(n^{-1}) \qquad \bar{x} \ne E(\bar{X}). \tag{2.46}$$

Here,

$$\begin{aligned} \hat{w}_n &= \sqrt{n}\hat{w} = \text{sgn}(\hat{s})\sqrt{2n\{\hat{s}\bar{x} - K(\hat{s})\}} \\ \hat{u}_n &= \sqrt{n}\hat{u} = \hat{s}\sqrt{nK''(\hat{s})} \end{aligned} \tag{2.47}$$

and

$$K'(\hat{s}) = \bar{x}$$

is the saddlepoint equation. This expression may be derived directly from (2.44) or by replacing random variable X with $\bar{X}$ in (1.21). Thus (2.46) offers no greater generality than has already been expressed through the basic formula in (1.21).

In settings where $\{X_i\}$ are lattice variables, the same asymptotics applies with the appropriate continuity corrections. For integer lattice variables, these asymptotics apply to the previously discussed corrections for $\sum_i X_i$ whose support is also the integer lattice. However, when the mean is computed, the support of $\bar{X}$ is the n^{-1}-lattice and the asymptotics now apply to the appropriate continuity corrections given in section 2.4.4 below.

(4) The $O(n^{-1})$ term in (2.46) applies to values of $\bar{x}$ in *large deviation sets*, that is sequences of sets for which $\bar{x} - \mu$ remains bounded as $n \to \infty$, where $\mu = E(\bar{X})$. The asymptotic rate becomes the higher value $O(n^{-3/2})$ when considered over sequences of *sets of bounded central tendency,* or sets of $\bar{x}$ for which $\sqrt{n}(\bar{x} - \mu)$ remains bounded with increasing n. In either case, the $O(n^{-1})$ term in (2.46) in the continuous setting is shown in Daniels (1987) to be

$$O(n^{-1}) = -\frac{\phi(\hat{w}_n)}{n^{3/2}} \left\{ \frac{1}{\hat{u}} \left(\tfrac{1}{8}\hat{\kappa}_4 - \tfrac{5}{24}\hat{\kappa}_3^2 \right) - \frac{1}{\hat{u}^3} - \frac{\hat{\kappa}_3}{2\hat{u}^2} + \frac{1}{\hat{w}^3} \right\}$$
$$+ O(n^{-2}), \tag{2.48}$$

where $\hat{u}$ and $\hat{w}$ are $\hat{u}_n$ and $\hat{w}_n$ without the $\sqrt{n}$ factor as seen in (2.47), and $\hat{\kappa}_3$ and $\hat{\kappa}_4$ are given in (2.33). Computations of this term appear in section 2.4.1 where there is also some discussion about assessing the accuracy of the approximation.

(5) The second-order continuous saddlepoint CDF approximation for arbitrary random variable X with CGF K is given by (1.21) with the inclusion of (2.48) using $n = 1$. This second-order approximation preserves all the properties of the first-order CDF approximation presented in section 2.1 including invariance under linear transformation and symmetry when present in the true CDF. In the lattice setting, the second-order term for $\widehat{\Pr}_2$ in (1.32) is given in Kolassa (1994, §5.5).

(6) Distributions are increasing functions with limiting values of 0 and 1 in the left and right tails. In most of our examples, these same properties carry over to the saddlepoint CDF approximation but there are some exceptions. Examples for which $\hat{F}$ is decreasing or has the wrong limit in the tails are given in Exercise 17 and in section 16.5. Exercise 16 provides some understanding of this in the context of a sample mean. The exercise consists of showing that the derivative of the Lugannani and Rice approximation is the second-order saddlepoint density to order $O(n^{-3/2})$ when $\bar{x}$ stays in sets of bounded central tendency. More specifically,

$$\frac{d}{d\bar{x}}\hat{F}(\bar{x}) = \hat{f}(\bar{x})\left\{1 + \frac{1}{n}\left(\tfrac{1}{8}\hat{\kappa}_4 - \tfrac{5}{24}\hat{\kappa}_3^2\right) + O\left(n^{-3/2}\right)\right\}. \tag{2.49}$$

If, for small n, the term in curly brackets should happen to be negative, then $\hat{F}(\bar{x})$ would be decreasing. The practical situations in which this occurs are not common due to the extreme accuracy exhibited by the second-order density approximation that also preserves relative error over large deviation regions.

The result in (2.49) might have been anticipated before doing the actual differentiation. Over sets of bounded central tendency (which are compact), $\hat{F}(\bar{x})$ has error $O(n^{-3/2})$ for $F(\bar{x})$ and is smooth. Therefore, passing to derivatives, it is reasonable to expect that $\hat{F}'(\bar{x})$ should also have the same error for $F'(\bar{x}) = f(\bar{x})$ on such sets. For $\hat{F}'(\bar{x})$ to determine $f(\bar{x})$ to this order, it must include the second-order correction term of order $O(n^{-1})$.

2.4 Further topics

2.4.1 Mellin transforms

The Mellin transform of a positive random variable X is defined as the MGF of $\ln X$ or

$$m(s) = E(X^s) = \int_0^\infty x^s f(x)\,dx. \tag{2.50}$$

In most of our applications, this integral is convergent over a neighborhood of 0 which is the domain of the Mellin transform. This definition differs from the ordinary definition of the Mellin transform of f which has traditionally been defined as $E(X^{s-1})$. Due to the awkwardness in dealing with this traditional definition, the saddlepoint discussion has been built around the more convenient form given in (2.50).

The Mellin transform itself is less important than the concept of working with the MGF of $\ln X$ in place of the MGF of X. Many distributions which lack easily computed MGFs have simple Mellin transforms. The more common examples include the gamma, beta, folded Student-t ($|X|$ where $X \sim$ Student-t), F, and Weibull. Several of these distributions are heavier-tailed and lack MGFs convergent for $s > 0$. The MGFs for the transformation $\ln X$, however, converge over maximal neighborhoods of 0 that are open so that saddlepoint methods are available for application. In the computations, $\Pr(X \leq x)$ is replaced with $\Pr(\ln X \leq \ln x)$ and the CDF approximation (1.21) is applied to the Mellin transform using the saddlepoint equation

$$\frac{m'(\hat{s})}{m(\hat{s})} = \ln x.$$

2.4 Further topics

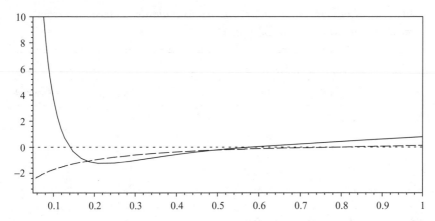

Figure 2.1. Percentage relative errors vs. α when approximating Pr{Beta$(\alpha, 0.8) < 0.4$} using the saddlepoint approximation from the Mellin transform (*dashed*) and that in (1.26) (*solid*).

Distributions of products of random variables with tractable Mellin transforms may be approximated by using such saddlepoint methods. If independent variables $\{X_i\}$ have Mellin transforms $\{m_i(s)\}$, then the product $\prod_{i=1}^n X_i$ has Mellin transform $\prod_{i=1}^n m_i(s)$ to which these saddlepoint methods may be very simply applied.

Examples

Beta (α, β)   The MGF of $\ln X$ when $X \sim$ Beta(α, β) is

$$M_{\ln X}(s) = E(X^s) = \frac{\Gamma(\alpha+s)\,\Gamma(\alpha+\beta)}{\Gamma(\alpha)\,\Gamma(\alpha+\beta+s)} \qquad s \in (-\alpha, \infty).$$

Computation of the saddlepoint CDF approximation requires the use of $(\ln \Gamma)'$ and $(\ln \Gamma)''$, the di- and tri-gamma functions, which are available, for example, in Maple. Figure 2.1 plots the relative error of this new approximation (dashed line) in the context described in figure 1.7 and includes the error for the previous approximation (solid line) as determined by (1.26). The approximation based on the Mellin transform is more accurate for most of the values of α and particularly for small α. The relative error of the Mellin-based approximation appears to stay bounded as $\alpha \to 0$ with value -6.61 at $\alpha = 10^{-5}$.

Generalized variance   Suppose W is a $(k \times k)$ sample covariance matrix having a Wishart (n, Σ) distribution with n degrees of freedom and mean $n\Sigma$ that is positive definite. The pivotal quantity

$$D = \frac{|W|}{|\Sigma|} \sim \prod_{i=1}^k \chi^2_{n-i+1}, \qquad (2.51)$$

where $\{\chi^2_{n-i+1}\}$ are independent chi square variables. Exact computation of this distribution is intractable however the Mellin transform of (2.51) is easily expressed as a ratio of products of gamma functions.

Butler *et al.* (1992) assess the numerical accuracy of Lugannani and Rice approximation $\hat{F}$ using the distribution of $\ln D$. Table 2.1 has been reproduced from their paper. Column

Table 2.1. *Empirical coverages for various quantiles determined from three approximations to the CDF of D.*

	$k = 5$, $n = 6$			$k = 20$, $n = 20$*	
Prob.	Sad. App.	Hoel	Normal	Sad. App.	Normal
.995	.995	.999	.999	.995	.999
.990	.990	.997	.997	.990	.997
.975	.975	.989	.988	.975	.988
.950	.950	.972	.966	.950	.967
.900	.900	.930	.914	.901	.916
.100	.101	.068	.104	.103	.102
.050	.051	.029	.060	.052	.060
.025	.025	.013	.036	.026	.037
.010	.010	.004	.019	.010	.020
.005	.005	.002	.012	.005	.013

Note: * The Hoel approximation is not applicable in this instance.

"Sad. App." shows the accuracy attained by $\hat{F}$ for the distribution of $\ln D$. Consider, for example, the top entry 0.995 in the Sad. App. column with $k = 5$ and $n = 6$. This entry is a simulation-based approximation for

$$\Pr\{\ln D \leq \hat{F}^{-1}(0.995)\},$$

the true coverage of the 99.5 percentile as determined by inverting the saddlepoint CDF $\hat{F}$. The simulation records the value 0.995 as the proportion of times that variate $\ln D$ falls below the specified 99.5 saddlepoint percentile $\hat{F}^{-1}(0.995)$ in 2 million simulations of D. The empirical coverage is correct to the accuracy shown.

Consider the application in which the saddlepoint approximation is used to determine a centered 95% confidence interval for $\ln |\Sigma|$ with 2.5% probability in each tail. Taking $k = 5$ and $n = 6$, then the 95% confidence interval for $\ln |\Sigma|$, set using the percentiles of $\hat{F}$, has coverage estimated to be 97.5% − 2.5% = 95.0% using the simulation of 2 million repetitions from the table.

The two other approximations in table 2.1 are described in Johnson and Kotz (1972, p. 198). The Hoel and normal procedures fit gamma and normal distributions to $k^{-1} \ln D$. The table clearly suggests that the saddlepoint approximation is better and virtually exact.

Wilks' likelihood ratio statistic $\Lambda_{k,m,n}$ This is the likelihood ratio statistic for testing hypotheses in MANOVA. Suppose under the null hypothesis that the error sum of squares matrix W_e is a $(k \times k)$ Wishart (n, Σ) and the hypothesis sum of squares matrix W_h is a $(k \times k)$ Wishart (m, Σ). Then, according to Anderson (2003, chap. 8),

$$\Lambda_{k,m,n} = |W_e + W_h|^{-1} |W_e| \sim \prod_{i=1}^{k} \beta_i \left\{ \tfrac{1}{2}(n - i + 1), \tfrac{1}{2}m \right\}$$

where $\{\beta_i\}$ are independent beta variates with the indicated degrees of freedom.

Table 2.2. *Values of the Mellin-based saddlepoint approximation (Sad. App.) for Wilks' $\Lambda_{k,m,n}$ and Box's approximation evaluated at the true quantiles as listed for $\ln \Lambda_{k,m,n}$.*

k	m	n	Exact	Sad. App.	Box	Quantile
3	5	7	.100	.0995	.0816	−3.1379
			.050	.0497	.0382	−3.5294
			.025	.0248	.0178	−3.8960
			.010	.0099	.0064	−4.3543
			.005	.0050	.0045	−4.6840
5	5	9	.100	.0997	.0762	−4.3200
			.050	.0499	.0350	−4.7486
			.025	.0249	.0160	−5.1454
			.010	.0099	.0056	−5.6357
			.005	.0049	.0025	−5.9902
16	9	25	.100	.1006	.0449	−8.5360
			.050	.0496	.0179	−8.9138
			.025	.0252	.0075	−9.2372
			.010	.0099	.0022	−9.6392
			.005	.0050	.0009	−9.9103

Lugannani and Rice approximation (1.21) using the Mellin transform of $\Lambda_{k,m,n}$ was considered by Srivastava and Yau (1989) and Butler *et al.* (1992a). Table 2.2 is reproduced from the latter's table 3. The table entries are values of the Mellin-based saddlepoint (Sad. App.) and Box's (1949) CDF approximation evaluated at the exact percentiles for $\ln \Lambda_{k,m,n}$ listed in the last column with associated probabilities in the Exact column. For most practical purposes, the saddlepoint computations are as good as exact. Additional considerations of this test, including saddlepoint approximation to its power function, are given in sections 11.1–11.3.

Likelihood ratio test statistics used in MANOVA Many of these tests have null distributions characterized as the product of independent betas. Included among such tests are those for block independence, sphericity, intraclass correlation, and homogeneity across strata; see Anderson (2003, chap. 9–10) and Morrison (1990, chap. 7). Details of the *p*-values for these tests as well as saddlepoint approximation to some of the power functions are examined later in sections 11.2–11.3.

Accuracy

Saddlepoint approximation to the CDFs of positive random variables using the Mellin transform approach is usually extremely accurate. When direct saddlepoint approximation is also possible, as with the beta example, the Mellin transform approach is usually more

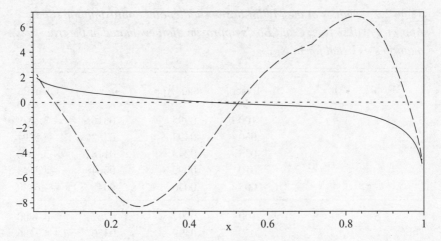

Figure 2.2. Relative error comparisons of the CDF approximation for $X \sim$ Beta $(2, 1)$ using direct approximation (*dashed*) and the Mellin-based approximation (*solid*).

accurate. We suggest that the reason for this is may be quite simple: for many of the positive random variables $X > 0$ commonly arising in applications, $\ln X$ is often has a more "normal" looking density than X. If one accepts this idea, then an explanation for the greater accuracy follows simply from examining the derivation of the saddlepoint density. The initial step uses Laplace's approximation to approximate either M_X, for direct approximation, or $M_{\ln X}$, for the Mellin transform approach. If the Laplace approximation is more accurate in determining $M_{\ln X}$ than M_X, then its saddlepoint density will be more accurate. Also its integral using the Temme approximation is likely to be more accurate. The two transforms are

$$M_X(s) = \int_0^\infty e^{sx} f(x)\, dx \qquad M_{\ln X}(s) = \int_0^\infty e^{s \ln x} f(x)\, dx,$$

where f is the density of X. Upon transformation $y = \ln x$, the Mellin integral becomes

$$M_{\ln X}(s) = \int_{-\infty}^\infty e^{sy} \{e^y f(e^y)\} dy, \qquad (2.52)$$

where the expression in curly brackets is the density of $Y = \ln X$. If this density has a more "normal" shape in y than $f(x)$ has in x, then the Mellin transform approach is likely to be more accurate for CDF saddlepoint approximation.

Consider, for example, $X \sim$ Beta $(\alpha, 1)$ with density

$$f(x) = \alpha x^{\alpha-1} \qquad x \in (0, 1). \qquad (2.53)$$

A simple calculation determines the density of $Y = -\ln X$ as Exponential (α). Further transformation as $Z = \ln Y = \ln\{-\ln(X)\}$ shows that Z has the distribution of $-G - \ln \alpha$ where $G \sim$ Gumbel $(0, 1)$ as in section 1.1. For $\alpha = 2$, the CDF of the density in (2.53) is $F(x) = x^2$ so we might easily determine relative accuracies. Direct saddlepoint approximation for the CDF of X yields the relative error plot in figure 2.2 with the dashed line. When the probabilities are instead approximated by way of the CDF of $Y = -\ln X$ and plotted on the x-scale, the relative errors are given as the solid curve and are much

2.4 Further topics

Table 2.3. *Second-order correction terms to* $\Pr\{\text{Beta}(\alpha, 0.8) < 0.4\}$ *using direct and Mellin transform approximations.*

α	Direct	Mellin
0.05	.142	$0.0^3 824$
0.1	.0786	$0.0^3 696$
0.2	.0380	$-0.0^4 600$
0.4	.0124	$-0.0^2 153$
0.6	$.0^2 301$	$-0.0^2 245$
0.8	$-.0^2 123$	$-0.0^2 291$
0.9	$-.0^2 240$	$-0.0^2 302$

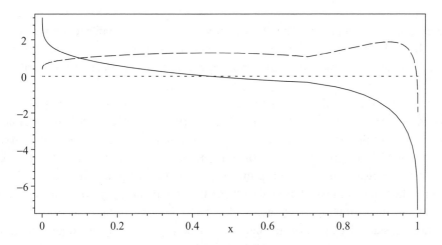

Figure 2.3. Relative error comparisons of the CDF approximation for $X \sim \text{Beta}(2, 1)$ using the Mellin-based approximation (*solid*) for the CDF of $-\ln X$ and that for $Z = \ln\{-\ln(X)\}$ (*dashed*).

smaller. Relative accuracy in the distribution tails is further improved by performing the approximation in terms of the CDF of Z as shown in figure 2.3. The increased accuracy supports our contention that the logarithm transformation in each instance helps by making the true density closer to a normal shape and hence improving upon accuracy for the Lugannani and Rice approximation.

Accuracy assessment In all of the examples, the accuracy of the saddlepoint approximations has been judged by making comparisons with the exact density or CDF. When exact computation is not possible, how can the accuracy of the various approximations be assessed? For example, what assessment procedure can demonstrate that the Mellin transform approach to the beta CDF is more accurate than direct approximation when the incomplete beta function is not computed? The next higher-order term in the saddlepoint expansion provides some evidence for this as may be seen in table 2.3. The entries are the $O(n^{-1})$ terms in (2.48) for the two approaches with $n = 1$ and the parameters α and

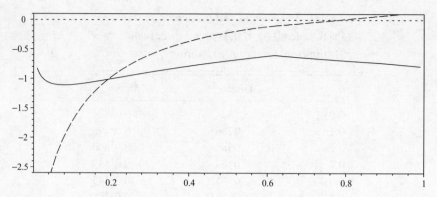

Figure 2.4. Percentage relative errors vs. α when approximating $\Pr\{\text{Beta}(\alpha, 0.8) < 0.4\}$ using the first-order saddlepoint approximation from the Mellin transform (*dashed*) and also including its second-order correction term (*solid*).

β controlling the asymptotics. These correction terms correctly indicate the greater accuracy of the Mellin transform approximation for smaller α. However, some of the correction terms adjust the approximations in the wrong direction as may be seen by comparison with figure 2.1.

There is also the possibility of improved approximation through the inclusion of the second-order term as suggested by some of the entries in the Mellin column of table 2.3. An assessment of this benefit is shown in figure 2.4 which compares the percentage relative errors of the Mellin approach both with and without the correction term. The correction terms do not appear to help except in the very difficult settings in which $\alpha < 0.2$. More generally, the benefit of this additional term in the saddlepoint expansions can vary considerably among the various applications. While it has not been especially successful here, the correction substantially improves the numerical accuracy for singly and doubly noncentral F distributions as shown in sections 12.1.2–12.1.5.

2.4.2 Central limit connections

Taylor expansion of both the saddlepoint density and CDF approximation for $\bar{X}$ shows the connection of these approximations to central limit theorems. First consider the behavior of the saddlepoint for fixed value of the the standardized variable $z = \sqrt{n}\,(\bar{x} - \mu)/\sigma$ where $\mu = K'(0) = E(\bar{X})$ and $\sigma^2 = K''(0)$. Fixing z restricts $\bar{x}$ to regions of bounded central tendency, e.g. where the central limit effect may be seen. If $\hat{s}$ is the saddlepoint for $\bar{x}$, then

$$\frac{\sigma z}{\sqrt{n}} = \bar{x} - \mu = K'(\hat{s}) - K'(0) = \hat{s} K''(0) + \tfrac{1}{2} \hat{s}^2 K^{(3)}(0) + O(\hat{s}^3) \qquad (2.54)$$

by Taylor expansion of $K'(\hat{s})$ about 0. If only the first term on the right side of (2.54) is retained, then one can solve for $\hat{s}$ as the value $\hat{s} = z/\sqrt{n}\sigma$. More generally, the solution when retaining the additional terms is

$$\hat{s} = \frac{z}{\sqrt{n}\sigma} + O(n^{-1}) \qquad (2.55)$$

as $n \to \infty$ if the standardized value of $\bar{x}$ is held fixed. Values for $\hat{w}_n^2/2$ can be expanded about $\hat{s} = 0$ as

$$\begin{aligned}
\tfrac{1}{2}\hat{w}_n^2 &= n\hat{s}\bar{x} - nK(\hat{s}) \\
&= n\hat{s}\bar{x} - n\left\{\hat{s}\mu + \tfrac{1}{2}\hat{s}^2\sigma^2 + O(n^{-3/2})\right\} \\
&= \tfrac{1}{2}z^2 + O(n^{-1/2}).
\end{aligned}$$

The middle line is obtained by Taylor expansion of $K(\hat{s})$ about 0 and the last line follows when $\bar{x} - \mu$ and $\hat{s}$ are expressed in terms of z using (2.54) and (2.55). Since $\mathrm{sgn}(\hat{w}_n) = \mathrm{sgn}(z)$, it follows that

$$\hat{w}_n = z + O(n^{-1/2}).$$

Also

$$\begin{aligned}
\hat{u}_n &= \hat{s}\sqrt{nK''(\hat{s})} \\
&= z + O(n^{-1/2}),
\end{aligned}$$

from (2.55). Transforming from (2.25), the saddlepoint density for variable $Z = \sqrt{n}(\bar{X} - \mu)/\sigma$ is

$$\hat{f}(z) = \frac{\sigma}{\sqrt{n}}\hat{f}(\bar{x}) = \sqrt{\frac{K''(0)}{2\pi K''(\hat{s})}}\exp\left(-\tfrac{1}{2}\hat{w}_n^2\right) \qquad (2.56)$$

$$= \phi(z) + O(n^{-1/2})$$

which is the local limit result. By the invariance property in Proposition 2.1.4, the saddlepoint CDF of Z at z is

$$\begin{aligned}
\hat{F}(z) &= \Phi(\hat{w}_n) + \phi(\hat{w}_n)\left(\frac{1}{\hat{w}_n} - \frac{1}{\hat{u}_n}\right) \\
&= \Phi(z) + \phi(z)O(n^{-1/2})
\end{aligned}$$

which is the central limit result.

The derivations have shown that Taylor expansions of the saddlepoint density and CDF along with the removal of the higher-order expansion terms leads to the central limit approximations. That these higher-order terms are crucial to the inherent accuracy of the saddlepoint approximation may be deduced from the relative inaccuracy seen when using central limit approximations.

The greater accuracy of the saddlepoint approximation may be further understood by considering that saddlepoint approximations make greater use of the information provided about the distribution. Central limit approximations use only the first two cumulants of the distribution whereas saddlepoint methods use knowledge of all the cumulants as expressed through values of the CGF over its entire convergence region. Also the saddlepoint method is not restricted to a particular shape such as the normal shape required with normal limits. In summary, saddlepoint methods use much more information about the distribution and consequently return much greater accuracy.

2.4.3 Asymptotic order with the Lugannani and Rice approximation

The asymptotic order of terms in the saddlepoint CDF expansion of (2.46) is now explained for fixed z, or when $\bar{x}$ remains in a set of bounded central tendency. In this case,

$$\hat{s} = O(n^{-1/2}) \qquad \hat{u}_n = z + O(n^{-1/2}) = \hat{w}_n,$$

so the leading term $\Phi(\hat{w}_n)$ is $O(1)$. The second term is

$$\phi(\hat{w}_n) \left(\frac{1}{\hat{w}_n} - \frac{1}{\hat{u}_n} \right) = O(n^{-1/2}) \qquad (2.57)$$

and a lower-order correction term. To show (2.57), write

$$\left(\frac{1}{\hat{w}_n} - \frac{1}{\hat{u}_n} \right) = \frac{1}{\sqrt{n}} \left(\frac{1}{\hat{w}} - \frac{1}{\hat{u}} \right)$$

and note that $\hat{s} \to 0$ as $n \to \infty$. Then, according to (2.76) in the appendix of section 2.5.2,

$$\left(\frac{1}{\hat{w}} - \frac{1}{\hat{u}} \right) = O(1)$$

so the order must be $O(n^{-1/2})$.

The same sort of arguments may be used to show that the next higher-order error term in (2.48) is $O(n^{-3/2})$ in this setting. Individual terms are not of this order, however the combination of terms in the curly braces is $O(1)$ so overall it is $O(n^{-3/2})$ as shown by Daniels (1987).

2.4.4 Continuity-corrected approximations on a δ-lattice

Suppose that X has support on the δ-lattice $L_\delta = \{\gamma, \gamma \pm \delta, \gamma \pm 2\delta, \ldots\}$. Saddlepoint probabilities for X, according to the lattice convention, would make the computations in terms of the variable $Y = \delta^{-1}(X - \gamma)$ supported on the integer lattice. In most practical applications, this description is sufficient. However, for completeness and also for theoretical reasons, a compilation of saddlepoint formulas is given that may be applied directly to the CGF of X. These formulas return the same computations as would occur when using the lattice convention but spare the user from basing them on the CGF for Y.

For small $\delta > 0$, the discretization may be sufficiently fine that the distribution is approximately continuous. In this instance, one might suspect that the use of continuity corrections would matter very little; this result is shown below. In this instance, as $\delta \to 0$, the continuity-corrected approximations converge to the uncorrected approximations that assume continuity.

For $x \in L_\delta$ the continuity-corrected survival function approximations are based on the CGF of X, denoted as K, that admits saddlepoint $\hat{s}$, solving $K'(\hat{s}) = x$, and admits $\tilde{s}$, solving

$$K'(\tilde{s}) = x^- = x - \tfrac{1}{2}\delta.$$

2.4 Further topics

First continuity correction

The Lugannani and Rice expression provides the survival function approximation $\widehat{\Pr}_1(X \geq x)$, with $x \neq E(X)$, if its inputs are

$$\hat{w} = \text{sgn}(\hat{s})\sqrt{2\{\hat{s}x - K(\hat{s})\}} \tag{2.58}$$

$$\tilde{u}_1 = \delta^{-1}(1 - e^{-\delta\hat{s}})\sqrt{K''(\hat{s})}. \tag{2.59}$$

Second continuity correction

The Lugannani and Rice expression provides the survival function approximation $\widehat{\Pr}_2(X \geq x)$ in both tails with inputs

$$\tilde{w}_2 = \text{sgn}(\tilde{s})\sqrt{2\{\tilde{s}x^- - K(\tilde{s})\}} \tag{2.60}$$

and

$$\tilde{u}_2 = 2\delta^{-1}\sinh\left(\tfrac{1}{2}\delta\tilde{s}\right)\sqrt{K''(\tilde{s})}. \tag{2.61}$$

When $\delta = 1$, these continuity-corrected expressions agree with their counterparts (1.27) and (1.32) in section 1.2.3.

Consistency as $\delta \to 0$

Both continuity-corrected results approach the continuous uncorrected results as $\delta \to 0$. For example, in (2.60) and (2.61),

$$\lim_{\delta \to 0}(\tilde{w}_2, \tilde{u}_2) = (\hat{w}, \hat{u}),$$

as given in (1.22), since $x^- \to x$, and $\tilde{s} \to \hat{s}$, so that

$$\lim_{\delta \to 0} 2\delta^{-1}\sinh\left(\tfrac{1}{2}\delta\tilde{s}\right) = \hat{s}.$$

This is certainly a property that would be expected to hold. It reflects on the continuity of the approximations shown in response to small changes in the underlying distributions under approximation.

2.4.5 A statistical interpretation of the saddlepoint CDF

The probability approximated by $\hat{F}(x)$ may be interpreted as a *p*-value in an hypothesis test. This idea extends into the discussion of tests of parameters in exponential families given in section 5.4.

The CGF of X with density f is defined in such a way that

$$1 = \int_{-\infty}^{\infty} e^{sx-K(s)} f(x)\, dx \qquad s \in (a, b); \tag{2.62}$$

that is, so that the positively valued integrand in (2.62) integrates to 1. The integrand, as such, is a density indexed by parameter s which can be denoted by

$$f(x;s) = e^{sx-K(s)} f(x) \qquad s \in (a,b) \tag{2.63}$$

and is called the *s-tilted density*. The class of densities in (2.63) may be considered as an artificially constructed class of densities called a *regular exponential family* in which the *true* density has the index $s = 0$.

Suppose the true value of s is not known and consider a test of

$$H_0 : s = 0 \quad \text{vs.} \quad H_1 : s < 0. \tag{2.64}$$

If the observed value of X is x, then the maximum likelihood estimate (MLE) of s maximizes the function $f(x;s)$. The critical value in this maximization is $\hat{s}$ which also solves the saddlepoint equation $x = K'(\hat{s})$. By uniqueness, the MLE and saddlepoint must therefore be the same. The likelihood ratio statistic is

$$-2\ln \frac{f(x;0)}{f(x;\hat{s})} = 2\{\hat{s}x - K(\hat{s})\} = \hat{w}^2.$$

The previous section indicated that $\hat{w} \simeq z$ in the context of the asymptotics in which x is a sample mean. Elaborating further in this context with x replaced by random value X, then the distribution of $\hat{w} = \text{sgn}(\hat{s})|\hat{w}|$, the *signed root of the likelihood ratio statistic*, is asymptotically Normal $(0, 1)$. The one-sided likelihood ratio test for the hypotheses indicated in (2.64) rejects H_0 for small values of $\hat{w}$. The p-value to first order is $\Phi(\hat{w})$, the leading term in the saddlepoint CDF. The second term involves

$$\hat{u} = (\hat{s} - 0)\sqrt{K''(\hat{s})}$$

which is the maximum likelihood statistic standardized by its asymptotic standard deviation estimated as $1/\sqrt{K''(\hat{s})}$, where

$$K''(\hat{s}) = -\left. \frac{\partial^2 \ln f(x;s)}{\partial s^2} \right|_{s=\hat{s}}$$

is the observed Fisher information in the exponential family. The second term in the saddlepoint CDF is noticeably smaller than $\Phi(\hat{w})$ in practical computations and provides a second-order correction to this first-order p-value.

The interpretation to be gained from this setup is that $\hat{F}(x)$ is a higher-order approximation to the true p-value $\Pr(X \leq x; s = 0)$ using datum x. The p-value concerns an hypothesis setup in which the known truth $H_0 : s = 0$ is contrasted with an alternative $H_1 : s < 0$ expressly created through the tilting of the true density as in (2.63).

2.4.6 Normalization

Previous examples have demonstrated that a normalized saddlepoint density or mass function can often achieve greater accuracy than its unnormalized counterpart. This improved accuracy does come with a cost which is the greater computational demand in determining the normalization constant. Normalization also provides an alternative to the adoption of the lattice convention for achieving uniquely defined saddlepoint mass functions. Recall that

2.4 Further topics

without the lattice convention, the summing of (1.13) could yield any positive total value simply by varying the size of the lattice span; uniqueness in the computation is attained by setting the lattice span to be 1. Routine normalization with any size lattice span provides an alternative which is computationally invariant under linear transformation and often more accurate than its unnormalized lattice convention counterpart.

Asymptotic arguments have also been given in Durbin (1980a) and Reid (1988) to reinforce the improved accuracy with normalization. In approximating the density of $\bar{X}$ at $\bar{x}$, when $\bar{x}$ remains in a set of bounded central tendency, then

$$f(\bar{x}) = \bar{f}(\bar{x}) \{1 + O(n^{-3/2})\}$$

when $\bar{f}$ is the normalized saddlepoint density. The effect of normalization is to improve the relative order from $O(n^{-1})$ to $O(n^{-3/2})$. For $\bar{x}$ in large deviation sets and outside of sets with bounded central tendency, the relative order remains as $O(n^{-1})$. See Durbin (1980a) for detailed arguments.

Example: Logistic (0, 1) density

The logistic density, which is symmetric about 0 with scale parameter 1, has density

$$f(x) = \frac{e^{-x}}{(1+e^{-x})^2} \qquad x \in (-\infty, \infty)$$

and MGF

$$M(s) = \Gamma(1+s)\Gamma(1-s) = \frac{\pi s}{\sin(\pi s)} \qquad s \in (-1, 1). \tag{2.65}$$

The latter expression for M is not defined at $s = 0$ but can be defined by continuity as 1. The saddlepoint equation

$$K'(\hat{s}) = 1/\hat{s} - \pi \cot(\pi \hat{s}) = x \qquad x \in (-\infty, \infty) \setminus \{0\}$$

$$K'(0) = 0$$

is obtained from differentiation of (2.65); the second equation follows by noting that both $x = 0$ and $K'(0)$ are expressions for the mean of the density. The saddlepoint solution $\hat{s}(x)$ approaches asymptote $\hat{s} = -1$ as $x \to -\infty$ and asymptote $\hat{s} = 1$ when $x \to \infty$. Figure 2.5 shows a comparative plot of the true density f (solid line) with the unnormalized saddlepoint density $\hat{f}$ (dotted line), and the normalized saddlepoint density $\bar{f}$ (dashed line). The portion of the plot with $x < 0$ is the mirror image by symmetry. The normalization constant for $\bar{f}$ was computed by numerically integrating $\hat{f}$ giving 0.89877. The normalized saddlepoint approximation $\bar{f}$ is very accurate and $\hat{f}$ constructs the proper shape for f but not the correct scaling of that shape.

The accuracy of $\bar{f}$ is confirmed in the probability computations of table 2.4. Exact computations in the F-column use the CDF

$$F(x) = (1 - e^{-x})^{-1} \qquad x \in (-\infty, \infty)$$

and saddlepoint computations are based on numerical integration $d\hat{s}$.

Table 2.4. *Effect of normalization on probability approximation using the saddlepoint density of the Logistic* (0, 1).

Probability computed from:	F	$\bar{f}$	$\hat{f}$	$\hat{s}$-range
$\Pr(-1 < X < 1)$	.4621	.4534	.4075	±0.2871
$\Pr(-2 < X < 2)$	.7616	.7513	.6753	±0.5000
$\Pr(-3 < X < 3)$	.9051	.8983	.8073	±0.6357
$\Pr(-4 < X < 4)$	.9640	.9607	.8634	±0.7208

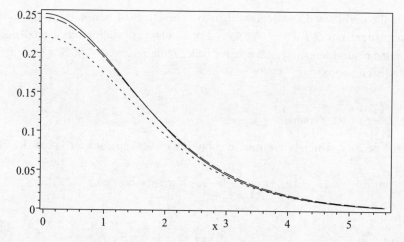

Figure 2.5. $\hat{f}(x)$ (*dots*), $\bar{f}(x)$ (*dashes*), and $f(x)$ (*solid*) vs. x for the logistic density.

Chapter 1 presented two examples for which the normalized saddlepoint density $\bar{f}$ is exactly f. In the case of the normal density $\hat{f}(x) \equiv f(x)$ while $\hat{f}(x) \propto f(x)$ for all x in the gamma case. For this to hold, one might suspect that the second-order correction term in (2.32) should not depend on $\hat{s}$. The reader is asked to verify this lack of dependence for the normal and gamma distributions in Exercise 15. Another distribution for which $\hat{f} \equiv f$ is the inverse Gaussian density in (13.17). This density is a non-regular case in which its MGF converges on a half-open neighborhood of 0 of the form $(-\infty, b]$, for some $b \geq 0$. Daniels (1980) and Blæsild and Jensen (1985) have shown the converse result: that these three distributions are the only one dimensional distributions for which $\bar{f}$ is exact.

2.5 Appendix

2.5.1 Asymptotic Laplace expansions

Derivation of (2.5) and (2.15)

A more careful development of the Laplace expansion is given. This leads to the second term in the expansion and the means for computing further terms.

Suppose that the integrand $\exp\{-ng(x)\}$ is sufficiently concentrated on (c, d) that the range of integration can be replaced by $(-\infty, \infty)$ without incurring much error. Taylor expansion of $g(x)$ about $\hat{x}$ yields

$$g(x) = g(\hat{x}) + \tfrac{1}{2} g''(\hat{x})(x - \hat{x})^2 + \sum_{i=3}^{\infty} \frac{g^{(i)}(\hat{x})}{i!}(x - \hat{x})^i$$

so that

$$\int_{-\infty}^{\infty} e^{-ng(x)} dx = e^{-ng(\hat{x})} \int_{-\infty}^{\infty} e^{-\tfrac{1}{2} ng''(\hat{x})(x-\hat{x})^2}$$

$$\times \exp\left\{ -n \sum_{i=3}^{\infty} \frac{g^{(i)}(\hat{x})}{i!}(x - \hat{x})^i \right\} dx.$$

The substitution $z = \sqrt{ng''(\hat{x})}(x - \hat{x})$ expresses the integral as a normal expectation of the form

$$\frac{e^{-ng(\hat{x})}}{\sqrt{ng''(\hat{x})}} \int_{-\infty}^{\infty} e^{-\tfrac{1}{2} z^2} \exp\left\{ -\sum_{i=3}^{\infty} n^{-\tfrac{1}{2}(i-2)} \frac{\hat{k}_i z^i}{i!} \right\} dz \tag{2.66}$$

where

$$\hat{k}_i = \frac{g^{(i)}(\hat{x})}{\{g''(\hat{x})\}^{i/2}} \quad i = 3, 4, \ldots$$

Let y be the summation expression in the curly brackets of (2.66). Further expansion as $\exp(-y) = 1 - y + \tfrac{1}{2} y^2 + O(y^3)$ yields

$$\sqrt{\frac{2\pi}{ng''(\hat{x})}} e^{-ng(\hat{x})} \int_{-\infty}^{\infty} \phi(z) \left\{ 1 - y + \tfrac{1}{2} y^2 + O(y^3) \right\} dz. \tag{2.67}$$

Subject to an integrability condition (Barndorff-Nielsen and Cox, 1989, §6.12), integration can be performed term-by-term within the curly brackets. The term 1 gives the leading term in the expansion. In the term $-y$, the odd powers ($i = 3, 5, \ldots$) of z provide odd moments of a standard normal which are 0. Also in $-y$, only the $i = 4$ term is of smaller order than $O(n^{-2})$. Squaring y and integrating term-by-term proceeds in the same way: terms with odd powers of z vanish and the only term of smaller order than $O(n^{-2})$ is the square of the $i = 3$ term in y. Therefore, to order $O(n^{-2})$, the integral is

$$\sqrt{\frac{2\pi}{ng''(\hat{x})}} e^{-ng(\hat{x})} \left[1 + \frac{1}{n} \left\{ -\frac{\hat{k}_4}{4!} E(Z^4) + \frac{1}{2} \left(\frac{\hat{k}_3}{3!}\right)^2 E(Z^6) \right\} \right] \tag{2.68}$$

where $Z \sim \text{Normal}(0, 1)$. Its fourth and sixth moments are $E(Z^4) = 3$ and $E(Z^6) = 15$; see Kendall and Stuart (1969, Eq. 3.16). Expression (2.68) reduces to (2.15).

Derivation of (2.20)

The derivation consists of a quite tedious computation of the third and fourth order derivatives of $\gamma(y) = g\{H^{-1}(y)\}$. The first derivative is

$$\gamma' = \frac{dg}{dx} \frac{dx}{dy},$$

where $x = H^{-1}(y)$. Chain rule differentiation gives

$$\gamma'' = \frac{d^2g}{dx^2}\left(\frac{dx}{dy}\right)^2 + \frac{dg}{dx}\frac{d^2x}{dy^2}, \tag{2.69}$$

where

$$\frac{d^2x}{dy^2} = \frac{d(1/h(x))}{dy} = \frac{d(1/h(x))}{dx}\frac{dx}{dy} = -\frac{h'(x)}{h(x)^3}.$$

At $\hat{y} = H(\hat{x})$, this is

$$\gamma''(\hat{y}) = \frac{g''(\hat{x})}{h(\hat{x})^2}. \tag{2.70}$$

Differentiation of (2.69) yields

$$\gamma''' = 3\frac{d^2g}{dx^2}\left(\frac{dx}{dy}\right)\left(\frac{d^2x}{dy^2}\right) + \frac{d^3g}{dx^3}\left(\frac{dx}{dy}\right)^3 + \frac{dg}{dx}\frac{d^3x}{dy^3}, \tag{2.71}$$

which gives

$$\gamma'''(\hat{y}) = -3\frac{g''(\hat{x})h'(\hat{x})}{h(\hat{x})^4} + \frac{g'''(\hat{x})}{h(\hat{x})^3}. \tag{2.72}$$

Further differentiation of (2.71) gives

$$\gamma''''(\hat{y}) = 4\frac{g''(\hat{x})}{h(\hat{x})}\left\{-\frac{h''(\hat{x})}{h(\hat{x})^4} + 3\frac{h'(\hat{x})^2}{h(\hat{x})^5}\right\} + 3g''(\hat{x})\frac{h'(\hat{x})^2}{h(\hat{x})^6}$$

$$- 6\frac{g'''(\hat{x})h'(\hat{x})}{h(\hat{x})^5} + \frac{g''''(\hat{x})}{h(\hat{x})^4}. \tag{2.73}$$

The three derivative expressions in (2.70), (2.72), and (2.73) are combined as in (2.15) to give an expression that simplifies to (2.20) after some tedious computation.

2.5.2 The removable singularities in the Lugannani and Rice approximations

The apparent singularity at the mean for the continuous distribution approximation in (1.21) is shown to be removable and its value is derived under continuity. This also leads to the removable singularity values for the continuity-corrected versions in (1.27), (1.32), (1.37), and (1.39). The method of derivation was given in Daniels (1987, §4) and entails writing $\hat{w}$ as an expansion in $\hat{u}$ so that the limiting value of $\hat{w}^{-1} - \hat{u}^{-1}$ may be determined as $\hat{s} \to 0$.

Write $\hat{w}^2$ as

$$\tfrac{1}{2}\hat{w}^2 = \hat{s}x + \{K(0) - K(\hat{s})\},$$

where, of course, $K(0) = 0$. Taylor expansion about $\hat{s}$ of the difference in curly brackets gives

$$\hat{w}^2 = \hat{s}^2 K''(\hat{s}) - \tfrac{1}{3}K'''(\hat{s})\hat{s}^3 + O(\hat{s}^4)$$
$$= \hat{u}^2 - \tfrac{1}{3}\hat{k}_3\hat{u}^3 + O(\hat{s}^4),$$

where $\hat{\kappa}_3$ is the third standardized cumulant evaluated at $\hat{s}$ as given in (2.33). Then

$$\hat{w} = \hat{u}\sqrt{1 - \tfrac{1}{3}\hat{\kappa}_3\hat{u} + O(\hat{s}^2)}$$
$$= \hat{u}\left\{1 - \tfrac{1}{6}\hat{\kappa}_3\hat{u} + O(\hat{s}^2)\right\} \tag{2.74}$$

after using $\sqrt{1-z} = 1 - z/2 + O(z^2)$ as $z \to 0$. Now write

$$\frac{1}{\hat{w}} - \frac{1}{\hat{u}} = \frac{1}{\hat{u}}\left(\frac{\hat{u}}{\hat{w}} - 1\right)$$
$$= \tfrac{1}{6}\hat{\kappa}_3 + O(\hat{s}) \tag{2.75}$$

from (2.74). From (2.75) we determine that

$$\lim_{\hat{s}\to 0}\left(\frac{1}{\hat{w}} - \frac{1}{\hat{u}}\right) = \lim_{\hat{s}\to 0}\tfrac{1}{6}\hat{\kappa}_3 = \frac{1}{6}\frac{K'''(0)}{\{K''(0)\}^{3/2}}. \tag{2.76}$$

This limit along with the fact that $\hat{w} \to 0$ as $\hat{s} \to 0$ determine the value in (1.21) by continuity.

For the continuity-corrected approximations, computations of the limits require that we replace $\hat{u}$ in (2.75) with $\tilde{u}_1$ and $\tilde{u}_2$. For $\widehat{\Pr}_1$ in (1.27),

$$\frac{1}{\hat{w}} - \frac{1}{\tilde{u}_1} = \frac{1}{\hat{u}}\left(\frac{\hat{u}}{\hat{w}} - \frac{\hat{s}}{1 - e^{-\hat{s}}}\right)$$
$$= \frac{1}{\hat{u}}\left\{\left(\frac{\hat{u}}{\hat{w}} - 1\right) + \left(1 - \frac{\hat{s}}{1 - e^{-\hat{s}}}\right)\right\}. \tag{2.77}$$

A Taylor expansion of the second term of (2.77) in curly braces is $-\hat{s}/2 + O(\hat{s}^2)$. When substituted in, this leads to

$$\lim_{\hat{s}\to 0}\left(\frac{1}{\hat{w}} - \frac{1}{\tilde{u}_1}\right) = \frac{1}{6}\frac{K'''(0)}{\{K''(0)\}^{3/2}} - \frac{1}{2\sqrt{K''(0)}},$$

and the value at the mean μ in (1.27). The same argument gives the value in the left tail as in (1.37).

The same approach leads to the limit for the $\widehat{\Pr}_2$ in (1.32). In this case

$$\lim_{\tilde{s}\to 0}\left(\frac{1}{\tilde{w}_2} - \frac{1}{\tilde{u}_2}\right) = \lim_{\tilde{s}\to 0}\frac{1}{\tilde{u}_3}\left\{\left(\frac{\tilde{u}_3}{\tilde{w}_2} - 1\right) + \left(1 - \frac{\tilde{s}}{2\sinh(\tilde{s}/2)}\right)\right\}$$
$$= \frac{1}{6}\frac{K'''(0)}{\{K''(0)\}^{3/2}} + \lim_{\tilde{s}\to 0}\frac{1}{\tilde{s}\sqrt{K''(\tilde{s})}}\left\{\frac{1}{24}\tilde{s}^2 + O(\tilde{s}^4)\right\}$$
$$= \frac{1}{6}\frac{K'''(0)}{\{K''(0)\}^{3/2}}.$$

What prevents the inclusion of an additional limiting term here, as occurs with the first continuity-correction, is the expansion $2\sinh(\tilde{s}/2) = \tilde{s} + O(\tilde{s}^3)$ which lacks a $O(\tilde{s}^2)$ term.

2.6 Exercises

1. Prove the equivariance property in (2.2) for first- and second-order saddlepoint densities. Also show that the continuous CDF approximation in (1.21) and its second-order adjustment in (2.48) are invariant under linear transformation.
2. Suppose that X has a density or mass function that is symmetric about μ.
   (a) For continuous X prove the continuous CDF approximation in (1.21) satisfies
   $$\hat{F}(x) = 1 - \hat{F}(2\mu - x) \qquad x \in \mathcal{I}_X.$$
   (b) Show that this symmetry is maintained when the second-order correction term in (2.48) is included.
   (c) If X has an integer lattice mass function symmetric about μ, prove that
   $$\widehat{\Pr}_i(X \leq k) = \widehat{\Pr}_i(X \geq 2\mu - k) \qquad k^- < \mu;\ k \in \mathcal{I}_X \tag{2.78}$$
   for $i = 2, 3$.
   (d) For the CDF in (1.37), show that
   $$\hat{F}_{1a}(k) = \widehat{\Pr}_1(X \geq 2\mu - k) \qquad k \leq \mu;\ k \in \mathcal{I}_X$$
   but that the same relationship cannot hold for $\hat{F}_1(k)$ specified in (1.29).
   (e) More generally, show that the continuity-corrected approximations $\widehat{\Pr}_2$ or $\widehat{\Pr}_3$ are invariant under the transformation $X \leftrightarrow -X$. Deduce from this that they are invariant under all 1-1 monotonic transformations.
3. Consider the asymptotic expansions for the Laplace approximation.
   (a) Derive the expression in (2.14).
   (b) Following the expansion arguments in section 2.5.1, show, by Taylor expansion of $h(x)$ about $\hat{x}$, that (2.19) is accurate to relative order $O(n^{-1})$.
   (c) Show, using the Taylor expansion arguments of part (b), that the $O(n^{-1})$ term is the same as that given in (2.20).
   (d) Fill in the details of the derivation of (2.20) as outlined in section 2.5.1.
4. Consider the higher-order expansion terms for $\Gamma(\alpha)$ in (2.17) and for $B(\alpha, \beta)$ in (2.18).
   (a) Derive these expressions.
   (b) Show that the use of (2.19) and (2.20) to determine expansions for $\Gamma(\alpha)$ and $B(\alpha, \beta)$, as set out in (2.21) and (2.22), gives the same expansions as in part (a).
5. The beta–binomial or Pólya mass function is the marginal distribution of X when defined conditionally as follows. Suppose $X|Y = y$ is Binomial (n, y) and Y is Beta (α, β). The marginal mass function is
   $$p(k) = \binom{n}{k} B(\alpha, \beta)^{-1} B(k + \alpha, n - k + \beta) \qquad k = 0, \ldots, n. \tag{2.79}$$
   Show that Laplace's approximation to $p(k)$ may be applied to give expressions of the form (2.79) in which the last Beta function is approximated by using (2.12).
6. Suppose that $X_1, \ldots, X_n$ is an i.i.d. sample from a Normal (μ, σ^2) distribution. If a conjugate normal–gamma prior is used on (μ, σ^{-2}) in which $\mu|\sigma^2 \sim$ Normal $(\mu_0, c_0\sigma^2)$ and $\sigma^{-2} \sim$ Gamma (α_0, β_0), then its posterior is again a normal–gamma distribution with updated parameters $(\mu_1, c_1, \alpha_1, \beta_1)$.
   (a) The marginal posterior distribution on μ is a translated and rescaled Student t distribution. Show that Laplace's approximation for marginalization gives a shape proportional to the true posterior t distribution.
   (b) The marginal posterior on σ^{-2} is gamma. Show that Laplace's approximation for marginalization gives the same gamma shape.

2.6 Exercises

7. Suppose the bivariate density of (X, Y) is Dirichlet (α, β, γ) with density

$$f(x, y) = \frac{\Gamma(\alpha + \beta + \gamma)}{\Gamma(\alpha)\Gamma(\beta)\Gamma(\gamma)} x^{\alpha-1} y^{\beta-1} (1 - x - y)^{\gamma-1}$$

over the simplex $\{(x, y) : x, y > 0; x + y < 1\}$. The marginal density of X is Beta $(\alpha, \beta + \gamma)$. Show that Laplace's approximation for marginalization gives a result proportional to this beta distribution.

8. The mean of a Gamma (α, β) distribution is α/β.
   (a) Show that Laplace's approximation for the mean is

   $$\frac{\alpha}{\beta} \frac{\hat{\Gamma}(\alpha)}{\Gamma(\alpha)} \left(1 + \frac{1}{12\alpha}\right).$$

   (b) Show that if (2.19) is applied instead with $h(x) = x$, the leading term is

   $$\frac{\alpha - 1}{\beta} \frac{\hat{\Gamma}(\alpha - 1)(\alpha - 1)}{\Gamma(\alpha)}.$$

   Also show the expression within curly braces in (2.20) is

   $$\frac{\alpha + 1/12}{\beta} \frac{\hat{\Gamma}(\alpha - 1)(\alpha - 1)}{\Gamma(\alpha)}.$$

   (c) Compare accuracies for the various approximations for $\alpha = 3/2$ and $\beta = 1$.

9. Consider the saddlepoint density and CDF approximations for the distribution of $\bar{X}$ in section 2.2.2.
   (a) Derive (2.25) from (1.4).
   (b) Show that

   $$-\frac{1}{n} \frac{\partial^2 \ln f(\bar{x}_s)}{\partial \bar{x}_s^2} \simeq \frac{1}{K''(s)} = O(1). \qquad (2.80)$$

   Deduce from (2.29) and (2.80) that $K'(s) = \bar{x}_s + O(n^{-1})$.

10. Derive the Temme approximation to the incomplete gamma function in (2.35).
    (a) Show that the substitution $w = \hat{w}(x)$ leads to

    $$h(w) = \frac{\hat{\Gamma}(\alpha)}{\Gamma(\alpha)} \frac{w}{(x - \alpha)/\sqrt{\alpha}} \qquad (2.81)$$

    where x is implicitly determined from w as $\hat{w}(x) = w$.
    (b) Prove $dw/dx > 0$ and, in doing so, show that

    $$h(0) = \frac{\hat{\Gamma}(\alpha)}{\Gamma(\alpha)} \lim_{u \to 0} \frac{\hat{w}(u)}{u} = \frac{\hat{\Gamma}(\alpha)}{\Gamma(\alpha)}$$

    where $u = (x - \alpha)/\sqrt{\alpha}$ and $\hat{w}(u)$ is $\hat{w}$ as a function of u (Example 3.19, Barndorff-Nielsen and Cox, 1989).
    (c) Compare this with the more general derivation of (1.21) in which $h(\hat{w}) = \hat{w}/\hat{u}$. Note that $h(w)$ in (2.81) is

    $$\frac{\hat{f}(x) \hat{w}(x)}{f(x) \hat{u}(x)}$$

    which effectively rewrites the original integral of the gamma density as the same integral of its saddlepoint density. This is the approach taken in the general derivation of (1.21) and explains why the approximations are related by (2.36).

(d) For $\alpha = 0.5$ plot the function $h_2(w)$ vs. w for $|w| < 2$. Use numerical integration to compute the error (2.45) of the Temme approximation.

11. Consider the Temme approximation to the incomplete beta function in (2.37).
    (a) Show the approximation is given in (2.39).
    (b) For $\alpha = 0.4$ and $\beta = 0.8$ plot the function $h_2(w)$ vs. w for $|w| < 2$. Use numerical integration to compute the error (2.45) of the Temme approximation.

12. Prove that $\lim_{\hat{s} \to 0} \hat{w}/\hat{s}$ exists and is $\sqrt{K''(0)}$ by Taylor expanding $K(\hat{s})$ and $x = K'(\hat{s})$ about $\hat{s} = 0$ in the expression for $\hat{w}$.

13. Derive the asymptotic expansion in (2.44) and verify the error term in (2.45).

14. Suppose X has a distribution of the form

$$X \sim \prod_{i=1}^{k} F_i(\alpha_i, \beta_i)$$

where $\{F_i\}$ are independent F-variables with the indicated degrees of freedom. Such distributions arise as the ratio of generalized variances in multivariate analysis as described in Mardia *et al.* (1979, thm. 3.7.2). Construct a saddlepoint approximation for the CDF of X and perform a simple computation with it. Verify its accuracy with simulation.

15. Verify that the correction term within the curly braces of (2.32) does not depend on $\hat{s}$ if $K(s)$ is the CGF of a normal or gamma distribution.

16. The idea underlying the derivation of $\hat{F}$ in (1.21) is to approximate the integral of saddlepoint density $\hat{f}$ with a Temme approximation. Thus, $\hat{F}$ is constructed from $\hat{f}$. Consider working in the opposite order and attempt to approximate f from $\hat{F}$ by differentiation.
    (a) Show that

    $$\frac{d\hat{w}}{dx} = \frac{\hat{s}}{\hat{w}} \qquad \frac{d\hat{u}}{dx} = \frac{1}{\sqrt{K''(\hat{s})}} + \frac{1}{2}\hat{s}\hat{k}_3$$

    and

    $$\hat{F}'(x) = \hat{f}(x)\left(1 + \frac{1}{\hat{u}^2} + \frac{\hat{k}_3}{2\hat{u}} - \frac{\hat{u}}{\hat{w}^3}\right).$$

    (b) In the asymptotic setting of a sample mean show that differentiation of (2.46) leads to

    $$\hat{F}'(\bar{x}) = \hat{f}(\bar{x})(1 + R/n),$$

    where

    $$R = \left(\frac{1}{\hat{u}^2} + \frac{\hat{k}_3}{2\hat{u}} - \frac{\hat{u}}{\hat{w}^3}\right) \tag{2.82}$$

    and the values of $\hat{w}$ and $\hat{u}$ are given in (2.47).

    (c) Compare R to (2.48) and note that the term in curly braces, which is known to be $O(1)$, is related to R according to

    $$O(1) = \frac{1}{\hat{u}}\left(\tfrac{1}{8}\hat{k}_4 - \tfrac{5}{24}\hat{k}_3^2\right) - \frac{R}{\hat{u}}.$$

    Since $\hat{u}^{-1} = O(\sqrt{n})$ on sets of bounded central tendency, infer that

    $$R = \left(\tfrac{1}{8}\hat{k}_4 - \tfrac{5}{24}\hat{k}_3^2\right) + O(n^{-1/2})$$

    so that $\hat{F}'(\bar{x})$ is the second-order saddlepoint density to order $O(n^{-3/2})$.

    (d) Apply this in estimating the Beta (α, β) density. What is the R term in this instance?

17. Consider the difficult problem of approximating the CDF for X that has a uniform density of height $1/2$ over the disconnected set $(-2, -1) \cup (1, 2)$.

(a) Show that the MGF of X is

$$M(s) = s^{-1}\{\sinh(2s) - \sinh(s)\} \qquad s \neq 0.$$

(b) Plot the saddlepoint CDF $\hat{F}$ and note that it is decreasing over the range $x \in (1.5, 1.6)$.

18. The Noncentral Beta (α, β, λ) distribution is defined as the distribution of $X = Y_\alpha/(Y_\alpha + Y_\beta)$ when $Y_\alpha \sim$ Noncentral $\chi^2(2\alpha, \lambda)$ (see Exercise 4, chapter 1) independently of $Y_\beta \sim$ Central $\chi^2(2\beta)$ with α and β assuming half-integer values.

(a) By defining Z_x as in (1.25), show that

$$\Pr(X \leq x) = \Pr(Z_x \leq 0)$$

can be saddlepoint approximated by using the CGF of Z_x given as

$$K_{Z_x}(s) = -\alpha \ln\{1 - 2s(1-x)\} + \frac{\lambda s(1-x)}{1 - 2s(1-x)} - \beta \ln(1 + 2sx).$$

(b) Furthermore, show using Maple that the saddlepoint solves the quadratic equation

$$a\hat{s}^2 + b\hat{s} + c = 0$$

where

$$a = -8x(x-1)^2(\beta + \alpha)$$
$$b = -2(x-1)(4\beta x + 4\alpha x + \lambda x - 2\alpha)$$
$$c = 2\alpha - 2\alpha x + \lambda - \lambda x - 2\beta x.$$

(c) Choose small values of α and β and a value for λ. Construct a small table of values for the CDF. Check the accuracy by computing the true values with specialized software or by integrating a truncated version of the true density given as

$$f(x) = \exp\left(-\tfrac{1}{2}\lambda\right)\left\{\sum_{i=0}^\infty \frac{1}{i!}\left(\frac{\lambda}{2}\right)^i \frac{x^{\alpha+i-1}}{B(\alpha+i,\beta)}\right\}(1-x)^{\beta-1}.$$

(d) The noncentral F distribution is the distribution of variable $Z = Y_\alpha/Y_\beta$. Values of its CDF are determined from the noncentral beta by $\Pr(Z \leq z) = \Pr\{X \leq z/(1+z)\}$.

19. The logit change in variables for the beta integral in section 2.2.1 increased the accuracy of Laplace's approximation for the beta function. For $X \sim$ Beta (α, β), another approximation to the incomplete beta uses the Lugannani and Rice approximation for $Y = \ln\{X/(1-X)\}$ as

$$F_{\alpha,\beta}(x) = \Pr(X \leq x) = \Pr\left(Y \leq \ln\frac{x}{1-x}\right).$$

(a) Compute the MGF of Y.

(b) For $x = 0.4$ and $\beta = 0.8$, compute the CDF approximations for both Y and $\ln X$. Table 2.5 gives, for the various values of α, the percentage relative errors in approximating $\Pr(X \leq 0.4)$ using the MGFs of $\ln\{X/(1-X)\}$ and $\ln X$.

Verify that the listed percentages are correct. This requires computing the true incomplete beta values either from software or by using the values listed in table 1.6. The relative errors are

$$100 \frac{\hat{F}_{\alpha,0.8}(0.4) - F_{\alpha,0.8}(0.4)}{\min\{F_{\alpha,0.8}(0.4), 1 - F_{\alpha,0.8}(0.4)\}}.$$

The errors are uniformly smaller when using the MGF of $\ln X$. Both methods, however, are considerably more accurate than the approximation from chapter 1 as may be seen by comparing table 2.5 to figure 2.1.

Table 2.5. *Percentage relative errors in saddlepoint CDF approximation when using logit and log transformation of the beta distribution from table 1.6.*

α	$\ln\{X/(1-X)\}$	$\ln X$
.1	−3.243	−1.722
.2	−1.881	−.9755
.4	−.7919	−.3665
.6	−.3580	−.1148
.8	−.2198	.001078
.9	−.1609	.007395

20. Consider the Mellin transform approach to the CDF of a Beta (α, β).
    (a) Show that the saddlepoint equation is
    $$\Psi(\alpha + \hat{s}) - \Psi(\alpha + \beta + \hat{s}) = \ln y, \qquad (2.83)$$
    where Ψ is the di-gamma function.
    (b) The di-gamma function has the recurrence formula
    $$\Psi(z+1) = \Psi(z) + \frac{1}{z},$$
    (see Abramowitz and Stegun 1970, §6.3.5). As a consequence, an explicit saddlepoint solution to (2.83) results when $\beta = 1$. Determine the solution and specify the CDF approximation.
    (c) Show that if $B \sim \text{Beta}(\alpha, 1)$ then B^{-1} has a Pareto distribution. Use the explicit solution in part (b) to determine a CDF approximation for this Pareto.
21. Suppose that X has support on the δ-lattice $L_\delta = \{\gamma, \gamma \pm \delta, \gamma \pm 2\delta, \ldots\}$.
    (a) Derive the continuity-corrected approximations for the CDF of X in (2.58) and (2.59).
    (b) Derive the continuity corrections in (2.60) and (2.61) for the second approximation.

3

Multivariate densities

The univariate saddlepoint density functions of (1.4) are generalized to provide density approximations in m dimensions. Also univariate mass functions on the integer lattice in (1.13) are generalized to higher dimensions. For these discrete settings, the lattice convention of section 2.1 is assumed to hold for each dimension so that the support of multivariate mass functions is a subset of the integer lattice in $\Re^m$ denoted as I^m.

3.1 Saddlepoint density and mass functions

3.1.1 Preliminaries

The development of approximations for multivariate density and mass functions proceeds in much the same manner as in the univariate setting since no conceptual differences exist. As in the univariate setting, the multivariate saddlepoint density and mass functions are expressions for approximating their true multivariate counterparts that are based on their multivariate MGFs defined on $\Re^m$.

Suppose $X = (X_1, \ldots, X_m)^T$ is a random vector with a nondegenerate distribution in $\Re^m$. If the components are all continuous, then let $f(x)$ be the density of X at x for all $x \in \Re^m$. When X consists entirely of lattice components, let $p(k)$ be the joint mass function for $k \in I^m$. The mixed setting, in which some of the components are continuous and the others are discrete lattice variates, can also be addressed by using the saddlepoint formulae of this chapter.

Let $\mathcal{X}$ be the support of random vector X and define its joint MGF in the continuous and lattice cases as

$$M(s) = E\{\exp(s^T X)\} = \begin{cases} \int_{\mathcal{X}} \exp(s^T x) f(x) \, dx \\ \sum_{k \in \mathcal{X} \subseteq I^m} \exp(s^T k) p(k) \end{cases}$$

for all values of $s = (s_1, \ldots, s_m) \in \Re^m$ such that the integral or sum converges. Various properties of M and CGF $K(s) = \ln M(s)$ are discussed in Barndorff-Nielsen and Cox (1989, §6.4) and Rockafellar (1970). Those that are important and pertinent to saddlepoint methods are listed below along with some assumptions that are made for this chapter.

(1) If $\mathcal{S}$ is the maximal convergence set for M in a neighborhood of $0 \in \Re^m$, then $\mathcal{S}$ is a convex set.

(2) Assume that $\mathcal{S}$ is open and refer to such a setting as *regular*. In such settings, all the moments of X are finite. The mean vector and covariance matrix of X are given by

$$E(X) = K'(0) = \left(\frac{\partial K}{\partial s_1}, \ldots, \frac{\partial K}{\partial s_m}\right)^T_{s=0}$$

and

$$\text{Cov}(X) = K''(0) = \left(\frac{\partial^2 K}{\partial s_i \partial s_j} : i, j = 1, \ldots, m\right)_{s=0}. \tag{3.1}$$

(3) If $\mathcal{I}_\mathcal{X} \subseteq \Re^m$ denotes the interior of the convex hull of $\mathcal{X}$, then $K' : \mathcal{S} \to \mathcal{I}_\mathcal{X}$ is a bijection (1-1 and onto).

(4) The CGF is a strictly convex function on $\mathcal{S}$ so that the matrices $\{K''(s) : s \in \mathcal{S}\}$ are all positive definite symmetric. This fact, in the particular case $s = 0$, assures that (3.1) is a nonsingular covariance matrix.

Simpler notation is used when dealing with bivariate distributions. In such instances the random vector is denoted as $(X, Y)^T$ with joint CGF $K(s, t)$ and joint density $f(x, y)$.

3.1.2 Saddlepoint density and mass functions

The saddlepoint density for continuous vector X is defined on $\mathcal{I}_\mathcal{X} \subseteq \Re^m$ as

$$\hat{f}(x) = \frac{1}{(2\pi)^{m/2} |K''(\hat{s})|^{1/2}} \exp\{K(\hat{s}) - \hat{s}^T x\}, \qquad x \in \mathcal{I}_\mathcal{X} \tag{3.2}$$

where saddlepoint $\hat{s}$ is the unique solution in $\mathcal{S}$ to the m-dimensional saddlepoint equation

$$K'(\hat{s}) = x. \tag{3.3}$$

If X is continuous with convex support $\mathcal{X}$, then $\mathcal{I}_\mathcal{X}$ is the interior of $\mathcal{X}$ and (3.2) is defined at all points of support except for the boundary points of $\mathcal{X}$. When $\mathcal{X}$ is not convex, then (3.2) is practically meaningful on $\mathcal{I}_\mathcal{X} \cap \mathcal{X}$; for the remaining values in $\mathcal{I}_\mathcal{X} \cap \mathcal{X}^c$ the saddlepoint density $\hat{f}$ in (3.2) is well-defined and provides positive weight even though the actual density value is zero. This is not really a problem as long as the set $\mathcal{I}_\mathcal{X} \cap \mathcal{X}^c$ can be identified and eliminated from consideration.

In the lattice setting, the saddlepoint mass function $\hat{p}(k)$, for $k \in \mathcal{I}_\mathcal{X}$, is expressed by using the same formula with k replacing x in both (3.2) and the saddlepoint equation (3.3). The saddlepoint formula is computable for any vector $x \in \mathcal{I}_\mathcal{X}$ regardless of whether $x \in I^m$, however it is only practically meaningful over the integer vectors $k \in \mathcal{I}_\mathcal{X} \cap \mathcal{X} \subseteq I^m$. The normalized versions of these expressions are denoted as $\bar{f}$ and $\bar{p}$. If $\mathcal{X}$ is not convex, then saddlepoint values at points in $\mathcal{I}_\mathcal{X} \setminus \mathcal{X}$ need to be excluded from the integration or summation during the normalization process.

3.1.3 Examples

Standard multivariate Normal$_m(0, I_m)$ density

Suppose the components of X are i.i.d. Normal $(0,1)$ with joint CGF $K(s) = s^T s/2$ for all $s \in \Re^m$. The saddlepoint equation is $K'(\hat{s}) = \hat{s} = x$, $\mathcal{X} = \Re^m$, and $K''(s) = I_m$.

Saddlepoint expression (3.2) reduces to

$$\hat{f}(x) = (2\pi)^{-m/2} \exp\left(-\tfrac{1}{2} x^T x\right) = \prod_{i=1}^{m} \phi(x_i) \qquad x \in \Re^m,$$

and exactly reproduces the true density.

Multivariate gamma density

Suppose $Z_1, \ldots, Z_n$ are a random sample of m-vectors from a multivariate Normal$_m(0, \Sigma)$ with Σ positive definite (denoted $\Sigma > 0$). The covariance sum of squares matrix

$$W = \sum_{i=1}^{n} Z_i Z_i^T$$

has a Wishart$_m(n, \Sigma)$ distribution and the diagonal entries of $W/2$ have a multivariate Gamma$_m(n/2, \Sigma)$ or a generalized Rayleigh distribution as described in Johnson and Kotz (1972, chap. 40, §3). Individual marginal densities for the multivariate Gamma are, of course, all Gamma. The components are independent when Σ is diagonal.

Consider the bivariate case in standardized variables so that $m = 2$ and Σ is a correlation matrix with off-diagonal correlation $\rho \in (-1, 1)$. For notational simplicity, denote the two variates as X and Y, and take $n = 2$ so the two marginal densities are Exponential (1). This special case is the bivariate exponential distribution introduced by Wicksell (1933) and Kibble (1941) and used by Downton (1970) in a reliability setting. Its density is

$$f(x, y) = (1 - \rho^2)^{-1} \exp\left(-\frac{x + y}{1 - \rho^2}\right) I_0 \left(\frac{2\sqrt{\rho^2 xy}}{1 - \rho^2}\right) \qquad x, y > 0 \qquad (3.4)$$

where I_0 denotes a Bessel I function of order 0. An alternative form of the density as an infinite expansion, that does not rely on special functions, is given in Johnson and Kotz (1972, chap. 40, Eq. 10.2). The reader is asked to derive the MGF in Exercise 1 as

$$M(s, t) = \{(1 - s)(1 - t) - \rho^2 st\}^{-1} \qquad (s, t) \in \mathcal{S} \qquad (3.5)$$

where $\mathcal{S}$ is the maximal neighborhood of convergence for (3.5) about $(s, t) = (0, 0)$. The boundary of $\mathcal{S}$ is determined by one of the pair of hyperbolic curves given by

$$\{(s, t) : (1 - s)(1 - t) - \rho^2 st = 0\}$$

and displayed in figure 3.1. In the (s, t)-plane, these two hyperbolas have common asymptotes $s = (1 - \rho^2)^{-1}$ and $t = (1 - \rho^2)^{-1}$ and a common 45 axis of symmetry given by the line $s = t$. The hyperbola to the lower left is denoted H_l and the one in the upper right is H_u. Hyperbola H_l is concave and $\mathcal{S}$ is the concave region of $\Re^2$ lying below and to the left of H_l; thus $\mathcal{S}$ is an open and convex set containing $(0, 0)$.

The correlation of X and Y is determined as ρ^2 by computing the Hessian of K at $(0, 0)$ using (3.5). When $\rho = 0$ the MGF separates in variables s and t to demonstrate that X and Y are independent in this instance. The left-hand side of figure 3.2 has two surface/contour plots displaying side and top views of this density with $\rho = 0.9$ and correlation $\rho^2 = 0.81$. The variables X and Y are exchangeable with $f(x, y) = f(y, x)$ for all $x, y > 0$. This feature may be seen in figure 3.2 through the reflective symmetry about the plane $y = x$.

78   *Multivariate densities*

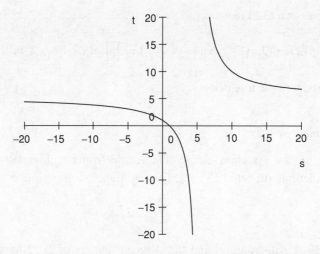

Figure 3.1.   Geometry showing the boundary of the convergence strip $\mathcal{S}$ for the bivariate exponential.

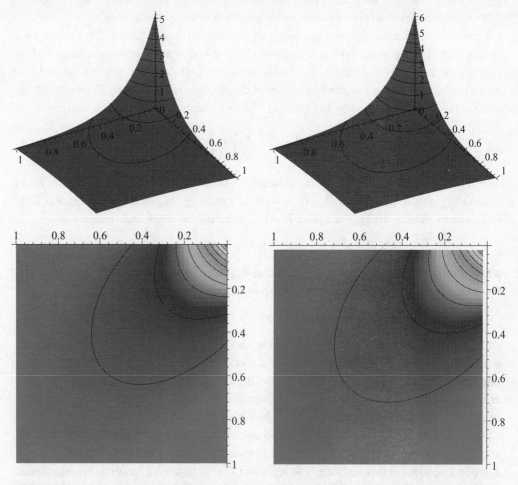

Figure 3.2.   Two views of the bivariate exponential density (left-hand side) with correlation 0.81 versus the same views for its saddlepoint density (right-hand side).

The unnormalized saddlepoint density is also shown on the right-hand side of figure 3.2. It is virtually identical in shape, and exhibits the same reflective symmetry. The symmetry of the saddlepoint density is easily explained by the exchangeability of the MGF in s and t; exchanging s and t along with x and y yields

$$\hat{s}(x, y) = \hat{t}(y, x)$$
$$\hat{t}(x, y) = \hat{s}(y, x)$$

and the saddlepoint expression for $\hat{f}(x, y)$ must be equal to that for $\hat{f}(y, x)$. The saddlepoint density appears to be slightly above the true density and needs to be normalized to rescale it to the height of the true density.

The geometry of the saddlepoint equations in $\mathfrak{R}^2$ is now described. Each equation in the pair of saddlepoint equations determines two distinct hyperbolic curves. One curve is in $\mathcal{S}$ and below H_l, while the another is above H_u. When both saddlepoint equations are considered simultaneously, the overlap yields two solutions: the saddlepoint solution in $\mathcal{S}$ that is below H_l and a bogus solution above H_u. Any search for the saddlepoint needs to be restricted to $\mathcal{S}$ since the root-seeking algorithm may otherwise find the bogus solution above H_u. Solving for $\hat{s}$ in terms of $\hat{t}$ in $\partial K/\partial \hat{s} = x$ yields

$$\hat{s} = \frac{1 - \hat{t}}{1 + \hat{t}(\rho^2 - 1)} - \frac{1}{x}. \tag{3.6}$$

The solution is kept in $\mathcal{S}$, by choosing $\hat{t}$ as the smaller root of the quadratic

$$y(1 - \rho^2)^2 \hat{t}^2 + (1 - \rho^2)(1 - \rho^2 - 2y)\hat{t} - (1 - \rho^2 - y + x\rho^2) = 0.$$

Upon further simplification, this is

$$\hat{t} = \frac{1}{1 - \rho^2} - \frac{1}{2y}\left\{1 + \sqrt{1 + 4xy\rho^2/(1 - \rho^2)^2}\right\}. \tag{3.7}$$

The saddlepoint equation is a bijective map from $\mathcal{S}$, below H_l, onto $(0, \infty)^2$. Outside of $\mathcal{S}$, it is also a bijection from the region above H_u onto $(0, \infty)^2$. The region in between H_l and H_u leads to negative values for M in (3.5). From this discussion, it is clear that bogus solutions to the saddlepoint equation can arise if solutions are allowed to stray outside of the region $\mathcal{S}$.

Additional understanding of the saddlepoint mapping $(\hat{s}, \hat{t}) \leftrightarrow (x, y)$ between $\mathcal{S}$ and $(0, \infty)^2$ is gained by considering the mapping near the boundaries. Consider passing clockwise along a curve in $\mathcal{S}$ which is very close and just below its upper boundary H_l. The image of this directed curve under the map K' is a curve on the distant horizon of $(0, \infty)^2$ moving from the y-axis to the x-axis. Now, complete the boundary of $\mathcal{S}$ by considering an arc passing clockwise from the southeast horizon around to the northwest horizon in the (s, t)-plane. The image of this portion in $(0, \infty)^2$ is an inward curve just above the x-axis moving toward the origin and then pulling upward alongside the y-axis to its horizon. More will be said about the geometry of this mapping when the conditional density approximation for Y given $X = x$ is considered in the next chapter.

Random variables X and Y are independent when $\rho = 0$, and the saddlepoint density also separates in this instance as we now show. In dealing with the joint MGF

with $\rho = 0$,
$$K(s, t) = -\ln(1-s) - \ln(1-t) \qquad s < 1, \, t < 1$$
and the two saddlepoint equations are solved separately as
$$\hat{s} = 1 - 1/x \qquad \hat{t} = 1 - 1/y.$$
Naturally, the same solution occurs in (3.6) and (3.7) by setting $\rho = 0$. This leads to the explicit bivariate density
$$\hat{f}(x, y) = \frac{1}{\hat{\Gamma}(1)^2} \exp(-x - y) \qquad x > 0, \, y > 0$$
which separates as the product of the two marginal saddlepoint densities given in (1.7).

Multinomial mass function

The Multinomial $(n; \pi_1, \ldots, \pi_m, \pi_{m+1})$ distribution for vector $X = (X_1, \ldots, X_m)^T$ with $\sum_{i=1}^{m+1} \pi_i = 1$ has mass function
$$p(k) = \Pr(X_1 = k_1, \ldots, X_m = k_m)$$
$$= \binom{n}{k_1, \ldots, k_{m+1}} \prod_{i=1}^{m+1} \pi_i^{k_i} \qquad k \in S^m, \tag{3.8}$$
where $k_{m+1} = n - \sum_{i=1}^m k_i$ and
$$S^m = \{k \in I^m : \sum_{i=1}^m k_i \le n, \, k_j \ge 0 \quad j = 1, \ldots, m\}.$$
The joint MGF of X is
$$M(s) = \left(\sum_{i=1}^m \pi_i e^{s_i} + \pi_{m+1}\right)^n \qquad s = (s_1, \ldots, s_m)^T \in \mathfrak{R}^m. \tag{3.9}$$

The validity of (3.9) is determined by noting that the mass function in (3.8) is the coefficient of the $\prod_{i=1}^{m+1} \pi_i^{k_i}$ term in the multinomial expansion of (3.9). The saddlepoint equations may be solved for any k in the interior of the convex hull of S^m ($k \in \mathcal{I}_\mathcal{X}$). The equations are
$$\frac{n\pi_i \exp(\hat{s}_i)}{\hat{\zeta} + \pi_{m+1}} = k_i \qquad i = 1, \ldots, m, \tag{3.10}$$
where $\hat{\zeta} = \sum_{i=1}^m \pi_i \exp(\hat{s}_i)$. Adding all m equations allows for the solution
$$\hat{\zeta} = \left(\frac{n}{k_{m+1}} - 1\right) \pi_{m+1}.$$
Further computation yields
$$\hat{s}_i = \ln\left(\frac{\pi_{m+1} k_i}{k_{m+1} \pi_i}\right) \qquad i = 1, \ldots, m.$$

Table 3.1. *Percentage relative errors of saddlepoint approximations $\hat{p}$ and $\bar{p}$ for the Trinomial $(10; 0.8, 0.15, 0.05)$ mass function.*

(k_1, k_2, k_3)	$p(k_1, k_2, k_3)$	% relative error $\hat{p}$	% relative error $\bar{p}$
(2, 7, 1)	$.0^4 1968$	13.4	.780
(3, 5, 2)	$.0^3 2449$	8.04	5.49
(4, 5, 1)	$.0^2 1960$	11.6	2.33
(6, 3, 1)	.03716	12.1	1.93
(7, 2, 1)	.08493	13.4	.778
(8, 1, 1)	.1132	17.8	3.09

The reader is asked to finish the saddlepoint derivation in Exercise 2 and, in particular, show that the Hessian matrix evaluated at $\hat{s}$ is

$$|K''(\hat{s})| = n^{-1} \prod_{i=1}^{m+1} k_i. \tag{3.11}$$

This leads to the saddlepoint mass function as

$$\hat{p}(k) = \widehat{\binom{n}{k_1, \ldots, k_{m+1}}} \binom{n}{k_1, \ldots, k_{m+1}}^{-1} p(k) \quad k \in S^m, \; k_i > 0 \; \forall i. \tag{3.12}$$

The first term is the multinomial coefficient with each factorial replaced with its Stirling's approximation as in (1.16).

The relationship between $\hat{p}$ and p in (3.12) suggests that the error of approximation is greatest when k has many component values of 1. Then, the errors in the Stirling approximations are at there worst and combine in a multiplicative manner. This is verified in the entries of table 3.1 which are mass probabilities for a Trinomial $(10; 0.8, 0.15, 0.05)$. The relative error decreases when there are fewer components with value 1. Normalization substantially improves upon the accuracy of $\hat{p}$. The probability that k is on the boundary in this example is 0.6882. The total sum of the $\hat{p}$-values off the boundary, needed to compute $\bar{p}$, is 0.3563.

Bivariate Poisson mass function

This distribution was introduced by Holgate (1964) as the joint distribution of

$$X = P_1 + P_3 \tag{3.13}$$
$$Y = P_2 + P_3$$

where P_1, P_2, and P_3 are independent Poisson variates with means λ, μ, and ν respectively. The distribution arises in a reliability context where it represents the number of failures in a two component system subjected to *fatal shocks*. Suppose X and Y count the total number of failures up to time t for the two types of components, where components that fail are immediately replaced to keep the system operational. Failures can occur individually or simultaneously as determined by three independent Poisson processes. The first and second processes, with rates λ and μ, govern individual failures of the x- and y-components respectively. The third process, with rate ν, induces fatal shocks for the simultaneous failure of both components. Such a fatal shock would result, for example, when both components are subjected to a common power surge of sufficient strength to damage both components. The mass function is determined from direct convolution of P_3 with the vector (P_1, P_2) as

$$p(i, j) = \sum_{k=0}^{\min(i,j)} \Pr(P_1 = i - k, P_2 = j - k) \Pr(P_3 = k)$$

$$= e^{-\lambda-\mu-\nu} \sum_{k=0}^{\min(i,j)} \frac{\lambda^{i-k} \mu^{j-k}}{(i-k)!(j-k)!} \frac{\nu^k}{k!} \qquad i \geq 0, \ j \geq 0.$$

The MGF is determined from (3.13) as

$$M(s, t) = \exp\{\lambda(e^s - 1) + \mu(e^t - 1) + \nu(e^{s+t} - 1)\} \quad (s, t) \in \Re^2. \tag{3.14}$$

The distribution has marginal distributions for X and Y that are Poisson $(\lambda + \nu)$ and Poisson $(\mu + \nu)$ respectively. The variables are independent when $\nu = 0$ as seen from either the construction or from (3.14).

The saddlepoint equations are

$$\begin{aligned} \lambda \hat{a} + \nu \hat{a} \hat{b} &= i & i > 0 \\ \mu \hat{b} + \nu \hat{a} \hat{b} &= j & j > 0 \end{aligned} \tag{3.15}$$

where $\hat{a} = \exp(\hat{s})$ and $\hat{b} = \exp(\hat{t})$. These equations admit an explicit solution, when (i, j) is not located on the boundary of the support, or when $i > 0$ and $j > 0$. The solution is

$$\hat{b} = (i - \lambda \hat{a})/\nu \hat{a},$$

where $\hat{a}$ is the larger root of

$$\lambda \nu \hat{a}^2 + \{\lambda \mu + \nu (j - i)\} \hat{a} - i \mu = 0.$$

Table 3.2 assesses the accuracy of the normalized saddlepoint mass function $\bar{p}$ in a setting that has failure rates of $\lambda = 1$, $\mu = 2$, and $\nu = 4$.

Normalization is simplified by the fact that the marginals are Poisson and the boundary probability is

$$d = \Pr(X = 0 \cup Y = 0) = e^{-(\lambda+\nu)} + e^{-(\mu+\nu)} - e^{-(\lambda+\mu+\nu)}.$$

Thus,

$$\bar{p}(i, j) = \frac{1-d}{\sum_{k=1}^{\infty} \sum_{l=1}^{\infty} \hat{p}(k, l)} \hat{p}(i, j) \qquad i > 0, \ j > 0$$

$$\simeq 0.9827 \hat{p}(i, j).$$

The percentage relative errors

$$R(i, j) = 100 \left\{ \frac{\bar{p}(i, j)}{p(i, j)} - 1 \right\} \qquad i > 0, \ j > 0$$

for the entries of table 3.2 are presented in table 3.3.

3.2 Development of the saddlepoint density

Table 3.2. *Values of the normalized saddlepoint mass function $\bar{p}$ (top) and its exact counterpart p (bottom) for the bivariate Poisson example.*

j	$i=1$	$i=2$	$i=3$	$i=6$	$i=9$
1	$.0^25905$	$.0^24977$	$.0^22321$	$.0^43560$	$.0^61028$
	$.0^25471$	$.0^24559$	$.0^22128$	$.0^43293$	$.0^79549$
2	$.0^29953$	.01562	.01156	$.0^33813$	$.0^51690$
	$.0^29119$	.01550	.01125	$.0^33673$	$.0^51633$
3	$.0^29284$	.02313	.02786	$.0^22338$	$.0^41693$
	$.0^28511$	.02249	.02817	$.0^22292$	$.0^41659$
5	$.0^22899$	.01276	.03093	.02090	$.0^35183$
	$.0^22675$	.01228	.03044	.02118	$.0^35153$
7	$.0^33748$	$.0^22365$	$.0^28778$	.03301	$.0^23932$
	$.0^33474$	$.0^22281$	$.0^28596$	.03359	$.0^23977$
10	$.0^55810$	$.0^45332$	$.0^33018$	$.0^26017$	$.0^26769$
	$.0^55404$	$.0^45159$	$.0^32960$	$.0^26009$	$.0^26928$

Table 3.3. *Percentage relative errors for the approximations in table 3.2.*

j	$i=1$	$i=2$	$i=3$	$i=6$	$i=9$
1	7.939	9.148	9.085	8.098	7.621
2	9.148	.7769	2.814	3.801	3.451
3	9.085	2.814	−1.101	2.018	2.050
5	8.385	3.896	1.605	−1.292	.5859
7	7.907	3.674	2.114	−1.706	−1.133
10	7.525	3.355	1.955	.1403	−2.311

The correlation of X and Y is easily determined from (3.14) or (3.13) as

$$\frac{\nu}{\sqrt{(\lambda+\nu)(\mu+\nu)}} = \frac{4}{\sqrt{30}} \simeq 0.730.$$

3.2 Development of the saddlepoint density

3.2.1 Laplace's approximation

Laplace's method in section 2.2.1 can be generalized to approximate m-dimensional integrals over a region $D \subseteq \Re^m$ as given in (3.16). Suppose that $g(\cdot)$ is a smooth function mapping D into $\Re$ with a unique minimum at $\hat{x}$ lying in the interior of D. Then,

$$I = \int_D e^{-g(x)} dx \simeq \frac{(2\pi)^{m/2}}{|g''(\hat{x})|^{1/2}} \exp\{-g(\hat{x})\} \qquad (3.16)$$

where g'' is the $(m \times m)$ Hessian matrix of second derivatives for g. Generally there are two important factors that determine the accuracy of the approximation: (i) the peakedness of integrand $\exp\{-g(\cdot)\}$ at $\hat{x}$ in $\Re^m$ and (ii) the degree to which $g(\cdot)$ is "quadratic-looking" in a neighborhood of $\hat{x}$. A tacit assumption when using this approximation is that the major contribution to the integral lies inside D and well away from its boundary. The approximation uses only local properties of g at $\hat{x}$ which include its value and its Hessian.

Example: Dirichlet $(\alpha_1, \ldots, \alpha_m, \alpha_{m+1})$ *integral*

This is an m-dimensional integral defined as

$$I = \frac{1}{\Gamma(\alpha.)} \prod_{i=1}^{m+1} \Gamma(\alpha_i) = \int_D \left(\prod_{i=1}^{m+1} y_i^{\alpha_i - 1} \right) dy, \qquad (3.17)$$

where $y = (y_1, \ldots, y_m)^T$, $y_{m+1} = 1 - \sum_{i=1}^m y_i$,

$$D = \left\{ y \in \Re^m : y_i > 0 \; \forall i, \; \sum_{i=1}^m y_i < 1 \right\}$$

is the simplex in $\Re^{m+1}$, $\alpha_i > 0$ for each i, and $\alpha. = \sum_{i=1}^{m+1} \alpha_i$. If the integration is performed in variable y, as stated in (3.17), then Laplace's approximation in (3.16) yields

$$\hat{I}_y = \frac{1}{(\alpha. - m - 1)^{m+1} \hat{\Gamma}(\alpha. - m - 1)} \prod_{i=1}^{m+1} (\alpha_i - 1) \hat{\Gamma}(\alpha_i - 1) \qquad (3.18)$$

as the reader is asked to show in Exercise 5. An interior minimum of g is obtained only when $\alpha_i > 1$ for all i; thus the approximation cannot be justified if $\alpha_i < 1$ for some i. Reasonable accuracy is achieved in the numerator by the fact that $(\alpha_i - 1)\hat{\Gamma}(\alpha_i - 1) \simeq \Gamma(\alpha_i)$, however, substantial inaccuracy may be anticipated in the denominator where $(\alpha. - m - 1)^{m+1} \hat{\Gamma}(\alpha. - m - 1)$ must approximate $\Gamma(\alpha.)$.

A more accurate approximation results from a change in the variable of integration. Take

$$dx = \prod_{i=1}^m dx_i = y_{m+1}^{-1} \prod_{i=1}^m y_i^{-1} dy_i = \left(\prod_{i=1}^{m+1} y_i^{-1} \right) dy, \qquad (3.19)$$

which expresses the Jacobian for the additive logistic transformation described in Aitchison (1986, §6.2) and given by

$$x_i = \ln(y_i / y_{m+1}) \qquad i = 1, \ldots, m.$$

The inverse transformation is

$$y_i = \frac{\exp(x_i)}{1 + \sum_{j=1}^m \exp(x_j)} \qquad i = 1, \ldots, m,$$

and the Dirichlet integral may be written as

$$I = \int_{\Re^m} \exp\left\{ \sum_{i=1}^m \alpha_i x_i - \alpha. \ln\left(1 + \sum_{j=1}^m e^{x_j}\right) \right\} dx.$$

The maximum of the integrand occurs at

$$\hat{x}_i = \ln(\alpha_i / \alpha_{m+1}) \qquad i = 1, \ldots, m \qquad (3.20)$$

for any setting in which I is defined, or when $\alpha_i > 0$ for all i. In terms of the original y_i-values, this maximum would occur at $\hat{y}_i = \alpha_i / \alpha.$. The Hessian matrix is the $(m \times m)$

3.2 Development of the saddlepoint density

matrix with components

$$g_{ij} = \frac{\partial^2 g}{\partial x_i \partial x_j} = \alpha. \, e^{x_i} (1+\xi)^{-1} \{\delta_{ij} - e^{x_j}(1+\xi)^{-1}\}, \tag{3.21}$$

where $\xi = \sum_{i=1}^{m} \exp(x_i)$ and δ_{ij} is the indicator function that $i = j$. The Hessian matrix evaluated at $\hat{x}$ has determinant $\alpha_{\cdot}^{-1} \prod_{i=1}^{m+1} \alpha_i$. Upon simplification, Laplace's approximation yields

$$\hat{I}_x = \frac{1}{\hat{\Gamma}(\alpha.)} \prod_{i=1}^{m+1} \hat{\Gamma}(\alpha_i), \tag{3.22}$$

which is clearly more accurate than the previous approximation $\hat{I}_y$ in (3.18). The numerical accuracy of $\hat{I}_x$ is tabulated for some selected examples in section 3.4.4.

Proof of (3.16). At critical value $\hat{x}$, the smooth function g has Taylor expansion

$$g(x) = g(\hat{x}) + \tfrac{1}{2}(x - \hat{x})^T g''(\hat{x})(x - \hat{x}) + R,$$

where the remainder R is assumed to be negligible. Retaining the quadratic portion of g only, then the right-hand side of (3.16) is approximately the multivariate normal integral

$$I \simeq e^{-g(\hat{x})} \int_D \exp\left\{-\tfrac{1}{2}(x-\hat{x})^T g''(\hat{x})(x-\hat{x})\right\} dx$$
$$\simeq e^{-g(\hat{x})} (2\pi)^{m/2} |g''(\hat{x})|^{-1/2}.$$

$\square$

The proof helps to clarify why the additive logistic transformation improves the accuracy of the approximation for the Dirichlet integral. This transformation is a bijection mapping the vector y in the simplex $D \subset \Re^{m+1}$ onto $x \in \Re^m$. Like previous changes of variable for this approximation, the integral in y is over a bounded set while the integration in new variable x is over the entire Euclidean space. The transformation has the effect of converting a Dirichlet vector into one whose density is more closely multivariate normal. This explains very simply why approximation $\hat{I}_x$ is more accurate than $\hat{I}_y$. More specifically, suppose $\{G_{\alpha_i} : i = 1, \ldots, m+1\}$ are independent gammas which underlie the Dirichlet with $G_{\alpha_i} \sim$ Gamma $(\alpha_i, 1)$. Then $Y_i = G_{\alpha_i} / \sum_{j=1}^{m+1} G_{\alpha_j}$ and

$$X_i = \ln G_{\alpha_i} - \ln G_{\alpha_{m+1}} \qquad i = 1, \ldots, m.$$

Clearly the joint density of $\{X_i\}$ should approximate a multivariate normal density better than the density of $\{Y_i\}$. The integration dx is a setting for which the Laplace derivation would be expected to achieve high accuracy.

Properties of Laplace's approximation

Invariance The Dirichlet example shows that the approximation is not invariant under all 1-1 transformations of the variable of integration. However, for nonsingular linear transformations of the form $y = Ax + b$, where A is $(m \times m)$ and nonsingular and $b \in \Re^m$, the approximation is invariant as the reader is asked to show in Exercise 9.

An alternative Laplace Suppose an additional factor $h(x)$ is allowed in the integrand of (3.16). Assume that $h(x)$ is a smooth and positively valued function on D. Then

$$\int_D e^{-g(x)} h(x)\, dx \simeq \frac{(2\pi)^{m/2}}{|g''(\hat{x})|^{1/2}} \exp\{-g(\hat{x})\} h(\hat{x}). \tag{3.23}$$

As an alternative to (3.23), the change of variables $dy = h(x)\, dx$ could be used along with (3.16) to give the same result to first order. Also, any higher order expansion of the integral in (3.23) in variable x is identical to the same order as the expansion in (3.28) using variable y. The proof of this is exactly the same as that used in the univariate setting.

Example Consider the Dirichlet $(\alpha_1, \ldots, \alpha_m, \alpha_{m+1})$ integral with

$$h(x) = \left(\prod_{i=1}^{m+1} x_i\right)^{-1} \qquad g(x) = -\sum_{i=1}^{m+1} \alpha_i \ln x_i.$$

The approximation from (3.23) leads to expression $\hat{I}_x$ in (3.22) as the reader is asked to show in Exercise 7.

Iterated Laplace The m-dimensional Laplace approximation in (3.16) might have been performed differently by first writing the integral as an iterated integral and performing two successive Laplace approximations, one nested inside the other. If the outer Laplace approximation is performed in a certain manner, then the iterated Laplace approximation agrees with the overall single Laplace approximation as indicated below. The formal statement of this property, when it applies to the true integrals, is Fubini's theorem.

Proposition 3.2.1 *Iterated Laplace approximation agrees with single-step Laplace approximation when the outer approximation is based upon (3.23) and takes $h(\cdot)$ as the Hessian factor determined from the inner approximation. Thus, Fubini's theorem continues to apply for Laplace approximations under these conditions.*

The result has been previously noted in Barndorff-Nielsen and Wood (1998).

Proof. Suppose x is partitioned into the two sets of components $x_a = (x_1, \ldots, x_p)$ and $x_b = (x_{p+1}, \ldots, x_m)$. Fubini's theorem states that

$$\int e^{-g(x)} dx = \int \left\{ \int e^{-g(x_a, x_b)} dx_a \right\} dx_b, \tag{3.24}$$

when the iterated regions of integration have been properly determined. Proposition 3.2.1 contends that this relationship continues to hold when integration is replaced by an appropriate Laplace approximation. If the inner integral of (3.24) is approximated as in (3.16), then the right side becomes

$$(2\pi)^{p/2} \int |g''_{aa}\{\hat{x}_a(x_b), x_b\}|^{-1/2} \exp\left[-g\{\hat{x}_a(x_b), x_b\}\right] dx_b, \tag{3.25}$$

where $g''_{aa} = \partial^2 g / \partial x_a \partial x_a^T$ and $\hat{x}_a = \hat{x}_a(x_b)$ denotes the minimum of $g(x_a, x_b)$ in x_a holding x_b fixed and written as an implicit function of x_b. The integral in (3.25) is approximated as in (3.23) taking $|g''_{aa}\{\hat{x}_a(x_b), x_b\}|^{-1/2}$ as the h-factor. Minimization of $g\{\hat{x}_a(x_b), x_b\}$ in x_b

3.2 Development of the saddlepoint density

leads to $\{\hat{x}_a(\hat{x}_b), \hat{x}_b\} = \hat{x}$, the overall minimum. Furthermore, it can be shown that

$$|g''(\hat{x})| = |g''_{aa}(\hat{x})| \times \left| \frac{\partial^2}{\partial x_b \partial x_b^T} g\{\hat{x}_a(x_b), x_b\} \right|_{x_b = \hat{x}_b} \qquad (3.26)$$

as indicated in Exercise 10. Putting this all together leads to an expression that is the single-step Laplace approximation as given in (3.16). □

The iterated Laplace approximation does not agree with the single-step Laplace approximation when the outer approximation is performed differently such as, for example, when the factor $|g''_{aa}|^{-1/2}$ is taken into the exponent as part of the outer g term.

Separability Consider situations in which a multivariate integral factors as the product of two integrals of lower dimension.

Proposition 3.2.2 *Laplace approximation of a multivariate integral preserves any separability found in the exact integration.*

Proof. Partition $x = (x_a, x_b)$ into two sets of components as previously described in (3). Now suppose that g separates additively in the subsets x_a and x_b as

$$g(x) = g_a(x_a) + g_b(x_b). \qquad (3.27)$$

Let $D_a = \{x_a : x \in D\}$ and $D_b = \{x_b : x \in D\}$ be the ranges of these separate components, and furthermore, assume that $D = D_a \times D_b$, where $\times$ denotes the Kronecker product. The conditions assure that the true integral I factors as $I_a \cdot I_b$, where

$$I_a = \int_{D_a} \exp\{-g_a(x_a)\} dx_a,$$

and I_b is defined similarly. Exercise 11 asks the reader to prove that the Laplace approximation (3.16) factors accordingly as $\hat{I} = \hat{I}_a \cdot \hat{I}_b$, where hats denote the Laplace approximations with

$$\hat{I}_a = (2\pi)^{p/2} |g''_a(\hat{x})|^{-1/2} \exp\{-g_a(\hat{x})\}.$$

□

Under appropriate conditions, Proposition 3.2.2 applies more generally to a partition of components that includes three or more subsets. The only apparent restriction is that the variable of integration for the Laplace approximation is the variable in which the exact separation occurs. Nonlinear 1-1 transformation of (x_a, x_b) to a nonseparable variable leads to a different approximation because the approximation is not invariant to such a transformation.

Remarks on asymptotics for Laplace approximations

(1) The approximation is often stated as an asymptotic expansion as a parameter $n \to \infty$. Typically $g(x)$ is replaced with $ng(x)$ in the expansion so that

$$\int_D e^{-ng(x)} dx = \frac{(2\pi)^{m/2}}{n^{m/2} |g''(\hat{x})|^{1/2}} \exp\{-ng(\hat{x})\}\{1 + O(n^{-1})\}, \qquad (3.28)$$

where the $O(n^{-1})$ term depends on arrays of third and fourth derivatives of g evaluated at $\hat{x}$. These are best described using the notation

$$\hat{g}_{ijk} = \left.\frac{\partial^3 g}{\partial x_i \partial x_j \partial x_k}\right|_{x=\hat{x}} \qquad \hat{g}_{ijkl} = \left.\frac{\partial^4 g}{\partial x_i \partial x_j \partial x_k \partial x_l}\right|_{x=\hat{x}}$$

for $i, j, k, l \in \{1, \ldots, m\}$. Also, denote $(\hat{g}^{ij}) = \{g''(\hat{x})\}^{-1}$ as the inverse of the Hessian. Then,

$$O(n^{-1}) = n^{-1}\left\{-\tfrac{1}{8}\hat{\kappa}_4 + \tfrac{1}{24}\left(2\hat{\kappa}_{23}^2 + 3\hat{\kappa}_{13}^2\right)\right\} + O(n^{-2}) \tag{3.29}$$

where

$$\begin{aligned}
\hat{\kappa}_4 &= \hat{g}_{ijkl}\hat{g}^{ij}\hat{g}^{kl} \\
\hat{\kappa}_{23}^2 &= \hat{g}_{ijk}\hat{g}_{rst}\hat{g}^{ir}\hat{g}^{js}\hat{g}^{kt} \\
\hat{\kappa}_{13}^2 &= \hat{g}_{ijk}\hat{g}_{rst}\hat{g}^{ij}\hat{g}^{kr}\hat{g}^{st}
\end{aligned} \tag{3.30}$$

are four- and six-dimensional sums over all values $i, j, k, l, r, s, t \in \{1, \ldots, m\}$ using the Einstein summation notation, appropriate for lattice sums as described in Barndorff-Nielsen and Cox (1989, §5.3) or McCullagh (1987, §1.2). The Einstein notation simply means that we interpret

$$\hat{g}_{ijkl}\hat{g}^{ij}\hat{g}^{kl} = \sum_{i=1}^{m}\sum_{j=1}^{m}\sum_{k=1}^{m}\sum_{l=1}^{m} \hat{g}_{ijkl}\hat{g}^{ij}\hat{g}^{kl}.$$

Essentially, the convention allows for the removal of a summation sign when its associated index appears as both a subscript and superscript in the added terms; this is clearly the case in the three expressions of (3.30). Examples of these higher order computations are given in section 3.4.4 where they are shown to improve upon the leading term approximation, particularly in higher dimensions.

(2) Proposition 3.2.2 considers the separability of a multivariate integral that factors into the product of lower dimensional integrals and also states that the first-order Laplace approximation factors accordingly. Now, it is possible to show that the higher order terms in the expansion also separate. When g separates as in (3.27), the Taylor expansion of g also separates as two separate Taylor expansions for g_a and g_b; this means the entire Laplace expansion separates in x_a and x_b. When taken to order $O(n^{-1})$, however, the approximations are not quite separable because smaller terms of order $O(n^{-2})$ are needed for complete factorability. To make this more explicit, suppose that $O_a(n^{-1})$ and $O_b(n^{-1})$ denote the respective second-order terms in the two separate Laplace expansions. Then the product of the two second-order expansions is

$$\hat{I}_a\left\{1 + O_a(n^{-1})\right\} \cdot \hat{I}_b\left\{1 + O_b(n^{-1})\right\} = \hat{I}_a\hat{I}_b\left\{1 + O(n^{-1}) + O(n^{-2})\right\},$$

where $O(n^{-1}) = O_a(n^{-1}) + O_b(n^{-1})$ and $O(n^{-2})$ includes the factor $O_a(n^{-1}) \times O_b(n^{-1})$. Therefore, to order $O(n^{-1})$, the m-dimensional expansion of g includes all but the last term so it is not quite separable. Working from the m-dimensional expansion of g, the $O(n^{-1})$ term may be seen as having the form $O_a(n^{-1}) + O_b(n^{-1})$. Indeed, all three higher order correction terms in (3.30) separate in an additive manner. For example, the nonzero terms of $\hat{\kappa}_{23}^2$ are

$$\hat{\kappa}_{23}^2 = (\hat{g}_a)_{ijk}(\hat{g}_a)_{rst}\,\hat{g}_a^{ir}\hat{g}_a^{js}\hat{g}_a^{kt} + (\hat{g}_b)_{ijk}(\hat{g}_b)_{rst}\,\hat{g}_b^{ir}\hat{g}_b^{js}\hat{g}_b^{kt}$$
$$= \hat{\kappa}_{a23}^2 + \hat{\kappa}_{b23}^2,$$

where the summation of terms involving g_a is over indices in $\{1, \ldots, p\}$, and the sum with g_b is over $\{p+1, \ldots, m\}$. This separation for all three terms in $O(n^{-1})$ assures that it is $O_a(n^{-1}) + O_b(n^{-1})$.

(3) A discussion of the approximation for large or increasing values of m is deferred to section 3.4.4.

3.2.2 Derivation of the saddlepoint density

The multivariate density is derived in the same manner as used with the univariate density but with attention for the multivariate aspects of the derivation. The derivation below works with the setting in which $n = 1$, so that f is the density of a single random vector.

Proof of (3.2). The multivariate CGF of f is

$$\exp\{K(s)\} = \int_{\mathcal{X}} \exp\{s^T x + \ln f(x)\}\,dx \qquad s \in \mathcal{S}.$$

With fixed s, Laplace's approximation to the right-hand side, taking $g(s, x) = -s^T x - \ln f(x)$, is

$$\exp\{K(s)\} \simeq (2\pi)^{m/2}\,|g''(s, x_s)|^{-1/2}\exp(s^T x_s)f(x_s), \qquad (3.31)$$

where x_s minimizes $g(s, x)$ over x. Of course the presumption in using Laplace is that $g''(s, x_s) = \partial^2 g/\partial x \partial x^T|_{x=x_s}$ is positive definite. The minimum x_s is a critical value of g that satisfies

$$0 = s + \frac{\partial}{\partial x_s}\ln f(x_s). \qquad (3.32)$$

Expression (3.32) defines an implicit relationship between s and root x_s, and the $(m \times m)$ derivative matrix

$$\frac{\partial s}{\partial x_s^T} = -\frac{\partial^2}{\partial x_s \partial x_s^T}\ln f(x_s) = g''(s, x_s) > 0, \qquad (3.33)$$

or positive definite. Solving for $\ln f(x_s)$ in (3.31) gives

$$\ln f(x_s) \simeq K(s) - s^T x_s - \tfrac{1}{2}m\ln(2\pi) + \tfrac{1}{2}\ln|g''(s, x_s)|. \qquad (3.34)$$

Now, suppose, as in the univariate setting, that the last term in (3.34) is roughly constant in x_s so its derivative with respect to x_s can be ignored. Then

$$\frac{\partial}{\partial x_s}\ln f(x_s) \simeq \frac{\partial s^T}{\partial x_s}\{K'(s) - x_s\} - s$$

and

$$0 = s + \frac{\partial}{\partial x}\ln f(x)\bigg|_{x=x_s} \iff 0 = \frac{\partial s^T}{\partial x_s}\{K'(s) - x_s\}.$$

From (3.33), $\partial s^T/\partial x_s$ is a nonsingular matrix, so s and x_s are related through the saddlepoint equation $K'(s) = x_s$ to the accuracy considered. The Hessian in Laplace's approximation is now

$$g''(s, x_s) = -\frac{\partial^2}{\partial x_s \partial x_s^T} \ln f(x_s) = \frac{\partial s}{\partial x_s^T} = \left(\frac{\partial x_s}{\partial s^T}\right)^{-1} = K''(s)^{-1}, \quad (3.35)$$

where the latter equality comes from differentiating the saddlepoint equation. Substitution of (3.35) into (3.34) gives the saddlepoint relationship in (3.2). For the $n > 1$ setting in which the density of a sample mean is approximated, the order of all approximations that have been made, and hence the final density approximation, is $O(n^{-1})$. □

Remarks on asymptotics

(1) The saddlepoint density approximation for $\bar{X}$, the mean of i.i.d. variables $X_1, \ldots, X_n$ with multivariate CGF $K(s)$, is

$$\hat{f}(\bar{x}) = \frac{n^{m/2}}{(2\pi)^{m/2} |K''(\hat{s})|^{1/2}} \exp\{nK(\hat{s}) - n\hat{s}^T \bar{x}\} \quad \bar{x} \in \mathcal{I}_\mathcal{X}, \quad (3.36)$$

where saddlepoint $\hat{s}$ is the unique solution to

$$K'(\hat{s}) = \bar{x},$$

for $\bar{x} \in \mathcal{I}_\mathcal{X}$, the interior of the convex hull of the support of X_1. According to its derivation, this approximation is accurate to asymptotic order $O(n^{-1})$.

(2) The second-order saddlepoint expansion of $f(\bar{x})$ is

$$f(\bar{x}) = \hat{f}(\bar{x})\{1 + O(n^{-1})\},$$

where

$$O(n^{-1}) = n^{-1}\left\{\tfrac{1}{8}\hat{\kappa}_4 - \tfrac{1}{24}\left(2\hat{\kappa}_{23}^2 + 3\hat{\kappa}_{13}^2\right)\right\} + O(n^{-2}). \quad (3.37)$$

The three new terms in (3.37) are expressed in Einstein summation notation as

$$\begin{aligned}\hat{\kappa}_4 &= \hat{K}_{ijkl}\hat{K}^{ij}\hat{K}^{kl} \\ \hat{\kappa}_{23}^2 &= \hat{K}_{ijk}\hat{K}_{rst}\hat{K}^{ir}\hat{K}^{js}\hat{K}^{kt} \\ \hat{\kappa}_{13}^2 &= \hat{K}_{ijk}\hat{K}_{rst}\hat{K}^{ij}\hat{K}^{kr}\hat{K}^{st}.\end{aligned} \quad (3.38)$$

The subscripted $\hat{K}$ terms in (3.38) are derivatives of $\hat{K}$ with respect to the subscripted components evaluated at $\hat{s}$. For example,

$$\hat{K}_{ijk} = \left.\frac{\partial^3 K(s)}{\partial s_i \partial s_j \partial s_k}\right|_{s=\hat{s}}.$$

The superscripted terms are components of

$$(\hat{K}^{ij}) = K''(\hat{s})^{-1},$$

the inverse of the Hessian evaluated at $\hat{s}$. Note that, in the univariate case $m = 1$, the second-order correction term (3.37) agrees with its univariate counterpart in (2.32). A derivation of this term is given in McCullagh (1987, §6.3).

3.3 Properties of multivariate saddlepoint densities

Some of the elementary properties of saddlepoint densities are listed below. Some informal understanding of the proofs is given, while formal justifications are left as exercises.

3.3.1 Equivariance

Proposition 3.3.1 *Multivariate saddlepoint densities are equivariant under nonsingular linear transformation. More specifically, suppose $Y = AX + b$ is a nonsingular $(m \times m)$ linear transformation of random vector X. Then, the saddlepoint density of Y, denoted $\hat{f}_Y$, is the usual Jacobian transformation of the saddlepoint density for X, or*

$$\hat{f}_Y(y) = ||A||^{-1} \hat{f}_X\{A^{-1}(y - b)\}, \tag{3.39}$$

where $||A|| = \text{abs}\{\det(A)\}$.

The property would also hold for lattice variables, as discussed in section 2.1.1, were it not for the adoption of the lattice convention.

Example. Suppose $X \sim \text{Normal}_m(0, I_m)$ and $Y = \Sigma^{1/2} X + \mu$, where $\Sigma^{1/2}$ is a nonsingular square root of $\Sigma > 0$. Following (3.39), then

$$\begin{aligned}\hat{f}_Y(y) &= |\Sigma|^{-1/2} (2\pi)^{-m/2} \exp\left[-\tfrac{1}{2}\{\Sigma^{-\frac{1}{2}}(y-\mu)\}^T \{\Sigma^{-\frac{1}{2}}(y-\mu)\}\right] \\ &= (2\pi)^{-m/2} |\Sigma|^{-1/2} \exp\left\{-\tfrac{1}{2}(y-\mu)^T \Sigma^{-1}(y-\mu)\right\},\end{aligned}$$

which agrees with the true $\text{Normal}_m(\mu, \Sigma)$ density.

Proposition 3.3.2 *A consequence of the first equivariance property is that relative errors in saddlepoint density approximation are invariant to linear transformation and, in particular, to location and scale changes.*

3.3.2 Independence

Proposition 3.3.3 *Saddlepoint densities and mass functions separate or factor into subsets of variables if and only if the corresponding true densities separate into the same variable subsets.*

Example. See the discussion for the bivariate exponential example of section 3.1.3.

3.3.3 Marginalization

Partition the continuous variable $X = (X_a, X_b)$ so that $X_a = (X_1, \ldots, X_p)^T$ and $X_b = (X_{p+1}, \ldots, X_m)^T$. The marginal density of X_a is

$$f(x_a) = \int f(x_a, x_b) dx_b. \tag{3.40}$$

If the true joint density in (3.40) is replaced with its joint saddlepoint density $\hat{f}(x_a, x_b)$, then what are the circumstances under which Laplace's approximation to marginalization

$\int \hat{f}(x_a, x_b)dx_b$ yields $\hat{f}(x_a)$, the marginal saddlepoint density of X_a? In other words, when are $\hat{f}(x_a)$ and $\hat{f}(x_a, x_b)$ related as in (3.40) but with the marginalization integral replaced by Laplace's approximation?

Proposition 3.3.4 *The marginal saddlepoint density $\hat{f}(x_a)$ is Laplace's approximation to $\int \hat{f}(x_a, x_b)dx_b$ from (3.23) when the h factor is chosen to be the Hessian term of $\hat{f}(x_a, x_b)$.*

The proof is deferred to chapter 4 where it is stated as Exercise 23.

3.3.4 Symmetry

Proposition 3.3.5 *Symmetries possessed by density/mass functions are generally also reproduced in their saddlepoint density/mass function counterparts. An incomplete list of such symmetries would include exchangeable, antipodal, spherical, and elliptical symmetries.*

Exchangeable symmetry for the distribution of random vector X would require that the distribution remain unchanged when the components of X are permuted; that is, PX must have the same distribution as X, where P is an $(m \times m)$ permutation matrix defined as any one of the $m!$ permutations of the rows of I_m. Antipodal symmetry refers to property in which $-X$ has the same distribution as X. Spherical symmetry requires that OX have the same distribution as X for any $(m \times m)$ orthogonal matrix O. Note that all permutation matrices and the matrix $-I_m$ are orthogonal matrices and that $\{P\}$ and $\{I_m, -I_m\}$ are proper subgroups of the orthogonal group $\{O\}$; such symmetries are therefore "weaker" than spherical symmetry. Symmetry in which densities remain constant over elliptical surfaces of the form

$$\{x : (x - \mu)^T \Sigma^{-1} (x - \mu) = \text{const.}\}, \tag{3.41}$$

for some μ and $\Sigma > 0$, is equivalent to spherical symmetry for the transformed variable $y = \Sigma^{-1/2}(x - \mu)$.

Examples. The normal and bivariate exponential examples of section 3.1.3 illustrate exchangeable, antipodal, and spherical symmetries. The normal example above, used to show equivariance under linear transformation, provides an example in which elliptical symmetry is maintained in the saddlepoint density. Further examples are provided in the next section.

Proof of proposition 3.3.5. These symmetries exist in the saddlepoint density/mass functions because they manifest in the saddlepoint vectors as a result of the symmetry of the CGFs. For example, the equivariant property of the saddlepoint density results because the saddlepoint for variable Y at $y = Ax + b$ is linear in the saddlepoint for X at x, thus allowing the equivariance to pass through to the saddlepoint expressions. Also, the preservation of elliptical symmetry as in (3.41) results from the preservation of spherical symmetry in the variable $y = \Sigma^{-1/2}(x - \mu)$ in combination with the equivariance property when converted back to the original variable x. □

3.4 Further examples

3.4.1 Bootstrap distribution of the sample mean

The saddlepoint approximation to this bootstrap distribution was given by Davison and Hinkley (1988). Suppose that $w_1, \ldots, w_n$ are the values of an i.i.d. sample from a population with unknown mean μ. The sample mean $\bar{w}$ provides an estimate of the population mean μ and the estimation error $\bar{w} - \mu$ can be assessed using bootstrap resampling. Such resampling involves drawing $W_1^*, \ldots, W_n^*$ as a random sample of the data with replacement, so $\{W_i^*\}$ are i.i.d. with each W_i^* having a uniform distribution over the data values, or

$$\Pr\{W_i^* = w_j\} = 1/n \qquad j = 1, \ldots, n. \tag{3.42}$$

The values $W_1^*, \ldots, W_n^*$ are a *resample* of the data as a population, with population mean $\bar{w}$. Percentage points for the distribution of $\bar{W} - \mu$ can be approximated using the corresponding percentage points for the distribution of $\bar{W}^* - \bar{w}$ induced from the randomization in (3.42). A saddlepoint approximation to the distribution of $\bar{W}^* - \bar{w}$ given $\{w_i\}$, or equivalently $n\bar{W}^*$, is based on the underlying multinomial distribution in the randomization. Let X_i be the number of times w_i is selected in resampling. Then $n\bar{W}^*$ has the same distribution as

$$Y = \sum_{i=1}^{n} w_i X_i \tag{3.43}$$

where $X = (X_1, \ldots, X_{n-1}) \sim$ Multinomial $(n; n^{-1}, \ldots, n^{-1})$. A simple calculation, using either the MGF of W_1^*, or the joint MGF of X in (3.9) in connection with representation (3.43), shows the MGF for Y is

$$M(s) = E(e^{sY}) = \hat{M}(s)^n \tag{3.44}$$

where

$$\hat{M}(s) = E(e^{sW_1^*}) = \frac{1}{n} \sum_{i=1}^{n} e^{sw_i}$$

is the empirical MGF of an individually sampled W_i^*.

The CDF of $\bar{W}^* - \bar{w}$ can be approximated by using the MGF in (3.44) in conjunction with the Lugannani and Rice expression in (1.21). Davison and Hinkley (1988) used the sample of $n = 10$ values

9.6, 10.4, 13.0, 15.0, 16.6, 17.2, 17.3, 21.8, 24.0, 33.8

and determined the approximate percentage points for $\bar{W}^* - \bar{w}$ given in table 3.4. The support for $\bar{W}^* - \bar{w}$ is $[-8.27, 15.93]$. The "Exact" values are the empirical percentage points based on 50,000 bootstrap resamplings of $\bar{W}^* - \bar{w}$. Normal approximation uses a Normal $(0, s^2/n)$ where s^2 is the sample variance of $\{w_i\}$. The Fisher–Cornish approximation is discussed in Kendall and Stuart (1969, §6.25).

A subtle issue concerning the support of $\bar{W}^* - \bar{w}$ has been overlooked in the saddlepoint application. The distribution of $\bar{W}^* - \bar{w}$ has been computed in table 3.4 as if it were continuous but its support is clearly not on an interval. The support is on a δ-lattice of increment $\delta = 0.01$. The mass values, however, are rough in the sense that they fluctuate

Table 3.4. *Approximations for the percentage points of the resampling distribution of $\bar{W}^* - \bar{w}$. Exact percentiles were estimated by simulating 50,000 bootstrap resamplings.*

Probability	Exact	Saddlepoint	Normal	Two-term Fisher–Cornish
.0001	−6.34	−6.31	−8.46	−6.51
.01	−4.42	−4.43	−5.29	−4.63
.05	−3.34	−3.33	−3.74	−3.48
.10	−2.69	−2.69	−2.91	−2.81
.20	−1.86	−1.86	−1.91	−1.95
.80	1.80	1.80	1.91	1.87
.90	2.87	2.85	2.91	3.00
.95	3.73	3.75	3.74	3.97
.99	5.47	5.48	5.29	5.89
.9999	9.33	9.12	8.46	10.2

wildly between neighboring points leaving many gaps with lattice points of mass zero. This roughness is smoothed over with both the saddlepoint CDF and density function, which is a point discussed by Feuerverger (1989). If continuity correction were to be used in this context, it would result in only a very minor change in the saddlepoint results of the table. This is because as $\delta \to 0$, the continuity-corrected CDF approximations on the δ-lattice approach those in the table that assume continuity. See section 2.4.4 for a discussion of this.

3.4.2 *Durbin–Watson statistic and the Dirichlet bootstrap*

The Durbin–Watson (1950, 1951) statistic is used to test for serial correlation in the residual vector $\hat{\varepsilon}^T = (\hat{\varepsilon}_1, \ldots, \hat{\varepsilon}_t)$ from a least squares regression. The statistic is

$$D = \frac{\sum_{i=2}^{t}(\hat{\varepsilon}_i - \hat{\varepsilon}_{i-1})^2}{\sum_{i=1}^{t}\hat{\varepsilon}_i^2} = \frac{\hat{\varepsilon}^T A \hat{\varepsilon}}{\hat{\varepsilon}^T \hat{\varepsilon}},$$

and is a ratio of quadratic forms, for an appropriated defined $(t \times t)$ matrix A (Johnson and Kotz, 1986, vol. II Durbin–Watson Test). The support of D is approximately bounded between 0 and 4 and it has a mean value of about 2 under the null hypothesis that the first-order serial correlation in the true error sequence $\{\varepsilon_i\}$ is zero. Canonical reduction of the residual error into independent degrees of freedom allows the null distribution to take the form

$$D \sim \frac{\sum_{i=1}^{n} w_i \chi_i^2}{\sum_{i=1}^{n} \chi_i^2} = \sum_{i=1}^{n} w_i D_i \qquad (3.45)$$

where $\{\chi_i^2\}$ are the i.i.d. χ_1^2 canonical variables, n is the degrees of freedom in the residuals and $\{w_i\}$ are fixed and known eigenvalues of a matrix determined from the design matrix of the linear model. As expressed in (3.45), the distribution is a weighted sum of exchangeable

Table 3.5. *Approximations to the percentage points of the resampling distribution of $D - \bar{w}$. Exact percentiles were estimated by simulating 100,000 bootstrap resamplings.*

Probability	Exact	Saddlepoint
.0001	−5.94	−6.15
.01	−4.07	−4.07
.05	−3.01	−3.04
.10	−2.45	−2.47
.20	−1.72	−1.74
.80	1.62	1.61
.90	2.73	2.72
.95	3.73	3.72
.99	5.75	5.73
.9999	9.68	10.60

Dirichlet $(1/2, \ldots, 1/2)$ components $\{D_i = \chi_i^2 / \sum_{j=1}^n \chi_j^2\}$ with support on the simplex in $\Re^n$.

Similar distribution theory occurs with the Dirichlet or Bayesian bootstrap introduced by Rubin (1981) as an alternative to the previously discussed multinomial bootstrap. Taking $\{w_i\}$ as the data, the distributional representation for this bootstrap is a weighted sum of Dirichlet $(1, \ldots, 1)$ components as specified on the right side of (3.45). Saddlepoint approximations to such distributions were introduced in Booth and Butler (1990), Paolella (1993), and Lieberman (1994b) with further discussion appearing in Butler and Paolella (1998a, 1998b). The CDF of D may be determined from the individual $\{\chi_i^2\}$ or gamma variables that underlie the Dirichlet. Suppose more generally that $(D_1, \ldots, D_{n-1})$ is Dirichlet $(\alpha_1, \ldots, \alpha_n)$ as constructed from independent $\{G_i\}$ with $G_i \sim$ Gamma $(\alpha_i, 1)$. Then

$$\Pr(D \leq d) = \Pr\left\{Z_d = \sum_{i=1}^n (w_i - d) G_i \leq 0\right\}. \tag{3.46}$$

The Lugannani and Rice approximation may now be applied to the CDF of Z_d. The details of this are specified in Exercise 16.

Table 3.5 has been reproduced from Booth and Butler (1990) and provides percentage points for the Dirichlet bootstrap using the bootstrap data of section 3.4.1.

3.4.3 *Rotationally symmetric density*

Feller (1971, p. 523) considers the bivariate density for variables (X, Y) given by

$$f(x, y) = (2\pi)^{-1} \alpha^2 \exp\left(-\alpha \sqrt{x^2 + y^2}\right) \qquad (x, y) \in \Re^2, \tag{3.47}$$

for $\alpha > 0$. The density is rotationally symmetry and has MGF

$$M(s, t) = \{1 - \alpha^{-2}(s^2 + t^2)\}^{-3/2} \qquad s^2 + t^2 < \alpha^2,$$

as shown in Feller. Because of the equivariance property, there is no loss of generality in taking scale parameter $\alpha = 1$, so the convergence set of M is the open unit disk $\mathcal{S} = \{(s, t) : s^2 + t^2 < 1\}$. The saddlepoint equations admit explicit solution in Euclidean coordinates but their description is more interesting in polar coordinates. Suppose that (r, θ) and $(\hat{\rho}, \hat{\theta})$ are the polar coordinates for (x, y) and $(\hat{s}, \hat{t})$ respectively. Then the saddlepoint equations are

$$\hat{\theta} = \theta$$
$$\hat{\rho} = (2r)^{-1} \left(\sqrt{9 + 4r^2} - 3\right).$$

These equations show that $K' = (\partial K / \partial x, \partial K / \partial y)$ maps points on the circle of radius $\hat{\rho} < 1$ in $\mathcal{S}$ to points on the circle of radius $r \in (0, \infty)$ in $\Re^2$ in a manner that preserves angular coordinates. The circle with $r = \infty$ is the image of the boundary circle $\hat{\rho} = 1$. The saddlepoint density is most easily expressed in polar coordinates as

$$\hat{f}(x, y) = (2\pi)^{-1} \tfrac{1}{3}(1 + \hat{\rho}^2)^{-1/2} \exp\left(-\tfrac{1}{2}\sqrt{9 + 4r^2} - 3\right) \quad (x, y) \in \Re^2. \tag{3.48}$$

Note that this is the density of (X, Y) expressed in terms of $r^2 = x^2 + y^2$ and not the density of (R, Θ), the polar coordinates for (X, Y). The saddlepoint density is not exact, as in the normal example, but possesses the rotational symmetry of the true density.

The joint density of (R, Θ) is determined from Jacobian transformation of (3.47) as

$$f(r, \theta) = \alpha^2 r \exp(-\alpha r) \cdot \frac{1}{2\pi} = f(r) f(\theta) \qquad r > 0, \theta \in (0, 2\pi).$$

The variables are independent, with $R \sim \text{Gamma}(2, \alpha)$ and $\Theta \sim \text{Uniform}(0, 2\pi)$. Direct saddlepoint approximation to $f(r, \theta)$ yields

$$\hat{f}(r, \theta) = \hat{f}(r) \cdot \hat{f}(\theta) = \hat{\Gamma}(2) f(r) \cdot \hat{f}(\theta). \tag{3.49}$$

where the saddlepoint for $\hat{f}(\theta)$ is not explicit. This may be seen by noting that the MGF of Θ is

$$M(s) = \frac{e^{2\pi s} - 1}{2\pi s} \qquad s \neq 0,$$

and is defined by continuity to be 1 at $s = 0$. Jacobian transformation of (3.49) to approximate the density of (X, Y) yields

$$\check{f}(x, y) = r^{-1} \hat{f}(r, \theta) = \hat{\Gamma}(2) \alpha^2 \exp(-\alpha r) \hat{f}(\theta). \tag{3.50}$$

This expression is not the same as (3.48), because the saddlepoint for $\hat{f}(\theta)$ is not explicit. In addition, it does not capture the rotational symmetry of the density because $\hat{f}(\theta)$ is dependent on the angle θ. The diagram below shows the two routes used in finding an approximation to $f(x, y)$: the upper route is direct saddlepoint approximation yielding (3.48), while the lower route yields (3.50).

$$\begin{array}{ccccc} (X, Y) & \Longrightarrow & & & \hat{f}(x, y) \\ \downarrow & & & & \Downarrow \\ (R, \Theta) & \Longrightarrow & \hat{f}(r, \theta) & \longrightarrow & \check{f}(x, y). \end{array}$$

The symbols $\longrightarrow$ and $\Longrightarrow$ indicate Jacobian density transformation and saddlepoint approximation respectively. The following general principle may be deduced:

Table 3.6. *Exact Dirichlet integral values I, Laplace's approximation to first-order $\hat{I}_x$, and the percentage relative error for the various dimensions m and values of parameter α.*

m	$\alpha_1, \ldots, \alpha_{m+1}$	I	$\hat{I}_x$	% relative error
2	.5 (1, 1, 1)	6.283	4.189	50.0
3	.5 (1, 1, 1, 1)	9.870	5.568	77.3
4	.5 (1, …, 1)	13.16	6.317	108
5	.5 (1, …, 1)	15.50	6.348	144
10	.5 (1, …, 1)	10.36	1.946	432
50	.5 (1, …, 1)	$.0^{11}1541$	$.0^{15}6180$	2.49×10^5
100	.5 (1, …, 1)	$.0^{38}2976$	$.0^{45}5552$	5.36×10^8
2	.25 (1, 2, 3)	8.886	5.513	61.2
3	.25 (1, 2, 3, 4)	5.924	3.316	78.6
4	.25 (1, …, 5)	1.614	.8370	92.8
5	.25 (1, …, 6)	.1796	.08765	105

Proposition 3.4.1 *A saddlepoint approximation sandwiched in between a nonlinear nonsingular transformation and its inverse mapping, does not generally yield the same approximation as direct saddlepoint approximation.*

When sandwiched between linear transformations, however, the approximations must be the same by the equivariance property. Exercises 13 and 14 ask the reader to consider an example of rotational symmetry in $\Re^3$.

3.4.4 High dimensional Laplace approximation

For fixed dimension m, the integrand in (3.28), which is the asymptotic version of Laplace's approximation, grows more concentrated about its maximum $\hat{x}$ as n increases in value. Within this context, the approximation also achieves an accuracy that generally increases with the value of n. Alternatively, consider the situation in which m is large relative to n, or perhaps both m and n increase together.

This situation is easily examined by considering the accuracy of Laplace's approximation for the Dirichlet integral in various dimensions. The various values of $\{\alpha_i\}$ play the role of n, and all must grow if the asymptotics are to be valid. The setting is particularly convenient for judging the accuracy since exact values can be determined. Table 3.6 shows that the accuracy declines quite dramatically with increasing m for settings in which the α-values remain fixed and do not increase with m.

Second-order Laplace approximations are considerably better than first order. For the setting in which m increases with the value of n, the additional correction term $O(n^{-1})$, given in (3.30), may not have the indicated order in n when m is no longer fixed. Before such computations can be made, the requisite second, third and fourth derivatives need to be computed and evaluated at $\hat{x}$. Second-order derivatives from (3.21) are expressed as

$$g_{ij} = \alpha \cdot \left\{ e^{x_i} (1+\xi)^{-1} \delta_{ij} - e^{x_i + x_j} (1+\xi)^{-2} \right\}, \tag{3.51}$$

where $\xi = \sum_{i=1}^{m} \exp(x_i)$ and δ_{ij} is the indicator that $i = j$. When evaluated at $\hat{x}$, then

$$\hat{y}_i = e^{\hat{x}_i}/(1+\hat{\xi}) = \alpha_i/\alpha., \tag{3.52}$$

and

$$\hat{g}_{ij} = \alpha.(\hat{y}_i \delta_{ij} - \hat{y}_i \hat{y}_j). \tag{3.53}$$

Values of $(\hat{g}^{ij})$, the inverse of the Hessian at $\hat{x}$, are determined by inversion as

$$\hat{g}^{ij} = \alpha_{m+1}^{-1}\left(1 + \frac{\alpha_{m+1}}{\alpha_i}\delta_{ij}\right). \tag{3.54}$$

Third cumulants result from differentiation of (3.51) using the product rule, treating δ_{ij} as a constant, and using the obvious fact that $\partial x_j/\partial x_i = \delta_{ij}$; this gives, after simplification,

$$g_{ijk} = \alpha.\left[e^{x_i}(1+\xi)^{-1}\delta_{ijk} - \left\{e^{x_i+x_k}\delta_{ij} + e^{x_i+x_j}(\delta_{ik}+\delta_{jk})\right\}\right. \\ \left. \times (1+\xi)^{-2} + 2e^{x_i+x_j+x_k}(1+\xi)^{-3}\right]. \tag{3.55}$$

At the maximum in (3.52), this becomes

$$\hat{g}_{ijk} = \alpha.[\hat{y}_i\delta_{ijk} - \{\hat{y}_i\hat{y}_j(\delta_{ik}+\delta_{jk}) + \hat{y}_i\hat{y}_k\delta_{ij}\} \\ + 2\hat{y}_i\hat{y}_j\hat{y}_k]. \tag{3.56}$$

Differentiation of (3.55) followed by evaluation at the maximum gives

$$\hat{g}_{ijkl} = \alpha.\left[\hat{y}_i\delta_{ijkl} - \{\hat{y}_i\hat{y}_j(\delta_{ik}+\delta_{jk})(\delta_{il}+\delta_{jl}) + \hat{y}_i\hat{y}_k\delta_{ij}(\delta_{il}+\delta_{kl}) \right. \\ \left. + \hat{y}_i\hat{y}_l\delta_{ijk}\} + 2\{\hat{y}_i\hat{y}_j\hat{y}_k(\delta_{il}+\delta_{jl}+\delta_{kl}) + \hat{y}_i\hat{y}_j\hat{y}_l(\delta_{ik}+\delta_{jk}) \right. \\ \left. + \hat{y}_i\hat{y}_k\hat{y}_l\delta_{ij}\} - 6\hat{y}_i\hat{y}_j\hat{y}_k\hat{y}_l\right]. \tag{3.57}$$

The second to fourth derivative patterns in (3.53), (3.56), and (3.57) are exactly those of the corresponding cumulants for the multinomial distribution given in McCullagh (1987, Problem 2.16). McCullagh uses a particular kind of bracket notation to express these patterns. In his notation, we can write

$$\hat{g}_{ijk} = \alpha.(\hat{y}_i\delta_{ijk} - \hat{y}_i\hat{y}_j\delta_{ik}\,[3] + 2\hat{y}_i\hat{y}_j\hat{y}_k),$$

where $\hat{y}_i\hat{y}_j\delta_{ik}\,[3]$ is an index pattern representing the three terms in the curly brackets of (3.56). The fourth derivative is written as

$$\hat{g}_{ijkl} = \alpha.\{\hat{y}_i\delta_{ijkl} - \hat{y}_i\hat{y}_j(\delta_{ik}\delta_{jl}\,[3] + \delta_{jkl}\,[4]) \\ + 2\hat{y}_i\hat{y}_j\hat{y}_k\delta_{il}\,[6] - 6\hat{y}_i\hat{y}_j\hat{y}_k\hat{y}_l\}.$$

Here $\hat{y}_i\hat{y}_j(\delta_{ik}\delta_{jl}\,[3] + \delta_{jkl}\,[4])$ denotes the seven terms of the two indicated types that are included in the first curly brackets of (3.57). The term $2\hat{y}_i\hat{y}_j\hat{y}_k\delta_{il}\,[6]$ stands for the six terms of the indicated pattern that occur within the second curly bracket. Further discussion about this type of notation can be found in McCullagh (1987, chaps. 1 and 2) and Barndorff-Nielsen and Cox (1989, chap. 5).

The agreement that exists between our derivative patterns and those of the multinomial cumulants leads to explicit formulas for the components of the second-order correction

3.4 Further examples 99

Table 3.7. *A continuation of table 3.6 that compares the second-order approximations $\hat{I}_{x2}$ and $\hat{I}_{xsm}$ to the exact values I and shows their percentage relative errors.*

m	I	$\hat{I}_{x2}$	$\hat{I}_{xsm}$	% error $\hat{I}_{x2}$	% error $\hat{I}_{xsm}$
2	6.283	6.050	6.533	−3.71	3.98
3	9.870	9.048	10.40	−8.33	5.37
4	13.16	11.37	14.06	−13.6	6.84
5	15.50	12.46	16.70	−19.6	7.74
10	10.36	5.483	11.99	−47.1	15.7
50	$.0^{11}1541$	$.0^{14}587$	$.0^{11}303$	−100.	96.6
100	$.0^{38}2976$	$.0^{44}999$	$.0^{37}113$	−100.	280.
2	8.886	8.575	9.608	−3.50	8.12
3	5.924	5.368	6.156	−9.39	3.92
4	1.614	1.355	1.554	−16.0	−3.72
5	.1796	.1578	.1952	−12.1	8.69

terms in Laplace's approximation. Using Problem 9 in McCullagh (1987), it can be shown that

$$\hat{\kappa}_4 = \sum_{i=1}^{m+1} \alpha_i^{-1} - \alpha_.^{-1}(m^2 + 4m + 1) \tag{3.58}$$

$$\hat{\kappa}_{23}^2 = \hat{\kappa}_4 + \alpha_.^{-1} m(m+1)$$

$$\hat{\kappa}_{13}^2 = \hat{\kappa}_4 + 2\alpha_.^{-1} m.$$

Suppose O is the second-order correction term or the $O(n^{-1})$ term in (3.29). When computed in this context however, $n = 1$ and

$$O = \frac{1}{12} \left(\sum_{i=1}^{m+1} \alpha_i^{-1} - \alpha_.^{-1} \right), \tag{3.59}$$

as determined from the reduction in (3.58).

Two second-order approximations can now be computed by combining the first-order approximation in (3.22) with correction term (3.59). The first approximation simply includes one more term in the Laplace expansion and is denoted as

$$\hat{I}_{x2} = \hat{I}_x(1 + O).$$

The second approximation is

$$\hat{I}_{xsm} = \hat{I}_x e^O,$$

which is the modification in Shun and McCullagh (1995). The two approximations are related according to the Taylor expansion $e^O \simeq 1 + O + \cdots$ Both second-order approximations improve on the accuracy of $\hat{I}_x$ as may be seen by extending table 3.6 to include these two additional approximations. The values in the rows of table 3.7 should be compared with values in the corresponding rows of table 3.6. Approximation $\hat{I}_{xsm}$ is generally better than $\hat{I}_{x2}$ which is considerably better than $\hat{I}_x$. In judging these relative errors, note that they are computed as $100(\hat{I}_{x2}/I - 1)$ and are restricted to the range $(-100, \infty)$ since the lower bound for $\hat{I}_{x2}/I$ is 0. For cases $m = 50$ and 100, such errors can be deceiving since $\hat{I}_{xsm}$ is clearly the better approximation despite $\hat{I}_{x2}$ having one of its relative errors smaller. Approximation $\hat{I}_{xsm}$, unlike $\hat{I}_{x2}$, appears to capture the correct order as a power of 10.

The asymptotic behavior of these correction terms can be described as $m \to \infty$ by considering the special case of exchangeability in which $\alpha_i \equiv \alpha$ for $i = 1, \ldots, m+1$. In this setting

$$O = \alpha^{-1} \frac{m(m+2)}{12(m+1)},$$

which is of order $O(1)$ if m is chosen to be of order $O(\alpha)$ when $\alpha \to \infty$. By comparison with the example of Shun and McCullagh (1995), their correction term is $O(1)$ when m is $O(\sqrt{n})$ as $n \to \infty$. In their context, the dimension must grow at a slower rate if it is to achieve a bounded error correction.

3.4.5 High dimensional saddlepoint mass function

The Multinomial $(n; \pi_1, \ldots, \pi_m, \pi_{m+1})$ distribution has the saddlepoint mass function given in (3.12). This expression is exact, apart from the substitution of Stirling's approximation for the combinatoric. With large values of m, however, Stirling's approximation and the saddlepoint density can be quite inaccurate. The inclusion of the second-order saddlepoint corrections in (3.37) and (3.38) is quite effective in compensating for the inaccuracy of the first-order saddlepoint mass function.

Computations for the second to fourth cumulants of the multinomial are identical to those of the previous example and result in the same cumulant patterns. Suppose the second-order saddlepoint approximation at $k = (k_1, \ldots, k_m)$, with $k_{m+1} = n - \sum_{i=1}^{m} k_i$, is

$$p(k) \simeq \hat{p}(k)(1 + O),$$

where $p(k)$ and $\hat{p}(k)$ are given in (3.12). Since the cumulant patterns are the same as in the Laplace example, the three correction terms $\hat{\kappa}_4$, $\hat{\kappa}_{23}^2$, and $\hat{\kappa}_{13}^2$ in (3.38) must have the form expressed in (3.58), but with k_i replacing α_i for $i = 1, \ldots, m+1$ and n replacing α. The correction term O is therefore -1 times expression (3.59) with similar replacements, or

$$O = -\frac{1}{12}\left(\sum_{i=1}^{m+1} k_i^{-1} - n^{-1}\right).$$

Table 3.8 lists the values for six multinomial mass functions along with their first-order and two second-order saddlepoint approximations

$$\hat{p}_2(k) = \hat{p}(k)(1 + O)$$
$$\hat{p}_{2m}(k) = \hat{p}(k)e^O.$$

The latter approximation has been suggested in McCullagh (1987, §6.3) and shows remarkably greater accuracy than $\hat{p}_2$. Each mass function of dimension m has $m + 1$ possible outcomes, has been assumed exchangeable with $\pi_i \equiv (m+1)^{-1}$, and has $n = 2(m+1)$ trials. In each instance, the density has been evaluated at the vector $k = (2, \ldots, 2)$.

3.5 Multivariate CDFs

Table 3.8. *Multinomial density values comparing exact p, first-order saddlepoint $\hat{p}$, and the two second-order approximations $\hat{p}_2$ and $\hat{p}_{2m}$.*

m	$p(k)$	$\hat{p}(k)$	$\hat{p}_2(k)$	$\hat{p}_{2m}(k)$
3	.0384	.0448	.0379	.0384
5	$.0^2 344$	$.0^2 438$	$.0^2 331$	$.0^2 343$
10	$.0^5 674$	$.0^4 106$	$.0^5 577$	$.0^5 672$
20	$.0^{10} 196$	$.0^{10} 467$	$.0^{11} 593$	$.0^{10} 195$
50	$.0^{27} 287$	$.0^{26} 236$	$-.0^{26} 266$	$.0^{27} 282$
100	$.0^{55} 169$	$.0^{53} 110$	$-.0^{55} 353$	$.0^{55} 164$

Table 3.9. *Percentage relative errors of the approximations from table 3.8.*

m	% error $\hat{p}$	% error $\hat{p}_2$	% err. $\hat{p}_{2m}$
3	16.8	−1.48	−.130
5	27.3	−3.67	−.195
10	57.0	−14.4	−.358
20	138.	−69.8	−.682
50	723.	−1025.	−1.65
100	6404.	−309.	−3.24

Table 3.9 compares the percentage relative errors of the three approximations, again confirming that $\hat{p}_{2m}$ is considerably more accurate than $\hat{p}_2$ or $\hat{p}$ for both low and high dimensions.

3.5 Multivariate CDFs

Converting a multivariate saddlepoint density $\hat{f}(x)$ into a multivariate probability $\Pr(X \in A)$ for a wide range of sets $A \subset \Re^{m_y}$ is a very important computational problem in probability. When A is a rectangular set of the form $A = \prod_{i=1}^{m}(-\infty, x_i]$, then $\Pr(X \in A)$ is the multivariate CDF. For $m = 2$ Wang (1990) has provided analytical approximations. Kolassa (2003) provides a general approach for right tail event $A = \prod_{i=1}^{m}(x_i, \infty)$ with arbitrary m. Unfortunately, many practical settings demand computations over sets A that are not rectangular and, outside of specific classes of sets, it appears exceedingly difficult to construct analytical methods that can accurately accommodate arbitrary sets.

Analytical approximations to some commonly occurring multivariate sampling CDFs have been given in Butler and Sutton (1998). These multivariate CDFs include the continuous Dirichlet distribution and three discrete multivariate distributions: the multinomial, multivariate hypergeometric, and multivariate Pólya.

To address the more general problem, Butler *et al.* (2005) have provided a Monte Carlo method that integrates and normalizes the saddlepoint density using importance sampling.

3.6 Exercises

1. (a) Derive the bivariate exponential MGF in (3.5) as a marginal MGF of a (2×2) Wishart MGF.
   (b) Show that the saddlepoint equations have the explicit solution given in (3.6) and (3.7).
2. (a) In the trinomial setting with $m = 2$, show that the Hessian matrix at $\hat{s}$ is

$$K''(\hat{s}) = n(\hat{\zeta} + \pi_3)^{-2} \begin{bmatrix} \pi_1 e^{\hat{s}_1}(\pi_3 + \pi_2 e^{\hat{s}_2}) & -\pi_1 \pi_2 e^{(\hat{s}_1 + \hat{s}_2)} \\ -\pi_1 \pi_2 e^{(\hat{s}_1 + \hat{s}_2)} & \pi_2 e^{\hat{s}_1}(\pi_3 + \pi_1 e^{\hat{s}_1}) \end{bmatrix}$$

where $\hat{\zeta} = \pi_1 \exp(\hat{s}_1) + \pi_2 \exp(\hat{s}_2)$. Show that its determinant is $k_1 k_2 k_3/n$, which is the expression in (3.11).
   (a) Verify (3.11) for general m.
   (b) Verify (3.12) for general m.
   (c) Check the accuracy of one of the entries in table 3.1.
3. The multivariate logistic distribution is described in Johnson and Kotz (1972, chap. 42, §5) and has the density

$$f(x) = m! \left(1 + \sum_{i=1}^{m} e^{-x_i}\right)^{-(m+1)} \exp\left(-\sum_{i=1}^{m} x_i\right) \qquad x_i > 0 \ \forall_i.$$

Its MGF is

$$M(s) = \left\{\prod_{i=1}^{m} \Gamma(1 - s_i)\right\} \Gamma\left(1 + \sum_{i=1}^{m} s_i\right) \qquad s \in \mathcal{S},$$

where $\mathcal{S} = \{s \in \Re^m : \sum_{i=1}^{m} s_i > -1, \ s_i < 1 \ \forall_i\}$.
   (a) Verify that the marginal distributions are univariate logistic.
   (b) For the bivariate case of $m = 2$, construct a surface plot of the true density, the saddlepoint density, and its percentage relative error.
   (c) Normalize the saddlepoint density numerically. Repeat the surface plots of part (b) using the normalized saddlepoint density instead.
4. The bivariate Negative Binomial $(n; \pi_1, \pi_2, \pi_3)$ mass function arises from independent trials that lead to only one of the three outcomes A, B, or C with probabilities π_1, π_2, and π_3 respectively. Suppose the independent trials are stopped when there are n occurrences of outcome C, a situation referred to as inverse sampling. Let X and Y count the number of occurrences of A and B.
   (a) Show that the joint mass function of X and Y is

$$p(i, j) = \frac{\Gamma(n + i + j)}{i! j! \Gamma(n)} \pi_1^i \pi_2^j \pi_3^n \qquad i \geq 0, \ j \geq 0.$$

   (b) Derive the joint MGF as

$$M(s, t) = \pi_3^n (1 - \pi_1 e^s - \pi_2 e^t)^{-n} \qquad (s, t) \in \mathcal{S}$$

where

$$\mathcal{S} = \{(s, t) : \pi_1 e^s + \pi_2 e^t < 1\}.$$

Show that $\mathcal{S}$ is a convex open set so there is regularity.
   (c) Show that the marginal distributions are Negative Binomial (n, θ), as in Exercise 9 of chapter 1, where $\theta = \pi_3/(\pi_1 + \pi_3)$ for X and $\theta = \pi_3/(\pi_2 + \pi_3)$ for Y.
   (d) Show the saddlepoint for $i > 0$ and $j > 0$ is

$$\hat{s} = -\ln \frac{\pi_1(n + i + j)}{i} \qquad \hat{t} = -\ln \frac{\pi_2(n + i + j)}{j} \qquad i > 0, \ j > 0.$$

(e) Deduce that

$$K''(\hat{s}, \hat{t}) = n^{-1} \begin{bmatrix} i(n+i) & ij \\ ij & j(n+j) \end{bmatrix}$$

$|K''(\hat{s}, \hat{t})| = ij(n+i+j)/n$, and

$$\hat{p}(i, j) = \frac{\hat{\Gamma}(n+i+j)}{\hat{i}!\hat{j}!\hat{\Gamma}(n)} \pi_1^i \pi_2^j \pi_3^n \qquad i > 0, \; j > 0.$$

(f) From the MGF, show that the correlation is

$$\sqrt{\frac{\pi_1 \pi_2}{(1-\pi_1)(1-\pi_2)}}$$

which is always positive.

5. Show that Laplace's approximation to the Dirichlet integral leads to the approximation in (3.18).
   (a) Show that $\hat{y}_i = a_i/a_.$, where $a_i = \alpha_i - 1 > 0$ and $a_. = \alpha_. - m - 1$.
   (b) Show the Hessian matrix evaluated at $\hat{y}$ is $(\hat{g}_{ij})$ where

$$\hat{g}_{ij} = a_.^2 \left(a_{m+1}^{-1} + \delta_{ij} a_i^{-1} \right).$$

   (c) Show the Hessian determinant reduces to

$$|g''(\hat{x})| = a_.^{2m+1} \left(\prod_{i=1}^{m+1} a_i \right)^{-1}.$$

   (d) Upon simplification, show that the final form is given in (3.18).

6. Derive Laplace's approximation to the Dirichlet integral in (3.22) using the additive logistic change of variable.
   (a) Show that the change of variable has the Jacobian specified in (3.19).
   (b) Show the maximum occurs at (3.20).
   (c) Show the Hessian matrix is given in (3.21) and has the specified determinant when evaluated at $\hat{x}$.
   (d) Show the approximation simplifies to (3.22).

7. Show that the approximation to the Dirichlet integral from (3.23) is $\hat{I}_x$ as in (3.22).

8. The Dirichlet–Multinomial distribution is a mixture distribution arising in the following manner. Suppose the conditional distribution of $Y = (Y_1, \ldots, Y_m)|X = x = (x_1, \ldots, x_m)$ is Multinomial $(n; x, x_{m+1})$ with $x_{m+1} = 1 - \sum_{i=1}^m x_i$. In addition, let the marginal density of X be Dirichlet $(\alpha_1, \ldots, \alpha_m, \alpha_{m+1})$ with density

$$f(x) = \frac{1}{\Gamma(\alpha_.)} \prod_{i=1}^{m+1} \left\{ \Gamma(\alpha_i) x_i^{\alpha_i - 1} \right\} \qquad x \in D,$$

where $x_{m+1} = 1 - \sum_{i=1}^m x_i$, D is the simplex in $\Re^{m+1}$, and $\alpha_. = \sum_{i=1}^{m+1} \alpha_i$. The marginal distribution of Y is Dirichlet–Multinomial $(n; \alpha_1, \ldots, \alpha_m, \alpha_{m+1})$ with mass function

$$p(k) = \Pr(Y_1 = k_1, \ldots, Y_m = k_m)$$

$$= \binom{n}{k_1, \ldots, k_m, k_{m+1}} \frac{\Gamma(\alpha_.)}{\Gamma(n+\alpha_.)} \prod_{i=1}^{m+1} \frac{\Gamma(k_i + \alpha_i)}{\Gamma(\alpha_i)},$$

where $k_{m+1} = n - \sum_{i=1}^m k_i$. From context, it should be apparent that this mass function is the predictive Bayesian distribution of counts in n future multinomial trials whose vector parameter of probabilities X has a Dirichlet posterior. The mass function also arises in other contexts where it is referred to as the multivariate Pólya distribution. If $\hat{p}(k)$ is the marginalization performed using the additive logistic change of variables from x to y, show that

$$\hat{p}(k) = \binom{n}{k_1, \ldots, k_m, k_{m+1}} \frac{\Gamma(\alpha.)}{\hat{\Gamma}(n+\alpha.)} \prod_{i=1}^{m+1} \frac{\hat{\Gamma}(k_i + \alpha_i)}{\Gamma(\alpha_i)}.$$

9. Show that Laplace's approximation is invariant to nonsingular linear transformation of the variable of integration.
10. Show that (3.26) holds.
    (a) In particular, since $\hat{x}_a = \hat{x}_a(x_b)$ solves

    $$g'_a\{\hat{x}_a(x_b), x_b\} = 0,$$

    use implicit differentiation to show that

    $$\frac{\partial \hat{x}_a}{\partial x_b^T} = -g''_{aa}(\hat{x}_a, x_b)^{-1} g''_{ab}(\hat{x}_a, x_b) \qquad (3.60)$$

    where

    $$g'' = \begin{pmatrix} g''_{aa} & g''_{ab} \\ g''_{ba} & g''_{bb} \end{pmatrix}.$$

    (b) Show that

    $$\frac{\partial}{\partial x_b} g\{\hat{x}_a(x_b), x_b\} = g'_b\{\hat{x}_a(x_b), x_b\}$$

    and

    $$\frac{\partial^2}{\partial x_b \partial x_b^T} g\{\hat{x}_a(x_b), x_b\} = g''_{ba}(\hat{x}_a, x_b) \frac{\partial \hat{x}_a}{\partial x_b^T} + g''_{bb}(\hat{x}_a, x_b). \qquad (3.61)$$

    Combining (3.60) and (3.61), gives

    $$\left| \frac{\partial^2}{\partial x_b \partial x_b^T} g\{\hat{x}_a(x_b), x_b\} \right|_{x_b = \hat{x}_b} = \left| g''_{bb}(\hat{x}) - g''_{ba}(\hat{x}) g''_{aa}(\hat{x})^{-1} g''_{ab}(\hat{x}) \right|.$$

    Expression (3.26) is now

    $$|g''(\hat{x})| = |g''_{aa}(\hat{x})| \times |g''_{bb}(\hat{x}) - g''_{ba}(\hat{x}) g''_{aa}(\hat{x})^{-1} g''_{ab}(\hat{x})|,$$

    or the block determinant identity given in Press (1972, §2.6.2).
11. Show that the multivariate Laplace approximation separates as the product of marginal Laplace approximations under the appropriate conditions described in Proposition 3.2.2 of section 3.2.1.
12. Consider the accuracy of the two density approximations in section 3.4.3.
    (a) Normalize the approximations determined from (3.48) and (3.50). Compare their surface plots with the true density in (3.47).
    (b) Tabulate the relative errors of the two approximations over some common values and decide which of the normalized approximations is better.
    (c) The density approximation in (3.50) was computed from (3.49) by Jacobian transformation. Suppose that in determining (3.49), $\Theta \sim$ Uniform $(-\pi, \pi)$ is used instead of $\Theta \sim$ Uniform $(0, 2\pi)$. Does this change the approximation in (3.50)? If so, in what way?
13. The distribution of a random vector (X, Y, Z) with rotational symmetry in $\mathfrak{R}^3$ can be characterized in terms of its random polar coordinates (R, Θ, Ψ), where $R > 0$ is the radius, $\Theta \in (0, 2\pi)$ is the longitude, and Ψ is the polar distance or latitude. The characterization assumes that the

components of (R, Θ, Ψ) are mutually independent with a uniform density on Θ, density

$$f(\psi) = \tfrac{1}{2} \sin \psi \qquad \psi \in (0, \pi)$$

for Ψ, and density $g(r)$ for R on $r > 0$.

(a) Using $dxdydz = r^2 \sin \psi \, dr d\psi d\theta$, derive the density $f(x, y, z)$ in terms of g.

(b) Feller (1971, p. 524) expresses the characteristic function of (X, Y, Z) as

$$\phi(s, t, u) = E \exp\{i(sX + tY + uZ)\}$$
$$= \int_0^\infty (r\rho)^{-1} \sin(r\rho) g(r) dr,$$

where $\rho = \sqrt{s^2 + t^2 + u^2}$. Using

$$\sin a = \frac{e^{ia} - e^{-ia}}{2i},$$

show that the joint MGF is

$$M(s, t, u) = \phi(-is, -it, -iu)$$
$$= \int_0^\infty (r\rho)^{-1} \sinh(r\rho) g(r) dr, \qquad (3.62)$$

when g is chosen so the integral in (3.62) converges in a neighborhood of zero.

(c) Suppose $g(r) = re^{-r}$ for $r > 0$ so that $R \sim \text{Gamma}(2, 1)$. Show that

$$M(s, t, u) = \left(1 - \rho^2\right)^{-1} \qquad \rho < 1. \qquad (3.63)$$

(d) Determine the saddlepoint density of (X, Y, Z) from (3.63).

(e) Determine the saddlepoint density of the random polar coordinates (R, Θ, Ψ). Transform to Euclidean coordinates and compare your answer with that from part (d). Which is more accurate?

14. Repeat all parts of Exercise 13 using a density for R that is χ_3^2. A simple explicit expression should result in (3.62).

15. Consider the bootstrap example of section 3.4.1.

(a) Show that the MGF for the bootstrap resampling distribution of $\sum_{i=1}^n W_i^*$ is given in (3.44) by direct computation. Confirm this result by also computing it from the distributional representation in (3.43) using the MGF of the Multinomial distribution.

(b) In approximating

$$\Pr\left(\bar{W}^* \le w\right) = \Pr\left(\sum_{i=1}^n W_i^* \le nw\right),$$

show that the saddlepoint equation is

$$\frac{\sum_{i=1}^n w_i e^{\hat{s} w_i}}{\sum_{i=1}^n e^{\hat{s} w_i}} = w. \qquad (3.64)$$

Show (3.64) admits a solution if and only if $w \in (\min w_i, \max w_i)$.

(c) Confirm one of the percentiles in table 3.4.

16. Derive the saddlepoint expressions for the distribution of the Durbin–Watson test statistic in (3.45).

(a) Define $l_1 \le \ldots \le l_n$ as the ordered values of $\{w_i - d\}$. Show that d is in the interior of the support of D if and only if $0 \in (l_1, l_n)$, the interior of the support of Z_d.

(b) For d as in (a), show that the CGF of Z_d is

$$K_d(s) = -\frac{1}{2} \sum_{i=1}^n \ln(1 - 2sl_i) \qquad s \in \left(\frac{1}{2l_1}, \frac{1}{2l_n}\right). \qquad (3.65)$$

(c) Determine expressions for the Lugannani and Rice approximation. Choose some weights and compute a small table of CDF values.

(d) Suppose that D has distribution

$$D \sim \frac{\sum_{i=1}^n w_i G_i}{\sum_{i=1}^n G_i},$$

where $\{G_i\}$ are independently distributed Gamma $(\alpha_i, 1)$. Show that the CDF for D is computed as above but that Z_d now has the CGF

$$K_d(s) = -\sum_{i=1}^n \alpha_i \ln(1 - s l_i) \qquad s \in (l_1^{-1}, l_n^{-1}).$$

Show the saddlepoint $\hat{s}$ solves

$$0 = \sum_{i=1}^n \frac{\alpha_i l_i}{1 - \hat{s} l_i} \qquad \hat{s} \in (l_1^{-1}, l_n^{-1}). \tag{3.66}$$

The degenerate case in which $\Pr(D = w) = 1$ occurs when $w_i \equiv w$; the saddlepoint approximations are not meaningful in this setting.

4

Conditional densities and distribution functions

Saddlepoint approximations for conditional densities and mass functions are presented that make use of two saddlepoint approximations, one for the joint density and another for the marginal. In addition, approximations for univariate conditional distributions are developed. These conditional probability approximations are particularly important because alternative methods of computation, perhaps based upon simulation, are likely to be either very difficult to implement or not practically feasible. For the roles of conditioning in statistical inference, see Reid (1995).

4.1 Conditional saddlepoint density and mass functions

Let (X, Y) be a random vector having a nondegenerate distribution in $\Re^m$ with $\dim(X) = m_x$, $\dim(Y) = m_y$, and $m_x + m_y = m$. With all components continuous, suppose there is a joint density $f(x, y)$ with support $(x, y) \in \mathcal{X} \subseteq \Re^m$. For lattice components on I^m, assume there is a joint mass function $p(j, k)$ for $(j, k) \in \mathcal{X} \subseteq I^m$. All of the saddlepoint procedures discussed in this chapter allow both X and Y to mix components of the continuous and lattice type.

Approximations are presented below in the continuous setting by using symbols f, x, and y. Their discrete analogues simply amount to replacing these symbols with p, j, and k. We shall henceforth concentrate on the continuous notation but also describe the methods as if they were to be used in both the continuous and lattice settings. Any discrepancies that arise for the lattice case are noted.

The conditional density of Y at y given $X = x$ is defined as

$$f(y|x) = \frac{f(x, y)}{f(x)} \qquad (x, y) \in \mathcal{X} \tag{4.1}$$

and 0 otherwise, where f is used generically to denote the (conditional or marginal) density of the variable(s) within. A natural approach to approximating (4.1) is to use two separate saddlepoint approximations for $f(x, y)$ and $f(x)$. Denoting such approximations with the inclusion of hats, then

$$\hat{f}(y|x) = \frac{\hat{f}(x, y)}{\hat{f}(x)} \qquad (x, y) \in \mathcal{I}_\mathcal{X} \tag{4.2}$$

defines a *double-saddlepoint density or mass function*. Symbol $\hat{f}$ is used generically to indicate a saddlepoint density/mass function and $\mathcal{I}_\mathcal{X}$ denotes the interior of the convex hull of the support $\mathcal{X}$. The idea of using two saddlepoint approximations to recover a conditional mass function was introduced by Daniels (1958) with its full elaboration, as in (4.2), presented in Barndorff-Nielsen and Cox (1979).

Double-saddlepoint density (4.2) may be expressed in terms of the joint CGF of (X, Y) as follows. Let $K(s, t)$ denote the joint CGF for $(s, t) \in \mathcal{S} \subseteq \Re^m$ where $\mathcal{S}$ is open and the components s and t are associated with X and Y respectively. Then, the saddlepoint density for the numerator is given in (3.2) as

$$\hat{f}(x, y) = (2\pi)^{-m/2} \left|K''(\hat{s}, \hat{t})\right|^{-1/2} \exp\{K(\hat{s}, \hat{t}) - \hat{s}^T x - \hat{t}^T y\} \tag{4.3}$$

for $(x, y) \in \mathcal{I}_\mathcal{X}$. Here, the m-dimensional saddlepoint $(\hat{s}, \hat{t})$ solves the set of m equations

$$K'(\hat{s}, \hat{t}) = (x, y), \tag{4.4}$$

where K' is the gradient with respect to both components s and t, and K'' is the corresponding Hessian. The denominator saddlepoint density is determined from the marginal CGF of X given as $K(s, 0)$ and defined on the subplane $\{s : (s, 0) \in \mathcal{S}\} = \mathcal{S}_0 \subseteq \Re^{m_x}$. Since $\mathcal{S}$ does not contain its boundary points in $\Re^m$, $\mathcal{S}_0$ also does not in the lower dimensional subspace $\Re^{m_x}$. Hence, it too is an open set relative to the lower dimensional subspace. Thus the setting is regular and the marginal saddlepoint density is

$$\hat{f}(x) = (2\pi)^{-m_x/2} \left|K''_{ss}(\hat{s}_0, 0)\right|^{-1/2} \exp\left\{K(\hat{s}_0, 0) - \hat{s}_0^T x\right\} \tag{4.5}$$

for values of x such that $(x, y) \in \mathcal{I}_\mathcal{X}$, where $\hat{s}_0$ is the m_x-dimensional saddlepoint for the denominator that solves

$$K'_s(\hat{s}_0, 0) = x. \tag{4.6}$$

Here K'_s denotes the gradient with respect to component s only and K''_{ss} denotes its corresponding Hessian. Putting (4.3) and (4.5) together, gives a convenient computational form for the double-saddlepoint density as

$$\hat{f}(y|x) = (2\pi)^{-m_y/2} \left\{\frac{\left|K''(\hat{s}, \hat{t})\right|}{\left|K''_{ss}(\hat{s}_0, 0)\right|}\right\}^{-1/2} \tag{4.7}$$
$$\times \exp\left[\{K(\hat{s}, \hat{t}) - \hat{s}^T x - \hat{t}^T y\} - \{K(\hat{s}_0, 0) - \hat{s}_0^T x\}\right].$$

The approximation is well-defined as long as $(x, y) \in \mathcal{I}_\mathcal{X}$ for fixed x, which places y in a set we denote as $\mathcal{Y}_x$. Exercise 1 asks the reader to show that $\mathcal{Y}_x \subseteq \Re^{m_y}$ is an open convex set and that $\mathcal{Y}_x \supseteq \mathcal{I}_\mathcal{X}$, the interior of the convex hull of the conditional support of Y given $X = x$. Within $\mathcal{Y}_x$, the approximation is practically meaningful if y is an interior point of the conditional support for Y given $X = x$.

Converting a conditional saddlepoint density $\hat{f}(y|x)$ into a conditional probability $\Pr(Y \in A|x)$ for a wide range of sets $A \subset \Re^{m_y}$ is a very difficult problem. The problem can be addressed by using the Monte Carlo integration methods in Butler *et al.* (2005) that make use of importance sampling. While the conditioning makes these methods rather slow in execution, alternative approaches appear to be even slower and more difficult to implement.

4.1.1 Examples

Independence of X and Y

In the full generality of the independence assumption,

$$\hat{f}(y|x) = \frac{\hat{f}(x,y)}{\hat{f}(x)} = \frac{\hat{f}(x)\hat{f}(y)}{\hat{f}(x)} = \hat{f}(y),$$

the marginal saddlepoint density of Y. The double-saddlepoint density of Y given x is the same as the marginal or single-saddlepoint density of Y.

This might also have been explained in terms of the reduction of expression (4.7). Under independence, the joint CGF separates as

$$K(s,t) = K(s,0) + K(0,t)$$

so the numerator saddlepoint equations are $K'_s(\hat{s}, 0) = x$ and $K'_t(0, \hat{t}) = y$. The first of these equations is identical to the denominator saddlepoint equation $K'_s(\hat{s}_0, 0) = x$; therefore, by uniqueness of saddlepoint solutions, $\hat{s} = \hat{s}_0$. Using the same argument, the second of these equations reveals that $\hat{t}$ is the saddlepoint in the marginal saddlepoint density $\hat{f}(y)$. These facts, when used to reduce (4.7), also lead to the single-saddlepoint density $\hat{f}(y)$.

Binomial (n, θ)

Suppose $Z_i \sim$ Poisson (λ_i) independently for $i = 1, 2$. The conditional mass function of $Y = Z_1$ given $X = Z_1 + Z_2 = n \geq 2$ is Binomial (n, θ) with $\theta = \lambda_1/(\lambda_1 + \lambda_2)$. The double-saddlepoint mass function uses CGF

$$K(s,t) = K_{Z_1}(s+t) + K_{Z_2}(s)$$
$$= \lambda_1(e^{s+t} - 1) + \lambda_2(e^s - 1)$$

defined on $(s, t) \in \Re^2$. When approximating

$$p(k|n) = \Pr(Y = k | X = n) = \binom{n}{k} \theta^k (1-\theta)^{n-k},$$

the numerator saddlepoint is

$$(\hat{s}, \hat{t}) = \left(\ln \frac{n-k}{\lambda_2}, \ln \frac{k\lambda_2}{(n-k)\lambda_1} \right) \qquad k = 1, \ldots, n-1$$

with Hessian

$$K''(\hat{s}, \hat{t}) = \begin{pmatrix} n & k \\ k & k \end{pmatrix}.$$

The denominator saddlepoint is

$$\hat{s}_0 = \ln \frac{n}{\lambda_1 + \lambda_2}.$$

Together, these results reduce to

$$\hat{p}(k|n) = \widehat{\binom{n}{k}} \theta^k (1-\theta)^{n-k} \qquad k = 1, \ldots, n-1 \qquad (4.8)$$

Table 4.1. *Characterizations of the various distributions that are the result of conditioning the independent variables in the left column on their sum.*

Z_i distribution	$Y \mid X = n$
Poisson (λ_i)	Multinomial $(n; \lambda_1/\lambda., \ldots, \lambda_{m+1}/\lambda.)^*$
Binomial (n_i, θ)	Hypergeometric $(n_1, \ldots, n_{m+1}; n)$
Negative Binomial (n_i, θ)	Pólya $(n_1, \ldots, n_{m+1}; n)$
Gamma (α_i, β)	Dirichlet $(\alpha_1, \ldots, \alpha_{m+1})$ when $n = 1$

Note: * Here, $\lambda. = \sum_{i=1}^{m+1} \lambda_i$.

which is exact apart from Stirling's approximation for the combinatoric. For $n = 2$, only the value $k = 1$ leads to a lattice point $(2, 1)$ in $\mathcal{I}_\mathcal{X} = \{(n, k) : 0 < k < n\}$ for the double-saddlepoint mass function; thus $n \geq 2$ is required.

The double-saddlepoint approximation in (4.8) agrees with the single-saddlepoint approximation for Binomial (n, θ) given in (1.19). None the less, these two approximations have been derived from completely different theoretical perspectives. The double-saddlepoint approximation is based on a conditional distributional characterization of the Binomial and is determined from the associated joint CGF; the single-saddlepoint approximation derives directly from the univariate CGF. Additional situations are given later in which both the single- and double-saddlepoint approaches may be used. Quite often the two approaches are seen to agree.

Conditioning on sums

Certain multivariate distributions with dependence may be characterized by starting with some independent components and conditioning on their sum. Four such distributions are summarized below. Suppose $Z_1, \ldots, Z_{m+1}$ are independent and take $Y = (Z_1, \ldots, Z_m)^T$ and $X = \sum_{i=1}^{m+1} Z_i$. The various mass/density functions that may be characterized as the conditional distribution of $Y \mid X = n$ are listed in table 4.1. All these distributional characterizations are easily derived. For the conditional characterization of the multivariate Pólya, see Johnson and Kotz (1986, vol. VII, Pólya distribution, multivariate, Eq. 3). In each of these instances, the double-saddlepoint mass/density function reproduces its true counterpart apart from the use of Stirling's approximation in place of the factorial or gamma functions involved. For the Dirichlet density, this means that the normalized saddlepoint density is exact.

Saddlepoint approximations to the multivariate CDFs of the four distributions listed on the right side of table 4.1 have been given in Butler and Sutton (1998).

Bivariate exponential

Let (X, Y) have this bivariate density, as given in (3.4), with $\rho = 0.9$. Since the marginal density of X is Exponential (1), the exact conditional density of Y given $X = 1.5$ is easily determined from (3.4). The double-saddlepoint density is the ratio of the joint saddlepoint

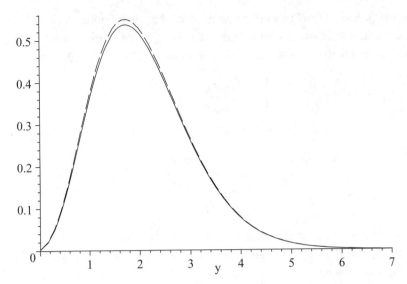

Figure 4.1. The conditional density $f(y|x)$ (*solid*) and double-saddlepoint density $\hat{f}(y|x)$ (*dashed*) of a bivariate exponential with $\rho = 0.9$ and $x = 1.5$.

density, as determined in section 3.1.3, and the saddlepoint approximation to Exponential (1), which is exact apart from Stirling's approximation to $\Gamma(1) = 1$. Figure 4.1 compares the exact conditional density (solid line) with its double-saddlepoint approximation (dashed).

4.1.2 Properties

Equivariance

Proposition 4.1.1 *Double-saddlepoint density $\hat{f}(y|x)$ maintains equivariance under non-singular linear transformation of variable Y as occurs with the true density $f(y|x)$.*

Symmetry

Proposition 4.1.2 *The various symmetries of multivariate conditional densities and mass functions for variable Y given $X = x$ are preserved in their conditional double-saddlepoint counterparts.*

These symmetries concern exchangeability, as well as antipodal, spherical, and elliptical symmetries all of which relate to group invariances. The arguments for showing that such symmetries hold in the conditional setting simply amount to restricting the group invariance to the y-components in $\hat{f}(x, y)$. The subgroup action is now on component y alone and leaves the value of x fixed.

4.1.3 Remarks on asymptotics

(1) The double-saddlepoint density approximation is often stated in the context of the conditional density of sample means. Suppose $(X_1, Y_1), \ldots, (X_n, Y_n)$ are i.i.d. m-vectors

with support $\mathcal{X}$ and CGF $K(s, t)$ convergent on $\mathcal{S}$. The distribution of $(\bar{X}, \bar{Y})$ has support in the convex hull of $\mathcal{X}$ and its joint CGF $nK(s/n, t/n)$ converges on $(s, t) \in n\mathcal{S}$. In the continuous setting with $(\bar{x}, \bar{y}) \in \mathcal{I}_\mathcal{X}$, the double-saddlepoint density of $\bar{Y}$ at $\bar{y}$ given $\bar{X} = \bar{x}$ is

$$\hat{f}(\bar{y}|\bar{x}) = \left(\frac{n}{2\pi}\right)^{m_y/2} \left\{\frac{|K''(\hat{s}, \hat{t})|}{|K''_{ss}(\hat{s}_0, 0)|}\right\}^{-1/2} \qquad (4.9)$$
$$\times \exp\left[n\left\{K(\hat{s}, \hat{t}) - \hat{s}^T \bar{x} - \hat{t}^T \bar{y}\right\} - n\left\{K(\hat{s}_0, 0) - \hat{s}_0^T \bar{x}\right\}\right],$$

where $(\hat{s}, \hat{t}) \in \mathcal{S}$ is the numerator saddlepoint solving

$$K'_s(\hat{s}, \hat{t}) = \bar{x} \qquad (4.10)$$
$$K'_t(\hat{s}, \hat{t}) = \bar{y},$$

and $\hat{s}_0$ is the denominator saddlepoint solving $K'_s(\hat{s}_0, 0) = \bar{x}$.

The saddlepoint density of $\bar{Y}$ given $\bar{X} = \bar{x}$ that results from using the joint CGF of $(\bar{X}, \bar{Y})$, expressed as $nK(s/n, t/n)$, in expression (4.7) agrees with (4.9). However, the saddlepoints associated with using expression (4.7) are not quite the same as those in (4.10). Solving the saddlepoint equation (4.4) with K replaced by $nK(s/n, t/n)$ and (x, y) by $(\bar{x}, \bar{y})$ provides saddlepoints $(\check{s}, \check{t})$ that relate to the roots of (4.10) according to $(\check{s}, \check{t}) = n(\hat{s}, \hat{t}) \in n\mathcal{S}$. Likewise, the denominator saddlepoint is $(\check{s}_0, 0) = n(\hat{s}_0, 0) \in n\mathcal{S}$.

With the equivalence of these two expressions, it is clear that (4.9) offers no further generalization of the basic $n = 1$ formula in (4.7). The presentation will therefore continue to work with the simpler $n = 1$ formula.

In the lattice setting, the conditional mass function is given by (4.9) but with the leading power of n as $n^{-m_y/2}$ and not $n^{m_y/2}$ (see Barndorff-Nielsen and Cox, 1979 and Daniels, 1954, §8). The reason for this difference has to do with whether a Jacobian term is included when converting the density of a vector sum to that of the vector mean.

(2) As a statement of approximation, (4.9) is often expressed as the leading term of an asymptotic expansion as $n \to \infty$. For $(\bar{x}, \bar{y}) \in \mathcal{I}_\mathcal{X}$,

$$f(\bar{y}|\bar{x}) = \hat{f}(\bar{y}|\bar{x})\{1 + O(n^{-1})\}$$

where

$$O(n^{-1}) = \frac{1}{n}(O_N - O_D) + O(n^{-2})$$

and the terms O_N and O_D are the second-order correction terms for the saddlepoint densities in the numerator and denominator respectively. To show this, write

$$\frac{f(\bar{x}, \bar{y})}{f(\bar{x})} = \frac{\hat{f}(\bar{x}, \bar{y})\{1 + n^{-1}O_N + O(n^{-2})\}}{\hat{f}(\bar{x})\{1 + n^{-1}O_D + O(n^{-2})\}} \qquad (4.11)$$
$$= \hat{f}(\bar{y}|\bar{x})\{1 + n^{-1}(O_N - O_D) + O(n^{-2})\}$$

when the denominator term in curly braces has been expanded as a Taylor series.

(3) Normalization of the double-saddlepoint density has the same effect on the asymptotic error as it does with the single-saddlepoint density. Let $(\bar{X}, \bar{Y})$ have mean (μ_x, μ_y). If $(\bar{x}, \bar{y})$ remains in a large deviation set as $n \to \infty$, so that $(\bar{x} - \mu_x, \bar{y} - \mu_y)$ remains in a fixed

compact subset of its support, then, in both the continuous and lattice cases,

$$f(\bar{y}|\bar{x}) = \bar{f}(\bar{y}|\bar{x})\{1 + O(n^{-1})\},$$

where $\bar{f}$ is the normalized density. Over sets of bounded central tendency in which $\sqrt{n}(\bar{x} - \mu_x, \bar{y} - \mu_y)$ stays bounded, the $O(n^{-1})$ error improves to $O(n^{-3/2})$.

4.2 Conditional cumulative distribution functions

A double-saddlepoint CDF approximation for Y given $X = x$ when $\dim(Y) = 1$ has been given in Skovgaard (1987). Using the notation of the previous section, suppose that

$$F(y|x) = \Pr(Y \leq y | X = x)$$

is the true CDF.

4.2.1 Continuous distributions

If Y is a continuous variable for which $F(y|x)$ admits a density, then the Skovgaard (1987) approximation is

$$\hat{F}(y|x) = \Phi(\hat{w}) + \phi(\hat{w})\left(\frac{1}{\hat{w}} - \frac{1}{\hat{u}}\right) \qquad \hat{t} \neq 0 \qquad (4.12)$$

where

$$\hat{w} = \mathrm{sgn}(\hat{t})\sqrt{2\left[\{K(\hat{s}_0, 0) - \hat{s}_0^T x\} - \{K(\hat{s}, \hat{t}) - \hat{s}^T x - \hat{t} y\}\right]}$$
$$\hat{u} = \hat{t}\sqrt{|K''(\hat{s}, \hat{t})| / |K''_{ss}(\hat{s}_0, 0)|}, \qquad (4.13)$$

and the saddlepoints $(\hat{s}, \hat{t})$ and $\hat{s}_0$ are determined as in (4.4) and (4.6). Expression (4.12) is meaningful so long as $(x, y) \in \mathcal{I}_\mathcal{X}$. Because both sets of saddlepoints are within $\mathcal{S}$, the joint convergence region of $K(s, t)$, both Hessians in $\hat{u}$ are positive definite and the square roots in $\hat{u}$ are well defined. To see that $\hat{w}$ is well defined as a root of a positive number, note that

$$K(\hat{s}, \hat{t}) - \hat{s}^T x - \hat{t} y = \inf_{(s,t) \in \mathcal{S}} \{K(s, t) - s^T x - ty\}$$
$$K(\hat{s}_0, 0) - \hat{s}_0^T x = \inf_{\{s:(s,0) \in \mathcal{S}\}} \{K(s, 0) - s^T x\}$$

and the minimum at the top must be the smaller of the two.

If the value (x, y) leads to the saddlepoint solution $\hat{t} = 0$, then $\hat{s} = \hat{s}_0$ and $\hat{w} = 0 = \hat{u}$ so that (4.12) is undefined. This singularity is removable however, and the approximation may be defined by continuity to be the expression in equation (2.11) of Skovgaard (1987). This expression is complicated by the fact that it involves the lattice of third derivatives for K. The computation of such values can often be avoided by using the following numerical approximation. Perform the saddlepoint computations at $y \pm 10^{-8}$ and compute $\hat{F}(y - 10^{-8}|x)$ and $\hat{F}(y + 10^{-8}|x)$ using 30 digit accuracy in a package such as Maple. If these two values are sufficiently close, then the computation is essentially determined by continuity. Some examples are provided later.

The double-saddlepoint density in (4.7) with dim(Y) = 1 may be written in terms of $\hat{w}$ for later use. From (4.13),

$$\hat{f}(y|x) = \left\{ \frac{|K''(\hat{s}, \hat{t})|}{|K''_{ss}(\hat{s}_0, 0)|} \right\}^{-1/2} \phi(\hat{w}). \tag{4.14}$$

4.2.2 Lattice distributions

If the support of Y is the integer lattice, then the continuity corrections to (4.12), as introduced in Skovgaard (1987), should be used to achieve the greatest accuracy. These two continuity corrections find their motivation in the inversion theory approach originally developed in Daniels (1987). Based on these corrections, two survival approximations are recommended below for the reasons that have been discussed in sections 1.2.5–1.2.7.

These approximations presume a conditioning variable X which may consist of any mixture of lattice and continuous components. However, to ease the burden in notation, the approximations are presented below as if the conditioning variable X consists entirely of lattice variables.

First continuity correction

Suppose $(\hat{s}, \hat{t})$ is the solution to $K'(\hat{s}, \hat{t}) = (j, k)$ required for the numerator saddlepoint with $\hat{t} \neq 0$. Then

$$\widehat{\Pr}_1(Y \geq k | X = j) = 1 - \Phi(\hat{w}) - \phi(\hat{w})\left(\frac{1}{\hat{w}} - \frac{1}{\tilde{u}_1}\right) \qquad \hat{t} \neq 0 \tag{4.15}$$

where

$$\hat{w} = \operatorname{sgn}(\hat{t})\sqrt{2\left[\{K(\hat{s}_0, 0) - \hat{s}_0^T j\} - \{K(\hat{s}, \hat{t}) - \hat{s}^T j - \hat{t}k\}\right]}$$

$$\tilde{u}_1 = (1 - e^{-\hat{t}})\sqrt{|K''(\hat{s}, \hat{t})|/|K''_{ss}(\hat{s}_0, 0)|}, \tag{4.16}$$

and $\hat{s}_0$ solves $K'_s(\hat{s}_0, 0) = j$.

In Exercise 3, an alternative conditional saddlepoint CDF expression $\hat{F}_{1a}(k| X = j)$ is developed for $\hat{t} < 0$ which is the left-tail analogue of (1.37). This expression is not recommended for general applications since probabilities computed from it are not compatible with (4.15) in that

$$\hat{F}_{1a}(k| X = j) \neq 1 - \widehat{\Pr}_1(Y \geq k+1| X = j).$$

Second continuity correction

If $k^- = k - 1/2$ is the offset value of k and $(\tilde{s}, \tilde{t})$ is the offset saddlepoint solving

$$K'(\tilde{s}, \tilde{t}) = \left(j, k - \tfrac{1}{2}\right)$$

with $\tilde{t} \neq 0$, then

$$\widehat{\Pr}_2(Y \geq k | X = j) = 1 - \Phi(\tilde{w}_2) - \phi(\tilde{w}_2)\left(\frac{1}{\tilde{w}_2} - \frac{1}{\tilde{u}_2}\right) \qquad \tilde{t} \neq 0 \tag{4.17}$$

4.2 Conditional cumulative distribution functions

where

$$\tilde{w}_2 = \operatorname{sgn}(\tilde{t})\sqrt{2\left[\left\{K(\hat{s}_0, 0) - \hat{s}_0^T j\right\} - \left\{K(\tilde{s}, \tilde{t}) - \tilde{s}^T j - \tilde{t}k^-\right\}\right]}$$

$$\tilde{u}_2 = 2\sinh(\tilde{t}/2)\sqrt{|K''(\tilde{s}, \tilde{t})| / |K''_{ss}(\hat{s}_0, 0)|}. \tag{4.18}$$

Saddlepoint $\hat{s}_0$ is unchanged.

Following the development in section 1.2.5, the continuity correction $k^+ = k + 1/2$ may be used in the left tail to produce a CDF approximation $\hat{F}_2(k|X=j)$ that is similar to (1.39). There is no need for such an expression however, since it leads to

$$\hat{F}_2(k|X=j) = 1 - \widehat{\Pr}_2(Y \geq k+1|X=j) \tag{4.19}$$

and so (4.17) suffices for all computation with this correction.

4.2.3 Examples

Beta (α, β)

Suppose that $Z_1 \sim$ Gamma $(\alpha, 1)$ independently of $Z_2 \sim$ Gamma $(\beta, 1)$. The conditional distribution of $Y = Z_1$ given $X = Z_1 + Z_2 = 1$ is Beta (α, β) and the Skovgaard approximation can be used to approximate the CDF. This approximation turns out to be analytically the same as the single-saddlepoint approximation in (1.26) of section 1.2.2 that was obtained by expressing the probability as a linear combination of Gammas. See Exercise 4.

An alternative approach uses the conditional Mellin transform to approximate the CDF of $\ln\{Z_1/(Z_1 + Z_2)\}$ given $Z_1 + Z_2 = 1$. Exercise 5 explores this development.

Binomial (n, θ)

If $Z_1 \sim$ Poisson (λ) independently of $Z_2 \sim$ Poisson (μ), then $Y = Z_1$ given $X = Z_1 + Z_2 = n$ has a Binomial (n, θ) distribution with $\theta = \lambda/(\lambda + \mu)$. In this context, the two continuity-corrected Skovgaard approximations agree analytically with their single-saddlepoint counterparts given in (1.33) and (1.34) of section 1.2.3. See Exercise 6.

Hypergeometric $(M, N; n)$

Let $Z_1 \sim$ Binomial (M, θ) independently of $Z_2 \sim$ Binomial (N, θ). The conditional distribution of $Y = Z_1$ given $X = Z_1 + Z_2 = n$ is Hypergeometric $(M, N; n)$ with mass function

$$p(k) = \binom{M}{k}\binom{N}{n-k} \Big/ \binom{M+N}{n} \qquad k = k_l, \ldots, k_u, \tag{4.20}$$

where $k_l = \max(0, n - N)$ and $k_u = \min(M, n)$. For $k \in [k_l + 1, k_u - 1]$, the two continuity-corrected Skovgaard approximations may be computed. The form of these solutions is most easily presented using the notation of a 2×2 contingency table as given by Skovgaard (1987). In this regard, consider the notational change in table 4.2 in which parameters of the hypergeometric on the left are replaced with $\{l_{ij} : i, j = 1, 2\}$, the entries for a 2×2 table on the right.

Table 4.2. *The left table is notation connected with the hypergeometric mass function in (4.20) while the right table provides a simpler notation.*

k	$M-k$	M		l_{11}	l_{12}	$l_{1\cdot}$
$n-k$	$N-n+k$	N	$\equiv$	l_{21}	l_{22}	$l_{2\cdot}$
n	$M+N-n$	$M+N$		$l_{\cdot 1}$	$l_{\cdot 2}$	$l_{\cdot\cdot}$

For the first approximation, the numerator saddlepoints are

$$\hat{s} = \ln \frac{l_{21}(1-\theta)}{l_{22}\theta} \qquad \hat{t} = \ln \frac{l_{11}l_{22}}{l_{12}l_{21}} \qquad (4.21)$$

and the denominator saddlepoint is

$$\hat{s}_0 = \ln \frac{l_{\cdot 1}(1-\theta)}{l_{\cdot 2}\theta}.$$

If we let the right (left) tail refer to values of $k = l_{11}$ for which $\hat{t} > 0$ ($\hat{t} < 0$), then it follows from (4.21) that l_{11} is in the right (left) tail accordingly as

$$l_{11} > \frac{l_{21}l_{1\cdot}}{l_{2\cdot}} \iff l_{11} > \frac{l_{\cdot 1}l_{1\cdot}}{l_{\cdot\cdot}} = E(Y|X=n).$$

Thus it works out here that $\hat{t} > 0$ if and only if $k = l_{11} > E(Y|X)$ but this is not always the case when using the Skovgaard approximation. In this setting, the first continuity correction with $k = l_{11}$ has the inputs

$$\hat{w} = \operatorname{sgn}(\hat{t})\sqrt{2\left(\sum_{i,j} l_{ij}\ln l_{ij} - \sum_{i} l_{i\cdot}\ln l_{i\cdot} - \sum_{j} l_{\cdot j}\ln l_{\cdot j} + l_{\cdot\cdot}\ln l_{\cdot\cdot}\right)}$$

$$\tilde{u}_1 = \left(1 - \frac{l_{12}l_{21}}{l_{11}l_{22}}\right)\sqrt{\frac{l_{11}l_{12}l_{21}l_{22}l_{\cdot\cdot}}{l_{1\cdot}l_{2\cdot}l_{\cdot 1}l_{\cdot 2}}}. \qquad (4.22)$$

The determination of $\tilde{u}_1$ in (4.22) is based on the Hessian computations

$$|K''(\hat{s},\hat{t})| = \frac{l_{11}l_{12}l_{21}l_{22}}{l_{1\cdot}l_{2\cdot}} \qquad |K''_{ss}(\hat{s}_0,0)| = \frac{l_{\cdot 1}l_{\cdot 2}}{l_{\cdot\cdot}}.$$

The second approximation uses the continuity correction $k^- = l_{11}^- = l_{11} - 1/2$ as shown in table 4.3. Since the row and column totals of table 4.3 need to remain fixed, the adjustment in cell (1, 1) must be offset by changing all the other cell entries as shown in the table. The consequences of these corrections on the second approximation are as follows. The numerator saddlepoint $(\tilde{s}, \tilde{t})$ is given by the expressions in (4.21) but with l_{ij}^--values replacing l_{ij}-values. The denominator saddlepoint $\hat{s}_0$ is unchanged. The values for $\tilde{w}_2$ and $\tilde{u}_2$ are

$$\tilde{w}_2 = \operatorname{sgn}(\tilde{t})\sqrt{2\left(\sum_{i,j} l_{ij}^- \ln l_{ij}^- - \sum_{i} l_{i\cdot}\ln l_{i\cdot} - \sum_{j} l_{\cdot j}\ln l_{\cdot j} + l_{\cdot\cdot}\ln l_{\cdot\cdot}\right)}$$

$$\tilde{u}_2 = \left(l_{11}^- l_{22}^- - l_{12}^- l_{21}^-\right)\sqrt{\frac{l_{\cdot\cdot}}{l_{1\cdot}l_{2\cdot}l_{\cdot 1}l_{\cdot 2}}}. \qquad (4.23)$$

4.2 Conditional cumulative distribution functions

Table 4.3. *Notation indicating the continuity correction of the second survival approximation as applied to the hypergeometric distribution.*

$l_{11}^- = l_{11} - \frac{1}{2}$	$l_{12}^- = l_{12} + \frac{1}{2}$	$l_{1\cdot}$
$l_{21}^- = l_{21} + \frac{1}{2}$	$l_{22}^- = l_{22} - \frac{1}{2}$	$l_{2\cdot}$
$l_{\cdot 1}$	$l_{\cdot 2}$	$l_{\cdot\cdot}$

Table 4.4. *Tail probabilities for $H \sim$ Hypergeometric (M, N, n) along with three continuity-corrected approximations.*

M, N, n	k	Exact $\Pr(H \geq k)$	Double-saddlepoint $\widehat{\Pr}_1$	$\widehat{\Pr}_2$	Single-saddlepoint single-$\widehat{\Pr}_2^\dagger$
6, 6, 6	6	$.0^2 1082$	undefined	$.0^3 9756$	$.0^2 1159$
	5	.04004	.04592	.03928	.04014
	4	.2835	.2986	.2828	.2837
	3	.7165	.7303*	.7172	.7163
	2	.95996	.96404	.96071	.95986
	1	.99892	.99920	.99902	.99884
30, 50, 10	4	.5600	.5628	.5605	.5602
20, 10, 10	7	.5603	.5677	.5600	.5603
	9	.06205	.06669	.06140	.06191
100, 200, 20	12	.01047	.01054	.01046	.01046
	17	$.0^5 1249$	$.0^5 1281$	$.0^5 1242$	$.0^5 1244$

Note: † Denotes the single-saddlepoint $\widehat{\Pr}_2$ approximation.

This second approximation is the same expression as (5.5) of Skovgaard (1987) but derived from a different point of view. Skovgaard's development uses four i.i.d. Poisson (1) variables as the entries in a 2 × 2 table. He then conditions entry (1, 1) on the row and column sums to create a hypergeometric random variable that conditions one variable on three others. The approach here differs in that it conditions one variable on only one other. The two binomial variables in our approach may be thought of as those arising at an intermediate step in which the four Poissons are conditioned to have fixed row totals. While both approaches clearly accommodate the Skovgaard approximation procedure, it is not at all clear why the two procedures should lead to exactly the same analytical expressions. An explanation for this based on the idea of cuts is given in Booth and Butler (1990). Exercise 7 asks the reader to replicate the approach used by Skovgaard and to verify that it leads to continuity-corrected approximations that agree with (4.22) and (4.23).

Table 4.4 provides some numerical examples of these approximations and compares them to the exact values and the single-saddlepoint Lugannani and Rice approximation

Table 4.5. *Survival probabilities for P having various Pólya distributions.*

M, N, n	k	$\Pr(P \geq k)$	$\widehat{\Pr}_1(P \geq k)$	$\widehat{\Pr}_2(P \geq k)$	single-$\widehat{\Pr}_2$
6, 9, 6	6	.01192	undefined	.01139	.01105
	5	.07043	.07549	.06990	.06855
	4	.2167	.2227	.2166	.2153
	3	.4551	.4609	.4560	.4572
	2	.7233	.7272	.7254	.7285
	1	.9225	.9238	.9258	.9274
20, 60, 20	5	.5686	.5691	.5692	.5697

with sinh correction denoted as single-$\widehat{\Pr}_2$ in the last column. The first six rows contain tail probabilities for a Hypergeometric $(6, 6, 6)$ distribution for $k \geq 1$. Because $M = N$, the true mass function is symmetric about its mean 3. The single and double approximations with sinh corrections preserve this symmetry since

$$\widehat{\Pr}_2(H \leq k) = \widehat{\Pr}_2(H \geq 6 - k) \qquad k = 1, \ldots, 6 \tag{4.24}$$

whether computed marginally or conditionally. The starred entry has been computed at the mean where the first approximation has a removable singularity. The computation was made by changing $l_{11} = 3$ to the values 3 ± 10^{-8} and offsetting the other 2×2 table entries accordingly; the two computations agreed to over 10 significant digits. Close examination reveals that single-$\widehat{\Pr}_2$ is the most accurate with $\widehat{\Pr}_2$ only slightly worse. However even the worst approximation $\widehat{\Pr}_1$ would be considered highly accurate were we not already spoilt by the accuracy of the other two. The symmetries connected with using the two sinh corrections may be seen in the table by noting that the entries for $k = 3$ and 4 add to 1; the corresponding sum for the $\widehat{\Pr}_1$ correction shows its lack of symmetry. Exercise 3 asks the reader to show that the conditional CDF approximation $\hat{F}_{1a}$, defined in (4.85), also preserves the symmetry of this example.

Pólya $(M, N; n)$

Let $Z_1 \sim$ Negative Binomial (M, θ) independently of $Z_2 \sim$ Negative Binomial (N, θ). The conditional distribution of $Y = Z_1$ given $X = Z_1 + Z_2 = n$ is Pólya $(M, N; n)$ with mass function

$$p(k) = \binom{n}{k} \frac{\Gamma(M+k)\Gamma(N+n-k)}{\Gamma(M+N+n)} \frac{\Gamma(M+N)}{\Gamma(M)\Gamma(N)} \qquad k = 0, \ldots, n. \tag{4.25}$$

The two continuity-corrected Skovgaard approximations for the survival function may be computed and the development is similar to that of the hypergeometric. The details are discussed in Exercise 8. Table 4.5 compares exact computation with these two approximations as well as the single-saddlepoint Lugannani and Rice approximation with sinh correction denoted by single-$\widehat{\Pr}_2$. The double-saddlepoint $\widehat{\Pr}_2$ is clearly more accurate than both single-$\widehat{\Pr}_2$ and the other double-saddlepoint $\widehat{\Pr}_1$. The uniform mass function on $\{0, \ldots, m\}$ is a special case and is the Pólya $(1, 1, m)$ mass function. Exercise 9 considers this example and table 4.7 reveals the numerical accuracy achieved by the two continuity-corrected approximations when $m = 9$. Double-saddlepoint approximation $\widehat{\Pr}_2$ in table 4.7 achieves the greatest accuracy. When compared to all the single-saddlepoint approximations evaluated in table 1.11, it still achieves the greatest accuracy.

When both single- and double-saddlepoint approximations may be used, one might suppose offhand that the single-saddlepoint methods must be more accurate since they require less approximation. The Pólya and uniform examples provide counterexamples to this offhand assessment. In fact, an intuitive argument that favors greater accuracy for the double-saddlepoint method might be this: because the double-saddlepoint method affords greater flexibility by fitting two saddlepoints as opposed to one, it has the capability of achieving greater accuracy.

4.2.4 Proof of the Skovgaard approximation

The proof follows the same approach used to derive the Lugannani and Rice approximation and, like the previous development, only applies in the continuous setting. The idea is to approximate the right tail probability as the integral of the double-saddlepoint density. This, in turn, is approximated by using the Temme procedure in (2.34).

Proofs of the continuity corrections in the lattice case are best understood by using the inversion theory approach found in Skovgaard (1987).

Proof of (4.12). An approximation to $F(z|x)$ is the integral of the double-saddlepoint density given in (4.14) or

$$\int_{-\infty}^{z} \hat{f}(y|x)\,dy = \int_{-\infty}^{z} \left\{ \frac{|K''(\hat{s}, \hat{t})|}{|K''_{ss}(\hat{s}_0, 0)|} \right\}^{-1/2} \phi(\hat{w})\,dy \qquad (4.26)$$

where $\hat{w}$ is given in (4.13). A change of variables from dy to $d\hat{w}$ allows for the use of the Temme approximation. First, however, the change in variables requires the Jacobian in the mapping $y \leftrightarrow \hat{w}$.

Lemma 4.2.1 *The mapping $y \leftrightarrow \hat{w}$ is smooth and monotonic increasing with derivative*

$$\frac{dy}{d\hat{w}} = \begin{cases} \hat{w}/\hat{t} & \text{if } \hat{t} \neq 0 \\ \sqrt{|K''(\hat{s}_0, 0)| \,/\, |K''_{ss}(\hat{s}_0, 0)|} & \text{if } \hat{t} = 0. \end{cases}$$

Proof. Rewrite $\hat{w}$ as

$$\tfrac{1}{2}\hat{w}^2 = \{K(\hat{s}_0, 0) - \hat{s}_0^T x\} - \{K(\hat{s}, \hat{t}) - \hat{s}^T x - \hat{t} y\} \qquad (4.27)$$

and differentiate implicitly to get

$$\hat{w}\frac{d\hat{w}}{dy} = -\frac{d\hat{s}^T}{dy}\hat{K}'_s - \frac{d\hat{t}}{dy}\hat{K}'_t + \frac{d\hat{s}^T}{dy}x + \hat{t} + \frac{d\hat{t}}{dy}y$$

$$= \hat{t}, \qquad (4.28)$$

where $(\hat{K}'_s, \hat{K}'_t)$ is $K'(\hat{s}, \hat{t}) = (x, y)$. Since $\operatorname{sgn}(\hat{w}) = \operatorname{sgn}(\hat{t})$, then $dy/d\hat{w} > 0$ for $\hat{t} \neq 0$. From (4.28), the limiting value as $\hat{t} \to 0$ is

$$\lim_{\hat{t}\to 0} \frac{dy}{d\hat{w}} = \lim_{\hat{t}\to 0} \frac{\hat{w}}{\hat{t}}.$$

The limit may be determined by expressing (4.27) in terms of $\hat{t}$ as

$$\tfrac{1}{2}\hat{w}^2 = g(0) - g(\hat{t}), \tag{4.29}$$

where we define

$$g(t) = K(\hat{s}_t, t) - \hat{s}_t^T x - ty \tag{4.30}$$

with $\hat{s}_t$ as the unique solution to the first m_x saddlepoint equations

$$K_s'(\hat{s}_t, t) = x$$

for fixed value of t. Taylor expansion of $g(0)$ in (4.29) about $\hat{t}$ yields

$$\tfrac{1}{2}\hat{w}^2 = -g'(\hat{t})\hat{t} + \tfrac{1}{2}g''(\hat{t})\hat{t}^2 + O(\hat{t}^3). \tag{4.31}$$

In order to evaluate the right side, the reader is asked in Exercise 11 to determine that $g'(\hat{t}) = 0$ and

$$g''(\hat{t}) = \hat{K}_{tt}'' - \hat{K}_{ts}''(\hat{K}_{ss}'')^{-1}\hat{K}_{st}'' = \frac{|\hat{K}''|}{|\hat{K}_{ss}''|}. \tag{4.32}$$

Combining (4.32) and (4.31), we get

$$\lim_{\hat{t}\to 0}\left(\frac{\hat{w}}{\hat{t}}\right)^2 = \lim_{\hat{t}\to 0}\frac{|\hat{K}''|}{|\hat{K}_{ss}''|} = \frac{|K''(\hat{s}_0, 0)|}{|K_{ss}''(\hat{s}_0, 0)|} \tag{4.33}$$

and the lemma follows. □

The change in variables $y \leftrightarrow \hat{w}$ in (4.26), with $\hat{w}_z$ as the image of z under this mapping, gives

$$\int_{-\infty}^{\hat{w}_z}\left\{\frac{|K''(\hat{s}, \hat{t})|}{|K_{ss}''(\hat{s}_0, 0)|}\right\}^{-1/2}\frac{\hat{w}}{\hat{t}}\phi(\hat{w})d\hat{w} = \int_{-\infty}^{\hat{w}_z}h(\hat{w})\phi(\hat{w})d\hat{w}, \tag{4.34}$$

where $h(\hat{w})$ is defined to assure equality in (4.34). The Temme approximation in (2.34) can now be applied with the assurance from Lemma 19 that $h(0) = \lim_{\hat{t}\to 0}h(\hat{w}) = 1$. This gives

$$\Phi(\hat{w}_z) + \phi(\hat{w}_z)\frac{1 - h(\hat{w}_z)}{\hat{w}_z}$$

which is (4.12). □

4.2.5 Properties

Invariance

Proposition 4.2.2 *Probability approximation for continuous Y is invariant to linear transformation of Y when using either the Skovgaard approximation or integration of the double-saddlepoint density.*

Proposition 4.2.3 *In linear transformation of lattice variable Y, all the approximations are invariant to increasing linear transformation. For decreasing linear transformation* however, $\widehat{\Pr}_2$, using the Skovgaard approximation, and the summed double-saddlepoint mass function are invariant but $\widehat{\Pr}_1$ is not.

Example Consider the Binomial (23, 0.2) mass function from section 2.1.3 as well as its continuity-corrected Skovgaard approximations mentioned as the second example of section 4.2.3. These two Skovgaard approximations are equivalent to their continuity-corrected Lugannani and Rice counterparts in both tails and therefore inherit their invariance properties. As such, $\widehat{\Pr}_2$ is invariant to the transformation $X \to 23 - X$ whereas the first approximation $\widehat{\Pr}_1$ is not.

Symmetry

Proposition 4.2.4 *Let the conditional distribution of Y given $X = x$ be symmetric for almost every value of x. Furthermore, suppose that the conditional mean is linear in x or $E(Y|x) = \varsigma + \tau^T x$, for fixed values of ς and τ. Under such conditions, the Skovgaard approximation is also symmetric in its continuous version and in its second continuity-corrected version $\widehat{\Pr}_2$. The first correction is not symmetric but its alternative version, defined in (4.85) and (4.84), does preserves symmetry.*

Without any loss in generality, take $\varsigma = 0$ and assume that the point of symmetry is $E(Y|x) = \tau^T x$. This is possible since the approximations are invariant to translation as mentioned in Propositions 4.2.2 and 4.2.3. In the continuous setting for Y, the symmetry for the Skovgaard approximation is specified as

$$\hat{F}(\tau^T x - y|x) = 1 - \hat{F}(\tau^T x + y|x) \qquad (4.35)$$

for all values of $y > 0$. The proof is somewhat long and is therefore deferred to section 4.5.1 of the appendix.

In the lattice setting for Y, the symmetry relationship in (4.35) requires that y assume either integer of half-integer values. This symmetry holds for the second continuity-correction since the sinh function is odd, as the reader is asked to show in Exercise 14.

Examples

Beta (α, α) This distribution is symmetric with mean $1/2$. The Skovgaard approximation is the same as its Lugannani and Rice counterpart and therefore exhibits the underlying symmetry.

Binomial $(10, 1/2)$ In this example of section 2.1.2, symmetry was exhibited by the second continuity-corrected Lugannani and Rice approximation, which is known to be the same as its Skovgaard counterpart constructed from i.i.d. Poisson variables. The mean in this context is linear in 10, the value conditioned upon. The first continuity-corrected Skovgaard approximation $\widehat{\Pr}_1$, and its alternative version defined in (4.85) and (4.84), agree with their single-saddlepoint counterparts; thus the numerical work in section 2.1.2 demonstrates that the former is not symmetric while the alternative version does preserve symmetry.

Hypergeometric $(6, 6, 6)$ The distribution is symmetric with a mean that is linear in the conditioned value. Table 4.4 shows that $\widehat{\Pr}_2$ preserves symmetry whereas $\widehat{\Pr}_1$ does not.

Uniform mass function The symmetry of a uniform mass function over the set $\{0, \ldots, 9\}$ is reflected in table 4.7 by both continuity-corrected approximations $\widehat{\Pr}_2$ and the alternate version $\hat{F}_{1a}$, as defined in (4.85) and (4.84).

Singularity in the Skovgaard approximation

Proposition 4.2.5 *The removable singularity in $\hat{F}(y|x)$ occurs when $\hat{t} = 0$ but generally this is not at the value $y = E(Y|x)$. However, under the condition that $E(Y|X = x)$ is linear in x, the singularity must occur at the conditional mean.*

If $E(Y|x) = \varsigma + \tau^T x$, then it is shown in section 4.5.2 of the appendix that

$$\hat{t}(x, y) = 0 \quad \Leftrightarrow \quad y = E(Y|X = x), \tag{4.36}$$

where $\hat{t}(x, y)$ indicates the dependence of $\hat{t}$ on (x, y).

Examples The result applies to every example considered so far in this chapter including Binomial (n, θ), Hypergeometric (M, N, n), Pólya (M, N, n), the conditionals for the bivariate exponential, and Beta (α, β). Linearity of the conditional mean in the bivariate exponential is left as Exercise 15.

4.2.6 Remarks on asymptotics

(1) The Skovgaard approximation, as it applies to the sample means $(\bar{X}, \bar{Y})$ of i.i.d. vectors $(X_1, Y_1), \ldots, (X_n, Y_n)$, is stated in terms of the CGF of (X_1, Y_1) denoted by $K(s, t)$ (see Remark 1 of section 4.1.3). Using the saddlepoints described in (4.10), the approximation to $\Pr(\bar{Y} \leq \bar{y} | \bar{X} = \bar{x})$, for $(\bar{x}, \bar{y}) \in \mathcal{I}_\mathcal{X}$, has

$$\hat{w}_n = \sqrt{n}\hat{w} \qquad \hat{u}_n = \sqrt{n}\hat{u} \tag{4.37}$$

in the continuous setting where $\hat{w}$ and $\hat{u}$ are the expressions in (4.13) based on K, the CGF of (X_1, Y_1).

For the discrete case, where $\bar{Y}$ has support on a δ/n-lattice, the obvious analogues to those in section 2.4.4 apply for continuity correction. When using the second correction, the cutoff is offset by using $\bar{y}^- = \bar{y} - \delta/(2n)$.

(2) The asymptotic accuracy of the approximation based upon (4.37), as the leading term of an expansion, was given by Skovgaard (1987) and is similar to that of the Lugannani and Rice approximation discussed in section 2.3.2. If $(\bar{x}, \bar{y})$ remains in a large deviation set as $n \to \infty$, then

$$F(\bar{y}|\bar{x}) = \hat{F}(\bar{y}|\bar{x}) + O(n^{-1})$$

in both the continuous and lattice cases. Over sets of bounded central tendency, the $O(n^{-1})$ error improves to $O(n^{-3/2})$.

An expression for the next higher-order term was derived by Kolassa (1996) as

$$O(n^{-1}) = -\frac{\phi(\hat{w}_n)}{n^{3/2}} \left\{ \frac{1}{\hat{u}}(O_N - O_D) - \frac{1}{\hat{u}^3} - \frac{\hat{\lambda}_3}{2\hat{t}\hat{u}} + \frac{1}{\hat{w}^3} \right\} + O(n^{-2}), \tag{4.38}$$

where $O_N - O_D$ is the second-order correction term for the double-saddlepoint density and

$$\hat{\lambda}_3 = \hat{K}_{ijk}\hat{K}^{ij}\hat{K}^{km} \qquad (4.39)$$

in the Einstein notation introduced in (3.38). Notice that the m in term $\hat{K}^{km}$ is held fixed and represents the last component corresponding to variable $\bar{y}$. The term in curly brackets is $O(1)$ as $n \to \infty$ on sets of bounded central tendency even though the individual terms diverge. On such sets, the $O(n^{-2})$ improves to $O(n^{-5/2})$.

(3) The derivative of the Skovgaard approximation yields an expression which is the double-saddlepoint density with its second-order correction to order $O(n^{-3/2})$. This order requires that $(\bar{x}, \bar{y})$ stay in sets of bounded central tendency. More specifically

$$\hat{F}'(\bar{y}|\bar{x}) = \hat{f}(\bar{y}|\bar{x})\left\{1 + \frac{1}{n}(O_N - O_D) + O(n^{-3/2})\right\}. \qquad (4.40)$$

The derivation of this is given in section 4.5.3 of the appendix.

In the majority of applications, (4.40) is positive, however in unusual settings (4.40) can be negative so that $\hat{F}(\bar{y}|\bar{x})$ is decreasing. Also there is no guarantee that the extreme tail limits are 0 and 1 but again this is unusual.

As with the derivative of the Lugannani and Rice approximation, the form of (4.40) should not be surprising. Considering that $\hat{F}(\bar{y}|\bar{x})$ is accurate to order $O(n^{-3/2})$ for $F(\bar{y}|\bar{x})$, we expect its derivative to maintain the same accuracy for $F'(\bar{y}|\bar{x})$. In doing so, it has to contain the $O(n^{-1})$ correction term as in (4.40).

4.3 Further examples: Linear combinations of independent variables

The third example of section 4.1.1 provides a rich source of interesting examples of linear combinations. Suppose that $Z_1, \ldots, Z_n$ are independent variables with the distributions in table 4.1. Consider the conditional distribution of Y given X where these variables are defined as

$$Y = \sum_{i=1}^{n} w_i Z_i \qquad X = \sum_{i=1}^{n} Z_i \qquad (4.41)$$

with $\{w_i\}$ as any fixed collection of weights.

4.3.1 Bootstrap distribution of the sample mean

The distribution theory concerns a linear combination of components for a multinomial distribution. Suppose that $Z_i \sim$ Poisson (λ_i) for $i = 1, \ldots, n$. The conditional distribution of Y given $X = n$, as defined in (4.41), has the same distribution as $\sum_{i=1}^{n} w_i M_i$ where $(M_1, \ldots, M_{n-1}) \sim$ Multinomial $(n; \{\pi_i\})$ with $\pi_i = \lambda_i / \sum_j \lambda_j$. In the exchangeable setting in which all values of λ_i are equal, the conditional distribution is the bootstrap distribution of a resampled sum of data with the weights as the data values (see section 3.4.1). Unequal values of λ_i would be equivalent to performing non-uniform resampling using sampling probabilities proportional to $\{\lambda_i\}$.

In the most general setting, the joint CGF of X and Y is easily determined for use with the Skovgaard approximation. In this instance, the double-saddlepoint Skovgaard approximation is the same analytically as the single-saddlepoint Lugannani and Rice based on a linear combination of Multinomial $(n; \{\pi_i\})$. We ask the reader to show this equivalence in Exercise 16. The equivalence was first shown in Booth and Butler (1990) in the exchangeable setting. In this context, the Skovgaard approximation to the bootstrap distribution is the same as that developed by Davison and Hinkley (1988) and previously discussed in section 3.4.1 and Exercise 15 of chapter 3.

4.3.2 Combinations of Dirichlet components

Suppose that $Z_i \sim$ Gamma $(\alpha_i, 1)$ independently in (4.41). The conditional distribution of Y given $X = 1$ is the same as that for $\sum_{i=1}^{n} w_i D_i$ where $(D_1, \ldots, D_{n-1}) \sim$ Dirichlet $(\alpha_1, \ldots, \alpha_n)$ with support on the simplex in $\Re^n$. The Skovgaard approximation can be shown to agree analytically with its single-saddlepoint counterpart from section 3.4.2 and Exercise 16 of chapter 3. Exercise 17 has the reader explore this development in the manner in which the result was discovered. There are many steps for showing agreement and the exercise leads the reader through those needed to show that the $\hat{w}$ values of both approximations are the same. Showing that the $\hat{u}$ values agree is rather difficult and the arguments for that are given in section 4.5.4 of the appendix.

The Durbin–Watson statistic and the Dirichlet bootstrap arise in the special situations in which the Dirichlet components are exchangeable with $\alpha_i \equiv \alpha$. It was in this exchangeable context that the equivalence between the Skovgaard and Lugannani and Rice approximations was first shown by Paolella (1993) and Butler and Paolella (1998a). The extension of this equivalence to any $\{\alpha_i\}$ however should not be surprising for the following reason. Suppose $\{\alpha_i\}$ lie on a δ-lattice. As a result of this and the infinite divisibility of the Gamma variables, both X and Y may be expressed as (weighted) sums of i.i.d. Gamma $(\delta, 1)$ variables which is the exchangeable context in which the equivalence was first known to hold. Since this argument holds for $\{\alpha_i\}$ lying on any δ-lattice no matter how fine, then it should be possible to show it holds for any $\{\alpha_i\}$. This argument is outlined in Exercise 17 and section 4.5.4 of the appendix.

4.3.3 Bayesian posterior for the mean of an infinite population

Suppose a characteristic of interest in an infinite population assumes a finite number of distinct values $w_1, \ldots, w_m$. If $\pi_1, \ldots, \pi_m$ are the unknown population frequencies of these values, then $\mu = \sum_i w_i \pi_i$ is the average population characteristic. In a random sample of size n, the sufficient statistics would be the frequency counts in the m categories or $\{n_i\}$. Starting with a conjugate Dirichlet $(\alpha_1^0, \ldots, \alpha_m^0)$ prior on $\{\pi_i\}$, the posterior on $\{\pi_i\}$ is Dirichlet $(\alpha_1^1, \ldots, \alpha_m^1)$ with update $\alpha_i^1 = \alpha_1^0 + n_i$. The posterior distribution for the population mean μ given $\{n_i\}$ is

$$\mu \sim \sum_{i=1}^{m} w_i \pi_i \qquad \{\pi_i\} \sim \text{Dirichlet}\left(\alpha_1^1, \ldots, \alpha_m^1\right).$$

Saddlepoint approximation to this posterior distribution for μ is the same as in the previous example.

4.3.4 Bayesian posterior on the mean of a finite population

Assume the same setup as in previous example but now let the population be finite rather than infinite. Suppose the population size N is known and the population frequencies are $\{N_i\}$ with $\sum_{i=1}^{m} N_i = N$. Take $\mathbf{N} = (N_1, \ldots, N_{m-1})^T$ as the unknown parameter. A random sample of size n has a multivariate Hypergeometric $(N_1, \ldots, N_m; n)$ likelihood with sufficient statistic $\mathbf{n} = (n_1, \ldots, n_{m-1})^T$. The conjugate prior on $\mathbf{N}$ is multivariate Pólya $(\alpha_1^0, \ldots, \alpha_m^0; N)$, which the reader is asked to show in Exercise 18. Accordingly, the posterior on $\mathbf{N}$ is $\mathbf{n} + \mathbf{P}$ where $\mathbf{P} \sim$ Pólya $(\alpha_1^1, \ldots, \alpha_m^1; N - n)$, where the update is $\alpha_i^1 = \alpha_i^0 + n_i$. The posterior on the population mean $\mu = N^{-1} \sum_{i=1}^{m} w_i N_i$ given $\{n_i\}$ is

$$\mu \sim N^{-1} \sum_{i=1}^{m} w_i(n_i + P_i) \quad \{P_i\} \sim \text{Pólya}\left(\alpha_1^1, \ldots, \alpha_m^1; N - n\right). \qquad (4.42)$$

This posterior is the constant translation $N^{-1} \sum_i w_i n_i$ of a linear combination of multivariate Pólya components to which the Skovgaard approximation may be applied by using $Z_i \sim$ Negative Binomial $(\alpha_i, 1/2)$ as in (4.41).

Bayesian inference in finite populations with $m = 2$ distinct values and a univariate Pólya posterior was developed by Jeffreys (1961, III, §3.2). A very general framework, within which the multivariate Pólya is a special case, was given by Hill (1968). Note that the uniform prior on $\{N_i\}$, or the so-called Bose–Einstein statistics (Feller, 1968, §II.5), is a special case of the Pólya with $\alpha_i^0 \equiv 1$.

4.3.5 Sampling distribution of the mean in finite population sampling

Consider the alternative frequentist inference in the last example. A random sample of the population of size n yields category frequencies $\{H_i\}$ for characteristics $\{w_i\}$ in which $\{H_1, \ldots, H_{m-1}\} \sim$ Hypergeometric $(N_1, \ldots, N_m; n)$. The sample mean $\hat{\mu}$ is an unbiased estimator of population mean μ with distribution

$$\hat{\mu} \sim \frac{1}{n} \sum_{i=1}^{m} w_i H_i. \qquad (4.43)$$

A Skovgaard approximation to the CDF of $\hat{\mu}$ may be constructed by taking $Z_i \sim$ Binomial $(N_i, 1/2)$ as in (4.41).

4.3.6 Permutation and Wilcoxon rank sum tests

The two sample permutation test and the Wilcoxon rank sum test are nonparametric tests that compare the location of the treatment group with that of the control group. Suppose that the data are m i.i.d. observations from treatment and n from control. The data are summarized by pooling the responses and ordering them from smallest to largest values as $w_1, \ldots, w_{m+n}$. A sequence of binary variables $z_1, \ldots, z_{m+n}$ is also used to identify those responses connected to the treatment group. In particular $z_i = 1$ (0) if response w_i is in the treatment (control) group. A randomization test is based on the permutation distribution for $Y = \sum_{i=1}^{m+n} w_i Z_i$ given $X = \sum_{i=1}^{m+n} Z_i = m$. With $y = \sum_{i=1}^{m+n} w_i z_i$, the randomization p-value is the computation $\Pr(Y \leq y | X = m)$ that assumes equal probability for all the $\binom{m+n}{m}$

Table 4.6. *Hours of pain relief for two drugs (Lehmann and D'Abrera, 1975, p. 37).*

Hours	0.0	2.1	2.3	2.5	2.8	3.1	3.3	4.2
Tr.	1	1	1	1	1	0	0	0
Hours	4.4	4.5	4.7	4.8	4.9	5.8	6.6	6.8
Tr.	1	0	0	1	0	0	1	0

distinct orderings of the indicator values of $\{z_i\}$. This uniform distribution is constructed by assuming that $Z_1, \ldots Z_{m+n}$ are i.i.d. Bernoulli (p) for arbitrary $p \in (0, 1)$ and conditioning on $X = m$.

Table 4.6 contains the data for comparing two drugs with $m = 8 = n$. The approximate two-sided p-value using approximation (4.12) without continuity correction was given in Booth and Butler (1990, table 1) as 0.097 and compares to an exact value 0.102 found in table 2 of Robinson (1982). Robinson also reports an exact 95% confidence interval as $(-0.30, 3.26)$ while Butler and Booth invert their test to determine the approximate interval $(-0.31, 3.25)$.

The Wilcoxon rank sum test is the same test but with the scores replaced by the ranks. Thus w_i is replaced with i so that $Y = \sum_{i=1}^{m+n} i Z_i$ which attains value 51. The exact one-sided p-value is 0.0415 and is determined from table B of Lehmann and D'Abrera (1975). This value is doubled to give p-value 0.0830 for the two-sided test since the null distribution is symmetric about its mean 68. By comparison, the continuity-corrected one-sided p-value $\widehat{\Pr}_2$ from (4.17) is 0.0414 and, due to its preservation of symmetry, it also may be doubled to give 0.0828 for the two-sided approximation. The other continuity correction $\widehat{\Pr}_1$ achieves the less accurate p-value 0.0320 and cannot be doubled for the two-sided test. The continuous formula yields the value 0.0368 and may also be doubled. Such values, however, are not approximations for the p-value but rather the mid-p-value, which is a topic discussed in section 6.1.4.

4.4 Further topics

4.4.1 Central limit connections

Under quite general conditions the distribution of $(\bar{X}, \bar{Y})$ is multivariate normal to order $O(n^{-1/2})$. This is expressed through the conditional density and CDF of $\bar{Y}$ given $\bar{X} = \bar{x}$ which are normal to the same order. Since the double-saddlepoint density and CDF reproduce the true conditional density and CDF to the higher order $O(n^{-1})$, one can expect that such central limit approximations would result from an expansion of the double-saddlepoint approximations to the lower order $O(n^{-1/2})$. This is shown below using Taylor expansions of the double-saddlepoint approximations. In particular, the double-saddlepoint density and CDF are shown to agree with their asymptotic conditional normal density and CDF to order $O(n^{-1/2})$. The effect of the Taylor expansions is to diminish the second-order $O(n^{-1})$ accuracy of the double-saddlepoint density and the higher third-order $O(n^{-3/2})$ accuracy of the CDF approximation to the first-order accuracy $O(n^{-1/2})$ of the central limit approximations on sets of bounded central tendency.

4.4 Further topics

Suppose the mean of $(\bar{X}, \bar{Y})$ is (μ_x, μ_y) and the covariance in block form is

$$\mathrm{Cov}\begin{pmatrix} \bar{X} \\ \bar{Y} \end{pmatrix} = \frac{1}{n} K''(0,0) = \frac{1}{n}\Sigma = \frac{1}{n}\begin{pmatrix} \Sigma_{xx} & \Sigma_{xy} \\ \Sigma_{yx} & \Sigma_{yy} \end{pmatrix},$$

where Σ_{xy} is $(m_x \times m_y)$ etc. The standardized value of $(\bar{X}, \bar{Y})$ converges weakly to a Normal$_m(0, I_m)$. The result is used practically by assuming that the limit has been reached so that $(\bar{X}, \bar{Y})$ is normal with its true mean and covariance. The asymptotic conditional distribution of $\bar{Y}$ given $\bar{X} = \bar{x}$ is the appropriate condition normal distribution given as

$$\bar{Y} | \bar{X} = \bar{x} \sim \mathrm{Normal}_{m_y}(\mu_{y.x}, \Sigma_{yy.x}), \tag{4.44}$$

where

$$\mu_{y.x} = \mu_y + \Sigma_{yx}\Sigma_{xx}^{-1}(\bar{x} - \mu_x)$$

and

$$\Sigma_{yy.x} = \Sigma_{yy} - \Sigma_{yx}\Sigma_{xx}^{-1}\Sigma_{xy}$$

are the conditional mean and covariance in the normal limit.

Taylor expansion of saddlepoint quantities leads to the local limit result

$$\hat{f}(\bar{y}|\bar{x}) = \mathcal{N}_{m_y}(\bar{y}; \mu_{y.x}, n^{-1}\Sigma_{yy.x}) + O(n^{-1/2}) \tag{4.45}$$

where $\mathcal{N}_m(\bar{y}; \mu, \Sigma)$ denotes a Normal$_m(\mu, \Sigma)$ density at $\bar{y}$. Also, with dim $Y = 1$,

$$\hat{F}(\bar{y}|\bar{x}) = \Phi\left\{\sqrt{n/\Sigma_{yy.x}}(\bar{y} - \mu_{y.x})\right\} + O(n^{-1/2}). \tag{4.46}$$

Proof of (4.45) and (4.46). The first step in the general development is to determine an expansion for saddlepoint $\hat{t}$ for use in a later expansion of $\hat{w}_n$. Standardize $(\bar{x}, \bar{y})$ in terms of (z_x, z_y) by defining

$$\frac{1}{\sqrt{n}}\begin{pmatrix} z_x \\ z_y \end{pmatrix} = \begin{pmatrix} \bar{x} - \mu_x \\ \bar{y} - \mu_y \end{pmatrix} = \begin{Bmatrix} K'_s(\hat{s}, \hat{t}) - K'_s(0,0) \\ K'_t(\hat{s}, \hat{t}) - K'_t(0,0) \end{Bmatrix}.$$

Taylor expansion of the gradient on the right side about $(0, 0)$ gives

$$\frac{1}{\sqrt{n}}\begin{pmatrix} z_x \\ z_y \end{pmatrix} = K''(0,0)\begin{pmatrix} \hat{s} \\ \hat{t} \end{pmatrix} + Q, \tag{4.47}$$

where Q represents the quadratic and higher order terms in the expansion. For fixed values of (z_x, z_y), it is determined from (4.47) that

$$\begin{pmatrix} \hat{s} \\ \hat{t} \end{pmatrix} = \frac{1}{\sqrt{n}}\Sigma^{-1}\begin{pmatrix} z_x \\ z_y \end{pmatrix} + O(n^{-1})$$

as $n \to \infty$. The expansion in $\hat{t}$ simplifies as follows. Use Σ^{yx} for the (y, x) block of Σ^{-1}, etc. so that

$$\hat{t} = \frac{1}{\sqrt{n}}(\Sigma^{yx} z_x + \Sigma^{yy} z_y) + O(n^{-1})$$

$$= \frac{1}{\sqrt{n}}\left(-\Sigma_{yy.x}^{-1}\Sigma_{yx}\Sigma_{xx}^{-1} z_x + \Sigma_{yy.x}^{-1} z_y\right) + O(n^{-1})$$

$$= \Sigma_{yy.x}^{-1}(\bar{y} - \mu_{y.x}) + O(n^{-1}) \tag{4.48}$$

where, in determining the second expression, we have used the block inversion expressions in Press (1972, §2.6.1).

It is now possible to show (4.45) by working with the saddlepoint density in terms of $\hat{w}_n$ as in (4.14). First write the expression for $\hat{w}_n$ in (4.37) in terms of a multivariate analogue of the function g, defined in (4.30). Define

$$g(t) = K(\hat{s}_t, t) - \hat{s}_t^T \bar{x} - t^T \bar{y}$$

with $\hat{s}_t$ as the unique solution to the first m_x saddlepoint equations $K'_s(\hat{s}_t, t) = \bar{x}$ for fixed t. Express $\hat{w}_n$ as

$$\hat{w}_n^2/(2n) = g(0) - g(\hat{t}).$$

A multivariate Taylor expansion, comparable to that in (4.31) and (4.32), gives

$$\hat{w}_n^2/(2n) = \hat{t}^T (\hat{K}''_{tt} - \hat{K}''_{ts} \hat{K}''^{-1}_{ss} \hat{K}''_{st}) \hat{t}/2 + O(n^{-3/2}). \tag{4.49}$$

The convergence expression

$$\hat{K}''_{tt} - \hat{K}''_{ts} \hat{K}''^{-1}_{ss} \hat{K}''_{st} = \Sigma_{yy.x} + O(n^{-1/2}), \tag{4.50}$$

as well as the expansion of $\hat{t}$ in (4.48) are substituted into (4.49) to give

$$\tfrac{1}{2}\hat{w}_n^2 = \tfrac{1}{2} n (\bar{y} - \mu_{y.x})^T \Sigma_{yy.x}^{-1} (\bar{y} - \mu_{y.x}) + O(n^{-1/2}). \tag{4.51}$$

The same arguments as used above show that $\hat{s}_0 = O(n^{-1/2})$ so the remaining factor in saddlepoint density (4.9) is

$$\frac{|K''(\hat{s}, \hat{t})|}{|K''_{ss}(\hat{s}_0, 0)|} = \frac{|\Sigma|}{|\Sigma_{xx}|} + O(n^{-1/2})$$

$$= |\Sigma_{yy.x}| + O(n^{-1/2}). \tag{4.52}$$

Substituting expressions (4.52) and (4.51) into the double-saddlepoint density in (4.14) shows (4.45).

To show (4.46) in the case dim $\bar{Y} = 1$, use $\text{sgn}(\bar{y} - \mu_{y.x}) = \text{sgn}(\hat{t})$ so that (4.51) becomes

$$\hat{w}_n = \sqrt{n/\Sigma_{yy.x}}(\bar{y} - \mu_{y.x}) + O(n^{-1/2}).$$

The leading term in the Skovgaard approximation is now

$$\Phi\left\{\sqrt{n/\Sigma_{yy.x}}(\bar{y} - \mu_{y.x})\right\} + O(n^{-1/2})$$

as in (4.46). At this point it only remains to show that the second term in the approximation is $O(n^{-1/2})$. Skovgaard (1987) showed that the singularity at $\hat{t} = 0$ is removable so that

$$\lim_{\hat{t} \to 0} \left(\frac{1}{\hat{w}} - \frac{1}{\hat{u}}\right) = O(1). \tag{4.53}$$

On sets of bounded central tendency, $\hat{t} \to 0$ at the rate $O(n^{-1/2})$. Therefore, using (4.53),

$$\phi(\hat{w}_n) \frac{1}{\sqrt{n}} \left(\frac{1}{\hat{w}} - \frac{1}{\hat{u}}\right) = O(n^{-1/2}).$$

4.4 Further topics

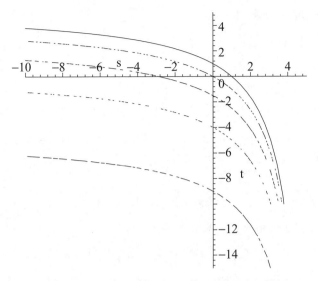

Figure 4.2. Saddlepoint manifolds $\mathcal{S}_x$ in coordinates (s, t) associated with $\hat{f}(y|x)$ for fixed values of x. Starting in the lower left, the curves have $x = 0.1$ (*dashed*), 0.2 (*dotted*), 0.3 (*dot-dash*), and 1 (*dot-dot-dash*). The solid line is the boundary $\mathcal{S}$.

4.4.2 Saddlepoint geometry

The geometry of saddlepoints for a conditional density is developed using the bivariate exponential density in figure 4.1 as an example. Conditional on x, the collection of saddlepoints $(\hat{s}, \hat{t})$ associated with $\hat{f}(y|x)$ forms a smooth manifold in the interior of $\mathcal{S} \subseteq \Re^m$ which we denote as $\mathcal{S}_x$. Distinct values of x yield distinct manifolds and, as x varies over all its possibilities, these manifolds accumulate so that $\cup_x \mathcal{S}_x = \mathcal{S}$.

This partition of $\mathcal{S}$ may be seen in figure 4.2 using the bivariate exponential example with $\rho = 0.9$. The upper right boundary of $\mathcal{S}$ is the solid hyperbolic curve

$$(1-s)(1-t) - \rho^2 st = 0.$$

Starting in the bottom left, the curves $\mathcal{S}_{0.1}$, $\mathcal{S}_{0.2}$, $\mathcal{S}_{0.3}$, and $\mathcal{S}_{1.0}$ are shown as the appropriately coded lines. The solid boundary curve and all the dotted lines approach the horizontal asymptote $t = (1 - \rho^2)^{-1} = 5.26$ to the far upper left and the vertical asymptote $s = (1 - \rho^2)^{-1}$ to the far bottom right. As y increases from 0 to ∞, the saddlepoint $(\hat{s}, \hat{t})$ traces out the curves beginning in the bottom right and moving toward the upper left in a counter-clockwise direction. This can be seen algebraically by examining the parametric curves determining the saddlepoint equations (3.6) and (3.7).

Certain characteristics of figure 4.2 must hold more generally which we now investigate in a regular setting in which $\mathcal{S}$ is open. Note that $\partial \hat{t}/\partial y > 0$ for all values of (x, y). This must hold generally when $m_y = 1$ and, for $m_y > 1$, $\partial \hat{t}/\partial y^T$ must be a positive definite matrix. To show this write

$$K'(\hat{s}, \hat{t}) = \begin{pmatrix} x \\ y \end{pmatrix}$$

and differentiate both sides $\partial/\partial y^T$ to get

$$\begin{pmatrix} \hat{K}''_{ss} & \hat{K}''_{st} \\ \hat{K}''_{ts} & \hat{K}''_{tt} \end{pmatrix} \frac{\partial}{\partial y^T} \begin{pmatrix} \hat{s} \\ \hat{t} \end{pmatrix} = \begin{pmatrix} 0 \\ I_{m_y} \end{pmatrix}.$$

Solving,

$$\partial \hat{t}/\partial y^T = (\hat{K}'')^{tt} = \left(\hat{K}''_{tt} - \hat{K}''_{ts} \hat{K}''^{-1}_{ss} \hat{K}''_{st} \right)^{-1} \tag{4.54}$$

using the block inversion expressions in Press (1972, §2.6.1). This must be positive definite and consequently each diagonal element $\partial \hat{t}_i/\partial y_i > 0$. In the case $m_y = 1$, this expression is

$$\partial \hat{t}/\partial y = |\hat{K}''_{ss}|/|\hat{K}''| \tag{4.55}$$

by Cramer's rule. The change of $\hat{s}$ with y is

$$\partial \hat{s}/\partial y^T = -\left(\hat{K}''_{ss}\right)^{-1} \hat{K}''_{st} \, \partial \hat{t}/\partial y^T.$$

In our example, this value was negative but components can be negative or positive.

The following generalizations may be deduced from the above in a regular setting. As $y_i \to \infty \, (-\infty)$, or the upper (lower) limit of its support, while holding x and the other y-components fixed, $\hat{t}_i$ increases (decreases) monotonically eventually achieving a positive (negative) value. The pair $(\hat{s}, \hat{t})$ approaches the boundary of $\mathcal{S}$.

4.4.3 Normalizing double-saddlepoint densities

With the proper change of variable, the normalization constant can be computed without the need to solve for all components in the numerator saddlepoint equation. This is similar to the manner in which a saddlepoint solution was avoided while normalizing the single-saddlepoint density. We suppose that the true density $f(y|x)$ is non-zero over $\mathcal{Y}_x = \{y : (x, y) \in \mathcal{I}_\mathcal{X}\}$; if the conditional support is a proper subset of $\mathcal{Y}_x$ then the integration must be so restricted. The normalization constant is

$$c = \int_{\mathcal{Y}_x} \hat{f}(y|x) \, dy.$$

Changing the variable of integration from dy to $d\hat{t}$ requires some care since both components of $(\hat{s}, \hat{t})$ vary with the value of y. With x fixed, the value of $(\hat{s}, \hat{t})$ lies in the m_y-dimensional manifold $\mathcal{S}_x$ and either y or $\hat{t}$ may be used to index points on the manifold. The Jacobian is given in (4.54) so

$$c = \int_{\mathcal{T}_x} \hat{f}(y|x) |\partial y/\partial \hat{t}^T| d\hat{t} = \int_{\mathcal{T}_x} \hat{f}(y|x) |K''(\hat{s}, \hat{t})|/|K''_{ss}(\hat{s}, \hat{t})| d\hat{t}, \tag{4.56}$$

where both $\hat{f}(y|x)$ and $\partial y/\partial \hat{t}^T$ depend on $\hat{t}$ through both $\hat{s}$ and $\hat{t}$. The region of integration is $\mathcal{T}_x = \{\hat{t} : (\hat{s}, \hat{t}) \in \mathcal{S}_x\}$ when the support is $\mathcal{Y}_x$. Numerical computation of (4.56), would use a grid of $\hat{t}$-values which, along with the fixed value x, determined a grid of $(\hat{s}, \hat{t})$ pairs which, in turn, determine a grid of y-values to pair with the fixed x. Thus, each grid point $(x, \hat{t})$ determines a pair $(\hat{s}, \hat{t})$ that may require a root search when finding $\hat{s}$. The final determination of (x, y) from $(\hat{s}, \hat{t})$ is explicit. Examples that make use of such numerical integration are given in Butler *et al.* (2005).

4.4.4 Conditional single-saddlepoint approximations

Some conditional distributions have explicit conditional CGFs to which a single-saddlepoint approximation may be applied. Some examples of this setting are considered below. For other conditional distributions, the conditional CGF may not be very tractable or simple to compute but a double-saddlepoint approximation may be possible based on the joint CGF of the variables involved. Still in other situations, both single- and double-saddlepoint approximation may be used. Some examples of this last situation include the binomial distribution and applications to linear combinations of multinomial or Dirichlet components where the single- and double-saddlepoint approximations turn out to be the same.

Examples that have simple double-saddlepoint approximations but for which the true conditional CGF is difficult to compute include the beta distribution, whose MGF is the confluent hypergeometric function, the hypergeometric, whose MGF is the Gauss hypergeometric function, and the Pólya distribution whose MGF is also expressed by using the Gauss hypergeometric function. Single-saddlepoint approximations for some very simple cases of these distributions were presented in section 4.2.3 but their more general usage with larger parameters may become difficult and computationally intractable. We shall return to reconsider some of these distributions in chapter 10 using sequential saddlepoint methods.

There is one general setting in which a conditional MGF can be determined from its joint MGF by using calculus. This occurs when the conditioning variables X are nonnegative and integer valued. For simplicity, let X be one-dimensional and suppose we would like to compute a probability for Y given that $X = i$. The conditional MGF of Y given $X = i$ can often be computed from its joint MGF, or equivalently, the *probability-moment generating function* (P-MGF) of X and Y. The P-MGF is defined as

$$M_p(z, t) = E\left(z^X e^{t^T Y}\right) \qquad (z, t) \in \mathcal{S}_p, \tag{4.57}$$

where, in a regular setting, $\mathcal{S}_p$ is an open neighborhood of $(1, 0)$. Expression $M_p(z, t)$ is simply $M(\ln s, t)$, where M is the joint MGF and $\mathcal{S}_p$ as the preimage of $\mathcal{S}$ under the mapping $(z, t) \to (\ln s, t)$. The form of $M_p(z, t)$ is particularly convenient for computing the conditional MGF since the conditional MGF of Y given $X = i$ is

$$M(t | X = i) = \frac{\partial^i M_p(z, t)/\partial z^i |_{z=0}}{\partial^i M_p(z, t)/\partial z^i |_{z=0, t=0}}. \tag{4.58}$$

Proof of (4.58). Take the expectation in (4.57) conditionally as

$$M_p(z, t) = \sum_{i=0}^{\infty} E\left(e^{t^T Y} | X = i\right) z^i \Pr(X = i)$$

so that

$$\frac{1}{i!} \frac{\partial^i}{\partial z^i} M_p(z, t) \Big|_{z=0} = E\left(e^{t^T Y} | X = i\right) \Pr(X = i). \tag{4.59}$$

Expression (4.59) at $t = 0$ is $\Pr(X = i)$, so the ratio of (4.59) divided by its value at $t = 0$ is the conditional MGF of Y. □

Examples

Bivariate Poisson This mass function was introduced in section 3.1.3 in the context of modelling fatal shocks for a two component system. From the joint MGF in (3.14),

$$M_P(z, t) = \exp\{-(\lambda + \mu + \nu) + \mu e^t\} \exp\{z(\lambda + \nu e^t)\}.$$

Upon differentiation $\partial^i / \partial z^i$, (4.58) is computed as

$$M(t|X = i) = \exp\{\mu(e^t - 1)\} \times \left\{ \frac{\lambda}{\lambda + \nu} + \frac{\nu}{\lambda + \nu}(e^t - 1) \right\}^i. \qquad (4.60)$$

In this instance, one can see that (4.60) must be the correct answer. From the construction of the bivariate Poisson in (3.13), the conditional mass function of Y given $X = i$ must be the convolution of P_2, having a Poisson (μ) distribution, and $P_3 | P_1 + P_3 = i$, which is an independent Binomial $(i, \nu/(\nu + \lambda))$. The convolution is apparent from the two factors of (4.60) which are their MGFs.

Bivariate negative binomial This example was introduced in Exercise 4 of chapter 3 and is left for consideration as Exercise 21.

Note that variable Y may be continuous or discrete but that X must be a lattice variable on the nonnegative integers. Variable X might also be a vector of nonnegative lattice variables. Exercise 22 asks the reader to determine the generalization of (4.58) in this instance.

When X is a continuous variable, there is no longer any simple means for determining the conditional MGF of Y given $X = x$ from its joint MGF. In such instances, the double-saddlepoint approximation is a useful procedure. In chapter 10, an alternative method is introduced that approximates the conditional MGF $\hat{M}(t|x)$ using an initial saddlepoint approximation. The subsequent use of $\hat{M}(t|x)$, as the input for an additional single-saddlepoint approximation, in determining probabilities for Y given $X = x$ is referred to as the *sequential saddlepoint approximation*.

4.5 Appendix

4.5.1 Symmetry of the Skovgaard approximation

Suppose the conditional distribution of Y given $X = x$ is symmetric for almost every value of x and $E(Y|x) = \tau^T x$. The symmetry in (4.35) follows as we now show.

Derivation of (4.35)

Define $M_0(t|x)$ as the conditional MGF of $Y - \tau^T x$ given $X = x$. Since this distribution is symmetric about 0, its MGF is an even function of t. The joint MGF of (X, Y) is now

$$\begin{aligned}
M(s, t) &= E\big[E\{e^{(s+t\tau)^T X + t(Y - \tau^T X)} \big| X\}\big] \\
&= E\{e^{(s+t\tau)^T X} M_0(t|X)\} \\
&= E\{e^{(s+t\tau)^T X} M_0(-t|X)\} \\
&= E\big[E\{e^{(s+t\tau)^T X - t(Y - \tau^T X)} \big| X\}\big] \\
&= M(s + 2t\tau, -t). \qquad (4.61)
\end{aligned}$$

Property (4.61) allows for a relationship to be determined between $(\hat{s}, \hat{t})$, the saddlepoint for $\hat{F}(\tau^T x + y | x)$ in the right tail, and $(\check{s}, \check{t})$, the saddlepoint for $\hat{F}(\tau^T x - y | x)$ on the left tail. In particular, an identity for the derivative of its CGF is first determined from (4.61) as

$$K'_t(s, t) = 2\tau^T K'_s(s + 2t\tau, -t) - K'_t(s + 2t\tau, -t). \tag{4.62}$$

After rearrangement and using

$$K'_s(s + 2t\tau, -t) = K'_s(s, t) \tag{4.63}$$

then

$$K'_t(s + 2t\tau, -t) = 2\tau^T K'_s(s, t) - K'_t(s, t). \tag{4.64}$$

In the right tail, $(\hat{s}, \hat{t})$ solves

$$\begin{pmatrix} K'_s(\hat{s}, \hat{t}) \\ K'_t(\hat{s}, \hat{t}) \end{pmatrix} = \begin{pmatrix} x \\ \tau^T x + y \end{pmatrix} \tag{4.65}$$

to give values for $\hat{w}$ and $\hat{u}$. In the left tail, $(\check{s}, \check{t})$ solves

$$\begin{pmatrix} K'_s(\check{s}, \check{t}) \\ K'_t(\check{s}, \check{t}) \end{pmatrix} = \begin{pmatrix} x \\ \tau^T x - y \end{pmatrix} \tag{4.66}$$

to give values $(\check{w}, \check{u})$. The two equations in (4.65) may be put together to form the bottom equation in (4.66) so that

$$\begin{aligned} \tau^T x - y &= 2\tau^T x - (\tau^T x + y) \\ &= 2\tau^T K'_s(\hat{s}, \hat{t}) - K'_t(\hat{s}, \hat{t}) \\ &= K'_t(\hat{s} + 2\hat{t}\tau, -\hat{t}) \end{aligned} \tag{4.67}$$

from (4.64). The top equation in (4.65) may be reexpressed using (4.63) as

$$x = K'_s(\hat{s}, \hat{t}) = K'_s(\hat{s} + 2\hat{t}\tau, -\hat{t}).$$

Together with the equation determined in (4.67) we get the left tail saddlepoint equations but expressed instead in terms of $\hat{s}$ and $\hat{t}$. Since these equations must admit a unique solution, we are able to determine the relationship between $(\hat{s}, \hat{t})$ and $(\check{s}, \check{t})$ as

$$\begin{aligned} \check{s} &= \hat{s} + 2\hat{t}\tau \\ \check{t} &= -\hat{t}. \end{aligned} \tag{4.68}$$

The denominator saddlepoints $\hat{s}_0$ and $\check{s}_0$ are determined by taking $\hat{t} = 0 = \check{t}$ in the top expressions of (4.65) and (4.66). They are the same equation so $\hat{s}_0 = \check{s}_0$. It is now a simple exercise to show that $\hat{w} = -\check{w}$ by using (4.61), and (4.68).

Showing $\hat{u} = -\check{u}$ would appear to be simple from (4.68), however finding the relationships between the Hessians is more involved. Further differentiation of (4.62) gives the following Hessian terms, where we use $\hat{K}''_{st}$ to denote the vector $K''_{st}(\hat{s}, \hat{t})$, etc.

$$\begin{aligned} K''_{ss}(\check{s}, \check{t}) &= \hat{K}''_{ss} \\ K''_{st}(\check{s}, \check{t}) &= 2\hat{K}''_{ss}\tau - \hat{K}''_{st} \\ K''_{tt}(\check{s}, \check{t}) &= 4\tau^T \hat{K}''_{ss}\tau - 4\tau^T \hat{K}''_{st} + \hat{K}''_{tt}. \end{aligned} \tag{4.69}$$

Then, using block determinant computation as in Press (1972, Eq. 2.6.8),

$$|K''(\check{s},\check{t})| = |K''_{ss}(\check{s},\check{t})|\{K''_{tt}(\check{s},\check{t}) - K''_{st}(\check{s},\check{t})^T K''_{ss}(\check{s},\check{t})^{-1} K''_{st}(\check{s},\check{t})\}$$
$$= |\hat{K}''_{ss}|\{\hat{K}''_{tt} - \hat{K}''_{ts}(\hat{K}''_{ss})^{-1}\hat{K}''_{st}\}$$
$$= |K''(\hat{s},\hat{t})|,$$

where the second line is determined by substituting the expressions of (4.69) into the first line. It now follows that $\hat{u} = -\check{u}$.

The symmetry in (4.35) follows from $(\hat{w}, \hat{u}) = -(\check{w}, \check{u})$. □

4.5.2 The removable singularity in the Skovgaard approximation

The singularity of $\hat{F}(y|x)$ is shown to occur at $y = E(Y|x)$ when $E(Y|x) = \varsigma + \tau^T x$.

Derivation of (4.36)

Define the conditional MGF of Y given $X = x$ as

$$M(t|x) = E(e^{tY}|X = x)$$

so that the joint MGF is

$$M(s,t) = E\{e^{s^T X} M(t|X)\}. \tag{4.70}$$

For fixed x, consider the y-value for which $\hat{t} = 0$. Using the form of (4.70), this y-value solves the saddlepoint equations

$$x = M(\hat{s},0)^{-1} E\{X e^{\hat{s}^T X} M(0|X)\}$$
$$= M(\hat{s},0)^{-1} E(X e^{\hat{s}^T X}) \tag{4.71}$$

and

$$y = M(\hat{s},0)^{-1} E\{e^{\hat{s}^T X} M'(0|X)\}$$
$$= M(\hat{s},0)^{-1} E\{e^{\hat{s}^T X}(\varsigma + \tau^T X)\}$$
$$= \varsigma + \tau^T M(\hat{s},0)^{-1} E(e^{\hat{s}^T X} X). \tag{4.72}$$

From (4.71), this last term is $\varsigma + \tau^T x$ and y is this value when $\hat{t} = 0$. By the uniqueness of saddlepoints, $\hat{t}$ must be 0 when $y = \varsigma + \tau^T x$.

4.5.3 Differentiation of the Skovgaard approximation

Differentiation of this approximation is shown to lead to the second-order double-saddlepoint density to order $O(n^{-3/2})$.

Derivation of (4.40)

Take $\hat{w}_n = \sqrt{n}\hat{w}$ and $\hat{u}_n = \sqrt{n}\hat{u}$. Straightforward differentiation shows that

$$\hat{F}'(\bar{y}|\bar{x}) = \phi(\hat{w}_n)\left\{\sqrt{n}\frac{\partial \hat{w}}{\partial \bar{y}}\left(\frac{\hat{w}}{\hat{u}} - \frac{1}{n\hat{w}^2}\right) + \frac{1}{\sqrt{n}\hat{u}^2}\frac{\partial \hat{u}}{\partial \bar{y}}\right\}.$$

4.5 Appendix

Using $\partial \hat{w}/\partial \bar{y} = \hat{t}/\hat{w}$ then

$$\hat{F}'(\bar{y}|\bar{x}) = \phi(\hat{w}_n)\left(\frac{\sqrt{n}\hat{t}}{\hat{u}} - \frac{\hat{t}}{\sqrt{n}\hat{w}^3} + \frac{1}{\sqrt{n}\hat{u}^2}\frac{\partial \hat{u}}{\partial \bar{y}}\right). \tag{4.73}$$

The derivative of $\hat{u}$ from its definition is

$$\frac{\partial \hat{u}}{\partial \bar{y}} = \frac{\partial \hat{t}}{\partial \bar{y}}\sqrt{\frac{|\hat{K}''|}{|K''_{ss0}|}} + \frac{\hat{t}}{2}\sqrt{\frac{|K''_{ss0}|}{|\hat{K}''|}} \times \frac{|\hat{K}''|}{|K''_{ss0}|}\text{tr}\{(\hat{K}'')^{-1}\hat{L}\} \tag{4.74}$$

where $K''_{ss0} = K''_{ss}(\hat{s}_0, 0)$ and the last term of (4.74) is the factor $\partial|\hat{K}''|/\partial \bar{y}$ as specified in Press (1972, §2.14.2). The matrix $\hat{L}$ has (i, j) entry

$$\hat{L}_{ij} = \sum_{k=1}^{m-1}\hat{K}_{ijk}\frac{\partial \hat{s}_k}{\partial \bar{y}} + \hat{K}_{ijm}\frac{\partial \hat{t}}{\partial \bar{y}}. \tag{4.75}$$

To determine $\partial(\hat{s}, \hat{t})/\partial \bar{y}$, differentiate the saddlepoint equation to get

$$\frac{\partial}{\partial \bar{y}}\begin{pmatrix}\hat{s}\\\hat{t}\end{pmatrix} = (\hat{K}'')^{-1}\begin{pmatrix}0\\1\end{pmatrix} = (\hat{K}^{1m}, \ldots, \hat{K}^{mm})^T. \tag{4.76}$$

From (4.76), we also determine that

$$\partial \hat{t}/\partial \bar{y} = \hat{K}^{mm} = |\hat{K}''_{ss}|/|\hat{K}''| \tag{4.77}$$

from Cramer's rule. Substitute (4.76) into (4.75) and, after some algebra,

$$\text{tr}\{(\hat{K}'')^{-1}\hat{L}\} = \hat{\lambda}_3,$$

as given in (4.39). Now, (4.74) reduces to

$$\frac{\partial \hat{u}}{\partial \bar{y}} = \frac{|\hat{K}''_{ss}|}{\sqrt{|\hat{K}''||K''_{ss0}|}} + \frac{1}{2}\hat{u}\hat{\lambda}_3. \tag{4.78}$$

Substitution of (4.78) into (4.73) gives the final expression

$$\hat{F}'(\bar{y}|\bar{x}) = \sqrt{n}\phi(\hat{w}_n)\sqrt{\frac{|K''_{ss0}|}{|\hat{K}''|}}\left\{1 + \frac{1}{n}\left(\frac{1}{\hat{u}^2}\frac{|\hat{K}''_{ss}|}{|K''_{ss0}|} + \frac{\hat{\lambda}_3}{2\hat{t}} - \frac{\hat{u}}{\hat{w}^3}\right)\right\}$$
$$= \hat{f}(\bar{y}|\bar{x})\left\{1 + \frac{1}{n}\left(\frac{1}{\hat{u}^2} + \frac{\hat{\lambda}_3}{2\hat{t}} - \frac{\hat{u}}{\hat{w}^3}\right) + O(n^{-3/2})\right\}, \tag{4.79}$$

which has the same form as the derivative of the Lugannani and Rice expression given in (2.82) of Exercise 16 in chapter 2.

It remains to show that the correction term in the curved braces of (4.79) is the second-order correction term $O_N - O_D$ of the double-saddlepoint density to order $O(n^{-3/2})$. The term in (4.79) is a part of the $O(1)$ term in (4.38) as expressed in

$$O(1) = \frac{1}{\hat{u}}\left\{(O_N - O_D) - \left(\frac{1}{\hat{u}^2} + \frac{\hat{\lambda}_3}{2\hat{t}} - \frac{\hat{u}}{\hat{w}^3}\right)\right\}. \tag{4.80}$$

Since $\hat{u} = O(n^{-1/2})$, then substitution of (4.80) into (4.79) gives the final result.

4.5.4 Linear combination of Dirichlet components

In this context, we show that the $\hat{u}$ value for the double-saddlepoint Skovgaard approximation agrees with its counterpart in the single-saddlepoint approximation.

A direct comparison of these two analytical expressions produces what seems to be very different numerical quantities. We show, however, that they are the same analytically using an indirect approach that makes use of the relationship between $\check{t}$ and $\hat{s}$. Exploiting the fact that $\check{t} = \alpha_\cdot \hat{s}$ as described in Exercise 17, differentiate both sides of this equality with respect to cutoff d to determine that

$$|K''(\check{s},\check{t})| = \alpha_\cdot^{-3} K''_d(\hat{s}), \tag{4.81}$$

where $K(s, t)$ refers to the bivariate CGF used in double-saddlepoint approximation, and $K_d(s)$ is given in (3.65) for the single-saddlepoint approximation. Then

$$\check{t}\sqrt{\frac{|K''(\check{s},\check{t})|}{K''_{ss}(\check{s}_0,0)}} = \alpha_\cdot \hat{s}\sqrt{\frac{\alpha_\cdot^{-3} K''_d(\hat{s})}{\alpha_\cdot^{-1}}} = \hat{s}\sqrt{K''_d(\hat{s})},$$

and the $\hat{u}$ values are the same.

Derivation of (4.81)

The computation of $\partial\check{t}/\partial d$ is determined by (4.55) as

$$\frac{\partial\check{t}}{\partial d} = \frac{\check{K}''_{ss}}{|K''(\check{s},\check{t})|},$$

where

$$\check{K}''_{ss} = \sum_{i=1}^n \frac{\alpha_i}{(1-\check{s}-\check{t}w_i)^2} = \frac{1}{\alpha_\cdot^2}\sum_{i=1}^n \frac{\alpha_i}{(1-\hat{s}l_i)^2}. \tag{4.82}$$

The derivative $\partial\hat{s}/\partial d$ is determined by differentiating the single-saddlepoint equation $K'_d(\hat{s}) = 0$ to get

$$0 = \sum_{i=1}^n \frac{\partial K'_d(\hat{s})}{\partial l_i}\frac{\partial l_i}{\partial d} + K''_d(\hat{s})\frac{\partial\hat{s}}{\partial d}$$

$$= -\sum_{i=1}^n \frac{\alpha_i}{(1-\hat{s}l_i)^2} + K''_d(\hat{s})\frac{\partial\hat{s}}{\partial d}$$

$$= -\alpha_\cdot^2 \check{K}''_{ss} + K''_d(\hat{s})\frac{\partial\hat{s}}{\partial d} \tag{4.83}$$

from (4.82). Solving for $\partial\hat{s}/\partial d$ in (4.83) and setting it equal to $\alpha_\cdot^{-1}\partial\check{t}/\partial d$ leads to (4.81).

4.6 Exercises

1. Suppose that (X, Y) has a joint density or mass function on $\Re^m$ or I^m with support $\mathcal{X}$ and let $\mathcal{I}_\mathcal{X}$ denote the interior of the convex hull of $\mathcal{X}$.
    (a) Show that $\mathcal{Y}_x = \{y \in \Re^{m_y} : (x, y) \in \mathcal{I}_\mathcal{X}\} \subseteq \Re^{m_y}$ is an open convex set and that $\mathcal{Y}_x \supseteq \mathcal{I}_x$, the interior of the convex hull of the conditional support of Y given $X = x$.

(b) Give an example in which $\mathcal{Y}_x \supset \mathcal{I}_x$ and $\mathcal{Y}_x \cap \mathcal{I}_x^c$ is not a set of measure zero in $\Re^{m_y}$. What implications does this have concerning normalization of the conditional density/mass function?

2. Determine the specific expressions of the double-saddlepoint mass and density functions resulting from the characterizations in table 4.1.

3. Suppose Y is a lattice variable.
   (a) Derive a conditional saddlepoint CDF expression for Y given $X = j$ which is the left tail analogue of (1.37). Show it has the form
   $$\hat{F}_{1a}(k|\,X=j) = \Phi(\hat{w}) + \phi(\hat{w})\left(\frac{1}{\hat{w}} - \frac{1}{\tilde{u}_{1a}}\right) \qquad \hat{t} < 0 \qquad (4.84)$$
   where
   $$\hat{w} = \mathrm{sgn}(\hat{t})\sqrt{2\left[\{K(\hat{s}_0, 0) - \hat{s}_0^T j\} - \{K(\hat{s}, \hat{t}) - \hat{s}^T j - \hat{t}k\}\right]}$$
   $$\tilde{u}_{1a} = (e^{\hat{t}} - 1)\sqrt{|K''(\hat{s}, \hat{t})| / |K''_{ss}(\hat{s}_0, 0)|},$$
   $(\hat{s}, \hat{t})$ is the solution to $K'(\hat{s}, \hat{t}) = (j, k)$, and $\hat{s}_0$ solves $K'_s(\hat{s}_0, 0) = j$.
   (b) Show numerically that the CDF approximation
   $$\begin{cases} \hat{F}_{1a}(k|\,X=j) & \text{if } \hat{t} < 0 \\ 1 - \widehat{\Pr}_1(Y \geq k+1|\,X=j) & \text{if } \hat{t} > 0 \end{cases} \qquad (4.85)$$
   preserves the symmetry in (4.24) for the hypergeometric example with $(M, N, n) = (6, 6, 6)$ in table 4.4.

4. Suppose $Z_1 \sim$ Gamma $(\alpha, 1)$ independently of $Z_2 \sim$ Gamma $(\beta, 1)$. Show that the Skovgaard approximation for the conditional CDF of Z_1 given $Z_1 + Z_2 = 1$ is analytically the same as the approximation in (1.26) of section 1.2.2.

5. A conditional Mellin transform approach can be used to approximate the CDF of $\ln B$, for $B \sim$ Beta (α, β). If $B = Z_1/(Z_1 + Z_2)$ as in Exercise 4, then
   $$\Pr(B \leq y) = \Pr(\ln Z_1 \leq \ln y|\, Z_1 + Z_2 = 1)$$
   where the Skovgaard approximation may be applied to the latter probability.
   (a) Show that the joint MGF of $(X, Y) = (Z_1 + Z_2, \ln Z_1)$ is
   $$M(s, t) = \frac{\Gamma(\alpha + t)}{\Gamma(\alpha)}(1 - s)^{-(\alpha + \beta + t)} \qquad s < 1, t > -\alpha. \qquad (4.86)$$
   (b) Show that the two equations for determining $(\hat{s}, \hat{t})$ may be reduced to the root of a single equation.
   (c) Plot the resulting double-saddlepoint density for $\ln B$ and compare it to the true density.
   (d) Compare the single equation from part (b) with the single-saddlepoint equation from the unconditional Mellin transform approach given in (2.83).
   (e) Compare the percentage relative error of the conditional Mellin transform approach using the Skovgaard approximation with that of the unconditional Mellin approach using the Lugannani and Rice approximation.

6. Let $Z_1 \sim$ Poisson (λ) independently of $Z_2 \sim$ Poisson (μ). Show that the two continuity-corrected Skovgaard approximations for Z_1 given $Z_1 + Z_2 = n$, namely $\widehat{\Pr}_1$ and $\widehat{\Pr}_2$, agree analytically with their single-saddlepoint counterparts given in (1.33) and (1.34) of section 1.2.4.

7. Let $\{L_{ij} : i, j = 1, 2\}$ be the entries in a 2×2 contingency table that are i.i.d. Poisson (1). In addition, suppose the row sums are fixed at values $l_{1\cdot}$ and $l_{2\cdot}$ and the first column sum is fixed at $l_{\cdot 1}$.

(a) Derive the true conditional mass function of L_{11} given $L_{1\cdot} = l_{1\cdot}$, $L_{2\cdot} = l_{2\cdot}$, and $L_{\cdot 1} = l_{\cdot 1}$.

(b) Determine the double-saddlepoint density approximation to this conditional mass function.

(c) Derive the two continuity-corrected Skovgaard approximations and show that they agree with the expressions in (4.22) and (4.23) (Skovgaard, 1987, §5).

8. Let $Z_1 \sim$ Negative Binomial (M, θ) independently of $Z_2 \sim$ Negative Binomial (N, θ). The conditional distribution of $Y = Z_1$ given $X = Z_1 + Z_2 = n$ is Pólya $(M, N; n)$ with mass function given in (4.25). The Skovgaard approximation for the CDF may be computed and the development is very similar to that with the hypergeometric. It is again convenient to change the notation to that of a 2×2 contingency table:

k	M	$M + k$		l_{11}	l_{12}	$l_{1\cdot}$
$n - k$	N	$N + n - k$	$\equiv$	l_{21}	l_{22}	$l_{2\cdot}$
n	$M + N$	$M + N + n$		$l_{\cdot 1}$	$l_{\cdot 2}$	$l_{\cdot\cdot}$

First continuity correction:

(a) Show that the saddlepoints for $\widehat{\Pr}_1$ when approximating

$$\Pr(Y \geq k \mid X = n)$$

are

$$\hat{s} = \ln\left\{\frac{l_{21}}{l_{2\cdot}(1 - \theta)}\right\} \qquad \hat{t} = \ln\left(\frac{l_{11} l_{2\cdot}}{l_{1\cdot} l_{21}}\right)$$

and

$$\hat{s}_0 = \ln\left\{\frac{l_{\cdot 1}}{l_{\cdot\cdot}(1 - \theta)}\right\}.$$

(b) Determine that $\hat{w}$ is the same expression as given in (4.22) while $\tilde{u}_1$ is

$$\tilde{u}_1 = \left(1 - \frac{l_{21} l_{1\cdot}}{l_{11} l_{2\cdot}}\right) \sqrt{\frac{l_{21} l_{1\cdot} l_{11} l_{2\cdot} l_{\cdot 2}}{l_{22} l_{12} l_{\cdot 1} l_{\cdot\cdot}}}. \tag{4.87}$$

(c) Show that this Pólya distribution can also be expressed as the marginal distribution of P where $P|\theta \sim$ Binomial (n, θ) and $\theta \sim$ Beta (M, N). From this derive the mean as

$$E(P) = l_{12} \frac{l_{\cdot 1}}{l_{\cdot 2}}.$$

After further algebra, show that

$$\hat{t} \leq 0 \Leftrightarrow l_{11} \leq E(P).$$

Second continuity correction:

(d) The right-tail continuity correction $k^- = k - 1/2 = l_{11}^-$ induces the following change in notation:

$l_{11}^- = l_{11} - \frac{1}{2}$	l_{12}	$l_{1\cdot}^- = l_{1\cdot} - \frac{1}{2}$
$l_{21}^- = l_{21} + \frac{1}{2}$	l_{22}	$l_{2\cdot}^- = l_{2\cdot} + \frac{1}{2}$
$l_{\cdot 1}$	$l_{\cdot 2}$	$l_{\cdot\cdot}$

Table 4.7. *Double-saddlepoint approximations for a uniform distribution on* $\{0, \ldots, 9\}$.

k	$\Pr(U \le k)$	$\widehat{\Pr}_1 (U \le k)$	$\widehat{F}_{1a} (k)$	$\widehat{\Pr}_2(U \le k)$
0	.1	.1006	undefined	.0990
1	.2	.1984	.2179	.2015
2	.3	.2964	.3128	.3016
3	.4	.3943	.4100	.4009
4	.5	.4922	.5078	.5*
k	$\Pr(U \ge k)$	$\widehat{\Pr}_1 (U \ge k)$		$\widehat{\Pr}_2 (U \ge k)$
5	.5	.5078		.5*
6	.4	.4100		.4009
7	.3	.3128		.3016
8	.2	.2179		.2015
9	.1	undefined		.0990

The value $\tilde{w}_2$ is now the value in (4.22) with $\{l_{ij}, l_{i\cdot}\}$ replaced with these corrected values. Show also that

$$\tilde{u}_2 = (l_{11}^- l_{2\cdot}^- - l_{1\cdot}^- l_{21}^-) \sqrt{\frac{l_{\cdot 2}}{l_{22} l_{12} l_{\cdot 1} l_{\cdot \cdot}}}. \tag{4.88}$$

9. (a) Show that the Pólya $(1, 1, m)$ mass function is the special case that puts uniform mass on the values $\{0, \ldots, m\}$.
   (b) Let U denote the variable with a uniform mass function on $\{0, \ldots, 9\}$. Compute the two continuity-corrected approximations to the CDF of U. Verify table 4.7.
   (c) Note the symmetry displayed by $\widehat{F}_{1a}(k)$ and $\widehat{\Pr}_2$ as a result of the underlying symmetry of the mass function. Since the mean is 4.5, the starred entries are removable singularities for $\widehat{\Pr}_2$ and were computed at 4.5 ± 10^{-8}. Compare the accuracy of these methods with the single-saddlepoint approximations in table 1.11. Note that overall the double-saddlepoint $\widehat{\Pr}_2(U \ge k)$ is the most accurate.
10. The sequence of steps below motivates a double-saddlepoint approximation to the central t_{n-1}-distribution.
    (a) If $Z_1, \ldots, Z_n$ are i.i.d. Normal (μ, σ^2), show that the t-statistic for testing $\mu = 0$ may be written as

    $$T = \sqrt{1 - n^{-1}} \frac{Y}{\sqrt{X - n^{-1}Y^2}} = g(X, Y),$$

    where $Y = \sum_i Z_i$ and $X = \sum_i Z_i^2$.
    (b) For the questions in parts (b)–(e), suppose the null setting in which $\mu = 0$. Show that T is independent of X. (Note that T is scale invariant.)
    (c) Assume for simplicity that $\sigma = 1$. By part (b),

    $$\Pr(T \le z) = \Pr\{g(X, Y) \le z | X = 1\}$$
    $$= \Pr(Y \le y | X = 1) \tag{4.89}$$

    for a suitably defined y. Find y and compute the joint CGF of (X, Y).

(d) Show that the double-saddlepoint approximation for the computation in (4.89) leads to explicit saddlepoints and determine the Skovgaard approximation to the CDF of the t_{n-1}-distribution.

(e) Demonstrate the accuracy of the approximation with a small table for the t_1 or Cauchy distribution.

(f) Suppose we didn't assume $\mu = 0$. Could this argument be used to approximate the CDF of the non-central t_{n-1}? Why or why not?

11. Show the following results needed in the derivation of the Skovgaard approximation.

    (a) Differentiate $g(t)$ in (4.30) to show that
    $$g'(t) = K'_t(\hat{s}_t, t) - y$$
    and $g'(\hat{t}) = 0$.

    (b) Use implicit differentiation to show that
    $$\frac{\partial \hat{s}_t}{\partial t} = -K''_{ss}(\hat{s}_t, t)^{-1} K''_{st}(\hat{s}_t, t),$$
    and
    $$g''(t) = K''_{ts}(\hat{s}_t, t) \frac{\partial \hat{s}_t}{\partial t} + K''_{tt}(\hat{s}_t, t).$$
    Combine these results to show that $g''(\hat{t})$ has the value in (4.32).

    (c) Consider an alternative to the approach used in the text for determining $\lim_{\hat{t} \to 0} \hat{w}/\hat{t}$. Start with the expression for $\hat{w}$ in (4.13) and use bivariate Taylor expansions for $K(\hat{s}, \hat{t})$ and $y = K'_t(\hat{s}, \hat{t})$ about $(\hat{s}_0, 0)$ to compute an expansion for the ratio $\hat{w}/\hat{t}$. Combine terms and use the fact that
    $$\frac{\partial \hat{s}}{\partial \hat{t}} = -K''_{ss}(\hat{s}, \hat{t})^{-1} K''_{st}(\hat{s}, \hat{t}),$$
    to derive $\lim_{\hat{t} \to 0} \hat{w}/\hat{t}$ as the value in (4.33).

12. (a) Prove that $\hat{f}(y|x)$ is equivariant to nonsingular linear transformation of the y-variable.

    (b) Show that the various Skovgaard approximations, both continuous and continuity-corrected, are invariant to increasing linear transformation of variable Y.

    (c) Show that all versions of the Skovgaard approximation except the first continuity-corrections are invariant to decreasing linear transformation. The binomial distribution provides a counterexample to the invariance of the first continuity-correction.

13. The two right-tail continuity corrections of section 4.2.2 were introduced by Skovgaard (1987) using an inversion theory approach to the approximation and continuity correction. For the second correction, derive a left-tail approximation for Y given $X = j$ from the right-tail approximation of $-Y$ given $X = j$ and denote it as $\hat{F}_2(k|X = j)$. Verify that it satisfies (4.19).

14. (a) Prove that the second continuity-corrected Skovgaard approximation preserves the inherent symmetry of the true underlying conditional distribution when this symmetry holds for almost every x-value and $E(Y|x) = \varsigma + \tau^T x$ for fixed ς and τ.

    (b) Prove that the alternate version of the first continuity-corrected Skovgaard approximation, defined in (4.85) and (4.84), preserves symmetry under the same conditions specified in part (a).

15. Prove that $E(Y|x)$ is linear in x when (X, Y) is bivariate gamma. As a consequence, $\hat{t} = 0$ when $y = E(Y|x)$. (Hint: Construct $X = \sum_i X_i^2$ and $Y = \sum_i Y_i^2$ from i.i.d. bivariate normals $\{(X_i, Y_i)\}$. Compute $E(Y_i^2|x_i)$ and determine that $E(Y|x_1, \ldots, x_n)$ depends on $\{x_i\}$ only through $\sum_i x_i^2$.)

16. (a) Derive the single-saddlepoint CDF approximation to $\Pr(Z \leq z)$ where $Z = \sum_{i=1}^{n} w_i M_i$, with $(M_1, \ldots, M_{n-1}) \sim$ Multinomial $(n; \{\pi_i\})$. Show that the single-saddlepoint $\hat{s}$ solves

$$\frac{z}{n} = \frac{\sum_{i=1}^{n} \pi_i w_i e^{\hat{s} w_i}}{\sum_{i=1}^{n} \pi_i e^{\hat{s} w_i}} \qquad \frac{z}{n} \in (\min_i w_i, \max_i w_i).$$

(b) Suppose that $Y = \sum_{i=1}^{n} w_i P_i$ and $X = \sum_{i=1}^{n} P_i$ where $\{P_i\}$ are independent Poisson (λ_i) variables. Take $\{\lambda_i\}$ proportional to $\{\pi_i\}$ so that

$$\Pr(Z \leq z) = \Pr(Y \leq z | X = n). \tag{4.90}$$

Derive the double-saddlepoint Skovgaard approximation to the right side of (4.90). If $(\check{s}, \check{t})$ and $\check{s}_0$ are the numerator and denominator saddlepoints respectively, show that $\check{s}_0 = \ln(n/\lambda_.)$, where $\lambda_. = \sum_{i=1}^{n} \lambda_i$, and

$$z = \sum_i \lambda_i w_i \exp(\check{s} + \check{t} w_i) \tag{4.91}$$

$$n = \sum_i \lambda_i \exp(\check{s} + \check{t} w_i).$$

(c) Take ratios of the two equations in (4.91) to show that $\check{t} = \hat{s}$. Denote the single-saddlepoint inputs to Lugannani and Rice as $(\hat{u}, \hat{w})$ and the double-saddlepoint inputs to Skovgaard as $(\check{u}, \check{w})$. Show that $(\check{u}, \check{w}) = (\hat{u}, \hat{w})$.

17. Show that the double-saddlepoint Skovgaard approximation for the linear combination of Dirichlet components in (3.45) agrees with the single-saddlepoint approximation discussed in Exercise 16 of chapter 3.

(a) Suppose that $Y = \sum_{i=1}^{n} w_i G_i$ and $X = \sum_{i=1}^{n} G_i$ where $\{G_i\}$ are independently distributed Gamma $(\alpha_i, 1)$ Show that the joint CGF of X and Y is

$$K(s, t) = -\sum_{i=1}^{n} \alpha_i \ln(1 - s - t w_i) \qquad (s, t) \in \mathcal{S},$$

where

$$\mathcal{S} = \{(s, t) : s < \min(1 - t \min_i w_i, \ 1 - t \max_i w_i)\}.$$

If $d \in (\min_i w_i, \max_i w_i)$, show that the saddlepoints $(\check{s}, \check{t})$ for determining $\Pr(Y \leq d | X = 1)$ satisfy

$$d = \sum_i \frac{\alpha_i w_i}{1 - \check{s} - \check{t} w_i} \tag{4.92}$$

$$1 = \sum_i \frac{\alpha_i}{1 - \check{s} - \check{t} w_i}. \tag{4.93}$$

(b) Take $n = 3$, $\alpha_i = i$ for $i = 1, 2, 3$, $w_1 = 2$, $w_2 = 4$, $w_3 = 5$, and $d = 3$. Compute

$$(\check{s}, \check{t}) = (8.018169813, -4.339389938). \tag{4.94}$$

Show that this results in the value of $(\hat{w}, \hat{u})$ given by

$$(\check{w}, \check{u}) = (-2.23058677, -2.86870248). \tag{4.95}$$

(c) For the example in (b) show that the single-saddlepoint approximation from Exercise 16 of chapter of 3 has saddlepoint

$$\hat{s} = -0.7232316563.$$

Show that the resulting single-saddlepoint value of $(\hat{w}, \hat{u})$ agrees with (4.95) to the number of decimals displayed.

(d) Note from the computational work that $\check{t} = 6\hat{s}$ or, more generally, $\check{t} = \alpha.\hat{s}$ where $\alpha. = \sum_i \alpha_i$. To determined the relationship of $(\check{s}, \check{t})$ to $\hat{s}$, subtract $d \times$ equation (4.93) from (4.92) so that

$$0 = \sum_{i=1}^{n} \frac{\alpha_i(w_i - d)}{1 - \check{s} - \check{t}w_i}$$
$$= \sum_{i=1}^{n} \frac{\alpha_i(w_i - d)}{1 - \frac{\check{t}}{1-\check{s}-\check{t}d}(w_i - d)}. \quad (4.96)$$

Expression (4.96) is the single-saddlepoint equation in (3.66) that yields a unique root; therefore this relationship must be

$$\hat{s} = \frac{\check{t}}{1 - \check{s} - \check{t}d}. \quad (4.97)$$

(e) Together, between the deduced relationship (4.97) and the computationally observed relationship $\check{t} = \alpha.\hat{s}$, the relationship is

$$\check{t} = \alpha.\hat{s} \quad (4.98)$$
$$\check{s} = 1 - \alpha.(1 + \hat{s}d),$$

a fact determined by Paolella (1993) and Butler and Paolella (1998a) in the special exchangeable case. Show that the values in (4.98) solve (4.92) and (4.93). Show also that these values are in $\mathcal{S}$. Now deduce from both of these results that the values in (4.98) must be the double-saddlepoint values. Verify these expressions with the numerical example used in part (b).

(f) Using (4.98), show that the single-saddlepoint $\hat{w}$ value is the same as that for the double-saddlepoint approximation.

(g) Showing equality for the $\hat{u}$ values is considerably more difficult. The details are given in section 4.5.4 of the appendix.

18. Suppose $w_1, \ldots, w_m$ are the values of a population characteristic which occur with frequencies $\{N_i : i = 1, .., m\}$ where the total $\sum_{i=1}^{m} N_i = N$ is known. Take $(N_1, \ldots, N_{m-1})^T = \mathbf{N}$ as the unknown parameter. Suppose that a random sample of size n without replacement yields frequencies $\{n_i\}$. Denote $(n_1, \ldots, n_{m-1})^T = \mathbf{n}$ as the sufficient statistic.

(a) A multivariate Pólya $(\alpha_1^0, \ldots, \alpha_m^0; N)$ prior on $\mathbf{N} = (N_1, \ldots, N_{m-1})$ has mass function

$$\Pr(\mathbf{N} = k) = \left\{ \frac{\Gamma(N + \alpha_.^0)}{\Gamma(\alpha_.^0)N!} \right\}^{-1} \prod_{i=1}^{m} \left\{ \frac{\Gamma(k_i + \alpha_i^0)}{\Gamma(\alpha_i^0)k_i!} \right\}$$

defined on $\{k_i \geq 0, \sum_{j=1}^{m-1} k_j \leq N\}$ with $k_m = N - \sum_{i=1}^{m-1} k_i$ and $\alpha_.^0 = \sum_i \alpha_i^0$. Note that this is the same mass function as the Dirichlet–Multinomial distribution in Exercise 8 of chapter 3. Show that it combines with data $\mathbf{n}$ to give a posterior distribution on $\mathbf{N} - \mathbf{n}$ which is Pólya $(\alpha_1^1, \ldots, \alpha_m^1; N - n)$ with update $\alpha_i^1 = \alpha_1^0 + n_i$.

(b) Derive a Skovgaard saddlepoint approximation for the posterior on the population mean in (4.42).

(c) Create a small population and sample to implement the approximation using a uniform prior.

(d) Determined a 90% posterior confidence range for the population mean. This may be determined by inverting the Skovgaard approximation to find the 5th and 95th percentage points of its posterior.

19. Suppose the sampling framework of Exercise 18. A random sample without replacement yields category frequencies $\{H_i\} \sim$ Hypergeometric $(N_1, \ldots, N_m; n)$.

4.6 Exercises

(a) Derive the results needed to approximate the CDF of the unbiased estimator of population mean $\hat{\mu}$ given in (4.43).

(b) Use the same data as in Exercise 18 to compute the sampling distribution of $\hat{\mu}$. Unlike the Bayesian approach, this approximate sampling distribution cannot be used directly in determining a confidence interval since it depends on the values of $\{N_i\}$.

(c) Replace each N_i in the Skovgaard approximation with NH_i/n, its unbiased estimator. Using this "Plug-in" approximation, determine its 5th and 95th percentage points and compare with the interval from Exercise 18.

(d) Discuss the merits and demerits of "plugging in" estimates of the unknown parameters in saddlepoint approximations.

20. Suppose that (X, Y) has the bivariate logistic distribution of Exercise 3 of chapter 3.

    (a) Plot the boundaries of S as in figure 4.2. For a half dozen values of x, also plot the saddlepoint manifolds $\{S_x\}$ created by fixing the value $X = x$ in the double-saddlepoint density.

    (b) Unlike the bivariate exponential setting, region S is now bounded. For fixed x, describe $(\hat{s}, \hat{t}) \in S_x$ as $y \to 0$ or $y \to \infty$. Does the plot agree with the geometry discussed in section 4.4.2?

21. Suppose (X, Y) has the bivariate negative binomial mass function of Exercise 4 of chapter 3.

    (a) Compute the conditional MGF of Y given $X = i$.

    (b) Is it possible to determine the true conditional distribution from the manner in which the bivariate distribution was constructed? If so, can you recognize the conditional MGF in part (a)?

22. Suppose $M_p(z, t)$ is the joint P-MGF of $X = (X_1, X_2)$ and Y where both components of X are nonnegative and integer-valued.

    (a) Derive a generalization of (4.58) for the conditional MGF of Y given $X = (i, j)$.

    (b) Generalize this further by determining the most general expression when $\dim X = r$.

23. Suppose that $\hat{f}(x, y)$ is the joint saddlepoint density of (X, Y) computed from the joint CGF $K(s, t)$. Let $\hat{f}(x)$ be the marginal saddlepoint density of X computed from $K(s, 0)$, the CGF of X. Prove that Laplace's approximation to

$$\int \hat{f}(x, y)\,dy,$$

leads to $\hat{f}(x)$ when (3.23) is applied with the h factor as $\left|K''(\hat{s}, \hat{t})\right|^{-1/2}$ when integrating dy.

(a) Write

$$\int \hat{f}(x, y)\,dy = \int \hat{f}(x, y)\left\|\frac{\partial y}{\partial \hat{t}}\right\|d\hat{t} = \int \hat{f}(x, y)\frac{|\hat{K}''|}{|\hat{K}''_{ss}|}d\hat{t}. \quad (4.99)$$

Define

$$g(x, \hat{t}) = \hat{s}^T x + \hat{t}^T y - K(\hat{s}, \hat{t})$$

as the exponential term in $\hat{f}(x, y)$ with $\hat{s}$ and y as implicit functions of $\hat{t}$. Show that $\partial g(x, \hat{t})/\partial \hat{t} = 0$ if and only if $\hat{t} = 0$.

(b) Show

$$\frac{\partial^2}{\partial \hat{t} \partial \hat{t}^T} g(x, \hat{t})|_{\hat{t}=0} = K''_{tt}(\hat{s}_0, 0) - K''_{ts}(\hat{s}_0, 0)K''_{ss}(\hat{s}_0, 0)^{-1}K''_{st}(\hat{s}_0, 0)$$

so that

$$\left|\frac{\partial^2}{\partial \hat{t} \partial \hat{t}^T} g(x, \hat{t})|_{\hat{t}=0}\right| = |K''(\hat{s}_0, 0)|/|K''_{ss}(\hat{s}_0, 0)|. \quad (4.100)$$

Table 4.8. *Responses of two groups to audio analgesia along with group membership indicators.*

Resp.	−10.5	−3.1	−3.0	0.0	.9	1.7	3.1	3.3	5.0
Grp.	1	1	1	1	0	1	0	0	1
Resp.	5.4	5.6	5.7	7.6	7.7	10.6	13.3	17.9	
Grp.	0	0	0	0	1	0	0	0	

(c) Complete the proof using (4.99) and (4.100). Note that the h factor is now $|K''(\hat{s}, \hat{t})|^{1/2}$ when integrating $d\hat{t}$.

24. The data in table 4.8 are responses to an audio analgesia for two distinct groups of persons given in Lehmann and D'Abrera (1975, p. 92).

    (a) For the two sample permutation test, show that the uncorrected approximation (4.12) leads to approximate p-value $p = 0.011$ as reported in Booth and Butler (1990). Robinson (1982, table 2) reports an exact value of $p = 0.012$.

    (b) Replicate the approximate 95% confidence interval (1.92, 13.56) from Booth and Butler. The exact confidence interval from Robinson is (1.88, 13.52).

    (c) For the Wilcoxon rank sum test, use (4.15) and (4.17) to determine continuity-corrected p-values.

    (d) Invert the two continuity-corrected approximations to determine 95% confidence intervals.

    (e) Invert (4.12) which does not use the continuity correction. Of the three intervals in parts (d) and (e), which do you think has the most accurate coverage? Why?

5

Exponential families and tilted distributions

Exponential families provide the context for many practical applications of saddlepoint methods including those using generalized linear models with canonical links. More traditional exponential family applications are considered in this chapter and some generalized linear models are presented in chapter 6. All these settings have canonical sufficient statistics whose joint distribution is conveniently approximated using saddlepoint methods. Such approximate distributions become especially important in this context because statistical methods are almost always based on these sufficient statistics. Furthermore, these distributions are most often intractable by other means when considered to the degree of accuracy that may be achieved when using saddlepoint approximations. Thus, in terms of accuracy, both in this context and in the many others previously discussed, saddlepoint methods are largely unrivaled by all other analytical approximations.

The subclass of *regular exponential families* of distributions was introduced in section 2.4.5 by "tilting" a distribution whose MGF converges in the "regular" sense. Tilting combined with Edgeworth expansions provide the techniques for an alternative derivation of saddlepoint densities that is presented in this chapter. This alternative approach also gives further understanding as to why the saddlepoint density achieves its remarkable accuracy.

Both unconditional and conditional saddlepoint approximations are considered for the canonical sufficient statistics in a regular exponential family. The computation of p-values for various likelihood ratio tests and the determination of some uniformly most powerful unbiased tests are some of the practical examples that are considered.

5.1 Regular exponential families

5.1.1 Definitions and examples

Let z be the data and denote Z as the random variable for the data before observation. Suppose the distribution of Z is a member of a parametric class of densities or mass functions of the form

$$f(z;\xi) = \exp\{\theta(\xi)^T x(z) - c(\xi) - d(z)\}$$

for $\xi \in \Xi$. Then the parametric class $\{f(\cdot\,;\xi) : \xi \in \Xi\}$ is an *exponential family* of distributions. The canonical sufficient statistic is $x = x(z)$ of dimension m and $\theta = \theta(\xi)$ is the canonical parameterization of ξ which ranges over $\Theta = \{\theta(\xi) : \xi \in \Xi\} \subseteq \Re^m$. The exponential family is *full* if Θ includes all possible values of θ for which the parametric class could be defined; more formally, the family is full if $\Theta = \tilde{\Theta}$ where

$$\tilde{\Theta} = \left\{ \theta : \int \exp\{\theta^T x(z) - d(z)\} dz < \infty \right\}.$$

The set $\tilde{\Theta}$ is convex as may be shown by using Hölder's inequality. A *regular exponential family* is a family that is full and for which Θ is open in $\Re^m$. Alternatively, Θ might be a differentiable manifold in $\Re^m$ in which case it is a *curved exponential family* as discussed in section 7.3. A thorough treatment of the properties for all these classes is given in Barndorff-Nielsen (1978, 1982) and Lehmann (1986).

Most of the distributional families so far considered provide examples of regular exponential families. From among the univariate distributions, these include Binomial (n, ξ),[1] Normal (μ, σ^2), Gamma (α, β), Poisson (λ), Negative Binomial (n, ξ), Beta (α, β) with one parameter fixed, and Pareto (α) where α is a scale parameter. Parametric families that are not exponential families include the Gumbel (μ, σ^2) and Logistic (μ, σ^2), where the parameters are location and scale parameters. Families of multivariate densities that are regular exponential families include Normal$_m(\mu, \Sigma)$, Wishart$_m(n, \Sigma)$ for fixed n, Multinomial $(n; \pi_1, \ldots, \pi_m + 1)$ for fixed n, and Dirichlet $(\alpha_1, \ldots, \alpha_m + 1)$ with a linear constraint on the $\{\alpha_i\}$. Multivariate families that are not exponential families include the multivariate gamma and bivariate Poisson families.

For a regular exponential family, $\xi \leftrightarrow \theta$ are 1-1 and the notation may be simplified by working with the canonical parameterization and expressing the exponential family as

$$f(z; \theta) = \exp\{\theta^T x(z) - c(\theta) - d(z)\}, \tag{5.1}$$

where $c(\theta)$ is $c\{\xi(\theta)\}$. If z denotes the data, then the *likelihood of θ* (or of ξ through θ) is any function that conveys the shape of (5.1) in terms of θ; formally, the likelihood is (5.1) or

$$\mathcal{L}(\theta) \stackrel{\theta}{\propto} \exp\{\theta^T x(z) - c(\theta)\}, \tag{5.2}$$

which differs only by the constant factor $\exp\{-d(z)\}$.

5.1.2 Properties

Important facts about regular exponential families that are relevant to saddlepoint approximation are given below. Some of the easier proofs are left as exercises. A thorough discussion of all the intricacies of exponential families is given in Barndorff-Nielsen (1978). Random variable X below refers to the canonical sufficient statistic in (5.1).

The CGF of canonical sufficient statistic X

The CGF of X for $\theta \in \Theta$ is

$$K(s) = c(s + \theta) - c(\theta) \qquad s \in \mathcal{S}_\theta \tag{5.3}$$

---

[1] Parametric dependence in this family is through ξ only and n is assumed fixed. The same applies to the negative binomial family.

where $\mathcal{S}_\theta = \{s : s + \theta \in \Theta\}$. This is a regular setting since $\mathcal{S}_\theta$ is always an open neighborhood of 0 when $\theta \in \Theta$. The mean and covariance of X are

$$E(X; \theta) = c'(\theta) \qquad \text{Cov}(X; \theta) = c''(\theta) \qquad (5.4)$$

where $c'(\theta)$ is the $m \times 1$ gradient and $c''(\theta)$ is the $m \times m$ Hessian.

Product families

Let $Z_1, \ldots, Z_n$ be i.i.d. with a density that is a member of the regular exponential family in (5.1). Then, the joint density of $Z = (Z_1, \ldots, Z_n)^T$ is a regular exponential family with sufficient statistic $X_n = \sum_{i=1}^n x(Z_i)$ that has CGF $nK(s)$.

The Fisher (1934) factorization theorem

The factorization theorem states informally that the density of sufficient statistic X, given as $f_X(x; \theta)$, is also the likelihood of θ. This occurs because $f_X(x; \theta)$ reproduces the density of the raw data in terms of its shape in θ. Suppose that dim X < dim Z. Then we may express this as

$$\mathcal{L}(\theta) \stackrel{\theta}{\propto} f_Z(z; \theta) = f_X(x; \theta) g(z|x). \qquad (5.5)$$

In the discrete case, $g(z|x)$ is the conditional mass function of Z given $X = x$. For the continuous case it may be called the conditional density of Z given $X = x$ but it is not an ordinary density of the type we have been discussing since it is a density[2] on the surface $\{z : x(z) = x\}$. The marginal density of X is found by integrating over the surface, with dS as the surface area element, so that

$$f_X(x; \theta) = \int_{\{z:x(z)=x\}} f_Z(z; \theta) \, dS$$
$$= \exp\{\theta^T x - c(\theta) - d_1(x)\} \qquad (5.6)$$

with

$$e^{-d_1(x)} = \int_{\{z:x(z)=x\}} e^{-d(z)} dS.$$

The density of X is a member of a regular exponential family as may be seen from the form of (5.6). The computation also demonstrates that $g(z|x)$ is not dependent on θ. From (5.6), the conditional density of Z given $X = x$ is

$$g(z|x) = \exp\{-d(z) + d_1(x)\} 1_{\{z:x(z)=x\}}$$

which does not depend on θ and which is now integrated dS.

---

[2] For smooth $x(\cdot)$, $g(z|x) dS$ is a density on the surface $x(z) = x$ and integrates to 1 when dS is the surface area element. Conditonal probabilities for Z given x require that the surface integral be computed using differential forms as described in Buck (1965, chap. 7).

Specification of canonical parameters

There is some latitude in the designation of the canonical parameter in the exponential family. Any set of linearly independent combinations of the canonical parameter θ can be made canonical in the following way. Suppose A^T is an $r \times m$ matrix of rank $r < m$, and let $\eta = A^T \theta$ be r such linear combinations. Let B^T to be any $m \times m$ nonsingular matrix whose first r rows are those in A^T. Then the canonical parameter and sufficient statistic can be modified as

$$\theta^T x = (B^T \theta)^T B^{-1} x = \eta^T y_1 + \vartheta^T y_2,$$

where y_1 and y_2, are the upper and lower component vectors of $B^{-1} x$ and the new canonical sufficient statistics, and ϑ is the lower vector of $B^T \theta$.

Marginal families

For notational purposes, let (X, Y) replace X above as the canonical sufficient statistic. If the joint distribution of (X, Y) is a member of a regular exponential family, then the marginal of X is not necessarily a member of an exponential family.

Example: Multivariate Gamma This parametric family of distributions is not an exponential family as may be seen in the special bivariate exponential setting of section 3.1.3 where the density is expressed in terms of a Bessel function. The distribution is the marginal for the diagonal of a Wishart distribution whose parametric class of densities is a regular exponential family.

Maximum likelihood estimation

In a regular exponential family, the log-likelihood

$$l(\theta) = \ln \mathcal{L}(\theta) = \theta^T x - c(\theta)$$

is a strictly concave function over $\theta \in \Theta$. Let the support of X be $\mathcal{X}$ and denote the interior of its convex hull as $\mathcal{I}_\mathcal{X}$. Then the maximum likelihood estimator (MLE) of θ exists if and only if $x \in \mathcal{I}_\mathcal{X}$. If the MLE $\hat{\theta}$ exists, then it is the unique solution in Θ to the MLE equation

$$c'(\hat{\theta}) = x. \tag{5.7}$$

Example: Poisson (λ) This mass function provides a simple example of the need to consider solutions with x in $\mathcal{I}_\mathcal{X}$. The canonical parameter is $\theta = \ln \lambda$ and

$$c(\theta) = \lambda = e^\theta$$

upon reparametrization. The support is $\mathcal{X} = \{0, 1, \ldots\}$ and $\mathcal{I}_\mathcal{X} = (0, \infty)$. The MLE equation is

$$e^{\hat{\theta}} = k,$$

which has a unique solution for $k \in (0, \infty)$ but no solution on the boundary at 0.

Fisher information

Define the *observed Fisher information* for $m \times 1$ parameter ξ as the $m \times m$ matrix

$$j(\xi) = -\frac{\partial^2 l(\xi)}{\partial \xi \partial \xi^T}$$

and the *expected Fisher information* as

$$i(\xi) = E\{j(\xi)\}.$$

For the canonical parameterization in a regular exponential family, $\xi = \theta$ so that

$$j(\theta) \equiv i(\theta) = c''(\theta)$$

follows from direct computation. If τ is any 1-1 smooth reparametrization of θ in this regular exponential family, then

$$i(\tau) = \frac{\partial \theta^T}{\partial \tau} c''(\theta) \frac{\partial \theta}{\partial \tau^T} \tag{5.8}$$

and

$$j(\tau) = i(\tau) - \sum_{i=1}^{m} \{x_i - c'_i(\theta)\} \frac{\partial^2 \theta_i}{\partial \tau \partial \tau^T}, \tag{5.9}$$

where $x^T = (x_1, \ldots, x_m)$, $c(\theta)^T = \{c_1(\theta), \ldots, c_m(\theta)\}$, and $\theta^T = (\theta_1, \ldots, \theta_m)$. At the MLE $\hat{\tau}$, the observed and expected information are the same, or

$$j(\hat{\tau}) = i(\hat{\tau}).$$

Proof of (5.9). Using the chain rule, compute

$$\frac{\partial l}{\partial \tau_i} = \frac{\partial \theta^T}{\partial \tau_i} x - c'(\theta)^T \frac{\partial \theta}{\partial \tau_i}$$

$$= \frac{\partial \theta^T}{\partial \tau_i} \{x - c'(\theta)\}$$

and

$$-\frac{\partial^2 l}{\partial \tau_i \partial \tau_j} = \frac{\partial \theta^T}{\partial \tau_i} c''(\theta) \frac{\partial \theta}{\partial \tau_j} - \frac{\partial^2 \theta^T}{\partial \tau_i \partial \tau_j} \{x - c'(\theta)\},$$

which is (5.9). Expression (5.8) results by taking its expectation. At the MLE, the term $x - c'(\hat{\theta})$ vanishes so that $j(\hat{\tau}) = i(\hat{\tau})$. $\square$

Example: Mean parameterization This parametrization of θ is $\mu = E(X; \theta) = c'(\theta)$. The mapping $\mu \leftrightarrow \theta$ is a smooth bijection and an important reparametrization in inference considerations for exponential families. The expected Fisher information concerning μ is

$$i(\mu) = \frac{\partial \theta^T}{\partial \mu} c''(\theta) \frac{\partial \theta}{\partial \mu^T} = c''(\theta)^{-1} = i(\theta)^{-1}.$$

At the MLE,

$$j(\hat{\mu}) = i(\hat{\mu}) = i(\hat{\theta})^{-1} = j(\hat{\theta})^{-1}. \tag{5.10}$$

Conditional families

Let (X, Y) be the canonical sufficient statistics and $\theta = (\chi, \psi) \in \Theta$ the conforming canonical parameters. The regular exponential family of (X, Y) is

$$f(x, y; \theta) = \exp\{\chi^T x + \psi^T y - c(\chi, \psi) - d_1(x, y)\}. \tag{5.11}$$

The conditional distribution of Y given $X = x$ is a member of a regular exponential family of the form

$$f(y|x; \psi) = \exp\{\psi^T y - c(\psi|x) - d_1(x, y)\}, \tag{5.12}$$

where

$$e^{c(\psi|x)} = \int_{\{y:(x,y)\in\mathcal{S}\}} \exp\{\psi^T y - d_1(x, y)\} dy, \tag{5.13}$$

and $\mathcal{S}$ is the joint support of (X, Y). The conditional CGF of Y given $X = x$ is

$$K_{Y|x}(t) = c(t + \psi|x) - c(\psi|x). \tag{5.14}$$

Example: Dirichlet $(\alpha_1, \ldots, \alpha_{m+1})$ With a linear constraint on $\{\alpha_i\}$, this class of densities is a regular exponential family. It may be constructed from another exponential family within which it represents the conditional joint density of m canonical sufficient variables given another single canonical component. Let $Z = (Z_1, \ldots, Z_{m+1})$ be "data" with independent components $Z_i \sim$ Gamma (α_i, β). Take $\xi = (\{\alpha_i\}, \beta)$ so the density of Z is

$$f(z; \xi) = \exp\left[-\beta z. + \sum_{i=1}^{m+1}\{(\alpha_i - 1)\ln z_i - \ln \Gamma(\alpha_i)\} + \alpha. \ln \beta\right]. \tag{5.15}$$

A convenient choice for representing the linear constraint on $\{\alpha_i\}$ is to fix the value of α_{m+1}. Take $x = z.$ and $y^T = (\ln z_1, \ldots, \ln z_m)$ with $\chi = -\beta$ and $\psi^T = (\alpha_1 - 1, \ldots, \alpha_m - 1)$. With α_{m+1} fixed, (x, y) is canonical sufficient and (χ, ψ) is the canonical parameter. The conditional distribution of Y given $X = 1$ is that of $(\ln D_1, \ldots, \ln D_m)$ where $(D_1, \ldots, D_m)$ is the Dirichlet vector. Through Jacobian transformation of the Dirichlet, this conditional density may be determined as

$$f(y| X = 1; \psi) = \frac{\Gamma(\alpha.)}{\prod_{i=1}^{m+1} \Gamma(\alpha_i)} \exp\{\psi^T y + 1^T y\}$$

and has the form of (5.12). The conditional CGF of $Y|X = 1$ is given in (5.14) in terms of (5.13) which is

$$e^{-c(\psi|x)} = \frac{\Gamma(\alpha.)}{\prod_{i=1}^{m+1} \Gamma(\alpha_i)};$$

thus from (5.14),

$$E\left(e^{t^T Y}| X = 1\right) = \frac{\Gamma(\alpha.)}{\Gamma(t. + \alpha.)} \prod_{i=1}^{m} \frac{\Gamma(t_i + \alpha_i)}{\Gamma(\alpha_i)}$$

with $t^T = (t_1, \ldots, t_m)$ and $t. = \sum_{i=1}^{m} t_i$.

5.2 Edgeworth expansions

These expansions are introduced without a great deal of formality and concern for whether the detailed conditions are met for their existence. Our purpose here is to use them to motivate an alternative development of saddlepoint approximations and not to provide formal derivations. This new approach provides some additional understanding of saddlepoint densities and also some reasons for why they are able to achieve greater accuracy.

5.2.1 Tchebycheff–Hermite polynomials

If first and higher derivatives of $\phi(z)$, the standard normal density, are computed, the chain rule assures that $\phi(z)$ remains as a factor in each successive derivative. In addition, each derivative has a polynomial factor that depends on the order of the derivative. Apart from a sign term, these polynomial factors define the collection of Tchebycheff–Hermite polynomials. Formally,

$$\frac{d^k}{dz^k}\phi(z) = (-1)^k H_k(z)\phi(z),$$

where $\{H_k(z) : k = 0, 1, \ldots\}$ forms the sequence of Tchebycheff–Hermite polynomials. A partial list is compiled for later use and is easily derived by hand or with Maple as

$$\begin{aligned}
H_0(z) &= 1 & H_4(z) &= z^4 - 6z^2 + 3 \\
H_1(z) &= z & H_5(z) &= z^5 - 10z^3 + 15z \\
H_2(z) &= z^2 - 1 & H_6(z) &= z^6 - 15z^4 + 45z^2 - 15 \\
H_3(z) &= z^3 - 3z & H_7(z) &= z^7 - 21z^5 + 105z^3 - 105z.
\end{aligned} \quad (5.16)$$

A more complete discussion of these polynomials is found in Kendall and Stuart (1969, §6.14) or Barndorff-Nielsen and Cox (1989, §1.6). Two important properties are emphasized that are relevant to saddlepoint approximation. First,

$$\begin{aligned}
H_i(0) &= 0 \quad \text{if} \quad i \text{ is odd} \\
H_i(0) &\neq 0 \quad \text{if} \quad i \text{ is even.}
\end{aligned} \quad (5.17)$$

The first seven polynomials in (5.16) demonstrate this property. Secondly, the polynomials are orthogonal under normal weighting, or

$$\int_{-\infty}^{\infty} H_i(z) H_j(z) \phi(z) \, dz = \begin{cases} 0 & \text{if } i \neq j \\ i! & \text{if } i = j. \end{cases} \quad (5.18)$$

The integral in (5.18) has the interpretation as a weighted inner product of $H_i(z)$ and $H_j(z)$. The collection of functions $\{H_i(z)\phi(z)\}$ form a basis for the standard densities $f(z)$ with mean 0 and variance 1 when $f^2(z)$ is integrable.[3] Subject to certain strong conditions,[4]

$$f(z) = \sum_{i=0}^{\infty} c_i H_i(z) \phi(z) \quad (5.19)$$

---

[3] Not all densities are squared integrable. For example, a χ_1^2 density is not. Densities that are not squared integrable are necessarily unbounded.

[4] The density $f(z)$ must be continuous and satisfy Cramér's conditions which specify that (a) f is of bounded variation over any finite interval, and (b) the integral

$$\int_{-\infty}^{\infty} f(z) e^{z^2/4} dz$$

must converge. See Kendall and Stuart (1969, §6.22) or Barndorff-Nielsen and Cox (1989, §1.6). With $f^2(z)$ integrable, the expansion is L_2-convergent. The additional conditions guarantee that it is also pointwise convergent.

for some constants $\{c_i\}$ which are the coordinates of f with respect to the basis elements. Each coordinate, like a Fourier coefficient, is determined by taking the inner product (without a weight function) of basis component $H_i(z)$ and $f(z)$ as

$$c_i = \frac{1}{i!} \int_{-\infty}^{\infty} H_i(z) f(z) dz. \tag{5.20}$$

To see this is true, substitute the expression for $f(z)$ in (5.19) into (5.20). The orthogonality of $\{H_i(z)\}$ in (5.18) then reduces the right side to c_i.

5.2.2 Gram–Charlier series Type A

This series expansion, for standardized density f with mean 0 and variance 1, is given by (5.19) when the coefficients $\{c_i\}$ are expressed in terms of the moments of f. To re-express $\{c_i\}$, substitute the polynomial expressions for $\{H_i\}$ into (5.20). Coefficient c_i is therefore a linear function of the first i moments of f, where the ith moment is

$$\rho_i = \int_{-\infty}^{\infty} z^i f(z) \, dz$$

with $\rho_1 = 0$ and $\rho_2 = 1$. Substituting the linear expression of c_i in terms of $\{\rho_1, \ldots, \rho_i\}$ into (5.19) gives the Gram–Charlier series of type A for f of the form

$$f(z) = \phi(z) \left\{ 1 + \tfrac{1}{6}\rho_3 H_3(z) + \tfrac{1}{24}(\rho_4 - 3) H_4(z) + \cdots \right\}. \tag{5.21}$$

The Gram–Charlier series for a nonstandard density $f(x)$ with mean μ and variance σ^2 is determined from Jacobian transformation $x = \sigma z + \mu$ as

$$f(x) = \frac{1}{\sigma} \phi(z) \left\{ 1 + \tfrac{1}{6}\rho_3 H_3(z) + \tfrac{1}{24}(\rho_4 - 3) H_4(z) + \cdots \right\}, \tag{5.22}$$

where, for simplicity, we have retained $z = (x - \mu)/\sigma$ as the standardized value of x. The $\{\rho_i\}$ are now the standardized moments of $X \sim f$ defined as

$$\rho_i = E \left(\frac{X - \mu}{\sigma} \right)^i.$$

5.2.3 Edgeworth series

The Edgeworth series of f is determined by replacing the standardized moments of the Gram–Charlier series with the associated cumulants. Define the standardized cumulants of f as

$$\kappa_i = \frac{K^{(i)}(0)}{\sigma^i},$$

where $K^{(i)}(0)$ is the ith cumulant of $X \sim f$. The first i standardized cumulants of X are 1-1 with the first i standardized moments $\{\rho_1, \ldots, \rho_i\}$ for any i. Thus, coefficients $\{c_1, \ldots, c_i\}$ in (5.19) may now be rewritten in terms of $\{\kappa_1, \ldots, \kappa_i\}$; the first seven such coefficients are

$$
\begin{aligned}
&c_0 = 1 && c_5 = \kappa_5/120 \\
&c_1 = c_2 = 0 && c_6 = (\kappa_6 + 10\kappa_3^2)/720 \\
&c_3 = \kappa_3/6 && c_7 = (\kappa_7 + 35\kappa_4\kappa_3)/5040.
\end{aligned}
\qquad (5.23)
$$

The series expansion in (5.19) using these coefficients is referred to as the Edgeworth expansion of type A. Terms up to $i = 6$ suffice in determining the second-order correction terms of the saddlepoint density as we shall see below.

When the cumulants of X are known, the density f can be approximated with a truncated version of the Edgeworth expansion in (5.19) using only the first few coefficients from (5.23). When cumulants are unknown and must be estimated from data, estimates of $\{\kappa_i\}$ can be used to determine the coefficients in (5.23) so a density estimate is determined. In both instances the approximations can be negative and probabilities computed from them can also be negative or greater than 1. This cannot happen using (normalized) saddlepoint density approximations and is an important advantage of the latter procedures.

Suppose we want the Edgeworth expansion of $X = \bar{X}$, the sample mean of $X_1, \ldots, X_n$ that are i.i.d. with CGF K, mean μ, and variance σ^2. This expansion is now easily determined by using the $n = 1$ setting above but applied to the CGF of $\bar{X}$ given as

$$K_{\bar{X}}(s) = nK(s/n). \qquad (5.24)$$

The standardized cumulants of $\bar{X}$ are determined from (5.24) as

$$\frac{K_{\bar{X}}^{(i)}(0)}{\{K_{\bar{X}}^{(2)}(0)\}^{i/2}} = \frac{n^{1-i}K^{(i)}(0)}{\{n^{-1}K^{(2)}(0)\}^{i/2}} = n^{1-i/2}\kappa_i \qquad i \geq 3.$$

From this, it is apparent that the third and fourth standardized cumulants are $O(n^{-1/2})$ and $O(n^{-1})$ respectively with higher cumulants of decreasingly smaller order. The nonzero coefficients are

$$
\begin{aligned}
&c_0 = 1 && c_5 = \frac{\kappa_5}{120n^{3/2}} \\
&c_3 = \frac{\kappa_3}{6\sqrt{n}} && c_6 = \frac{\kappa_6}{720n^2} + \frac{\kappa_3^2}{72n} \\
&c_4 = \frac{\kappa_4}{24n} && c_7 = \frac{1}{5040}\left(\frac{\kappa_7}{n^{5/2}} + \frac{35\kappa_4\kappa_3}{n^{3/2}}\right).
\end{aligned}
\qquad (5.25)
$$

The expansion at value $\bar{x}$ in terms of its standardized value $z = \sqrt{n}\,(\bar{x} - \mu)/\sigma$ is determined from (5.19) and (5.25). The expansion that includes terms to order $O(n^{-2})$ is

$$f_{\bar{X}}(\bar{x}) = \frac{\sqrt{n}}{\sigma}\phi(z)\left[1 + \frac{\kappa_3}{6\sqrt{n}}H_3(z) + \frac{1}{n}\left\{\frac{\kappa_4}{24}H_4(z) + \frac{\kappa_3^2}{72}H_6(z)\right\}\right.$$
$$\left. + \frac{1}{n^{3/2}}\left\{\frac{\kappa_5}{120}H_5(z) + \frac{\kappa_4\kappa_3}{144}H_7(z)\right\} + O(n^{-2})\right]. \qquad (5.26)$$

This expression to order $O(n^{-3/2})$ has been the starting point in the derivation of saddlepoint densities from Edgeworth expansions in Barndorff-Nielsen and Cox (1979, Eq. 2.1) and Reid (1988, Eq. 5). A similar development is found in the next section.

Remarks on (5.26)

(1) The leading term of the expansion is the local limit theorem in which the density of $\bar{X}$ is equal to its asymptotic normal density to order $O(n^{-1/2})$.

(2) The Edgeworth expansion, truncated to include only the $O(n^{-1/2})$ term with $H_3(z)$, is usually less accurate in practical applications than the saddlepoint density in (2.25) of the same asymptotic order $O(n^{-1})$. When allowed to include the terms of order $O(n^{-1})$, so both κ_3 and κ_4 must be computed in (5.26), the Edgeworth expansion can be competitive in accuracy with the first-order saddlepoint density in the middle of the distribution. The value $\bar{x}$ is said to be close to the middle when its standardized value z is small in magnitude. Increasingly larger values of $|z|$, however, lead to deteriorating accuracy for the Edgeworth expansion. This may be understood by noting that the Hermite polynomials in the Edgeworth expansion are unbounded with increasing values of $|z|$. Relative error for the saddlepoint density, by contrast, is generally uniformly bounded to order $O(n^{-1})$ over all values of z including those z-values in the tails.

(3) At $\bar{x} = \mu$, the leading term of the Edgeworth expansion, which is the limiting normal density for $\bar{X}$, achieves the higher order asymptotic accuracy $O(n^{-1})$. With $\bar{x} = \mu$, then $z = 0$ and the odd-ordered Hermite polynomials are all zero. The Edgeworth expansion has $H_4(0) = 3$ and $H_6(0) = -15$ and is

$$f_{\bar{X}}(\mu) = \frac{1}{\sigma}\sqrt{\frac{n}{2\pi}} \left\{ 1 + \frac{1}{n}\left(\tfrac{1}{8}\kappa_4 - \tfrac{5}{24}\kappa_3^2\right) + O(n^{-2}) \right\}. \tag{5.27}$$

Elimination of the odd-ordered Hermite polynomials leads to an expansion at μ that is in powers of n^{-1} and not the usual powers of $n^{-1/2}$. This higher order asymptotic accuracy suggests that the expansion may achieve greater numerical accuracy at the mean.

(4) When $\bar{x} \neq \mu$ the expansion is in powers of $n^{-1/2}$ except when the density is symmetric. Then all odd cumulants are zero and the expansion is again in powers of n^{-1} as may be seen from (5.26).

Examples

Gumbel $(0, 1)$ Take $n = 1$ and $X \sim$ Gumbel $(0, 1)$ with CGF $K(s) = \ln \Gamma(1-s)$ as introduced in section 1.1.3. Figure 5.1 plots the Gumbel density (solid) along with the Edgeworth approximations in (5.26). Two approximations are plotted. The first includes $O(n^{-1/2})$ terms (dotted) and the second has the additional $O(n^{-1})$ terms (dashed).

Both are quite inaccurate in comparison to the saddlepoint density in figure 1.3 which shows very little graphical difference.

Uniform $(-1, 1)$ Let $X_1, \ldots, X_n$ be i.i.d. Uniform $(-1, 1)$. The density for $\bar{X}$ is given by Daniels (1954) and Feller (1971, I.9, Eq. 9.7) as

$$f(\bar{x}) = \frac{n^n}{2^n(n-1)!} \sum_{i=0}^{n} (-1)^i \binom{n}{i} \left\langle 1 - \bar{x} - \frac{2i}{n} \right\rangle^{n-1} \qquad |\bar{x}| \leq 1 \tag{5.28}$$

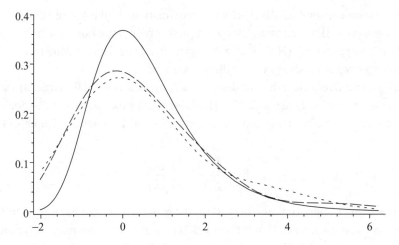

Figure 5.1. The Gumbel density (*solid*), and Edgeworth approximations to order $O(n^{-1})$ (*dots*), and $O(n^{-3/2})$ (*dashed*).

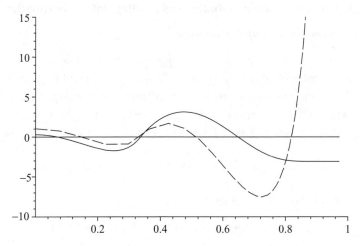

Figure 5.2. Percentage relative errors in approximating the density for the average of 3 Uniform $(-1, 1)$ variables with the normalized saddlepoint approximation (*solid*) and its Edgeworth expansion (*dashed*) to order $O(n^{-3/2})$.

where $\langle x \rangle = x$ for $x \geq 0$ and $= 0$ for $x < 0$. The density is symmetric so odd cumulants κ_3 and κ_5 are 0. The standardized z-value here is $z = \sqrt{3n}\bar{x}$ and $\kappa_4 = -6/5$. The Edgeworth expansion is

$$f(\bar{x}) = \sqrt{3n}\phi(z)\left\{1 - \frac{1}{20n}H_4(z) + O(n^{-2})\right\}. \tag{5.29}$$

With the density symmetric, the plots need only show the right portion over $(0, 1)$. Using $n = 3$, figure 5.2 shows the relative error for the Edgeworth approximation to order $O(n^{-2})$ given in (5.29) (dotted) and the normalized saddlepoint density (solid).

The dashed line continues upward unbounded as it approaches boundary value 1. The normalized approximation has been used because the relative error of the unnormalized density ranges from 3% to 9% over the range $(0, 1)$. Numerical integration determines the normalization constant as 1.061. Both approximations in the figure show comparable accuracy when $|\bar{x}| < 0.6$ even though the asymptotic order of the Edgeworth approximation is $O(n^{-2})$ as compares with $O(n^{-3/2})$ for the normalized saddlepoint density. For $|\bar{x}| > 0.6$ the Edgeworth expansion shows diminishing accuracy.

It should be intuitively clear that the density for $\bar{X}$ approaches 0 at the edges of its support; this may be shown formally using (5.28). The Edgeworth expansion in (5.29) does not tend to 0 and so its relative error must approach ∞ as $|\bar{x}| \to 1$. By contrast, Daniels (1954, §5) shows

$$\lim_{|\bar{x}| \to 1} \frac{\hat{f}(\bar{x})}{f(\bar{x})} = \frac{\Gamma(n)}{\hat{\Gamma}(n)}$$

where $\hat{\Gamma}$ is Stirling's approximation in (1.8) for the gamma function. For $n = 3$, this leads to a relative error approaching 2.81% for $\hat{f}(\bar{x})$ as $|\bar{x}| \to 1$. For the normalized version $\bar{f}(\bar{x})$, this error approaches -3.10% as confirmed numerically in figure 5.2.

5.3 Tilted exponential families and saddlepoint approximations

5.3.1 Tilting and saddlepoint approximation

Exponential tilting was introduced for density approximation by Esscher (1932) and further developed as a general probabilistic method by Daniels (1954) and Barndorff-Nielsen and Cox (1979). The method provides a means for creating a regular exponential family from a distribution whose CGF converges in the regular sense. Let X_0 be a random variable for such a density (mass) function f with CGF $K(s)$ that converges on (a, b). Then

$$e^{K(s)} = \int_{-\infty}^{\infty} e^{sx} f(x) dx \qquad s \in (a, b). \tag{5.30}$$

Rearrangement of (5.30) shows that the function

$$f(x; s) = \exp\{sx - K(s)\} f(x) \qquad s \in (a, b) \tag{5.31}$$

is a density for each $s \in (a, b)$. This collection of densities defines a *tilted regular exponential family* indexed by s. Density $f(x; s)$ is the s-*tilted density* and X_s is used to denote a random variable with this density. The mean and variance of canonical sufficient X_s are

$$E(X_s) = K'(s) \tag{5.32}$$
$$\text{Var}(X_s) = K''(s).$$

Note that $\text{Var}(X_s) = K''(s) > 0$ for all $s \in (a, b)$ and this fact is consistent with the convexity of K. The ith standardized cumulant of X_s is

$$\kappa_i(s) = \frac{K^{(i)}(s)}{\{K''(s)\}^{i/2}} \qquad i \geq 3. \tag{5.33}$$

A common setting in which saddlepoint methods are especially useful occurs when a distribution's CGF $K(s)$ is known but its associated density $f(x)$ is either unknown or intractable. Direct Edgeworth approximation for $f(x)$ has been discussed in the previous section as a method for approximation. The Esscher (1932) tilting method is an indirect Edgeworth expansion that consists of two steps: (i) First $f(x)$ is written in terms of $f(x; s)$

using (5.31); then (ii) $f(x;s)$ is Edgeworth expanded for a judicious choice of $s \in (a, b)$. We say indirect because the Edgeworth expansion is not for $f(x)$ directly when $s = 0$, but for this specially chosen s-tilted member of the exponential family. Step (i) is to specify that

$$f(x) = \exp\{K(s) - sx\} f(x;s) \qquad \forall s \in (a, b). \tag{5.34}$$

Step (ii) entails a choice for s such that the Edgeworth approximation for $f(x;s)$ is as accurate as possible. The third remark of the previous section suggests that the value s should be chosen so that the Edgeworth expansion of $f(x;s)$ is at its mean. This requires that the fixed argument value x be the mean of X_s with density $f(\cdot;s)$. Formally, this amounts to choosing $s = \hat{s}$ to solve

$$E(X_{\hat{s}}) = K'(\hat{s}) = x$$

or the saddlepoint equation. The Edgeworth expansion at the mean given in (5.27), with $n = 1$, $z = 0$, and cumulants determined from (5.33), is

$$f(x;\hat{s}) \simeq \frac{1}{\sqrt{2\pi K''(\hat{s})}} \left[1 + \left\{\tfrac{1}{8}\kappa_4(\hat{s}) - \tfrac{5}{24}\kappa_3^2(\hat{s})\right\}\right]. \tag{5.35}$$

The notation $\hat{\kappa}_i = \kappa_i(\hat{s})$ is used and (5.35) is substituted into (5.34) to give

$$f(x) = \exp\{\hat{s}x - K(\hat{s})\} f(x;\hat{s})$$
$$\simeq \frac{1}{\sqrt{2\pi K''(\hat{s})}} \exp\{K(\hat{s}) - \hat{s}x\} \left\{1 + \left(\tfrac{1}{8}\hat{\kappa}_4 - \tfrac{5}{24}\hat{\kappa}_3^2\right)\right\},$$

which is the second-order saddlepoint density.

The asymptotic development arises by taking $X_0 = \bar{X}$, the sample mean of n i.i.d. variables with CGF K convergent on (a, b). The density of $\bar{X}$ at $\bar{x}$ or $f(\bar{x})$ is tilted to create the regular exponential family

$$f(\bar{x};s) = \exp\{ns\bar{x} - nK(s)\} f(\bar{x}) \qquad s \in (a, b),$$

where $\bar{X}_s \sim f(\bar{x};s)$. The mean of $\bar{X}_s$ is $K'(s)$ and the optimal tilt is the saddlepoint tilt $\hat{s}$ that solves $K'(\hat{s}) = \bar{x}$. The ith standardized cumulant for $\bar{X}_s$ is $n^{1-i/2}\kappa_i(s)$ where $\kappa_i(s)$ is given in (5.33). The same argument is applied to give the saddlepoint expansion

$$f(\bar{x}) = \hat{f}(\bar{x}) \left\{1 + \frac{1}{n}\left(\tfrac{1}{8}\hat{\kappa}_4 - \tfrac{5}{24}\hat{\kappa}_3^2\right) + O(n^{-2})\right\}, \tag{5.36}$$

where $\hat{f}(\bar{x})$ is the saddlepoint density in (2.25).

5.3.2 Comments on tilting and saddlepoint accuracy

(1) The saddlepoint density at x is the indirect Edgeworth expansion at the $\hat{s}$-tilted density chosen so that the mean of this tilted density is the value x. This $\hat{s}$-tilt is "optimal" in the sense that the Edgeworth expansion used has $z = 0$ so it can achieve its greatest accuracy. In the sample mean setting, the tilted expansion in (5.36) is in powers of n^{-1} rather than the powers of $n^{-1/2}$ which are characteristic of the Edgeworth expansion when $z \neq 0$.

(2) As x varies over $\mathcal{I}_\mathcal{X}$, the tilting parameter varies over (a, b); thus, the set of saddlepoint density values $\{\hat{f}(x) : x \in \mathcal{I}_\mathcal{X}\}$ depends on the entire parametric collection of Edgeworth

expansions $\{f_{\text{Edge}}(\cdot\,;s) : s \in (a,b)\}$, where $f_{\text{Edge}}(\cdot\,;s)$ is the Edgeworth expansion for the density of X_s.

(3) Multivariate Edgeworth expansions along with tilting arguments can be used to derive the multivariate saddlepoint density in (3.2), and the asymptotic expansion for the density of a multivariate mean in (3.36) and (3.37). Expansions for conditional densities can also be approached from this point of view. Barndorff-Nielsen and Cox (1989) and McCullagh (1987) provide extensive accounts.

(4) Distributional approximations result when the cumulative integral of the density in (5.34) is expressed in terms of the cumulative integral of the tilted density on its right hand side. With $\hat{s}$-tilting, Robinson (1982) and Daniels (1987) have derived a continuous CDF approximation that differs from the Lugannani and Rice approximation and is a little less accurate in practice. The approximation for the CDF of $\bar{X}$ is

$$1 - \hat{F}(\bar{x}) = \exp\left(\frac{\hat{u}_n^2 - \hat{w}_n^2}{2}\right)\left[\{1 - \Phi(\hat{u}_n)\}\left\{1 - \frac{\hat{\kappa}_3 \hat{u}_n^3}{6\sqrt{n}} + O_1(n^{-1})\right\} \right.$$
$$\left. + \phi(\hat{u}_n)\left\{\frac{\hat{\kappa}_3}{6\sqrt{n}}(\hat{u}_n^2 - 1) + O_2(n^{-1})\right\}\right], \tag{5.37}$$

where $\hat{u}_n$, and $\hat{w}_n$ are given in (2.47). The approximation has a different form than the Lugannani and Rice approximation and is slightly more difficult to compute since it requires the standardized cumulant $\hat{\kappa}_3$. Daniels (1987) derives the $O(n^{-1})$ terms in this expression using a different approach based on inversion theory. He shows that

$$O_1(n^{-1}) = \frac{1}{n}\left(\frac{\hat{\kappa}_4 \hat{u}_n^4}{24} + \frac{\hat{\kappa}_3^2 \hat{u}_n^6}{72}\right) \tag{5.38}$$
$$O_2(n^{-1}) = -\frac{1}{n}\left\{\frac{\hat{\kappa}_4}{24}(\hat{u}_n^3 - \hat{u}_n) + \frac{\hat{\kappa}_3^2}{72}(\hat{u}_n^5 - \hat{u}_n^3 + 3\hat{u}_n)\right\}$$

and that the accuracy of (5.37) improves to $O(n^{-3/2})$ when these are included. Practical applications demonstrate that these additional correction terms improve the approximation's accuracy so it is comparable with that of the Lugannani and Rice approximation. The disadvantage of this approximation over Lugannani and Rice is that it requires computation of the third and fourth cumulants. However, once such cumulants have been computed, the next term of the Lugannani and Rice expansion in (2.48) could also have be computed. Thus the fair comparison using four cumulants should be between the resulting second-order Lugannani and Rice approximation and (5.37) coupled with (5.38).

5.4 Saddlepoint approximation in regular exponential families

5.4.1 Saddlepoint densities for canonical sufficient statistics

Suppose that X is the m-dimensional sufficient statistic for the regular exponential family in (5.1). Take θ as the true value of the canonical parameter so the CGF of X is $c(s+\theta) - c(\theta)$. According to the Fisher factorization theorem, the density of X is proportional to the likelihood function; however, the integration needed to extract the correct likelihood factor is difficult to perform and requires integration in z-space on the m-dimensional manifold defined by holding $x(z)$ fixed. Saddlepoint approximations offer an attractive simple alternative. The saddlepoint equation in approximating $f(x;\theta)$ is

$$c'(\hat{s}+\theta)=x,$$

whose unique solution is the MLE $\hat{\theta}=\hat{s}+\theta\in\Theta$, so that

$$\hat{s}=\hat{\theta}-\theta. \tag{5.39}$$

The resulting saddlepoint density is

$$\hat{f}(x;\theta)=(2\pi)^{-m/2}|c''(\hat{\theta})|^{-1/2}\exp\{c(\hat{\theta})-c(\theta)-(\hat{\theta}-\theta)^T x\}$$
$$=(2\pi)^{-m/2}|j(\hat{\theta})|^{-1/2}\frac{\mathcal{L}(\theta)}{\mathcal{L}(\hat{\theta})} \tag{5.40}$$

when the exponential has been expressed in terms of the likelihood ratio $\mathcal{L}(\theta)/\mathcal{L}(\hat{\theta})$ and $c''(\hat{\theta})$ is the Fisher information and therefore denoted as $j(\hat{\theta})$.

The same expression applies in the discrete setting with $x\in I^m$, the integer lattice in $\Re^m$. Determination of the mass function for such a discrete sufficient statistic would require summing over $\{z:x(z)=x\in I^m\}$.

Expression (5.40) shows that the saddlepoint density of X depends only on the likelihood function and various aspects of it. The complicated integration needed to determine the exact density requires integration of $\exp\{-d(z)\}$ over the manifold $\{z:x(z)=x\}$ but $d(z)$ never enters directly into the saddlepoint density, only indirectly through its interrelationship with function $c(\cdot)$.

Tilting provides another perspective on this saddlepoint density. By the factorization theorem, the ratio of densities for the canonical sufficient statistic agrees with the likelihood ratio or

$$\frac{f(x;\theta)}{f(x;\theta+s)}=\frac{\mathcal{L}(\theta)}{\mathcal{L}(\theta+s)},$$

since the $\exp\{-d(z)\}$ term cancels in the ratio. Then

$$f(x;\theta)=\frac{\mathcal{L}(\theta)}{\mathcal{L}(\theta+s)}f(x;\theta+s)\qquad\forall s\in\mathcal{S}_\theta$$
$$=\frac{\mathcal{L}(\theta)}{\mathcal{L}(\theta+\hat{s})}f(x;\theta+\hat{s}), \tag{5.41}$$

where $\hat{s}$ is the optimal tilt determined by making x the mean of density $f(\cdot;\theta+\hat{s})$, e.g. $\hat{s}$ is chosen to solve

$$c'(\theta+\hat{s})=x.$$

The tilt in this setting is from θ to $\theta+\hat{s}=\hat{\theta}$ whereas, when $\theta=0$ in the previous subsection, the tilt was from 0 to $\hat{s}$. Edgeworth expansion at the mean in the univariate setting with $m=1$ gives

$$f(x;\theta+\hat{s})=\{2\pi j(\hat{\theta})\}^{-1/2}\left\{1+\left(\tfrac{1}{8}\hat{\kappa}_4-\tfrac{5}{24}\hat{\kappa}_3^2\right)+\cdots\right\}, \tag{5.42}$$

where

$$\hat{\kappa}_i=\frac{c^{(i)}(\hat{\theta})}{\{c''(\hat{\theta})\}^{i/2}}=\frac{-l^{(i)}(\hat{\theta})}{\{-l''(\hat{\theta})\}^{i/2}}\qquad i\geq 3.$$

The use of l emphasizes the dependence on the likelihood function alone. With $m > 1$, a multivariate Edgeworth expansion, as in Barndorff-Nielsen and Cox (1979), leads to

$$f(x; \theta + \hat{s}) = (2\pi)^{-m/2} |j(\hat{\theta})|^{-1/2} \left[1 + \left\{\tfrac{1}{8}\hat{\kappa}_4 - \tfrac{1}{24}\left(2\hat{\kappa}_{23}^2 + 3\hat{\kappa}_{13}^2\right)\right\} + \cdots \right], \qquad (5.43)$$

where terms like $\hat{\kappa}_4$ are computed as in (3.38) with $c(\cdot)$ in place of $K(\cdot)$ and with evaluations at the argument value $\hat{\theta}$. Combining the tilt in (5.41) with Edgeworth expansion of the saddlepoint tilted density in (5.43) gives the tilted multivariate saddlepoint density to second order as

$$f(x; \theta) = \hat{f}(x; \theta)\left[1 + \left\{\tfrac{1}{8}\hat{\kappa}_4 - \tfrac{1}{24}\left(2\hat{\kappa}_{23}^2 + 3\hat{\kappa}_{13}^2\right)\right\} + \cdots \right], \qquad (5.44)$$

where the leading factor is in (5.40). This expansion depends only on the likelihood and its derivatives at $\hat{\theta}$ or $\{l(\hat{\theta}), l'(\hat{\theta}), l''(\hat{\theta}), \ldots\}$. Since $l(\cdot)$ is an analytic function over Θ, its collection of derivatives at $\hat{\theta}$ determine the Taylor coefficients of $l(\theta)$ and hence the entire likelihood function over $\theta \in \Theta$; therefore the expansion depends only on the likelihood function. This should not be surprising however, because the density of X is completely determined from the likelihood function by the factorization theorem. A surprise would occur if something else besides the likelihood function was required in the saddlepoint expansion to determine the true density of X.

Examples

Poisson (λ) Suppose $Z_1, \ldots, Z_n$ is an i.i.d. sample with this mass function. The canonical sufficient statistic is $X = \sum_i Z_i$ and $\theta = \ln \lambda$ is the canonical parameter. The saddlepoint expansion of the mass function of X is given in (5.44) as

$$p(k) = \frac{1}{k!}(n\lambda)^k e^{-n\lambda} \left\{1 - \frac{1}{12k} + O(k^{-2})\right\} \qquad k \geq 1. \qquad (5.45)$$

This also agrees with the known saddlepoint expansion of $X \sim$ Poisson $(n\lambda)$ whose saddlepoint mass function is given in (1.17) and which to second-order may be deduced from (2.32).

Other previous examples In the Poisson (λ) product family above, the distribution for the canonical sufficient statistic X is a regular saddlepoint setting for a previously considered example. The saddlepoint density of X has been previously computed from its CGF but the saddlepoint density may now be determined from the likelihood function of its product exponential family structure as expressed in (5.44). The same applies to some other i.i.d. product family examples previously considered. Suppose $Z_1, \ldots, Z_n$ is an i.i.d. sample from any of the following: Binomial $(1, \xi)$, Normal $(\mu, 1)$, Gamma (α, β) for fixed α, Negative Binomial $(1, \xi)$, Normal$_m(\mu, I_m)$, or Multinomial $(1; \pi_1, \ldots, \pi_{m+1})$. Then the saddlepoint mass/density function of canonical sufficient $X = \sum_i Z_i$ in the product family, as determined from (5.44), agrees with the previously derived saddlepoint densities for these distributions. Consideration of these examples is left as an exercise.

Normal (μ, σ^2) If $Z_1, \ldots, Z_n$ is an i.i.d. sample from this density, then the product family has canonical sufficient statistics $X = \sum_i Z_i^2$ and $Y = \sum_i Z_i$ and canonical parameter

5.4 Saddlepoint approximation in regular exponential families

$\theta^T = (\chi, \psi)$ with $\chi = -1/(2\sigma^2)$ and $\psi = \mu/\sigma^2$. The joint density of (X, Y) is

$$f(x, y; \chi, \psi) = \frac{(-\chi)^{n/2}}{\sqrt{n\pi}\,\Gamma\left(\frac{n-1}{2}\right)}(x - y^2/n)^{\frac{1}{2}(n-3)} \exp\left(\chi x + \psi y + \frac{n\psi^2}{4\chi}\right) \quad (5.46)$$

with support constraints $x > 0$ and $x - y^2/n > 0$.

The likelihood of the exponential family is

$$\mathcal{L}(\chi, \psi) = \exp\left[\chi x + \psi y - n\left\{-\frac{\psi^2}{4\chi} - \frac{1}{2}\ln(-\chi)\right\}\right]. \quad (5.47)$$

From this, simple computations show $|j(\chi, \psi)| = n^2/\{4(-\chi)^3\}$. The invariance property of MLEs determines estimates for the canonical parameters as $\hat\chi = -1/(2\hat\sigma^2)$ and $\hat\psi = \hat\mu/\hat\sigma^2$, where $\hat\mu = y/n$ and $\hat\sigma^2 = x/n - (y/n)^2$ are the usual MLEs of mean and variance. The saddlepoint density in (5.40) works out to be

$$\hat f(x, y; \chi, \psi) = f(x, y; \chi, \psi)\,\frac{\Gamma\left(\frac{n-1}{2}\right)\sqrt{\frac{n}{2}}}{\hat\Gamma\left(\frac{n}{2}\right)}, \quad (5.48)$$

as the reader is asked to show in Exercise 8.

Gamma (α, β) Suppose $Z_1, \ldots, Z_n$ is an i.i.d. sample from this density. The canonical sufficient statistics are $X = \sum_i \ln Z_i$ and $Y = \sum_i Z_i$ with corresponding canonical parameters α and $-\beta$. The joint density of (X, Y) is rather intractable since its determination requires integration over an $(n - 2)$-dimensional manifold in $(0, \infty)^n$. The likelihood is

$$\mathcal{L}(\alpha, \beta) = \exp[\alpha x - \beta y - n\{\ln \Gamma(\alpha) - \alpha \ln \beta\}].$$

The MLEs for α and β are not explicit and solve

$$\Psi(\hat\alpha) - \ln \hat\beta = x/n \quad (5.49)$$
$$\hat\alpha/\hat\beta = y/n \quad (5.50)$$

where $\Psi = (\ln \Gamma)'$. The joint saddlepoint density from (5.40) is

$$\hat f(x, y; \alpha, \beta) = (2\pi)^{-1}\frac{\hat\beta}{n\sqrt{\Psi'(\hat\alpha)\hat\alpha - 1}}\frac{\mathcal{L}(\alpha, \beta)}{\mathcal{L}(\hat\alpha, \hat\beta)}, \quad (5.51)$$

where $\hat\alpha$ and $\hat\beta$ are implicit functions of (x, y) through the MLE equations in (5.49) and (5.50). The joint support is

$$\mathcal{S} = \{(x, y) : y > 0 \quad \text{and} \quad x/n \le \ln(y/n)\}. \quad (5.52)$$

The second inequality is the order relationship of the geometric mean to the arithmetic mean for the data.

5.4.2 Conditional saddlepoint densities for canonical sufficient statistics

Suppose the canonical sufficient statistic (X, Y) that is conformable with canonical parameters $(\chi, \psi) \in \Theta$ as in (5.11) with dim $X = m_x$ and dim $Y = m_y$. The conditional density of Y given $X = x$ is also a regular exponential family indexed by canonical parameter ψ. Its double-saddlepoint density is based on the joint CGF of (X, Y) given as

$$K(s, t) = E\{\exp(s^T X + t^T Y)\}$$
$$= c(s + \chi, t + \psi) - c(\chi, \psi),$$

and the marginal CGF of X which is $K(s, 0)$. The double-saddlepoint density in (4.7) has a numerator saddlepoint for $\hat{f}(x, y)$ that solves

$$c'(\hat{\chi}, \hat{\psi}) = c'(\hat{s} + \chi, \hat{t} + \psi) = (x, y)$$

for $(\hat{\chi}, \hat{\psi}) \in \Theta$, where $c'(\chi, \psi) = \partial c / \partial(\chi, \psi)$. The MLE $(\hat{\chi}, \hat{\psi})$ is related to the saddlepoint according to

$$(\hat{s}, \hat{t}) = (\hat{\chi} - \chi, \hat{\psi} - \psi). \tag{5.53}$$

The denominator saddlepoint $\hat{s}_0$ that determines $\hat{f}(x)$ solves

$$c'_\chi(\hat{\chi}_\psi, \psi) = c'_\chi(\hat{s}_0 + \chi, \psi) = x$$

where the left expression is free to vary over $\{\hat{\chi}_\psi : (\hat{\chi}_\psi, \psi) \in \Theta\}$ and c'_χ denotes $\partial c(\chi, \psi) / \partial \chi$. The saddlepoint is therefore related to $\hat{\chi}_\psi$, the constrained MLE of χ holding ψ fixed, by

$$\hat{s}_0 = \hat{\chi}_\psi - \chi. \tag{5.54}$$

This determination of saddlepoints in (5.53) and (5.54) allows the double-saddlepoint density to be expressed entirely in terms of likelihood quantities as

$$\hat{f}(y|x; \psi) = (2\pi)^{-m_y/2} \left\{ \frac{|j(\hat{\chi}, \hat{\psi})|}{|j_{xx}(\hat{\chi}_\psi, \psi)|} \right\}^{-1/2} \frac{\mathcal{L}(\hat{\chi}_\psi, \psi)}{\mathcal{L}(\hat{\chi}, \hat{\psi})}, \tag{5.55}$$

where the observed Fisher information matrix j has been partitioned as

$$j(\chi, \psi) = \begin{pmatrix} j_{\chi\chi} & j_{\chi\psi} \\ j_{\psi\chi} & j_{\psi\psi} \end{pmatrix}.$$

Note that the approximation in (5.55) does not depend on parameter χ but only ψ, as occurs with the true conditional density.

Examples

Normal $(0, \sigma^2)$ *continuation* The joint density of the canonical sufficient statistics in (5.46) admits a conditional density for Y given $X = x$ that is difficult to normalize in general. In the special case when $\mu = 0$ so that $\psi = 0$, this integration is explicit and the conditional density is

$$f(y|x; \psi = 0) = \frac{\Gamma\left(\frac{n}{2}\right)}{\sqrt{n\pi x}\,\Gamma\left(\frac{n-1}{2}\right)} \left(1 - \frac{y^2}{nx}\right)^{(n-3)/2} \qquad |y| < \sqrt{nx}. \tag{5.56}$$

5.4 Saddlepoint approximation in regular exponential families

Jacobian transformation of (5.56) shows that $W = Y/\sqrt{X}$ is independent of X and hence provides its marginal distribution. The additional transformation

$$T = \sqrt{1 - n^{-1}} \frac{W}{\sqrt{1 - W^2/n}}, \tag{5.57}$$

as used in Exercise 10 of chapter 4, leads to the ordinary T-statistic having a central t_{n-1}-density.

For $\psi = 0$, the double-saddlepoint approximation in (5.55) uses the MLEs $\hat{\chi}$ and $\hat{\psi}$ given above as well as the MLE of χ determined by constraining $\psi = 0$ or $\hat{\chi}_0 = -1/(2\hat{\sigma}_0^2) = -n/(2x)$. It is left as an exercise to show that

$$\hat{f}(y|x; \psi = 0) = \frac{1}{\sqrt{2\pi x}} \left(1 - \frac{y^2}{nx}\right)^{(n-3)/2} \tag{5.58}$$

$$= f(y|x; \psi = 0) \frac{\Gamma\left(\frac{n-1}{2}\right) \sqrt{\frac{n}{2}}}{\Gamma\left(\frac{n}{2}\right)}. \tag{5.59}$$

A simpler approach in deriving (5.59) is to first use the fact that $X/\sigma^2 \sim \chi_n^2$ when $\psi = 0$; thus the marginal density approximation to X differs by Stirling's approximation in the factor $\Gamma(n/2)$ as seen in (1.9). Then, by using (5.48),

$$\hat{f}(y|x; \psi = 0) = \frac{\hat{f}(x, y; \chi, \psi = 0)}{\hat{f}(x; \psi = 0)}$$

$$= \frac{f(x, y; \chi, \psi = 0) \Gamma\left(\frac{n-1}{2}\right) \sqrt{\frac{n}{2}} / \hat{\Gamma}\left(\frac{n}{2}\right)}{f(x; \psi = 0) \Gamma\left(\frac{n}{2}\right) / \hat{\Gamma}\left(\frac{n}{2}\right)},$$

which agrees with (5.59).

Jacobian transformation of the conditional saddlepoint density in (5.58) from $Y|x$ to $T|x$, according to (5.57), leads to a saddlepoint approximation to the T_{n-1} density $\hat{f}_{n-1}$ which is exact upon normalization. The relationship with the true density f_{n-1} is

$$\hat{f}_{n-1}(t) = f_{n-1}(t) \frac{\Gamma\left(\frac{n-1}{2}\right) \sqrt{\frac{n}{2}}}{\Gamma\left(\frac{n}{2}\right)},$$

which follows directly from the relationship in (5.59).

Normal (μ, σ^2) *continuation* For $\mu \neq 0$ the saddlepoint development for Y given x of the previous example is easily extended, however the exact distribution theory requires numerical normalization. Fixing $\psi \neq 0$, then the following expressions can be easily derived and are used in subsequent examples:

$$\hat{\chi}_\psi = -\frac{n}{4x}\left(1 + \sqrt{1 + 4x\psi^2/n}\right) = -\frac{n}{4x}(1 + R_\psi),$$

where R_ψ is the so-defined square root,

$$\frac{\mathcal{L}(\hat{\chi}_\psi, \psi)}{\mathcal{L}(\hat{\chi}, \hat{\psi})} = \left(1 - \frac{y^2}{nx}\right)^{n/2} \left(\frac{1 + R_\psi}{2}\right)^{n/2} \exp\left\{\psi y + \frac{n}{2}(1 - R_\psi)\right\} \tag{5.60}$$

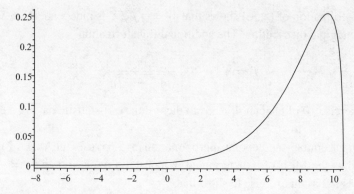

Figure 5.3. Plot of the conditional density and normalized saddlepoint density for Y given x with $\psi = 1/2$.

and

$$\left\{ \frac{|j(\hat{\chi}, \hat{\psi})|}{|j_{\chi\chi}(\hat{\chi}_\psi, \psi)|} \right\}^{-1/2} = \frac{1}{\sqrt{x}} \left(1 - \frac{y^2}{nx}\right)^{-3/2} \frac{2}{1+R_\psi} \sqrt{1 + \frac{4x\psi^2}{n(1+R_\psi)}}. \quad (5.61)$$

Consider a sample of size $n = 4$ from a Normal $(2, 4)$ distribution so that we may take $\psi = 1/2$, $\chi = -1/8$. Condition on the value $x = 28$. The true conditional density and the normalized saddlepoint approximation in (5.55) are overlaid in figure 5.3 and exhibit no graphical difference. Indeed the relative error is typically $O(10^{-5})\%$ over the entire range $|y| < \sqrt{nx} - 10^{-3} = 10.582$ and $O(10^{-3})\%$ at $y = 10.583$, which is 5×10^{-5} below the support boundary.

Gamma (α, β) continuation The conditional saddlepoint density of Y given $X = x$ requires that we solve for the constrained MLE of α for fixed β. This is the root $\hat{\alpha}_\beta$ for which

$$\Psi(\hat{\alpha}_\beta) = \ln \beta + \frac{x}{n}.$$

Note that $\hat{\alpha}_\beta$ does not vary with y. The conditional density from (5.55) is

$$\hat{f}(y|x; \beta) = (2\pi)^{-1/2} \left[\frac{n\{\Psi'(\hat{\alpha})\hat{\alpha} - 1\}/\hat{\beta}^2}{\Psi'(\hat{\alpha}_\beta)} \right]^{-1/2}$$
$$\times \exp\left[(\hat{\alpha}_\beta - \hat{\alpha})x - (\beta - \hat{\beta})y - n\left\{\ln \frac{\Gamma(\hat{\alpha}_\beta)}{\Gamma(\hat{\alpha})} + \hat{\alpha} \ln \hat{\beta} - \hat{\alpha}_\beta \ln \beta \right\} \right],$$

where $\hat{\alpha}$ and $\hat{\beta}$ are implicit functions of (x, y). The exponent simplifies somewhat using the saddlepoint equation identities $\hat{\beta}y = n\hat{\alpha}$ and

$$-\hat{\alpha}x - n\hat{\alpha}\ln\hat{\beta} = -n\hat{\alpha}\Psi(\hat{\alpha})$$

so that

$$\hat{f}(y|x; \beta) = (2\pi n)^{-1/2} \left\{ \frac{\Psi'(\hat{\alpha})\hat{\alpha} - 1}{\Psi'(\hat{\alpha}_\beta)} \right\}^{-1/2} \times \frac{n\hat{\alpha}}{y} \quad (5.62)$$
$$\times \exp\left[-\beta y + n\hat{\alpha}\{1 - \Psi(\hat{\alpha})\} - n \ln \frac{\Gamma(\hat{\alpha}_\beta)}{\Gamma(\hat{\alpha})} + \hat{\alpha}_\beta (x + n \ln \beta) \right].$$

5.4 Saddlepoint approximation in regular exponential families

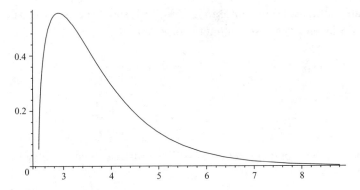

Figure 5.4. The normalized saddlepoint density $\bar{f}(y|x;\beta = 1)$ vs. y for the Gamma $(\alpha, \beta = 1)$ example.

Only two quantities change with the value of y in (5.62): y and $\hat{\alpha}$. By allowing $\hat{\alpha}$ to range over $(0, \infty)$, values of y can be traced out through the maximum likelihood relationship

$$y = y(\hat{\alpha}) = n\hat{\alpha}/\hat{\beta} = n\hat{\alpha}\exp\{x/n - \Psi(\hat{\alpha})\}. \tag{5.63}$$

The plot of this normalized conditional density is the parametric curve

$$\{(y, \bar{f}(y|x;\beta))\} : y = y(\hat{\alpha}), \quad \hat{\alpha} \in (0, \infty)\}$$

plotted in figure 5.4. The figure is based on $n = 4$ data points simulated from a Gamma $(1, 1)$ which yielded $z_1 = 1.01$, $z_2 = 0.132$, $z_3 = 2.79$, and $z_4 = 0.391$. For this data, the value of X conditioned upon is $x = -1.928$. The support constraint on $Y|x$, determined from (5.52), requires that $y > 2.470$. The normalization constant is the rather large value 1.313 and has been computed using the change of variable $dy = y'(\hat{\alpha})d\hat{\alpha}$ from (5.63). The true conditional density is not shown since, by comparison, it is very difficult to compute.

5.4.3 Saddlepoint CDF for a canonical sufficient statistic

Suppose the canonical sufficient statistic X has dimension $m = 1$. In this instance, the Lugannani and Rice approximation is explicit in terms of the MLE for canonical parameter θ. This leads to an explicit CDF approximation for the distribution of $\hat{\theta}(X)$ since X and $\hat{\theta}(X)$ have a monotonic increasing relationship through the MLE equation $c'(\hat{\theta}) = x$.

Let $\hat{\theta} = \hat{\theta}(x)$ denote the MLE for cutoff x. The approximation for

$$\Pr(X \leq x; \theta) = \Pr\{\hat{\theta}(X) \leq \hat{\theta}\}$$

is given as in (1.21) by taking

$$\hat{w} = \text{sgn}(\hat{\theta} - \theta)\sqrt{-2\ln\frac{\mathcal{L}(\theta)}{\mathcal{L}(\hat{\theta})}} \tag{5.64}$$

$$\hat{u} = (\hat{\theta} - \theta)\sqrt{j(\hat{\theta})}$$

when X has a continuous density.

Appropriate continuity correction as in (2.58)–(2.61) should be used when the distribution of X lies on a δ-lattice. The first continuity correction is

$$\widehat{\Pr}_1(X \geq x) = 1 - \Phi(\hat{w}) - \phi(\hat{w})\left(\frac{1}{\hat{w}} - \frac{1}{\tilde{u}_1}\right) \qquad \hat{\theta} \neq \theta \qquad (5.65)$$

with $\hat{w}$ given in (5.64) and

$$\tilde{u}_1 = \delta^{-1}[1 - \exp\{-\delta(\hat{\theta} - \theta)\}]\sqrt{j(\hat{\theta})}. \qquad (5.66)$$

A description of the second correction $\widehat{\Pr}_2(X \geq x)$ is based on the offset likelihood defined as

$$\mathcal{L}_-(\theta) = \exp\{\theta x^- - c(\theta)\},$$

where $x^- = x - \delta/2$ is the offset value of x. If $\tilde{\theta}$ is the root of $c'(\tilde{\theta}) = x^-$, then the inputs to the Lugannani and Rice expression, as in (5.65), are

$$\tilde{w}_2 = \text{sgn}(\tilde{\theta} - \theta)\sqrt{-2\ln\frac{\mathcal{L}_-(\theta)}{\mathcal{L}_-(\tilde{\theta})}} \qquad (5.67)$$

$$\tilde{u}_2 = 2\delta^{-1}\sinh\left\{\frac{\delta(\tilde{\theta} - \theta)}{2}\right\}\sqrt{j(\tilde{\theta})}. \qquad (5.68)$$

Applications of these continuity corrections can be found in chapter 6.

A statistical interpretation for the saddlepoint CDF as a p-value in an hypothesis test can be given that generalizes the interpretation given in section 2.4.5. Suppose the true canonical parameter is $\theta + s$ and consider testing

$$H_0 : s = 0 \quad \text{vs.} \quad H_1 : s < 0 \qquad (5.69)$$

for θ the fixed hypothesized value. The quantity $\hat{w}$ is the signed root of the likelihood ratio statistic that has an approximate Normal $(0, 1)$ distribution under H_0; $\hat{u}$ is the asymptotically studentized MLE which has the same approximate distribution. The rejection region of the test for p-value computation is $\{X \leq x\}$ or $\{\hat{\theta}(X) \leq \hat{\theta}\}$. The term $\Phi(\hat{w})$ is a first-order normal approximation to this p-value and the Lugannani and Rice approximation represents a third-order refinement on sets of bounded central tendency.

5.4.4 Testing a renewal process is a Poisson process: A continuation of Gamma (α, β)

Renewal processes are often used to model the interarrival times of component failures for a system as a sequence of i.i.d. random variables. Suppose we assume a renewal process has a Gamma (α, β) interarrival time. A test that such a renewal process is a Poisson process with exponential interarrival times compares $H_0 : \alpha = 1$ with $H_1 : \alpha > 1$ if the test is one-sided. Parameter β is a nuisance parameter so the appropriate conditional test is based upon the distribution of X given y.

Distribution theory

The distribution of X given y has an interesting structure which leads to a simple form for the conditional test. First transform from (x, y) to $(\hat{\alpha}, y)$. An expression for $\hat{\alpha}$ in terms of x

5.4 Saddlepoint approximation in regular exponential families

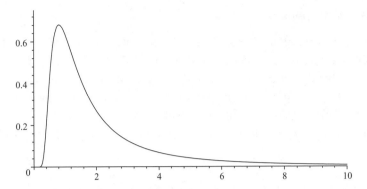

Figure 5.5. The unnormalized saddlepoint density for $\hat{\alpha}$ in (5.73) with $\alpha = 1$ and $n = 4$.

and y is determined by substituting $\hat{\beta}$ from (5.50) into (5.49) so that

$$x = n\{\Psi(\hat{\alpha}) - \ln(n\hat{\alpha}/y)\} \quad (5.70)$$

or

$$u = \sum_{i=1}^{n} \ln(z_i/z.) = n\{\Psi(\hat{\alpha}) - \ln(n\hat{\alpha})\} \quad (5.71)$$

when expressed in terms of the original data in which $x = \sum_i \ln z_i$ and $z. = \sum_i z_i = y$. The vector $\{d_i = z_i/z.\}$ is an exchangeable Dirichlet with common parameter α and is independent of $y = z.$. Since $\hat{\alpha}$ is a function of $\{d_i\}$, it too is independent of y. This fact is discussed in Barndorff-Nielsen (1978, p. 183) and appears to have been first noted by Cox and Lewis (1966).

The joint saddlepoint density of $(\hat{\alpha}, y)$ also separates in the two variables to reflect the independence of its true joint density. The Jacobian in transforming from $(x, y) \to (\hat{\alpha}, y)$ is determined from (5.70) as

$$\frac{\partial x}{\partial \hat{\alpha}} = n\{\Psi'(\hat{\alpha}) - 1/\hat{\alpha}\}.$$

The joint saddlepoint density is

$$\hat{f}(\hat{\alpha}, y; \alpha, \beta) = \hat{f}(\hat{\alpha}; \alpha)\hat{f}(y; \alpha, \beta) \quad (5.72)$$

where $\hat{f}(y; \alpha, \beta)$ is the saddlepoint approximation to the Gamma $(n\alpha, \beta)$ density of Y given in (1.7), and

$$\hat{f}(\hat{\alpha}; \alpha) = \frac{1}{2\pi} \frac{\hat{\Gamma}(n\alpha)}{\Gamma(\alpha)^n n^{n\alpha-1}} \sqrt{\hat{\alpha}\Psi'(\hat{\alpha}) - 1} \quad (5.73)$$
$$\times \exp\left[n\{-\alpha \ln \hat{\alpha} + (\alpha - \hat{\alpha})\Psi(\hat{\alpha}) + \hat{\alpha} + \ln \Gamma(\hat{\alpha})\}\right].$$

A plot of this density with $\alpha = 1$ and $n = 4$ is shown in figure 5.5.

A conditional test of $H_0 : \alpha = 1$ that treats β as a nuisance parameter may be based on the distribution theory of $\hat{\alpha}$ or u given y. A discussion of the various saddlepoint approximations to deal with these conditional probabilities is deferred to section 5.4.5. An alternative unconditional approach for this problem may be based on the concept of marginal likelihood which is now introduced.

Marginal likelihood approach

An unconditional *marginal likelihood* approach uses only a portion of the likelihood function. Consider the parametrization $(\alpha, n\alpha/\beta)$ with MLE $(\hat{\alpha}, z.)$. The MLE $\hat{\alpha}$ is a function of the exchangeable Dirichlet vector $\{d_i = z_i/z.\}$ which is independent of $z.$. The density for this Dirichlet vector defines the marginal likelihood of shape parameter α as discussed in Fraser (1968, 1979) or Cox and Hinkley (1974)). The likelihood factors into its marginal likelihood by transforming the data $(z_1, \ldots, z_n) \to (d_1, \ldots, d_{n-1}, z.)$. The Jacobian is $z_.^{n-1}$ so that

$$\prod_{i=1}^{n} f(z_i; \alpha, \beta) = f(d_1, \ldots, d_{n-1}; \alpha) f(z.; \alpha, \beta) \left| \frac{\partial(z_1, \ldots, z_n)}{\partial(d_1, \ldots, d_{n-1}, z.)} \right|^{-1}$$

$$= \frac{\Gamma(n\alpha)}{\Gamma(\alpha)^n} \exp\left\{ (\alpha - 1) \sum_{i=1}^{n} \ln d_i \right\} \qquad (5.74)$$

$$\times \frac{\beta^{n\alpha}}{\Gamma(n\alpha)} z_.^{n\alpha-1} e^{-\beta z.} \times z_.^{1-n}. \qquad (5.75)$$

The middle expression (5.74) is the marginal likelihood used for inference about α. It is a regular exponential family with canonical sufficient statistic $u = \sum_{i=1}^{n} \ln d_i$ that has CGF determined from the exponential structure as

$$K_U(s) = n \ln \frac{\Gamma(\alpha + s)}{\Gamma(\alpha)} - \ln \frac{\Gamma\{n(\alpha + s)\}}{\Gamma(n\alpha)} \qquad s > -\alpha.$$

The unused portion of the likelihood in (5.75) consists of the density of $z.$ and the Jacobian factor. With two unknown parameters appearing in this portion of the likelihood, it is difficult to extract further information about α from this factor. Consequently, the term is ignored under the supposition that it contains little information beyond what has already been extracted from the marginal likelihood.

As an example, consider aircraft 13 of the air-conditioning data in Cox and Lewis (1966). The data are $z_1, \ldots, z_{16}$ representing 16 waiting times to failure with $y = z. = 1312$ and $x = 65.5$ so that $u = -49.37$. The p-value in the test of the null hypothesis above is computed as

$$\Pr\{U > -49.37; \alpha = 1\} \simeq 0.09885$$

using the Lugannani and Rice approximation. The maximum marginal likelihood estimate used in this computation is $\hat{\alpha}_M = 1 + \hat{s} = 1.653$.

Marginal likelihood ratio test

This test compares $H_0 : \alpha = 1$ versus $H_1 : \alpha \neq 1$ using a two-sided test that rejects for large and small values of U. The p-value may be computed by determining u_{LM}, the left tail value of u such that the marginal likelihood ratio statistic at u_{LM} has the same value as at $u_{UM} = -49.37$, or the value 0.3690. Suppose that $\mathcal{L}_M(\alpha)$ is the marginal likelihood in (5.74). If s is the saddlepoint for the observed value of U given as $u = K'_U(s)$, then $s = \alpha - 1$ and marginal likelihood ratio for testing $\alpha = 1$ may be expressed in terms

5.4 Saddlepoint approximation in regular exponential families

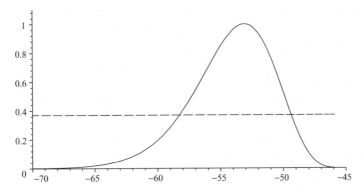

Figure 5.6. The marginal likelihood ratio Λ_M vs. u (*solid*). The dashed line has height 0.369 which is the attained value for the marginal likelihood ratio.

of s as

$$\Lambda_M(s) = \frac{\mathcal{L}_M(1)}{\mathcal{L}_M(\alpha)} = \frac{\Gamma(n)}{\frac{\Gamma\{n(1+s)\}}{\Gamma(1+s)^n} \exp\{s K'_U(s)\}}.$$

The parametric plot of $\{(K'_U(s), \Lambda_M(s)) : s > -1\}$ is shown in figure 5.6 and plots Λ_M versus u along with a dotted line of height 0.3690 for the attained value of Λ_M.

We determine numerically that $u_{LM} = -58.26$, a value also confirmed from the plot. The figure demonstrates that the mapping $u \to \Lambda_M$ is two-to-one reflecting the two sided nature of the marginal likelihood ratio test. The saddlepoint approximation to the p-value of this test is

$$\widehat{\Pr}\{\Lambda_M < 0.3690\} = \widehat{\Pr}\{U < -58.26\} + \widehat{\Pr}\{U > -49.37\}$$
$$= 0.06412 + 0.09885 = 0.1630$$

and has been computed from the CGF of U using $\alpha = 1$. This is the p-value of the *marginal* likelihood ratio test and not the ordinary likelihood ratio test which entails using the additional factor (5.75) in the likelihood ratio.

Likelihood ratio test

This test statistic always depends on the data through a sufficient statistic such as $(\hat{\alpha}, y)$. In this particular setting however, the dependence further reduces to depending only on $\hat{\alpha}$ as the reader is asked to show in Exercise 12. The MLE $\hat{\alpha}$ itself depends only on u through the relationship $u = u(\hat{\alpha})$ in (5.71). Thus the likelihood ratio Λ depends only on u and is independent of y making conditional computation unnecessary. The likelihood ratio as a function of $\hat{\alpha}$ is

$$\Lambda(\hat{\alpha}) = \left\{\frac{n\Gamma(\hat{\alpha})}{(n\hat{\alpha})^{\hat{\alpha}}}\right\}^n \exp\left[\{n - u(\hat{\alpha})\}(\hat{\alpha} - 1)\right] \qquad (5.76)$$

and is a two-to-one mapping from $\hat{\alpha} \to \Lambda$. The mapping $\hat{\alpha} \leftrightarrow u$ is monotone increasing as may be seen by computing

$$\frac{\partial u}{\partial \hat{\alpha}} = n\{\Psi'(\hat{\alpha}) - 1/\hat{\alpha}\}$$

and deducing that this is positive for all $\hat{\alpha}$ as a result of differentiating expression (6.3.21) in Abramowitz and Stegun (1970). The mapping $u \to \Lambda$ is therefore also two-to-one. A p-value computation for Λ now follows by using the same approach applied for marginal likelihood. The two-to-one mapping is inverted so that p-value probabilities may be expressed in term of events for U.

Consider the same two-sided test for the aircraft data. The MLE is $\hat{\alpha} = 1.746$ and leads to likelihood ratio $\Lambda = 0.2825$. The second value of $\hat{\alpha}$ that gives this Λ-value is $\hat{\alpha}_L = 0.6404$. The two values of u associated with this Λ-value are $u_L = -59.65$ and $u_U = -49.37$. The saddlepoint approximation to the distribution of U determines an approximate p-value as

$$\widehat{\Pr}\{\Lambda < 0.2825\} = \widehat{\Pr}\{U < -59.65\} + \widehat{\Pr}\{U > -49.37\}$$
$$= 0.03286 + 0.09885 = 0.1317. \tag{5.77}$$

Brute force simulation has also been used to determine this p-value and provides a 95% confidence interval as 0.1310 ± 0.00131. The simulation was stopped when the half-width of the 95% confidence interval achieved 1% relative error. The saddlepoint approximation is well inside the confidence interval. By comparison, the asymptotic χ_1^2 test has the p-value 0.1118.

5.4.5 Conditional saddlepoint CDF for a canonical sufficient statistic

If dim $Y = 1$, the double-saddlepoint density of Y given $X = x$ in (5.55) has an associated double-saddlepoint CDF approximation. For continuous Y, it is a simple exercise to show that the Skovgaard approximation in (4.12) is based on

$$\hat{w} = \text{sgn}(\hat{\psi} - \psi)\sqrt{-2\ln\frac{\mathcal{L}(\hat{\chi}_\psi, \psi)}{\mathcal{L}(\hat{\chi}, \hat{\psi})}} \tag{5.78}$$

$$\hat{u} = (\hat{\psi} - \psi)\sqrt{\frac{|j(\hat{\chi}, \hat{\psi})|}{|j_{\chi\chi}(\hat{\chi}_\psi, \psi)|}}.$$

If Y has δ-lattice support, the continuity corrections in section 4.2.2 should be used and modified as in (2.58)–(2.61) to account for $\delta \neq 1$. For the conditional survival function of Y given $X = x$, the first continuity correction is

$$\widehat{\Pr}_1(Y \geq y | X = x) = 1 - \Phi(\hat{w}) - \phi(\hat{w})\left(\frac{1}{\hat{w}} - \frac{1}{\tilde{u}_1}\right) \qquad \hat{\psi} \neq \psi$$

where $\hat{w}$ is given in (5.78) and

$$\tilde{u}_1 = \delta^{-1}\{1 - e^{-\delta(\hat{\psi}-\psi)}\}\sqrt{\frac{|j(\hat{\chi}, \hat{\psi})|}{|j_{\chi\chi}(\hat{\chi}_\psi, \psi)|}}. \tag{5.79}$$

For the second correction, let $y^- = y - \delta/2$, and define

$$\mathcal{L}_-(\chi, \psi) = \exp\{\chi^T x + \psi y^- - c(\chi, \psi)\}$$

5.4 Saddlepoint approximation in regular exponential families

as the offset likelihood function. With $(\tilde{\chi}, \tilde{\psi})$ as the root of $c'(\tilde{\chi}, \tilde{\psi}) = (x, y^-)$, then the conditional survival function when $\tilde{\psi} \neq \psi$ has the inputs

$$\tilde{w}_2 = \text{sgn}(\tilde{\psi} - \psi)\sqrt{-2\ln\frac{\mathcal{L}_-(\hat{\chi}_\psi, \psi)}{\mathcal{L}_-(\tilde{\chi}, \tilde{\psi})}} \tag{5.80}$$

$$\tilde{u}_2 = 2\delta^{-1}\sinh\{\delta(\tilde{\psi} - \psi)/2\}\sqrt{\frac{|j(\tilde{\chi}, \tilde{\psi})|}{|j_{\chi\chi}(\hat{\chi}_\psi, \psi)|}}. \tag{5.81}$$

Applications of these corrections are considered in chapter 6.

There is also a statistical interpretation for the Skovgaard formula as an approximate p-value in an hypothesis test. The canonical parameter is taken to be $(\chi, \psi + s)$ and the test (5.69) is under consideration which states that ψ is the true value of the interest parameter with arbitrary nuisance parameter χ. Value $\hat{w}$ is the signed root of the likelihood ratio test. The first-order replacement of $j(\hat{\chi}_\psi, \psi)$ for $j(\hat{\chi}, \hat{\psi})$ in $\hat{u}$ makes $\hat{u}$ the conditionally studentized MLE. This may be seen by noting that the asymptotic conditional variance of $\hat{\psi}$ given $X = x$ is estimated by

$$j_{\psi\psi \cdot \chi}^{-1}(\hat{\chi}_\psi, \psi) = \frac{|j_{\chi\chi}(\hat{\chi}_\psi, \psi)|}{|j(\hat{\chi}_\psi, \psi)|} = \left.\left(j_{\psi\psi} - j_{\psi\chi}j_{\chi\chi}^{-1}j_{\chi\psi}\right)^{-1}\right|_{\hat{\chi}_\psi, \psi}.$$

Both $\hat{w}$ and $\hat{u}$ are asymptotically standard normal under the appropriate standard conditions and the Skovgaard expression provides a third-order correct p-value approximation for asymptotics that are applicable on sets of bounded central tendency.

Examples

Normal $(0, \sigma^2)$ continuation The conditional CDF approximation of $\Pr\{Y \leq y | X = x\}$ is determined from (5.78) to be

$$\hat{w} = \text{sgn}(y)\sqrt{-n\ln\left(1 - \frac{y^2}{nx}\right)} = \text{sgn}(\hat{\mu})\sqrt{-n\ln\frac{\hat{\sigma}^2}{\hat{\sigma}^2 + \hat{\mu}^2}}$$

$$\hat{u} = \frac{y}{\sqrt{x}}\sqrt{1 - \frac{y^2}{nx}} = \frac{\sqrt{n}\hat{\mu}\hat{\sigma}}{\hat{\sigma}^2 + \hat{\mu}^2}.$$

According to (5.57), the T_{n-1} statistic is a monotonic function of $Y/\sqrt{X}$ and therefore independent of X. Thus we may condition on $X = 1$ so that

$$\Pr\{T_{n-1} \leq \tau\} = \Pr\{T_{n-1} \leq \tau | X = 1\}$$

$$= \Pr\left\{Y \leq \frac{\tau\sqrt{n}}{\sqrt{\tau^2 + n - 1}} \bigg| X = 1\right\}.$$

This leads to a Skovgaard approximation for the T_{n-1} distribution which uses values

$$\hat{w} = \text{sgn}(\tau)\sqrt{n\ln\left(1 + \frac{\tau^2}{n-1}\right)} \tag{5.82}$$

$$\hat{u} = \frac{\tau\sqrt{n(n-1)}}{\tau^2 + n - 1}.$$

172 *Exponential families and tilted distributions*

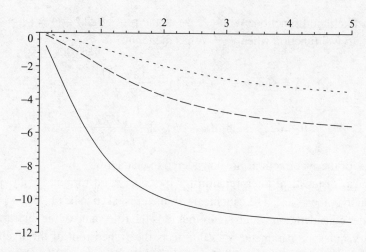

Figure 5.7. Percentage relative errors of the Skovgaard approximation for the right tail probabilities of the T_1 (*solid*), T_3 (*dashed*), and T_5 (*dotted*) distributions.

Consider a simple example with three degrees of freedom and $\tau = 2$. Then $\Pr\{T_3 \geq 2\} = 0.0697$ and the saddlepoint approximation is 0.0670 with a relative error of -3.76%.

Figure 5.7 plots the right-tail percentage relative error

$$100 \left(\frac{1 - \hat{F}(\tau)}{\Pr(T_{n-1} > \tau)} - 1 \right) \qquad \text{vs.} \qquad \tau \in (0, 5)$$

for $n = 2, 4, 6$ and degrees of freedom 1, 3, and 5. As the degrees of freedom increase further the relative errors continue to decrease to 0. From (5.82),

$$\lim_{n \to \infty} \hat{w} = \tau = \lim_{n \to \infty} \hat{u},$$

so the approximation becomes $\Phi(\tau)$ as $n \to \infty$, which is the exact result.

Normal (μ, σ^2) continuation When $\mu \neq 0$, the conditional power attained by the T test of $H_0 : \mu = 0$ can be computed from the Skovgaard approximation. Using the relationship of T_{n-1} to $W = Y/\sqrt{X}$ in (5.57) and conditioning on $X = x$ with $\psi \neq 0$,

$$\Pr\{T_{n-1} \leq \tau | X = x; \psi\} = \Pr\left\{ Y \leq \frac{\tau \sqrt{nx}}{\sqrt{\tau^2 + n - 1}} \bigg| X = x; \psi \right\}. \qquad (5.83)$$

The cutoff for Y given in (5.83) is y, its attained value, and the computation proceeds using $\hat{w}$ and $\hat{u}$ in (5.78) based on the likelihood ratio and Fisher information given in (5.60) and (5.61).

Take the previous example in which $n = 4$ and a T_3 statistic attains the value $\tau = 2$. The exact and saddlepoint p-values are 0.0697 and 0.0670 respectively when testing $H_0 : \psi = 0$ vs. $H_1 : \psi > 0$. Some sufficient statistics that lead to T_3 assuming the value $\tau = 2$ are a sample mean of 2 and a sample standard deviation of 2; thus $y = 8$ and $x = 28$. The attained conditional power of the T_3 test when $\psi = 1/2$ is approximated as

$$\Pr\{T_3 \geq \tau | X = 28; \psi = \tfrac{1}{2}\} = \Pr\{Y \geq 8 | X = 28; \psi = \tfrac{1}{2}\}$$
$$\simeq 0.552. \qquad (5.84)$$

5.4 Saddlepoint approximation in regular exponential families

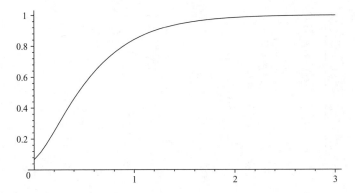

Figure 5.8. The conditional power approximation $\widehat{\Pr}\{T_3 \geq 2|\, X = 28;\psi\}$ for the T_3 test vs. $\psi > 0$.

Figure 5.8 shows a plot of the saddlepoint approximation for this attained conditional power versus $\psi > 0$.

This is not the power of the T_3 test obtained from the noncentral T_3 distribution with noncentrality parameter $\lambda = \sqrt{4}\mu/\sigma$; it is the power conditional on $X = 28$. This attained conditional power serves as a *post-data* estimate of the attained sensitivity of the T_3 test to alternative values of ψ. By contrast, the unconditional noncentral T_3 with noncentrality λ serves as a *pre-data* design criterion.

The conditional power function in figure 5.8 can be modified to give an estimate of the unconditional power function. Figure 5.9 provides a plot of the noncentral power

$$\beta(\lambda) = \Pr(T_3 \geq 2; \lambda)$$

vs. noncentrality parameter λ as the solid line. The dashed line is its estimate using the data with $n = 4$. The basic idea is to consider that the true conditional power function is an unbiased estimate of the unconditional power function as noted in Lehmann (1986, §4.4, p. 151). In fact, by the Lehmann–Scheffé lemma, it is a UMVUE for the power. The saddlepoint approximation to the attained conditional power function in terms of $(X, Y) = (28, 8)$ provides the conditional power estimate. An adjustment is also necessary because the plot in figure 5.8 is vs. $\psi = \mu/\sigma^2$ whereas the unconditional power is vs.

$$\lambda = \sqrt{n}\mu/\sigma = \sqrt{n}\sigma\psi.$$

The adjustment is to use

$$\hat{\beta}(\lambda) = \widehat{\Pr}\{Y \geq 8|\, X = 28; \psi = \lambda/(\sqrt{n}\tilde{\sigma}) = \lambda/4\}$$

where $\tilde{\sigma} = 2$ is the sample standard deviation. The estimate in figure 5.9 is quite good for $\lambda \in (0, 2.5)$ even with only three degrees of freedom for $\tilde{\sigma}$ as a estimate of σ. The conditional power estimate in (5.84) with $\psi = 1/2$ estimates the true power 0.5569 at $\lambda = 2$ and is very close to the crossing point in the plot. The true power function was computed using expressions (9) and the one just below (10) in Johnson and Kotz (1970, §31.5, p. 205). In the generalized context of multiparameter exponential families of section 6.2, conditional power functions are again reconsidered as estimates for their associated unconditional power functions.

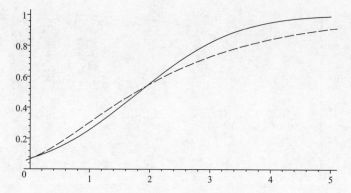

Figure 5.9. The estimate $\hat{\beta}(\lambda)$ (*dotted*) of the unconditional power $\beta(\lambda)$ (*solid*) versus λ for the one-sided t_3-test.

Gamma (α, β) continuation: Testing a renewal process is a Poisson process The optimal test of $H_0 : \alpha = 1$ versus $H_1 : \alpha > 1$ rejects for large values of $X = \sum_{i=1}^{n} \ln Z_i$ and its p-value is computed conditional on $Y = Z_{.} = z_{.}$ which removes distributional dependence on β. The p-value for this optimal conditional test turns out to be the p-value for the one-sided marginal likelihood test. The conditional p-value, computed with $\alpha = 1$, is

$$\Pr\{X > x | Z_{.} = z_{.}; \alpha\} = \Pr\{X - n \ln Z_{.} > x - n \ln z_{.} | Z_{.} = z_{.}; \alpha\}$$
$$= \Pr\{U > u | Z_{.} = z_{.}; \alpha\} \tag{5.85}$$

where $U = \sum_i \ln(Z_i/Z_{.})$ and u is the observed value of U. Since U is independent of $Z_{.}$ in (5.85), the conditional p-value is the value $\Pr\{U > u; \alpha = 1\}$ that has been approximated as 0.09885 using the Lugannani and Rice approximation for the distribution of U.

The Skovgaard approximation provides an alternative approximation for the left hand side of (5.85). Exercise 11 asks the reader to show that this approximation leads to 0.09847 as the one-tailed p-value which differs from the Lugannani and Rice approximation in the third significant digit.

The marginal likelihood ratio test was previously considered as a two-sided test that also includes the left tail for which $U < u_L = -58.26$. The Lugannani and Rice computation for this marginal left-tail probability was 0.06412. The equivalent left-tail probability in terms of the conditional distribution of X given $Z_{.} = z_{.} = 4.323$ requires a Skovgaard approximation for the equivalent event that $X < 56.61$. This yields 0.06439 which differs in the third significant digit.

Two sample gamma comparison: Testing equal shape Consider two independent Gamma samples such that $Z_{i1}, \ldots, Z_{in_i}$ is an i.i.d. sample from a Gamma (α_i, β_i) for $i = 1, 2$. The test of equal shape parameters $H_0 : \alpha_1 = \alpha_2$ versus $H_1 : \alpha_1 < \alpha_2$ can be based either on the marginal likelihood for (α_1, α_2) or by working directly with the exponential family likelihood as in the previous example. The reparametrization $\alpha_1 = \alpha$ and $\alpha_2 = \alpha + \psi$ may be used to transform the test to $H_0 : \psi = 0$ versus $H_1 : \psi > 0$.

The marginal likelihood is based on the joint density of the two independent Dirichlet vectors $\{D_{ij} = Z_{ij}/Z_{i.}\}$. With canonical parameter (α, ψ), the associated canonical sufficient

5.4 Saddlepoint approximation in regular exponential families

statistic is $(U_1 + U_2, U_2)$, where $U_i = \sum_{j=1}^{n_i} \ln Z_{ij}$. The conditional test rejects for large values of U_2 and p-values are computed conditional on $u_1 + u_2$.

Suppose samples one and two are the air conditioner failure times on board aircraft 2 and 13 respectively for the data of Cox and Lewis (1966). The data are $n_1 = 23$, $u_1 = -85.93$ and $n_2 = 16$, $u_2 = -49.36$. Exercise 14 asks the reader to compute the p-value from the marginal likelihood as

$$\Pr\{U_2 > -49.36 | u_1 + u_2; \psi = 0\} \simeq 0.10127. \tag{5.86}$$

The maximum marginal likelihood estimates are $\hat{\alpha}_M = 0.9327$ and $\hat{\psi}_M = 0.7227$ and the constrained MLE is $\hat{\alpha}_{0M} = 1.1270$.

Direct computation using the Skovgaard approximation with the four-parameter exponential family provides a second approximation. If the canonical parameter is $(\alpha, \psi, -\beta_1, -\beta_2)$ the corresponding sufficient statistic is $(X_1 + X_2, X_2, Y_1, Y_2)$ where

$$X_i = \sum_{j=1}^{n_i} \ln Z_{ij} \qquad Y_i = \sum_{j=1}^{n_i} Z_{ij}. \tag{5.87}$$

Exercise 14 asks the reader to show that the true conditional p-value of the optimal test in this context is the same as that determined from the marginal likelihood approach, or

$$\Pr\{X_2 > x_2 | x_1 + x_2, y_1, y_2; \psi\} = \Pr\{U_2 > u_2 | u_1 + u_2; \psi\} \tag{5.88}$$

for all ψ. The Skovgaard approximation to the left side of (5.88) results in a p-value of 0.10133 which differs from the approximation used with marginal likelihood in the fourth significant digit. The MLEs are $\hat{\alpha} = 0.9655$, $\hat{\psi} = 0.7802$, $\hat{\beta}_1 = 0.01009$, and $\hat{\beta}_2 = 0.02129$. The constrained MLEs with $\psi = 0$ are $\hat{\alpha}_0 = 1.175$, $\hat{\beta}_{10} = 0.01228$, and $\hat{\beta}_{20} = 0.01433$.

Two sample gamma comparison: Testing equal scales Suppose equal shape parameters are assumed for the previous example and we want to further test equality of the scale parameters. The gamma parameters are $\alpha_1 = \alpha_2 = \alpha$, $\beta_1 = \beta$, and $\beta_2 = \beta + \psi$ and the hypothesis test is $H_0 : \psi = 0$ versus $H_1 : \psi \neq 0$.

The attained values of the canonical sufficient statistics associated with canonical parameters $\alpha, -\beta$, and $-\psi$ are

$$x_1 = \sum_{ij} \ln z_{ij} = 156.6 \qquad x_2 = z_{..} = 3513 \qquad y = z_{2.} = 1312.$$

With such data, the conditional p-value of the likelihood ratio test can be determined by adding its contributions from each tail of Y. The MLE for ψ is determined as $\hat{\psi} = 0.002051$. The one-sided test versus $H_1 : \psi > 0$ rejects for small values of Y and the Skovgaard approximation for the p-value in this case is

$$\widehat{\Pr}\{Y < 1312 | x_1, x_2; \psi = 0\} = 0.3169.$$

To obtain the other contribution to the conditional p-value of the likelihood ratio test, the value y_U must be determined such that the likelihood ratio statistic using (x_1, x_2, y_U) is the same as that with (x_1, x_2, y), or likelihood ratio value 0.8776. Since $\hat{\psi} > 0$, MLE value $\hat{\psi}_U$ based on (x_1, x_2, y_U) must be negative and the corresponding rejection region for $H_0 : \psi = 0$

versus $H_1 : \psi < 0$ is for large values of Y. Numerical solution yields $y_U = 1572.6$ with $\hat{\psi}_U = -0.001977$. The right tail portion needed for the conditional p-value is

$$\widehat{\Pr}\{Y > 1572.6 | x_1, x_2; \psi = 0\} = 0.3042.$$

This gives a two-sided p-value of 0.6211 for the likelihood ratio test.

5.4.6 Likelihood ratio, UMP, and UMP unbiased tests

One parameter regular exponential families

Consider hypothesis testing for a single canonical parameter θ of dimension 1 with canonical sufficient statistic X. In one-sided testing, the uniformly most powerful (UMP) test of $H_0 : \theta = 0$ versus $H_1 : \theta > 0$ among tests of level α rejects for large values of X (Lehmann, 1986, §3.3). The cutoff of the test is chosen to assure that the level is α. In the discrete setting, randomization of the decision on the boundary is needed to achieve this exact level however, for simplicity, we ignore this complication here. The p-value associated with this test is $\Pr\{X \geq x; \theta = 0\}$ which may be approximated using the Lugannani and Rice approximation. The attained power function is the plot of $\Pr\{X \geq x; \theta\}$ versus $\theta \geq 0$ and is approximated in the same manner.

In two-sided testing of $H_0 : \theta = 0$ versus $H_1 : \theta \neq 0$, optimality for a test among those of level α only occurs when the level α tests are further restricted to those that are *unbiased*. A level α test is said to be unbiased if its power function exceeds level α for all $\theta \neq 0$. The UMP test among the class of unbiased level α tests rejects for $X \notin (x_l, x_u)$ where the cutoffs are determined so the level is α and the power function has slope 0 at $\theta = 0$ (Lehmann, 1986, §4.2). If $\beta(\theta; x_l, x_u)$ is the power function of the test with rejection region $(-\infty, x_l) \cup (x_u, \infty)$, then x_l and x_u are chosen to solve

$$\beta(0; x_l, x_u) = \alpha$$
$$\left. \frac{\partial}{\partial \theta} \beta(\theta; x_l, x_u) \right|_{\theta=0} = 0. \tag{5.89}$$

The derivative of this power function is an analytic function of θ over Θ for any regular exponential family (Lehmann, 1986, §2.7, thm. 9). Such smoothness makes the condition (5.89) a necessary condition that a test be unbiased since otherwise the power does not have a local minimum at $\theta = 0$ and dips below α on one side.

The practical limitations of finding cutoffs that satisfy the constraints in (5.89) have limited the usage of such tests. Saddlepoint methods, however, are useful for determining such rejection regions as well as for computing p-values and attained power functions associated with the test. The next example addresses these problems.

Example: Testing a renewal process is a Poisson process The test of $H_0 : \alpha = 1$ versus $H_1 : \alpha \neq 1$ has the framework of a one-parameter exponential family when inference is based on the marginal likelihood of α given in (5.74). The rejection region is $U \notin (u_l, u_u)$ and, for p-value determination, one of the boundary values is the observed value $u = -49.37$. In this particular case, since $\hat{\alpha}_M > 1$, $u_u = -49.37$ and u_l must be determined numerically so the test conforms to the structure (5.89) that assures an unbiasedness property. The Lugannani and Rice approximation, denoted as $\widehat{\Pr}$, can be used to solve for the approximate value of

5.4 Saddlepoint approximation in regular exponential families

Table 5.1. *Acceptance regions and saddlepoint probabilities for the two-side UMP unbiased test of $H_0 : \alpha = 1$ when the right cutoff is the listed value of u.*

u	(u_l, u_u)	$\widehat{\Pr}\{U \leq u_l\}$	$\widehat{\Pr}\{U \geq u_u\}$	p-value
-49.37	$(-58.27, -49.37)$	.06400	.09885	.1629
-47	$(-64.68, -47)$	$.0^2 0832$	$.0^2 4249$	$.0^2 6332$
-45	$(-79.08, -45)$	$.0^6 1278$	$.0^6 4276$	$.0^6 5554$

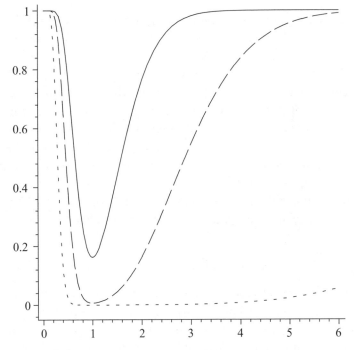

Figure 5.10. Attained power functions versus α for UMP unbiased tests of $H_0 : \alpha = 1$ vs. $H_1 : \alpha \neq 1$ when $u = -49.37$ (*solid*), -47 (*dashed*), and -45 (*dotted*).

u_l as the root of

$$\frac{\hat{\partial}}{\hat{\partial}\alpha} \widehat{\Pr}\{u_l < U \leq u | \alpha\}\Big|_{\alpha=1} = 0. \tag{5.90}$$

The notation $\hat{\partial}/\hat{\partial}\alpha$ indicates a numerical derivative of increment $\hat{\partial}\alpha = 10^{-8}$ carrying 30 digit computational accuracy in Maple.

To better understand the nature of this test and the unbiased condition, three values for u have been considered: the observed value -49.37 and the more extreme values -47 and -45. Table 5.1 gives the acceptance regions and left- and right-tail probabilities that result from imposing the unbiasedness condition of (5.90). The last column of p-values suggests that the first value is insignificant while the other more extreme values are quite significant. The null mean of U is -53.09 and well within the acceptance regions. The three attained power curve estimates associated with table 5.1 are shown in figure 5.10. The unbiasedness of the tests may be seen from the flat slope of the power functions at $\alpha = 1$.

Table 5.2. *Acceptance regions and saddlepoint probabilities for the two-sided marginal likelihood ratio test of $H_0 : \alpha = 1$ when the right cutoff is the listed value of u.*

u	(u_{LM}, u_{UM})	$\widehat{\Pr}\{U \leq u_{LM}\}$	$\widehat{\Pr}\{U \geq u_{UM}\}$	p-value
-49.37	$(-58.26, -49.37)$	.06412	.09885	.1630
-47	$(-64.67, -47)$	$.0^2 2092$	$.0^2 4249$	$.0^2 6340$
-45	$(-79.06, -45)$	$.0^6 1295$	$.0^6 4276$	$.0^6 5571$

The marginal likelihood ratio test turns out to be almost exactly the same test as the UMP unbiased test. Table 5.2 displays acceptance regions and significance probabilities for this test that admit direct comparison with table 5.1. All the entries are nearly identical which suggests that the marginal likelihood ratio test is essentially UMP unbiased in this context. The close agreement of these tests is especially striking when one considers the manner in which the left edges of the acceptance regions have been determined. For each value of u, the values of u_l for the UMP unbiased tests have been chosen to assure that the power function has slope zero at $\alpha = 1$. The corresponding values of u_{LM}, for the marginal likelihood ratio test, have been chosen so the marginal likelihood at u_{LM} in the left tail agrees with the attained marginal likelihoods in the right tail. The fact that $u_l \simeq u_{LM}$ is somewhat surprising and confirms the somewhat intuitive notion about marginal likelihood: that it contains all recoverable information about α in the full likelihood.

Multiparameter regular exponential families

Suppose a regular exponential family has canonical parameter $\theta = (\chi, \psi)$ and canonical sufficient statistic (x, y). An optimal hypothesis test for $H_0 : \psi = 0$ versus $H_1 : \psi > 0$ in the dim $\psi = 1$ case has a simple form. The UMP unbiased test of level α rejects for large values of Y as described in Lehmann (1986, §4.4) and has conditional p-value $\Pr\{Y > y | x; \psi = 0\}$ that may be approximated quite simply by using the Skovgaard approximation.

When testing the two-sided alternative $H_1 : \psi \neq 0$, the UMP test in the class of level α tests that are *unconditionally unbiased* rejects for $Y \notin (y_l, y_u)$. The values $y_l < y_u$ are determined so the conditional level of the test is α and the conditional power at $\psi = 0$ has a slope of 0. If we denote the conditional power function connected to rejection region $Y \notin (y_l, y_u)$ as

$$\beta(\psi; y_l, y_u | x) = \Pr\{Y \notin (y_l, y_u) | x; \psi\}, \qquad (5.91)$$

then the UMP unbiased level α test has acceptance region (y_l, y_u) solving

$$\beta(0; y_l, y_u | x) = \alpha$$

$$\left. \frac{\partial}{\partial \psi} \beta(\psi; y_l, y_u | x) \right|_{\psi=0} = 0. \qquad (5.92)$$

The same method for approximating (y_l, y_u) can be applied as in the single parameter setting except now the Skovgaard approximation is used to approximate the conditional power function in (5.92).

The likelihood ratio test of level α rejects for $Y \notin (y_L, y_U)$ where the rejection region is chosen to solve the alternative equations

$$\beta(0; y_L, y_U | x) = \alpha$$
$$\hat{w}(x, y_L) = -\hat{w}(x, y_U). \tag{5.93}$$

The notation $\hat{w}(x, y_L)$ indicates the $\hat{w}$-value in the Skovgaard approximation resulting from (x, y_L) as the canonical sufficient statistic. The different rejection regions determined from (5.92) and (5.93) are due to the second equation. In certain symmetric settings these equations can agree so the tests are equivalent, however in most settings they will differ. It would be quite interesting to explore just how different these two tests actually are; perhaps the likelihood ratio test, considered conditional on $X = x$, and the UMP unbiased test are most often very close as was seen in the previous unconditional example. Condition (5.92) is a necessary condition for a test to be *conditionally unbiased,* but is no longer a necessary condition for unconditional unbiasedness as in the one-parameter setting. For unconditional unbiasedness the condition need not hold for every value of x but only when averaged over the distribution of X. In situations where the likelihood ratio test differs from the UMP unbiased test, the former test cannot be conditionally unbiased. This is because it satisfies (5.93) and not (5.92) so its conditional power function does not achieve a local minimum at the point null hypothesis. Often, however, it is unconditionally unbiased.

The reader is asked to make further comparisons between likelihood ratio tests and UMP unbiased tests in the exercises.

Inference about ψ in the dim $\psi > 1$ setting is difficult to approach analytically since the rejection region for Y may be a nonrectangular region $B \subset \Re^{m_y}$. In the rectangular setting, Kolassa (2004) has provided analytic approximations for $\Pr(Y_i > y_i : i = 1, \ldots, m_y | X = x; \psi)$. Alternatively, Butler *et al.* (2005) address the more general computation using importance sampling to integrate the double-saddlepoint density $\hat{f}(y|x; \psi)$ and to approximate $\Pr(Y \in B | X = x; \psi)$ for any region $B \subset \Re^{m_y}$. The main advantage of this saddlepoint/simulation approach is its simplicity and generality; for example, B may be the rejection region for a likelihood ratio test.

5.5 Exercises

1. (a) Prove that the CGF of the sufficient statistic for the regular exponential family in (5.1) is as given in (5.3).
   (b) Show that the first two cumulants are given by (5.4).
2. Show that the joint density of i.i.d. variables from a regular exponential family form another regular exponential family.
3. (a) Show that the Multinomial $(n; \pi_1, \ldots, \pi_{m+1})$ mass function, for n fixed, is a regular exponential family.
   (b) Derive the same mass function as a conditional mass function in an $(m+1)$-dimensional regular exponential family.
   (c) Determine the conditional CGF from the conditional characterization as given in (5.14).
4. Assess the accuracy of the tilted CDF approximation in (5.37) using a continuous univariate example that has been previously discussed. In particular:

(a) Tabulate values for the Lugannani and Rice approximations, with and without the higher order correction term (2.48), along with the $O(n^{-1})$ and $O(n^{-3/2})$ approximations from (5.37) and (5.38). Compare these to the exact values.

(b) Plot the four curves for percentage relative errors.

5. Suppose $Z_1, \ldots, Z_n$ is an i.i.d. sample from any of the following families: Binomial $(1, \xi)$, Normal $(\mu, 1)$, Gamma (α, β) for fixed α, Negative Binomial $(1, \xi)$, Normal$_m(\mu, I_m)$, or Multinomial $(1; \pi_1, \ldots, \pi_{m+1})$. Pick one of these settings and derive the saddlepoint mass/density function of the canonical sufficient statistic $X = \sum_i Z_i$ in the product family, based on the characteristics of the exponential family as in (5.44). Verify that it agrees with the previously derived saddlepoint density for this distribution.

6. Suppose $Z_1, \ldots, Z_n$ is an i.i.d. sample from a Multinomial $(1; \pi_1, \ldots, \pi_{m+1})$.

(a) Derive the first-order saddlepoint mass function of the canonical sufficient $X = \sum_{i=1}^n Z_i$ in the product family, based on characteristics of the exponential family. Verify that it agrees with (3.12).

(b) If $X^T = (X_1, \ldots, X_{m-1}, Y)$ denotes the components of X, then the conditional mass function of Y given $x_1, \ldots, x_{m-1}$ is Binomial $(n - x_., \omega)$ where $x_. = \sum_{i=1}^{m-1} x_i$ and $\omega = \pi_m / (\pi_m + \pi_{m+1})$. Derive the double-saddlepoint mass function using the exponential family characteristics and verify that it is exact up to Stirling's approximation for the combinatoric.

(c) Compute the continuity-corrected Skovgaard approximations for the CDF of the binomial in part (b) using the conditional setting of the exponential family. Verify that they agree with the single-saddlepoint approximations in (1.33) and (1.34).

7. Suppose $Z \sim$ Beta $(\alpha, 1)$ for $\alpha > 0$. The canonical statistic is $X = \ln Z$ with α as the canonical parameter.

(a) Compute the saddlepoint density of X for some value of α.

(b) Use Jacobian transformation to approximate the beta density of Z from this saddlepoint density. Plot the true beta density, its saddlepoint approximation, and a normalized approximation.

(c) Determine the Lugannani and Rice approximation for X based on the exponential family characteristics. Verify that it is the same as the approximation in section 2.4.1 based on the Mellin transform approach.

(d) In Exercise 20 of chapter 2, the reader was asked to show that Z^{-1} has a Pareto distribution and also to find a CDF approximation. Does the approximation resulting from (c) agree with the approximation determined in chapter 2?

8. Let $Z_1, \ldots, Z_n$ be an i.i.d. sample from a Normal (μ, σ^2).

(a) Derive the true density of $X = \sum_i Z_i^2$ and $Y = \sum_i Z_i$ in (5.46) starting from the known distributions of the sample mean and variance.

(b) Verify the relationship of the saddlepoint density to the true density given by (5.48).

(c) With $\mu = 0$, show that the conditional density of Y given $X = x$ is as given in (5.56).

(d) Make the transformation to $W = Y/\sqrt{X}$ and determine that the conditional density of W given $X = x$ is free of x. From this marginal density, derive the central t_{n-1} density for T as determined from W in (5.57).

(e) Derive the conditional saddlepoint density of Y given $X = x$ given in (5.58).

9. Derive the Skovgaard approximation for the T_{n-1} distribution as given in (5.82).

10. Consider inference for σ^2 in the normal sample of Exercise 8.

(a) Derive the true conditional density of X given $Y = y$, which depends only on $\chi = -1/(2\sigma^2)$.

(b) Determine the corresponding conditional saddlepoint density.

(c) Using part (a), show that the conditional density of $X - Y^2/n$ given $Y = y$ is free of y so that $X - Y^2/n$ is independent of Y.

(d) Repeat part (c) but use the saddlepoint density from part (b) instead. Is the relationship between the true marginal density and the transformed saddlepoint density of $X - Y^2/n$ what you would expect?

11. Let $Z_1, \ldots, Z_n$ be an i.i.d. sample from a Gamma (α, β) distribution.
    (a) Determine the saddlepoint density of $X = \sum_i \ln Z_i$ given $Y = \sum_i Z_i$.
    (b) Based on the four data points used to construct figure 5.4, plot the normalized conditional saddlepoint density of X given $Y = 4.323$ with $n = 4$ and $\alpha = 1$.
    (c) Numerically integrate this density to approximate

    $$P\{X > -1.928 | Y = 4.323; \alpha = 1\}, \tag{5.94}$$

    the p-value in testing $H_0 : \alpha = 1$ versus $H_1 : \alpha > 1$.
    (d) Use the Skovgaard approximation to approximate (5.94) and compare this answer to that of part (c).

12. Consider the likelihood ratio test that a renewal process is a Poisson process as specified in (5.76).
    (a) Derive (5.76) and plot $\Lambda(\hat{\alpha})$ versus $\hat{\alpha}$ to determine that the mapping $\hat{\alpha} \to \Lambda(\hat{\alpha})$ is two-to-one.
    (b) Plot $u = u(\hat{\alpha})$ versus $\hat{\alpha}$ to verify that $u \leftrightarrow \hat{\alpha}$ is monotone increasing.
    (c) Confirm the validity of the likelihood ratio p-value given in (5.77).
    (d) Plot a saddlepoint estimate of the attained power function of the test in a neighborhood of $\alpha = 1$.

13. Consider the Skovgaard approximations for testing that a renewal process is a Poisson process as described in (5.85).
    (a) For the aircraft data, show that the optimal conditional test of $H_0 : \alpha = 1$ versus $H_0 : \alpha > 1$ has a p-value whose Skovgaard approximation is 0.09847.
    (b) Show that the Skovgaard approximation for the two-sided marginal likelihood ratio test is

    $$\widehat{\Pr}\{\Lambda_M < 0.3690\} = \widehat{\Pr}\{X < 56.61 \mid y\} + \widehat{\Pr}\{X > 65.5 \mid y\}$$
    $$= 0.06439 + 0.09847 = 0.1628$$

    where $y = 1312$ and $\alpha = 1$.
    (c) Show that the same computation as in (b), using the Lugannani and Rice approximation to $\Pr\{U > u; \alpha = 1\}$, gives 0.1630.
    (d) Derive the Skovgaard approximation to the p-value of likelihood ratio test Λ and compare it to the previously determined value of 0.1317 using saddlepoint approximation to U.

14. Consider the test of equal shape parameters in the two sample gamma setting of section 5.4.5. Suppose samples of size n_1 and n_2 from populations that are Gamma (α, β_1) and Gamma $(\alpha + \psi, \beta_2)$. There are two approaches to the one-sided test of $H_0 : \psi = 0$ versus $H_1 : \psi > 0$.
    (a) The first approach restricts attention to the marginal likelihood of (α, ψ) that is an exponential family with canonical sufficient statistic $(U_1 + U_2, U_2)$. The optimal test derived from the marginal likelihood rejects for large U_2 conditional on $u_1 + u_2$. Derive numerical expressions to carry out these computations and verify the values of the maximum marginal likelihood estimates and p-value in (5.86). Show the two equations that determine $(\hat{\alpha}_M, \hat{\psi}_M)$ can be reduced to solving the single equation

    $$n_1 \{\Psi(\hat{\alpha}_M) - \Psi(n_1 \hat{\alpha}_M)\} = u_1. \tag{5.95}$$

    (b) The second approach starts with the full likelihood which is a four parameter exponential family with canonical parameter $(\alpha, \psi, \beta_1, \beta_2)$ and canonical sufficient statistic $(X_1 + X_2, X_2, Y_1, Y_2)$ as defined in (5.87). Determine numerical expressions for using the Skovgaard approximation and verify the MLEs in the text and the p-value of 0.10133. Show

that the four equations that determine $(\hat{\alpha}, \hat{\psi}, \hat{\beta}_1, \hat{\beta}_2)$ can be reduced to solving the single equation

$$n_1 \{\Psi(\hat{\alpha}) - \ln(n_1\hat{\alpha}/y_1)\} = x_1. \tag{5.96}$$

(c) Prove that the two theoretical p-values, under consideration for approximation in parts (a) and (b), are really the same as stated in (5.88).

(d) The extremely close p-values of 0.10127 and 0.10133 from parts (a) and (b) suggest the need to take a closer look at the MLE equations in (5.95) and (5.96). Show that the first-order approximation $\Psi(z) \simeq \ln z$ given in (6.3.18) of Abramowitz and Stegun (1970) makes these equations equivalent. Likewise show that the same approximation, when applied to the second equation from (a) that determines $\hat{\psi}$, makes that equation equivalent to the second equation in (b) that determines its value for $\hat{\psi}$.

15. For the two-sample gamma setting of Exercise 14, consider the two-sided test of equal shape parameters with $H_0 : \psi = 0$ versus $H_1 : \psi \neq 0$.
    (a) Derive the conditional p-value of the marginal likelihood ratio test using the exponential family structure of the marginal likelihood.
    (b) Derive the conditional p-value of the likelihood ratio test using the exponential family structure of the full likelihood.
    (c) Determine the p-value for the UMP unbiased test.

16. The three tests in Exercise 15 are best compared by determining their rejection regions for a common level of α.
    (a) Using the aircraft data with $\alpha = 0.05$ or 0.10, compute rejection regions for the three tests that incorporate the appropriate conditioning needed for the optimality of the various tests.
    (b) Compare the three tests by plotting their conditional power functions.

17. Consider the two-sided test of equal scale parameters in a two sample gamma setting with equal shape parameters.
    (a) Show, in particular, the three equations that determine the MLE $(\hat{\alpha}, \hat{\beta}, \hat{\psi})$ can be reduced to solving the single equation

    $$n_1 \left\{\Psi(\hat{\alpha}) - \ln \frac{n_1\hat{\alpha}}{z_{1\cdot}}\right\} + n_2 \left\{\Psi(\hat{\alpha}) - \ln \frac{n_2\hat{\alpha}}{z_{2\cdot}}\right\} = y$$

    for common shape estimate $\hat{\alpha}$.
    (b) Determine the rejection region for the likelihood ratio test at level $\alpha = 0.1$.
    (c) Compute the rejection region of the UMP unbiased test of level $\alpha = 0.1$. Is it close to the rejection region in part (b)?
    (d) Compare the conditional power functions of these two tests by plotting them on a common graph.

6

Further exponential family examples and theory

Statistical inference is considered in five practical settings. In each application, saddlepoint approximations offer an innovative approach for computing p-values, mid-p-values, power functions, and confidence intervals. As case studies, these five settings motivate additional considerations connected with both the saddlepoint methodology and theory that are used for making statistical inference. Each additional topic is addressed as it arises through its motivating example.

The first application concerns logistic regression, with an emphasis on the determination of LD50. p-value computation, test inversion to determine confidence intervals, and the notion of mid-p-values are considered. Saddlepoint treatment of prospective and retrospective analyses are also compared.

The second application deals with common odds ratio estimation. Both single- and double-saddlepoint methods are applicable and these two approaches are compared with several examples. Power function computations and their properties are discussed in connection with both the single- and double-saddlepoint approaches.

Time series data may also be analyzed and an example dealing with an autoregression of Poisson counts is given.

The final two applications are concerned with modeling nonhomogeneous Poisson processes, and inference with data that has been truncated. The former example reconsiders the Lake Konstanz data analyzed in Barndorff-Nielsen and Cox (1979) and the latter deals with the truncation of binomial counts that occur with genetic data.

The general use of saddlepoint methods for testing purposes in exponential families was introduced in Davison (1988), however these applications had already been anticipated much earlier in the discussion by Daniels (1958) concerning the logistic setting as well as in Barndorff-Nielsen and Cox (1979). Test inversion to determine confidence intervals was first considered in Pierce and Peters (1992).

6.1 Logistic regression and LD50 estimation

6.1.1 Example: Aphids

Quantal response data for determining the effect of nicotine on the mortality of aphids are presented in Morgan (1992, p. 95) and a small portion of this data is reproduced in table 6.1.

At nicotine level l_i for $i = 1, \ldots, I = 6$, the number of deaths d_i among n_i aphids receiving this dose level are recorded. The aphid deaths are assumed to be independent

Table 6.1. *Mortalities among aphids subjected to various doses of nicotine.*

Dose l_i	Total n_i	Dead d_i
0.01	4	0
0.02	5	1
0.03	3	1
0.04	5	2
0.06	5	4
0.08	4	4

Bernoulli (π_i) trials with π_i as the probability of death for an aphid with dose l_i. The model specifies that $D_i|l_i \sim$ Binomial (n_i, π_i) with π_i as a function of l_i. The canonical parameter in the Binomial is the logit of π_i which, for simplicity, is assumed to be linear in the dose, so that

$$\ln \frac{\pi_i}{1-\pi_i} = l_i \chi + \eta. \tag{6.1}$$

This is the so-called canonical link function relating π_i to l_i and leads to an exponential family with canonical sufficient statistics

$$x = \sum_{i=1}^{I} l_i d_i = 0.69 \qquad y = d. = \sum_{i=1}^{I} d_i = 12,$$

and canonical parameter $\theta = (\chi, \eta) \in \Re^2 = \Theta$. The exponential family has a likelihood of the form

$$\mathcal{L}(\chi, \eta) = \exp\{\chi x + \eta d. - c(\chi, \eta)\} \tag{6.2}$$

with

$$c(\chi, \eta) = -\sum_{i=1}^{I} n_i \ln(1-\pi_i) = \sum_{i=1}^{I} n_i \ln\{1 + \exp(l_i \chi + \eta)\},$$

which determines the joint CGF of X and Y.

The MLE for the logistic fit is $(\hat{\chi}, \hat{\eta}) = (87.68, -3.661)$ with asymptotic standard deviations, computed from $c''(\hat{\chi}, \hat{\eta})^{-1}$, as 32.22 and 1.359 respectively. The constrained MLE with $\chi = 0$ is $\hat{\eta}_0 = -0.1542$. The data in table 6.1 clearly reveal an increasing death rate with dosage level. This is confirmed by the test comparing $H_0 : \chi = 0$ versus $H_1 : \chi > 0$; its p-values from the Skovgaard approximations, with and without continuity corrections as described in §5.4.5, are the very significant values

$$\widehat{\Pr}(X \geq 0.69 | Y = 62; \chi = 0) = \begin{cases} 0.0^3 1609 & \text{if uncorrected} \\ 0.0^3 2442 & \text{if } \widehat{\Pr}_1 \\ 0.0^3 2344 & \text{if } \widehat{\Pr}_2. \end{cases} \tag{6.3}$$

The two continuity corrections are nearly the same and are about 50% larger than the uncorrected approximation. By comparison, the (unconditional) log-likelihood ratio and

Table 6.2. 95% confidence intervals for LD50 from p-value inversion using the three saddlepoint approximations described in section 6.1.5.

	uncorrected	$\widehat{\Pr}_1$	$\widehat{\Pr}_2$
lower	.02894	.03123	.03117
upper	.05803	.06252	.06200

Wald tests attain the values $\chi_1^2 = 14.20$ and 7.406 respectively with asymptotic p-values 0.0^31645 and 0.0^26501. The likelihood ratio p-value is very close to the uncorrected saddlepoint p-value for this example.

Exact p-value computations are known to lead to p-values that are larger than the corresponding asymptotic approximations (Routledge, 1994, §1). As may be seen in (6.3), the continuity-corrected saddlepoint approximations are also larger and exhibit the same trait due to their accuracy.

An LD-50 dose level is a value of l, λ say, whose associated probability of death is $1/2$, i.e.

$$0 = \ln \frac{\pi}{1-\pi} = \lambda \chi + \eta.$$

The MLE for LD50 dosage is $\hat{\lambda} = -\hat{\eta}/\hat{\chi} = 0.04176$, a quite reasonable estimate when judged from the data in table 6.1. A 95% confidence interval for the true value $\lambda = -\eta/\chi$ can be computed by inverting 2.5% level one-sided hypothesis tests over a grid of values for λ. An hypothesis test, for a fixed λ-value under consideration, is constructed by changing the model parameterization so that the parameter $\lambda\chi + \eta = \psi_\lambda$ is canonical in the exponential family. Then, the test of hypothesis becomes a test that $\psi_\lambda = 0$. Details of this analysis are given in Chapman et al. (2007) and are partially reproduced below in section 6.1.5. The method, for the aphid data, determines the confidence intervals in table 6.2.

It turns out that the interval produced without continuity correction is "better centered" than the others. This means that the undershoot and overshoot coverage probabilities tend to be closer to the target value of 2.5% in each tail which provides for a more centered 95% confidence interval. This is discussed in section 6.1.4 where the subject of mid-p-values is also introduced. Before dealing with LD50 however, a general discussion of the logistic model is needed.

6.1.2 Logistic regression and significance tests

A general logistic dose response relationship using m independent variables or dosage components admits data as $D_i \sim \text{Binomial}(n_i, \pi_i)$ for cases $i = 1, \ldots, I$ with link function

$$\ln \frac{\pi_i}{1-\pi_i} = l_i^T \chi. \tag{6.4}$$

Here $l_i^T = (l_{i1}, \ldots, l_{im})$ sets the individual dosage levels for case i and, more generally, consists of the independent variables in the regression relationship with $\chi^T = (\chi_1, \ldots \chi_m)$

as the regression slopes. This general model includes the aphid example if η is included in the χ vector as an independent variable representing location. Suppose that L is the $I \times m$ matrix whose rows are $\{l_i^T\}$; then

$$L^T = (l_1 \ \ldots \ l_I) = (l_{ji}) \tag{6.5}$$

is $m \times I$. Assume that L consists of fixed values either set to certain levels by design or, when random, fixed and conditioned upon for inference purposes. Let $d^T = (d_1, \ldots d_I)$ contain the binomial responses. The canonical parameter is χ with corresponding canonical sufficient statistic

$$x = (x_1, \ldots, x_m)^T = L^T d = \sum_{i=1}^{I} l_i d_i.$$

A p-value for the test of an individual regression component such as χ_1 is now a routine double-saddlepoint computation applied to the conditional distribution of $X_1 | x_2, \ldots, x_m$. No theoretical issues arise when the m independent variable values in L lie on regular δ-lattices. Should X_1 lie on such a δ-lattice, the continuity corrections of section 5.4.5 generally provide greater accuracy; however there are questions about the relevance of exact p-values in this context as discussed in section 6.1.4. Without such lattice support, formal justification for the use of saddlepoint methods must rest upon the assumption that the lattice increments are small enough that the continuous formulas can provide approximate p-values.

The CGF of X is $c(s + \chi) - c(\chi)$ where

$$c(\chi) = -\sum_{i=1}^{I} n_i \ln(1 - \pi_i) = \sum_{i=1}^{I} n_i \ln\left\{1 + \exp\left(l_i^T \chi\right)\right\},$$

with π_i denoting a function of χ given as

$$\pi_i = \pi_i(\chi) = \frac{\exp\left(l_i^T \chi\right)}{1 + \exp\left(l_i^T \chi\right)}.$$

It is simpler to maintain the π_i notation and so the discussion will conform to this by denoting MLEs as $\hat{\pi}_i = \pi_i(\hat{\chi})$. Correspondingly, if $\hat{\mu}^T = (n_1 \hat{\pi}_1, \ldots, n_I \hat{\pi}_I)$ is the MLE for the mean of d, then the numerator saddlepoint solves the m equations

$$x = c'(\hat{\chi}) = L^T \hat{\mu} = \sum_{i=1}^{I} l_i n_i \hat{\pi}_i.$$

If $\hat{D} = \text{diag}\{n_i \hat{\pi}_i (1 - \hat{\pi}_i)\}$ is the MLE for the covariance of d, then the Hessian is

$$c''(\hat{\chi}) = L^T \hat{D} L = \sum_{i=1}^{I} n_i \hat{\pi}_i (1 - \hat{\pi}_i) l_i l_i^T. \tag{6.6}$$

This is the observed Fisher information matrix for the canonical parameter χ so that the asymptotic variance of $\hat{\chi}_i$, used in its normal approximation confidence interval, is the ith diagonal element of $(L^T \hat{D} L)^{-1}$. Simple modifications to these equations give the denominator saddlepoint.

6.1.3 Confidence interval theory from p-value inversion

Confidence intervals may be obtained by inverting the *p*-value computations just described in section 6.1.2. Suppose that (X, Y) is canonical sufficient with associated canonical parameter (χ, ψ). where ψ is a scalar interest parameter. An upper confidence interval for ψ of the form (ψ_L, ∞) is constructed from data (x, y) using the appropriate tail probability associated with the one-sided test of $H_0 : \psi = \psi_0$ versus $H_1 : \psi > \psi_0$. The probability is determined from the conditional distribution of $Y|X = x$ that does not depend on the nuisance parameter χ. Insignificance of ψ_0 at a 2.5% level is the indicator that the value ψ_0 belongs in this 97.5% one-sided confidence interval. Thus, the boundary value ψ_L solves

$$\widehat{\Pr}(Y \geq y | X = x; \psi_L) = 0.025. \tag{6.7}$$

The other one-sided test with alternative $H_1 : \psi < \psi_0$ determines a 97.5% lower interval $(-\infty, \psi_U)$ in which the upper edge must solve

$$\widehat{\Pr}(Y \leq y | X = x; \psi_U) = 0.025. \tag{6.8}$$

The intersection of these two one-sided intervals is (ψ_L, ψ_U), a centered 95% two-sided confidence interval.

This method may be applied, for example, to the aphid data taking $x = 1^T d = d.$ as the sufficient statistic associated with location vector 1. Exercise 1 asks the reader to determine 95% confidence intervals for χ and η in this manner. Note that ψ_L is determined graphically by plotting the increasing function

$$\widehat{\Pr}(Y \geq y | X = x; \psi) \quad \text{vs.} \quad \psi \tag{6.9}$$

and noting the appropriate cutoff at the inverse value of the height 0.025. Likewise ψ_U is determined by plotting the the left side of (6.8) versus ψ as a decreasing function and inverting it at a height of 0.025.

Further understanding of these confidence procedures is obtained by investigating the pivotal nature of the $\hat{w}$ argument in the Lugannani and Rice and Skovgaard approximations. Consider first the Lugannani and Rice approximation in the one parameter setting of section 5.4.3 where θ is canonical and X is canonical sufficient and attains the value x. When checking whether θ is in the confidence interval, the dominant term for the tail probability $\widehat{\Pr}(X \geq x; \theta)$ is $1 - \Phi\{\hat{w}(\theta)\}$, with $\hat{w}(\theta)$ given in (5.64) and often referred to as the *signed likelihood ratio statistic* of the test. As discussed in sections 2.4.5 and 4.4.1, $\hat{w}(\theta)$ has an approximate Normal $(0, 1)$ distribution with large samples; thus the distribution of $1 - \Phi\{\hat{w}(\theta)\}$ is roughly uniform on $(0, 1)$ as a probability integral transform. The signed likelihood ratio statistic is also a decreasing function in θ as the reader is asked to show in Exercise 2. Therefore, $1 - \Phi\{\hat{w}(\theta)\}$ is an approximate pivot that is increasing in θ. Since the additional correction term for the Lugannani and Rice approximation is often much smaller, it may be presumed that its inclusion does not substantially alter this monotonicity; the details of such power monotonicity are discussed in section 6.2.2. Thus $\hat{w}(\theta)$ is an approximate Normal $(0, 1)$ pivot whose pivotal accuracy is improved when using the additional correction afforded by the Lugannani and Rice approximation. This motivates the method that sets the lower confidence value θ_L as the solution to

$$\widehat{\Pr}(X \geq x; \theta_L) = 0.025,$$

where $\widehat{\Pr}$ is the Lugannani and Rice approximation. The upper confidence value is similarly motivated and determined by solving

$$\widehat{\Pr}(X \leq x; \theta_U) = 0.025 \qquad (6.10)$$

where (6.10) is presumed to be decreasing in θ.

In the Skovgaard setting, the confidence interval is for interest parameter ψ in the presence of nuisance parameter χ. The leading term of the approximation is $1 - \Phi\{\hat{w}(\psi)\}$ which is also increasing in ψ since $\hat{w}(\psi)$ is decreasing in ψ; a proof of this is outlined in Exercise 3. Monotonicity of the power function determined from the Skovgaard approximation is discussed in section 6.2.4. Thus the Skovgaard approximation provides a refined conditional pivot that improves upon the first-order pivot which assumes $\hat{w}(\psi) \sim$ Normal $(0, 1)$.

6.1.4 Mid-p-values and their inversion

In the discrete setting of Y, an exact one-sided significance test for ψ is likely to be overly conservative in the sense that the actual type I error rate is smaller than the nominal level for the test. For example, consider the nominal 5% level test that is usually implemented by computing the p-value and rejecting H_0 if $p < 0.05$. Often this rejection rule is overly conservative when p is the exact p-value. This means that p tends to be somewhat larger than it should be if a type I error rate of 5% is to be maintained. This causes the test to accept H_0 too often. One remedy for this is to deflate the p-value slightly by computing mid-p-values as discussed in Agresti (1992), Routledge (1994), and Kim and Agresti (1995).

A mid-p-value is the p-value deflated by half the mass at the cutoff y. When testing $H_0 : \psi = \psi_0$ versus $H_1 : \psi > \psi_0$, it may be specified as

$$\text{mid-}\Pr(Y \geq y \mid X = x; \psi_0) = \tfrac{1}{2}\{\Pr(Y \geq y|X = x; \psi_0) + \Pr(Y \geq y + \delta|X = x; \psi_0)\}$$

if the support of Y is on the δ-lattice. There are a number of ways to approximate this value. The most obvious approach uses two continuity-corrected saddlepoint approximations at y and $y + \delta$. A different method of approximation however, has proved to be not only simpler but also more accurate. This entails using the uncorrected continuous version of the Skovgaard formula and treating Y as if it were a continuous variable. This approach requires only a single-saddlepoint computation at y. Exercise 4 considers an alternative approximation that proves to be less reliable.

When testing with alternative $H_1 : \psi < \psi_0$, the p-value $\Pr(Y \leq y|X = x; \psi_0)$ is likewise reduced by half the probability of y and the mid-p-value is

$$\text{mid-}\Pr(Y \leq y \mid X = x; \psi_0) = \tfrac{1}{2}\{\Pr(Y \leq y|X = x; \psi_0) + \Pr(Y \leq y - \delta|X = x; \psi_0)\}.$$

The deflation to the mid-p-value causes the test to reject H_0 more often and thus decreases the over-conservatism and increases the type I error rate.

An understanding of why the continuous Lugannani and Rice and Skovgaard formulas approximate mid-p-values requires some consideration of Fourier inversion formulas for CDFs. Suppose X is absent from consideration, and $\Pr(Y \leq y; \psi_0)$ is the p-value. If the CDF of Y has a step discontinuity at y, then it is a well-known fact that exact Fourier inversion of the MGF of Y to determine the CDF at y does not actually yield the CDF at y; it rather gives the average of its values at y^+ and y^- or the mid-p-value. See theorem 6.2.1 of

6.1 Logistic regression and LD50 estimation

Chung (2001) as concerns univariate CDF inversion and Theorem 10.7b of Henrici (1977) as concerns the general inversion of Laplace transforms. Then, by the very nature of the exact inversion, the approximate inversion, performed by using the continuous Lugannani and Rice formula, provides a natural and direct route to approximating the mid-p-value. This property also extends to the setting in which there is conditioning on $X = x$ so that the continuous Skovgaard formula provides a direct approximation for the mid-p-values given above.

If pairs of continuity corrections are used to approximate mid-p-values, then these left- and right-tail expressions are related to one another. The relationship is

$$\text{mid-}\widehat{\Pr}(Y \leq y | X = x; \psi) = 1 - \text{mid-}\widehat{\Pr}(Y \geq y | X = x; \psi) \tag{6.11}$$

when $\widehat{\Pr}$ is either the uncorrected Skovgaard approximation or the first $\widehat{\Pr}_1$ or second $\widehat{\Pr}_2$ continuity correction. This result easily follows from (1.29) or (1.36). Thus a single formula suffices for expressing the mid-p-value concept in both tails.

Example

For the aphid data, the mid-p-values given below may be compared with the p-value entries of (6.3).

$$\text{mid-}\widehat{\Pr}(X \geq 0.69 | Y = 62; \chi = 0) = \begin{cases} 0.0^3 1609 & \text{if uncorrected} \\ 0.0^3 1754 & \text{if mid-}\widehat{\Pr}_1 \\ 0.0^3 1678 & \text{if mid-}\widehat{\Pr}_2. \end{cases} \tag{6.12}$$

The mid-p-values are smaller than their ordinary p-value counterparts as expected. All three approximations are very similar but the uncorrected approximation provides the simplest approximation.

Mid-p-value inversion

When Y is discrete, confidence intervals for ψ, constructed by inverting the p-values of one-sided tests, may also be too conservative and wide. In such instances, the overly wide confidence intervals attain coverage that often exceeds 95% even though the nominal coverages have been set at 95%. A remedy for this is to invert mid-p-values rather than p-values.

To see that the use of mid-p-values shortens confidence intervals, suppose that ψ_L is the left edge of the confidence interval from exact p-value inversion; thus

$$\Pr(Y \geq y | X = x; \psi_L) = 0.025.$$

Now, with mid-p-value deflation, $\Pr(Y \geq y + \delta | X = x; \psi_L) < 0.025$ so that

$$\text{mid-}\Pr(Y \geq y | X = x; \psi_L) < 0.025. \tag{6.13}$$

To achieve equality in (6.13), ψ_L must be increased which shortens the confidence interval on the left side.

A 95% confidence interval is computed by solving the equations

$$\text{mid-}\widehat{\Pr}(Y \geq y | X = x; \psi_L) = 0.025 = 1 - \text{mid-}\widehat{\Pr}(Y \geq y | X = x; \psi_U) \tag{6.14}$$

Table 6.3. *95% confidence intervals for LD50 using mid-p-value inversion with the various continuity corrections.*

	uncorrected p-value	mid-$\widehat{\mathrm{Pr}}_1$	mid-$\widehat{\mathrm{Pr}}_2$
lower	.02894	.03208	.03214
upper	.05803	.06035	.05994

for $\psi_L < \psi_U$, where the latter equation has made use of (6.11). Thus all three of the proposed mid-p inversion methods may be implemented by solving the two equations in (6.14). These equations are easily solved using quasi-Newton root-finders in the various programming languages and are routine computational exercises.

Example

For the aphid data, the mid-p-value confidence intervals for LD50 of level 95% are given in table 6.3. The theory for their determination is given in section 6.1.5.

6.1.5 LD50 confidence intervals

The logistic model of the aphid example may be extended to include m multiple dose component levels. The methodological development for the determination of LD50 confidence regions is presented below with further discussion in Chapman *et al.* (2007).

Suppose the multiple dose levels are summarized in the $I \times m$ matrix L with $m \times 1$ regression vector χ. Let η be a separate location variable so that the ith case has the link

$$\ln \frac{\pi_i}{1 - \pi_i} = l_i^T \chi + \eta$$

with inverse relationship

$$\pi_i(\chi, \eta) = \frac{\exp\left(l_i^T \chi + \eta\right)}{1 + \exp\left(l_i^T \chi + \eta\right)}. \tag{6.15}$$

The canonical sufficient statistics are $x = L^T d$ and $d_.$. Let $\lambda^T = (\lambda_1, \ldots \lambda_m)$ be one of a collection of dose levels achieving LD50, i.e.

$$0 = \ln \frac{\pi}{1 - \pi} = \lambda^T \chi + \eta = \psi_\lambda.$$

A confidence region for λ is derived from the theory above by creating an exponential family within which parameter ψ_λ is canonical. This is achieved by adding and subtracting $\lambda^T \chi d_.$ so that (χ, ψ_λ) becomes canonical. The adding and subtracting is equivalent to a linear transformation of (χ, η) using the $(m+1) \times (m+1)$ matrix B_λ^T defined as

$$B_\lambda^T \begin{pmatrix} \chi \\ \eta \end{pmatrix} = \begin{pmatrix} I_m & 0 \\ \lambda^T & 1 \end{pmatrix} \begin{pmatrix} \chi \\ \eta \end{pmatrix} = \begin{pmatrix} \chi \\ \psi_\lambda \end{pmatrix}.$$

Then,
$$B_\lambda^{-1}\begin{pmatrix}x\\d.\end{pmatrix}=\begin{pmatrix}I_m & -\lambda\\0^T & 1\end{pmatrix}\begin{pmatrix}x\\d.\end{pmatrix}=\begin{pmatrix}x-\lambda d.\\d.\end{pmatrix}=\begin{pmatrix}z_\lambda\\d.\end{pmatrix}.$$

The likelihood remains unchanged as
$$\mathcal{L}(\chi,\eta)=\exp\{\chi^T x+\eta d.-c(\chi,\eta)\} \tag{6.16}$$
$$=\exp\{\chi^T z_\lambda+\psi_\lambda d.-c(\chi,\psi_\lambda-\lambda^T\chi)\}=\mathcal{L}_\lambda(\chi,\psi_\lambda),$$

where
$$c(\chi,\eta)=-\sum_{i=1}^{I}n_i\ln\{1-\pi_i(\chi,\eta)\}=\sum_{i=1}^{I}n_i\ln\left\{1+\exp\left(l_i^T\chi+\eta\right)\right\}.$$

If $(\hat{\chi},\hat{\eta})$ is the MLE, then the space of MLEs for λ consists of
$$\{\hat{\lambda}:\hat{\psi}_{\hat{\lambda}}=\hat{\lambda}^T\hat{\chi}+\hat{\eta}=0\} \tag{6.17}$$

which is an $(m-1)$-dimensional subplane with gradient $\hat{\chi}$. A 95% confidence region in $\Re^m$ is formed by finding the collection of λ-values satisfying the following two constraints:
$$0.025<\text{mid-}\widehat{\Pr}(Y\geq d.|Z_\lambda=z_\lambda;\psi_\lambda=0)<0.975. \tag{6.18}$$

Numerical computation

Uncorrected *p*-value computation using the continuous Skovgaard formula is recommended to keep the computations simple. The computations required for (6.18) depend on λ in a very complicated way: through all the inputs to the saddlepoint approximation including the statistic z_λ upon which the coverage is conditioned.

The computational burden in finding these confidence bands is minimal however. As λ varies for confidence interval determination, the MLE $(\hat{\chi},\hat{\psi}_\lambda)$ is an explicit function of $(\hat{\chi},\hat{\psi}_0=\hat{\eta})$ for each λ; thus there is no new root-finding required for the numerator saddlepoint with each new λ-value. Only the single root $\hat{\chi}_{0\lambda}$ in the denominator saddlepoint needs to be found with each new λ-value. This also is not difficult when the incremental change in λ, or $\Delta\lambda$, is small as it would be over a fine grid of λ-values. The search for the root $\hat{\chi}_{0,\lambda+\Delta\lambda}$ at $\lambda+\Delta\lambda$ may be started at $\hat{\chi}_{0\lambda}$, the root for the previous value λ, and located after only a few Newton iterations. See Exercise 5 for further details.

Example: Aphids

A plot of the survival probabilities used to determine the confidence interval from the continuous Skovgaard approximation is shown in figure 6.1. The figure plots the mid-*p* survival probabilities versus λ for the aphid data. Inversions of the horizontal lines at heights 0.025 and 0.975 provide the 95% confidence interval (0.02894, 0.05803) for λ.

6.1.6 Prospective and retrospective analyses

The aphid data is a prospective study. With such sampling schemes, each aphid is selected, exposed to a certain dosage of nicotine, and then monitored over time until its outcome,

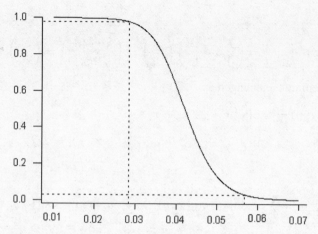

Figure 6.1. Plots of (6.18) versus λ to determined a 95% confidence interval for LD50.

dead or alive, can be recorded in response to the dosage. An alternative sampling scheme is retrospective sampling such as would occur when sampling from pre-existing historical records. In the context of the aphid data, this would amount to supposing that the data in table 6.1 were obtained by sampling previous data recorded in these historical records. The main presumption here is that some researcher has already summarized these experiments and has provided an historical list of sacrificed aphids listing the nicotine dose and the response, dead or alive, for each aphid. If the historical list is sampled using a scheme for which the sampling probability of an aphid is not dependent on the administered dosage, then the sampling is said to be retrospective sampling.

In retrospective sampling, suppose that ζ and $1 - \zeta$ are the respective probabilities of sampling dead and alive aphids from the historical list. If the linear logistic model is used to fit the retrospective data set, then McCullagh and Nelder (1989, §4.3.3) have shown that the likelihood is the same as for prospective sampling except for a translation of the canonical intercept parameter. For the aphid data, the model is the same as (6.1) except that the intercept parameter η is now replaced by $\eta^* = \eta + \ln\{\zeta/(1 - \zeta)\}$. The slope parameter χ in the retrospective model is the same as that for the prospective model. These same equivalences apply more generally for any linear logistic model.

If data from a prospective study have exponential likelihood $\mathcal{L}(\eta, \chi)$, then the same data from a retrospective study would have the tilted likelihood $\mathcal{L}(\eta^*, \chi)$. It follows that conditional inference for the slope χ treating η^* as a nuisance parameter is identical for either model. The conditional saddlepoint approximations for the corresponding canonical sufficient statistics are also identical so that saddlepoint inference is also identical. Inference that concerns the intercept parameter, such as LD50 estimation, does however depend on whether the sample is prospective or retrospective; see Chapman *et al.* (2007) for details.

Saddlepoint methods in retrospective sampling have also been discussed in Wang and Carroll (1999) from a different perspective. Their approach is to develop density approximations for the MLEs directly from an estimating equation perspective, a subject to be considered in chapter 12. By taking this approach, the authors are forced to deal with random covariates $\{l_i\}$ and use this randomness to justify continuous densities for the MLEs. One of their main conclusions is that the marginal densities for the slope estimates are the same to higher asymptotic order regardless of whether the data are obtained from prospective or retrospective sampling.

The main conceptual difficulty with the Wang–Carroll approach is that it violates the strong likelihood principle; see Berger and Wolpert (1988). From the perspective of conditional inference, the sampled random covariates are ancillary statistics whose marginal distribution is irrelevant in making inference about the slope coefficients. Therefore, methods which do not condition on the ancillary covariates violate the conditionality principle and hence the strong likelihood principle; see Birnbaum (1962) and Bjørnstad (1996). A Bayesian perspective is similar if a flat improper prior is chosen for the location parameter η. Since the two likelihoods agree so $\mathcal{L}(\eta, \chi) = \mathcal{L}(\eta^*, \chi)$ and the priors on η and η^* are the same, then the marginal posteriors on χ must be the same.

6.2 Common odds ratio in 2 × 2 tables

6.2.1 With $1 : m_i$ matching cases

Such data occur when comparing a treatment to a control in a setting for which the responses are Bernoulli trials. The situation is made more difficult by the presence of covariates that may confound the comparison. In such instances, the covariate effects may be controlled by matching each treatment subject to control subjects that have the same covariate value.

Suppose there are n treatment subjects and let π_{1i} be the success probability of the ith treatment subject expressed as

$$\ln \frac{\pi_{1i}}{1 - \pi_{1i}} = \chi_i + \psi. \tag{6.19}$$

Here, χ_i is the effect due to the covariates and ψ is a treatment effect common to all treatment subjects. The response is $X_{1i} \sim$ Bernoulli (π_{1i}), the indicator of success. This treatment subject is matched with m_i control subjects who share the same, or approximately the same, covariates as measured with the composite effect χ_i; let their common success probability π_{2i} be

$$\ln \frac{\pi_{2i}}{1 - \pi_{2i}} = \chi_i. \tag{6.20}$$

Now $X_{2i} \sim$ Binomial (m_i, π_{2i}) records the number of matched control successes. The experimental results are summarized as a set of n two by two tables with row and column totals given as

	Successes	Failures	Total
Treatment	x_{1i}	$1 - x_{1i}$	1
Control	x_{2i}	$m_i - x_{2i}$	m_i
Total	$x_{\cdot i}$	$m_i + 1 - x_{\cdot i}$	$m_i + 1$

(6.21)

for $i = 1, \ldots, n$.

Parameter ψ measures the treatment effect and its relationship to the common odds ratio φ is

$$\frac{\pi_{1i}}{1 - \pi_{1i}} \div \frac{\pi_{2i}}{1 - \pi_{2i}} = e^{\psi} = \varphi \tag{6.22}$$

Table 6.4. *The crying baby data in which x_{1i} and x_{2i} record the number of babies that are not crying.*

Day $i =$	1	2	3	4	5	6	7	8	9
x_{1i}	1	1	1	0	1	1	1	1	1
x_{2i}	3	2	1	1	4	4	5	4	3
m_i	8	6	5	6	5	9	8	8	5
Day $i =$	10	11	12	13	14	15	16	17	18
x_{1i}	0	1	1	1	1	1	1	0	1
x_{2i}	8	5	8	5	4	4	7	4	5
m_i	9	6	9	8	5	6	8	6	8

for all i. The exponential family structure has canonical parameter (χ, ψ) with $\chi = (\chi_1, \ldots, \chi_n)^T$ and corresponding canonical sufficient statistic (X, Y), where $X = (X_{\cdot 1}, \ldots, X_{\cdot n})^T$ and

$$Y = X_{1 \cdot} = \sum_{i=1}^{n} X_{1i}.$$

Testing $\psi = 0$ or equivalently $\varphi = 1$

A conditional test of $H_0 : \psi = 0$ that is free of the nuisance parameter χ is based on the conditional null distribution of Y given $X = x$. Davison (1988) first considered using the appropriate double-saddlepoint approximation to determine the p-value of this test. The p-value and power computation for the test will be introduced later in section 6.2.3.

The structure of this distribution is best understood with reference to the table in (6.21). When conditioning on vector $X = x$, the X_{1i} component of Y depends only on conditioning component $X_{\cdot i}$ with $X_{1i} | X_{\cdot i} = x_{\cdot i} \sim$ Bernoulli $\{\hat{\pi}_i\} = B_i$. Here,

$$\hat{\pi}_i = x_{\cdot i}/(m_i + 1)$$

is the null estimate of $\pi_{1i} = \pi_{2i}$. Therefore, the conditional null distribution is characterized as a sum of independent Bernoulli variables

$$Y|X = x \sim \sum_{i=1}^{n} B_i \qquad (6.23)$$

and is said to be Bernoulli decomposable; see Quine (1994).

A single-saddlepoint approximation for the distribution in (6.23) is now based on its CGF

$$K(s) = \sum_{i=1}^{n} \ln\{\hat{\pi}_i(e^s - 1) + 1\}. \qquad (6.24)$$

This approach has been recommended in Bedrick and Hill (1992). While the distribution theory has arisen through conditioning, the conditional distribution may be approximated directly using a single-saddlepoint approximation as previously suggested in section 4.4.4.

The crying baby data in table 6.4, taken from Cox (1970, p. 71), and also used in Davison (1988), provides an interesting example for discussion. The covariate in this instance is

6.2 Common odds ratio in 2 × 2 tables

Table 6.5. *P-value approximations in testing $\varphi = 1$ vs. $\varphi > 1$ for the crying baby data.*

Saddlepoint Type	Exact	Uncorrected	$\widehat{\Pr}_1$	$\widehat{\Pr}_2$
Single	0.045	0.02428	0.04564	0.04474
Double	0.045	0.02399	0.04737	0.04264

the time variable "Day" which leads to 18 tables. The one-sided test of $H_0 : \varphi = 1$ versus $H_1 : \varphi > 1$ has the *p*-values listed in table 6.5 when using the various continuity-corrected and uncorrected saddlepoint approximations. Single-saddlepoint approximations have been derived directly from (6.24) whereas the double-saddlepoint approximations have been determined from the methods suggested later in section 6.2.3.

The exact *p*-value was computed as 0.045 in Cox (1970). A high degree of accuracy is seen using both the single- and double-saddlepoint methods with continuity correction.

Power function

The attained conditional power function for the test $H_0 : \psi = 0$ versus $H_1 : \psi > 0$ is the plot of $\Pr(Y \geq y | X = x; \psi)$ as a function of ψ where x and y are the attained values of X and Y. At $\psi = 0$ or $\varphi = 1$, this attained power is the *p*-value and its value for $\varphi \neq 1$ reveals the rejection probability of the testing procedure for alternative values of φ. For this situation in which there are canonical nuisance parameters, conditional tests have been used and the attained conditional power function provides a uniformly minimum variance unbiased estimator of the unconditional power function. Exercise 6 asks the reader to show that the noncentral distribution of $Y | X = x$, with $\psi \neq 0$, remains of the form in (6.23) but with $B_i \sim \text{Bernoulli } \{\hat{\pi}_{i\varphi}\}$ where

$$\hat{\pi}_{i\varphi} = \frac{\hat{\pi}_i \varphi}{\hat{\pi}_i \varphi + (1 - \hat{\pi}_i)} = \frac{\hat{\pi}_i e^{\psi}}{\hat{\pi}_i e^{\psi} + (1 - \hat{\pi}_i)} = \hat{\pi}_{i\psi}. \tag{6.25}$$

The noncentral CGF is now given in (6.24) but with $\{\hat{\pi}_{i\varphi} = \hat{\pi}_{i\psi}\}$ replacing $\{\hat{\pi}_i\}$. Two notations have been defined in (6.25) because both have importance. On the practical side, the plotting of power should be versus common odds ratio φ; however, insight into the saddlepoint theory for power computation is best understood in terms of the canonical parameter ψ.

In saddlepoint approximation to the power function, simplification occurs when determining the nonnull saddlepoints denoted as $\hat{s}^{\psi}$ when $\psi \neq 0$. If $K_{\psi}(s)$ denotes the noncentral CGF, then the saddlepoint equation may be written as

$$x_1 = K'_{\psi}(\hat{s}^{\psi}) = \sum_{i=1}^{n} \frac{\hat{\pi}_i \exp(\hat{s}^{\psi} + \psi)}{\hat{\pi}_i \exp(\hat{s}^{\psi} + \psi) + 1 - \hat{\pi}_i}. \tag{6.26}$$

The null saddlepoint solution $\hat{s}^0$ occurs in (6.26) when $\psi = 0$; therefore the noncentral saddlepoint is related to the null saddlepoint according to

$$\hat{s}^{\psi} = \hat{s}^0 - \psi. \tag{6.27}$$

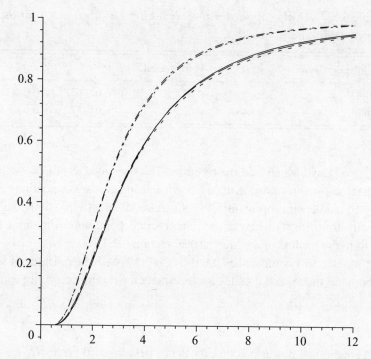

Figure 6.2. Single-saddlepoint approximations for attained power, using $\widehat{\Pr}_1$ (*dot-dashed*) and $\widehat{\Pr}_2$ (*dot-dot-dashed*) and for mid-p power using no correction (*solid*), mid-$\widehat{\Pr}_1$ (*dashed*), and mid-$\widehat{\Pr}_2$ (*dotted*).

When plotting the power function over a grid of values for $\varphi = e^{\psi}$, a separate saddlepoint root search for each grid value of ψ is unnecessary. The power function becomes an explicit function of $\hat{s}^0$ and ψ as a result of (6.27). The important practical consequence is that the entire power function plot is provided explicitly once the single null saddlepoint $\hat{s}^0$ solution is found. The uncorrected inputs for the Lugannani and Rice tail probability are

$$\hat{w}(\psi) = \text{sgn}(\hat{s}^0 - \psi)\sqrt{2\{(\hat{s}^0 - \psi)x_{1\cdot} - K_{\psi}(\hat{s}^0 - \psi)\}}$$
$$\hat{u}(\psi) = (\hat{s}^0 - \psi)\sqrt{K''_{\psi}(\hat{s}^0 - \psi)}.$$

The continuity corrections are likewise explicit and it should be clear that relationship (6.27) continues to hold when the saddlepoint equation is offset to $x_{1\cdot} - 1/2 = x_{1\cdot}^-$ for use with the second correction.

Figure 6.2 plots single-saddlepoint approximations for the attained conditional power and mid-p-power functions versus odds ratio $\varphi = e^{\psi}$ when testing $H_0 : \varphi = 1$ versus $\varphi > 1$. The two higher plots are $\widehat{\Pr}_1(Y \geq 15 | X = x; \varphi)$ (dot-dashed) and $\widehat{\Pr}_2(Y \geq 15 | X = x; \varphi)$ (dot-dot-dashed) and approximate the larger attained power. The lower three plots are approximations for mid-p-power using the uncorrected p-value (solid), mid-$\widehat{\Pr}_1$ (dashed), and mid-$\widehat{\Pr}_2$ (dotted). The three plots are all very close together and this once again reinforces the point that the uncorrected saddlepoint approximation is a very simple and accurate way to determine mid-p-value power functions.

6.2 Common odds ratio in 2 × 2 tables

Table 6.6. *Attained conditional power computations for the single- and double-saddlepoint approximations.*

		$\widehat{\Pr}_1$		$\widehat{\Pr}_2$	
φ	Exact	Single	Double	Single	Double
2	.3097	.3137	.3226	.3089	.3009
3	.5619	.5676	.5810	.5610	.5543
4	.7240	.7297	.7439	.7231	.7213
6	.8808	.8852	.8967	.8802	.8857
8	.9417	.9448	.9531	.9413	.9491

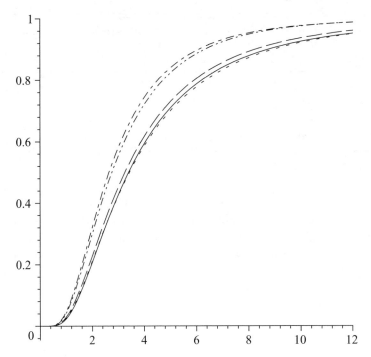

Figure 6.3. Double-saddlepoint approximations for the power and mid-p power plots displayed in figure 6.2 with the same denotations.

The corresponding double-saddlepoint approximations, to be introduced in section 6.2.3, are shown in figure 6.3 using similar denotations. All are approximately correct however the various approximations for the same power function show slightly greater variability than in the single-saddlepoint setting.

A closer numerical comparison of the p-values in figures 6.2 and 6.3 with the exact power computations, taken from Strawderman and Wells (1998), is shown in table 6.6. The $\widehat{\Pr}_2$ approximation with the single-saddlepoint approximation is the most accurate and its double-saddlepoint counterpart is also very close. It appears, based on table 6.6 and the two figures, that the double-saddlepoint $\widehat{\Pr}_1$ and mid-p-$\widehat{\Pr}_1$ approximations are the least accurate and are responsible for the variability of approximations seen in figure 6.3. Overall,

Table 6.7. *95% confidence intervals for the odds ratio φ from p-value inversion using the crying baby data.*

Saddlepoint type		exact	uncorrected	$\widehat{\mathrm{Pr}}_1$	$\widehat{\mathrm{Pr}}_2$
Single	lower	0.865	1.01	0.861	0.865
	upper	21.4	15.5	20.0	21.5
Double	lower	0.865	1.01	0.853	0.876
	upper	21.4	15.6	16.8	19.1

Table 6.8. *95% confidence intervals for the odds ratio φ from mid-p-value inversion using the crying baby data.*

Saddlepoint type		p-value uncorrected	mid-p-value $\widehat{\mathrm{Pr}}_1$	mid-p-value $\widehat{\mathrm{Pr}}_2$	K&A$_1$	K&A$_2$
Single	lower	1.01	0.971	0.977	1.04	1.01
	upper	15.5	16.1	17.0	14.9	11.1
Double	lower	1.01	0.961	0.987		
	upper	15.6	14.0	15.5		

the single-saddlepoint approximations are only slightly better than their double-saddlepoint counterparts. Note that all the approximations achieve entirely acceptable accuracy in this example. Both the single- and double-saddlepoint approximations should continue to maintain such accuracy in other practical settings but the simplicity of the single-saddlepoint makes it preferable with $1:m_i$ matching.

Confidence interval for $\varphi = e^\psi$

First consider the inversion of ordinary power functions represented by the higher graphs in figures 6.2 and 6.3. These inversions at height 0.025 determine the lower confidence limits listed in table 6.7 when using both single- and double-saddlepoint approximations. It is apparent from the figures that inversion of the uncorrected saddlepoint approximations, representing mid-p-power inversion, leads to the largest value 1.01 and least conservative among the lower cutoffs as seen in the table. Likewise, inversion of the uncorrected left-tail approximation leads to the least conservative values 15.5 and 15.6 among the upper cutoffs. The exact bands in table 6.7 have been taken from Kim and Agresti (1995). The $\widehat{\mathrm{Pr}}_2$ intervals are extremely accurate for both the single- and double-saddlepoint methods. The $\widehat{\mathrm{Pr}}_1$ method is slightly less accurate.

The inversion of mid-p-values leads to the confidence intervals in table 6.8. These intervals have shorter length and tend to have less conservative coverage. Kim and Agresti (1995) have provided interval K&A$_1$ determined by inverting the separate one-sided tests but using a modified p-value similar to the mid-p-value. They also suggest K&A$_2$ determined by inverting a two-sided test with modified p-value. They argue that K&A$_2$ often leads to shorter confidence intervals.

6.2.2 Single-saddlepoint power functions and confidence intervals

The ease with which power computations may be performed, as seen with the previous example, extends quite generally to power computations in any regular exponential family. In any test of a scalar canonical parameter, the power function is an explicit computation once the null saddlepoint value has been determined. Thus, only one saddlepoint solution for the p-value is needed to be able to determine the entire power function.

Consider a one parameter exponential family as given in (5.1) with likelihood (5.2). Let $\hat{\theta}$ be the MLE for the canonical parameter based on canonical sufficient x. If $\hat{s}^0$ denotes the saddlepoint when $\theta = 0$, then $\hat{s}^0 = \hat{\theta}$. The saddlepoint at x when $\theta \neq 0$ is denoted by $\hat{s}^\theta$ and, according to (5.39), is

$$\hat{s}^\theta = \hat{\theta} - \theta = \hat{s}^0 - \theta. \tag{6.28}$$

Thus, the saddlepoint with $\theta \neq 0$ is given in terms of the null saddlepoint $\hat{s}^0$ when testing $H_0 : \theta = 0$. It is this obvious and simple relationship that is responsible for the explicit power function computation based only on the MLE $\hat{\theta} = \hat{s}^0$. The power at θ using an uncorrected Lugannani and Rice approximation, or $\widehat{\Pr}(X \geq x; \theta)$, has the inputs

$$\hat{w}(\theta) = \text{sgn}(\hat{s}^\theta)\sqrt{-2\ln\frac{\mathcal{L}(\theta)}{\mathcal{L}(\hat{\theta})}} = \text{sgn}(\hat{\theta} - \theta)\sqrt{-2\ln\frac{\mathcal{L}(\theta)}{\mathcal{L}(\hat{\theta})}}$$

$$\hat{u}(\theta) = (\hat{\theta} - \theta)\sqrt{j(\hat{\theta})} \tag{6.29}$$

where $j = \partial^2 \ln \mathcal{L}/\partial\theta^2$.

When using mid-p-power functions with X on the δ-lattice, then only two saddlepoints are needed when computing mid-$\widehat{\Pr}_1$: the MLE $\hat{\theta}$ determined from the observed value x, and the MLE based on the offset value $x + \delta$. Likewise when computing mid-$\widehat{\Pr}_2$ power, only two MLEs are required: $\tilde{\theta}$, the null saddlepoint or MLE at $x - \delta/2$ which leads to non-null saddlepoint $\tilde{s}^\theta = \tilde{\theta} - \theta$ for $\theta \neq 0$; and the MLE at the additional offset value $x + \delta/2$. Thus, when inverting mid-p-$\widehat{\Pr}_i$ for $i = 1, 2$, only two saddlepoints are need to determine the confidence interval.

The power function approximations have properties that mimic their exact counterparts when the parameter space is $\Theta = (a, b)$ and the exponential family is regular. The saddle-point power computation $\widehat{\Pr}(X \geq x; \theta)$ below refers to the uncorrected Lugannani and Rice approximation so either X is continuous or the test uses mid-p-values. Proofs for these results are outlined in Exercises 9 and 10.

(1) Generally, the power function approximation $\widehat{\Pr}(X \geq x; \theta)$ is monotonic increasing in θ. A sufficient condition for this at θ is

$$\frac{\hat{w}^3(\theta)}{\hat{u}(\theta)} > 1. \tag{6.30}$$

Since $\text{sgn}(\hat{w}) = \text{sgn}(\hat{u})$, and $\hat{w}(\theta)$ roughly approximates $\hat{u}(\theta)$ for all θ in many standard applications, this inequality should hold quite generally. The plot in figure 6.2 clearly

Table 6.9. *Attained conditional powers in the extremes of $\Theta = (0, \infty)$.*

φ	$\widehat{\Pr}_1(\varphi)$	$\widehat{\Pr}_2(\varphi)$	$\widehat{\Pr}(\varphi)$
100	0.9^558	0.9^500	0.9^389
0.01	$.0^{22}393$	$.0^{22}383$	$.0^{23}664$

demonstrates its increasing nature. Monotonicity of the true power function has been shown in Lehmann (1986, §3.3).

(2) The limiting powers at either extreme exist and are

$$\lim_{\theta \to a} \widehat{\Pr}(X \geq x; \theta) = 0 \quad \text{and} \quad \lim_{\theta \to b} \widehat{\Pr}(X \geq x; \theta) = 1. \quad (6.31)$$

This replicates the behavior of the true tail probabilities as discussed in Barndorff-Nielsen (1978, §9.8 (viii)). Extreme power computations that approach these limits can be shown using the computations for figure 6.2. These computations remain numerically stable well into the limits because the power function is explicit. Table 6.9 provides such values for $\varphi = 100$ and $1/100$.

6.2.3 With $m_{1i} : m_{2i}$ matching cases

Many data sets used to infer common odds ratios involve $1 : m_{2i}$ matching; the ith individual in treatment is matched according to relevant covariates with m_{2i} subjects in the control group. This often occurs because treatment responses are expensive to acquire while controls are inexpensive. Furthermore, the treatment subjects are generally chosen to have distinctly different covariate values so that the treatment/control comparisons accommodate the full range of covariate values expressed in the population under study.

Still there are settings in which two or more treatment subjects share a common covariate effect χ_i. Conditioning on all the individual 2×2 row and column totals would overcondition the inference. This is understood by considering the canonical sufficient statistics in the exponential family: all 2×2 tables for treatment individuals sharing a common covariate effect χ_i must be added to form a single 2×2 table. This combination forms the minimal sufficient statistic for the nuisance parameter χ_i. Suppose, after combination, the ith covariate level has m_{1i} treatment subjects matched with m_{2i} controls and data are summarized as

	Successes	Failures	Total
Treatment	x_{1i}	$m_{1i} - x_{1i}$	m_{1i}
Control	x_{2i}	$m_{2i} - x_{2i}$	m_{2i}
Total	$x_{\cdot i}$	$m_{\cdot i} - x_{\cdot i}$	$m_{\cdot i}$

(6.32)

for $i = 1, \ldots, n$. The test for the common odds ratio value of $\varphi = 1$ uses statistic $Y = X_1.$ and conditions on the column totals $X = (X_{\cdot 1}, \ldots, X_{\cdot n})^T$. This results in the null distribution theory

$$Y|X = x \sim \sum_{i=1}^{n} H_i,$$

where $x = (x_{\cdot 1}, \ldots, x_{\cdot n})^T$ and $H_i \sim$ Hypergeometric $(m_{1i}, m_{2i}, x_{\cdot i})$ as specified in section 4.2.3.

Single-saddlepoint approach

A conceptually simple approach for approximating the null distribution is to use a single-saddlepoint approximation. The null MGF is the product of MGFs of $\{H_i\}$ or

$$M(s) = \prod_{i=1}^{n} \binom{m_{\cdot i}}{x_{\cdot i}}^{-1} \sum_{j=L_i}^{U_i} \binom{m_{1i}}{j} \binom{m_{2i}}{x_{\cdot i} - j} e^{sj} \tag{6.33}$$

where

$$L_i = \max(0, x_{\cdot i} - m_{2i}) \qquad U_i = \min(m_{1i}, x_{\cdot i}).$$

The null saddlepoint $\hat{s}^0$ is easily determined from (6.33) as the solution to $M'(\hat{s}^0)/M(\hat{s}^0) = y$. With $\hat{s}^0$ determined, the power function approximation is an explicit function that is best described in terms of the tilted exponential family structure. For $\psi = \ln \varphi \neq 0$, the joint noncentral density of $\{H_i\}$ is an exponential family with likelihood

$$\mathcal{L}(\psi) = \exp\{y\psi - c_\cdot(\psi)\}, \tag{6.34}$$

where

$$c_\cdot(\psi) = \sum_{i=1}^{n} c_i(\psi) = \sum_{i=1}^{n} \ln \left\{ \sum_{j=L_i}^{U_i} \binom{m_{1i}}{j} \binom{m_{2i}}{x_{\cdot i} - j} e^{\psi j} \right\}. \tag{6.35}$$

The null saddlepoint is the value $\hat{s}^0 = \hat{\psi}$ maximizing (6.34) and the nonnull saddlepoint is $\hat{s}^\psi = \hat{\psi} - \psi$. The mid-$p$-power function is approximated by using the inputs such as those in (6.29) with ψ replacing θ.

Strawderman and Wells (1998) advocated a modification to this approach. They used the fact that the probability generating function (PGF) for the hypergeometric, or (6.33) as a function of $t = \exp(s)$, has all its roots real-valued. An initial computation of these roots along with a factorization of the PGFs involved were used to reduce the computational effort when computing the power function over a grid of ψ values.

An approach that is arguably equally efficient avoids finding all these roots. The only root-finding actually required when determining the power function is an initial determination of $\hat{\psi}$, through differentiation of (6.35) in its summation form. Once determined, all noncentral saddlepoints are explicit, although many values of $c_\cdot(\psi)$ need to be evaluated through summation. An even more efficient approach performs the evaluations of $c_\cdot(\psi)$ and $c'_\cdot(\psi)$ using Gauss hypergeometric functions as described in Booth and Butler (1998).

As an example, consider the penicillin data in Kim and Agresti (1995) given in table 6.10. The single-saddlepoint p-values were computed by Strawderman and Wells (1998) and are reproduced in table 6.11 along with double-saddlepoint p-values that we now introduce.

Table 6.10. *Three informative 2 × 2 tables that show the effectiveness of no time delay with penicillin injection.*

Penicillin level	Delay	Cured	Died
1/4	none	3	3
1/4	1.5 hrs.	0	6
1/2	none	6	0
1/2	1.5 hrs.	2	4
1	none	5	1
1	1.5 hrs.	6	0

Table 6.11. *p-value approximations in testing $\varphi = 1$ vs. $\varphi > 1$ for the penicillin data.*

Type	Exact	Uncorrected	$\widehat{\Pr}_1$	$\widehat{\Pr}_2$
Single	0.0200	$0.0^2 7735$	0.02131	0.01992
Double	0.0200	$0.0^2 7656$	0.02214	0.02054

Double saddlepoint approach

This approach, as compares with single-saddlepoint methods, involves a lot less computational effort and returns a (mid) p-value with little loss in accuracy. This approach is recommended when most of the values $\{m_{1i}\}$ exceed 1. It was initially suggested in Davison (1988) and further simplified in Booth and Butler (1998) to the point where the saddlepoint is determined as the solution to a single equation in one unknown. Upon determining the solution to this equation, the double-saddlepoint approximation to the power function is also explicit and requires no further root-finding. Any concerns about the accuracy of this approach, as initially expressed in Bedrick and Hill (1992) and continued in Strawderman and Wells (1998), has been exaggerated. The crying baby data set, chosen by Strawderman and Wells because of the sparseness of the tables, serves as a case in point as does the penicillin data in table 6.11. These accuracy issues are addressed below after first presenting the method.

A double-saddlepoint approximation is derived for the distribution of $Y = X_1.$, the total in cell (1, 1) over all tables, given $X = (X_{.1}, \ldots, X_{.n})^T$, the column totals for all tables. The likelihood assumes that

$$X_{1i} \sim \text{Binomial}(m_{1i}, \pi_{1i}) \quad X_{2i} \sim \text{Binomial}(m_{2i}, \pi_{2i})$$

with the π-values specified in (6.19) and (6.20) and constrained so that the common odds is given in (6.22). The canonical parameters associated with Y and X^T are $\psi = \ln \varphi$ and $\chi^T = (\chi_1, \ldots, \chi_n)$. Details of the derivation are outlined in Exercise 7 and only the ideas are presented here. The standard theory results in $n + 1$ saddlepoint equations. Those n equations which take derivatives in χ may be solved explicitly in χ when ψ is held fixed.

These solutions are more easily expressed using the reparameterization

$$a_i = e^{\chi_i} = \frac{\pi_{2i}}{1 - \pi_{2i}}.$$

If $\hat{\chi}_\varphi = (\hat{\chi}_{1\varphi}, \ldots, \hat{\chi}_{n\varphi})$ denotes the MLE of χ holding the odds ratio $\varphi = e^\psi$ fixed, then the corresponding constrained MLEs of $\{a_i\}$ are related by

$$\hat{a}_i(\varphi) = \exp(\hat{\chi}_{i\varphi}). \tag{6.36}$$

For fixed φ, the constrained MLEs in (6.36) form the larger roots of the quadratics $A_i a_i^2 + B_i a_i + C_i$ where

$$A_i = \varphi(m_{\cdot i} - x_{\cdot i}) \tag{6.37}$$
$$B_i = (m_{1i} - x_{\cdot i})\varphi + (m_{2i} - x_{\cdot i})$$
$$C_i = -x_{\cdot i}.$$

Substituting $\hat{a}_i(\varphi)$ into the remaining saddlepoint equation that differentiates in ψ leads to $\hat{\varphi}$, the MLE of φ, as the solution to

$$x_{1\cdot} = \hat{\varphi} \sum_{i=1}^n \frac{m_{1i}\hat{a}_i(\hat{\varphi})}{1 + \hat{a}_i(\hat{\varphi})\hat{\varphi}}. \tag{6.38}$$

This is the only equation that needs to be solved in order to obtain the double-saddlepoint power function. The likelihood maximized in χ holding ψ fixed, or $\mathcal{L}(\hat{\chi}_\psi, \psi)$, is conveniently expressed in terms of φ as

$$\mathcal{L}(\varphi) = \exp\left\{ \sum_{i=1}^n x_{\cdot i} \ln \hat{a}_i(\varphi) + x_{1\cdot} \ln \varphi - c(\varphi) \right\} \tag{6.39}$$

where

$$c(\varphi) = \sum_{i=1}^n [m_{1i} \ln\{1 + \hat{a}_i(\varphi)\varphi\} + m_{2i} \ln\{1 + \hat{a}_i(\varphi)\}].$$

The uncorrected inputs for the Skovgaard approximation that determine the power function are

$$\hat{w}(\varphi) = \text{sgn}(\hat{\varphi}/\varphi - 1)\sqrt{-2 \ln \frac{\mathcal{L}(\varphi)}{\mathcal{L}(\hat{\varphi})}} \tag{6.40}$$

$$\hat{u}(\varphi) = \ln(\hat{\varphi}/\varphi) \sqrt{\prod_{i=1}^n \left\{ \frac{H_{1i}(\hat{\varphi}) + H_{2i}(\hat{\varphi})}{H_{1i}(\varphi) + H_{2i}(\varphi)} \right\} \times \sum_{j=1}^n \frac{H_{1j}(\hat{\varphi})H_{2j}(\hat{\varphi})}{H_{1j}(\hat{\varphi}) + H_{2j}(\hat{\varphi})}}$$

where

$$H_{1i}(\varphi) = \frac{m_{1i}\hat{a}_i(\varphi)\varphi}{\{1 + \hat{a}_i(\varphi)\varphi\}^2} \qquad H_{2i}(\varphi) = \frac{m_{2i}\hat{a}_i(\varphi)}{\{1 + \hat{a}_i(\varphi)\}^2}.$$

In order to use the second continuity correction, the solution in (6.38) must be offset by replacing the left side with $x_{1\cdot} - 1/2$ to get the offset MLE $\tilde{\varphi}$. Details of all these derivations are outlined in Exercise 7.

Figure 6.4 shows a plot of various approximations to the power function for the test of $\varphi = 1$ versus $\varphi > 1$ using the penicillin data. The two solid lines, that are virtually

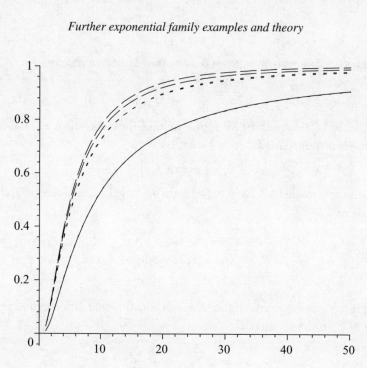

Figure 6.4. Single- and double-saddlepoint approximations for the attained conditional power function using the penicillin data. Shown are the pairs of uncorrected (*solid*), $\widehat{\Pr}_1$ (*dashed*), and $\widehat{\Pr}_2$ (*dotted*).

indistinguishable, represent the uncorrected approximations for the single- and double-saddlepoint approximations. Likewise, the virtually coincident dotted lines are the pair of second continuity-corrections, while the diverging dashed lines are the pair of first continuity-corrections. This discrepancy replicates the same difference that was seen in the power plots for the crying baby data.

In both data sets, the single- and double-saddlepoint approximations offered virtually no practical difference when used with the uncorrected and second continuity-corrected approximations. The computational simplicity of the double-saddlepoint method makes it preferable for this setting.

The issue of double-saddlepoint accuracy

Accuracy of the double-saddlepoint approximation in the common odds setting has been questioned largely as a result of some very discrepant simulation errors reported in Bedrick and Hill (1992). These errors were especially large with 1 : 1 matching. Further questions about its accuracy were subsequently raised in Strawderman and Wells (1998) and the combined results have given the impression that it is not an accurate method in the most difficult settings of this and other examples. For the most demanding settings in which the number of informative 2×2 tables is small, it is not as accurate as the single-saddlepoint but it still provides quite acceptable accuracy in most practical settings as shown below.

Consider 1 : 1 matching as used by Bedrick and Hill with row totals of $m_{1i} = 1 = m_{2i}$. A table is informative when it contributes to the randomness in the test statistic and this only occurs when $x_{\cdot i} = 1$. Assume n such informative tables. The true distribution of the test statistic from (6.23) and (6.25) is Binomial $\{n, \varphi/(\varphi + 1)\}$. Exercise 8 provides the explicit form of the double-saddlepoint approximation $\widehat{\Pr}_2$ in this instance. Table 6.12 compares the

6.2 Common odds ratio in 2 × 2 tables

Table 6.12. *Double-saddlepoint p-value computations using $\widehat{\Pr}_2$ as compared to the true p-values with n = 25 informative 2 × 2 tables.*

$x_1.$ =	14	17	20	23
$\Pr(X_1. \geq x_1.\|n=25)$	.3450	.05388	$.0^2 2039$	$.0^5 9716$
$\widehat{\Pr}_2(X_1. \geq x_1.\|n=25)$	.3341	.04592	$.0^2 2261$	$.0^3 1515$

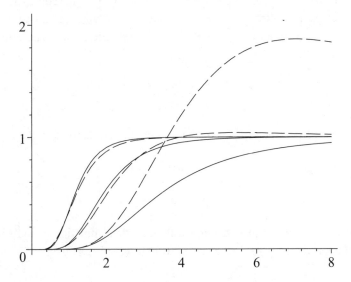

Figure 6.5. Some exact power functions (*solid*) with corresponding double-saddlepoint approximations (*dashed*) using n = 25 informative 2 × 2 tables with 1 : 1 matching.

exact and double-saddlepoint *p*-values ($\varphi = 1$) computed with $n = 25$ using the various cutoff values of $x_1.$ denoted in the table columns.

In the practically important settings with $x_1.$ ranging up to 20, the *p*-value approximations are quite acceptable but show deteriorating relative error in the extreme tail. The source of inaccuracy, recorded in the simulations of Bedrick and Hill (1992), must have occurred with very large values of $x_1. > 20$ for which the *p*-value is already quite small (< 0.002).

To further confirm this, plots of the true attained power functions (solid) for the cases with $x_1. = 14, 17$, and 20 are displayed from left to right in figure 6.5. The respective double-saddlepoint $\widehat{\Pr}_2$ approximations are shown as dashed lines. With $x_1. = 14$ the Skovgaard power function approximation is quite accurate, however with $x_1. = 17$ it slightly exceeds value 1 on $\varphi \in (4, 8)$ where it is also slightly decreasing. For $x_1. = 20$ the curve is highly inaccurate, well above the value 1, and clearly decreasing with large φ. From a practical point of view, however, $x_1. = 20$ is already quite far into the right tail of the null distribution and the pathology seen in the figure is therefore a matter of lesser practical relevance. Despite the appearances in figure 6.5, the power is asymptotically 1 or

$$\lim_{\varphi \to \infty} \widehat{\Pr}_2(X_1. \geq x_1.|n; \varphi) = 1, \tag{6.41}$$

as the reader is asked to show in Exercise 8.

6.2.4 Double-saddlepoint power functions and confidence intervals

The explicitness of power computations for conditional tests in exponential families has some generality beyond the common odds setting. Suppose the sufficient statistics X and Y have the regular exponential family structure in (5.11) with canonical parameters χ and ψ. In the test of $\psi = 0$ versus $\psi > 0$, the conditional power function is the tail probability $\Pr(Y \geq y | X = x; \psi)$ as a function of ψ. Once the p-value has been determined, saddlepoint approximation to this conditional power function is explicit under the condition that $\hat{\chi}_\psi$, the MLE of χ for fixed ψ, is an explicit function of ψ. Thus, one saddlepoint solution for the p-value provides the entire power function without further root finding.

To show this, suppose that $(\hat{\chi}, \hat{\psi})$ is the MLE from data (x, y). The null saddlepoint assuming parameter $(\chi, 0)$ is

$$\hat{s}^0 = \hat{\chi} - \chi$$
$$\hat{t}^0 = \hat{\psi}.$$

Assuming alternative parameter (χ, ψ), the numerator saddlepoint of the attained power function is

$$\hat{s}^\psi = \hat{\chi} - \chi = \hat{s}^0$$
$$\hat{t}^\psi = \hat{\psi} - \psi = \hat{t}^0 - \psi.$$

Thus the numerator saddlepoint is explicit following determination of the MLE. The denominator saddlepoint $\hat{s}_0^\psi = \hat{\chi}_\psi - \psi$ is only explicit when $\hat{\chi}_\psi$ is the explicit solution to

$$c'_\chi(\hat{\chi}_\psi, \psi) = x.$$

Thus, if $\hat{\chi}_\psi$ is explicit, then the whole conditional power function is determined from the MLE without any further root finding. This is the situation with the common odds ratio where $\{\hat{a}_i(\varphi)\}$ are explicit as the roots of quadratic equations. The uncorrected inputs into the Skovgaard approximation in (4.12) that determine conditional power are

$$\hat{w}(\psi) = \text{sgn}(\hat{\psi} - \psi) \sqrt{-2 \ln \frac{\mathcal{L}(\hat{\chi}_\psi, \psi)}{\mathcal{L}(\hat{\chi}, \hat{\psi})}}$$

$$\hat{u}(\psi) = (\hat{\psi} - \psi) \sqrt{\frac{|j(\hat{\chi}, \hat{\psi})|}{|j_{\chi\chi}(\hat{\chi}_\psi, \psi)|}}. \qquad (6.42)$$

The usual adjustments for discrete Y are given in section 5.4.5.

The conditional power function approximations have properties that mimic their exact counterparts. The conditional distribution of Y given $X = x$ is a one-parameter regular exponential family with $\psi \in (a_x, b_x)$, see §5.1.2. As such, the conditional power is monotonic increasing in ψ and approaches 0 and 1 as $\psi \to a_x$ and $\psi \to b_x$ respectively. Sufficient conditions to guarantee that the uncorrected Skovgaard approximation $\widehat{\Pr}(Y \geq y | X = x; \psi)$ is also monotonic increasing at ψ are more complicated than for Lugannani and Rice; see Exercise 11. It is however simple to show that

$$\lim_{\psi \to a_x} \widehat{\Pr}(Y \geq y | X = x; \psi) = 0 \quad \text{and} \quad \lim_{\psi \to b_x} \widehat{\Pr}(Y \geq y | X = x; \psi) = 1. \qquad (6.43)$$

6.2.5 Pascual sampling

Suppose the common odds setting but consider sampling the treatment and control groups using Pascual sampling. Rather than fixing the row totals as occurs in Binomial sampling, Pascual sampling fixes the number of failures m_{1i} and m_{1i} in each group and lets the resultant number of successes x_{1i} and x_{2i} vary according to the Pascal or Negative Binomial distribution. Operationally such data would occur when the overall aim is to limit the number of failures in both groups either due to costs or their more dire consequences. See Martz and Waller (1982, §7.1.2) who discuss the use of Pascual sampling in life testing. As with binomial sampling, control subjects are chosen to match the treatment subjects in terms of some relevant covariates.

All the previous considerations of this section that pertain to binomial sampling may be extended to deal with Pascual sampling. Suppose there are n tables of the form

	Successes	Failures	Total
Treatment	x_{1i}	m_{1i}	$x_{1i} + m_{1i}$
Control	x_{2i}	m_{2i}	$x_{2i} + m_{2i}$
Total	$x_{\cdot i}$	$m_{\cdot i}$	$x_{\cdot i} + m_{\cdot i}$

(6.44)

where X_{1i} has the mass function

$$\Pr(X_{1i} = k) = \binom{k + m_{1i} - 1}{k} \pi_{1i}^k (1 - \pi_{1i})^{m_{1i}} \qquad k = 0, 1, \ldots,$$

and likewise for X_{2i}. The canonical links are

$$\ln(\pi_{1i}) = \chi_i + \psi \qquad \text{and} \qquad \ln(\pi_{2i}) = \chi_i$$

and differ from those using binomial sampling as in (6.19) and (6.20).

The test that $\psi = 0$ is based on the conditional distribution of

$$Y = X_{1\cdot} = \sum_{i=1}^{n} X_{1i}$$

given $x = (x_{\cdot 1}, \ldots, x_{\cdot n})^T$. The appropriate distribution theory may be approached either by using single- or double-saddlepoint approximations. In the null setting the single-saddlepoint approximation uses the distributional representation

$$Y = \sum_{i=1}^{n} P_i$$

where P_i are independent Pólya $(m_{1i}, m_{2i}; x_{\cdot 1})$. The double-saddlepoint approach uses the exponential family structure and conditions Y on x in order to rid the analysis of the nuisance parameters $\chi = (\chi_1, \ldots, \chi_n)^T$. The solution to the system of equations for finding joint MLEs of ψ and χ leads to explicit MLEs for χ with fixed ψ; thus all the development of this section continues to hold in this setting. The development of these results is considered in Exercises 12–14.

Table 6.13. *Month counts of poliomyelitis for a six year span.*

	Jan	Feb	Mar	Apr	May	Jun	Jul	Aug	Sep	Oct	Nov	Dec
1970	0	1	0	0	1	3	9	2	3	5	3	5
1971	2	2	0	1	0	1	3	3	2	1	1	5
1972	0	3	1	0	1	4	0	0	1	6	14	1
1973	1	0	0	1	1	1	1	0	1	0	1	0
1974	1	0	1	0	1	0	1	0	1	0	0	2
1975	0	1	0	1	0	0	1	2	0	0	1	2

6.3 Times series analysis of truncated count data

Table 6.13 lists the monthly number of cases of poliomyelitis reported by the U.S. Center for Disease Control during the years 1970–1975.

The data are a portion of a 14 year data set previously analyzed and tabulated in Zeger (1988). The data have been further considered by Fahrmeir and Tutz (1994) and also by Fokianos (2001) who used truncated Poisson regression models. Following the analysis of the last author, a generalized linear model is fit using a canonical link that includes regression on time, lag-1 autoregression, as well as annual frequencies. The data have 72 dependent responses that are the counts from February 1970 up to and including the count for January 1976 (not shown) which is 0.

The example introduces several new topics for discussion which include (i) the consideration of data whose dependence is modelled as a Markov process and whose model structure leads to a time heterogenous generalized linear model; and (ii) the introduction of truncation into the model.

Let y_t denote the polio count at time $t = 2, \ldots, 73$ ranging from February 1970 up to and including January 1976. An untruncated model might assume that $Y_t \sim$ Poisson (λ_t) where

$$\ln \lambda_t = \alpha + \beta t/10^3 + \gamma y_{t-1} + \delta \sin(2\pi t/12) + \zeta \cos(2\pi t/12). \quad (6.45)$$

Instead we follow the approach taken in Fokianos (2001) and fit a truncated Poisson (λ_t) in which the counts for y_t cannot exceed 14, the largest value seen in the table. The likelihood in this instance is

$$\prod_{t=2}^{73} p(y_t | y_{t-1}; \theta) \stackrel{\theta}{\propto} \exp\left[\sum_{t=2}^{73} \{y_t \ln \lambda_t - \ln c_{14}(\lambda_t)\}\right]$$

where

$$c_m(\lambda) = \sum_{k=0}^{m} \frac{\lambda^k}{k!}.$$

The autoregressive likelihood is that of an exponential family in which $\theta = (\alpha, \beta, \gamma, \delta, \zeta)^T$ and the canonical sufficient statistic is $x = (x_1, \ldots, x_5) = \sum_{t=2}^{73} z_t y_t$ where

$$z_t = (z_{t1}, \ldots, z_{t5}) = \{1, t/10^3, y_{t-1}, \sin(2\pi t/12), \cos(2\pi t/12)\}.$$

Table 6.14. *Significance values for the regression terms in the Poisson time series.*

Nonzero in alternate hypothesis	Null hypothesis	$\hat{w}$	$\hat{u}$	χ_1^2 deviance test	Skovgaard p-value
$\alpha, \beta, \gamma, \delta, \zeta$	$\zeta = 0$	$.0^2 1849$	$.0^2 1849$	.9985	.9938
$\alpha, \beta, \gamma, \delta$	$\delta = 0$	-4.347	-3.790	$.0^4 139$	$.0^4 160$
α, β, γ	$\gamma = 0$	2.714	2.956	$.0^2 665$	$.0^2 604$
α, β, γ	$\beta = 0$	-4.233	-3.779	$.0^4 231$	$.0^4 260$
α, γ	$\gamma = 0$	3.742	4.294	$.0^3 182$	$.0^3 157$
α, β	$\beta = 0$	-4.956	-4.782	$.0^6 720$	$.0^6 748$

Maximum likelihood estimation is based on solving the five equations

$$\sum_{t=2}^{73} z_t y_t = \sum_{t=2}^{73} \frac{1}{c_{14}(\lambda_t)} \frac{\partial}{\partial \theta} c_{14}(\lambda_t) = \sum_{t=2}^{73} z_t \lambda_t \frac{c_{13}(\lambda_t)}{c_{14}(\lambda_t)}$$

and leads to

$$\hat{\theta} = (\hat{\alpha}, \hat{\beta}, \hat{\gamma}, \hat{\delta}, \hat{\zeta}) = (1.000, -25.39, 0.04496, -0.6579, 0.0^3 2621).$$

p-values for model selection are shown in table 6.14. Starting with all five variables (including location), the standard χ_1^2 deviance test that the regression coefficient of the cosine term $\zeta = 0$ gives $p = 0.9985$ as shown in the table. The Skovgaard p-value is 0.9938 and is determined by first noting that $\hat{\zeta} > 0$ and then computing the saddlepoint approximation to $2\Pr(Z_5 > z_5 | z_1, \ldots, z_4; \zeta = 0)$ to provide a two-sided p-value (the opposite inequality would be use if $\hat{\zeta}$ were < 0). The continuous formula has been used to allow for the expression of mid-p-values in testing as previously discussed in section 6.1.4. The two sets of p-values are quite comparable for all the tests as seen in the table. The inputs into the Skovgaard approximation are indicated as $\hat{w}$ and $\hat{u}$ and the deviance value is $\hat{w}^2$.

The agreement in the p-values from the continuous Skovgaard approximation and the deviance test might have been expected when using this somewhat larger sample size. This is because the larger sample size allows for the expression of the asymptotic normal approximations for the MLEs and sufficient statistics as derived in Fokianos (2001).

Notwithstanding the larger sample sizes, the agreement in the first two rows of the table is mildly surprising since it is difficult to theoretically justify using the Skovgaard p-values in these two instances. Justification is formally lacking since the support of X_4 and X_5 is not on a regularly spaced lattice. Indeed the (z_{t4}, z_{t5}) coordinates are the coordinates for $\sqrt[12]{1}$, the 12 complex roots of 1, which are equally spaced on the unit circle and which, when projected down onto the coordinate axes, tend to congregate near ± 1. The other sufficient statistics X_1, X_2, and X_3 are lattice variables which formally justifies the last four tests that express mid-p-value significance levels.

6.4 Exponential families of Markov processes

This example provides a non-normal model of a Markov process whose likelihood function is a regular exponential family. When used with a canonical link function, then the canonical

Table 6.15. *Years for major freezes in Lake Konstanz during AD 875 to 1974.*

875	895						
928							
1074	1076						
1108							
1217	1227	1277					
1323	1325	1378	1379	1383			
1409	1431	1435	1460	1465	1470	1479	1497
1512	1553	1560	1564	1565	1571	1573	
1684	1695						
1763	1776	1788	1796				
1830	1880						
1963							

sufficient statistics are linear in the data. The distributions of these sufficient statistics may be determined by using saddlepoint approximations since their joint CGF is specified by the likelihood shape.

Durbin (1980a) pointed out the usefulness of this approach with dependent data and provided a circular autoregression model with lag 1 and normal errors as an example. Exercise 15 considers the particulars for this example. A more general treatment of autoregressive dependence with normal models was considered by Daniels (1956). He developed several different circularly defined time series models whose associated autoregressive statistics have distributions that are amenable to saddlepoint methods. Extensions to noncircularly defined autoregressive models have been considered in Butler and Paolella (1998b).

Data from a Markov process $\{Z_t\}$ generates an exponential family if the density of Z_t given z_{t-1} has the form

$$f(z_t|z_{t-1};\theta) = \exp\{\theta^T x(z_t, z_{t-1}) - c_t(\theta, z_{t-1}) - d(z_t, z_{t-1})\}. \tag{6.46}$$

In such instances, standard saddlepoint approximations may be used to approximate the distribution of the sufficient statistic $\sum_t x(z_t, z_{t-1})$.

6.4.1 Example: Gamma process

Suppose that $Z_t|z_{t-1} \sim$ Gamma $\{\alpha(\theta, t, z_{t-1}), \beta(\theta, t, z_{t-1})\}$ where $\alpha(\theta, t, z_{t-1})$ and $\beta(\theta, t, z_{t-1})$ are both linear in θ. Then the likelihood has the form given in (6.46) in which θ is the canonical parameter.

6.4.2 Example: Nonhomogeneous Poisson process

Barndorff-Nielsen and Cox (1979) analyzed the record of major freezes of Lake Konstanz from AD 875 to 1974 as reproduced in table 6.15. The table shows a histogram that represents a rough approximation to the hazard rate of the process. With a peak hazard during the fifteenth century, they suggest that the hazard function should be modelled as quadratic and proceed to test the validity of this hypothesis.

6.4 Exponential families of Markov processes

Table 6.16. *Hypothesis tests concerning the hazard rate function for major freezes of Lake Konstanz.*

Nonzero in alternate hypothesis	Null hypothesis	$\hat{w}$	$\hat{u}$	χ_1^2 deviance test	Skovgaard p-value
α, β, γ	$\gamma = 0$	-2.524	-1.871	.01160	.01617
α, β	$\beta = 0$	.8311	.8288	.4059	.4078

In scaling the data, they took the year of the first major freeze AD 875 as the origin for time $t = 0$ and let a single time unit be $1974 - 875 = 1099$ years; accordingly the $n = 37$ freezes that are shown occurred during the time interval $[0, 1)$.

Their time-dependent Poisson process used a log-linear intensity function $\lambda_0(t) = \exp(\alpha + \beta t)$ and they compared it to a log-quadratic fit

$$\lambda(t) = \exp(\alpha + \beta t + \gamma t^2).$$

Starting at time t, the survival function of the interarrival time is

$$\bar{F}(u; t) = \exp\left\{-\int_t^{t+u} \lambda(w) dw\right\}$$

and has the density

$$g(u; t) = -\frac{d}{du} \bar{F}(u; t) = \lambda(t + u) \exp\left\{-\int_t^{t+u} \lambda(w) dw\right\}. \quad (6.47)$$

Denoting the times for major freezes as $0 = t_0, \ldots, t_{36}$, then the likelihood in terms of $\theta = (\alpha, \beta, \gamma)$ and (6.47) is

$$\mathcal{L}(\theta) = \lambda(0) \times \prod_{i=1}^{36} g(t_i - t_{i-1}; t_{i-1}) \times \exp\left\{-\int_{t_{36}}^1 \lambda(t) dt\right\}$$

$$= \lambda(0) \times \prod_{i=1}^{36} \left[\lambda(t_i) \exp\left\{-\int_{t_{i-1}}^{t_i} \lambda(t) dt\right\}\right] \times \exp\left\{-\int_{t_{36}}^1 \lambda(t) dt\right\},$$

where the last term is the probability that no freeze occurred from 1963 to 1974. The likelihood has exponential family form

$$\mathcal{L}(\theta) = \exp(37\alpha + \beta x_1 + \gamma x_2) \exp\left\{-\int_0^1 \exp(\alpha + \beta t + \gamma t^2) dt\right\}$$

where 37, $x_1 = \sum_{i=1}^{36} t_i = 19.958$ and $x_2 = \sum_{i=1}^{36} t_i^2 = 12.739$ are the values for the three sufficient statistics.

Tests for the fit of the linear and quadratic hazard rates are shown in table 6.16. The respective MLEs when fitting the quadratic, linear, and constant functions are

$$(\hat{\alpha}, \hat{\beta}, \hat{\gamma}) = (2.067, 7.108, -6.374)$$
$$(\hat{\alpha}_0, \hat{\beta}_0) = (3.364, 0.4746)$$
$$\hat{\alpha}_1 = 3.611.$$

The p-values shown in the table are two sided; the one-sided values are obtained by taking half and are $p = 0.008085$ for the Skovgaard p-value and $p = 0.00580$ for the signed deviance test.

Comparable single-saddlepoint p-values were reported by Barndorff-Nielsen and Cox (1979) as $p = 0.008594$ (normal approximation), 0.008370 (one correction term), and 0.008180 (two correction terms). The last value with two correction terms is quite close to the Skovgaard double-saddlepoint p-value. The idea of their single-saddlepoint approximation is to proceed in two steps: (i) replace the double-saddlepoint density approximation by a conditional density approximation that constrains the numerator saddlepoint and denominator saddlepoint to be equal; and (ii) integrate this density to obtain a normal approximation that is reported as their first p-value. Two further refinements result when additional correction terms in the tilted Edgeworth expansions are retained. The values of these refinements push the single-saddlepoint approximation closer to the double-saddlepoint approximation.

The test for linear hazard versus constant hazard is not significant when the quadratic term is excluded. When the quadratic term is included, the linear term is significant when using the Wald test. The MLE $\hat{\beta} = 7.108$ has standard error 2.964 so that $z = 2.398$ has significance $p = 0.002468$.

6.5 Truncation

Truncation of the support for datum z that underlies an exponential family leads to another exponential family with a different normalization function. Suppose Z has the exponential family of (5.1) and the circumstances under which Z is obtained restricts its value to the set R. Then the new exponential family for this truncated Z is

$$f_R(z; \theta) = \exp\{\theta^T x(z) - c_R(\theta) - d(z)\}$$

where

$$c_R(\theta) = \int_R \exp\{\theta^T x(z) - d(z)\} dz$$

is the normalization function that is convergent over some set $\theta \in \Theta_R$. Allowing for such truncation and choosing $\theta \in \Theta_R$, then the CGF of sufficient statistic $X = X(Z)$ is $c_R(s + \theta) - c_R(\theta)$ which is defined for $s + \theta \in \Theta_R$. Thus, if Θ_R is an open set, then the exponential family is regular and saddlepoint methods may be readily applied to approximate the distribution of X to accommodate the truncation.

Binomial sampling example

Truncation issues arise in studying genetic deficiencies as concerns "ascertainment bias." Consider estimating the probability p that a child of normal parents, who are able to have deficient children, actually is deficient. For such studies it is common to obtain data through the deficient children so the data are the siblings of deficient children with normal parents.

Given that one child in a family has the deficiency, the conditional probability that the family has k deficient out of n children is

$$p(k; n, p) = \binom{n}{k} \frac{p^k (1-p)^{n-k}}{1 - (1-p)^n} \qquad k = 1, \ldots, n. \tag{6.48}$$

Severe ascertainment bias occurs if the truncation of value $k = 0$ represented in (6.48) is ignored in the data when estimating p.

Suppose that the data are four families with (k, n)-values $(1, 4)$, $(1, 4)$, $(2, 6)$, and $(2, 6)$. The total number of deficient children in the four families is sufficient and attains the value $x = 6$. The exponential family has likelihood

$$\mathcal{L}(\theta) = \exp\{6\theta - c_0(\theta)\}$$

where

$$c_0(\theta) = 2 \ln\{(1 + e^\theta)^4 - 1\} + 2 \ln\{(1 + e^\theta)^6 - 1\}$$

and the canonical parameter is $\theta = \ln\{p/(1-p)\}$. Consider testing the hypotheses $H_0 : p = 1/4$ ($\theta = -\ln 3$) versus $H_1 : p < 1/4$ ($\theta < -\ln 3$). The exact p-value is computed as $\Pr(X \le 6) = 0.5171$; the continuity-corrected approximations are $\widehat{\Pr}_1(X \le 6) = 0.5153$ and $\widehat{\Pr}_2(X \le 6) = 0.5152$ and quite accurate even with such small sample sizes.

6.6 Exercises

1. Consider the aphid data in table 6.1.
   (a) Determine separate 95% confidence intervals for χ and η using first-order asymptotics for their MLEs. Compare these intervals to those computed by inverting the p-values in (6.7) and (6.8) using the uncorrected procedure as well as the various continuity corrections based on $\widehat{\Pr}_1$ and $\widehat{\Pr}_2$.
   (b) Now compute confidence intervals by inverting the various mid-p-value estimates and compare the resulting intervals to those of part (a).
2. Consider the Lugannani and Rice approximation in the one parameter setting of section 5.4.3 in which θ is canonical. Prove that the signed likelihood ratio statistic

$$\hat{w}(\theta) = \operatorname{sgn}(\hat\theta - \theta) \sqrt{-2 \ln \frac{\mathcal{L}(\theta)}{\mathcal{L}(\hat\theta)}}$$

$$= \operatorname{sgn}(\hat\theta - \theta) \sqrt{-2\{c(\hat\theta) - c(\theta) - (\hat\theta - \theta)x\}}$$

is decreasing in θ. Implicitly differentiate $\hat{w}(\theta)^2 / 2$ in θ to get that

$$\hat{w}(\theta) \frac{\partial \hat{w}}{\partial \theta} = c'(\theta) - x.$$

Replace x with $c'(\hat\theta)$ and use

$$\operatorname{sgn}\{\hat{w}(\theta)\} = -\operatorname{sgn}\{c'(\theta) - c'(\hat\theta)\},$$

to show that $\partial \hat{w}/\partial \theta < 0$ for all $\theta \ne \hat\theta$.

3. Consider the Skovgaard approximation in the multiparameter setting of section 5.4.5 in which ψ is canonical. Prove that the signed likelihood ratio statistic

$$\hat{w}(\psi) = \text{sgn}(\hat{\psi} - \psi)\sqrt{-2\ln\frac{\mathcal{L}(\hat{\chi}_\psi, \psi)}{\mathcal{L}(\hat{\chi}, \hat{\psi})}}$$

is decreasing in ψ.

(a) Use implicitly differentiation on

$$\tfrac{1}{2}\hat{w}(\psi)^2 = \{c(\hat{\chi}_\psi, \psi) - \hat{\chi}_\psi^T x - \psi y\} - \{c(\hat{\chi}, \hat{\psi}) - \hat{\chi}^T x - \hat{\psi} y\}$$

to get that

$$\hat{w}\frac{\partial \hat{w}}{\partial \psi} = c'_\psi(\hat{\chi}_\psi, \psi) - y = c'_\psi(\hat{\chi}_\psi, \psi) - c'_\psi(\hat{\chi}, \hat{\psi}). \tag{6.49}$$

(b) Show that $c'_\psi(\hat{\chi}_\psi, \psi)$ is monotonic increasing in ψ by computing

$$\frac{\partial}{\partial \psi}c'_\psi(\hat{\chi}_\psi, \psi) = c''_{\psi\psi}(\hat{\chi}_\psi, \psi) - c''_{\psi\chi}(\hat{\chi}_\psi, \psi)c''_{\chi\chi}(\hat{\chi}_\psi, \psi)^{-1}c''_{\chi\psi}(\hat{\chi}_\psi, \psi)$$

$$= \left.\left(j_{\psi\psi} - j_{\psi\chi}j_{\chi\chi}^{-1}j_{\chi\psi}\right)\right|_{\hat{\chi}_\psi, \psi}$$

$$= \frac{|j(\hat{\chi}_\psi, \psi)|}{|j_{\chi\chi}(\hat{\chi}_\psi, \psi)|} = j_{\psi\psi\cdot\chi}(\hat{\chi}_\psi, \psi), \tag{6.50}$$

the asymptotic conditional variance of $\hat{\psi}$ given $X = x$ as described in section 5.4.5. Since (6.50) is positive, and

$$c'_\psi(\hat{\chi}_{\hat{\psi}}, \hat{\psi}) = c'_\psi(\hat{\chi}, \hat{\psi}),$$

then

$$\text{sgn}(\hat{w}) = \text{sgn}(\hat{\psi} - \psi) = -\text{sgn}\{c'_\psi(\hat{\chi}_\psi, \psi) - c'_\psi(\hat{\chi}_{\hat{\psi}}, \hat{\psi})\}$$

compares the left and right side of (6.49). Thus $\partial \hat{w}/\partial \psi < 0$ for all $\psi \neq \hat{\psi}$.

4. An alternative saddlepoint approximation for the mid-p-value computes $\widehat{\text{Pr}}_1(Y \geq y|X = x; \psi)$ and removes half the (double-) saddlepoint density at y. If Y is an integer-lattice variable with $\delta = 1$, then the (double-) saddlepoint density has the form

$$\phi(\hat{w})/\sqrt{\hat{j}} \tag{6.51}$$

with $\hat{j}$ as the observed (conditional) information. Half of (6.51) should be removed from the tail approximation in this instance. For the setting in which Y has δ-lattice support, it is necessary to remove δ^{-1} times half of (6.51) in order to have the mass probability at y conform to the lattice convention described in section 2.1.1.

(a) In the general case, show that the resulting mid-$\widehat{\text{Pr}}_1$ tail approximation is

$$\text{mid-}\widehat{\text{Pr}}_1(Y \geq y|X = x; \psi) = 1 - \Phi(\hat{w}) - \phi(\hat{w})\left(\frac{1}{\hat{w}} - \frac{1}{\tilde{u}_{R1}}\right) \tag{6.52}$$

where

$$\tilde{u}_{R1} = \frac{2\delta\{1 - e^{-\delta(\hat{\psi}-\psi)}\}}{2\delta^2 - 1 + e^{-\delta(\hat{\psi}-\psi)}}\sqrt{\hat{j}}. \tag{6.53}$$

Note that $\tilde{u}_{R1}$ is no more difficult to compute than $\tilde{u}_1$ since both are determined from $\hat{\psi}$ and $\hat{j}$.

(b) Show that a Taylor expansion of $\tilde{u}_{R1}$ in $\hat{t} = \hat{\psi} - \psi$ leads to

$$\tilde{u}_{R1} = \left\{ \hat{t} + \hat{t}^2 \frac{1-\delta^2}{2\delta} + O(\hat{t}^3) \right\} \sqrt{\hat{j}} = \hat{t}\sqrt{\hat{j}} + O(\hat{t}^2), \tag{6.54}$$

which is the uncorrected approximation to order $O(\hat{t}^2)$. Curiously enough, the higher order of $O(\hat{t}^3)$ is achieved in (6.54) when $\delta = 1$.

(c) Show that the left tail expression, appropriate with alternatively small values of ψ, is

$$\text{mid-}\widehat{\Pr}_1(Y \leq y | X = x; \psi) = \Phi(\hat{w}) + \phi(\hat{w}) \left(\frac{1}{\hat{w}} - \frac{1}{\tilde{u}_{L1}} \right), \tag{6.55}$$

where

$$\tilde{u}_{L1} = \frac{2\delta\{1 - e^{-\delta(\hat{\psi}-\psi)}\}}{2\delta^2 + 1 - e^{-\delta(\hat{\psi}-\psi)}} \sqrt{\hat{j}}$$

$$= \left\{ \hat{t} - \hat{t}^2 \frac{1+\delta^2}{2\delta} + O(\hat{t}^3) \right\} \sqrt{\hat{j}} = \hat{t}\sqrt{\hat{j}} + O(\hat{t}^2).$$

(d) A virtue of this mid-p-value approximation is that it requires only a single-saddlepoint determination at y. Unfortunately it is an unreliable approximation when used to compute smaller tail probabilities. Show that (6.52) leads to a negative mid-p-value approximation when applied to the aphid data as in (6.12).

5. Derive the tail probability computation in (6.18) when no continuity correction is used.

(a) Show that the MLEs for $\mathcal{L}_\lambda(\chi, \psi_\lambda)$ are $\hat{\chi}$ and $\hat{\psi}_\lambda = \lambda^T \hat{\chi} + \hat{\eta}$ where $(\hat{\chi}, \hat{\eta})$ maximizes $\mathcal{L}(\chi, \eta)$. Also show that $(\hat{\chi}, \hat{\eta})$ solves the $m+1$ equations

$$\sum_{i=1}^{I} \binom{l_i}{1} n_i \pi_i (\hat{\chi}, \hat{\eta}) = \binom{x}{d.}, \tag{6.56}$$

where $\pi_i(\chi, \eta)$ is given in (6.15).

(b) Show that the constrained MLE of χ with $\psi_\lambda = 0$ is $\hat{\chi}_{0\lambda}$ that solves

$$\sum_{i=1}^{I} (l_i - \lambda) \{ n_i \pi_{i\lambda}(\hat{\chi}_{0\lambda}) - d_i \} = 0,$$

where

$$\pi_{i\lambda}(\chi) = \frac{\exp\{(l_i - \lambda)^T \chi\}}{1 + \exp\{(l_i - \lambda)^T \chi\}}.$$

(c) For $\hat{\psi}_\lambda \neq 0$, the approximate survival function in (6.18) has the inputs

$$\hat{w}_\lambda = \text{sgn}(\hat{\psi}_\lambda) \sqrt{-2 \ln \frac{\mathcal{L}(\hat{\chi}_{0\lambda}, -\lambda^T \hat{\chi}_{0\lambda})}{\mathcal{L}(\hat{\chi}, \hat{\eta})}} \tag{6.57a}$$

$$\hat{u}_\lambda = \hat{\psi}_\lambda \sqrt{\frac{|c''(\hat{\chi}, \hat{\eta})|}{\tilde{h}_{0\lambda}}}, \tag{6.57b}$$

where $\mathcal{L}(\chi, \eta)$ is the likelihood in (6.16). Prove that the Hessian of $\ln \mathcal{L}_\lambda(\chi, \psi_\lambda)$ at $(\hat{\chi}, \hat{\psi}_\lambda)$ is related to the Hessian of $\ln \mathcal{L}(\chi, \eta)$ at $(\hat{\chi}, \hat{\eta})$ through the relationship

$$-\frac{\partial^2 \ln \mathcal{L}_\lambda(\chi, \psi_\lambda)}{\partial(\chi, \psi_\lambda)^2} = \frac{\partial^2 c(\chi, \psi_\lambda - \chi\lambda)}{\partial(\chi, \psi_\lambda)^2} = B_\lambda^{-1} c''(\chi, \eta) \left(B_\lambda^{-1} \right)^T.$$

Thus,
$$\left| -\frac{\partial^2 \ln \mathcal{L}_\lambda(\hat{\chi}, \hat{\psi}_\lambda)}{\partial(\hat{\chi}, \hat{\psi}_\lambda)^2} \right| = \left| B_\lambda^{-1} c''(\hat{\chi}, \hat{\eta})(B_\lambda^T)^{-1} \right| = |c''(\hat{\chi}, \hat{\eta})|$$

is the appropriate entry in (6.57b). Show also that
$$c''(\chi, \eta) = \sum_{i=1}^{I} n_i \pi_i(\chi, \eta)(\{1 - \pi_i(\chi, \eta)\}) \begin{pmatrix} l_i l_i^T & l_i \\ l_i^T & 1 \end{pmatrix}.$$

(d) The Hessian value $\hat{h}_{0\lambda}$ is the determinant of the $m \times m$ upper left block of the Hessian from part (c) but evaluated at $\hat{\chi}_{0\lambda}$. As such, show that
$$\hat{h}_{0\lambda} = \left| (I_m \ 0) B_\lambda^{-1} c''(\hat{\chi}_{0\lambda}, -\lambda^T \hat{\chi}_{0\lambda}) (B_\lambda^{-1})^T \begin{pmatrix} I_m \\ 0^T \end{pmatrix} \right|$$
$$= \left| (I_m \ -\lambda) c''(\hat{\chi}_{0\lambda}, -\lambda^T \hat{\chi}_{0\lambda}) \begin{pmatrix} I_m \\ -\lambda^T \end{pmatrix} \right|. \quad (6.58)$$

6. Prove that the noncentral distribution of the test statistic for a common odds ratio in section 6.2.1 has the form in (6.23) but with $B_i \sim$ Bernoulli $\{\hat{\pi}_{i\psi}\}$, where
$$\hat{\pi}_{i\psi} = \frac{\hat{\pi}_i}{\hat{\pi}_i + (1 - \hat{\pi}_i) e^{-\psi}}.$$

7. Derive the double-saddlepoint approximations for the power function when testing the common odds ratio.

(a) Using the logistic link functions, show that the likelihood in terms of $\varphi = e^\psi$ and $a = (a_1, \ldots, a_n)^T$, with $a_i = e^{\chi_i}$, may be expressed as
$$\mathcal{L}(a, \varphi) = \exp\left\{ \sum_{i=1}^{n} x_{\cdot i} \ln a_i + x_{1\cdot} \ln \varphi - c(a, \varphi) \right\}$$

where
$$c(a, \varphi) = \sum_{i=1}^{n} \{m_{1i} \ln(1 + a_i \varphi) + m_{2i} \ln(1 + a_i)\}.$$

(b) With φ fixed, determine that the MLE of a_i is the root of the quadratic in (6.37). The likelihood must be differentiated with respect to χ_i, the canonical parameter, and the chain rule helps to keep computations to a minimum. To determine that the larger root is the MLE, consider the situation in which the canonical sufficient statistics are at their joint mean. Show in this setting that the positive root is the only possible solution. Why does this argument suffice for determining the correct root?

(c) Determine that the solution to (6.38) provides the MLE of φ. Now verify the expression for $\hat{w}(\varphi)$ in (6.40).

(d) Verify the expression for $\hat{u}(\varphi)$ in (6.40). Show that the required information matrix, when evaluated at the MLE of a with φ held fixed, has the form
$$j(\chi, \psi) = \begin{pmatrix} H_{11} + H_{21} & 0 & 0 & H_{11} \\ 0 & \ddots & 0 & \vdots \\ 0 & 0 & H_{1n} + H_{2n} & H_{1n} \\ H_{11} & \cdots & H_{1n} & \sum_{i=1}^{n} H_{1i} \end{pmatrix}$$

where $H_{1i} = H_{1i}(\varphi)$, etc. Use the block diagonal structure of $j(\chi, \psi)$ to figure its determinant and complete the verification of $\hat{u}(\varphi)$ in (6.40).

8. Suppose 1 : 1 matching is used to test the common odds ratio and that there are n informative tables. Show that the saddlepoint is explicit and derive the explicit attained power function for the double-saddlepoint approximation.
   (a) Show that $a_i(\varphi) = 1/\sqrt{\varphi}$ for each i.
   (b) Determine that the offset MLE from (6.38) using $x_{1.} - 1/2$ in place of $x_{1.}$ is
   $$\tilde{\varphi} = \left(\frac{x_{1.} - 1/2}{n - x_{1.} + 1/2} \right)^2.$$
   (c) Show that the offset likelihood function is
   $$\mathcal{L}_-(\varphi) = \varphi^{x_{1.} - (n+1)/2} \left(2 + \sqrt{\varphi} + 1/\sqrt{\varphi} \right)^{-n}$$
   and that
   $$\tilde{w}_2(\varphi) = \operatorname{sgn}\left(\frac{\tilde{\varphi}}{\varphi} - 1 \right) \sqrt{-2 \ln \frac{\mathcal{L}_-(\varphi)}{\mathcal{L}_-(\tilde{\varphi})}}.$$
   (d) Show that
   $$\tilde{u}_2(\varphi) = 2 \sinh\left(\tfrac{1}{2} \ln \frac{\tilde{\varphi}}{\varphi} \right) \sqrt{\frac{n}{2}} \frac{\tilde{\varphi}^{(n+1)/4}}{\varphi^{n/4}} \frac{(1 + \sqrt{\varphi})^n}{(1 + \sqrt{\tilde{\varphi}})^{n+1}}.$$
   (e) Show that $\tilde{w}_2(\varphi) \to -\infty$ and $\tilde{u}_2(\varphi) \to -\infty$ as $\varphi \to \infty$. Use this to show that (6.41) holds.
9. Consider the power function approximation for a single parameter exponential family that uses the Lugannani and Rice approximation with the inputs (6.29). Prove that a sufficient condition for this approximation to be monotonic at θ is that $\hat{w}^3(\theta)/\hat{u}(\theta) > 1$. Show that
$$\frac{\partial}{\partial \theta} \widehat{\Pr}(X \geq x; \theta) = \phi\{\hat{w}\} \left\{ \frac{1}{\hat{w}^2} \frac{\partial \hat{w}}{\partial \theta} \left(1 - \frac{\hat{w}^3}{\hat{u}} \right) + \frac{1}{(\hat{\theta} - \theta)^2 \sqrt{j(\hat{\theta})}} \right\}.$$
When combined with the results of the implicit differentiation in Exercise 2, conclude that this derivative is positive under the specified conditions.
10. Prove (6.31) for a regular exponential family with parameter space $\Theta = (a, b) \subseteq (-\infty, \infty)$.
    (a) If $\mathcal{L}(\theta)$ is the likelihood for the exponential family, show that
    $$\lim_{\theta \uparrow b} \mathcal{L}(\theta) = 0 = \lim_{\theta \downarrow a} \mathcal{L}(\theta). \tag{6.59}$$
    (b) From (6.59) deduce that (6.31) is true.
11. For the approximation of conditional power using the Skovgaard approximation with inputs (6.42), show that the limits in (6.43) hold. In addition, show that conditions sufficient to guarantee that
$$\frac{\partial}{\partial \psi} \widehat{\Pr}(Y \geq y | X = x; \psi) > 0$$
are more complicated than for the one-parameter setting in Exercise 9.
12. (a) Derive details of the single-saddlepoint approximation for testing common odds ratio with Pascual sampling as described in section 6.2.5.
    (b) Apply these methods to the crying baby data in table 6.4 assuming that $\{x_{2i}\}$ are pre-set values (which, of course, they are not). Derive p-values using the uncorrected Lugannani and Rice approximation and the $\widehat{\Pr}_1$ and $\widehat{\Pr}_2$ corrections and compare with the entries in table 6.5.
    (c) Compute the various power functions and compare to those in figure 6.2.
    (d) Determine 95% confidence intervals for the common odds ratio using mid-p-values and compare to those in table 6.8.

13. (a) Derive details of the double-saddlepoint approximation for testing common odds ratio with Pascual sampling as described in section 6.2.5. Show that $\hat{\chi}_\psi$, the MLE of χ for fixed ψ, is explicit in ψ. As a result, show that the MLE $\hat{\psi}$ can be determined as the root of a single equation in one unknown.
    (b) Apply these methods to the crying baby data in table 6.4 as described in Exercise 12 and compute all the quantities described in 12(b)–12(d).
    (c) Compare the relative values of the single- and double-saddlepoint methods. Taking values of the single-saddlepoint approximations as the gold standard, describe the adequacy of the double-saddlepoint approximations in this context and for this data set.
14. In Exercises 12 and 13, suppose that Pascual sampling is used for the treatment group and Binomial sampling is used for control. Describe how the test of common odds ratio might proceed under these circumstances. How could saddlepoint methods be used?
15. Suppose that $z = (z_1, \ldots, z_n)$ is generated by the circular autoregression

$$z_t = \rho z_{t-1} + \varepsilon_t \qquad t = 1, \ldots, n,$$

where $z_0 = z_n$ and $\{\varepsilon_t\}$ are i.i.d. Normal $(0, \sigma^2)$.
    (a) Show that the likelihood function is

$$f(z; \rho, \sigma^2) = \frac{1 - \rho^n}{(2\pi\sigma^2)^{n/2}} \exp\left\{ -\frac{(1 + \rho^2)x_0 - 2\rho x_1}{2\sigma^2} \right\}$$

where

$$x_0 = \sum_{t=1}^n z_t^2 \quad \text{and} \quad x_1 = \sum_{t=1}^n z_t z_{t-1}.$$

   (b) If $|\rho| < 1$, then we may ignore the exponentially small term ρ^n. Show in this instance that approximate MLEs are

$$\hat{\rho} = \frac{x_1}{x_0} \quad \text{and} \quad \hat{\sigma}^2 = \frac{1}{n}\{(1 + \hat{\rho}^2)x_0 - 2\hat{\rho}x_1\}.$$

   (c) Ignoring the term ρ^n, derive a saddlepoint density for (x_0, x_1). Transform and determine the saddlepoint density of the MLEs $(\hat{\rho}, \hat{\sigma}^2)$.
   (d) Marginalize the density in part (c) to find an approximate density for the MLE $\hat{\rho}$.
16. Suppose the distribution of $Z|\lambda$ is Poisson (λ) and $\lambda \sim$ Gamma (α, β). Then the marginal mass function of Z is Negative Binomial $\{\alpha, \beta/(\beta + 1)\}$ as described in Exercise 9 of chapter 1. The additional dispersion due random λ can be useful when modelling discrete data and provides an alternative to modelling count data as Poisson. Use the negative binomial to model the counts for the poliomyelitis data in table 6.13. For a suitable value of m, fit a canonical link in which $Z_t|z_{t-1} \sim$ Negative Binomial (m, p_t) with $\ln p_t$ defined by the right side of (6.45).

7

Probability computation with p^*

The p^* density was introduced in Barndorff-Nielsen (1980, 1983) and has been prominently featured as an approximation for the density of the maximum likelihood estimate (MLE). Its development from saddlepoint approximations and its role in likelihood inference for regular exponential families are discussed in section 7.1. Section 7.2 considers group transformation models, such as linear regression and location and scale models. In this setting the p^* formula provides the conditional distributions of MLEs given a maximal ancillary statistic and agrees with expressions developed by Fisher (1934) and Fraser (1979). In curved exponential families, the p^* formula also approximates the conditional distribution of MLEs given certain approximate ancillaries. This development is explored in section 7.3 with a more detailed discussion given for the situation in which the ancillaries are affine.

7.1 The p^* density in regular exponential families

In the context of a regular exponential family, the p^* density is the normalized saddlepoint density for the MLE. In its unnormalized form in the continuous setting, p^* is simply a Jacobian transformation removed from the saddlepoint density of the canonical sufficient statistic. Consider the regular exponential family in (5.1) with canonical parameter $\theta \in \Theta \subseteq \Re^m$ and canonical sufficient statistic X. Statistic X is also the MLE $\hat{\mu}$ for the mean parameterization $\mu = c'(\theta)$ as discussed in section 5.1.2. Therefore the saddlepoint density for X, as given in (5.40), is also the density for MLE $\hat{\mu}$ and may be reexpressed as

$$\hat{f}(\hat{\mu}; \theta) = (2\pi)^{-m/2} |j(\hat{\theta})|^{-1/2} \frac{\mathcal{L}(\theta)}{\mathcal{L}(\hat{\theta})}, \qquad (7.1)$$

where both terms $\mathcal{L}(\hat{\theta})$ and $j(\hat{\theta})$ are viewed as implicit functions of $\hat{\mu} = x$ through the canonical MLE $\hat{\theta}$. Jacobian transformation of the density from $\hat{\mu}$ to $\hat{\theta}$ in this setting requires the factor

$$\left| \frac{\partial \hat{\mu}}{\partial \hat{\theta}^T} \right| = |c''(\hat{\theta})| = |j(\hat{\theta})| \qquad (7.2)$$

so that (7.1) and (7.2) together yield the unnormalized saddlepoint density for $\hat{\theta}$ as

$$p^\dagger(\hat{\theta}; \theta) = (2\pi)^{-m/2} |j(\hat{\theta})|^{1/2} \frac{\mathcal{L}(\theta)}{\mathcal{L}(\hat{\theta})}. \qquad (7.3)$$

When normalized, expression (7.3) is the p^* formula for the density of the canonical parameter MLE $\hat{\theta}$. This normalization replaces the leading constant with $\varkappa(\theta)$, a constant that depends on θ. If τ is any other 1-1 smooth reparameterization of θ, then Jacobian transformation yields

$$p^*(\hat{\tau};\theta) = p^*(\hat{\theta};\theta)\left|\frac{\partial\hat{\theta}}{\partial\hat{\tau}^T}\right| = \varkappa(\theta)\left|\frac{\partial\hat{\theta}^T}{\partial\hat{\tau}}j(\hat{\theta})\frac{\partial\hat{\theta}}{\partial\hat{\tau}^T}\right|^{1/2}\frac{\mathcal{L}(\theta)}{\mathcal{L}(\hat{\theta})}$$

$$p^*(\hat{\tau};\tau) = \varkappa(\tau)|j(\hat{\tau})|^{1/2}\frac{\mathcal{L}(\tau)}{\mathcal{L}(\hat{\tau})} \tag{7.4}$$

where $\mathcal{L}(\tau)$ and $\varkappa(\tau)$ are now the likelihood and normalization constant reparametrized in terms of τ. Expression (7.4) is the p^* formula for the density of an arbitrary parametrization of the MLE $\hat{\tau}$. Note that $p^*(\hat{\tau};\tau)$ and $p^*(\hat{\theta};\theta)$ have exactly the same structural form as given in (7.3), a property that is referred to as parameterization invariance. Writing $\theta = \theta(\tau)$ and $x = x(\hat{\tau})$ to make explicit the naturally occurring implicit relationships, then, from (5.40), the likelihood terms in (7.4) should be taken as standing for

$$\mathcal{L}(\tau) = \exp\left[\theta(\tau)x(\hat{\tau}) - c\{\theta(\tau)\}\right] \tag{7.5}$$

$$j(\tau) = \frac{\partial^2}{\partial\tau\partial\tau^T}c(\theta) = \frac{\partial\theta^T}{\partial\tau}c''\{\theta(\tau)\}\frac{\partial\theta}{\partial\tau^T}.$$

Formula (7.4) expresses the saddlepoint approximation for the distribution of the MLE using an economy of notation whose meaning becomes apparent from the likelihood notation in (7.5). With striking simplicity the formula embodies the Fisherian idea that the exact distribution of $\hat{\tau}$ is characterized by its likelihood shape; the presence of the likelihood ratio and $j(\cdot)$ is a continual reminder of this fact.

In a single dimension, the mapping $\hat{\tau} \leftrightarrow \hat{\theta}$ must be monotonic in order to be a bijection. In this instance the CDF of $\hat{\tau}$ is specified by the CDF of $\hat{\theta}$ as given in section 5.4.3.

The discrete setting does not entail Jacobian transformation, hence $p^*(\hat{\tau})$ is the expression (7.1) with $\hat{\theta}$ viewed as an implicit function of $\hat{\tau}$. Generally the mass function of $\hat{\tau}$ is not distributed on a lattice and it is perhaps simpler to refer matters back to the original mass function of X; see Severini (2000b).

7.1.1 Examples

Gamma (α, β)

Let $Z_1, \ldots, Z_n$ be a random sample from this distribution. The canonical sufficient statistics $X = \sum_{i=1}^n \ln Z_i$ and $Y = \sum_{i=1}^n Z_i$ are also the MLEs for the mean parameterization expressed in (5.49) and (5.50) as

$$\tau_1 = E(X) = n\{\Psi(\alpha) - \ln\beta\}$$
$$\tau_2 = E(Y) = n\alpha/\beta.$$

Therefore, $p^\dagger(\hat{\tau}_1, \hat{\tau}_2) = \hat{f}(x, y; \alpha, \beta)$ as given in (5.51). With the mixed parameterization (α, τ_2), then $p^\dagger(\hat{\alpha}, \hat{\tau}_2)$ is given in (5.72).

7.1 The p* density in regular exponential families

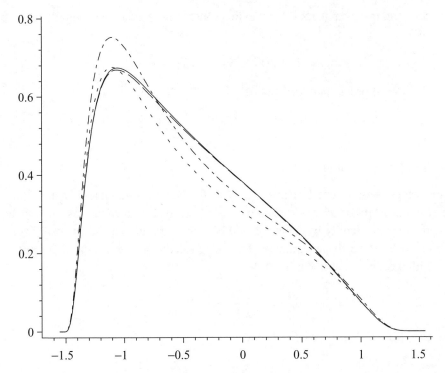

Figure 7.1. Exact density (*solid*), p^* density (*dotted*), normalized p^* (*dot-dashed*), and second-order p^* density (*dashed*) of $\hat{\theta}$, the MLE for the $HBS\,(\theta)$ distribution.

Hyperbolic secant (HBS) distribution

This distribution was studied by Jørgensen (1997) as an exponential dispersion model with quadratic variance function. A special case involves random variable X with the density

$$f(x;\theta) = \frac{e^{\theta x} \cos\theta}{2\cosh(\pi x/2)} \qquad x \in (-\infty, \infty)$$

for $\theta \in (-\pi/2, \pi/2)$, a distribution which is called the HBS (θ) density. With $\theta = 0$ the density is symmetric and for $\theta \neq 0$ it is skewed. The MLE is found to be $\hat{\theta} = \arctan x \in (-\pi/2, \pi/2)$. The $p^\dagger$ formula is

$$p^\dagger(\hat{\theta};\theta) = \frac{\cos\theta \sec^2\hat{\theta}}{\sqrt{2\pi}} \exp\{(\theta - \hat{\theta})\tan\hat{\theta}\} \qquad \hat{\theta} \in (-\pi/2, \pi/2).$$

The true density is

$$f(\hat{\theta};\theta) = \frac{\cos\theta \sec^2\hat{\theta}}{2\cosh(\pi\tan\hat{\theta}/2)} \exp(\theta\tan\hat{\theta}) \qquad \hat{\theta} \in (-\pi/2, \pi/2).$$

A comparison of these densities is shown in figure 7.1 for $\theta = -0.7$. Plotted are the true density (solid), the $p^\dagger$ density (dotted) and p^* (dot-dashed) using the normalization constant 0.8988. Neither of these saddlepoint densities accurately approximates the peculiar shape of this density. To improve upon accuracy, $p^\dagger$ is adjusted to include the second-order correction term and the result is shown as the dashed line. The correction term provides a remarkable improvement to $p^\dagger$ and makes it virtually exact.

The correction is given in (5.42) and in this context the standardized cumulants simplify to

$$\hat{\kappa}_3(\hat{\theta}) = 2\sin(\hat{\theta}) \qquad \hat{\kappa}_4(\hat{\theta}) = 2\{2\sin^2(\hat{\theta}) + 1\}$$

so that the second-order approximation is

$$p^\dagger(\hat{\theta};\theta)\left\{1 + \tfrac{1}{8}\hat{\kappa}_4(\hat{\theta}) - \tfrac{5}{24}\hat{\kappa}_3^2(\hat{\theta})\right\}. \tag{7.6}$$

Planck's radiation formula

In statistical physics, the energy density $f(x)$ vs. frequency x for radiation that is in thermal equilibrium is given by Planck's formula. As a density, this formula may be embedded in a two parameter family by letting $\beta > 0$ be a temperature dependent scale parameter and letting $\alpha > 0$ be a shape parameter. Planck's formula has the specific shape parameter $\alpha = 3$ within the parametric family of densities

$$f(x;\alpha,\beta) = \frac{\beta^{\alpha+1} x^\alpha}{C(\alpha+1)(e^{\beta x} - 1)} \qquad x > 0$$

where $C(\alpha+1) = \Gamma(\alpha+1)\zeta(\alpha+1)$ and

$$\zeta(\alpha+1) = \sum_{j=1}^{\infty} j^{-(\alpha+1)}$$

is Riemann's zeta function. See Johnson and Kotz (1970, §6.1).

With $X \sim f(x;\alpha,\beta)$ and β as a fixed number, α is the canonical parameter and $\ln(\beta x)$ is canonical sufficent with CGF

$$K(s) = \kappa(s + \alpha + 1) - \kappa(\alpha + 1)$$

where $\kappa(z) = \ln C(z)$. The MLE $\hat{\alpha}$ solves the equation

$$\kappa'(\hat{\alpha} + 1) = \ln(\beta x) \tag{7.7}$$

with

$$\kappa'(z) = \Psi(z) + \frac{\zeta'(z)}{\zeta(z)},$$

and Ψ as the digamma function. The Fisher information is $\kappa''(\hat{\alpha}+1)$ with

$$\kappa''(z) = \Psi'(z) + \frac{\zeta''(z)}{\zeta(z)} - \left\{\frac{\zeta'(z)}{\zeta(z)}\right\}^2$$

so that

$$\begin{aligned}
p^\dagger(\hat{\alpha};\alpha) &= \frac{1}{\sqrt{2\pi}}\sqrt{\kappa''(\hat{\alpha}+1)}\,\frac{f(x;\alpha,\beta)}{f(x;\hat{\alpha},\beta)} \\
&= \frac{1}{\sqrt{2\pi}}\sqrt{\kappa''(\hat{\alpha}+1)}\,\frac{C(\hat{\alpha}+1)}{C(\alpha+1)}\exp\{(\alpha-\hat{\alpha})\kappa'(\hat{\alpha}+1)\}.
\end{aligned}$$

This expression depends on β only through the maximum likelihood equation in (7.7).

7.1 The p^* density in regular exponential families

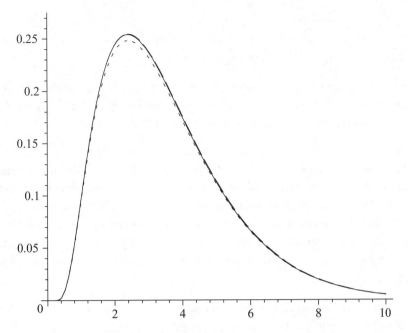

Figure 7.2. The exact density (*solid*) of $\hat{\alpha}$, the $p^{\dagger}$ density (*dotted*), and the p^* density (*dot-dashed*) from Planck's radiation formula.

Figure 7.2 compares $p^{\dagger}(\hat{\alpha}; 3)$ (dotted) with the true density (solid) and the p^* density (dashed) where the latter requires division by the value $\int p^{\dagger}(\hat{\alpha}; 3) d\hat{\alpha} \simeq 0.9754$. The p^* density is virtually indistinguishable from the true density. The accuracy is simply uncanny when one considers that the sample size here is 1.

7.1.2 Likelihood ratio statistics and Bartlett correction

One parameter models

The p^* formula provides a density approximation for the null distribution of the log-likelihood ratio statistic. Consider first a one parameter exponential family with canonical parameter θ and canonical sufficient statistic $X = \sum_{i=1}^{n} Z_i$ that has the CGF $n\{c(\theta + s) - c(\theta)\}$. The log-likelihood ratio statistic for testing that $\theta = \theta_0$ is

$$\lambda_n = -2 \ln \frac{\mathcal{L}(\theta_0)}{\mathcal{L}(\hat{\theta})} = -2x(\theta_0 - \hat{\theta}) + 2n\{c(\theta_0) - c(\hat{\theta})\}, \tag{7.8}$$

with $\hat{\theta}$ solving $nc'(\hat{\theta}) = x$.

The propositions given below are due to Barndorff-Nielsen and Cox (1979, §6.3, 1984).

Proposition 7.1.1 *The p^* density for λ_n under the null hypothesis is a χ_1^2 density to order $O(n^{-1})$ and given by the leading term on the right side of (7.9). An expansion for the true density is*

$$p(\lambda) = \frac{1}{\sqrt{2\pi\lambda}} e^{-\lambda/2} \left\{ 1 + \frac{1-\lambda}{n} \vartheta(\theta_0) + O\left(n^{-3/2}\right) \right\} \tag{7.9}$$

with

$$\vartheta(\theta) = \frac{1}{8} \frac{c^{(4)}(\theta)}{\{c''(\theta)\}^2} - \frac{5}{24} \frac{\{c^{(3)}(\theta)\}^2}{\{c''(\theta)\}^3}.$$

The term $\vartheta(\hat{\theta})/n$ would be the second-order correction term for the saddlepoint density of X. The order $O(n^{-3/2})$ results from the assumption that $\hat{\theta} - \theta = O(n^{-1/2})$ thus this order applies for λ-values that remain in sets of bounded central tendency.

Proof. Details of the proof are sufficiently technical to be banished to the appendix in section 7.4.1. The ideas of the proof, however, are quite simple: Take a saddlepoint expansion of the density of X to second-order. Follow with a Jacobian transformation to determine the density of λ_n. Simple calculations determine the Jacobian however the transformation from $x \to \lambda_n$ has two difficulties. First it is 2 to 1 so it is necessary to sum contributions from both values of x that map into a given fixed λ_n. Furthermore the Jacobian is specified in terms of $\hat{\theta}$ and much of the hard work has to do with replacing expressions in $\hat{\theta}$ with their equivalents in terms of λ_n. □

Bartlett (1937) suggested that the asymptotic χ_1^2 distribution of λ_n could be improved upon by dividing λ_n by a factor $(1 - b/n)$ that equates the mean of $p_n = \lambda_n/(1 - b/n)$ with 1, the asymptotic degree of freedom. The mean of λ_n is determined from (7.9) as

$$E(\lambda_n) = \int_0^\infty \lambda p(\lambda) d\lambda = 1 + (1-3)\vartheta(\theta_0)/n + O(n^{-3/2})$$

so that $b = 2\vartheta(\theta_0)$. The factor $1 - 2\vartheta(\theta_0)/n$ is called the Bartlett correction for λ_n.

Proposition 7.1.2 *The density for p_n under the null hypothesis is*

$$f(p) = \frac{1}{\sqrt{2\pi p}} e^{-p/2} \{1 + O(n^{-3/2})\} \tag{7.10}$$

and χ_1^2 to order $O(n^{-3/2})$.

A proof of this is suggested in Exercise 4. Barndorff-Nielsen and Hall (1988) have shown that the $O(n^{-3/2})$ term in (7.10) is zero and the error is actually $O(n^{-2})$. For an example, see Exercise 5.

Vector parameter, composite hypotheses, and more general models

Similar propositions with slight modification apply with rather general models as described in Barndorff-Nielsen and Cox (1984). These model extensions have the regularity to assure that the asymptotic distribution theory is χ^2, but also go well beyond the single parameter and exponential model setting detailed above. Vector parameters are allowed along with null hypotheses that test a particular value for a subvector with the complimentary portion as a nuisance parameter. The models allow for both regular and curved exponential families (see section 7.3) where, in the latter setting, distributional results need to be made conditional on appropriately chosen (approximate) ancillaries as described in Barndorff-Nielsen (1986) and Barndorff-Nielsen and Wood (1998). Technically what is required in this latter setting dealing with curved exponential families is that the ancillaries must admit relative error

$O(n^{-3/2})$ in the p^* formula (see section 7.3.6 for details). Results of the propositions are also extendible to transformation models which form the next topic.

Exercise 6 provides an example.

7.2 Conditional inference and p^* in group transformation models

Linear regression models and special cases, such location and location/scale models, are examples of continuous group transformation models. The name derives from the fact that such continuous parametric families may be generated by affine groups. The theoretical development and properties of such models are explained in Lehmann (1986, chap. 6). Such a model admits an MLE $\hat{\theta}$ for the model parameter θ and also possesses ancillary statistics A whose distributions do not depend on the parameter. Most often A can be chosen so that $(\hat{\theta}, A)$ is a sufficient statistic. In this context, the two main results of this section can be stated quite informally.

The first important result concerns an observation of Fisher (1934) that the conditional density for $\hat{\theta}|A$ is determined entirely by the likelihood shape. This remarkable idea is very simple yet it has eluded the mainstream of statistical thought. An implicit understanding of his work appeared 34 years later in the group transformation models studied by Fraser (1968, 1979). However, it was 54 years later that the specific details of Fisher's paper were finally elucidated in Efron and Hinkley (1978). We present the details for these ideas below.

The second result is the recognition in Barndorff-Nielsen (1980) that the shape of this conditional density is exactly reproduced by the p^* formula. We summarize both of these results informally.

Proposition 7.2.1 *If $(\hat{\theta}, A)$ is a sufficient statistic in a group transformation model, then the exact shape of the conditional density for $\hat{\theta}|A$ is expressed in terms of the likelihood function through the p^* formula.*

The proposition is explained by addressing increasingly more complex models as was originally done in Fisher (1934).

7.2.1 Location model

Suppose a random sample of size n is drawn from the density $g(y - \mu)$ with $\mu \in \Re$. Denote the observed order statistics as $y_1 \leq y_2 \leq \cdots \leq y_n$ and collectively as the vector y. Since y is sufficient, its density is

$$f(y; \mu) = n! \prod_{i=1}^{n} g(y_i - \mu) \stackrel{\mu}{\propto} \mathcal{L}(\mu) \qquad 0 \leq y_1 \leq \cdots \leq y_n.$$

Transform $y \to (t, a^*)$ where $t = y_1$ and $a^* = (a_2^*, \ldots, a_n^*)$ is an ancillary with $a_i^* = y_i - y_1$ and $a_1^* = 0$. The joint density of (t, a^*) is

$$f(t, a^*; \mu) = n! g(t - \mu) \prod_{i=2}^{n} g(a_i^* + t - \mu). \tag{7.11}$$

Note that μ always appears in (7.11) in terms of $t - \mu$; hence $\hat{\mu} = t + \lambda(a^*)$ for some function λ. Now define a second set of "residual" ancillaries as

$$a_i = a_i^* - \lambda(a^*) \quad i = 1, \ldots, n \quad (7.12)$$
$$= y_i - \hat{\mu}$$

with $a = (a_1, \ldots, a_n)$. The Jacobian for $(t, a^*) \to (\hat{\mu}, a)$ is

$$\left| \frac{\partial(\hat{\mu}, a)}{\partial(t, a^*)} \right| = \begin{vmatrix} \partial\hat{\mu}/\partial t & - \\ 0 & \partial a/\partial a^{*T} \end{vmatrix} = \begin{vmatrix} 1 & - \\ 0 & q(a) \end{vmatrix} = |q(a)|$$

and dependent only on a. Thus the conditional density is

$$f(\hat{\mu}|a; \mu) = \varrho(a)^{-1} \prod_{i=1}^{n} g(a_i + \hat{\mu} - \mu) = \varrho(a)^{-1} \mathcal{L}(\mu) \quad (7.13)$$

for some function ϱ. Function ϱ is simply the normalization constant expressed by taking $\hat{p} = \hat{\mu} - \mu$ as a pivot and writing

$$\varrho(a) = \int_{-\infty}^{\infty} \prod_{i=1}^{n} g(a_i + \hat{p}) d\hat{p}.$$

The presentation confirms what has been suggested in Fisher's observation: that likelihood shape $\mathcal{L}(\mu)$ alone suffices in determining the graph for the density of $\hat{\mu}|a$. It is worth presenting the abstruse wording used by Fisher (1934) himself to convey this idea:

> if attention is confined to samples having a given configuration [value for a] the sampling distribution of $\hat{\mu}$ for a given μ is found from $\mathcal{L}(\mu)$ for a given $\hat{\mu}$, the probability curve in the first case being the mirror image of the likelihood curve in the second.

To understanding the meaning of this, denote the observed value of $\hat{\mu}$ as $\hat{\mu}_{\text{obs}}$ and treat it as the constant that locates the mode of $\mathcal{L}(\mu)$. Then the density of $\hat{p}|a$ is given as

$$f(p|a) = \varrho(a)^{-1} \mathcal{L}(\hat{\mu}_{\text{obs}} - p) \quad (7.14)$$

where the dependence of $\mathcal{L}(\mu)$ on a is suppressed. For any value $p > 0$ in the right tail of the "probability curve," the $\Pr(p < \hat{p} < p + dp|a) = f(p|a)dp$ "in the first case" is determined by (7.14), as $\varrho(a)^{-1} \mathcal{L}(\hat{\mu}_{\text{obs}} - p)$, and hence as a value to the left of the likelihood maximum $\hat{\mu}_{\text{obs}}$ "in the second case." Figure 7.3 illustrates this point geometrically. The left plot shows $\mathcal{L}(\mu)$ versus μ with a MLE of $\hat{\mu}_{\text{obs}} = -0.6949$. The right plot is obtained from the left plot by revolving it about the vertical axis and translating it so that the mode is at $d = 0$. The differential of probability

$$\Pr(-2 < \hat{p} < -2 + dp|a) = f(-2|a)dp$$

is shown for the density on the right. Its value determined from the likelihood in (7.14) is $\varrho(a)^{-1} \mathcal{L}(1.3051)dp$ and displayed on the left.

The survival function of $\hat{p}$ is computed by integrating (7.14) so that

$$\Pr(\hat{p} > p|a) = \frac{\int_p^{\infty} \mathcal{L}(\hat{\mu}_{\text{obs}} - t)dt}{\int_{-\infty}^{\infty} \mathcal{L}(\hat{\mu}_{\text{obs}} - t)dt} = \frac{\int_{-\infty}^{\hat{\mu}_{\text{obs}} - p} \mathcal{L}(\mu)d\mu}{\int_{-\infty}^{\infty} \mathcal{L}(\mu)d\mu} \quad (7.15)$$

7.2 Conditional inference and p* in group transformation models

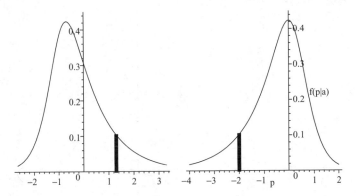

Figure 7.3. A plot of $\mathcal{L}(\mu)$ vs. μ (on the left) for the location model along with its reflected image (on the right) that plots the conditional density of pivot $\hat{p}|a$.

upon substitution with $\mu = \hat{\mu}_{\text{obs}} - t$. The ratio in (7.15) makes the distributional dependence on $\mathcal{L}(\mu)$ explicit. As one can clearly see, it also agrees with the fiducial or improper Bayesian posterior computation of probability for the event $\{\mu < \hat{\mu}_{\text{obs}} - p\}$ in which μ is random and the data are fixed.

The conditional density of $\hat{p}|a$ from p^*

For the location model, the p^* formula replicates the density in (7.14). In this instance both $\mathcal{L}(\hat{\mu})$ and $j(\hat{\mu})$ involve the vector $y - \hat{\mu}\mathbf{1}$ with $\mathbf{1} = (1, \ldots, 1)^T$ and therefore depend only on the ancillary a; both these terms drop out upon normalization. Thus $p^*(\hat{\mu}) \propto \mathcal{L}(\mu)$ and, upon transformation to $\hat{p} = \hat{\mu} - \mu$, and normalization the p^* formula for $\hat{p}$ is

$$p^*(p) = \frac{\mathcal{L}(\hat{\mu}_{\text{obs}} - p)}{\int_{-\infty}^{\infty} \mathcal{L}(\hat{\mu}_{\text{obs}} - p) dp} = f(p|a).$$

Example: Hyperbolic secant

Suppose $y_1 = -2$ and $y_2 = -1$ are the order statistics of a sample of size $n = 2$ from the density

$$g(y; \mu) = \frac{e^{-0.7(y-\mu)} \cos(-0.7)}{2 \cosh\{\pi(y - \mu)/2\}} \qquad y \in (-\infty, \infty).$$

The MLE is determined as $\hat{\mu}_{\text{obs}} = -0.6949$ and the ancillaries are $a_1 = -0.9852$ and $a_2 = 0.01418$. In Fisher's words, these ancillaries comprise the "configuration" since they, and not $\hat{\mu}_{\text{obs}}$, are the quantities responsible for determining the shape for the conditional density of $\hat{p} = \hat{\mu} - \mu$ given a. Figure 7.4 plots this density as in (7.14). Normalization gives $\varrho(a) = 0.1957$.

Example: Exponential

Suppose a random sample of size n is drawn from the density

$$g(y - \mu) = e^{-(y-\mu)} \qquad y > \mu.$$

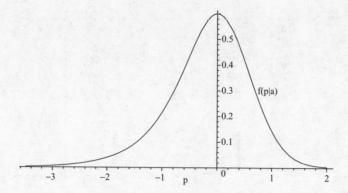

Figure 7.4. A plot of the conditional density of $\hat{p}$ given a as specified in (7.14).

Then $\hat{\mu} = y_1$, the smallest order statistic. Direct transformation shows that $\hat{\mu}$ and a^* are independent and that $\hat{p} \sim$ Exponential (n).

7.2.2 Location and scale model

Suppose a random sample of size n is drawn from the density $g\{(y-\mu)/\sigma\}/\sigma$ with $\theta = (\mu, \sigma) \in \Re \times (0, \infty)$. Let the components of y be the observed order statistics and transform $y \to (t_1, t_2, a^*)$ with $t_1 = y_1$, $t_2 = y_n - y_1$, and $a^* = (a_2^*, \ldots, a_{n-1}^*)$ with $a_i^* = (y_i - t_1)/t_2 \in (0, 1)$ and $a_1^* = 0$ and $a_n^* = 1$. Ancillary a^* determines a small square $[a^*, a^* + da^*] \subset \Re^{n-2}$ which Fisher called the *reference set*. The terminology expresses his early formulation of the conditionality principle that requires inferential probabilities be conditional on a^*. The transformation has Jacobian t_2^{n-2} computed as the determinant of the matrix $\partial y/\partial(t_1, t_2, a^{*T})$ which has zero entries above the diagonal. Thus

$$f(t_1, t_2, a^*; \theta) = \frac{t_2^{n-2}}{\sigma^n} \prod_{i=1}^n g\{(y_i - \mu)/\sigma\}.$$

The MLE $(\hat{\mu}, \hat{\sigma})$ is related to (t_1, t_2, a^*) according to

$$\hat{\mu} = t_1 + \mu(a^*)t_2 \tag{7.16}$$
$$\hat{\sigma} = \lambda(a^*)t_2 \tag{7.17}$$

where $\mu(a^*)$ and $\lambda(a^*)$ are some functions of a^*. To see this, let $a_+^* = (a_1^*, \ldots, a_n^*)^T$ and note that in vector form

$$\frac{1}{\sigma}(y - \mathbf{1}\mu) = \frac{1}{\sigma}\{(t_1 - \mu)\mathbf{1} + t_2 a_+^*\} = \frac{t_2}{\sigma}\left\{\frac{t_1 - \mu}{t_2}\mathbf{1} + a_+^*\right\}. \tag{7.18}$$

Based on the form of (7.18), likelihood maximization in μ must result in (7.16) and subsequent maximization over σ must give (7.17). The transformation $(t_1, t_2, a^*) \to (\hat{\mu}, \hat{\sigma}, a)$, with $a = (a_2, \ldots, a_{n-1})$, and

$$a_i = \{a_i^* - \mu(a^*)\}/\lambda(a^*) = (y_i - \hat{\mu})/\hat{\sigma} \qquad i = 1, \ldots, n$$

as the studentized residuals, has Jacobian $\lambda^{n-3}(a^*)$ and leads to the conditional density of $\hat{\mu}, \hat{\sigma} | a$ as

$$f(\hat{\mu}, \hat{\sigma} | a; \mu, \sigma) = \varrho(a)^{-1} \hat{\sigma}^{n-2} \prod_{i=1}^{n} g\{(\hat{\sigma} a_i + \hat{\mu} - \mu)/\sigma\} \quad (7.19)$$

where $\varrho(a)$ is its normalization constant.

In terms of $(\hat{\mu}, \hat{\sigma})$, expression (7.19) is proportional to $\hat{\sigma}^{-2} \mathcal{L}(\mu, \sigma)/\mathcal{L}(\hat{\mu}, \hat{\sigma})$. Therefore, to show that the p^* formula agrees with (7.19), it is only necessary to verify that the observed Fisher information $j(\hat{\mu}, \hat{\sigma}, a) \propto \hat{\sigma}^{-4}$ in $(\hat{\mu}, \hat{\sigma})$. In fact, straightforward computations show that $j(\hat{\mu}, \hat{\sigma}, a) = \hat{\sigma}^{-4} |D(a)|$ where

$$D(a) = \left\{ \sum_{i=1}^{n} h''(a_i) \right\} \left\{ -n + \sum_{i=1}^{n} a_i^2 h''(a_i) \right\} - \left\{ \sum_{i=1}^{n} a_i h''(a_i) \right\}^2$$

and $h(y) = \ln g(y)$.

Example: Normal

If $g(y) = \phi(y)$, then $a_i = (y_i - \bar{y})/\hat{\sigma}$ with $\hat{\sigma}^2 = \sum_{i=1}^{n}(y_i - \bar{y})^2/n$. Direct computation of (7.19) shows that $\bar{y}, \hat{\sigma}^2$, and $\{a_i : i = 1, \ldots, n\}$ form a mutually independent triple since the density of $\bar{y}, \hat{\sigma} | a$ does not depend on a and it also separates into $\bar{y}$ and $\hat{\sigma}$. The two factors reveal that $\bar{y} \sim \text{Normal}(\mu, \sigma^2/n)$ and that $\hat{\sigma}$ has a density that is consistent with $n\hat{\sigma}^2 \sim \sigma^2 \chi_{n-1}^2$.

Example: Weibull

Suppose the data are i.i.d. from a Weibull (α, β) distribution with density

$$f(t) = \frac{\beta}{\alpha}(t/\alpha)^{\beta-1} \exp\{-(t/\alpha)^{\beta}\} \quad y > 0.$$

The log transformation $y = \ln t$ changes the distribution to that of a left-skewed Gumbel (μ, σ) with density

$$f(y) = \frac{1}{\sigma} \exp\left\{(y - \mu)/\sigma - e^{(y-\mu)/\sigma}\right\} \quad -\infty < y < \infty,$$

where $\mu = \ln \alpha$ and $\sigma = 1/\beta$. Lawless (1973, 1974) considered conditional inference for α and β through location/scale estimates on the log scale. See Exercise 8 for a numerical example.

7.2.3 Linear model

Extensions to the general linear model are derived in Fraser (1979) and summarized below. Suppose that dependent $n \times 1$ vector y is no longer ordered but linearly related to $p - 1$ independent variables and an additional location variable expressed though the fixed $n \times p$ matrix X of rank p. Then

$$y = X\beta + \sigma e$$

with $\theta = (\beta, \sigma) \in \Re^p \times (0, \infty)$. Let $e = (e_1, \ldots, e_n)^T$ have the n-variate density $g(e)$. Fraser (1979, §6.1.4) shows that the density for the least squares estimates $b = (X^T X)^{-1} X^T y$ and $s = \|y - Xb\|$ given ancillary $a^* = (y - Xb)/s$ is

$$f(b, s | a^*; \theta) \stackrel{b,s}{\propto} \frac{s^{n-p-1}}{\sigma^n} g\left[\frac{1}{\sigma}\{sa^* + X(b - \beta)\}\right]. \tag{7.20}$$

Using arguments similar to those used in the location/scale model, the conditional density of the MLE $(\hat{\beta}, \hat{\sigma})$ given $a = (y - X\hat{\beta})/\hat{\sigma}$ is shown in Exercise 9 to be

$$f(\hat{\beta}, \hat{\sigma} | a; \theta) = \varrho(a)^{-1} \hat{\sigma}^{n-p-1} g\left[\frac{1}{\sigma}\{(\hat{\sigma} a + X(\hat{\beta} - \beta)\}\right]. \tag{7.21}$$

This is also shown to agree with the p^* formula in which $j(\hat{\beta}, \hat{\sigma}, a) = \hat{\sigma}^{-2(p+1)} K(a)$.

7.3 Approximate conditional inference and p^* in curved exponential families

Suppose x is the value for a canonical sufficient statistic in a curved exponential family indexed by the parameter ξ. Then 1-1 transformations $x \leftrightarrow (\hat{\xi}, a)$ may be determined in such a way that $\hat{\xi}$ is the MLE of ξ and a is an approximate ancillary statistic; details concerning the nature of a are given below. The main result to be presented here, a result that originated in Barndorff-Nielsen (1980, 1983) and Hinkley (1980b), concerns approximation to the true density for $\hat{\xi} | a$. Briefly and informally it states that this conditional density is approximated to higher asymptotic order by the p^* formula when computed using the likelihood of the curved exponential family. Barndorff-Nielsen (1980) has referred to the determination of the density for $\hat{\xi} | a$ as the *conditionality resolution* of the exponential family indexed by ξ.

Some of the technical notation required for discussing curved exponential families is presented in the first subsection. An affine ancillary a is introduced followed by a derivation of p^* as the approximate conditional distribution of $\hat{\xi} | a$ for this case in sections 7.3.2–7.3.4. The presentation is an adaptation of the proofs used in Hinkley (1980b) and Barndorff-Nielsen (1980) that provide some geometrical understanding of the problem as well as draw connections to the curvature of the model (Efron 1975) and the Efron-Hinkley (1978) ancillary. The final sections 7.3.5–7.3.6 describe generalizations of these results to likelihood ancillaries as well as some of the asymptotic and large deviation properties that have been shown to hold for the p^* approximation in Skovgaard (1990) and Barndorff-Nielsen and Wood (1998).

7.3.1 Curved exponential families

Let z be the data and $x = x(z)$ the $m \times 1$ canonical sufficient statistic for an exponential family of the form

$$f(z; \xi) = \exp\left\{\theta_\xi^T x - c\left(\theta_\xi\right) - d(z)\right\}. \tag{7.22}$$

The subscripting of θ_ξ is used to indicate the dependence of the $m \times 1$ canonical parameter on the lower dimensional vector $\xi \in \Xi \subset \Re^p$ with $p < m$ and Ξ as an open subset of $\Re^p$. The parametric class $\{f(\cdot; \xi) : \xi \in \Xi\}$ is referred to as an (m, p)-*curved exponential family* of distributions. The graph of $\{\theta_\xi : \xi \in \Xi\} \subset \Re^m$ is assumed to be a nonlinear differential

7.3 Approximate conditional inference and p^* in curved exponential families

manifold that possesses all its derivatives. If the manifold is linear, so that $\theta_\xi = B\xi$ for some $m \times p$ matrix B, then (7.22) reduces to a regular exponential family with sufficient statistic $B^T x$ and canonical parameter ξ. Geometry plays some importance in these families because the parametric class of distributions may be characterized through $\{\theta_\xi : \xi \in \Xi\} \subset \Re^m$, the graph that the canonical parameter traces out in canonical parameter space. If $\mu_\xi = E(X)$ denotes the mean parametrization, then the family is also characterized through the graph of $\{\mu_\xi : \xi \in \Xi\} \subset \Re^m$ in the mean parameter space. This follows as a result of the 1-1 correspondence $\theta_\xi \leftrightarrow \mu_\xi$.

Some of the calculus required for differentiating $l_\xi = \ln f(z;\xi)$ is now stated. All results follow from elementary application of the chain rule. Denote

$$\mathrm{Cov}(X;\xi) = c''(\theta_\xi) = \Sigma_\xi,$$

and let $\dot{\theta}_\xi = \partial \theta_\xi / \partial \xi^T$ be an $m \times p$ vector. Then

$$\dot{\mu}_\xi = \frac{\partial \mu_\xi}{\partial \xi^T} = \Sigma_\xi \dot{\theta}_\xi$$

is $m \times p$ and

$$\frac{\partial c(\theta_\xi)}{\partial \xi^T} = \mu_\xi^T \dot{\theta}_\xi.$$

The first derivative of l_ξ with respect to ξ^T is the $1 \times p$ vector

$$\dot{l}_\xi^T = \frac{\partial l_\xi}{\partial \xi^T} = (x - \mu_\xi)^T \dot{\theta}_\xi. \tag{7.23}$$

The $p \times p$ Hessian requires notation that allows for the expression of the second derivative matrix of a vector. For this, two methods are given. First one can write

$$\ddot{l}_\xi = \frac{\partial^2 l_\xi}{\partial \xi \partial \xi^T} = -\dot{\theta}_\xi^T \Sigma_\xi \dot{\theta}_\xi + D_\xi \tag{7.24}$$

with

$$(D_\xi)_{ij} = \left\{ (x - \mu_\xi)^T \frac{\partial^2 \theta_\xi}{\partial \xi_i \partial \xi_j} \right\}. \tag{7.25}$$

Alternatively D_ξ may be written as

$$D_\xi = \sum_{i=1}^m (x_i - \mu_{\xi i}) \frac{\partial^2 \theta_{\xi i}}{\partial \xi \partial \xi^T} = (x - \mu_\xi) \odot \ddot{\theta}_\xi,$$

where $\theta_{\xi i}$ indicates the ith component of vector θ_ξ and the the identity defines the meaning for the vector-matrix product operation $\odot$. From (7.24), the expected Fisher information is

$$i_\xi = -E(\ddot{l}_\xi) = \dot{\theta}_\xi^T \Sigma_\xi \dot{\theta}_\xi = \dot{\theta}_\xi^T \dot{\mu}_\xi. \tag{7.26}$$

The last expression $\dot{\theta}_\xi^T \dot{\mu}_\xi$ is often the simplest way of determining i_ξ in examples. From this, the relationship of the observed and expected information is

$$j_\xi = -\ddot{l}_\xi = i_\xi - D_\xi. \tag{7.27}$$

The MLE solves the p equations

$$0 = \dot{l}_\xi = \dot{\theta}_\xi^T (x - \mu_\xi), \qquad (7.28)$$

however, since $D_\xi \neq 0$, then

$$j_\xi = i_\xi - D_\xi \neq i_\xi$$

which differs from the equality that occurs in the regular exponential setting.

Further simplifications in notation result when considering the special case in which $p = 1$. Then, (7.24) becomes

$$\ddot{l}_\xi = -\dot{\theta}_\xi^T \Sigma_\xi \dot{\theta}_\xi + \ddot{\theta}_\xi^T (x - \mu_\xi) \qquad (7.29)$$

where $\ddot{\theta}_\xi = \partial^2 \theta_\xi / \partial \xi^2$ is $p \times 1$. Likewise

$$j_\xi = i_\xi - \ddot{\theta}_\xi^T (x - \mu_\xi). \qquad (7.30)$$

Let the first m derivatives of θ_ξ form the rows of the $m \times m$ matrix

$$\Delta_\xi = \begin{pmatrix} \dot{\theta}_\xi^T \\ \vdots \\ \partial^m \theta_\xi^T / \partial \xi^m \end{pmatrix}. \qquad (7.31)$$

Then further differentiation of (7.29) shows that

$$\text{Cov}(\dot{l}_\xi, \ddot{l}_\xi \ldots, \partial^m l_\xi / \partial \xi^m) = \Delta_\xi \Sigma_\xi \Delta_\xi.$$

7.3.2 Examples

The gamma hyperbola example was introduced by Fisher (1990, *Statistical Methods, Experimental Design, and Scientific Inference*, pp. 169–175) to motivate the need for considering ancillary statistic a along with the MLE $\hat{\xi}$. He emphasizes that $\hat{\xi}$ alone does not fully capture the information in the data and that there is a need to have an ancillary a in order to make $(\hat{\xi}, a)$ fully informative or, in our more precise statistical parlance, minimal sufficient. Fisher's example is a precursor to the development of the conditionality resolutions that the p^* formula provides and therefore provides a convenient example to motivate many of the theoretical points to be made.

Further examples that consider correlation, times series, and the Behrens–Fisher problems are also curved exponential families and will be used later to provide numerical examples for the theoretical developments.

Gamma hyperbola

Fisher (1990) suggested a simple example of a (2,1) curved exponential family in which $X_1 \sim$ Exponential (ξ) independent of $X_2 \sim$ Exponential $(1/\xi)$. The canonical parameter is $\theta_\xi = (\xi, 1/\xi)^T$ for $\xi > 0$ which traces out a hyperbola in $\Theta = (0, \infty)^2$. The mean parameter is $\mu_\xi = (1/\xi, \xi)^T$ which is the interchange of coordinates in θ_ξ and corresponds to a rotation of the hyperbola plot of the canonical parameter about the 45° line. The solid line in figure 7.5 is a plot of $\mu_\xi = (1/\xi, \xi)$ with $1/\xi$ along the horizontal axis. The same hyperbola traces

7.3 Approximate conditional inference and p* in curved exponential families

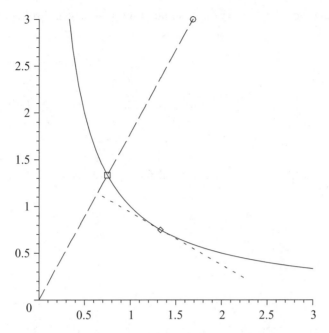

Figure 7.5. Plot of $1/\xi$ vs. ξ (*solid*) representing the graph of θ_ξ.

the canonical parameter curve but, because of the interchange of coordinates, ξ is the value plotted along the horizontal axis.

The lines in the figure show the geometry involved when determining the MLE $\hat{\xi} = \sqrt{x_2/x_1}$ from the data $x^T = (27/16, 3)$ which is indicated by the hollow circle. The MLE is $\hat{\xi} = 4/3$ and $\mu_{\hat{\xi}} = (3/4, 4/3)^T$ is the mean parameter shown as the square. Its reflection across the 45° line is $\theta_{\hat{\xi}} = (4/3, 3/4)^T$ denoted by a diamond. The geometry imposed on $\hat{\xi}$ by the maximum likelihood equation (7.28) may be seen through the two lines in the figure. Equation (7.28) requires that $\dot{\theta}_{\hat{\xi}}$, the tangent vector to the canonical parameter curve shown with the dotted line, must be perpendicular to $x - \mu_{\hat{\xi}}$ as shown with the dashed line.

This example offers a simple introduction to the affine ancillaries to be considered below. The geometry in determining $\hat{\xi}$ suggests that all data points lying on the dashed line lead to the same value for $\hat{\xi}$. In terms of curvilinear coordinates, we may think about $\hat{\xi}$ as the coordinate of x along the hyperbola "axis." A complementary coordinate to complete the sufficient statistic would be a measurement of position along the dashed line given as

$$\mathcal{A}_{\hat{\xi}} = \{y : \dot{\theta}_{\hat{\xi}}^T(y - \mu_{\hat{\xi}}) = 0\}$$

and called the *affine ancillary space*. Consider measuring the distance of x from $\mu_{\hat{\xi}}$ within this geometrical setting. A reasonable distance based on intuitive grounds is the non-Euclidean distance which "studentizes" $x - \mu_{\hat{\xi}}$ as

$$q^2 = (x - \mu_{\hat{\xi}})^T \Sigma_{\hat{\xi}}^{-1} (x - \mu_{\hat{\xi}}) \tag{7.32}$$
$$= 2\left(\sqrt{x_1 x_2} - 1\right)^2$$

after some computation. Note that the distribution of $X_1 X_2$ is invariant to scale changes and hence $X_1 X_2$ is exactly ancillary in this setting. The intuitive choice of an ancillary as a studentized distance has lead to an exact ancillary for this somewhat artificial but illustrative example.

Bivariate normal correlation

Suppose $(z_{11}, z_{12})^T, \ldots, (z_{n1}, z_{n2})^T$ are i.i.d. data from a Normal$_2$ $(0, R)$ with

$$R = \begin{pmatrix} 1 & \xi \\ \xi & 1 \end{pmatrix} \qquad |\xi| < 1.$$

The likelihood is a (2, 1) curved exponential family with canonical statistic having the components

$$x_1 = \frac{1}{2} \sum_{i=1}^n \left(z_{i1}^2 + z_{i2}^2 \right) \qquad x_2 = \sum_{i=1}^n z_{i1} z_{i2} \qquad (7.33)$$

and canonical parameter as

$$\theta^T = \frac{1}{1 - \xi^2} (-1 \quad \xi).$$

This example was first considered by Efron and Hinkley (1978) and Barndorff-Nielsen (1980).

ARMA (p, q) time series

The standard ARMA (p, q) times series model with normal errors is a curved exponential family. As an example, consider a simple stationary AR (1) model in which

$$Z_t = \alpha Z_{t-1} + \sigma \varepsilon_t \qquad t = 1, \ldots, T$$

with $|\alpha| < 1$, $\varepsilon_1, \ldots, \varepsilon_T$ as i.i.d. standard normal, and Z_0 fixed at z_0. The likelihood is a (3, 2) curved exponential family with canonical parameter and sufficient statistic

$$\theta^T = \frac{1}{\sigma^2} \left(-\frac{1}{2}, \alpha, -\frac{\alpha^2}{2} \right) \qquad \text{and} \qquad x = \sum_{t=1}^T \left(z_t^2, z_t z_{t-1}, z_{t-1}^2 \right).$$

Behrens–Fisher problem

Consider two independent samples with unequal population variances, i.e. $Z_{i1}, \ldots, Z_{in_i}$ are i.i.d. Normal (μ_i, σ_i^2) for $i = 1, 2$. With different population means, this is a (4, 4) exponential family in which the sample means and variances are the sufficient statistics. When testing the equality $\mu_1 = \mu_2 = \mu$, the model becomes a (4, 3) family under the null hypothesis and it is unclear how to determine suitable p-values.

7.3.3 Affine ancillaries

The computation of q^2 in (7.32) is a Mahalanobis distance of the type commonly considered in multivariate analysis. Below q is shown to be the value of the so-called Efron–Hinkley (1978) affine ancillary that was originally constructed by using the concept of the *curvature*

7.3 Approximate conditional inference and p^* in curved exponential families

of the exponential family. After briefly considering some properties of curvature, Fisher's gamma hyperbola example is reconsidered in light of these new concepts. All of this provides a prelude to the introduction of the affine ancillaries in a general (m, p) curved exponential family that rounds out the subsection.

(2, 1) Curved exponential families

The gamma hyperbola example is generalized to an arbitrary (2, 1) exponential setting. The transformation of $x - \mu_\xi$ by Δ_ξ, as given in (7.31), leads to

$$\Delta_\xi(x - \mu_\xi) = \begin{pmatrix} 0 \\ \ddot{\theta}_\xi^T(x - \mu_\xi) \end{pmatrix} = \begin{pmatrix} 0 \\ i_\xi - j_\xi \end{pmatrix}$$

by (7.30). Now reexpress

$$\begin{aligned} q^2 &= (x - \mu_\xi)^T \Delta_\xi^T \left(\Delta_\xi \Sigma_\xi \Delta_\xi^T \right)^{-1} \Delta_\xi (x - \mu_\xi) \\ &= \begin{pmatrix} 0 \\ i_\xi - j_\xi \end{pmatrix}^T \begin{pmatrix} \dot{\theta}_\xi^T \Sigma_\xi \dot{\theta}_\xi & \dot{\theta}_\xi^T \Sigma_\xi \ddot{\theta}_\xi \\ \ddot{\theta}_\xi^T \Sigma_\xi \dot{\theta}_\xi & \ddot{\theta}_\xi^T \Sigma_\xi \ddot{\theta}_\xi \end{pmatrix}^{-1} \begin{pmatrix} 0 \\ i_\xi - j_\xi \end{pmatrix} \\ &= (i_\xi - j_\xi)^2 \times \{(2, 2) \text{ element of the inverse}\} \\ &= (j_\xi - i_\xi)^2 \times [\mathrm{Var}(\dot{l}_\xi)/\{\mathrm{Var}(\dot{l}_\xi)\mathrm{Var}(\ddot{l}_\xi) - \mathrm{Cov}^2(\dot{l}_\xi, \ddot{l}_\xi)\}]_{\xi=\hat{\xi}}. \end{aligned}$$

The relationship of q^2 to the intrinsic curvature $\gamma_\xi \geq 0$ of the likelihood can now be made explicit. Define (see Efron, 1975 and Hinkley, 1980a)

$$\gamma_\xi^2 = \frac{1}{i_\xi^2} \mathrm{Var}(\ddot{l}_\xi) \left\{ 1 - \frac{\mathrm{Cov}^2(\dot{l}_\xi, \ddot{l}_\xi)}{\mathrm{Var}(\dot{l}_\xi)\mathrm{Var}(\ddot{l}_\xi)} \right\}$$

which, apart from the constant $1/i_\xi^2$, is the variance after regressing $\ddot{l}_\xi$ on $\dot{l}_\xi$. With this definition, then $q^2 = a_2^2$ where

$$a_2 = -\frac{j_\xi - i_\xi}{i_\xi \gamma_\xi}.$$

Statistic a_2 has been shown in Efron and Hinkley (1978) to be an approximate ancillary under quite general circumstances. The following relevant facts about curvature are easily derived.

(1) In a regular exponential family the curvature is identically zero. Otherwise the curvature must be nonzero for some values of ξ.
(2) For fixed ξ, γ_ξ does not change with reparametrization; thus it is an intrinsic property of a curved exponential family.
(3) If Z_1 has density $f(z; \xi)$ with curvature $\bar{\gamma}_\xi$, then, for $Z_1, \ldots, Z_n$ i.i.d., the likelihood $\prod_{i=1}^n f(z_i; \xi)$ has curvature $\gamma_\xi = \bar{\gamma}_\xi/\sqrt{n}$. For the latter setting,

$$\sqrt{n}\left(\frac{j_\xi}{i_\xi} - 1\right) \xrightarrow{\mathcal{D}} \text{Normal}(0; \bar{\gamma}_\xi^2) \qquad \text{as } n \to \infty. \qquad (7.34)$$

By using Slutsky's theorem, it follows that $a_2 \xrightarrow{\mathcal{D}}$ Normal $(0, 1)$ as $n \to \infty$ and a_2 is asymptotically ancillary. Note that (7.34) affords an interpretation for the per observation curvature as

$$\bar{\gamma}_\xi^2 \simeq \text{Var}\left(\sqrt{n} j_\xi / i_\xi\right).$$

Thus in models with large $|\bar{\gamma}_\xi|$, an inference which uses i_ξ in place of j_ξ may result in a substantial difference.

(4) For a group transformation model in which ξ is a location or scale parameter, γ_ξ is not dependent on ξ. In either case, a_2 is an exact ancillary.

(5) Ancillary a_2 is invariant to reparameterization of the model since q^2, as a Mahalanobis distance, is invariant.

Gamma hyperbola

The curvature and ancillary statistics for Fisher's model are now reconsidered. Reparameterize $\omega = -2 \ln \xi$ so that $\hat{\omega} = -2 \ln \hat{\xi} = \ln x_1 - \ln x_2$. Taking $a = \ln(x_1 x_2)$ as an ancillary, then the transformation $(x_1, x_2) \to (\hat{\omega}, a)$ leads to the joint density

$$f(\hat{\omega}, a; \omega) = \tfrac{1}{2} e^a \exp\left[-2e^{a/2} \cosh\{(\hat{\omega} - \omega)/2\}\right] \qquad (\hat{\omega}, a) \in \Re^2. \tag{7.35}$$

The conditional density of $\hat{\omega}$ given a is symmetric about location parameter ω and upon normalization is

$$f(\hat{\omega}|a; \omega) = \frac{1}{4 K_0(2e^{a/2})} \exp\left[-2e^{a/2} \cosh\{(\hat{\omega} - \omega)/2\}\right] \qquad \hat{\omega} \in \Re, \tag{7.36}$$

where $K_0(\cdot)$ denotes a Bessel K function. Using the properties of curvature, then γ_ω is not dependent on ω and also a_2 is an exact ancillary. It is easiest to show that $\gamma_\xi \equiv \gamma = 1/\sqrt{2}$ using the original likelihood in ξ. Additional computations based on (7.35) show that

$$a_2 = -\left(\frac{j_{\hat{\omega}}}{\gamma i_{\hat{\omega}}} - 1\right) = -\sqrt{2}\left(\sqrt{x_1 x_2} - 1\right). \tag{7.37}$$

(m, p) curved exponential families

The development of affine ancillaries in $(m, 1)$ exponential families follows from $(2, 1)$ in the rather straightforward manner indicated below that also reveals the geometry involved. The more general construction for (m, p) families is less transparent and uses the arguments in Barndorff-Nielsen (1980).

$(m, 1)$ *families* Following the notation for the $(2, 1)$ case, let $\mathcal{A}_\xi$ be the $(m - 1)$-dimensional hyperplane in the space of mean values for x that is orthogonal to $\dot{\theta}_\xi$ and passes through μ_ξ. In a $(2, 1)$ family, the value of a_2 with

$$a_2^2 = (x - \mu_\xi)^T \Sigma_\xi^{-1}(x - \mu_\xi)$$

represents the coordinate of $x - \mu_\xi$ within $\mathcal{A}_\xi$. The generalization to $(m, 1)$ families amounts to determining additional values $a_3, \ldots, a_m$ such that $a^T = (a_2, \ldots, a_m)$ is a coordinate for

7.3 Approximate conditional inference and p* in curved exponential families

$x - \mu_\xi$ within $\mathcal{A}_\xi$ and which consists of "standard independent" variables that satisfy

$$\|a\|^2 = (x - \mu_\xi)^T \Sigma_\xi^{-1} (x - \mu_\xi). \tag{7.38}$$

Gram–Schmidt orthogonalization achieves this decomposition in the following manner. First remove the constrained value $\dot{\theta}_\xi^T (x - \mu_\xi) = 0$ from (7.38) by transforming

$$\Delta_\xi (x - \mu_\xi) = \begin{pmatrix} 0 \\ u_\xi \end{pmatrix} \tag{7.39}$$

where

$$u_\xi = \begin{pmatrix} \ddot{\theta}_\xi^T (x - \mu_\xi) \\ \vdots \\ (\partial^m \theta_\xi / \partial \xi^m)^T (x - \mu_\xi) \end{pmatrix}_{\xi = \hat{\xi}} = \begin{pmatrix} \ddot{l}_\xi - E(\ddot{l}_\xi) \\ \vdots \\ \partial^m l_\xi / \partial \xi^m - E(\partial^m l_\xi / \partial \xi^m) \end{pmatrix}_{\xi = \hat{\xi}}$$

is the quantity to be further orthogonalized. Decompose

$$\Delta_\xi \Sigma_\xi \Delta_\xi = \begin{pmatrix} i_\xi & v_{21\xi}^T \\ v_{21\xi} & V_{22\xi} \end{pmatrix}$$

where $V_{22\xi} = \text{Cov}(u_\xi)$ so that the right side of (7.38) is

$$(x - \mu_\xi)^T \Delta_\xi (\Delta_\xi \Sigma_\xi \Delta_\xi)^{-1} \Delta_\xi (x - \mu_\xi)$$
$$= \begin{pmatrix} 0 \\ u_\xi \end{pmatrix}^T \begin{pmatrix} i_\xi & v_{21\xi}^T \\ v_{21\xi} & V_{22\xi} \end{pmatrix}^{-1} \begin{pmatrix} 0 \\ u_\xi \end{pmatrix}$$
$$= u_\xi^T V_\xi^{22} u_\xi$$

with $V_\xi^{22} = V_{22\xi} - v_{21\xi} v_{21\xi}^T / i_\xi$. The final step is to let Ω_ξ be a $(m-1) \times (m-1)$ lower triangular matrix such that $\Omega_\xi V_\xi^{22} \Omega_\xi^T = I_{m-1}$. The transformation $a = \Omega_\xi u_\xi$ produces the desired result with orthogonalized and standardized values assuring the equality in (7.38).

The use of lower triangular matrix Ω_ξ as $(V_\xi^{22})^{-1/2}$ means that a_j depends only on the first j derivatives $\ddot{l}_\xi, \ldots, \partial^j l_\xi / \partial \hat{\xi}^j$ which reflects the Gram–Schmidt orthogonalization. Furthermore, a_2 depends only on $\ddot{l}_\xi$ and therefore must be the Efron–Hinkley ancillary. The representation of u_ξ above in terms of the first m derivatives of l_ξ affords the following statistical interpretation of the affine ancillary construction: a_2 is the studentized residual in regressing $\ddot{l}_\xi$ on $\dot{l}_\xi$ evaluated at $\xi = \hat{\xi}$; a_3 is the studentized residual in regressing $\dddot{l}_\xi$ on $\dot{l}_\xi$ and $\ddot{l}_\xi$ and evaluating it at $\xi = \hat{\xi}$; etc.

(m, p) families The approach used in Barndorff-Nielsen (1980) is to consider "affine" transformations of x by determining a $(m-p) \times m$ matrix A_ξ such that $a_\xi = A_\xi (x - \mu_\xi)$ has mean zero and covariance I_{m-p}. Then it is reasonable to expect that a_ξ is weakly dependent on ξ since its first two moments are independent of ξ. Furthermore, upon substituting $\hat{\xi}$ for ξ, $a_{\hat{\xi}}$ should maintain this weak dependence so that $a_{\hat{\xi}}$ is an approximate affine ancillary. A second property not explicitly mentioned in Barndorff-Nielsen (1980), is that the estimating equation $\dot{\theta}_\xi^T (x - \mu_\xi)$, whose root determines $\hat{\xi}$, should be uncorrelated with a_ξ. The ancillary construction for $(m, 1)$ families above satisfies both of these properties as outlined in Exercise 10. This general approach also has a geometrical interpretation, however only the computational aspects of this general case are considered.

Let Ψ_ξ be any $m \times (m - p)$ matrix whose column vectors are orthogonal to the column vectors of $\dot\mu_\xi = \Sigma_\xi \dot\theta_\xi$ so that $\Psi_\xi^T \Sigma_\xi \dot\theta_\xi = 0$.

Lemma 7.3.1 *Let*

$$A_\xi = \left(\Psi_\xi^T \Sigma_\xi \Psi_\xi\right)^{-1/2} \Psi_\xi^T$$

and define

$$a = a_{\hat\xi} = A_{\hat\xi}(x - \mu_{\hat\xi}). \tag{7.40}$$

Then a is a $(m - p) \times 1$ affine ancillary statistic as required in the conditions stated above.

Proof. The lemma is a simple exercise. □

The lemma suggests a specific choice for an affine ancillary but it also disguises an entire class of ancillaries. The value of a may vary according to the choice of columns for Ψ_ξ; any basis for the space orthogonal to the column space of $\dot\mu_\xi$ may be used. Furthermore, any $(m - p) \times (m - p)$ orthogonal transformation of a results in a vector satisfying the same two conditions mentioned in the lemma. If a_ξ is allowed to have any covariance that does not depend on ξ and not just the identity matrix, then affine ancillaries include all linear transformations of a.

Any affine ancillary as in (7.40) has the property

$$a^T a = (x - \mu_{\hat\xi})^T \Sigma_{\hat\xi}^{-1}(x - \mu_{\hat\xi}). \tag{7.41}$$

which follows directly.

7.3.4 p^* as a conditionality resolution with affine ancillaries

Theorem 7.3.2 *For the continuous (m, p) curved exponential family in (7.22), let a be any affine ancillary as specified in lemma 7.3.1. Suppose also that the transformation $x \to (\hat\xi, a)$ is 1-1. Then the conditional density of $\hat\xi$ given a is approximated by the p^* density which provides a conditionality resolution for the curved exponential family.*

The proof is given at the end of this section.

The nature of the approximation requires some clarification. Using the notation in (7.25)–(7.27), let

$$p^\dagger(\hat\xi | a; \xi) = (2\pi)^{-p/2} |j_{\hat\xi}|^{1/2} \frac{\mathcal{L}(\xi)}{\mathcal{L}(\hat\xi)} \tag{7.42}$$

$$= (2\pi)^{-p/2} |i_{\hat\xi} - \{x(\hat\xi, a) - \mu_{\hat\xi}\} \odot \ddot\theta_{\hat\xi}|^{1/2} \frac{\exp\left\{\theta_\xi^T x(\hat\xi, a) - c(\theta_\xi)\right\}}{\exp\left\{\theta_{\hat\xi}^T x(\hat\xi, a) - c(\theta_{\hat\xi})\right\}}$$

where $x = x(\hat\xi, a)$ makes the dependence of x on $(\hat\xi, a)$ explicit. If

$$\varkappa(a; \xi)^{-1} = \int p^\dagger(\hat\xi | a; \xi) d\hat\xi,$$

then
$$p^*(\hat{\xi}|a;\xi) = \varkappa(a;\xi)|j_{\hat{\xi}}|^{1/2}\frac{\mathcal{L}(\xi)}{\mathcal{L}(\hat{\xi})} \quad (7.43)$$

is the normalized version of $p^\dagger$. In (7.42) note that $p^\dagger$ depends on $\hat{\xi}$ through $i_{\hat{\xi}}$, $\ddot{\theta}_{\hat{\xi}}$, $x(\hat{\xi}, a)$, $\mu_{\hat{\xi}}$ and $\theta_{\hat{\xi}}$. The simplicity and elegance of the $p^\dagger$ and p^* formulas as seen in (7.42) and (7.43) belie their underlying complexity and this discrepancy has been a source of misunderstanding and confusion about them. However, it will be seen that these expressions are quite easy to compute and extremely accurate in many applications.

Gamma hyperbola

Compare the true density of $\hat{\xi}|a$ with the $p^\dagger$ density approximation. The simpler ancillary $a = \sqrt{x_1 x_2}$ is a 1-1 function of the affine ancillary and so we may condition on this a instead. According to (7.36), the true density of $\hat{\xi}|a$ from the transformation $\hat{\xi} = e^{-\hat{\omega}/2}$ is

$$f(\hat{\xi}|a;\xi) = \frac{1}{2K_0(2e^{a/2})}\frac{1}{\hat{\xi}}\exp\{-e^{a/2}(\hat{\xi}/\xi + \xi/\hat{\xi})\} \quad \hat{\xi} > 0. \quad (7.44)$$

By comparison the $p^\dagger$ density is

$$p^\dagger(\hat{\xi}|a;\xi) = \frac{1}{2\hat{K}_0(2e^{a/2})}\frac{1}{\hat{\xi}}\exp\{-e^{a/2}(\hat{\xi}/\xi + \xi/\hat{\xi})\} \quad \hat{\xi} > 0, \quad (7.45)$$

and exact up to a normalization constant where

$$\hat{K}_0(z) = \sqrt{\frac{\pi}{2z}}e^{-z}$$

is given in Abramowitz and Stegun (1972, 9.7.2) as the leading term in an asymptotic expansion (as $|z| \to \infty$) of K_0. For the numerical setting previously given in which $a = -5\sqrt{2}/4$, the argument of $\hat{K}_0$ in (7.45) is quite small and leads to a ratio of normalization constants in which $p^\dagger/f$ is 0.5543.

The exactness of the p^* density in this setting should not be surprising. Rather than viewing x_1, x_2 as data, suppose the model is conditional on a and $\hat{\omega}$ is viewed as the lone observation. Then the density of $\hat{\omega}|a$ determines the likelihood and reflects a single observation from a location model with parameter ω. Then, according to the Fisher & Barndorff-Nielsen result in proposition 7.2.1, the p^* formula is exact. Another example of this occurs in Fisher's (1973, p. 140) normal circle example as developed in Exercise 11.

The next example provides a prototype for the sort of curved exponential family models in which the accuracy of $p^\dagger$ and p^* may be determined. This class of examples is distinguished by the fact that the exact joint density for the canonical sufficient statistic X can be determined and thus an exact conditionality resolution is possible through Jacobian transformation. Let $g(x;\xi)$ denote this density for X and suppose the mapping $x = x(\hat{\xi}, a)$ expresses a smooth relationship through the set of linear equations

$$\dot{\theta}_{\hat{\xi}}^T(x - \hat{\mu}_{\hat{\xi}}) = 0 \quad (7.46)$$
$$A_{\hat{\xi}}(x - \hat{\mu}_{\hat{\xi}}) = a.$$

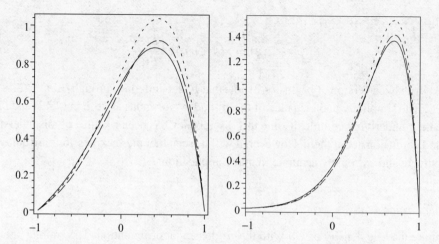

Figure 7.6. Plots of the true density $f(\hat{\xi}|a;\xi)$ (*solid*), $p^\dagger(\hat{\xi}|a;\xi)$ (*dotted*) and $p^*(\hat{\xi}|a;\xi)$ (*dashed*) when conditioning on $a = 0.2735$ for the settings in which $\xi = 1/4$ (left figure) and $\xi = 1/2$ (right figure).

Then the exact conditionality resolution is

$$f(\hat{\xi}|a;\xi) = \varrho(a;\xi) g\{x(\hat{\xi},a);\xi\} \left\| \frac{\partial x}{\partial(\hat{\xi},a)} \right\| \qquad (7.47)$$

where $\varrho(a;\xi)^{-1}$ is the appropriate normalization $d\hat{\xi}$.

Bivariate normal correlation

The joint density for X_1 and X_2 may be determined as a marginal distribution computed from the three dimensional Wishart density for $\sum_{i=1}^{n} Z_i Z_i^T$ where $Z_i = (Z_{i1}, Z_{i2})^T$. The appendix in section 7.4.2 provides the details and shows that

$$g(x_1, x_2; \xi) = \frac{(x_1^2 - x_2^2)^{n/2-1}}{2^{n-1} \Gamma(n/2)^2 (1-\xi^2)^{n/2}} \exp\left(-\frac{x_1 - \xi x_2}{1-\xi^2}\right) \qquad 0 < |x_2| < x_1. \qquad (7.48)$$

The dependence of x_1 and x_2 on $(\hat{\xi}, a)$ is expressed as

$$\begin{aligned} x_1 &= n + a\sqrt{n(1+\hat{\xi}^2)} \\ x_2 &= \hat{\xi}(2x_1 - n + n\hat{\xi}^2)/(1+\hat{\xi}^2). \end{aligned} \qquad (7.49)$$

The expression for x_2 derives from the estimating equation for $\hat{\xi}$ as given in (7.46) or it can be determined directly as the MLE of density (7.48). Also some longer computations provide the Jacobian with determinant

$$\left| \frac{\partial x}{\partial(\hat{\xi},a)} \right| = n \left(\sqrt{n(1+\hat{\xi}^2)} + \frac{2a}{1+\hat{\xi}^2} - 2a\hat{\xi} \right).$$

As a numerical example, take $n = 4$ with data $(-1, -1/2)$, $(0, 0)$, $(1, 1)$, and $(2, 3/2)$. Then $x_1 = 4\frac{3}{4}$, $x_2 = 4\frac{1}{2}$, $\hat{\xi} = 0.9379$, and $a = 0.2735$. Conditioning on this value of a, figure 7.6 compares $f(\hat{\xi}|a;\xi)$ (solid), $p^\dagger(\hat{\xi}|a;\xi)$ (dotted) and $p^*(\hat{\xi}|a;\xi)$ (dashed) for the

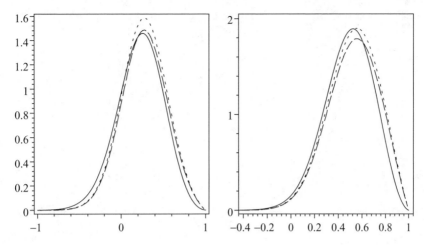

Figure 7.7. The same plots as in figure 7.6 but computed when conditioning on $a = 2.617$.

respective settings in which $\xi = 1/4$ and $1/2$. The accuracy is not the extreme accuracy we have become accustomed to but it is still quite good considering that $n = 4$ which is not much information about a covariance parameter.

A second numerical example looks at the accuracy when a is quite far from 0 when judged as a normal score. This example has $n = 4$ and $x_2 = 4\frac{1}{2}$ but doubles the value of x_1 so it is $x_1 = 9\frac{1}{2}$. Then, $\hat{\xi} = 0.3222$, and the large value $a = 2.617$ results. Figure 7.7 plots the same densities f, $p^\dagger$, and p^* as in the two previous figures for the respective settings in which $\xi = 1/4$ and $1/2$. The approximations experience some difficulty in the right tail in figure 7.7 (right).

Gamma exponential

A (2, 1) curved exponential family may be defined by supposing that $X_1 \sim$ Exponential (ξ) independently of $X_2 \sim$ Exponential (e^ξ). This family was considered in less detail in Butler (2003). To keep the analysis as simple as possible, suppose the data are $x_1 = 1$ and $x_2 = 2$. The MLE is

$$\hat{\xi} = \text{LambertW}(1/2) \simeq 0.3517$$

and solves an equation which, when rearranged, allows x_2 to be expressed in terms of $\hat{\xi}$ and x_1 as

$$x_2 = e^{-\hat{\xi}}(1/\hat{\xi} + 1 - x_1). \tag{7.50}$$

The affine ancillary, computed from (7.38), is

$$a^2 = (\hat{\xi}x_1 - 1)^2 + (e^{\hat{\xi}}x_2 - 1)^2 \simeq 3.817. \tag{7.51}$$

Fixing $a \simeq 1.954$ in (7.51) and substitution of (7.50) for the value of x_2 in (7.51) leads to two solutions for x_1 since we have not committed to the sign of a. The observed data are retrieved with the solution

$$x_1 = 1/\hat{\xi} - |a|/\sqrt{1 + \hat{\xi}^2}. \tag{7.52}$$

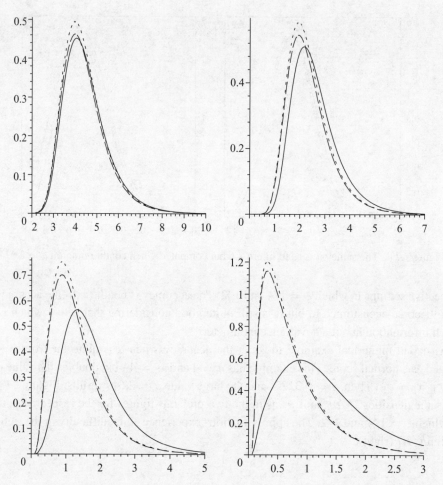

Figure 7.8. Densities for $\hat{\xi}$ in the Gamma Exponential example when conditioning on the affine ancillary $a = 1.954$. The plots reveal a range of accuracy from good to poor and show the exact density (*solid*), $p^\dagger(\hat{\xi}|a;\xi)$ (*dotted*), and $p^*(\hat{\xi}|a;\xi)$ (*dashed*) for $\xi = 4, 2, 1$, and $1/2$ respectively.

Having obtained expressions (7.50) and (7.52), the task of computing p^* as well as the exact conditionality resolution is now relatively simple. The success of p^* (dashed) and $p^\dagger$ (dotted) in approximating the exact density (solid) is revealed in figure 7.8 for $\xi = 4, 2, 1$, and $1/2$. The accuracy declines with the decrease in ξ-value particularly for $\xi \leq 1$.

The same data are used for each of the plots so the observed MLE $\hat{\xi} \simeq 0.3517$ and the observed ancillary $a \simeq 1.954$ apply for each plot. What is particularly interesting are the changes that occur in the normalization constant

$$f(a;\xi) = \int_0^\infty f(\hat{\xi}, a; \xi) d\hat{\xi}$$

as ξ ranges over 4, 2, 1, and 1/2. These values were computed using numerical integration that leads to the values 1.854, 1.240, 0.6689, and 0.3562 respectively. Note that they may be interpreted as the marginal probability rates for observing $a \simeq 1.954$ given the values of ξ. The plot of this marginal density value versus ξ is defined as the *marginal likelihood* of a. Its shape reveals the likelihood information about ξ provided in the observed value

7.3 Approximate conditional inference and p* in curved exponential families

of a and a plot is given in figure 7.10 later in the chapter. Note that as the chance of $a \simeq 1.954$ becomes less likely with the value of ξ, then so also deteriorates the accuracy of the p^* approximation. Unfortunately the dramatic change in these values also suggests that a is not quite as ancillary as has been assumed. What the evidence in this example does suggest is that the accuracy of the conditionality resolution based on the affine ancillary may deteriorate when the affine ancillary is not in a region of central tendency, e.g. if it is an unlikely value occurring in the tail of the density $f(a; \xi)$. This supposition would be consistent with theoretical properties of p^* and, in particular, with the last comment on asymptotics over large deviation regions given at the end of the chapter.

Proof of theorem 7.3.2. The proof is given in the asymptotic context in which x is the sum of n i.i.d. components. First the sufficiency of X is used to tilt its density from ξ to $\hat{\xi}$ so that

$$f(x; \xi) = \frac{\mathcal{L}(\xi)}{\mathcal{L}(\hat{\xi})} f(x; \hat{\xi}).$$

Were the exponential family regular so that $\hat{\xi}$ could be taken as canonical, then the multivariate Edgeworth expansion for $f(x; \hat{\xi})$ given in (5.43) could be used. However, the presence of curvature prevents usage of this particular version and requires the more general Edgeworth expression given in Barndorff-Nielsen and Cox (1979, Eq. 4.4) as

$$\begin{aligned}
f(x; \hat{\xi}) &= (2\pi)^{-m/2} |\Sigma_{\hat{\xi}}|^{-1/2} \exp\left\{-\tfrac{1}{2}(x - \mu_{\hat{\xi}})^T \Sigma_{\hat{\xi}}^{-1} (x - \mu_{\hat{\xi}})\right\}(1 + r_n) \\
&= (2\pi)^{-m/2} |\Sigma_{\hat{\xi}}|^{-1/2} \exp\left\{-\tfrac{1}{2} a^T a\right\}(1 + r_n)
\end{aligned} \tag{7.53}$$

using (7.41). Note that $x - \mu_{\hat{\xi}} = 0$ in a regular exponential family but with curvature only the p linear combinations $\hat{\theta}_{\hat{\xi}}^T (x - \mu_{\hat{\xi}})$ are zero. The error term r_n is

$$r_n = \frac{1}{6\sqrt{n}} (H^T \hat{\kappa})(a) + O(n^{-1})$$

where $(H^T \hat{\kappa})(a)$ denotes linear combinations of multivariate third-order Hermite polynomials whose argument values are a and which use weights that are cumulants of the $\hat{\xi}$-tilted distribution. Since $\hat{\xi} = \xi + O(n^{-1/2})$, then

$$\hat{\kappa} = \kappa(\hat{\xi}) = \kappa(\xi) + O(n^{-1/2}) = \kappa + O(n^{-1/2})$$

and the dependence of the error r_n on $\hat{\xi}$ has order $O(n^{-1})$ as expressed in

$$r_n = \frac{1}{6\sqrt{n}} (H^T \kappa)(a) + O(n^{-1}).$$

The second step in the proof considers the transformation $x \to (\hat{\xi}, a)$ applied to the density of X. The determinant of the Jacobian is

$$\left\| \frac{\partial x}{\partial (\hat{\xi}^T, a^T)} \right\| = |j_{\hat{\xi}}| |i_{\hat{\xi}}|^{-1/2} |\Sigma_{\hat{\xi}}|^{1/2} \tag{7.54}$$

and is sufficiently technical to be banished to the appendix in section 7.4.2. Combining (7.54) and (7.53) gives

$$f(\hat{\xi}, a; \xi) = (2\pi)^{-m/2} |j_{\hat{\xi}}| |i_{\hat{\xi}}|^{-1/2} \frac{\mathcal{L}(\xi)}{\mathcal{L}(\hat{\xi})} \exp\left\{-\tfrac{1}{2} a^T a\right\} (1 + r_n).$$

The relationship of j_ξ to i_ξ in (7.27) allows the expansion

$$|i_\xi| = |j_\xi + D_\xi| = |j_\xi| |I_m + j_\xi^{-1} D_\xi| \qquad (7.55)$$
$$= |j_\xi| \{1 + n^{-1/2} h_1(\xi, a) + O(n^{-1})\}$$

which is also justified in the appendix. Thus

$$f(\hat{\xi}, a; \xi) = (2\pi)^{-m/2} |j_\xi|^{1/2} \frac{\mathcal{L}(\xi)}{\mathcal{L}(\hat{\xi})} \exp\left\{-\tfrac{1}{2} a^T a\right\} \qquad (7.56)$$
$$\times (1 + r_n)\{1 + n^{-1/2} h_1(\xi, a) + O(n^{-1})\}.$$

From (7.57) the asymptotic order of $p^\dagger$ is $O(n^{-1/2})$ as given in (7.61). Since both of the error terms, r_n and $n^{-1/2} h_1(\xi, a)$, depend only on ξ and a and not $\hat{\xi}$, upon normalization the p^* achieves the order $O(n^{-1})$ as in (7.62). □

7.3.5 Likelihood ancillaries yielding conditionality resolutions

Historically the p^* formula for curved exponential families was introduced in Barndorff-Nielsen (1980) by conditioning upon the affine ancillaries as described above. Barndorff-Nielsen (1980, 1986) also suggested the use of ancillaries computed from likelihood ratio statistics and whose distributions are asymptotically ancillary. In a particular small sample example, Pedersen (1981) has shown that this latter ancillary is more distribution constant than its affine ancillary counterpart.

Assume a (m, p) curved exponential family with canonical parameter θ_ξ. Consider expanding the parameter space in such a way that the curved exponential family may be embedded in a regular exponential family. This is done by including a $(m - p)$ dimensional parameter η that complements ξ and makes $\theta \in \Re^m$ a smooth 1-1 transformation of the vector $(\xi, \eta) \in \Re^m$. Denoting $\theta = \theta(\xi, \eta)$, then it is also assumed that the curved exponential family is traced out by a particular value η_0 of the complementary variable so that $\theta_\xi = \theta(\xi, \eta_0)$ for all ξ.

As an example of this construction, consider the gamma hyperbola setting in which $\theta_\xi = (\xi, 1/\xi)$ for $\xi > 0$. Take $\theta = (\xi, \eta/\xi)$ so that the value $\eta_0 = 1$ assures the embedding $\theta_\xi = (\xi, 1/\xi)$ for the curved exponential family.

Now reconsider the general case but, for the sake of simplicity in discussion, suppose that $\eta = (\vartheta, \chi)$ has dimension 2. Write $\theta = \theta(\xi, \vartheta, \chi)$ and let the value $\eta_0 = (\vartheta_0, \chi_0)$ assure that the curved exponential family is traced out by the values of ξ. Then there are two canonical variates that underlie the likelihood ratio test of $H_0 : \eta = \eta_0$, a curved exponential family, versus $H_1 : \eta \neq \eta_0$, a full exponential family model. These canonical variates are the signed log likelihood ratio statistics that define the proposed likelihood ancillaries as

$$a_\chi^0 = \text{sgn}(\hat{\chi}_f - \chi_0) \sqrt{-2\left\{l(\hat{\xi}_{\chi_0}, \hat{\vartheta}_{\chi_0}, \chi_0) - l(\hat{\xi}_f, \hat{\vartheta}_f, \hat{\chi}_f)\right\}} \qquad (7.57)$$
$$a_\vartheta^0 = \text{sgn}(\hat{\vartheta}_{\chi_0} - \vartheta_0) \sqrt{-2\left\{l(\hat{\xi}_{\eta_0}, \eta_0) - l(\hat{\xi}_{\chi_0}, \hat{\vartheta}_{\chi_0}, \chi_0)\right\}}.$$

Here, the notation $\hat{\xi}_{\chi_0}, \hat{\vartheta}_{\chi_0}$ indicates that the MLE for (ξ, ϑ) is computed by holding parameter χ_0 fixed in the log likelihood $l(\xi, \vartheta, \chi_0)$. Also the subscript in $(\hat{\xi}_f, \hat{\vartheta}_f, \hat{\chi}_f)$ indicates the joint MLE under the full regular exponential family and the notation distinguishes $\hat{\xi}_f$ from

7.3 Approximate conditional inference and p* in curved exponential families

$\hat{\xi}$, the MLE under the true curved exponential family. Variables a_χ^0 and a_ϑ^0 are approximately normal for large sample sizes.

Theorem 7.3.3 *For the continuous (m, p) curved exponential family in (7.22), let a_χ^0 be a $(m - p)$-vector of likelihood ancillaries constructed as just described. Suppose also that the transformation $x \to (\hat{\xi}, a_\chi^0)$ is 1-1. Then the conditional density of $\hat{\xi}$ given a_χ^0 is approximated by the p^* density which provides a conditionality resolution for the curved exponential family.*

Proof. In the case in which the mapping $x \to (\hat{\xi}, a)$ to the affine ancillary is also 1-1, then $a \leftrightarrow a_\chi^0$ are 1-1. Jacobian transformation of the joint density expression for $(\hat{\xi}, a)$ in (7.57) shows that

$$f(\hat{\xi}, a_\chi^0; \xi) \simeq (2\pi)^{-m/2} |j_{\hat{\xi}}|^{1/2} \frac{\mathcal{L}(\xi)}{\mathcal{L}(\hat{\xi})} \times \exp\left\{-\tfrac{1}{2} a^T a\right\} \|\partial a^T / \partial a_\chi^0\|$$
$$= p^\dagger(\hat{\xi} | a_\chi^0; \xi) \times (2\pi)^{-(m-p)/2} \exp\left\{-\tfrac{1}{2} a^T a\right\} \|\partial a^T / \partial a_\chi^0\|$$

where $p^\dagger$ is more explicitly given as

$$p^\dagger(\hat{\xi} | a_\chi^0; \xi) = (2\pi)^{-p/2} |i_{\hat{\xi}} - \{x(\hat{\xi}, a_\chi^0) - \mu_{\hat{\xi}}\} \odot \ddot{o}_{\hat{\xi}}|^{1/2} \frac{\exp\left\{\theta_{\hat{\xi}}^T x(\hat{\xi}, a_\chi^0) - c(\theta_{\hat{\xi}})\right\}}{\exp\left\{\theta_{\hat{\xi}}^T x(\hat{\xi}, a_\chi^0) - c(\theta_{\hat{\xi}})\right\}}.$$
□

The choice of likelihood ancillaries in (7.57) for a particular curved exponential family is far from unique. In the particular parametric formulation above, the first ancillary was chosen to test $H_0 : \chi = \chi_0$ vs. $H_0 : \chi \neq \chi_0$, and the second to test $H_0' : \vartheta = \vartheta_0; \chi = \chi_0$ vs. $H_1' : \vartheta \neq \vartheta_0; \chi = \chi_0$. Reversing the order of the tests to first test $H_0 : \vartheta = \vartheta_0$ followed by the conditional test of $\chi = \chi_0$ leads to a different set of ancillaries. In general, the p-dimensional complementary parameter η determines $p!$ possible sets of likelihood ancillaries. On top of this, the choice of the complementary parameterization η is quite arbitrary and can lead to many different possibilities for likelihood ancillaries.

The next example shows that the likelihood ratio ancillary can be an exact ancillary and that its conditionality resolution may agree with the resolution based on an affine ancillary. The second example is a prototype for dealing with the computational aspects for conditionality resolutions based on likelihood ancillaries.

Gamma hyperbola

The likelihood ancillary constructed from (7.57) uses MLE $\hat{\xi} = \sqrt{x_2/x_1}$ when $\eta = 1$ and $(\hat{\xi}_f, \hat{\eta}_f) = \{1/x_1, 1/(x_1 x_2)\}$ otherwise. Then

$$a_\eta^0 = \text{sgn}(1 - a) 2\sqrt{a - \ln a - 1},$$

where $a = \sqrt{x_1 x_2}$. Likelihood ancillary a_η^0 is a smooth monotonic decreasing function of a mapping $(0, \infty)$ onto $(\infty, -\infty)$. Also the exact conditionality resolution is the same whether conditioning is based upon a, the affine ancillary in (7.37), or a_η^0.

Bivariate normal correlation

For this example the curved exponential family model may be expanded to suppose that the bivariate normal sample is from a Normal$_2$ $(0, \sigma^2 R)$ with the value $\sigma^2 = 1$ determining the (2,1) curved exponential family. The MLE under the alternative (2, 2) full exponential family is $(\hat{\xi}_f, \hat{\sigma}_f^2) = (1, \infty)$ and thus on the boundary of the parameter space. Indeed the likelihood function in terms of the parameters ξ and σ^2 is not continuous at $(1, \infty)$. In passing to the limit, any positive value can be obtained by taking a suitably chosen path. For this reason $l(1, \infty)$ is undefined and so also is the likelihood ancillary.

Gamma exponential

The completion of the (2, 1) exponential family to (2, 2) adds the parameter $\chi > 0$ so that $X_1 \sim$ Exponential (ξ) and $X_2 \sim$ Exponential (χe^ξ). The likelihood ancillary, based on the likelihood ratio test that $\chi = 1$, is the value

$$\tfrac{1}{2}(a_\chi^0)^2 = -\ln\hat{\xi} + 1/\hat{\xi} - 1 - (1 - \hat{\xi})x_1 - \ln x_1 - \ln(1/\hat{\xi} + 1 - x_1) \tag{7.58}$$

when its dependence on x_2 is replaced with x_1 using (7.50). The sign of a_χ^0 is

$$\operatorname{sgn}(\hat{\chi}_f - 1) = \operatorname{sgn}\left(e^{-1/x_1}/x_2 - 1\right) = \operatorname{sgn}(0.1839 - 1) = -1 \tag{7.59}$$

so that $a_\chi^0 \simeq -1.546$.

When fixing a_χ^0 rather than affine a, determinations of both the exact conditionality resolution and its p^* approximation become slightly more difficult computations. Consider first $p^\dagger$ computation. With a_χ^0 fixed and a grid of $\hat{\xi}$-values specified, the $p^\dagger$ formula requires the determination of each (x_1, x_2) pair associated with each $\hat{\xi}$-value on the grid. This can be done numerically by solving (7.58) for x_1 for all grid values of $\hat{\xi}$. Expression (7.58) admits two solutions and the smaller solution is the correct one for this example and leads to the negative sign of (7.59). Once x_1 has been determined, x_2 is explicitly given by (7.50). For fixed a_χ^0 and a grid of $(\hat{\xi}, x_1, x_2)$-triples, then $p^\dagger$ is computable.

The true joint density of $(\hat{\xi}, a_\chi^0)$ may also be computed the same way but with the additional complication of a Jacobian determination. If g is the joint density of $x = (x_1, x_2)$, then

$$f(\hat{\xi}, a_\chi^0; \xi) = g\{x(\hat{\xi}, a); \xi\} \, \|\partial(\hat{\xi}, a_\chi^0)/\partial x\|^{-1}. \tag{7.60}$$

The components of the Jacobian in (7.60) may be determined by using implicit differentiation applied to the MLE equation (7.50), which gives $\partial \hat{\xi}/\partial x$, and (7.58) which gives $\partial a_\chi^0/\partial x$. Since differentiation is implicit, the Jacobian depends on all four of the quantities involved but these are given on the grid of $(\hat{\xi}, x_1, x_2)$-triples. The numerical work above, has performed these Jacobian computations symbolically using Maple.

Figure 7.9 compares the exact conditionality resolutions $f(\hat{\xi}|a_\chi^0; \xi)$ (solid) with $p^\dagger(\hat{\xi}|a_\chi^0; \xi)$ (dotted) and $p^*(\hat{\xi}|a_\chi^0; \xi)$ (dashed) for the values $\xi = 4, 2, 1$, and $1/2$ respectively. These may be compared to similar plots in which affine ancillaries were conditioned upon in figure 7.8. The p^* and $p^\dagger$ expressions are considerably more accurate using likelihood ancillaries than when affine ancillaries were considered.

7.3 Approximate conditional inference and p* in curved exponential families 247

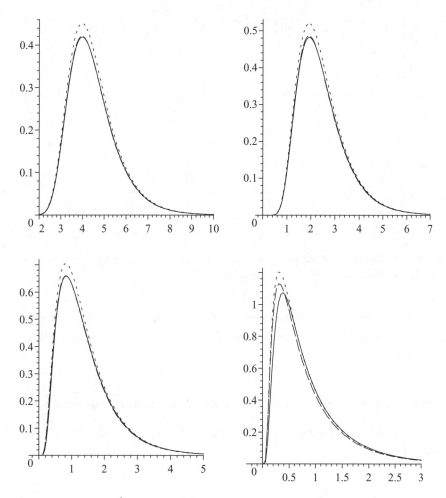

Figure 7.9. Densities for $\hat{\xi}$ when conditioning on the likelihood ancillary $a_\chi^0 = -1.546$. In each plot, $f(\hat{\xi}|a_\chi^0;\xi)$ (solid), $p^\dagger(\hat{\xi}|a_\chi^0;\xi)$ (dotted) and $p^*(\hat{\xi}|a_\chi^0;\xi)$ (dashed) are shown.

The normalization constants used in plotting the exact conditional density are values for the marginal density $f(a_\chi^0;\xi)$ of the likelihood ancillary at the observed value $a_\chi^0 \simeq -1.546$. These values are $0.1815, 0.1755, 0.1708$, and 0.1587 as ξ ranges over 4, 2, 1, and 1/2. Further values are given in the plot of the marginal likelihood of a_χ^0 as shown by the solid lines in figure 7.10. For comparison, the marginal likelihood of the affine ancillary a is plotted as the dashed line in figure 7.10 (right). Figure 7.10 (right) dramatically shows the superior ancillary behavior of the likelihood ancillary over the affine ancillary. The plot of the marginal likelihood of a_χ^0 is quite flat which demonstrates that a_χ^0 is very uninformative about the value of ξ, a trait expected of an ancillary statistic. These plots were computing by numerically integrating the joint density of $(\hat{\xi}, a_\chi^0)$ over the grid $\hat{\xi} \in \{0.02(0.04)9.98, 10\frac{1}{16}(\frac{1}{16})12, 12\frac{1}{8}(\frac{1}{8})16\}$. The fineness of the grid allowed for accurate determination of $f(a_\chi^0;\xi)$ down to $\xi = 0.1$. The upper grid limit of 16 allowed accurate determination of $f(a_\chi^0;\xi)$ up to $\xi = 12$.

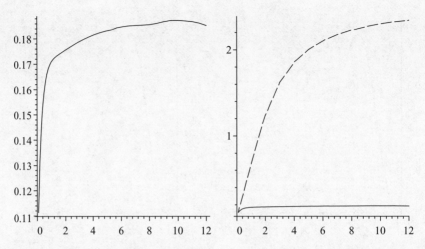

Figure 7.10. Marginal likelihood plots for $f(a_\chi^0; \xi)$ (*solid*) and $f(a; \xi)$ (*dashed*) versus ξ where a_χ^0 and a are the likelihood and affine ancillaries respectively.

The superior performance shown by the p^* formula when conditioning on a likelihood ancillary rather than affine, is consistent with its superior asymptotic properties that are described below.

7.3.6 Properties of p^* and $p^\dagger$ for affine, likelihood, and higher-order ancillaries

When fixing affine or likelihood ancillaries, the $p^\dagger$ and p^* formulas in (7.42) and (7.43) have the following properties:

Invariance and likelihood dependence

Both $p^\dagger$ and p^* are invariant to 1-1 smooth reparameterizations. This follows from the observation that all three of the factors in

$$\varkappa(a; \xi) \left\{ \frac{\mathcal{L}(\xi)}{\mathcal{L}(\hat{\xi})} \right\} \left(|j_{\hat{\xi}}|^{1/2} d\hat{\xi} \right)$$

are invariant.

Furthermore, the expressions are dependent only on the likelihood function $\mathcal{L}(\xi)$. This is consistent with the likelihood principle which states that the likelihood is fully informative about ξ. The likelihood information is extracted through the conditionality resolution embodied in the p^* formula as described in theorem 7.3.2.

Asymptotics over normal deviation regions

Consider the asymptotic setting in which $X = \sum_{i=1}^{n} Z_i$ where $\{Z_i\}$ are i.i.d. with the (m, p) curved exponential family

$$f(z; \xi) = \exp \left\{ \theta_\xi^T z - \bar{c}(\theta_\xi) - \bar{d}(z) \right\}.$$

The notation in (7.42) and (7.43) requires $c(\theta_\xi) = n\bar{c}(\theta_\xi)$ etc. Ancillary a may be either a likelihood or affine ancillary; both lead to the same error in this setting.

Over normal deviation regions for which $\hat{\xi} - \xi = O(n^{-1/2})$ and a is bounded, the true conditional density

$$f(\hat{\xi}|a;\xi) = p^\dagger(\hat{\xi}|a;\xi)\{1 + O(n^{-1/2})\} \tag{7.61}$$
$$= p^*(\hat{\xi}|a;\xi)\{1 + O(n^{-1})\}, \tag{7.62}$$

see Skovgaard (1990, p. 788) and Barndorff-Nielsen and Wood (1998, theorem 3.1). Furthermore,

$$\varkappa(a;\xi) = (2\pi)^{-p/2}\{1 + O(n^{-1})\} \tag{7.63}$$

where the $O(n^{-1})$ term is related to the Bartlett factor; see Barndorff-Nielsen and Cox (1984). According to (7.63), the $p^\dagger$ formula should be quite accurate without normalization when n is not too small.

Asymptotics over large deviation regions

In this setting, the likelihood ancillary has superior relative error properties. This is somewhat surprising because both the likelihood and affine ancillaries are first-order ancillaries, e.g. their distributions differ from their standard normal limit by a term that depends on ξ to order $O(n^{-1/2})$.

Consider the same asymptotic setting but on large deviation sets for which $\hat{\xi} - \xi = O(1)$ and likelihood or affine ancillary a is bounded. In the affine case, the errors of both $p^\dagger$ and p^* simply remain bounded in the sense that

$$0 < \frac{1}{B} < \liminf_{n\to\infty} \frac{p^*(\hat{\xi}|a;\xi)}{f(\hat{\xi}|a;\xi)} \le \limsup_{n\to\infty} \frac{p^*(\hat{\xi}|a;\xi)}{f(\hat{\xi}|a;\xi)} < B \tag{7.64}$$

for some $B > 0$. For the likelihood ancillary, the error in (7.62) is of order $O(n^{-1/2})$. These results were mentioned in Skovgaard (1990, p. 788) and appeared formally in theorem 3.1 of Barndorff-Nielsen and Wood (1998).

When a fails to stay bounded and grows $O(\sqrt{n})$, the error in (7.62) is $O(1)$ for the likelihood ancillary and relative error is preserved. However, with the affine ancillary, control of relative error is typically lost as noted in remark 3.6 of Barndorff-Nielsen and Wood (1998).

Asymptotics for higher-order ancillaries in curved exponential families

In some settings, such as in the gamma hyperbola example, the likelihood or affine ancillary is an exact ancillary. For other settings, these ancillaries may be second-order in the sense that they have a density that is standard normal with error term that is $O(n^{-1})$ in normal deviation regions (for which a is bounded). In either of these cases, the asymptotic results in (7.61) and (7.62) are the higher orders $O(n^{-1})$ and $O(n^{-3/2})$ respectively as discussed in Skovgaard (1990, p. 788) and Barndorff-Nielsen and Cox (1994).

Barndorff-Nielsen and Wood (1998) have also shown that a modification to the likelihood ancillary, denoted as a^*, can be used in a conditionality resolution to achieve the higher relative error accuracy achieved by saddlepoint densities. This order is $O(n^{-1}\|\hat{\xi} - \xi\|)$ in large deviation regions for $\hat{\xi} - \xi$ and a^*. Computations using a^* however are yet another order of difficulty greater that those above that have used a^0.

7.4 Appendix

7.4.1 Proofs for likelihood ratio propositions

Proof of proposition 7.1.1. The density for X to second order, as given in (5.40) and (5.42), is

$$f(x) = \frac{1}{\sqrt{2\pi n c''(\hat{\theta})}} e^{-\lambda_n/2} \left\{1 + \frac{1}{n}\vartheta(\hat{\theta}) + O(n^{-2})\right\}. \tag{7.65}$$

Replacing $\vartheta(\hat{\theta})/n$ in (7.65) with $\vartheta(\theta_0)/n$ incurs an error of order $O(n^{-3/2})$ in the curly braces when $\hat{\theta}$ remains on bounded sets of central tendency. Differentiating (7.8) gives the Jacobian for $x \to \lambda_n$ as

$$\frac{\partial \lambda_n}{\partial x} = -2(\theta_0 - \hat{\theta}) + 2\{x - nc'(\hat{\theta})\}\frac{\partial \hat{\theta}}{\partial x} = 2(\hat{\theta} - \theta_0). \tag{7.66}$$

This value is both positive and negative, depending on the value of x, and reveals a 2 to 1 transformation $\hat{\theta}(x) \to \lambda_n$. Summing over both terms yields the density

$$f(\lambda) = \frac{1}{2\sqrt{2\pi n}} e^{-\lambda/2} \sum_{\{\hat{\theta}:\hat{\theta} \to \lambda\}} \frac{1}{\sqrt{(\hat{\theta} - \theta_0)^2 c''(\hat{\theta})}} \left\{1 + \frac{1}{n}\vartheta(\theta_0) + O(n^{-3/2})\right\}. \tag{7.67}$$

What remains is to write the $\hat{\theta}$ expression on the right side in terms of λ_n. To do this, take (7.8) and expand $c(\hat{\theta})$ in the curly braces and also $x = nc'(\hat{\theta})$ about θ_0 to get

$$\lambda_n = n\left\{c''(\theta_0)(\hat{\theta} - \theta_0)^2 + \tfrac{2}{3}c'''(\theta_0)(\theta_0 - \hat{\theta})^3 + \tfrac{1}{4}c''''(\theta_0)(\theta_0 - \hat{\theta})^4 + O(n^{-5/2})\right\}$$
$$= nc''(\theta_0)(\hat{\theta} - \theta_0)^2 \left\{1 + \varsigma_3(\hat{\theta} - \theta_0) + \varsigma_4(\hat{\theta} - \theta_0)^2 + O(n^{-3/2})\right\}, \tag{7.68}$$

where ς_3 and ς_4 are constants defined by equality whose actual values ultimately play no role. The expansion in (7.68) needs to be inverted and we employ the method described in Barndorff-Nielsen and Cox (1989, §3.5). This method starts by removing the higher order terms in the expansion and solving to get the two approximate solutions

$$\hat{\theta} - \theta_0 = \pm\sqrt{\frac{\lambda_n}{nc''(\theta_0)}} = \pm\sqrt{q_n}$$

to order $O(n^{-1})$. The next step takes a trial solution for the inversion as the expansion

$$\hat{\theta} - \theta_0 = \pm\sqrt{q_n}\left\{1 + \eta_1\sqrt{q_n} + \eta_2 q_n + O(n^{-3/2})\right\}$$

and substitutes this into the right side of (7.68) with the purpose of determining values for the two coefficients η_1 and η_2. The resulting identity, leaving out unnecessary terms, is

$$q_n = q_n\{1 + \eta_1\sqrt{q_n} + O(n^{-1})\}^2\{1 \pm \varsigma_3\sqrt{q_n}(1 + \eta_1\sqrt{q_n})$$
$$+ \varsigma_4 q_n(1 + \eta_1\sqrt{q_n})^2 + O(n^{-3/2})\}.$$

7.4 Appendix

Retaining only the $O(n^{-1/2})$ term yields the equality

$$1 = 1 + \sqrt{q_n}(\pm \varsigma_3 + 2\eta_1) + O(n^{-1}).$$

To achieve equality, the coefficient of $\sqrt{q_n}$ must be zero so η_1 assumes the two values $\mp\varsigma_3/2$ for the two different inversions. We write these two inversions as

$$\hat{\theta} - \theta_0 = \pm\sqrt{q_n}\left\{1 \mp \eta_3\sqrt{q_n} + \eta_4 q_n + O(n^{-3/2})\right\} \tag{7.69}$$

where the actually values of η_3 and η_4 will not matter. The remaining factor required in (7.67) is $c''(\hat{\theta})$ which is expanded about θ_0 to get

$$c''(\hat{\theta}) = c''(\theta_0) + c'''(\theta_0)(\hat{\theta} - \theta_0) + O(n^{-1})$$
$$= c''(\theta_0)\left\{1 \pm \eta_5\sqrt{q_n} + O(n^{-1})\right\}.$$

Including this and (7.69) in the term required in (7.67) yields

$$\frac{1}{\sqrt{(\hat{\theta}-\theta_0)^2 c''(\hat{\theta})}} = \frac{1 \pm \eta_3\sqrt{q_n} - \eta_4 q_n + O(n^{-3/2})}{\sqrt{q_n c''(\theta_0)\left\{1 \pm \eta_5\sqrt{q_n} + O(n^{-1})\right\}}}$$
$$= \sqrt{\frac{n}{\lambda_n}}\left\{1 \pm \left(\eta_3 - \tfrac{1}{2}\eta_5\right)\sqrt{q_n} - \eta_6 q_n + O(n^{-3/2})\right\}.$$

Adding the two terms here cancels out the term preceded with $\pm$ and (7.67) becomes

$$f(\lambda) = \frac{1}{\sqrt{2\pi n}}e^{-\lambda/2}\sqrt{\frac{n}{\lambda}}\left\{1 - \eta_6\frac{\lambda}{nc''(\theta_0)} + O(n^{-3/2})\right\}$$
$$\times \left\{1 + \frac{1}{n}\vartheta(\theta_0) + O(n^{-3/2})\right\}$$
$$= \frac{1}{\sqrt{2\pi\lambda}}e^{-\lambda/2}\left\{1 + \frac{\eta_7\lambda + \vartheta(\theta_0)}{n} + O(n^{-3/2})\right\}. \tag{7.70}$$

The leading term of (7.70) is a χ_1 density. Integrating both sides yields

$$1 = 1 + \{\eta_7 + \vartheta(\theta_0)\}/n + O(n^{-3/2})$$

and equating powers of n shows that $\eta_7 = -\vartheta(\theta_0)$ to give (7.9). □

7.4.2 Derivations of p^* for curved exponential families

Derivation of the Jacobian in (7.54)

Let $b = \Psi_{\hat{\xi}}^T(x - \mu_{\hat{\xi}})$ so that $a = (\Psi_{\hat{\xi}}^T \Sigma_{\hat{\xi}} \Psi_{\hat{\xi}})^{-1/2}b$. The approach used in Barndorff-Nielsen is to break the transformation into two stages: first compute the Jacobian when transforming $x \to (\hat{\xi}, b)$ followed by $(\hat{\xi}, b) \to (\hat{\xi}, a)$. To deal with the first stage, take the derivative $\partial/\partial(\hat{\xi}^T, b^T)$ of the equality

$$\begin{pmatrix} \dot{\theta}_{\hat{\xi}}^T \\ \Psi_{\hat{\xi}}^T \end{pmatrix}(x - \mu_{\hat{\xi}}) = \begin{pmatrix} 0 \\ b \end{pmatrix}.$$

The product rule gives

$$\begin{pmatrix} \dot{\theta}_\xi^T \\ \Psi_\xi^T \end{pmatrix} \left\{ \frac{\partial x}{\partial(\hat{\xi}^T, b^T)} - (\dot{\mu}_\xi \ 0) \right\} + \begin{pmatrix} \ddot{\theta}_\xi \odot (x - \mu_\xi) & 0 \\ \dot{\Psi}_\xi^T \odot (x - \mu_\xi) & 0 \end{pmatrix} = \begin{pmatrix} 0 & 0 \\ 0 & I_{m-p} \end{pmatrix}$$

where, for example,

$$\{\dot{\Psi}_\xi^T \odot (x - \mu_\xi)\}_{ij} = \sum_{k=1}^m \frac{\partial (\Psi_\xi^T)_{ik}}{\partial \hat{\xi}_j} (x_k - \mu_{\xi k}).$$

The symmetry of $\ddot{\theta}_\xi$ assures that

$$\ddot{\theta}_\xi \odot (x - \mu_\xi) = (x - \mu_\xi) \odot \ddot{\theta}_\xi.$$

Thus, using (7.27) and $\Psi_\xi^T \dot{\mu}_\xi = 0$,

$$\begin{pmatrix} \dot{\theta}_\xi^T \\ \Psi_\xi^T \end{pmatrix} \frac{\partial x}{\partial(\hat{\xi}^T, b^T)} = \begin{pmatrix} \dot{\theta}_\xi^T \dot{\mu}_\xi - (x - \mu_\xi) \odot \ddot{\theta}_\xi & 0 \\ \Psi_\xi^T \dot{\mu}_\xi - \dot{\Psi}_\xi^T \odot (x - \mu_\xi) & I_{m-p} \end{pmatrix}$$

$$= \begin{pmatrix} j_\xi & 0 \\ -\dot{\Psi}_\xi^T \odot (x - \mu_\xi) & I_{m-p} \end{pmatrix} \quad (7.71)$$

and

$$\left\| \frac{\partial x}{\partial(\hat{\xi}^T, b^T)} \right\| = |j_\xi| \, \| \dot{\theta}_\xi \ \Psi_\xi \|^{-1}.$$

The Jacobian for $(\hat{\xi}, b) \to (\hat{\xi}, a)$ is

$$\left\| \frac{\partial b}{\partial a^T} \right\| = |\Psi_\xi^T \Sigma_\xi \Psi_\xi|^{1/2}.$$

The result follows if it can be shown that

$$\| \dot{\theta}_\xi \ \Psi_\xi \|^{-1} |\Psi_\xi^T \Sigma_\xi \Psi_\xi|^{1/2} = |i_\xi|^{-1/2} |\Sigma_\xi|^{1/2}.$$

To show this, use the orthogonality of Ψ_ξ and $\dot{\mu}_\xi$ as expressed in

$$\begin{pmatrix} \dot{\theta}_\xi^T \\ \Psi_\xi^T \end{pmatrix} \Sigma_\xi \left(\dot{\theta}_\xi \ \Psi_\xi \right) = \begin{pmatrix} i_\xi & 0 \\ 0 & \Psi_\xi^T \Sigma_\xi \Psi_\xi \end{pmatrix} \quad (7.72)$$

and the result follows upon taking determinants of both sides of (7.72).

Derivation of (7.55)

Perhaps the simplest argument for this result can be given for a $(m, 1)$ curved exponential family. Then

$$\frac{i_\xi}{j_\xi} = \frac{1}{1 + a_2 \gamma_\xi} = \frac{1}{1 + a_2 \bar{\gamma}_\xi / \sqrt{n}}$$

$$= 1 - a_2 \bar{\gamma}_\xi / \sqrt{n} + O(n^{-1})$$

$$= 1 - a_2 \bar{\gamma}_\xi / \sqrt{n} + O(n^{-1})$$

7.4 Appendix

upon replacing $\bar{\gamma}_{\hat{\xi}} = \bar{\gamma}_{\xi} + O(n^{-1/2})$. The same argument applies in general and is outlined in Barndorff-Nielsen (1980, §4). Denote the functional dependence of $|j_{\hat{\xi}}|$ on $\hat{\xi}$ and $a/\sqrt{n}$ by $|j_{\hat{\xi}}| = g(\hat{\xi}, a/\sqrt{n})$. Note that if $a = 0$ then $g(\hat{\xi}, 0) = |i_{\hat{\xi}}|$. Taylor expansion of g in its second argument gives

$$|j_{\hat{\xi}}| - |i_{\hat{\xi}}| = g'(\hat{\xi}, 0)a/\sqrt{n} + O(n^{-1}).$$

Then

$$\frac{|j_{\hat{\xi}}|}{|i_{\hat{\xi}}|} - 1 = \frac{g'(\hat{\xi}, 0)}{g(\hat{\xi}, 0)} a/\sqrt{n} + O(n^{-1})$$

$$= \frac{g'(\xi, 0)}{g(\xi, 0)} a/\sqrt{n} + O(n^{-1}).$$

Note that since $i_{\xi} = n\bar{\imath}_{\xi}$ with $\bar{\imath}_{\xi}$ as the per observation information, the ratio function g'/g, which does not grow with n, is required to cancel the growth of the determinants.

Conditionality resolution for bivariate normal correlation example

Suppose that $Z_1, \ldots, Z_n$ are i.i.d. Normal$_2(0, R)$ as described near (7.33). Define

$$u = \frac{1}{2}\sum_{i=1}^{n} z_{i1}^2$$

so that

$$W = \begin{pmatrix} w_{11} & w_{12} \\ w_{12} & w_{22} \end{pmatrix} = \begin{pmatrix} 2u & x_2 \\ x_2 & 2(x_1 - u) \end{pmatrix}$$

has a 2×2 Wishart $(n; R)$ density as given in Muirhead (1982, §3.2.1). Jacobian transformation leads to a joint density for (x_1, x_2, u) that has the form

$$f(x_1, x_2, u) = \frac{(1 - \xi^2)^{-n/2}}{2^{n-2}\Gamma(n/2)\Gamma\{(n-1)/2\}\sqrt{\pi}} \tag{7.73}$$

$$\times \left\{x_1^2 - x_2^2 - 4\left(u - \frac{x_1}{2}\right)^2\right\}_+^{(n-3)/2} \exp\left\{-\frac{x_1 - \xi x_2}{1 - \xi^2}\right\},$$

where $\{\cdot\}_+$ is the indicator function that the argument is positive. The marginal density is found by making the substitution

$$v = \frac{2(u - x_1/2)}{\sqrt{x_1^2 - x_2^2}}$$

so that the factor within $\{\cdot\}_+$ has the integral

$$\int_{\frac{1}{2}(x_1 - \sqrt{x_1^2 - x_2^2})}^{\frac{1}{2}(x_1 + \sqrt{x_1^2 - x_2^2})} \left\{x_1^2 - x_2^2 - 4\left(u - \frac{x_1}{2}\right)^2\right\}^{(n-3)/2} du \tag{7.74}$$

$$= \frac{1}{2}(x_1^2 - x_2^2)^{n/2 - 1} \int_{-1}^{1} (1 - v^2)^{(n-3)/2} dv$$

$$= \frac{1}{2}(x_1^2 - x_2^2)^{n/2 - 1} \frac{\Gamma\{(n-1)/2\}\Gamma(1/2)}{\Gamma(n/2)}$$

where the last integral is expressed in term of the beta function. The marginal density in (7.48) follows upon collecting the factors remaining from (7.73) and (7.74).

7.5 Exercises

1. Suppose that $X \sim HBS(\theta)$ as in section 7.1.1 with $\theta = -0.7$. Plot the first- and second-order Lugannani and Rice approximations for the CDF of $\hat{\theta}$ and graphically compare them to the true CDF computed using numerical integration. Does this first-order approximation show better accuracy than the first-order p^* density?

2. Let X have the density
$$f(x;\omega) = \frac{\ln \omega}{\omega - 1}\omega^x \qquad x \in (0,1) \tag{7.75}$$
for $\omega > 1$.
   (a) Compute the saddlepoint density for X and compare it graphically to (7.75) for $\omega = 1.1$ and 2.
   (b) Compute the p^* density for $\hat{\omega}$ along with its true density and compare them graphically for $\omega = 1.1$ and 2.
   (c) Suppose that X_1 and X_2 are i.i.d. with density $f(x;\omega)$ and $Y = X_1 + X_2$. Compute the saddlepoint density and the true density of Y and compare them graphically for $\omega = 1.1$ and 2.
   (d) Repeat the comparisons in (c) but for the p^* and true densities of $\hat{\omega}$.

3. Consider Planck's radiation density of section 7.1.2 and the CDF of $\hat{\alpha}$ when $\alpha = 3$.
   (a) Approximate this CDF using the Lugananni and Rice approximation and plot its relative error as compared to an approximation based on numerical integration.
   (b) Use the approximation in part (a) to determine the cutoff for a 5% level test of $\alpha = 3$ versus $\alpha > 3$. Also compute the approximate power curve of the test. How many saddlepoint equations need to be solved in order to compute the power curve?

4. For a single parameter model, prove proposition 7.1.2.
   (a) Determine the density of p_n from (7.9). Write
$$f(p) = \frac{1}{\sqrt{2\pi\lambda}} e^{-\lambda/2} h(p)$$
for some $h(p)$ and show through Taylor expansions that
$$h(p) = 1 + O(n^{-3/2})$$
as specified in (7.10).
   (b) Prove that $p_n = \lambda_n/(1 - 2b/n)$ is χ_1^2 to order $O(n^{-3/2})$ if and only if the density of λ_n has the form
$$f(\lambda) = \left(1 + \frac{b}{2n}\right)q_1(\lambda) - \frac{b}{2n}q_3(\lambda) + O(n^{-3/2})$$
where $q_i(\lambda)$ denotes a χ_i^2 density.

5. Suppose that $Z_1, \ldots, Z_n$ are i.i.d. $N(0, \sigma^2)$ and that only the value for $s^2 = \sum_{i=1}^n (Z_i - \bar{Z})^2 \sim \sigma^2 \chi_{n-1}^2$ has been observed.
   (a) In testing the null hypothesis $H_0 : \sigma^2 = 1$, show that the likelihood ratio statistic is
$$\lambda_n = -(n-1)\{\ln \tilde{\sigma}^2 - (\tilde{\sigma}^2 - 1)\} \tag{7.76}$$
where $\tilde{\sigma}^2 = s^2/(n-1)$.

(b) Show that

$$E(\lambda_n) = 1 + \frac{1}{3(n-1)} + O(n^{-3}). \qquad (7.77)$$

Hint: Use the MGF of $\ln s^2$ and the digamma expansion

$$\Psi(z) = \frac{\Gamma'(z)}{\Gamma(z)} = \ln z - \frac{1}{2z} - \frac{1}{12z^2} + O(z^{-4}).$$

This leads to the Bartlett corrected statistic as $p_n = \lambda_n / \{1 + 1/(3n)\}$.

(c) In Exercise 5(b), verify that the Bartlett correction is correct by computing $\vartheta(1)$, the second-order correction term to the saddlepoint density of s^2 when $\sigma^2 = 1$.

6. In Exercise 5, suppose that $Z_1, \ldots, Z_n$ are observed as i.i.d. $N(\mu, \sigma^2)$ with μ unknown.

(a) Show that the likelihood ratio test statistic for testing $H_0 : \sigma^2 = 1$ is

$$\lambda_n = -n\{\ln \hat{\sigma}^2 - (\hat{\sigma}^2 - 1)\} \qquad (7.78)$$

where $\hat{\sigma}^2 = s^2/n$. Note the similarity to (7.76).

(b) Compute

$$E(\lambda_n) = 1 + \frac{11}{6n} + O(n^{-2}).$$

Note that despite the similarity in test statistics (7.76) and (7.77), their Bartlett corrections are quite different.

7. Suppose a random sample of size n is drawn from the density

$$g(y; \mu) = e^{-(y-\mu)} \qquad y > \mu.$$

(a) Show that $\hat{\mu}$ is y_1, the smallest order statistic. Also show that $\hat{\mu}$ and $a^* = (a_2^*, \ldots a_n^*)$ with $a_i^* = y_i - y_1$ are independent.

(b) Deduce from (a) that $\hat{p} = \hat{\mu} - \mu \sim$ Exponential (n).

(c) Suppose a random sample of size n is drawn from the density

$$g(y; \mu, \sigma) = \frac{1}{\sigma} e^{-(y-\mu)/\sigma} \qquad y > \mu.$$

(d) Show that $\hat{\mu} = y_1$, the smallest order statistic, and $\hat{\sigma} = \bar{y} - y_1$.

(e) Using Fisher's expression for the conditional density of $\hat{\mu}$ and $\hat{\sigma}$ given $\{a_i\}$ with $a_i = (y_i - y_1)/(\bar{y} - y_1)$, show that $\hat{\mu}$, $\hat{\sigma}$ and $\{a_i\}$ form a mutually independent triple with $\hat{\mu} - \mu \sim$ Exponential (n/σ) and $\hat{\sigma} \sim$ Gamma $(n-1, n/\sigma)$.

8. Lieblein and Zelen (1956, p. 286) recorded the lifetimes of 23 ball bearings in millions of revolutions as

17.88	28.92	33.00	41.52	42.12	45.60
48.48	51.84	51.96	54.12	55.58	67.80
68.64	68.64	68.68	84.12	93.12	98.64
105.12	105.84	127.92	128.04	173.40	

Assume the data fit a Weibull (α, β) density as in section 7.2.2. The MLEs have been determined as $\hat{\alpha} = 81.99$ and $\hat{\beta} = 2.102$ by Thoman et al. (1969).

(a) Plot the marginal density of $(\hat{\mu} - \mu)/\hat{\sigma}$ given a. Check to see that it leads to the 95% interval for $\alpha = e^\mu$ of Lawless (1973, 1974) given as $(68.05, \infty)$.

(b) Plot the marginal density of $\hat{\sigma}/\sigma$ given a. Show that it leads to the 90% confidence interval for $\beta = 1/\sigma$ of Lawless (1973, 1974) given by (1.52, 2.59).

9. Consider the general linear model of section 7.2.3.
   (a) Use the form of the density for least square estimates in (7.20) to derive the density for the MLEs in (7.21).
   (b) Starting with the form of the p^* density, show that it leads to the same density for the MLEs.
   (c) In the special setting in which $\ln g(e) = \sum_{i=1}^{n} h(e_i)$, show that

$$j(\beta, \sigma, a) = \frac{-1}{\sigma^2} \sum_{i=1}^{n} \begin{bmatrix} h''(e_i)x_i x_i^T & \{h'(e_i) + e_i h''(e_i)\} x_i \\ \{h'(e_i) + e_i h''(e_i)\} x_i^T & 1 + 2e_i h'(e_i) + e_i^2 h''(e_i) \end{bmatrix}$$

where $X = (x_1, \ldots, x_p)$. Thus

$$|j(\hat{\beta}, \hat{\sigma}, a)| = \hat{\sigma}^{-2(p+1)} K(a),$$

for appropriate $K(a)$.

   (d) In the i.i.d. normal setting with $h(e_i) = \ln \phi(e_i)$, show that the triple $\hat{\beta}, \hat{\sigma}, a$ are mutually independent with $\hat{\beta} \sim N_p\{\beta, \sigma^2(X^T X)^{-1}\}$ and $n\hat{\sigma}^2 \sim \sigma^2 \chi^2_{n-p}$.

10. Consider affine ancillary $a = \Omega_\xi u_\xi$ constructed in a $(m, 1)$ curved exponential family as in (7.39). Show that a is an affine ancillary in the sense that it shares the same two properties as suggested in lemma 7.3.1 for the ancillary construction of Barndorff-Nielsen (1980). To begin, partition

$$\Delta_\xi = \begin{pmatrix} \dot{\theta}_\xi^T \\ \Delta_{\xi L} \end{pmatrix}$$

and define

$$Z_{1\xi} = \dot{\theta}_\xi^T (X - \mu_\xi)$$
$$Z_{2\xi} = \Omega_\xi \{\Delta_{\xi L}(X - \mu_\xi) - v_{21\xi} Z_{1\xi}/i_\xi\}.$$

The portion of $Z_{2\xi}$ within the curly braces is the linear regression of $\Delta_{\xi L}(X - \mu_\xi)$ on $Z_{1\xi}$. Show that $Z_{2\xi}$ has covariance I_{m-p} and is uncorrelated with $Z_{1\xi}$. Conclude that a is an affine ancillary as in lemma 7.3.1.

11. Suppose that $x = (x_1, x_2)^T$ is bivariate Normal$_2(\mu, I_2)$ with mean vector $\mu^T = (\cos \xi, \sin \xi)$ with $\xi \in (-\pi, \pi]$.
    (a) Transform to polar coordinates $(a, \hat{\xi})$ and show that

$$f(a, \hat{\xi}) = \frac{a}{2\pi} \exp\{-\tfrac{1}{2}(1 + a^2) + a \cos(\hat{\xi} - \xi)\} \qquad a > 0; \hat{\xi} \in (-\pi, \pi].$$

Using the fact that the Bessel I function has the integral representation (Abramowitz and Stegun, 1972, 9.6.16)

$$I_0(a) = \frac{1}{2\pi} \int_{-\pi}^{\pi} \exp\{a \cos(\hat{\xi} - \xi)\} d\hat{\xi}, \qquad (7.79)$$

determine that a is ancillary and that $\hat{\xi}|a$ is a location model with exact density

$$f(\hat{\xi}|a; \xi) = \frac{1}{2\pi I_0(a)} \exp\{a \cos(\hat{\xi} - \xi)\} \qquad \hat{\xi} \in (-\pi, \pi].$$

   (b) Conclude that the p^* density for $\hat{\xi}|a$ is exact. Derive the $p^\dagger$ density as

$$p^\dagger(\hat{\xi}|a; \xi) = \frac{1}{2\pi \hat{I}_0(a)} \exp\{a \cos(\hat{\xi} - \xi)\} \qquad \hat{\xi} \in (-\pi, \pi]$$

where
$$\hat{I}_0(a) = \frac{1}{\sqrt{2\pi a}} e^a$$

is the leading term in an asymptotic expansion (Abramowitz and Stegun, 1972, 9.7.1) for (7.79).

12. Suppose $X_1, \ldots, X_n$ are i.i.d. Normal (μ, μ^2).
    (a) Suggest a $(n-2)$ dimensional vector of ancillaries *outside* of the canonical sufficient statistic.
    (b) Suggest an exact ancillary *inside* the canonical sufficient statistic in that it is a function of the latter.
    (c) Compute the p^* density for $\hat{\mu}$ given a $(n-1)$ dimensional ancillary. Which ancillaries does it depend on?
    (d) Why must p^* be exact?
13. Suppose $X_1, \ldots, X_n$ are i.i.d. Gamma (α, α) with a mean of 1. Answer the same questions as in Exercise 12.
14. Consider the Behrens–Fisher problem as described in section 7.3.2.
    (a) Reparameterize in such a way that the $(4, 3)$ curved exponential family is embedded in a $(4, 4)$ full exponential family.
    (b) Determine an affine ancillary a and also the likelihood ancillary a^0 associated with the embedding created in part (a).
    (c) Now suppose that the data for populations 1 and 2 are $\{2, 3, 4\}$ and $\{1, 3, 5, 7, 9\}$ respectively. Compute the $p^\dagger$ density for the MLE of $(\mu, \sigma_1^2, \sigma_2^2)$ given the affine ancillary a. Normalize this to compute the p^* density for $\hat{\mu}$ given $\hat{\sigma}_1^2, \hat{\sigma}_2^2, a$.
    (d) Repeat part (c) but use the likelihood ratio ancillary a^0 in place of the affine ancillary. Compare the density to that of part (c).
    (e) The exact density for the MLE of $(\mu, \sigma_1^2, \sigma_2^2)$ given the affine ancillary a may be computed from the joint density of the canonical sufficient statistics. Use this approach to compute the true density for $\hat{\mu}$ given $\hat{\sigma}_1^2, \hat{\sigma}_2^2, a$ and compare this to the p^* approximation obtained in part (c).
    (f) Repeat part (e) but use the likelihood ratio ancillary a^0 in place of the affine ancillary. Compare the density to the p^* approximation obtained in part (d).
15. Consider the AR(1) time series model described in section 7.3.2. Starting with $z_0 = 1.2$, data were generated from the model with $\alpha = 1 = \sigma$ to give
$$(z_1, z_2, z_3) = (-1.4, -0.2, -0.76).$$
    (a) Reparameterize in such a way that the $(3, 2)$ curved exponential family is embedded in a $(3, 3)$ full exponential family.
    (b) Determine an affine ancillary a and also the likelihood ancillary a^0 associated with the embedding created in part (a).
    (c) Describe how one might go about computing the $p^\dagger$ density for the MLE of (α, σ^2) given the affine ancillary a. What are the difficulties that limit your ability to make this computation?
    (d) Repeat part (c) but use the likelihood ratio ancillary a^0 in place of the affine ancillary. Again, describe the difficulties involved.
16. Suppose that $X_1 \sim$ Exponential (ξ) independently of $X_2 \sim$ Exponential (ξ^2).
    (a) Show that the MLE for ξ is
$$\hat{\xi} = \frac{x_1}{4x_2}\left(\sqrt{1 + 24x_2/x_1^2} - 1\right). \quad (7.80)$$

(b) Show that the affine ancillary in (7.38) takes the value
$$a^2 = \hat{\xi} x_1(\hat{\xi} x_1 - 2) + \hat{\xi}^2 x_2(\hat{\xi}^2 x_2 - 2) + 2. \tag{7.81}$$

Note that it is also an exact ancillary due to being a function of the exact ancillary x_2/x_1^2.

(c) Compute the likelihood ancillary as
$$a^0 = 2(\hat{\xi} x_1 + \hat{\xi}^2 x_2) - 2\ln(\hat{\xi}^3 x_1 x_2) - 4$$

and note that it also is an exact ancillary.

(d) Show that both a and a^0 are monotone functions of x_2/x_1^2 when the appropriate sign of each is taken into account. Indeed for a given value of $\hat{\xi}$ and with fixed a^2 the values of x_1 and x_2 are constrained to an ellipse by (7.81) and a parabola by (7.80). The intersection of these two curves is two points with the choice of points determined by the sign of a. Plot the $p^\dagger$ and p^* densities for each and compare them.

(e) Derive the exact conditional density of $\hat{\xi}$ given ancillary $\tilde{a} = x_2/x_1^2$. (Hint: Determine an explicit form for the joint density of $(x_1/x_2, \tilde{a})$, use Maple to determine the normalization constant for the conditional density of x_1/x_2 given $\tilde{a}$, and transform to $\hat{\xi}$. Compare this plot with those of the p^* and $p^\dagger$ densities from part (d).

17. Suppose that $X_1 \sim$ Exponential (ξ) independently of $X_2 \sim$ Exponential $(\xi + \xi^2)$ and the data are $x_1 = 1$ and $x_2 = 2$.

    (a) Compute the $p^\dagger$ and p^* densities for $\hat{\xi}$ given an affine ancillary for several values of ξ.
    (b) Repeat part (a) but use a likelihood ancillary.
    (c) Derive the exact conditionality resolutions for part (a) and (b) and discuss the accuracy of the various approximations.
    (d) Can you construct an alternative exact ancillary?
    (e) When conditioning on the affine ancillary, is the success of the $p^\dagger$ and p^* densities related to the magnitude of $f(a; \xi)$, the marginal density of the ancillary at its observed value?

18. Assume the same model and data as in Exercise 17 but now make $X_2 \sim$ Exponential $(\xi + \sqrt{\xi})$. Answer all the same questions with this new model.

8

Probabilities with r^*-type approximations

Approximations to continuous univariate CDFs of MLEs in curved exponential and transformation families have been derived in Barndorff-Nielsen (1986, 1990, 1991) and are often referred to as r^* approximations. These approximations, along with their equivalent approximations of the Lugannani and Rice/Skovgaard form, are presented in the next two chapters. Section 8.2 considers the conditional CDF for the MLE of a scalar parameter given appropriate ancillaries. The more complex situation that encompasses a vector nuisance parameter is the subject of chapter 9.

Other approaches to this distribution theory, aimed more toward p-value computation, are also presented in section 8.5. Fraser and Reid (1993, 1995, 2001) and Fraser et al. (1999a) have suggested an approach based on geometrical considerations of the inference problem. In this approach, explicit ancillary expressions are not needed which helps to simplify the computational effort. Along these same lines, Skovgaard (1996) also offers methods for CDF approximation that are quite simple computationally. Specification of ancillaries is again not necessary and these methods are direct approximations to the procedures suggested by Barndorff-Nielsen above.

Expressions for these approximate CDFs involve partial derivatives of the likelihood with respect the parameter but also with respect to the MLE and other quantities holding the approximate ancillary fixed. The latter partial derivatives are called *sample space derivatives* and can be difficult to compute. An introduction to these derivatives is given in the next section and approximations to such derivatives, as suggested in Skovgaard (1996), are presented in appropriate sections.

8.1 Notation, models, and sample space derivatives

The notation used in chapter 7 is continued here. Let, for example, $\hat{\Xi}$ denote the MLE for ξ when considered as a random variable and use $p^{\dagger}(\hat{\xi}|a;\xi)$ in (7.42) as the approximation to the continuous density of $\hat{\Xi}$ at $\hat{\xi}$ given ancillary a and scalar parameter ξ. The models are assumed to be any of the models for which the p^* formula was found applicable in chapter 7; this includes, for example, transformation models in which a is an exact ancillary, and (m, p) curved exponential families in which a is an approximate $(m - p)$-dimensional affine ancillary.

In the discussion of transformation and curved exponential models in chapter 7, a specification of the ancillary statistic a assured that the log-likelihood function could be determined in terms the vector parameter ξ and $(\hat{\xi}, a)$, the minimal sufficient statistic. Denote this log-likelihood as $l = l(\xi; \hat{\xi}, a)$.

The $p \times 1$ gradient and $p \times p$ Hessian of l with respect to ξ are indicated by using the notation

$$l_{\xi;} = l_{\xi;}(\xi; \hat{\xi}, a) = \frac{\partial l(\xi; \hat{\xi}, a)}{\partial \xi}$$

$$l_{\xi\xi;} = l_{\xi\xi;}(\xi; \hat{\xi}, a) = \frac{\partial^2 l(\xi; \hat{\xi}, a)}{\partial \xi \partial \xi^T} = -j_\xi.$$

Often the semicolon is omitted when specifying these two quantities in the literature, particularly when there are no derivatives taken with respect to $\hat{\xi}$. This practice is not followed here because doing so would contribute to some confusion with the notation used for a curved exponential family. For example, in curved exponential families, $\ddot{\theta}_\xi$ refers to $\partial^2 \theta / \partial \xi \partial \xi^T$ and not to the first derivative of a mysterious quantity $\dot{\theta}$.

Partial derivatives with respect to $\hat{\xi}$ holding a fixed are referred to as sample space derivatives and there will be a need to consider the first two derivatives with respect to $\hat{\xi}$ denoted by

$$l_{;\hat{\xi}} = l_{;\hat{\xi}}(\xi; \hat{\xi}, a) = \frac{\partial l(\xi; \hat{\xi}, a)}{\partial \hat{\xi}^T}$$

$$l_{;\hat{\xi}\hat{\xi}} = l_{;\hat{\xi}\hat{\xi}}(\xi; \hat{\xi}, a) = \frac{\partial^2 l(\xi; \hat{\xi}, a)}{\partial \hat{\xi} \partial \hat{\xi}^T}.$$

Note that $l_{;\hat{\xi}}$ denotes a $1 \times p$ row vector whereas $l_{\xi;}$ is a $p \times 1$ column vector. It may seem excessively pedantic to emphasize that ξ (to the left of ;) and a (to the right) are held fixed in these computations. Forgetting to holding ξ fixed is a common mistake made when performing these computations in practice. Fixing a essentially makes a commitment to conditioning on a particular ancillary a. In addition, mixed derivatives will be needed such as $l_{\xi;\hat{\xi}} = \partial^2 l(\xi; \hat{\xi}, a)/\partial \xi \partial \hat{\xi}^T$, which is $p \times p$, and $l_{\xi\xi;\hat{\xi}}$ which is a three-dimensional lattice.

8.2 Scalar parameter approximations

The continuous CDF approximation for scalar MLE $\hat{\Xi}$ given ancillary a and scalar parameter ξ is given below. From Barndorff-Nielsen (1990), the CDF approximation for $\hat{\Xi}|a; \xi$ is

$$\widehat{\Pr}(\hat{\Xi} \leq \hat{\xi}|a; \xi) = \Phi(\hat{w}) + \phi(\hat{w})(1/\hat{w} - 1/\hat{u}) \qquad \hat{\xi} \neq \xi \tag{8.1}$$

where

$$\hat{w} = \text{sgn}(\hat{\xi} - \xi)\sqrt{-2\{l(\xi; \hat{\xi}, a) - l(\hat{\xi}; \hat{\xi}, a)\}} \tag{8.2}$$

$$\hat{u} = j_{\hat{\xi}}^{-1/2}\{l_{;\hat{\xi}}(\hat{\xi}; \hat{\xi}, a) - l_{;\hat{\xi}}(\xi; \hat{\xi}, a)\}. \tag{8.3}$$

The value of $\hat{w}$ is the familiar signed log-likelihood ratio while $\hat{u}$ is a new quantity specified in terms of the sample space derivatives. Factor $j_{\hat{\xi}} = -l_{\xi\xi;}(\hat{\xi}; \hat{\xi}, a)$ is the MLE for the Fisher information about ξ. An implicit assumption that is made when using (8.1) is that the mapping $\hat{\xi} \leftrightarrow \hat{w}$ is 1-1 and monotonic increasing. The derivation of (8.1) uses the method of Temme and is delayed until section 8.4.

8.2.1 r^* approximation

The current form of this approximation has evolved through the efforts of Barndorff-Nielsen (1986, 1990, 1991) and is remarkably simple. If

$$r^* = \hat{w} - \frac{1}{\hat{w}} \ln \frac{\hat{w}}{\hat{u}},$$

then the conditional CDF approximation simply treats r^* as Normal $(0, 1)$ or

$$\Pr(\hat{\Xi} \leq \hat{\xi} | a; \xi) \simeq \Phi(r^*). \tag{8.4}$$

In all the applications that the author has worked with or seen, the numerical differences between the computations in (8.1) and (8.4) are negligible. Typically when one is highly accurate, so also is the other. Likewise when one approximation does not work well, neither does the other. For an example in which they both fail, see the discussion of the inverse Gaussian distribution in section 16.3.1. One advantage of (8.1) however, is that it can produce a probability outside of $(0, 1)$ to demonstrate that it isn't working well; (8.4) cannot reveal its failure in this way.

As a result of their computational equivalence, the decision of whether to work with (8.1) or (8.4) is purely a matter of individual preference. Since the former approximation more readily reveals it faults and is also a saddlepoint approximation while the latter is not, the current discussion will naturally continue to give emphasis to the former approach.

Approximations (8.4) and (8.1) are related by Taylor expansion of the former as outlined in Exercise 1.

8.3 Examples

The examples below suggest that the sample space derivatives in (8.3) are simple to compute in transformation models but may not be so easily computed in curved exponential families. Section 8.5.1 discusses an approximation to the sample space derivative useful for the latter setting.

8.3.1 One-parameter exponential family

Suppose ξ is a 1-1 reparameterization of the canonical parameter θ in the exponential family with

$$l(\theta; x) = x\theta - c(\theta) - d(x). \tag{8.5}$$

In this case the expression (8.1) agrees with the Lugannani and Rice formula given in (5.64). If $\hat{w}(\xi)$ and $\hat{w}(\theta)$ are the $\hat{w}$ values in their respective parameterizations, then

$$\hat{w}(\xi) = \text{sgn}(\partial \hat{\xi}/\partial \hat{\theta}) \hat{w}(\theta). \tag{8.6}$$

The same relationship applies as concerns the respective $\hat{u}$ values. Thus the computation of (8.1) is invariant to the particular parameterization used and also preserves this invariance under decreasing transformations where the inequalities must be reversed. Thus no generality is lost in writing

$$\hat{u} = J_{\hat{\theta}}^{-1/2}\{l_{;\hat{\theta}}(\hat{\theta};\hat{\theta},a) - l_{;\hat{\theta}}(\theta;\hat{\theta},a)\}.$$

The difference

$$l(\hat{\theta};\hat{\theta},a) - l(\theta;\hat{\theta},a) = \{x\hat{\theta} - c(\hat{\theta})\} - \{x\theta - c(\theta)\}$$

can be differentiated $\partial/\partial\hat{\theta}$ but holding fixed those values of $\hat{\theta}$ that appear due to their occurrence to left of the semicolon in $l(\hat{\theta};\hat{\theta},a)$. This gives

$$\hat{u} = J_{\hat{\theta}}^{-1/2}\frac{\partial x}{\partial\hat{\theta}}(\hat{\theta} - \theta) = \sqrt{J_{\hat{\theta}}}(\hat{\theta} - \theta). \tag{8.7}$$

The invariance to parameterization also assures that $\hat{u}$ may be written in the mean parametrization $\hat{\mu} = x$ as

$$\hat{u} = J_{\hat{\mu}}^{-1/2}\{l_{;\hat{\mu}}(\hat{\theta};\hat{\theta},a) - l_{;\hat{\mu}}(\theta;\hat{\theta},a)\} = \sqrt{J_{\hat{\theta}}}(\hat{\theta} - \theta)$$

by direct computation. Thus $\hat{u}$ has an interpretation as a standardized score statistic.

8.3.2 Location model

Suppose the data $y_1, \ldots, y_n$ are i.i.d. from density $g(y - \mu)$ as described in (7.11). Then the approximate CDF of the MLE for μ at $\hat{\mu}$ given the vector ancillary a in (7.12) is given by the right side of (8.1) with

$$\hat{w} = \text{sgn}(\hat{\mu} - \mu)\sqrt{-2\sum_{i=1}^{n}\ln\frac{g(a_i + \hat{\mu} - \mu)}{g(a_i)}} \tag{8.8}$$

$$\hat{u} = -j_{\hat{\mu}}^{-1/2}\sum_{i=1}^{n}(\ln g)'(a_i + \hat{\mu} - \mu), \tag{8.9}$$

where $j_{\hat{\mu}} = -\sum_{i=1}^{n}(\ln g)''(a_i)$. Note in this instance that $l_{;\hat{\mu}}(\hat{\mu};,\hat{\mu},a) = 0$. The expression for $\hat{u}$ in (8.9) makes explicit its interpretation as a standardized score statistic. This particular expression was given earlier by Fraser (1990) and DiCiccio *et al.* (1990).

8.3.3 Gamma hyperbola

This distribution may be considered either as a (2, 1) curved exponential family (§ 7.3.2) or, with suitable reparameterization and conditioning on an ancillary, as the scale parameter model (7.44) or as the location model (7.36). Approximation (8.1) is considered from the first and last perspectives.

As a location model

Conditioning on the ancillary $a = \ln(x_1 x_2)$, exact computation of $\hat{w}$ and $\hat{u}$ in (8.2) and (8.3) can be based on the location model representation in parameterization ω. Here a is not

8.3 Examples

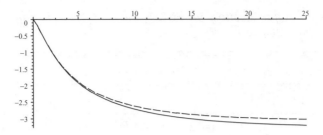

Figure 8.1. Percentage relative error of $\widehat{\Pr}(\hat{\Xi} > \hat{\xi}|a; \xi = 1)$ (*solid*) and $\Phi(r^*)$ (*dashed*) versus $\hat{\xi} \in (1, 25)$.

treated as data but a fixed number with a single datum $\hat{\omega}$. With a single datum, expressions (8.8) and (8.9) are not directly applicable and the computations are referred back to (8.2) and (8.3). The information is $j_{\hat{\omega}} = e^{a/2}/2$ and the exact input values for $\Pr(\hat{\Omega} \leq \hat{\omega}|a; \omega)$ are

$$\hat{w} = \text{sgn}(\hat{\omega} - \omega)\sqrt{4e^{a/2}\left[\cosh\{(\hat{\omega} - \omega)/2\} - 1\right]} \tag{8.10}$$

$$\hat{u} = \sqrt{2}e^{a/4}\sinh\{(\hat{\omega} - \omega)/2\}. \tag{8.11}$$

From direct computation

Using $\hat{\xi} = \sqrt{x_2/x_1}$, the log-likelihood is rewritten as

$$l(\xi; \hat{\xi}, a) = -x_1\xi - x_2/\xi = -e^{a/2}(\xi/\hat{\xi} + \hat{\xi}/\xi). \tag{8.12}$$

Also, in this parametrization, $j_{\hat{\xi}} = 2e^{a/2}/\hat{\xi}^2$. Direct differentiation of (8.12) leads to the input for computing $\Pr(\hat{\Xi} \leq \hat{\xi}|a; \xi)$ as

$$\hat{u} = \frac{1}{\sqrt{2}}e^{a/4}(\hat{\xi}/\xi - \xi/\hat{\xi}) = -\sqrt{2}e^{a/4}\sinh\{\hat{\omega} - \omega)/2\}.$$

Likewise $\hat{w}$ here is the negative of the $\hat{w}$ given in (8.10). The sign differences reflect the decreasing nature of the transformation $\omega = -2\ln\xi$.

Of course this equivalence is simply the result of the parameter invariance in the $(2, 1)$ exponential family under the transformation $\hat{\omega} = -2\ln\hat{\xi}$.

Numerical computations

Consider the numerical example from section 7.3.2 which conditions on $a = \ln(x_1x_2) = \ln(81/16) \simeq 1.622$. Figure 8.1 plots the percentage relative errors of $\widehat{\Pr}(\hat{\Xi} > \hat{\xi}|a; \xi = 1)$ in (8.1) (solid) and $\Phi(r^*)$ in (8.4) (dashed) for $\hat{\xi} \in (1, 25)$ or equivalently $0 > \hat{\omega} > -6.44$. The comparison has been made using numerical integration of the conditional density in (7.44) as the exact computation. For the range $\hat{\xi} \in (1/25, 1)$ or $6.44 > \hat{\omega} > 0$, the relative error plot is the mirror image of the plot about the vertical axis as a result of the symmetry of the exact distribution of $\hat{\omega}$ and the oddness of its approximation terms in (8.10). Within the plotted range from $\hat{\xi} \in (1/25, 25)$ the degree of relative accuracy is remarkable.

Some numerical comparisons of approximations for the tail probabilities $\Pr(\hat{\Xi} > \hat{\xi}|a; \xi = 1)$ are given in table 8.1 with accuracies that reflect the relative errors seen in figure 8.1.

Table 8.1. *Approximations for* $\Pr(\hat{\Xi} > \hat{\xi}|a; \xi = 1)$ *for the values of* $\hat{\xi}$ *listed in the columns.*

	$\hat{\xi}$-values						
Method	$1\frac{1}{10}$	$1\frac{1}{2}$	2	$2\frac{1}{2}$	3	6	15
Exact	.4178	.1873	.06223	.01974	.0²6138	.0⁵5218	.0¹⁴4279
(8.1)	.4177	.1868	.06187	.01956	.0²6067	.0⁵5106	.0¹⁴4151
(8.4)	.4177	.1867	.06187	.01956	.0²6067	.0⁵5108	.0¹⁴4156

As previously mentioned, there is virtually no difference in accuracy between (8.1), which uses the Lugannani and Rice format, and the normal approximation to r^* in (8.4).

8.3.4 $(m, 1)$ *curved exponential family*

The expression for $\hat{u}$ in (8.3) maintains its interpretation as a standardized score statistic in this context. Suppose that sufficient statistic x has log-likelihood

$$l(\xi; x) = \theta_\xi^T x - c(\theta_\xi) - d(x). \tag{8.13}$$

Then consider the data-dependent parameterization

$$\eta_{\hat{\xi}} - \eta_\xi = l_{;\hat{\xi}}(\hat{\xi}; x) - l_{;\hat{\xi}}(\xi; x) = (\theta_{\hat{\xi}} - \theta_\xi)^T \frac{\partial x}{\partial \hat{\xi}}. \tag{8.14}$$

The observed information for η is $j_{\hat{\eta}} = j_{\hat{\xi}}(\partial \hat{\xi}/\partial \hat{\eta})^2$. Direct differentiation of $\eta_\xi = l_{;\hat{\xi}}(\xi; x)$ gives

$$\frac{\partial \hat{\eta}}{\partial \hat{\xi}} = l_{\hat{\xi};\hat{\xi}}(\xi; x) = \dot{\theta}_\xi^T \frac{\partial x}{\partial \hat{\xi}}.$$

To determine this quantity, take the sample space derivative $\partial/\partial \hat{\xi}$ of the score equation $0 = l_{\hat{\xi};}(\hat{\xi}; x)$. Differentiating on both sides of the semicolon gives

$$l_{\hat{\xi};\hat{\xi}}(\hat{\xi}; x) = -l_{\hat{\xi}\hat{\xi};}(\hat{\xi}; x) = j_{\hat{\xi}}$$

so that $\partial \hat{\eta}/\partial \hat{\xi} = j_{\hat{\xi}}$. Thus

$$\hat{u} = j_{\hat{\xi}}^{-1/2}\{l_{;\hat{\xi}}(\hat{\xi}; x) - l_{;\hat{\xi}}(\xi; x)\} = \sqrt{j_{\hat{\eta}}}(\eta_{\hat{\xi}} - \eta_\xi) \tag{8.15}$$

is the standardized score. The form of $\hat{u}$ in (8.15) assures that (8.1) is invariant to monotone transformations of ξ.

Geometry in x-space

The geometry of $\hat{u}$ also reflects the fact that an ancillary a has been held fixed in its computation. Consider the geometrical meaning of the right side of (8.14). If $x \leftrightarrow (\hat{\xi}, a)$ is a conditionality resolution, then vector $\partial x/\partial \hat{\xi}$ points tangent to the manifold in x-space on which ancillary a has been held fixed. Parameter η_ξ is the inner product of θ_ξ with this tangent vector so it is the coordinate of the canonical parameter with respect to vector $\partial x/\partial \hat{\xi}$; thus $\eta_{\hat{\xi}} - \eta_\xi$ is the coordinate of the estimation error for θ_ξ with respect to tangent vector $\partial x/\partial \hat{\xi}$.

8.4 Derivation of (8.1)

Barndorff-Nielsen (1990) used the method of Temme in section 2.3.1 to integrate the $p^\dagger$ approximation over $(-\infty, \hat{\xi}]$. This proof is replicated below and is essentially the same argument used to derive the Lugannani and Rice approximation in section 2.3.2.

The integral

$$\int_{-\infty}^{\hat{\xi}} p^\dagger(z|a;\xi)dz = \frac{1}{\sqrt{2\pi}} \int_{-\infty}^{\hat{\xi}} |j_z|^{1/2} \exp\{l(\xi;z,a) - l(z;z,a)\}dz \tag{8.16}$$

can be put into the form for Temme's result by transforming for dz to $d\hat{w}$ where $\hat{w}$ depends on z through

$$\hat{w} = \text{sgn}(\hat{\xi} - \xi)\sqrt{-2\{l(\xi;z,a) - l(z;z,a)\}}.$$

By direct computation,

$$\frac{d\hat{w}}{dz} = \frac{\text{sgn}(\hat{\xi} - \xi)}{\sqrt{-2\{l(\xi;z,a) - l(z;z,a)\}}}[\{l_{\xi;}(z;z,a) + l_{;\xi}(z;z,a)\} - l_{;\xi}(\xi;z,a)]$$

$$= \frac{1}{\hat{w}}\{l_{;\xi}(z;z,a) - l_{;\xi}(\xi;z,a)\}. \tag{8.17}$$

The term $l_{\xi;}(z;z,a) = 0$ for all values of z since the score function at the MLE must always be 0. This leads to the following result.

Lemma 8.4.1 *The mapping $z \leftrightarrow \hat{w}$ is smooth with*

$$\frac{d\hat{w}}{dz} = \begin{cases} \hat{w}^{-1}\{l_{;\xi}(z;z,a) - l_{;\xi}(\xi;z,a)\} & \text{if } z \neq \xi \\ \sqrt{j_\xi} & \text{if } z = \xi. \end{cases}$$

Proof. The result follows from (8.17) except when $z = \xi$. The latter case is derived by passing to the limit as $z \to \xi$ so that $\hat{w} \to 0$. Then, using l'Hôpital's rule,

$$\lim_{z \to \xi} \frac{d\hat{w}}{dz} = \frac{l_{\xi;\xi}(z;z,a) + l_{;\xi\xi}(z;z,a) - l_{;\xi\xi}(\xi;z,a)}{d\hat{w}/dz}.$$

The second and third terms in the numerator cancel at $z = \xi$. The first term is found by differentiating the score equation

$$l_{\xi;}(\xi;\xi,a) = 0$$

with respect to $d\xi$ to get

$$l_{\xi;\xi}(\xi;\xi,a) = -l_{\xi\xi;}(\xi;\xi,a) = j_\xi. \tag{8.18}$$

Thus, at $z = \xi$,

$$\left(\frac{d\hat{w}}{dz}\right)^2_{z=\xi} = j_\xi$$

and the lemma follows. □

Note that the mapping $z \leftrightarrow \hat{w}$ is not assured of being 1-1 and monotonic increasing as occurred in the derivation of the Lugannani and Rice approximation. Increasing monotonicity, or $d\hat{w}/dz \geq 0$, is however assured if the condition

$$\text{sgn}(\hat{w}) = \text{sgn}\{l_{;\hat{\xi}}(z; z, a) - l_{;\hat{\xi}}(\xi; z, a)\}$$

holds for all $z \neq \xi$.

Implementing the change in variable to $d\hat{w}$ for (8.16) yields

$$\int_{-\infty}^{\hat{\xi}} |j_z|^{1/2} \frac{\hat{w}}{l_{;\hat{\xi}}(z; z, a) - l_{;\hat{\xi}}(\xi; z, a)} \phi(\hat{w}) d\hat{w} = \int_{-\infty}^{\hat{\xi}} h(\hat{w}) \phi(\hat{w}) d\hat{w}$$

where h is defined by the equality. Since, from Lemma 8.4.1,

$$\lim_{\hat{w} \to 0} h(\hat{w}) = 1,$$

then (8.1) follows directly from the Temme approximation in (2.34). □

8.5 Other versions of $\hat{u}$

The sample space derivatives of $\hat{u}$ in (8.3) have exact and approximate expressions that depend on the model structure as well as the ancillary that is held fixed. Thus, for example, fixing the likelihood ancillary usually leads to a different quantity than fixing the affine ancillary when both types of ancillaries are available. An elaboration of these exact expressions as well as their approximation as suggested by Skovgaard (1996) are given below.

From a much different perspective, Fraser *et al.* (1999a) have derived an expression that is different from $\hat{u}$ and denoted as $\breve{u}$ below. Rather than trying to approximate $\hat{u}$, they sought to determine the computation of a p-value through some fundamental considerations about what the proper statistical inference should be. The details of this approach require additional discussion and are presented last.

8.5.1 Skovgaard's approximate sample space derivatives

The derivation for this approximation is rather technical, and its details have therefore been banished to the appendix in section 8.8.1. Define

$$q(\xi_1, \xi_2; \xi_0) = E\left[\{l(\xi_1) - l(\xi_2)\}l_{\xi;}(\xi_0); \xi_0\right] \tag{8.19}$$

as the expectation over all the data (both $\hat{\Xi}$ and ancillary A) of the expression in (8.19) assuming parameter value ξ_0. Since the mean of $l_{\xi;}(\xi_0)$ is zero, the expected product is simply the covariance between $l(\xi_1) - l(\xi_2)$ and $l_{\xi;}(\xi_0)$.

In terms of q, the sample space derivative may be approximated as

$$l_{;\hat{\xi}}(\hat{\xi}; \hat{\xi}, a) - l_{;\hat{\xi}}(\xi; \hat{\xi}, a) \simeq q(\hat{\xi}, \xi; \hat{\xi}) i_\xi^{-1} j_\xi \tag{8.20}$$

where $i_\xi = E\{l_{\xi;}(\xi)^2; \xi\}$ is the expected Fisher information given ξ, and j_ξ is the observed Fisher information given ξ. In computing q, it is helpful to remember that the insertion of

One parameter exponential family

For a regular family, the log-likelihood is given in (8.5) and it is assumed that $\xi \leftrightarrow \theta$ is 1-1. The value of q in (8.20) is

$$E\left[\{l(\hat{\theta}) - l(\theta)\}l_{\xi;}(\hat{\xi}); \hat{\xi}\right]$$
$$= E\left[\{(\hat{\theta} - \theta)X - [c(\hat{\theta}) - c(\theta)]\}\{X - c'(\hat{\theta})\}\frac{\partial\hat{\theta}}{\partial\hat{\xi}}; \hat{\theta}\right] \quad (8.21)$$

where $\hat{\xi}, \hat{\theta},$ and θ are assume to be fixed and the expectation is with respect to the sufficient statistic X. The computation in (8.21) is

$$(\hat{\theta} - \theta)j_{\hat{\theta}}\partial\hat{\theta}/\partial\hat{\xi}$$

and the value of $\hat{u}$ with approximation (8.20) is

$$\check{u} = j_{\hat{\xi}}^{-1/2}(\hat{\theta} - \theta)j_{\hat{\theta}}\partial\hat{\theta}/\partial\hat{\xi} = \left\{j_{\hat{\theta}}\left(\frac{\partial\hat{\theta}}{\partial\hat{\xi}}\right)^2\right\}^{-1/2}(\hat{\theta} - \theta)j_{\hat{\theta}}\frac{\partial\hat{\theta}}{\partial\hat{\xi}}$$
$$= \sqrt{j_{\hat{\theta}}}(\hat{\theta} - \theta)$$

upon using the relationship of $j_{\hat{\xi}}$ and $j_{\hat{\theta}}$ in (5.8). Thus $\check{u}$ and $\hat{w}$ above agree with the appropriate entries in Lugannani and Rice given in (5.64). The use of the sample space approximations is exact in this case.

$(m, 1)$ curved exponential family

The same derivation incorporating approximation (8.20) in this context leads to

$$q(\hat{\xi}, \xi; \hat{\xi}) = (\theta_{\hat{\xi}} - \theta_{\xi})^T \Sigma_{\hat{\xi}}\dot{\theta}_{\hat{\xi}} = (\theta_{\hat{\xi}} - \theta_{\xi})^T \dot{\mu}_{\hat{\xi}}$$

so that $\hat{u}$ is approximately

$$\check{u} = \sqrt{j_{\hat{\xi}}i_{\hat{\xi}}^{-1}}\{(\theta_{\hat{\xi}} - \theta_{\xi})^T \Sigma_{\hat{\xi}}\dot{\theta}_{\hat{\xi}}\} = \sqrt{j_{\hat{\xi}}}(\dot{\theta}_{\hat{\xi}}^T \dot{\mu}_{\hat{\xi}})^{-1}(\theta_{\hat{\xi}} - \theta_{\xi})^T \dot{\mu}_{\hat{\xi}}. \quad (8.22)$$

The approximate sample space derivative is also invariant under 1-1 transformation of the index parameter ξ. If $\hat{\xi} \leftrightarrow \hat{\omega}$ is 1-1 and smooth, then, upon changing parameters, calculus shows that

$$j_{\hat{\xi}} = j_{\hat{\omega}}\left(\frac{\partial\hat{\omega}}{\partial\hat{\xi}}\right)^2 \qquad i_{\hat{\xi}} = i_{\hat{\omega}}\left(\frac{\partial\hat{\omega}}{\partial\hat{\xi}}\right)^2.$$

In the leading expression for $\check{u}$, the term $(\theta_{\hat{\xi}} - \theta_{\xi})^T \Sigma_{\hat{\xi}}$ is parameterization invariant and the remaining terms all contribute various powers of the factor $|\partial\hat{\omega}/\partial\hat{\xi}|$ that add to zero to leave the correct result in $\hat{\omega}$ along with $\text{sgn}(\partial\hat{\omega}/\partial\hat{\xi})$.

8.5.2 Exact sample space derivatives fixing affine ancillaries

If the choice is made to condition on an affine ancillary in a $(m, 1)$ curved exponential family, then an explicit expression for the sample space derivative can be determined. With a affine, the exact computation

$$l_{;\hat{\xi}}(\hat{\xi}; \hat{\xi}, a) - l_{;\hat{\xi}}(\xi; \hat{\xi}, a) = (\theta_{\hat{\xi}} - \theta_{\xi})^T \frac{\partial x}{\partial \hat{\xi}}, \tag{8.23}$$

requires the partial derivative of the sufficient statistic holding a fixed. This computation may be found in the first column of (7.71). For the case in which $\hat{\xi}$ is a scalar, this may be simplified to give

$$\hat{u} = j_{\hat{\xi}}^{-1/2}[\, j_{\hat{\xi}} - (x - \mu_{\hat{\xi}})^T \dot{\Psi}_{\hat{\xi}}](\dot{\theta}_{\hat{\xi}}\, \Psi_{\hat{\xi}})^{-1}(\theta_{\hat{\xi}} - \theta_{\xi}). \tag{8.24}$$

The main difficulty in implementing this computation is in determining the matrix Ψ_ξ as a function of ξ. Recall from lemma 7.3.1, section 7.3.3, that Ψ_ξ is any $m \times (m - 1)$ matrix whose columns are orthogonal to $\dot{\mu}_\xi$. This choice is never unique and, as seen in the gamma hyperbola example below, the choice can lead to varying degrees of accuracy.

8.5.3 Exact sample space derivatives in $(2, 1)$ curved exponential families fixing a likelihood ancillary

Conditioning upon a likelihood ancillary in this family leads to an exact expression for the sample space derivative that has been derived in Barndorff-Nielsen (1990). The derivation is outlined in Exercise 5 and results in

$$\hat{u} = \sqrt{j_{\hat{\xi}}}(1\ 0)(\dot{\theta}_{\hat{\xi}}\, \tilde{\theta} - \theta_{\hat{\xi}})^{-1}(\theta_{\hat{\xi}} - \theta_{\xi}), \tag{8.25}$$

where $\tilde{\theta}$ is the MLE for the full $(2, 2)$ exponential family. The likelihood ancillary in this instance is

$$a^2 = -2(\tilde{l} - \hat{l}) = 2\{(\tilde{\theta} - \theta_{\hat{\xi}})^T x - c(\tilde{\theta}) + c(\theta_{\hat{\xi}})\} \tag{8.26}$$

where $\tilde{l}$ is the likelihood maximized over two dimensional θ at $\tilde{\theta}$, and $\hat{l}$ is the maximized likelihood over the curved exponential family in ξ at $\hat{\xi}$.

8.5.4 The approach of Fraser, Reid, and Wu (1999a)

The standardized score statistic $\check{u}$ recommended from this approach is an evolution of many ideas in the work of Fraser and Reid (1993, 1995), Fraser *et al.* (1999a), and Fraser (2004). A simple description along with some motivation are presented. Basic computational aspects are considered first followed by some understanding of the procedure derived through various examples. Lastly, its relationship to $\hat{u}$ in (8.1) is considered.

Typically the approach requires a model structure that can be broken down into independent scalar observations. A generalization that allows for dependence and applies in curved exponential families settings is given below however. For the general independent setting, let $y = (y_1, \ldots, y_n)^T$ be the data of an independent sample with $Y_i \sim F_i(y; \xi)$ for scalar ξ and CDF F_i given. The ancillaries that are fixed include $\{a_i^s = F_i(y_i; \hat{\xi})\}$, the MLEs of the

8.5 Other versions of $\hat{u}$

probability integral transforms of the data. Fraser (2004) and earlier work fix $\{F_i(y_i; \xi)\}$, the collection of exact pivots instead, however see the comments further onward concerning their equivalence. Fixing $\{a_i^s\}$, then $v_i = \partial y_i / \partial \hat{\xi}$ is computed and as the "sensitivity" of y_i. The vector $v = (v_1, \ldots, v_n)^T$ records the change in the data as a result of a change in $\hat{\xi}$ when ancillaries $\{a_i^s\}$ are held fixed. Fraser (2004) calls v the "sensitivity vector" as it records the response direction in y-space to perturbations in $\hat{\xi}$ holding $\{a_i^s\}$ fixed. The sensitive direction points tangent to the locally defined ancillary manifold in y-space based on holding $\{a_i^s\}$ fixed which, of course, may not lead to a globally defined ancillary manifold. As the sensitive direction "responds" to shifts in the MLE value, it should be intuitively clear that the sensitive direction is useful in discriminating values for the true parameter ξ. With this in mind, let φ_ξ be the directional sample space derivative of the log-likelihood in the sensitive direction or

$$\varphi_\xi = \frac{d}{dv} l(\xi; y) = \frac{d}{dt} l(\xi; y + tv)|_{t=0} = \sum_{i=1}^{n} \frac{\partial l(\xi; y)}{\partial y_i} v_i. \qquad (8.27)$$

The value of φ_ξ depends on both the unknown parameter ξ and data y and may be considered a data-dependent parametrization that seeks to examine the change of the log-likelihood in the sensitive direction. The value of $\breve{u}$ is the signed and standardized MLE for this mixed parameterization φ_ξ or

$$\breve{u} = \text{sgn}(\hat{\xi} - \xi) \sqrt{j_{\hat{\varphi}}} |\varphi_{\hat{\xi}} - \varphi_\xi|, \qquad (8.28)$$

where $j_{\hat{\varphi}} = \partial^2 l / \partial \hat{\varphi}^2$ is the observed Fisher information about $\varphi_{\hat{\xi}}$.

The sophistication of tools and ideas needed in describing (8.28) belies its ease of computation and range of application. Its accuracy often matches that occurring with exact computation of $\hat{u}$ as seen in the examples below. Its limitation however is that it is a statistical rather than a probabilistic tool. It performs the statistical objectives of computing a p-value, attained conditional power as a function of parameter ξ, and the inversion of power leading to confidence intervals. Beyond this however, the computations may not necessarily be meaningful because they are only locally defined for y, the data observed. Whether they are locally or globally defined depends on whether the sensitive direction characterizes a globally defined ancillary manifold. The general theory of differential equations only guarantees a locally defined manifold; see theorem 1.1 of Ross (1974).

General independent observation model

Let $y = (y_1, \ldots, y_n)^T$ be the data observed for independent observations in which $Y_i \sim F_i(y; \xi)$. The sensitive direction in y-space is determined by computing the change in y_i holding approximate ancillary $a_i^s = F_i(y_i; \hat{\xi})$ fixed. Take the total differential of $F_i(y_i; \hat{\xi})$ with respect to both coordinates to get

$$0 = dF_i(y_i; \hat{\xi}) = \frac{\partial F_i(y_i; \hat{\xi})}{\partial y_i} dy_i + \frac{\partial F_i(y_i; \hat{\xi})}{\partial \hat{\xi}} d\hat{\xi}$$

so that

$$v_i = \partial y_i / \partial \hat{\xi} = -\frac{\partial F_i(y_i; \hat{\xi}) / \partial \hat{\xi}}{\partial F_i(y_i; \hat{\xi}) / \partial y_i}. \qquad (8.29)$$

Fraser and Reid (1993, 1995), Fraser *et al.* (1999a), and Fraser (2003, 2004) approach this differently but get the same direction. They fix the exact pivotal ancillaries $\{F_i(y_i; \xi)\}$ to determine the sensitive direction as a function of ξ and then substitute $\xi = \hat{\xi}$. Conceptually there may be a difference but not computationally. They contend that at the observed data y, the sensitive direction $v = (v_i, \ldots, v_n)^T$ is the tangent vector to a locally defined second order ancillary manifold in y-space. What is clear from (8.29) in the presentation above, is that v is tangent to the locally defined manifold fixing $\{a_i^s\}$.

Location model

Suppose the data $y_1, \ldots, y_n$ are i.i.d. from density $g(y - \mu)$. Ancillary a_i^s depends only the exact ancillary $y_i - \hat{\mu}$ so that $\partial y_i / \partial \hat{\mu} = 1$ and $v = \mathbf{1} = (1, \ldots, 1)^T$. Then the sensitive direction leads to the discriminating parameter

$$\varphi_\mu = \sum_{i=1}^{n} (\ln g)'(y_i - \mu)$$

which is the negative score function. Its Fisher information is

$$j_{\hat{\varphi}} = j_{\hat{\mu}} \left(\frac{\partial \hat{\varphi}}{\partial \hat{\mu}} \right)^{-2} = j_{\hat{\mu}}^{-1}$$

and

$$\check{u} = \operatorname{sgn}(\hat{\mu} - \mu) j_{\hat{\mu}}^{-1/2} \left| -\sum_{i=1}^{n} (\ln g)'(y_i - \mu) \right| = \hat{u}$$

as given in (8.9).

In this setting the ancillary manifold holding $\{a_i^s\}$ fixed is globally defined and given by

$$\{y : y - \hat{\mu}\mathbf{1} = a\} = \{c\mathbf{1} + a : c \in \Re\},$$

the linear space at a pointing in direction $\mathbf{1}$.

(m, 1) curved exponential family with independent sufficient statistics

Suppose that $(x_1, \ldots, x_m) = x^T$ are independent components of the canonical sufficient statistic and the log-likelihood is given in (8.13). Then

$$\varphi_\xi = \frac{\partial}{\partial v} l(\xi; x) = \{\theta_\xi - d'(x)\}^T v.$$

The observed information may be computed directly as

$$j_{\hat{\varphi}} = j_{\hat{\xi}} \left(\frac{\partial \hat{\varphi}}{\partial \hat{\xi}} \right)^{-2} = j_{\hat{\xi}} (\dot{\theta}_{\hat{\xi}}^T v)^{-2}. \tag{8.31}$$

This can be reduced if the directional derivative of the score equation

$$0 = l_{\xi;}(\hat{\xi}; x) = \dot{\theta}_{\hat{\xi}}^T (x - \mu_{\hat{\xi}})$$

is taken in the sensitive direction, e.g. holding $\{a_i^s\}$ fixed. Then

$$0 = l_{\xi\xi;}(\hat{\xi}; x) + l_{\xi;v}(\hat{\xi}; x) = -j_{\hat{\xi}} + \dot{\theta}_{\hat{\xi}}^T v \tag{8.32}$$

8.6 Numerical examples

and, combining (8.32) and (8.31) gives $j_{\hat{\varphi}} = j_{\hat{\xi}}^{-1} = j_{\check{\eta}}$ if $\check{\eta}_\xi = \theta_\xi^T v$. Thus $\check{u}$ has the form of a standardized score statistic

$$\check{u} = \text{sgn}(\hat{\xi} - \xi)\sqrt{j_{\check{\eta}}}|\check{\eta}_\xi - \check{\eta}_\xi|. \qquad (8.33)$$

The forms of $\check{u}$ and $\hat{u}$ in (8.15) are exactly the same except for the direction of the (locally) defined ancillary manifolds. The computation of $\hat{u}$ uses direction $\partial x/\partial \hat{\xi}$ computed holding either the likelihood or affine ancillary fixed; thus $\partial x/\partial \hat{\xi}$ specifies the direction of a globally defined ancillary manifold. By contrast, $\check{u}$ uses the sensitive direction tangent to a locally defined and sometimes globally defined ancillary. Note that $\check{u} = \hat{u}$ if $v \propto \partial x/\partial \hat{\xi}$ and the sensitive direction points in the same direction as a globally defined manifold.

(m, 1) curved exponential family with dependent sufficient statistics

Determining a sensitive direction becomes more difficult once dependence is allowed among the canonical sufficient statistics. An answer that deals with this may be inferred from appendix A of Fraser (2004) however. In theory, probability integral transforms for a sequence of conditional distributions are possible. With $\mathcal{X}_{i-1} = \{x_i, \ldots, x_{i-1}\}$, then $F_i(x|\mathcal{X}_{i-1}; \xi)$, the conditional CDF of X_i given $\mathcal{X}_{i-1}$, defines approximate ancillaries $a_i^s = F_i(x|\mathcal{X}_{i-1}; \hat{\xi})$ for $i \geq 2$. To implement this in practice, Skovgaard's approximation to F_i, as given in (5.78), may be used to define $\{a_i^s : i \geq 2\}$ and Lugannani and Rice in (5.64) approximates $F_1(x; \hat{\xi})$ to give a_1^s. This approach is examined numerically in the Bivariate Correlation example below.

8.6 Numerical examples

8.6.1 Gamma hyperbola

For this example, the approximate sample space derivative in (8.22) is exact in that it leads to the value $\hat{u}$ computed by fixing the exact ancillary. See Exercise 4.

When conditioning on likelihood ancillary (8.26), the value for $\hat{u}$ in (8.25) is also the exact value given previously.

The sensitive direction is computed as

$$v_1 = -x_1/\hat{\xi} \qquad v_2 = x_2/\hat{\xi}$$

and the directional derivative is

$$\varphi_\xi = \frac{\partial}{\partial t}\{-\xi(x_1 - x_1 t/\hat{\xi}) - \xi^{-1}1(x_2 + x_2 t/\hat{\xi})\}|_{t=0}$$
$$= (\xi x_1 - x_2/\xi)/\hat{\xi}.$$

Following the computations suggested in Exercise 4, then $\check{u} = \hat{u}$ in (8.25). This result should not be surprising for two reasons. First, the example is really a disguised location model for which the identity $\check{u} = \hat{u}$ always holds. Secondly, both of the "sensitive" direction ancillaries are given as

$$a_1^s = 1 - \exp(-\hat{\xi}x_1) = 1 - \exp\left(-\sqrt{x_1 x_2}\right) = 1 - \exp(-x_2/\hat{\xi}) = a_2^s$$

Table 8.2. *A comparison of approximation (8.1) using the sample space derivative expression (8.24) that conditions on two different affine ancillaries a. Values are for $\widehat{\Pr}(\hat{\Xi} > \hat{\xi}|a; \xi = 1)$ using the values of $\hat{\xi}$ listed in the columns.*

	$\hat{\xi}$-values						
Method	$1\tfrac{1}{10}$	$1\tfrac{1}{2}$	2	$2\tfrac{1}{2}$	3	6	15
(i) = (8.1)	.418	.187	.0619	.0196	$.0^2607$	$.0^5511$	$.0^{14}415$
(ii) = (8.34)	.470	.226	.0804	.0270	$.0^2877$	$.0^5870$	$.0^{14}821$

and are exactly ancillary. Holding them fixed specifies the exact globally defined ancillary manifold. Thus $v \propto \partial x / \partial \hat{\xi}$ and points tangent to the hyperbola on which $x_1 x_2$ is held fixed.

Suppose instead, the conditioning is on affine ancillaries. Then the value of $\hat{u}$ as well as its accuracy depend on the choice made for Ψ_ξ. In this example, Ψ_ξ is a vector that must be perpendicular to $\dot{\mu}_\xi = (-1/\xi^2, 1)$. Two simple choices are (i) $\Psi_\xi = (\xi, 1/\xi)$ and (ii) $\Psi_\xi = (1, 1/\xi^2)$ which differ by the factor ξ. Indeed all choices must be proportional since vectors orthogonal to $\dot{\mu}_\xi$ form a one-dimensional space. In case (i), the value for $\hat{u}$ agrees with the sample space derivative specified in (8.3) that fixes the exact ancillary; for case (ii),

$$\hat{u} = 2^{-1/2} e^{a/4} (\hat{\xi} - \xi) \{2/\hat{\xi} + e^{-a/2}(1/\xi - 1/\hat{\xi})\}. \tag{8.34}$$

Probabilities from the computation of (8.24) in cases (i) and (ii) are compared in table 8.2. An odd and somewhat troubling aspect of this development is that both cases have the same ancillary $a = \sqrt{2}(e^{a/2} - 1)$ but lead to different probabilities. In fact, when considering a (2, 1) family, there is always a unique affine ancillary defined by (7.40) in lemma 7.3.1 that is invariant to ξ-dependent scale changes in Ψ_ξ.

8.6.2 Bivariate correlation

This example has been continued from section 7.3.1 where it was used as an example for the p^* density. In that context, only the affine ancillaries were considered since the likelihood ancillary is undefined.

Conditioning on an affine ancillary

Suppose the data are $x_1 = 4\tfrac{3}{4}$, $x_2 = 4\tfrac{1}{2}$, and $n = 4$ as before. Then $\hat{\xi} = 0.9379$ and the affine ancillary $a = 0.2735$. Plots of p^* and $p^\dagger$ for $\xi = 1/4$ and $1/2$ have been given in figure 7.6.

In this example, the sample space derivatives of $\hat{u}$ in (8.3) can be computed directly as in (8.23) by differentiating the sufficient statistic expressions in (7.49) to get

$$\frac{\partial x_1}{\partial \hat{\xi}} = a\hat{\xi}\sqrt{\frac{n}{1+\hat{\xi}^2}} \qquad \frac{\partial x_2}{\partial \hat{\xi}} = n + \frac{2a\sqrt{n}}{(1+\hat{\xi}^2)^{3/2}}. \tag{8.35}$$

Table 8.3. (Normal correlation). A comparison of approximations for $\Pr(\hat{\Xi} > \hat{\xi}|a;\xi)$ versus "Exact" for $\xi = 1/4$ and $1/2$ where a is an affine ancillary.

Method $\xi = 1/4$	$\hat{\xi}$-values					
	.2	.4	.6	.8	.9	.95
Exact	.542	.370	.198	.0598	.0164	$.0^2430$
$\hat{u}$ in (8.1)	.587	.412	.228	.0709	.0198	$.0^2523$
$\check{u}$ in (8.22)	.590	.417	.231	.0715	.0199	$.0^2525$
$\int_{\hat{\xi}}^1 p^*(t)dt$	.574	.397	.216	.0659	.0182	$.0^2477$
$\int_{\hat{\xi}}^1 p^\dagger(t)dt$	.648	.448	.244	.0744	.0205	$.0^2539$
$1 - \Phi(\hat{w})$	.548	.352	.171	.0435	.0103	$.0^2237$
$\xi = 1/2$	.55	.7	.8	.9	.95	.98
Exact	.485	.287	.155	.0414	.0131	$.0^2223$
$\hat{u}$ in (8.1)	.516	.310	.170	.0524	.0146	$.0^2250$
$\check{u}$ in (8.22)	.519	.312	.170	.0525	.0146	$.0^2250$
$\int_{\hat{\xi}}^1 p^*(t)dt$	.504	.301	.163	.0501	.0139	$.0^2237$
$\int_{\hat{\xi}}^1 p^\dagger(t)dt$	.547	.326	.177	.0544	.0151	$.0^2257$
$1 - \Phi(\hat{w})$	.435	.235	.115	.0301	$.0^2725$	$.0^2106$

These derivatives determine expression (8.1) and lead to the numerical entries in the row (8.1) of table 8.3. They may be compared with the "Exact" value determined through numerical integration of the true conditional density of $\hat{\xi}|a$.

Row (8.22) uses Skovgaard's $\check{u}$-value despite the fact that we cannot condition on a likelihood ancillary. Most notably, this approach produces values quite close to those determined from exact sample space derivatives determined by fixing affine ancillaries. Indeed very little seems to be lost by using the much simpler expression (8.22).

One feature peculiar to this example is that Skovgaard's $\check{u}$-value may be obtained as an exact sample space derivative from (8.24). The value Ψ_ξ, as in (7.40) of lemma 7.3.1, must be chosen perpendicular to $\dot{\mu}_\xi$ where, in this instance, $\mu_\xi = n(1, \xi)^T$. Choosing $\Psi_\xi = (1, 0)^T$ results in an expression for $\hat{u}$ that agrees with $\check{u}$ in (8.22). This may help to explain why row "$\check{u}$ in (8.22)" and "$\hat{u}$ in (8.1)" are so similar.

The integrated p^* and $p^\dagger$ computations and the normal approximation have also been included for comparison. For these examples the saddlepoint computations do not display the level of accuracy we have become accustomed to. Indeed the normal approximation is most accurate at $\hat{\xi} = 0.2$ and 0.4 when $\xi = 1/4$. However, the saddlepoint approximations dominate for $\hat{\xi} \geq 0.6$ and dominate uniformly in $\hat{\xi}$ when $\xi = 1/2$. Interestingly (8.1), the consequence of using the Temme approximation in integrating $p^\dagger$, always yields a value that appears to be a compromise between integrating $p^\dagger$ and p^*. Curiously, $p^\dagger$ has not been normalized but the Temme argument seems to achieve normalization "on its own."

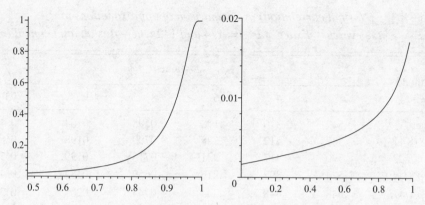

Figure 8.2. (Left): Right tail probability plot of the Fraser, Reid, and Wu approach for computing conditionally attained significance versus ξ. Similar plots, as described in (i) and (ii) just above, were graphically indistinguishable. (Right): Percentage relative error comparisons versus ξ of the Fraser *et al.* tail significance (numerator) versus the Skovgaard significance (denominator).

Conditioning on local ancillaries

The Fraser *et al.* (1999a) approximation uses (8.28) but cannot be computed for use in table 8.3 because it is only locally defined at the observed data. What can be computed for it is the p-value and associated attained significance level as a function of ξ. This attained significance function is interpreted conditionally on a locally defined ancillary that points in the sensitive direction.

Using independent components Following the approach suggested in Fraser (2004), transform the bivariate data $Z_i = (Z_{i1}, Z_{i2})^T$ to $W_{i1} = Z_{i1} + Z_{i2}$ and $W_{i2} = Z_{i1} - Z_{i2}$ which assures that all responses are independent. The likelihood does not change and is still a (2, 1) exponential family but the canonical sufficient statistics change and are now $Y_i = \sum_{i=1}^{4} W_{i1}^2$ for $i = 1, 2$ attaining values $37/2$ and $1/2$ respectively. They are independent with $\frac{1}{2}Y_1/(1+\xi)$ and $\frac{1}{2}Y_2/(1-\xi)$ both χ_4^2. Further details are found in Exercise 9.

Figure 8.2(left) plots the right tail p-value probability of Fraser *et al.* (1999a) versus ξ which is assumed to be conditional on a local ancillary pointing in the sensitive direction. Two other approaches to attained significance curves lead to graphically indistinguishable plots. With $\hat{\xi}$ and a as observed values, they include (i) the computation of $\Pr\{\hat{\Xi} > \hat{\xi}|a; \xi\}$ in (8.1) versus ξ using the exact sample space derivatives holding affine ancillary a fixed; and (ii) the same approximation but using Skovgaard's approximate sample space derivatives instead. The degree of similarity in values among these three approaches is uncanny but they are not the same expressions analytically. The percentage relative error between the Fraser, Reid, and Wu approach (in the numerator) and item (i) (in the denominator) increases monotonically from $0.0^4 103\%$ to $0.0^3 122\%$ as ξ increases over the range (0, 0.999). Figure 8.2(right) shows percentage relative error of the Fraser, Reid, and Wu tail significance (numerator) versus the Skovgaard method (denominator) with errors ranging between $0.0^2 2\%$ to 0.02%. A similar plot between approximation (8.1) (numerator) and the Skovgaard method (denominator) produced a graph indistinguishable from figure 8.2(right).

To underscore the remarkable equivalence in approaches, 90% confidence intervals were computed for each of the three methods. For Fraser *et al.* and the (i) and (ii) approaches, the respective intervals are $(0.6658, 0.9737)$, $(0.6658, 0.9737)$, and $(0.6657, 0.9737)$. Note that the first two agree to four significant digits.

The sensitive direction in (y_1, y_2)-space is $v = (9.546, -8.050)$ or $-40.14°$. Compare this to the tangent to the affine ancillary manifold. The tangent in (x_1, x_2)-space is given by evaluating the sample space derivatives in (8.35) since they were computed holding the affine ancillary fixed. This needs to be converted into a direction in (y_1, y_2)-space using the chain rule so that

$$\begin{pmatrix} \partial y_1/\partial \hat{\xi} \\ \partial y_2/\partial \hat{\xi} \end{pmatrix} = \begin{pmatrix} \partial y_1/\partial x_1 & \partial y_1/\partial x_2 \\ \partial y_2/\partial x_1 & \partial y_2/\partial x_2 \end{pmatrix} \begin{pmatrix} \partial x_1/\partial \hat{\xi} \\ \partial x_2/\partial \hat{\xi} \end{pmatrix}$$

$$= \begin{pmatrix} 2 & 2 \\ 2 & -2 \end{pmatrix} \begin{pmatrix} .3742 \\ 4.425 \end{pmatrix} = \begin{pmatrix} 9.598 \\ -8.101 \end{pmatrix}.$$

This direction of this vector is $-40.17°$, and remarkably close to the sensitive direction.

Using dependent components Suppose the sensitive direction is determined directly in terms of the dependent canonical sufficient statistics x_1 and x_2. Let $a_1^s = \hat{F}_1(x_1; \hat{\xi})$ with $\hat{F}_1$ as the Lugannani and Rice approximation (5.64) and $a_2^s = \hat{F}_2(x_2|x_1; \hat{\xi})$ with $\hat{F}_2$ as the Skovgaard approximation (5.78). These approximations are determined from the joint CGF

$$K(s, t) = -\tfrac{1}{2} n \ln\{(1 - s - \xi t)^2 - (\xi s + t)^2\}$$

that is derived from the characteristic function of the Wishart given in section 3.2.2 of Muirhead (1982). The sensitivities are readily computed numerically as

$$v_2 = -\frac{\hat{F}_2(x_2|x_1; \hat{\xi} + \varepsilon) - \hat{F}_2(x_2|x_1; \hat{\xi} - \varepsilon)}{\hat{F}_2(x_2 + \varepsilon|x_1; \hat{\xi}) - \hat{F}_2(x_2 - \varepsilon|x_1; \hat{\xi})}$$

for small ε with an analogous expression for v_1. The resulting sensitive directions in (x_1, x_2)- and (y_1, y_2)-space are

$$\begin{pmatrix} .3479 \\ 4.103 \end{pmatrix} \quad \text{and} \quad \begin{pmatrix} 8.901 \\ -7.509 \end{pmatrix}.$$

The angle for the latter vector is $-40.15°$ and differs only by a mere $0.01°$ from the angle computed using the independent decomposition approach. Since significance probabilities depend only on the angle expressed by v and not by its magnitude, the p-value and other significance values are essentially the same as occur with the independent component decomposition.

While the example has not been the greatest challenge for the method since the transformation from x to y is linear, the replacement of exact distributions with saddlepoint CDF approximations has also occurred. Little accuracy has been lost as a result of this replacement.

8.6.3 Gamma exponential

Both affine and likelihood ancillaries are considered in this example which has been continued from section 7.3.1.

Table 8.4. *(Gamma exponential). Approximations for* $\Pr(\hat{\Xi} > \hat{\xi}|a;\xi)$ *vs. "Exact" for* $\xi = 4$ *and* 2 *where a is an affine ancillary.*

Method $\xi = 4$	$\hat{\xi}$-values				
	4.5	5	5	7	8
Exact	.433	.264	.0865	.0267	.0²824
$\hat{u}$ in (8.1)	.400	.239	.0759	.0230	.0²702
$\check{u}$ in (8.22)	.386	.228	.0710	.0213	.0²645
$1 - \Phi(\hat{w})$	.287	.151	.0381	.0²954	.0²249
$\xi = 2$	2.5	3	4	5	6
Exact	.475	.280	.0844	.0245	.0²722
$\hat{u}$ in (8.1)	.364	.199	.0548	.0151	.0²432
$\check{u}$ in (8.22)	.335	.180	.0482	.0131	.0²375
$\int_{\hat{\xi}}^{1} p^*(t)dt$	.364	.199	.0542	.0148	.0²418
$\int_{\hat{\xi}}^{1} p^{\dagger}(t)dt$	.391	.214	.0583	.0159	.0²449
$1 - \Phi(\hat{w})$	.260	.124	.0267	.0²603	.0²147

Conditioning on an affine ancillary

Direct computation of sample space derivative $\partial x/\partial \hat{\xi}$ follows by using implicit differentiation on the expressions for x_1 and x_2 in (7.52) and (7.50). This leads to expression (8.1) in table 8.4. Comparisons of approximation (8.1) with the "Exact" computation suggest reasonably accurate with $\xi = 4$ but deterioration in accuracy at $\xi = 2$. The integrated p^* and $p^{\dagger}$ densities however make it clear that this deterioration is not due to the Temme method, but due to the underlying inaccuracy of p^* and $p^{\dagger}$ for the true conditionality resolution. Again the sample space derivative approximation using $\check{u}$ in (8.22) shows remarkable accuracy in tracking the results for $\hat{u}$ given in the line above.

Conditioning on a likelihood ancillary

In this example, the p^* and $p^{\dagger}$ formulas were shown to achieve considerably greater accuracy when conditioning was on the likelihood ancillary. The plot and table below show that this accuracy is maintained in the CDF approximations.

For $\xi = 2$, a plot of the percentage relative error of the right tail probability from (8.1) is shown in figure 8.3 as the solid line. The dashed line shows the same relative error using Skovgaard's approximate sample space derivatives in (8.22). The range for this plot has been chosen to assure that the "Exact" computations achieve sufficient accuracy when computed using the trapezoidal rule over the grid $\hat{\xi} \in \{0.02(0.04)9.98, 10\frac{1}{16}(\frac{1}{16})12, 12\frac{1}{8}(\frac{1}{8})16\}$.

Further numerical comparisons are made for $\xi = 1$ and $1/2$ in table 8.5 above. Of the four values of ξ considered, these two resulted in the least accuracy for the p^* formula however the table suggests that this inaccuracy is quite modest. The figure and table suggest that the computations with approximate sample space derivatives are also very accurate.

8.6 Numerical examples

Table 8.5. *(Gamma exponential). Approximations for* $\Pr(\hat{\Xi} > \hat{\xi}|a_\chi^0;\xi)$ *versus "Exact" for* $\xi = 1$ *and* $1/2$ *where* a_χ^0 *is a likelihood ancillary.*

Method	$\hat{\xi}$-values				
$\xi = 1$	2	3	4	5	6
Exact (Trapezoidal sum)	.197	.0561	.0162	.0²482	.0²148
$\hat{u}$ in (8.1)	.182	.0514	.0148	.0²442	.0²136
$\check{u}$ in (8.22)	.161	.0456	.0132	.0²399	.0²124
$1 - \Phi(\hat{w})$	.112	.0250	.0²599	.0²154	.0³422
$\xi = 1/2$	0.8	1.8	2.2	3	3.8
Exact (Trapezoidal sum)	.425	.0980	.0563	.0194	.0²697
$\hat{u}$ in (8.1)	.371	.0845	.0484	.0166	.0²595
$\check{u}$ in (8.22)	.325	.0713	.0411	.0143	.0²521
$1 - \Phi(\hat{w})$	.265	.0453	.0237	.0²696	.0²220

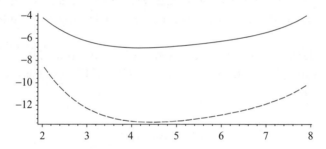

Figure 8.3. The percentage relative error of $\widehat{\Pr}(\hat{\Xi} > \hat{\xi}|a_\chi^0;\xi = 2)$ vs. $\hat{\xi} \in (2, 8)$ using exact (*solid*) sample space derivatives for fixed likelihood ancillary a_χ^0 and approximate sample space derivatives (*dashed*).

Sensitive direction and r^* connections

For this example, the ancillary direction is computed as $v^T = -(x_1/\hat{\xi}, x_2)$ which leads to the data dependent parameterization

$$\varphi_\xi = \xi x_1/\hat{\xi} + e^\xi x_2.$$

The standardized maximum likelihood departure value is

$$\check{u} = \text{sgn}(\hat{\xi} - \xi)|x_1(1 - \xi/\hat{\xi}) + x_2(e^{\hat{\xi}} - e^\xi)|\sqrt{j_{\hat{\xi}}}|x_1/\hat{\xi} + e^{\hat{\xi}} x_2|^{-1} \quad (8.36)$$

where

$$j_{\hat{\xi}} = 1/\hat{\xi}^2 + e^{\hat{\xi}} x_2.$$

Quite remarkably, it can be shown for any data (x_1, x_2), that $\check{u}$ is analytically the same as $\hat{u}$ in (8.3) computed by holding likelihood ancillary a_χ^0 fixed. This seems to suggest that the sensitive direction points tangent to the likelihood ancillary manifold or $\{(x_1, x_2) : a_\chi(x_1, x_2) = -1.546\}$. This is indeed the case. Implicit differentiation of (7.50) to determine

278   *Probabilities with r*-type approximations*

Table 8.6. *p-values for the various methods listed in the rows. "Exact" refers to trapezoidal summation for* $\Pr(\hat{\Xi} < \hat{\xi}|a_\chi^0; \xi)$ *and (8.36) accounts for the sensitive direction as well as* $\hat{u}$.

Method	$\xi = 1/2$	3/4	1	3/2	2
Exact (Trapezoidal sum)	.189	.0689	.0194	$.0^3 489$	$.0^5 120$
$(8.36) = \hat{u}$ in (8.1)	.238	.0864	.0239	$.0^3 583$	$.0^5 140$
$\check{u}$ in (8.22)	.259	.0990	.0289	$.0^3 796$	$.0^5 219$
Normal	.325	.130	.0392	$.0^2 112$	$.0^5 315$

$\partial x_2/\partial x_1$ holding a_χ^0 fixed requires the determination of $\partial \hat{\xi}/\partial x_1$ through (7.58). After long computations,

$$\partial x_2/\partial x_1 = \hat{\xi} x_2/x_1 = v_2/v_1, \qquad (8.37)$$

the direction of v. At the data this slope is 0.7035 with angle 35.13°.

Is this example merely a coincidence or are there greater generalities to these agreements? Some *p*-values using the various approaches are given in table 8.6 for the ξ-values listed in the columns.

The exact confidence interval for ξ determined by inverting $\Pr(\hat{\xi} < \hat{\xi}^0|a_\chi^0; \xi)$ gives $(0.0276, 0.664)$ while $\hat{u} = \check{u}$ in (8.36) gives $(0.0446, 0.717)$ and the use of $\check{u}$ in (8.22) gives $(0.0478, 0.748)$.

8.7 Properties

8.7.1 Equivariance

Both of the CDF approximations in (8.1) and (8.4) are invariant to monotonic increasing transformations of $\hat{\psi}$. They are also equivariant to monotonic decreasing transformations in that the inequality in the CDF expression before transformation gets reversed properly to determine the appropriate probability in the opposite tail after transformation. These properties all derive from the behavior of the $\hat{w}$ and $\hat{u}$ expressions is in (8.2) and (8.3).

The same properties are preserved when using Skovgaard's approximation in (8.20) in place of the exact sample space derivatives found in (8.1) and (8.4). Likewise, the attained significance function of Fraser, Reid and Wu using (8.28) is equivariant to monotonic transformation. See Exercise 12 for more details.

8.7.2 Symmetry

The gamma hyperbola example of section 8.2.2, when considered in its location parameter ω form, is an example in which the density of $\hat{\omega}$ given a is symmetric about ω. The values of $\hat{w}$ and $\hat{u}$ and $\check{u}$ all exhibited the characteristics needed to assure that the CDF approximations based on (8.1), (8.4), and (8.20) reflect such symmetry.

Preservation of symmetry may be shown to hold for any symmetric pivotal model such as with the gamma hyperbola. Suppose that x is sufficient and that there is a conditionality resolution $x \to (\hat{\omega}, a)$ such that the joint density of $(\hat{\omega} - \omega, a)$ does not depend on parameter ω and the conditional density of $\hat{\omega}$ given a is symmetric about ω. Then all the approximations based on (8.1), (8.4), (8.20), and (8.28) must preserve the symmetry of the conditionality resolution. The framework in which this may be proven is suggested in Exercise 13.

If the symmetry is not centered about the parameter of interest, then the extent to which symmetry is preserved appears difficult to determine in general.

8.7.3 Asymptotics for (8.1)

Consider the asymptotic setting in which $X = \sum_{i=1}^{n} Z_i$ where $\{Z_i\}$ are i.i.d. with $(m, 1)$ curved exponential family whose log-likelihood is given in (8.13). Let a^0 be a $(m-1)$-dimensional likelihood ancillary which is generally first-order ancillary or parameter-free to $O(n^{-1/2})$ in normal deviation regions.

In order to judge the relative error of the approximation (8.1), a more refined conditionality resolution has been introduced by Barndorff-Nielsen and Wood (1998, §2.4). Whereas a^0 is only a first-order ancillary, they have provided a constructive definition for a new $(m-1)$-dimensional ancillary a^* from a^0 that is a third-order ancillary. The means by which successive components of a^* are constructed from a^0 follows the manner by which r^* has been constructed from $\hat{w}$. The formulation of a^* is quite important from a theoretical point of view since fixing a^* rather than a^0 suggests the distribution of $\hat{\Xi}|a^*; \xi$ as a more highly refined conditionality resolution for which the goal of (8.1) is its approximation. Unfortunately a^* is virtually impossible to compute so its use seems limited to defining the theoretical quantity of interest.

Suppose that (8.1) is used and sample space derivatives are computed by fixing the likelihood ancillary a^0. Then (8.1) achieves saddlepoint accuracy for $\hat{\Xi}|a^*; \xi$ or

$$\Pr(\hat{\Xi} \leq \hat{\xi}|a^*; \xi) = \widehat{\Pr}(\hat{\Xi} \leq \hat{\xi}|a^0; \xi)\{1 + O(n^{-1}|\hat{\xi} - \xi|) + O(n^{-3/2})\} \qquad (8.38)$$

uniformly over $\hat{\xi} \in (\xi_0, \delta\sqrt{n})$ for fixed ξ_0 and $\delta > 0$ sufficiently small. The important point here, as emphasized by Barndorff-Nielsen and Wood (1998), is that the approximation is readily computed since it is based on fixed a^0, however it may be interpreted as being conditional on a^* since it reproduces this more refined conditionality resolution with the saddlepoint accuracy shown in (8.38).

If (8.1) is used and sample space derivatives are computed by fixing an affine ancillary a, then typically

$$\Pr(\hat{\Xi} \leq \hat{\xi}|a^*; \xi) = \widehat{\Pr}(\hat{\Xi} \leq \hat{\xi}|a; \xi)\{1 + O(n^{-1}\|a\|^2 |\hat{\xi} - \xi|) + O(n^{-1}|\hat{\xi} - \xi|)\}. \qquad (8.39)$$

See section 6 of Barndorff-Nielsen and Wood (1998) for precise details.

8.8 Appendix

8.8.1 Derivation of (9.24) with ξ as $p \times 1$ and (8.20) with $p = 1$

Define the $p \times 1$ vector

$$Q(\xi_1, \xi_0) = E\left[l(\xi_1)l_{\xi;}(\xi_0); \xi_0\right] \qquad (8.40)$$

so that $q(\xi_1, \xi_2; \xi_0) = Q(\xi_1, \xi_0) - Q(\xi_2, \xi_0)$.

Expand the left side of (9.24), the vector version of (8.20), so that

$$l_{;\hat{\xi}}(\hat{\xi};\hat{\xi},a) - l_{;\hat{\xi}}(\xi;\hat{\xi},a) \simeq (\hat{\xi}-\xi)^T l_{\xi;\hat{\xi}}(\hat{\xi};\hat{\xi},a) + \tfrac{1}{2}(\hat{\xi}-\xi)^T l_{\xi\xi;\hat{\xi}}(\hat{\xi};\hat{\xi},a)(\hat{\xi}-\xi) \quad (8.41)$$

is $1 \times p$. Here the meaning given to the last term is the $1 \times p$ computation

$$\tfrac{1}{2}\sum_{i=1}^{p}\sum_{j=1}^{p}(\hat{\xi}_i - \xi_i)(\hat{\xi}_j - \xi_j) l_{\xi_i\xi_j;\hat{\xi}}(\hat{\xi};\hat{\xi},a) = \tfrac{1}{2}\sum_{i=1}^{p}\sum_{j=1}^{p}(\hat{\xi}_i - \xi_i)(\hat{\xi}_j - \xi_j)\left.\frac{\partial^2 l_{;\hat{\xi}}(\xi;\hat{\xi},a)}{\partial \xi_i \partial \xi_j}\right|_{\xi=\hat{\xi}}.$$

The result follows by comparing the terms in (8.41) with the expansion

$$q(\hat{\xi},\xi;\hat{\xi})^T = \{Q(\hat{\xi},\hat{\xi}) - Q(\xi,\hat{\xi})\}^T \quad (8.42)$$
$$\simeq (\hat{\xi}-\xi)^T Q_1(\hat{\xi},\hat{\xi})^T + \tfrac{1}{2}(\hat{\xi}-\xi)^T Q_{11}(\hat{\xi},\hat{\xi})^T(\hat{\xi}-\xi),$$

where Q_1 and Q_{11} are the gradient and Hessian with respect to the first argument of Q. Note from the definition of Q in (8.40) that

$$Q_1(\xi_0, \xi_0)^T = E\left[l_{\xi;}(\xi_0) l_{\xi;}^T(\xi_0); \xi_0\right] = i_{\xi_0}$$

so that $Q_1(\hat{\xi},\hat{\xi}) = i_{\hat{\xi}}$. Likewise, for the three-dimensional array, it is shown below that

$$Q_{11}(\xi_0, \xi_0)^T = \left.\frac{\partial^2}{\partial \xi \partial \xi^T} Q(\xi, \xi_0)^T\right|_{\xi=\xi_0} \simeq l_{\xi\xi;\hat{\xi}}(\xi_0; \xi_0, a). \quad (8.43)$$

Thus the left side of (8.42) is approximately

$$q(\hat{\xi},\xi;\hat{\xi})^T \simeq (\hat{\xi}-\xi)^T i_{\hat{\xi}} + \tfrac{1}{2}(\hat{\xi}-\xi)^T l_{\xi\xi;\hat{\xi}}(\hat{\xi};\hat{\xi},a)(\hat{\xi}-\xi)$$
$$\simeq \left\{(\hat{\xi}-\xi)^T j_{\hat{\xi}} + \tfrac{1}{2}(\hat{\xi}-\xi)^T l_{\xi\xi;\hat{\xi}}(\hat{\xi};\hat{\xi},a)(\hat{\xi}-\xi)\right\} j_{\hat{\xi}}^{-1} i_{\hat{\xi}}$$
$$\simeq \left\{l_{;\hat{\xi}}(\hat{\xi};\hat{\xi},a) - l_{;\hat{\xi}}(\xi;\hat{\xi},a)\right\} j_{\hat{\xi}}^{-1} i_{\hat{\xi}}$$

from (8.41) which leads to (9.24).

To show (8.43), write

$$Q_{11}(\xi_0, \xi_0)^T = E\{l_{\xi\xi;}(\xi_0) l_{\xi;}^T(\xi_0); \xi_0\} \quad (8.44)$$

and expand its two terms as

$$l_{\xi\xi;}(\xi_0) = l_{\xi\xi;}(\xi_0; \hat{\xi}, a) \simeq l_{\xi\xi;}(\xi_0; \xi_0, a) + l_{\xi\xi;\hat{\xi}}(\xi_0; \xi_0, a)(\hat{\xi} - \xi_0) \quad (8.45)$$

and

$$l_{\xi;}(\xi_0) = l_{\xi;}(\xi_0; \hat{\xi}, a) \simeq l_{\xi;}(\xi_0; \xi_0, a) + l_{\xi;\hat{\xi}}(\xi_0; \xi_0, a)(\hat{\xi} - \xi_0)$$
$$\simeq l_{\xi;\hat{\xi}}(\xi_0; \xi_0, a)(\hat{\xi} - \xi_0) \simeq j_{\xi_0}(\hat{\xi} - \xi_0),$$

when the ancillary score term $l_{\xi;}(\xi_0; \xi_0, a)$ is ignored. The ancillary term $l_{\xi\xi;}(\xi_0; \xi_0, a)$ in (8.45) can also be ignored since, upon substitution of (8.45) into (8.44), it is multiplied by the score $l_{\xi;}^T(\xi_0)$ inside the expectation. Substituting both expansions into (8.44) and computing the expectation over $\hat{\Xi}$ with a fixed leads to

$$Q_{11}(\xi_0, \xi_0)^T \simeq E\left\{l_{\xi\xi;\hat{\xi}}(\xi_0; \xi_0, a)(\hat{\xi} - \xi_0)(\hat{\xi} - \xi_0)^T j_{\xi_0} | a; \xi_0\right\}$$
$$\simeq l_{\xi\xi;\hat{\xi}}(\xi_0; \xi_0, a) j_{\xi_0}^{-1} j_{\xi_0} = l_{\xi\xi;\hat{\xi}}(\xi_0; \xi_0, a).$$

8.8 Appendix

8.8.2 Derivation of (9.25)

The first step in showing

$$l_{\xi;\hat{\xi}}(\xi;\xi_0,a) \simeq S(\xi;\xi_0)^T i_{\xi_0}^{-1} j_{\xi_0} \qquad (8.46)$$

is to show that

$$S(\xi;\xi_0) - S(\xi_0;\xi_0) \simeq \{l_{\xi;\hat{\xi}}(\xi;\xi_0,a) - l_{\xi;\hat{\xi}}(\xi_0;\xi_0,a)\}^T. \qquad (8.47)$$

Use (9.26) to write the left side of (8.47) as expectations

$$E\left\{l_{\xi;}(\xi_0)l_{\xi;}(\xi)^T;\xi_0\right\} - E\left\{l_{\xi;}(\xi_0)l_{\xi;}(\xi_0)^T;\xi_0\right\}. \qquad (8.48)$$

Each of the terms $l_{\xi;}(\xi)^T = l_{\xi;}(\xi;\hat{\xi},a)^T$ and $l_{\xi;}(\xi_0)^T = l_{\xi;}(\xi_0;\hat{\xi},a)^T$ may be Taylor expanded in sample space derivatives about $\hat{\xi}_0$ to give the approximation

$$E\left[l_{\xi;}(\xi_0;\hat{\xi},a)\left\{l_{\xi;}(\xi;\xi_0,a) + l_{\xi;\hat{\xi}}(\xi;\xi_0,a)(\hat{\xi}-\xi_0)\right\}^T;\xi_0\right] \qquad (8.49)$$

$$- E\left[l_{\xi;}(\xi_0;\hat{\xi},a)\left\{l_{\xi;}(\xi_0;\xi_0,a) + l_{\xi;\hat{\xi}}(\xi_0;\xi_0,a)(\hat{\xi}-\xi_0)\right\}^T;\xi_0\right] \qquad (8.50)$$

Three reductions apply to (8.49) and (8.50). First, the score function $l_{\xi;}(\xi_0;\xi_0,a) = 0$ in (8.50). Secondly, sample space Taylor expansion of the score $0 = l_{\xi;}(\xi;\xi,a)$ leads to substituting

$$l_{\xi;}(\xi;\xi_0,a) \simeq -l_{\xi;\hat{\xi}}(\xi;\xi_0,a)(\xi-\xi_0)$$

into (8.49). Finally, Taylor expansion

$$l_{\xi;}(\xi_0;\hat{\xi},a) \simeq l_{\xi\xi;}(\hat{\xi};\hat{\xi},a)(\xi_0 - \hat{\xi}) = j_{\hat{\xi}}(\hat{\xi} - \xi_0)$$

allows for substitution for the leading terms in both (8.49) and (8.50). Upon substitution and rearrangement, (8.48) is approximately

$$E\left\{j_{\hat{\xi}}(\hat{\xi}-\xi_0)(\hat{\xi}-\xi_0)^T[l_{\xi;\hat{\xi}}(\xi;\xi_0,a) - l_{\xi;\hat{\xi}}(\xi_0;\xi_0,a)]^T;\xi_0\right\} \qquad (8.51)$$

$$- E\left\{j_{\hat{\xi}}(\hat{\xi}-\xi_0)(\xi-\xi_0)^T l_{\xi;\hat{\xi}}(\xi;\xi_0,a)^T;\xi_0\right\}. \qquad (8.52)$$

The first term in (8.51) is reduced by computing the expectation conditional upon a; thence its approximate value is

$$[l_{\xi;\hat{\xi}}(\xi;\xi_0,a) - l_{\xi;\hat{\xi}}(\xi_0;\xi_0,a)]^T.$$

In applications, the second term in (8.52) has the difference $\xi - \xi_0$ set to $\hat{\xi} - \hat{\xi}_\psi$ which is generally small; thus this term may be ignored altogether. This completes the motivation for (8.47).

Derivation of (8.46) begins with (8.47) and uses the multivariate version of (8.18) to give

$$l_{\xi;\hat{\xi}}(\xi;\xi_0,a) \simeq S(\xi;\xi_0)^T + l_{\xi;\hat{\xi}}(\xi_0;\xi_0,a) - S(\xi_0;\xi_0)^T$$
$$= S(\xi;\xi_0)^T + j_{\xi_0} - i_{\xi_0} = S(\xi;\xi_0)^T + i_{\xi_0}(i_{\xi_0}^{-1} j_{\xi_0} - I_p)$$
$$\simeq S(\xi;\xi_0)^T + S(\xi;\xi_0)^T(i_{\xi_0}^{-1} j_{\xi_0} - I_p) = S(\xi;\xi_0)^T i_{\xi_0}^{-1} j_{\xi_0}.$$

Replacement of i_{ξ_0} with $S(\xi;\xi_0)^T$ is justified on the grounds that $\xi - \xi_0$ is small in applications of the approximation.

8.9 Exercises

1. Show that Taylor expansion of functions Φ and ln in the r^* approximation of (8.4) leads to the saddlepoint approximation in (8.1).
2. Consider Planck's radiation density of section 7.1.1 and the CDF of $\hat{\alpha}$ when $\alpha = 3$.
   (a) Approximate this CDF using the r^* approximation in (8.4). Compare the accuracy with the Lugannani and Rice approximation used in Exercise 3 of chapter 7.
   (b) Consider the test of $\alpha = 3$ versus $\alpha > 3$ at a 5% level using the r^* approximation. Compute its power curve and compare this with the curve using the Lugannani and Rice approximation from Exercise 3 of chapter 7. How many saddlepoint equations need to be solved in order to compute the power curve based on r^*?
3. Consider the distribution with density (7.75) in Exercise 2 of chapter 7.
   (a) Using parameter values $\omega = 1.1$ and 2, compare the true CDF of $\hat{\omega}$ with the approximate CDFs determined by (8.1) and (8.4) using a small table of $\hat{\omega}$-values.
   (b) Plot their relative errors.
4. For the gamma hyperbola example, show that the following results hold.
   (a) The approximate sample space derivative in (8.20) is exact in that it leads to the value $\hat{u}$ computed by fixing the exact ancillary.
   (b) The parameterization from the sensitive direction leads to

   $$\varphi_{\hat{\xi}} - \varphi_{\xi} = x_1(\hat{\xi}/\xi - \xi/\hat{\xi}).$$

   When combined with

   $$j_{\hat{\varphi}} = j_{\hat{\xi}}(\partial\hat{\varphi}/\partial\hat{\xi})^{-2} = \sqrt{x_1 x_2}/(2x_1^2)$$

   then $\check{u} = \hat{u}$.
   (c) When conditioning on an affine ancillary with $\Psi_\xi = (\xi, 1/\xi)$, show that (8.24) leads to same exact result as in part (a).
   (d) When conditioning on an affine ancillary with $\Psi_\xi = (1, 1/\xi^2)$, show that (8.24) leads to the $\hat{u}$ value (8.34).
5. In a $(2, 1)$ curved exponential family, derive expression (8.25) by computing the sample space derivatives holding the likelihood ancillary in (8.26) fixed.
   (a) Compute derivatives of (8.26) with respect to $\hat{\xi}$ and a to get two equations.
   (b) Differentiate the MLE equation

   $$(x - \mu_{\hat{\xi}})^T \dot{\theta}_{\hat{\xi}} = 0$$

   to get two more equations which, when combine with part (a), give

   $$\begin{pmatrix} \dot{\theta}_{\hat{\xi}}^T \\ (\tilde{\theta}^T - \theta_{\hat{\xi}})^T \end{pmatrix} \begin{pmatrix} \dfrac{\partial x}{\partial \hat{\xi}} & \dfrac{\partial x}{\partial a} \end{pmatrix} = \begin{pmatrix} j_{\hat{\xi}} & 0 \\ 0 & a \end{pmatrix}. \qquad (8.53)$$

   (c) Solve (8.53) for $\partial x/\partial \hat{\xi}$ to show that $\hat{u}$ is as given in (8.25).
6. Consider the hyperbolic secant example of section 7.2.1 with location parameter μ and data $-2, -1$.
   (a) Compute approximations (8.1) and (8.4) for $\hat{\mu}|a$, the MLE for the location parameter given an ancillary.
   (b) Compare their accuracy with the exact CDF determined by numerically integrating the p^* density with a held fixed.
7. Exercise 11 of chapter 7 considers the normal circle example that may be considered both as a $(2, 1)$ curved exponential family and also as a location family.

(a) Determine (8.1) by first considering the model as a location model. Plot the resulting relative error of both (8.1) and (8.4) as compares to the exact conditional CDF determined by numerically integrating the p^* density.

(b) Determine the approximation (8.22) using (8.20) for the sample space derivatives and compare with the answers in part (a).

(c) Compute the exact sample space derivative using (8.25) and again compare results.

(d) Compute several CDF approximations based on (8.24) that use exact sample space derivatives but which condition on an affine ancillary. Do these results agree with any of the approximations in parts (a)–(c)?

8. Suppose that data $X_1, \ldots, X_n$ are derived as i.i.d. from density $g(x/\sigma)/\sigma$ for $\sigma > 0$.

(a) In this setting, derive a general expression for computing the conditional CDF of $\hat{\sigma}$ using (8.1).

(b) What do the expressions lead to when g is an Exponential (1)?

(c) Suppose that g is Normal (1, 1) and $g(x/\sigma)/\sigma$ is therefore a normal distribution with coefficient of variation that has value 1. What do the general expressions lead to in this case?

9. Derive the Fraser et al. (1999a) expression for $\tilde{u}$ in the bivariate correlation example.

(a) Compute the sensitive direction as
$$v = \{y_1/(1+\hat{\xi}), -y_2/(1-\hat{\xi})\}.$$

Hence, show that the directional derivative of log-likelihood leads to the data dependent parametrization
$$\varphi_\xi = -\frac{y_1}{4(1+\xi)(1+\hat{\xi})} + \frac{y_2}{4(1-\xi)(1-\hat{\xi})}.$$

(b) Complete the computation of $\tilde{u}$ by showing that
$$\varphi_{\hat{\xi}} = 4\hat{\xi}/(1-\hat{\xi}^2)$$

and
$$j_{\hat{\varphi}} = 16 j_{\hat{\xi}} \left\{ y_1(1+\hat{\xi})^{-3} + y_2(1-\hat{\xi})^{-3} \right\}^{-2}.$$

(c) For $y_1 = 37/2$ and $y_2 = 1/2$, verify that the 90% confidence interval for ξ is (0.6658, 0.9737).

10. Exercise 16 in chapter 7 introduced a (2, 1) curved exponential family derived from data in which $X_1 \sim$ Exponential (ξ) and independent of $X_2 \sim$ Exponential (ξ^2).

(a) Using Maple's symbolic computation or otherwise, compute the exact conditional CDF of $\hat{\xi}$ given $\tilde{a} = x_2^2/x_1$ by determining the cumulative integral of the density in terms of the erf function.

(b) Consider conditioning on an affine ancillary. In doing so, determine (8.1), and approximation (8.22) using (8.20). Compare the results to exact computations that may be deduced from part (a) for a range of ξ-values.

(c) Repeat part (b) but condition on a likelihood ancillary. Compare the accuracies of the conditionality resolutions determined here with those determined in part (b). What conclusion may be drawn?

(d) Compute $f(a; \xi)$, the marginal density of the affine ancillary at its observed value a, and plot its values versus ξ. Plot the marginal density of the likelihood ancillary and compare its plot with that of the affine ancillary. What do the plots reveal?

(e) Plot the attained significance probability function using the approach of Fraser et al. Compare it with (i) the same plot that uses exact sample space derivatives and which fixes the likelihood ancillary and (ii) the plot using Skovgaard's approximate sample space derivatives.

(f) Compute the sensitive direction and compare it to the tangent direction of the exact ancillary manifold defined by $\tilde{a}$ in part (a).

11. Assume the same model and data as in Exercise 10 but now make $X_2 \sim$ Exponential $(\xi + \xi^2)$. Answer all the same questions with this new model.

12. Prove that all of the CDF approximations are equivariant under smooth monotonic transformations of ξ.

    (a) Show this property holds for (8.1) and (8.4) and also when the sample space derivative in $\hat{u}$ is replaced by (8.20).

    (b) Show equivariance for the Fraser, Reid, and Wu significance function. In particular, show for a locally 1-1 transformation that the sensitive direction is invariant (up to a sign change). Show that the parameterization change, characterized by the directional derivative, is determined up to a scale factor so that $\breve{u}$ is equivariant.

13. Prove that all of the CDF approximations preserve symmetry when the sufficient statistic admits a symmetric pivotal model. Such a model has a conditionality resolution $(\hat{\omega}, a)$ such that the joint distribution of $(\hat{\omega} - \omega, a)$ does not depend on ω and the conditional density of $\hat{\omega}|a$ is symmetric about ω for any a. Show that the symmetry property holds for (8.1) and (8.4), when the sample space derivative in $\hat{u}$ is replace by (8.20), and also for the Fraser *et al.* significance function.

9

Nuisance parameters

In the previous chapter, ξ was a scalar interest parameter and there were no nuisance parameters in the model. This chapter extends the discussion to allow for vector nuisance parameters. The discussion and proofs are mostly adapted from the seminal paper by Barndorff-Nielsen (1991).

9.1 Approximation with nuisance parameters

Let $\xi^T = (\chi^T, \psi)$ consist of scalar interest parameter ψ and $(p-1) \times 1$ nuisance parameter χ. Inference for ψ centers about producing an approximate conditional CDF for the signed likelihood ratio test statistic for ψ defined as

$$\hat{w}_\psi = \text{sgn}(\hat{\psi} - \psi)\sqrt{-2\{l(\hat{\chi}_\psi, \psi) - l(\hat{\chi}, \hat{\psi})\}}. \tag{9.1}$$

With nuisance parameter lacking, the relationship between $\hat{\psi}$ and $\hat{w}_\psi$ was assumed to be 1-1 and monotonic and the conditional CDF of $\hat{w}_\psi$ readily converted into a CDF for $\hat{\psi}$. This is not always the case with nuisance parameters, a fact borne out in the examples and particularly evident with the normal linear regression example.

Our aim is to determine a CDF for $\hat{w}_\psi$ that is "largely" dependent on only the interest parameter ψ and with "minimal" dependence on nuisance parameter χ. Use $l(\hat{\chi}_\psi, \psi) = l(\hat{\chi}_\psi, \psi; \hat{\xi}, a)$ to simplify the notation, and let $\hat{\chi}_\psi$ be the constrained MLE for χ with ψ fixed.

Conceptually the approach to removing dependence on the nuisance parameter χ is simply to condition on $\hat{\chi}_\psi$ using tools that have already been put in place. Suppose $(\hat{\xi}, a)$ is a conditionality resolution with MLE $\hat{\xi}$, (approximate) ancillary a, and conditional density specified by using $p^\dagger(\hat{\xi}|a; \xi)$. With a fixed, the transformation $\hat{\xi} \to (\hat{\chi}_\psi, \hat{w}_\psi)$ brings the focus to considering $\hat{w}_\psi$ with the intent to condition upon $\hat{\chi}_\psi$. The computation and inclusion of a Jacobian for the transformation yields $p^\dagger(\hat{\chi}_\psi, \hat{w}_\psi|a; \xi)$. The dependence of this joint distribution on χ can be largely removed by conditioning on $\hat{\chi}_\psi$ and this is done by using its $p^\dagger$ formula. However, the $p^\dagger$ formula for $\hat{\chi}_\psi$ necessarily conditions on all available ancillary information and, since ψ is held fixed, this ancillary information not only includes a but also $\hat{w}_\psi$, a known approximate ancillary with respect to χ when ψ is held fixed. Division by $p^\dagger(\hat{\chi}_\psi|\hat{w}_\psi, a; \xi)$, the $p^\dagger$ for $\hat{\chi}_\psi$, yields the defining expression

$$p^\dagger(\hat{w}_\psi|a; \xi) = \frac{p^\dagger(\hat{\chi}_\psi, \hat{w}_\psi|a; \xi)}{p^\dagger(\hat{\chi}_\psi|\hat{w}_\psi, a; \xi)} \tag{9.2}$$

as a density that is "largely" independent of χ and mostly dependent on interest parameter ψ. A Temme type argument turns this density into the sought after conditional CDF of $\hat{w}_\psi | a; \xi$ with strong dependence on only the ψ component of ξ.

A common notation found in the literature indicates this CDF as $\hat{w}_\psi | a; \psi$ and simply uses ψ in place of ξ to indicate the strong dependence. Here this will be indicated by $\hat{\xi} | a; \star \psi$ where the $\star$ symbol means that the dependence rests largely on the value of ψ. Thus, the continuous CDF approximation for signed likelihood ratio statistic $\hat{W}_\psi$ given ancillary a, and scalar parameter ξ is

$$\widehat{\Pr}(\hat{W}_\psi \leq \hat{w}_\psi | a; \star \psi) = \Phi(\hat{w}_\psi) + \phi(\hat{w}_\psi)(1/\hat{w}_\psi - 1/\hat{u}) \qquad \hat{\psi} \neq \psi \qquad (9.3)$$

where $\hat{w}_\psi$ is given in (9.1) and $\hat{u}$ is the yet to be derived value

$$\hat{u} = \frac{\text{sgn}(\hat{\psi} - \psi)}{\sqrt{|j_{\chi\chi}(\hat{\xi}_\psi)||J_{\hat{\xi}}|}} \left\| \begin{array}{c} l_{\chi;\hat{\xi}}(\hat{\xi}_\psi; \hat{\xi}, a) \\ l_{;\hat{\xi}}(\hat{\xi}; \hat{\xi}, a) - l_{;\hat{\xi}}(\hat{\xi}_\psi; \hat{\xi}, a) \end{array} \right\|, \qquad (9.4)$$

with $\| \cdot \| = \text{abs}\{\det(\cdot)\}$. In (9.4), $J_{\hat{\xi}} = -l_{\xi\xi;}(\hat{\xi}; \hat{\xi}, a)$ is the $p \times p$ observed information matrix for ξ and $j_{\chi\chi} = -l_{\chi\chi;}(\hat{\xi}_\psi; \hat{\xi}, a)$ denotes its $(p-1) \times (p-1)$ subblock in component χ evaluated at $\hat{\xi}_\psi$. The sample differentiation $l_{;\hat{\xi}}$ is assumed to produce a p-dimensional row vector and $l_{\chi;\hat{\xi}}$ is a $(p-1) \times p$ matrix of derivatives. An implicit assumption that has been made is that the mapping $\hat{\xi} \leftrightarrow (\hat{\chi}_\psi, \hat{w}_\psi)$ is 1-1. Of course, the r^* version of this approximation is based on

$$r^* = \hat{w}_\psi - \frac{1}{\hat{w}_\psi} \ln \frac{\hat{w}_\psi}{\hat{u}} \sim N(0, 1) \qquad (9.5)$$

and is equivalent.

Expressions (9.3) and (9.5) do not depend on the particular form of the nuisance parameter χ in which they are implemented. For $\hat{w}_\psi$ this follows from the invariance property of the MLE. For $\hat{u}$ in (9.4) the result requires some technical computation that has been delegated to Exercise 1.

9.2 Examples

9.2.1 Regular exponential family

Canonical ψ

Suppose $\xi^T = \theta^T = (\chi^T, \psi)$ is the canonical parameter of a regular exponential family with corresponding sufficient statistic (x^T, y). Then (9.3) is the Skovgaard approximation for $\hat{\psi} | x; \psi$ or equivalently $y | x; \psi$ given in (5.78). In this relatively simple case, the CDF of $\hat{w}_\psi | x; \psi$ readily converts into the CDF of $\hat{\psi} | x; \psi$ because $\hat{w}_\psi \leftrightarrow \hat{\psi}$ is 1-1 and monotonic for fixed x.

In this setting the determinant in (9.4) simplifies considerably. First compute

$$l_{\chi;\hat{\theta}}(\hat{\theta}_\psi; \hat{\theta}) = \partial \{x - c'_\chi(\theta)\}/\partial \hat{\theta}^T = \partial x/\partial \hat{\theta}^T = j_{\chi,\theta}(\hat{\theta})$$

and

$$l_{;\hat{\theta}}(\hat{\theta}; \hat{\theta}) - l_{;\hat{\theta}}(\theta; \hat{\theta}) = (\hat{\theta} - \theta)^T j_{\hat{\theta}} = (\hat{\chi} - \chi)^T j_{\chi,\theta}(\hat{\theta}) + (\hat{\psi} - \psi) j_{\psi,\theta}(\hat{\theta})$$

9.2 Examples

so that

$$\hat{u} = \frac{\text{sgn}(\hat{\psi} - \psi)}{\sqrt{|j_{\chi\chi}(\hat{\theta}_\psi)||J_{\hat{\theta}}|}} \left\| \begin{matrix} j_{\chi,\theta}(\hat{\theta}) \\ (\hat{\chi} - \hat{\chi}_\psi)^T j_{\chi,\theta}(\hat{\theta}) + (\hat{\psi} - \psi) j_{\psi,\theta}(\hat{\theta}) \end{matrix} \right\|$$

$$= \frac{\text{sgn}(\hat{\psi} - \psi)}{\sqrt{|j_{\chi\chi}(\hat{\theta}_\psi)||J_{\hat{\theta}}|}} \left\| \begin{matrix} j_{\chi,\theta}(\hat{\theta}) \\ (\hat{\psi} - \psi) j_{\psi,\theta}(\hat{\theta}) \end{matrix} \right\|$$

$$= (\hat{\psi} - \psi)\sqrt{|J_{\hat{\theta}}|/|j_{\chi\chi}(\hat{\theta}_\psi)|}. \tag{9.6}$$

Noncanonical ψ

Often ψ is neither a component of the canonical parameter θ nor equivalent to a single component such as may occur through linear transformation. In this case, suppose it is possible to complete the parameter space by finding complementary parameter χ such that $\xi^T = (\chi^T, \psi)$ is 1-1 with θ. Then it is left as Exercise 1 to show that

$$\hat{u} = \text{sgn}(\hat{\psi} - \psi)\sqrt{\frac{|J_{\hat{\theta}}|}{|j_{\chi\chi}(\hat{\chi}_\psi)|}} \left\| \begin{matrix} \partial \theta^T/\partial \chi|_{\chi=\hat{\chi}_\psi} \\ (\hat{\theta} - \hat{\theta}_\psi)^T \end{matrix} \right\|. \tag{9.7}$$

In the special case in which ψ is canonical, then (9.7) reduces to (9.6). Furthermore, the value of $\hat{u}$ in (9.7) does not depend on the particular choice of the complementary parameter χ, a property that may be either shown directly or as emanating from the more general invariance of (9.4).

9.2.2 Curved exponential family

Let $\xi^T = (\chi^T, \psi)$ be the p-dimensional parameter for an (m, p) curved exponential family as in (7.22). The canonical parameter and sufficient statistic are θ_ξ and x respectively and

$$\hat{u} = \frac{\text{sgn}(\hat{\psi} - \psi)}{\sqrt{|J_{\hat{\xi}}||j_{\chi\chi}(\hat{\xi}_\psi)|}} \left\| \begin{pmatrix} \partial \theta_\xi^T/\partial \chi|_{\chi=\hat{\chi}_\psi} \\ (\hat{\theta} - \hat{\theta}_\psi)^T \end{pmatrix} \frac{\partial x}{\partial \hat{\xi}^T} \right\|. \tag{9.8}$$

The derivation is left as Exercise 1. In this setting, the $m \times p$ matrix $\partial x/\partial \hat{\xi}^T$ must be computed by fixing the appropriate $(m - p)$-dimensional ancillary; such computation may prove to be very difficult. In the regular exponential family setting, this sample space derivative is simple since there is no ancillary to fix; this leads to

$$\partial x/\partial \hat{\xi}^T = j_{\hat{\theta}}\, \partial \hat{\theta}/\partial \hat{\xi}^T$$

and (9.8) reduces to (9.7). Expression (9.8) is easily shown to be invariant to the particular nuisance parametrization χ used for implementation.

9.2.3 Independent regression model

The linear regression model in section 7.2.3 and Exercise 9 of chapter 7 allows explicit computation of the sample space derivatives in $\hat{u}$. Let $y_i = x_i^T \beta + \sigma e_i$ for $i = 1, \ldots, n$ be independent responses with $\beta^T = (\beta_1, \ldots, \beta_p)$ and $\xi^T = (\beta^T, \sigma)$. When necessary, let all cases be written together using the usual vector notation $y = X\beta + \sigma e$. Suppose the interest parameter is $\psi = \beta_p$, $\chi = (\beta_1, \ldots, \beta_{p-1}, \sigma)^T$ is a nuisance, and for simplicity maintain the use of original notation ψ and χ. In order to avoid awkward notation, the dimension of ξ needs to be taken here as $p + 1$ rather than the p used in the theoretic development.

If $\{e_i\}$ are i.i.d. with density g and $\ln g = h$, then the likelihood is

$$l(\xi) = -n \ln \sigma + \sum_{i=1}^{n} h(e_i). \tag{9.9}$$

The ancillaries to be held fixed are $\{a_i = (y_i - x_i^T \hat{\beta})/\hat{\sigma}\}$. Sample space derivatives of (9.9) require derivatives of $\{e_i\}$ which are easily computed by writing

$$e_i = \{a_i \hat{\sigma} + x_i^T(\hat{\beta} - \beta)\}/\sigma$$

so that

$$\partial e_i / \partial(\hat{\beta}^T, \hat{\sigma}) = (x_i^T, a_i)/\sigma.$$

Notation is also required to express maximization holding $\psi = \beta_p$ fixed. Under such constraints, denote the MLEs as $\hat{\xi}_\psi^T = (\hat{\beta}_\psi^T, \hat{\sigma}_\psi)$ and the residuals as $\hat{e}_\psi = (\hat{e}_{1\psi}, \ldots, \hat{e}_{n\psi})^T$ with

$$\hat{e}_{i\psi} = (y_i - x_i^T \hat{\beta}_\psi)/\hat{\sigma}_\psi.$$

Let $x_{i\backslash}$ be the $(p-1) \times 1$ vector obtained by removing the last component of x_i and denote the n-vector associated with β_p as $(x_{1,p}, \ldots, x_{n,p})$. Then an explicit form for $\hat{u}$ may be derived as

$$\hat{u} = \frac{\hat{\sigma}^{p+1} \operatorname{sgn}(\hat{\beta}_p - \beta_p)}{\hat{\sigma}_\psi^p \sqrt{|K_p(\hat{e}_\psi)||K_{p+1}(a)|}} \left| \frac{\hat{\rho}_{1\psi}}{\hat{\sigma}_\psi} \det \hat{D}_{1\psi} - \left(\frac{n}{\hat{\sigma}} + \frac{\hat{\rho}_{2\psi}}{\hat{\sigma}_\psi} \right) \det \hat{D}_{2\psi} \right| \tag{9.10}$$

where

$$\begin{pmatrix} \hat{\rho}_{1\psi} & \hat{\rho}_{2\psi} \end{pmatrix} = \sum_{i=1}^{n} h'(\hat{e}_{i\psi}) \begin{pmatrix} x_{i,p} & a_i \end{pmatrix} \tag{9.11}$$

are inner products of the score values $\{h'(\hat{e}_{i\psi})\}$ with components of x_p and a,

$$\hat{D}_{1\psi} = \sum_{i=1}^{n} h''(\hat{e}_{i\psi}) \begin{pmatrix} x_{i\backslash} \\ \hat{e}_{i\psi} \end{pmatrix} \begin{pmatrix} x_{i\backslash}^T & a_i \end{pmatrix} + \hat{\rho}_{2\psi} \iota_p \iota_p^T \tag{9.12}$$

$$\hat{D}_{2\psi} = \sum_{i=1}^{n} h''(\hat{e}_{i\psi}) \begin{pmatrix} x_{i\backslash} \\ \hat{e}_{i\psi} \end{pmatrix} x_i^T + \hat{\rho}_{1\psi} \iota_p \iota_p^T,$$

and ι_p is a $p \times 1$ vector consisting of 1 in its last row and zeros elsewhere. The notation $K_p(\hat{e}_\psi)$ and $K_{p+1}(a)$ indicate portions of the information determinants given respectively as

$$K_p(\hat{e}_\psi) = \det \left\{ \sum_{i=1}^{n} \begin{bmatrix} h''(\hat{e}_{i\psi}) x_{i\backslash} x_{i\backslash}^T & h''(\hat{e}_{i\psi}) \hat{e}_{i\psi} x_{i\backslash} \\ h''(\hat{e}_{i\psi}) \hat{e}_{i\psi} x_{i\backslash}^T & -1 + h''(\hat{e}_{i\psi}) \hat{e}_{i\psi}^2 \end{bmatrix} \right\} \tag{9.13}$$

$$K_{p+1}(a) = \det \left\{ \sum_{i=1}^{n} \begin{bmatrix} h''(a_i) x_i x_i^T & h''(a_i) a_i x_i \\ h''(a_i) a_i x_i^T & -1 + h''(a_i) a_i^2 \end{bmatrix} \right\}.$$

Further details are found in Exercise 2 where the reader is asked to derive these results as well as in Exercise 3 that deals with the special case of normal errors.

Normal linear regression

In this case, $\hat{w}_\psi$ and $\hat{u}$ reduce to

$$\hat{w}_\psi = \text{sgn}(t)\sqrt{n \ln\{1 + t^2/(n-p)\}} \tag{9.14}$$
$$\hat{u} = t\{1 + t^2/(n-p)\}^{-(p+1)/2}\sqrt{n/(n-p)}$$

where

$$t = \frac{\hat{\beta}_p - \beta_p}{\hat{\sigma}\sqrt{(X^T X)^{p,p}}}\sqrt{\frac{n-p}{n}} \tag{9.15}$$

is the ordinary t statistic for testing the value β_p.

The fact that $\hat{w}_\psi$ and $\hat{u}$ are based entirely on t suggests considering what might occur were a marginal likelihood to be used in place of the actual likelihood. Bartlett (1937) has advocated the use of a marginal likelihood ratio test instead of the usual likelihood ratio to adjust for degrees of freedom due to nuisance parameters. Taking the log-Student-t_{n-p} pivot as log-marginal likelihood l_M with all terms but β_p observed, then

$$\hat{w}_M = \text{sgn}(t)\sqrt{-2\{l_M(\beta_p) - l_M(\hat{\beta}_p)\}} \tag{9.16}$$
$$= \text{sgn}(t)\sqrt{(n-p+1)\ln\{1 + t^2/(n-p)\}}$$
$$\hat{u}_M = j_{\hat{\beta}_p}^{-1/2}\{l_{;\hat{\beta}_p}(\hat{\beta}_p) - l_{;\hat{\beta}_p}(\beta_p)\}$$
$$= t\{1 + t^2/(n-p)\}^{-1}\sqrt{(n-p+1)/(n-p)}$$

differ by replacing n by $n - p + 1$ and by using power -1 instead of $-(p+1)/2$. For small p, marginal and ordinary likelihood lead to roughly the same answers however this in not true for large values; see Kalbfleish and Sprott (1970, 1973) and Cox and Hinkley (1974, §2.1) as well as the discussion in the next subsection.

Simple linear regression with normal errors

This model fits a location parameter β_1 and an independent regression variable x_2 whose slope $\beta_2 = \psi$ is the interest parameter. Without any loss in generality, assume that the independent variables are orthogonal so $x_2^T 1 = 0$. The nuisance parameter is $\chi = (\beta_1, \sigma)$ and confidence intervals for slope β_2 are determined from (9.14) and (9.16) using the various degrees of freedom listed in table 9.1. Acceptable 95% confidence intervals result with as few as three degrees of freedom.

Some plots of the percentage relative error incurred when using (9.14) to compute right tail probabilities in place of exact Student-t probabilities are shown in Figure 9.1. The left plot suggests that the relative error achieves a limit under large deviations. See section 9.8.1 for statements of such results that are applicable to this setting.

Table 9.1. *Confidence ranges with level 95% determined from the CDF approximation for $\hat{w}_\psi$ given in (9.14) and (9.16) as compared to exact t intervals.*

d.f.	using (9.14)	using (9.16)	exact t interval
1	$|t| < 9.210$	$|t| < 11.25$	$|t| < 12.71$
3	$|t| < 3.012$	$|t| < 3.119$	$|t| < 3.182$
5	$|t| < 2.514$	$|t| < 2.550$	$|t| < 2.571$

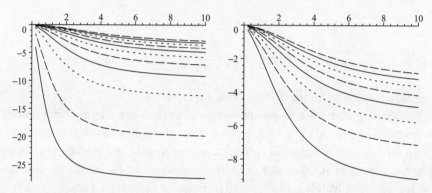

Figure 9.1. Percentage relative errors for right tail probabilities for various degrees of freedom (d.f.). From worst to best are d.f. = 1, 2, 4(2)20 for the left plot and d.f. = 6(2)20 on the right.

Implicit marginalization

This example illustrates that a marginalization process has occurred that leads to the usual t pivotal for inference about β_p. A deeper appreciation for the derivation of (9.3), and (9.2) in particular, may be gained by repeating this derivation in the context of this example. We use notation ψ in place of slope β_2.

The $p^\dagger$ expression for $\hat{\xi} = (\hat{\beta}_1, \hat{\psi}, \hat{\sigma})$ records the mutual independence of these three components which in turn are independent of the ancillary a. This is transformed to give $p^\dagger(\hat{\beta}_{1\psi}, \hat{\sigma}_\psi, \hat{w}_\psi; \xi)$ or $p^\dagger(\hat{\beta}_{1\psi}, \hat{\sigma}_\psi, t; \xi)$ since $\hat{w}_\psi$ is 1-1 with t. With orthogonality, these expressions are

$$\hat{\beta}_{1\psi} = \hat{\beta}_1 = \bar{y} \qquad t = \frac{(\hat{\psi} - \psi)}{\hat{\sigma}}\sqrt{x_2^T x_2 (n-2)/n} \qquad (9.17)$$
$$n\hat{\sigma}_\psi^2 = \|y - \bar{y}\mathbf{1} - \psi x_2\|^2 = n\hat{\sigma}^2 + (\hat{\psi} - \psi)^2 x_2^T x_2.$$

In this $p^\dagger$ density, $\hat{\beta}_{1\psi}$ expresses its independence of $(\hat{\sigma}_\psi, t)$ so that

$$p^\dagger(\hat{\beta}_{1\psi}, \hat{\sigma}_\psi, t; \xi) = p^\dagger(\hat{\beta}_{1\psi}; \xi) p^\dagger(\hat{\sigma}_\psi, t; \xi).$$

The computation

$$p^\dagger(t) = \frac{p^\dagger(\hat{\beta}_{1\psi}, \hat{\sigma}_\psi, t; \xi)}{p^\dagger(\hat{\beta}_{1\psi}, \hat{\sigma}_\psi | t; \xi)} = \frac{p^\dagger(\hat{\beta}_{1\psi}; \xi) p^\dagger(\hat{\sigma}_\psi, t; \xi)}{p^\dagger(\hat{\beta}_{1\psi}; \xi) p^\dagger(\hat{\sigma}_\psi | t; \xi)} \qquad (9.18)$$

makes use of t as a likelihood ancillary for the distribution of $\hat{\sigma}_\psi$. Indeed t is an exact ancillary with a Student-t_{n-p} distribution. In (9.18), the expression on the left now has $p^\dagger$ not depending on ξ or ψ since t is an exact parameter-free pivot.

9.3 Derivation of (9.3) and (9.4)

The $p^\dagger$ density for $\hat{\xi}|a;\xi$

$$p^\dagger(\hat{\xi}|a;\xi) = (2\pi)^{-p/2}|j_{\hat{\xi}}|^{1/2}\exp\{l(\xi) - l(\hat{\xi})\}$$

is transformed and rearranged to give

$$p^\dagger(\hat{w}_\psi, \hat{\chi}_\psi|a;\xi) = (2\pi)^{-p/2}|j_{\hat{\xi}}|^{1/2}|\exp\left\{l(\xi) - l(\hat{\xi}_\psi) - \hat{w}_\psi^2/2\right\} \left|\frac{\partial\hat{\xi}}{\partial(\hat{\chi}_\psi, \hat{w}_\psi)}\right|. \quad (9.19)$$

Inserting the value of the Jacobian determined in lemma 9.3.1 below, then expression (9.19) is

$$p^\dagger(\hat{\chi}_\psi, \hat{w}_\psi|a;\xi) = \frac{1}{(2\pi)^{p/2}}\frac{\hat{w}_\psi}{\hat{u}}|j_{\chi\chi}(\hat{\xi}_\psi)|^{1/2}\exp\left\{l(\xi) - l(\hat{\xi}_\psi) - \hat{w}_\psi^2/2\right\}.$$

According to (9.2), this is to be divided by

$$p^\dagger(\hat{\chi}_\psi|\hat{w}_\psi, a;\xi) = (2\pi)^{-(p-1)/2}|j_{\chi\chi}(\hat{\xi}_\psi)|^{1/2}\exp\{l(\xi) - l(\hat{\xi}_\psi)\}$$

so that

$$p^\dagger(\hat{w}_\psi|a;\xi) = (2\pi)^{-1/2}\frac{\hat{w}_\psi}{\hat{u}}\exp\left\{-\hat{w}_\psi^2/2\right\}. \quad (9.20)$$

The Temme type argument for integrating (9.20) now suffices in producing (9.3).

Lemma 9.3.1 *The Jacobian of the transformation* $(\hat{\chi}, \hat{\psi}) \longleftrightarrow (\hat{\chi}_\psi, \hat{w}_\psi)$ *is*

$$\frac{\partial(\hat{\chi}_\psi, \hat{w}_\psi)}{\partial(\hat{\chi}, \hat{\psi})} = \begin{pmatrix} \partial\hat{\chi}_\psi/\partial\hat{\chi}^T & \partial\hat{\chi}_\psi/\partial\hat{\psi} \\ \partial\hat{w}_\psi/\partial\hat{\chi}^T & \partial\hat{w}_\psi/\partial\hat{\psi} \end{pmatrix} = \begin{pmatrix} j_{\chi\chi}(\hat{\xi}_\psi)^{-1}l_{\chi;\hat{\xi}}(\hat{\xi}_\psi) \\ \{l_{;\hat{\xi}}(\hat{\xi}) - l_{;\hat{\xi}}(\hat{\xi}_\psi)\}/\hat{w}_\psi \end{pmatrix} \quad (9.21)$$

where the derivative to the left (right) of the semicolon produces a column (row) vector. The determinant is

$$\left|\frac{\partial(\hat{\chi}_\psi, \hat{w}_\psi)}{\partial(\hat{\chi}, \hat{\psi})}\right| = \frac{\hat{u}}{\hat{w}_\psi}\sqrt{|j_{\hat{\xi}}|/|j_{\chi\chi}(\hat{\xi}_\psi)|}. \quad (9.22)$$

Proof. Compute derivatives, or more simply differentials of

$$-\tfrac{1}{2}\hat{w}_\psi^2 = l(\hat{\xi}_\psi) - l(\hat{\xi})$$

to get

$$-\hat{w}_\psi d\hat{w}_\psi = \{l_{;\hat{\chi}}(\hat{\xi}_\psi) - l_{;\hat{\chi}}(\hat{\xi})\}d\hat{\chi} = \{l_{;\hat{\psi}}(\hat{\xi}_\psi) - l_{;\hat{\psi}}(\hat{\xi})\}d\hat{\psi}.$$

Solving for the sample space derivatives of $\hat{w}_\psi$ yields the bottom row of the matrix

$$\begin{pmatrix} j_{\chi\chi}(\hat{\xi}_\psi)^{-1}l_{\chi;\hat{\chi}}(\hat{\xi}_\psi) & j_{\chi\chi}(\hat{\xi}_\psi)^{-1}l_{\chi;\hat{\psi}}(\hat{\xi}_\psi) \\ \{l_{;\hat{\chi}}(\hat{\xi}) - l_{;\hat{\chi}}(\hat{\xi}_\psi)\}/\hat{w}_\psi & \{l_{;\hat{\psi}}(\hat{\xi}) - l_{;\hat{\psi}}(\hat{\xi}_\psi)\}/\hat{w}_\psi \end{pmatrix} \quad (9.23)$$

which is (9.21) when, for example, $l_{;\hat{\chi}}(\hat{\xi})$ and $l_{;\hat{\psi}}(\hat{\xi})$ are concatenated to give $l_{;\hat{\xi}}(\hat{\xi})$. The upper blocks of (9.21), or equivalently (9.23), are determined by taking differentials of the score

$$l_{\chi;}(\hat{\chi}_\psi, \psi; \hat{\chi}, \hat{\psi}, a) = 0$$

to get

$$l_{\chi\chi;}(\hat{\xi}_\psi)d\hat{\chi}_\psi + l_{\chi;\hat{\chi}}(\hat{\xi}_\psi)d\hat{\chi} = 0_{p-1}$$
$$l_{\chi\chi;}(\hat{\xi}_\psi)d\hat{\chi}_\psi + l_{\chi;\hat{\psi}}(\hat{\xi}_\psi)d\hat{\psi} = 0,$$

where 0_{p-1} is a $(p-1)$-vector of zeros. Solving these two systems of equations leads directly to the upper blocks in (9.23). The determinant in (9.22) follows from the definition of $\hat{u}$. □

9.4 Exact and approximate sample space derivatives

9.4.1 Skovgaard's approximate sample space derivatives

If ξ is $p \times 1$, then the covariance $q(\xi_1, \xi_2; \xi_0)$ defined in (8.19) is the $p \times 1$ vector of covariances for the likelihood difference $l(\xi_1) - l(\xi_2)$ with the gradient vector $l_{\xi;}(\xi_0)$ assuming parameter value ξ_0. Quite often these covariances are not difficult to compute. The function q allows the sample space derivative to be approximated as the matrix form of (8.20) which is

$$l_{;\hat{\xi}}(\hat{\xi}; \hat{\xi}, a) - l_{;\hat{\xi}}(\hat{\xi}_\psi; \hat{\xi}, a) \simeq q(\hat{\xi}, \hat{\xi}_\psi; \hat{\xi})^T i_\xi^{-1} j_\xi \quad (9.24)$$

where $i_\xi = E\{l_{\xi;}(\xi)l_{\xi;}(\xi)^T; \xi\}$ is the expected Fisher information for ξ, and j_ξ is the $p \times p$ observed Fisher information for ξ.

The expression for $\hat{u}$ in (9.4) contains the additional mixed derivative $l_{\chi;\hat{\xi}}(\hat{\xi}_\psi; \hat{\xi}, a)$ whose value may be approximated as the appropriate $(p-1) \times p$ block of

$$l_{\xi;\hat{\xi}}(\hat{\xi}_\psi; \hat{\xi}, a) \simeq S(\hat{\xi}_\psi; \hat{\xi})^T i_\xi^{-1} j_\xi \quad (9.25)$$

where

$$S(\xi; \xi_0) = E\{l_{\xi;}(\xi_0)l_{\xi;}(\xi)^T; \xi_0\} \quad (9.26)$$

is often a simple computation. Details for motivating this approximation are found in the appendix of section 8.8.2.

Regular exponential family

For this setting, these approximations return the exact sample space derivatives and lead to (9.6) for canonical ψ, and (9.7) for a non-canonical ψ.

(m, p) curved exponential family

Let $\xi^T = (\chi^T, \psi)$ be the p-dimensional parameter for an (m, p) curved exponential family as in (7.22). With canonical parameter θ_ξ and sufficient statistic x, the score is $l_{\xi;}(\xi) =$

$\dot{\theta}_\xi^T (x - \mu_\xi)$ with $\dot{\theta}_\xi^T = \partial \theta^T / \partial \xi$ and

$$q(\xi, \xi_1, \xi) = \text{Cov}[l_{\xi;}(\xi)\{l(\xi) - l(\xi_1)\}; \xi]$$
$$= E\{\dot{\theta}_\xi^T (x - \mu_\xi) x^T (\theta_\xi - \theta_{\xi_1}); \xi\}$$
$$= \dot{\theta}_\xi^T \Sigma_\xi (\theta_\xi - \theta_{\xi_1}) = \dot{\mu}_\xi^T (\theta_\xi - \theta_{\xi_1}).$$

Thus,

$$l_{;\hat{\xi}}(\hat{\xi}) - l_{;\hat{\xi}}(\hat{\xi}_\psi) \simeq (\hat{\theta} - \hat{\theta}_\psi)^T \dot{\mu}_\xi i_\xi^{-1} j_{\hat{\xi}}, \qquad (9.27)$$

where $i_\xi = \dot{\theta}_\xi^T \dot{\mu}_\xi$. Also, from (9.26),

$$S(\xi; \xi_0) = E\{\dot{\theta}_{\xi_0}^T (x - \mu_{\xi_0})(x - \mu_\xi)^T \dot{\theta}_\xi; \xi_0\}$$
$$= \dot{\theta}_{\xi_0}^T \Sigma_{\xi_0} \dot{\theta}_\xi = \dot{\mu}_{\xi_0}^T \dot{\theta}_\xi$$

so that

$$l_{\xi;\hat{\xi}}(\hat{\xi}_\psi) \simeq \dot{\theta}_{\hat{\xi}_\psi}^T \dot{\mu}_\xi i_\xi^{-1} j_{\hat{\xi}}. \qquad (9.28)$$

Substituting (9.27) and using the $l_{\chi;\hat{\xi}}$ portion of (9.28) along with some simplification leads to an approximate $\check{u}$ as

$$\check{u} = \frac{\text{sgn}(\hat{\psi} - \psi)}{|i_\xi|} \sqrt{\frac{|j_{\hat{\xi}}|}{|j_{\chi\chi}(\hat{\xi}_\psi)|}} \left\| \begin{pmatrix} \partial \theta^T / \partial \chi |_{\xi = \hat{\xi}_\psi} \\ (\hat{\theta} - \hat{\theta}_\psi)^T \end{pmatrix} \dot{\mu}_\xi \right\|. \qquad (9.29)$$

This expression is also invariant to the particular nuisance parametrization χ used for implementation.

9.4.2 The approach of Fraser, Reid and Wu (1999a)

This approach was discussed in section 8.5.4 in the case of a scalar parameter and provides an alternative standardized score statistic $\check{u}$ in the presence of nuisance parameters. For applications to linear and nonlinear regression analyses, see Fraser et al. (1999b). If independent data y are in n dimensions and ξ has dimension p, then an approximate ancillary statistic is conditioned upon to determine a p-dimensional mixed reparameterization φ. In the scalar case, φ is the directional derivative of the log-likelihood in a sensitive direction v pointing tangent to a locally defined ancillary space. Similarly, with nuisance parameters, a sensitive directional plane, spanned by the columns of $n \times p$ matrix V, forms a tangent space to a locally defined ancillary manifold. Now the new parameterization that responds to changes in $\hat{\xi}$ holding the ancillary fixed is the gradient of the log likelihood in the sensitive direction or $\varphi = \partial l(\xi; y)/\partial V$ evaluated at the observed $\hat{\xi}$ and y.

Having reduced to the p-dimensional parameter φ, further orthogonalization, aimed at inference for the interest parameter ψ, is performed as describe in their section 2.1 to determine the score statistic $\check{u}$. This leads to an expression for $\check{u}$ given in their expression (2.11) that has exactly the form of $\hat{u}$ in (9.4) but takes the sample space derivatives in the sensitive direction V rather than with respect to $\hat{\xi}$ holding ancillary a fixed.

For $\hat{u}$ in expression (9.4), the differentiation with respect to $\hat{\xi}$ holding a fixed is a directional derivative tangent to the ancillary manifold at the observed data. Therefore, if the column space of V is tangent to the ancillary manifold for a at the observed data, then $\breve{u} = \hat{u}$ as concluded by Fraser *et al.* (1999a). In many of their examples, $\breve{u}$ and $\hat{u}$ are sufficiently close as to provide probability computations agreeing to three significant digits but the expressions are analytically different. By continuity, the tangent plane to the locally defined ancillary must be very close but not exactly tangent to the likelihood ancillary. In section 8.6.3, the gamma exponential model provided an example in which the sensitive direction is exactly tangent to the likelihood ancillary at the data in the (2, 1) curved exponential family. Further clarification for the equivalence of $\breve{u}$ and $\hat{u}$ is needed. Also in curved exponential families, some understanding of the degree to which the sensitive direction points tangent to the likelihood ancillary manifold would be interesting; see the discussion of Fraser (2004).

9.5 Numerical examples

A wide range of examples that demonstrate the numerical accuracy of the r^* type approximations may be found in Pace and Salvan (1997), Severini (2000a), Bellio and Brazzale (1999, 2003), Sartori *et al.* (1999), Bellio (2003) and Bellio *et al.* (2000).

9.5.1 F distribution

Suppose $Y_1, \ldots, Y_m$ are i.i.d. Exponential (λ) and independently $Z_1, \ldots, Z_n$ are i.i.d. Exponential (μ). For an $M/M/1$ queue, the $\{y_i\}$ might be the observed interarrivals of tasks in a Poisson (λ) arrival process. The values of $\{z_i\}$ represent the observed service rates. Usually observations on a queue lead to values of m and n that are also random quantities, however suppose this aspect is ignored. If $\psi = \lambda/\mu < 1$, then the queue is stable and the value of the stability constant ψ is of interest. For example, over time, the queue length approaches a stationary distribution that has a Geometric (ψ) mass function.

Inference for ψ is generally based on the pivot

$$f = \hat{\psi}/\psi = \hat{\lambda}/(\hat{\mu}\psi) = (z/n)/(y/m)/\psi \sim F_{2n,2m}$$

where $y = \sum_{i=1}^{m} y_i$ and $z = \sum_{i=1}^{n} z_i$. Alternatively, approximation (9.3) may be used, and it will be seen that it reproduces the F pivot with remarkable accuracy. The likelihood is written in terms of parameters (μ, ψ) as

$$\mathcal{L}(\mu, \psi) = (\mu\psi)^m \mu^n \exp(-\mu\psi y - \mu z). \tag{9.30}$$

This results in a value for $\hat{w}_\psi$ that depends only on f and simply asserts the well-known fact that the likelihood ratio test of $H_0 : \lambda = \mu$ is an F test based on f. The MLEs involved are

$$\hat{\mu} = n/z \qquad \hat{\lambda} = m/y \qquad \hat{\psi} = \hat{\lambda}/\hat{\mu} \qquad \hat{\mu}_\psi = \frac{m+n}{\psi y + z}.$$

It is a simple exercise to show that the signed likelihood ratio may be written entirely in terms of f as

$$\hat{w}_\psi = \operatorname{sgn}(f-1)\sqrt{2(m+n)\{\ln(1 + nf/m) - \ln(1 + n/m)\} - 2n \ln f}. \tag{9.31}$$

9.5 Numerical examples

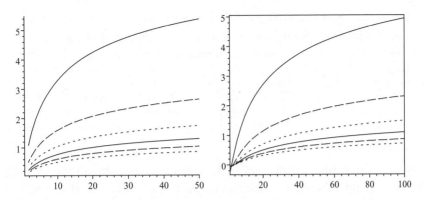

Figure 9.2. Percentage relative errors for left tail (left) and right tail (right) of the (9.3) approximation for the $F_{4m,2m}$ distribution. The plots, from worst to best, have $m = 1(1)6$.

The value of $\hat{u}$ is given directly by (9.7) since the likelihood is a (2, 2) regular exponential family in which ψ is not a canonical component. Take $\theta^T = (-\psi\mu, -\mu)$ and compute

$$\hat{u} = \text{sgn}(\hat{\psi} - \psi)\sqrt{\frac{(m/\hat{\lambda}^2)(n/\hat{\mu}^2)}{(m+n)/\hat{\mu}_\psi^2}} \left\| \begin{pmatrix} -\psi & -1 \\ \hat{\lambda} - \hat{\lambda}_\psi & \hat{\mu} - \hat{\mu}_\psi \end{pmatrix} \right\| \quad (9.32)$$
$$= (f-1)(1 + nf/m)^{-1}\sqrt{n(m+n)/m}.$$

The numerator under the square root in (9.32) has the computation of $|j_{\hat{\theta}}|$, the information for the canonical parameter. The denominator term has been computed as $-\partial^2 \ln \mathcal{L}(\mu, \psi)/\partial\mu^2|_{\mu=\hat{\mu}_\psi}$ thus holding ψ fixed. Holding λ fixed, leads to the wrong answer and is a common mistake when making these computations.

The marginal likelihood is the density of $\hat{\psi} \sim \psi F_{2n,2m}$ and leads to the exact same values for $\hat{w}_\psi$ and $\hat{u}$. In this instance the value $\hat{\psi}$ alone determines the likelihood as an expression of marginal likelihood. A reason for this equivalence is given in section 9.6. Once again (9.3) succeeds in capturing the true marginal distribution of $\hat{w}_\psi$ as it relates to the $F_{2n,2m}$ distribution.

Confidence intervals for ψ were constructed from (9.31) and (9.32) using $m = 1$, $n = 2$, $y = 2$, and $z = 3$ which leads to the values $\hat{\lambda} = 1/2$, $\hat{\mu} = 2/3$, and $\hat{\psi} = 3/4$. At level 95%, the approximate interval (centered in terms of probability) is (0.0183, 8.13) which compares to exact interval (0.0191, 7.99). For level 90% the intervals are (0.0377, 5.29) and (0.0390, 5.21) respectively.

Now consider the relative error in approximating the $F_{4m,2m}$ pivot with (9.3) through a sequence of numerical examples in which $m = 1(1)6$, $n = 2m$, and the MLEs are fixed at $\hat{\lambda} = 1/2$, $\hat{\mu} = 2/3$, and $\hat{\psi} = 3/4$ as used for the confidence intervals. Let $\hat{W}_{m\psi}$ denote the dependence of the random variable $\hat{W}_\psi$ on m. Figure 9.2 (left) plots the percentage relative error as given on the left of (9.33)

$$100 \left\{ \frac{\widehat{\Pr}(\hat{W}_{m\psi} \leq \hat{w}_{m\psi}; \psi)}{F_{4m,2m}(.75/\psi)} - 1 \right\} \quad 100 \left\{ \frac{\widehat{\Pr}(\hat{W}_{m\psi} > \hat{w}_{m\psi}; 1/\psi)}{1 - F_{4m,2m}(.75\psi)} - 1 \right\} \quad (9.33)$$

for $\psi \in (0.75, 50)$ thus assessing error in the left tail of $F_{4m,2m}$. Figure 9.2 (right) shows the relative error given on the right side of (9.33) for $\psi \in (4/3, 100)$ and shows error in the right tail of $F_{4m,2m}$.

Table 9.2. *Five trials of paired measurements of voltage and current across a load with fixed impedance as given in Hannig et al. (2003).*

Trial:	1	2	3	4	5
Voltage (μV)	5.007	4.994	5.005	4.990	4.999
Current (μA)	19.663	19.639	19.640	19.685	19.678

The error is quite small even for the $F_{4,2}$ distribution which shows the largest error. The range of probability coverage shown for $F_{4,2}$ and $F_{8,4}$ are $(0.0^3 848, 0.987)$ and $(0.0^5 352, 0.9^3 567)$ respectively. The plots suggest that large deviation errors are preserved. Theoretical statements to this effect are given in section 9.8.

9.5.2 Ratios of bivariate normal means

Suppose that $(y_1, z_1)^T, \ldots, (y_n, z_n)^T$ are i.i.d. data from a Normal$_2(\mu, \Sigma)$ with $\mu = (\mu_1, \mu_2)^T$ and arbitrary covariance pattern $\Sigma = (\sigma_{ij})$. A confidence interval for the ratio $\psi = \mu_2/\mu_1$ is required. The Fieller (1954) method is based on a t pivotal defined as

$$t_\psi = \frac{\sqrt{n-1}(\bar{z} - \psi \bar{y})}{\sqrt{\hat{\sigma}_{22} - 2\psi \hat{\sigma}_{12} + \psi^2 \hat{\sigma}_{11}}} \sim \text{Student-}t_{n-1} \qquad (9.34)$$

where $\hat{\Sigma} = (\hat{\sigma}_{ij})$ is the MLE rather than the unbiased estimate.

As an alternative to Fieller's method consider the use of (9.3). The likelihood function is a $(5, 5)$ regular exponential family in which the interest parameter is not canonical. Thus the value of $\hat{u}$ can be computed directly from (9.7) with a good deal of nasty algebra. A simpler derivation described below leads to

$$\hat{w}_\psi = \text{sgn}(t_\psi)\sqrt{n \ln\{1 + t_\psi^2/(n-1)\}} \qquad (9.35)$$
$$\hat{u} = t_\psi \{1 + t_\psi^2/(n-1)\}^{-1} \sqrt{n/(n-1)}.$$

Thus, the accuracy of (9.3) in reproducing Fieller's pivot is exactly the same as occurred in the normal linear regression model.

Consider the data in table 9.2 analyzed in Hannig *et al.* (2003). In order to determine an unknown fixed impedance ψ in a circuit, five pairs of measurements were made recording the voltage potential (μV) across the impedance and the current (μA) drawn. The ratio $\bar{z}/\bar{y}$ of average voltage to average current estimates the impedance in ohms as 0.254260 $\Omega = 254.260 \, \mu\Omega$.

The summary statistics for this data lead to a t pivotal of

$$t_\psi = \frac{\sqrt{4}(4.999 - 19.661\psi)}{\sqrt{0.0^3 3588\psi^2 - 2\psi(-0.0000432) + 0.0^4 412}}.$$

A comparison of confidence intervals in $\mu\Omega$ constructed from (9.3) and the exact method of the Fieller pivot are: $(253.612, 254.909)$ and $(253.604, 254.917)$ respectively at level 95%, and $(253.187, 255.336)$ and $(253.173, 255.350)$ respectively at level 99%. Both intervals are the same from a practical point of view.

Proof of (9.35)

Let ψ be an arbitrary number (not necessarily μ_2/μ_1) and transform the bivariate data using the transformation

$$\begin{pmatrix} x_{1i} \\ x_{2i} \end{pmatrix} = O_\psi \begin{pmatrix} y_i \\ z_i \end{pmatrix} \quad O_\psi = \begin{pmatrix} 1 & 0 \\ \psi & -1 \end{pmatrix}.$$

Then each pair $(x_{1i}, x_{2i})^T \sim \text{Normal}_2(\tau; \Omega)$ where

$$\tau = \begin{pmatrix} \tau_1 \\ \tau_2 \end{pmatrix} = \begin{pmatrix} \mu_1 \\ \psi\mu_1 - \mu_2 \end{pmatrix}$$

and $\Omega = O_\psi \Sigma O_\psi^T$. If parameter $\psi = \mu_2/\mu_1$, then $\tau_2 = 0$ in the transformed bivariate normal model. If $\psi \neq \mu_2/\mu_1$, then $\tau_2 \neq 0$. Thus, for a particular value of ψ, the likelihood ratio test that the mean ratio is ψ versus not ψ is equivalent to the likelihood ratio test that $\tau_2 = 0$ in the transformed bivariate model.

In the transformed data, $\{x_{1i}\}$ have mean $\tau_1 = \mu_1$ and variance $\omega_{11} = \sigma_{11}$ where $\Omega = (\omega_{ij})$. These parameters are unknown fixed parameters estimated by their sample mean and variance. The MLE for τ_2 is

$$\hat{\tau}_2 = n^{-1} \sum_{i=1}^n x_{2i} = \psi\bar{y} - \bar{z}$$

and the likelihood ratio test that $\tau_2 = 0$ is a t pivot based on the sample mean and variance of the $\{x_{2i}\}$ data. This t pivot reduces to the Fieller pivot. Since the likelihood is essentially dealing with the location parameter in a location/scale normal model, the relationship of the signed likelihood ratio test statistic $\hat{w}_\psi$ to the t pivot must be the same as that given in (9.14). Thus $\hat{w}_\psi$ takes the form in (9.35) with $p = 1$ upon recognizing that it is a location model with $n - 1$ degrees of freedom.

The structure of the likelihood again assures that the value of $\hat{u}$ is given in (9.14) with $n - 1$ degrees of freedom. The transformed likelihood factors into the product of marginal normal densities for $\{x_{2i}\}$ and the conditional normal densities of $\{x_{1i}|x_{2i}\}$. The former product depends only on τ_2 and ω_{22} which we call $\eta = (\omega_{22}, \tau_2)$. The latter product depends on location parameter $\tau_3 = \tau_1 - \beta\tau_2$, regression coefficient $\beta = \omega_{12}/\omega_{22}$, and the residual error $\omega_{1.2} = \omega_{11} - \omega_{12}^2/\omega_{22}$ collectively denoted as $\chi_o = (\tau_3, \beta, \omega_{1.2})^T$. The two sets of parameters η and χ_o are variation independent as a result of the L-independence of $\{x_{2i}\}$ and $\{x_{1i}|x_{2i}\}$ (Barndorff-Nielsen, 1978, §3.3) and the likelihood is completely separable in the two sets of parameters. The implication of this is that the collection of conditional densities $\{x_{1i}|x_{2i}\}$ and their associated orthogonal parameters may be ignored when computing $\hat{u}$. This idea is made precise in the next section. The consequence is that when determining $\hat{u}$, it suffices to only consider the set of location densities for $\{x_{2i}\}$ and such consideration leads to $\hat{u}$ as given in (9.14).

9.6 Variation independence, conditional likelihood, and marginal likelihood

These three concepts deal with the extent to which interest parameter ψ can be separated from nuisance parameter χ in the likelihood function. The expressions for $\hat{w}_\psi$ and $\hat{u}$ simplify for easier computation in the first two cases, and under restrictive conditions in the third case. These simpler expressions provide further insight into the nature of the approximation in (9.3) and how it deals with nuisance parameters.

Partition $\xi = (\chi_o, \chi_n, \psi)$ into two parts $\eta = (\chi_n, \psi)$ and χ_o and suppose that the likelihood is variation independent in η and χ_o. This means that (η, χ_o) varies over a rectangle (generally infinite) and $\partial^2 l / \partial \chi_o \partial \eta^T \equiv 0$ on that rectangle. From a Bayesian perspective, variation independence in the likelihood and independent priors imply posterior independence.

9.6.1 Variation independence

The next theorem is a slight generalization of the discussion in Barndorff-Nielsen (1991, §4).

Theorem 9.6.1 *If $\eta = (\chi_n, \psi)$ is variation independent of χ_o, then the values of $\hat{w}_\psi$ and $\hat{u}$ are the same whether computed from the entire likelihood based on parameter ξ or computed from the portion of the likelihood depending only on the reduced parameter η.*

Examples

F-ratio The ratio of exponential rates is $\psi = \lambda/\mu$. Transform the exponential sums $y \sim$ Gamma $(2m, \lambda)$ and $z \sim$ Gamma $(2n, \mu)$ to $z + \psi y \sim$ Gamma $(2m + 2n, \mu)$ and $\psi y/(z + \psi y) \sim$ Beta (m, n) where the beta variable is independent of its denominator sum. Since the distribution of $\hat{\psi}$ is $\psi F_{2n,2m}$ and determined only by the beta, the likelihood separates into the product of a gamma density, that depends on μ, and the density of $\hat{\psi} \sim \psi F_{2n,2m}$ that depends on ψ. Theorem 9.6.1 with $\chi_o = \mu$ and $\chi_n = \emptyset$ now explains the equivalence of the marginal likelihood results with those determined from the entire likelihood.

Interestingly enough, the value of $z + \psi y$ is not observable, however that does not exclude it from being used to separate the likelihood so that theorem 9.6.1 may be applied.

Fieller's pivot The justification for $\hat{w}_\psi$ and $\hat{u}$ in (9.35) has been based on theorem 9.6.1. In the transformed data, the likelihood separates into a regression model, based on the conditional densities $\{x_{i1}|x_{i2}\}$ with $\chi_o = (\tau_3, \beta, \omega_{2.1})$, and a location/scale model involving $\eta = (\omega_{22}, \tau_2)$ where $\chi_n = \omega_{22}$ and $\psi = \tau_2$. By theorem 9.6.1, the likelihood may be restricted to the location/scale model that is a special case of the regression in (9.14).

Normal linear regression At first sight, theorem 9.6.1 might appear to apply to this setting but it does not. Consider the simple linear regression with orthogonal independent variables; the more general case uses the same argument. The likelihood is a $(3, 3)$ regular exponential family and three mutually independent components constructed from (9.17) would appear to separate the likelihood:

$$n\hat{\sigma}_\psi^2 = n\hat{\sigma}^2 + (\hat{\psi} - \psi)^2 x_2^T x_2$$
$$\hat{\beta}_{1\psi} = \bar{y} \quad \text{and} \quad B = (\hat{\psi} - \psi)^2 x_2^T x_2 / (n\hat{\sigma}_\psi^2).$$

9.6 Variation independence, conditional likelihood, and marginal likelihood

The densities of $n\hat{\sigma}_\psi^2$ and $\hat{\beta}_{1\psi}$ depend on (β_1, σ^2) while $B \sim \text{Beta}(1, n-1)$ has no parameter dependence. Unfortunately the pivotal nature of B or equivalently t^2 is part of the difficulty. No single *statistic* can express the pivotal nature of the t^2 statistic determined from B as an expression for the marginal likelihood of ψ. Both $\hat{\psi}$ and $\hat{\sigma}^2$ are needed and this places σ^2 as a component of χ_o and also χ_n. Thus the marginal likelihood taken in (9.16) must generally lead to a different approximation than (9.14).

Proof of theorem 9.6.1

Only the value of $\hat{u}$ requires explanation. Consider the three blocks χ_o, χ_n and ψ for the matrix of sample space derivatives. Variation independence of χ_o and η assures that when ψ is fixed, $\hat{\chi}_{o\psi} = \hat{\chi}_o$. Separation also assures that the off-block information is zero, e.g.

$$j_{\chi_o \eta}(\hat{\xi}_\psi) = -l_{\chi_o \eta;}(\hat{\xi}_\psi; \hat{\xi}, a) = 0.$$

Further simplifications start with the MLE equations

$$l_{\chi_o;}(\hat{\chi}_o) = l_{\chi_o;}(\hat{\chi}_o, \hat{\eta}_\psi) = 0 \qquad (9.36)$$
$$l_{\chi_n;}(\hat{\eta}_\psi) = l_{\chi_n;}(\hat{\chi}_o, \hat{\eta}_\psi) = 0$$

in which for simplicity the dependence on MLE $\hat{\xi}$ and ancillary a to the right of the semicolon are suppressed. Since the likelihood separates in χ_o and η, $l_{\chi_o;}$ does not depend on $\hat{\eta}_\psi$ to the left of the semicolon; likewise $l_{\chi_n;}$ does not depend on $\hat{\chi}_o$ to the left of the semicolon. Differentiation in $\hat{\chi}_o$ leads to

$$l_{\chi_o;\hat{\chi}_o}(\hat{\chi}_o, \hat{\eta}_\psi) = j_{\chi_o \chi_o}(\hat{\chi}_o)$$
$$l_{\chi_n;\hat{\chi}_o}(\hat{\chi}_o, \hat{\eta}_\psi) = 0.$$

Differentiation of (9.36) in $\hat{\eta}$ shows $l_{\chi_o;\hat{\eta}}(\hat{\xi}_\psi) = 0$. In block form the matrix of sample space derivatives ordered in the sequence χ_o, χ_n, ψ is

$$\left\| \begin{matrix} j_{\chi_o \chi_o}(\hat{\chi}_o) & 0 & 0 \\ 0 & l_{\chi_n;\hat{\eta}} \\ - & \Delta l_{;\hat{\eta}} \end{matrix} \right\| = |j_{\chi_o \chi_o}(\hat{\chi}_o)| \times \left\| \begin{matrix} l_{\chi_n;\hat{\eta}} \\ \Delta l_{;\hat{\eta}} \end{matrix} \right\| \qquad (9.37)$$

where $l_{\chi_n;\hat{\eta}} = l_{\chi_n;\hat{\eta}}(\hat{\eta}_\psi)$ and $\Delta l_{;\hat{\eta}} = l_{;\hat{\eta}}(\hat{\eta}) - l_{;\hat{\eta}}(\hat{\eta}_\psi)$. Note that the second factor in (9.37) is the matrix of sample space derivatives for the portion of the likelihood depending on the reduced parameterization in η. The information determinants factor as

$$|j_{\chi\chi}(\hat{\xi}_\psi)| = |j_{\chi_o \chi_o}(\hat{\chi}_o)||j_{\chi_n \chi_n}(\hat{\eta}_\psi)|$$
$$|j_{\hat{\xi}}| = |j_{\chi_o \chi_o}(\hat{\chi}_o)||j_{\eta\eta}(\hat{\eta})|.$$

Thus $\hat{u}$ from the full parameterization ξ has the value

$$\hat{u} = \frac{\text{sgn}(\hat{\psi} - \psi)}{\sqrt{|j_{\chi_n \chi_n}(\hat{\eta}_\psi)||j_{\eta\eta}(\hat{\eta})|}} \left\| \begin{matrix} l_{\chi_n;\hat{\eta}}(\hat{\eta}_\psi) \\ l_{;\hat{\eta}}(\hat{\eta}) - l_{;\hat{\eta}}(\hat{\eta}_\psi) \end{matrix} \right\|$$

which is its value for the reduced parameterization. □

9.6.2 Conditional likelihood

Simpler computations for $\hat{u}$ exist under weaker conditions than have been given in theorem 9.6.1 and in particular when interest parameter ψ admits a conditional likelihood. Often $\hat{\xi}$ is more easily expressed in terms of a 1-1 transformation $\hat{\xi} \leftrightarrow (t, u)$ where t and u are themselves MLEs but for different parameters. Parameter ψ admits a conditional likelihood g if the likelihood, or equivalently the joint density of (t, u), factors as

$$f(t, u; \xi) = g(t|u; \psi) h(u; \xi). \tag{9.38}$$

The log conditional likelihood is $_c l(\psi; t|u) = \ln g(t|u; \psi)$.

Theorem 9.6.2 *(Conditional likelihood). If ψ has conditional likelihood $_c l(\psi) = {}_c l(\psi; t|u)$, then*

$$\hat{u} = \mathrm{sgn}(\hat{\psi} - \psi) |{}_c l_{;t}(\hat{\psi}) - {}_c l_{;t}(\psi)| \sqrt{|j_{uu}(\hat{\xi}_\psi)| / |j_{(u,t)}(\hat{\xi})|} \tag{9.39}$$

where

$$|j_{uu}(\hat{\xi}_\psi)| = |j_{\hat{\chi}\hat{\chi}}(\hat{\xi}_\psi)| \, |\partial \hat{\chi}_\psi / \partial u^T|^2$$
$$|j_{(u,t)}(\hat{\xi})| = |j_{\hat{\xi}}(\hat{\xi})| \, |\partial(\hat{\chi}, \hat{\psi}) / \partial(u, t)^T|^2$$

are the information transformations from $\hat{\chi}$ to u and $\hat{\xi}$ to (u, t) respectively.

Examples

Regular exponential family Suppose that (u, t) is the canonical sufficient statistic of an exponential family with corresponding canonical parameter $\theta = (\chi, \psi) = \xi$. Then the change of $_c l_{;t}$ as expressed in (9.39) is $|\hat{\psi} - \psi|$. Since (u, t) is the MLE of the mean parameterization, then $|j_{(u,t)}(\hat{\xi})| = |j_{\hat{\theta}}|^{-1}$ by relationship (5.10). Also, if τ is the mean of u,

$$|j_{uu}(\hat{\xi}_\psi)| = |j_{\tau\tau}(\hat{\xi}_\psi)| = |j_{\chi\chi}(\hat{\xi}_\psi)| \, |\partial \chi / \partial \tau^T|^2_{\xi = \hat{\xi}_\psi} = |j_{\chi\chi}(\hat{\xi}_\psi)|^{-1}$$

and

$$\hat{u} = (\hat{\psi} - \psi) \sqrt{|j_{\hat{\theta}}| / |j_{\chi\chi}(\hat{\theta}_\psi)|}$$

as given in (5.78).

Normal linear regression For the regression model of section 9.2.3, let the interest parameter be $\psi = \sigma$ with $\chi = \beta$. Since $\hat{\psi}$ and $\hat{\chi}$ are independent, the conditional likelihood g is determined as the marginal distribution of $t = n\hat{\psi}^2 \sim \sigma^2 \chi^2_{n-p}$. The normal density for $\hat{\chi}$ depends on χ, ψ and determines h. The value of $\hat{u}$, whether computed from (9.39) or from (9.4) is

$$\hat{u} = (\hat{\sigma}^2/\sigma^2 - 1)(\hat{\sigma}^2/\sigma^2)^{p/2} \sqrt{n/2}. \tag{9.40}$$

Proof of Theorem 9.6.2

Use the chain rule to determine that

$$(l_{;\hat{\chi}}, l_{;\hat{\psi}}) = \left\{ {}_c l_{;u}(\psi) + l_{;u}(\xi; u) \quad {}_c l_{;t}(\psi) \right\} \begin{pmatrix} \partial u / \partial \hat{\chi}^T & \partial u / \partial \hat{\psi} \\ \partial t / \partial \hat{\chi}^T & \partial t / \partial \hat{\psi} \end{pmatrix} \tag{9.41}$$

9.6 Variation independence, conditional likelihood, and marginal likelihood

where $l(\xi; u) = \ln h(u; \xi)$ and $_c l_{;u}(\psi) = {}_c l_{;u}(\psi; t|u)$. Denote the latter matrix in (9.41) as J and take the derivative in χ to get

$$l_{\chi;\hat{\xi}}(\hat{\xi}_\psi) = \{ l_{\chi;u}(\hat{\xi}_\psi; u) \quad 0_{m-1} \} J. \tag{9.42}$$

Differentiating the MLE equation $l_{\chi;}(\hat{\xi}_\psi; u) = 0$ with respect to u directly shows that

$$l_{\chi;u}(\hat{\xi}_\psi; u) = j_{\chi\chi}(\hat{\xi}_\psi) \partial \hat{\chi}_\psi / \partial u^T. \tag{9.43}$$

The bottom row in the matrix of sample space derivatives has a term that uses the abbreviated notation

$$l_{;t}(\hat{\psi}; \hat{\xi}) - l_{;t}(\psi; \hat{\xi}) = {}_c l_{;t}(\hat{\psi}) - {}_c l_{;t}(\psi) = \Delta_c l_{;t}.$$

Thus the matrix of sample space derivatives consists of an upper left $(m-1) \times (m-1)$ block for $l_{\chi;\hat{\chi}}$ and a lower right 1×1 block for $\Delta l_{;\hat{\psi}}$ that may be written as

$$\begin{pmatrix} l_{\chi;u}(\hat{\xi}_\psi; u) & 0_{m-1} \\ - & \Delta_c l_{;t} \end{pmatrix} J. \tag{9.44}$$

The determinant of (9.44) has factor $|\Delta_c l_{;t}|$ times the determinant of its cofactor times $||J||$. From (9.43), this leads to

$$\hat{u} = \frac{\operatorname{sgn}(\hat{\psi} - \psi)}{\sqrt{|j_{\chi\chi}(\hat{\xi}_\psi)| \, |j_{\hat{\xi}}|}} |\Delta_c l_{;t}| \, |j_{\chi\chi}(\hat{\xi}_\psi)| \, ||\partial \hat{\chi}_\psi / \partial u^T|| \, ||J||$$

$$= \operatorname{sgn}(\hat{\psi} - \psi) |\Delta_c l_{;t}| \sqrt{|j_{uu}(\hat{\xi}_\psi)| / |j_{(u,t)}(\hat{\xi})|} \tag{9.45}$$

upon transforming the information matrices.

9.6.3 Marginal likelihood

In this approach, the likelihood factors into two parts as in (9.38) but the marginal density of u isolates the interest parameter ψ instead of the conditional density. Thus if

$$f(t, u; \xi) = g(t|u; \xi) h(u; \psi), \tag{9.46}$$

then $_m l(\psi|u) = \ln h(u; \psi)$ is the log-marginal likelihood. Often the computation of $\hat{u}$ is not simplified by using the marginal likelihood structure. In order to get a simple form the auxiliary portion $l(\xi; t|u) = \ln g(t|u; \xi)$ must be such that $l_{\chi;\hat{\psi}}(\hat{\xi}_\psi; t|u) = 0$. Exercise 12 considers the particulars of this setting as well as some examples.

9.6.4 Accuracy with many nuisance parameters

When there are many nuisance parameters in the model, two conclusions may be drawn about the accuracy of approximation (9.3) or its simpler version in (9.39) based on conditional likelihood. First, if the nuisance parameters are variation independent, then no accuracy is lost as has been stated in theorem 9.6.1. However without this strict condition, then the examples and discussion have provided hints that this accuracy may be lost when there are a large number of nuisance parameters. The next example reveals the nature of the problem.

The Neyman and Scott (1948) problem

Suppose a one-way ANOVA has two observations per level and also an increasing number of levels. Thus let $y_{ij} = \mu_i + \sigma e_{ij}$ for $i = 1, .., m$ and $j = 1, 2$ with i.i.d. normal errors. Consider the behavior of (9.3) as $m \to \infty$.

Inference for $\psi = \mu_1$ The t pivot is $t = (\bar{y}_{i\cdot} - \mu_1)/\hat{\sigma}$ with

$$\hat{\sigma}^2 = \sum_{i=1}^{m}(y_{i1} - y_{i2})^2/(4m) \stackrel{\text{wp 1}}{\to} \sigma^2/2$$

and inconsistent. As $m \to \infty$, the values in (9.14) become

$$\hat{w}_\psi = \text{sgn}(t)\sqrt{2m \ln\{1 + t^2/m\}} \sim t\sqrt{2} \stackrel{\text{wp 1}}{\to} \sqrt{2}z$$
$$\hat{u} = t\{1 + t^2/m\}^{-(m+1)/2}\sqrt{2} \sim te^{-t^2/2}\sqrt{2} \stackrel{\text{wp 1}}{\to} \sqrt{2}ze^{-z^2/2} \quad (9.47)$$

where $z = \sqrt{2}(\bar{y}_{1\cdot} - \mu_1)/\sigma$ is a standard score. The limiting forms in (9.47) are interesting from the point of view of considering whether the relative error of the approximation is preserved with large m. For fixed value of $\bar{y}_{1\cdot}$ or z, the limiting percentage relative error $R(z)$ is uniformly bounded in $z \in \Re$ since

$$\lim_{z \to \pm\infty} R(z) = 100(1/\sqrt{2} - 1)\% \simeq -29.29\%.$$

Also it can be shown that $R(z)$ is symmetric about $z = 0$ with $0\% \geq R(z) > -29.3\%$ for all $z \geq 0$. Altogether this leads to the statement that the limiting relative error (as $m \to \infty$) is bounded between 0% and -29.3% with probability 1. The use of (9.3) for inference about μ_1 has not suffered badly. However note that the ordinary t pivot does not suffer too badly either since $t = (\bar{y}_{i\cdot} - \mu_1)/\hat{\sigma} \sim$ Student-t_{2m-1} and $t \stackrel{D}{\to} z \sim$ Normal $(0, 1)$ as $m \to \infty$. Exercise 13 outlines the arguments required for the above statements.

Inference for $\psi = \sigma^2$ The difficulty with nuisance parameters enters when using (9.3) to infer the variance. In this context

$$\hat{w}_\psi = \text{sgn}(\hat{\sigma}^2/\sigma^2 - 1)\sqrt{2m\{\hat{\sigma}^2/\sigma^2 - 1 - \ln(\hat{\sigma}^2/\sigma^2)\}} \quad (9.48)$$

and $\hat{u}$ is given in (9.40) as

$$\hat{u} = (\hat{\sigma}^2/\sigma^2 - 1)(\hat{\sigma}^2/\sigma^2)^{m/2}\sqrt{m}. \quad (9.49)$$

As $m \to \infty$, $\hat{w}_\psi \sim -\sqrt{2m}\sqrt{-1/2 + \ln 2}$ and $\hat{u} \sim -2^{-(m+1)/2}\sqrt{m}$. The relative error cannot stay bounded and all probability computations using (9.3) necessarily converge to 0. See Exercise 13 for details.

Solutions for many nuisance parameters

The Neyman–Scott example shows that the degree to which (9.3) is affected by an increasing number of nuisance parameters depends on the particular context. However, some general conclusions may be drawn.

9.6 Variation independence, conditional likelihood, and marginal likelihood

With many nuisance parameters, both (9.3) and its r^* version in (8.4) will tend to diminish in accuracy. Jensen (1992, lemma 2.1) has shown that the relative value of these two approximations is uniformly small over large deviation sets in which $r^* = O(\sqrt{n})$. These theoretical results combined with extensive numerical work point to very little difference in the ability of the two formulas to contend with large numbers of parameters.

When conditional likelihoods are used in place of ordinary likelihoods, the accuracy of (9.3) tends to improve considerably. This recommendation is consistent with Bartlett's (1937) recommendation to use conditional and marginal likelihood ratio tests instead of the ordinary likelihood ratio. If a conditional likelihood cannot be determined, then an approximate conditional likelihood is available. For example, a modified profile likelihood has been given in (3.2) of Barndorff-Nielsen (1983). His Example 3.5 shows that it yields the correct pivot $2m\hat{\sigma}^2 \sim \sigma^2 \chi_m^2$ for the Neyman–Scott problem. Sartori (2003) provides further support for modified profile likelihood.

In the conditional likelihood setting, the difficulties with $\hat{w}_\psi$ and $\hat{u}$ from the full likelihood are more transparent. For $\hat{w}_\psi$ the likelihood ratio is the conditional likelihood ratio times $h(u; \hat{\xi}_\psi)/h(u; \hat{\xi})$ and this latter factor can be substantial if $\dim(\xi)$ is large. Difficulties with the full likelihood $\hat{u}$ may be traced to the information determinants appearing in (9.39). If the exact conditional likelihood were used instead, then, according to (8.3), the conditional likelihood information for t or $\{-{}_cl_{;tt}(\hat{\psi}; t|u)\}^{-1/2}$ should be used in place of $\sqrt{|j_{uu}(\hat{\xi}_\psi)|/|j_{(u,t)}(\hat{\xi})|}$. However note the similarity of these two expressions. If in the latter, $\hat{\xi}$ is replaced by $\hat{\xi}_\psi$, then

$$|j_{(u,t)}(\hat{\xi}_\psi)| / |j_{uu}(\hat{\xi}_\psi)| = \left(j_{tt} - j_{tu} j_{uu}^{-1} j_{ut}\right)(\hat{\xi}_\psi) \tag{9.50}$$

and the right side of (9.50) is the asymptotic computation of conditional information for t given u implicitly assuming that the distribution of (u, t) is approximately normal. Thus there are two problems: the need to replace $\hat{\xi}$ with $\hat{\xi}_\psi$, and the use of the asymptotic rather than exact conditional likelihood information when computing $\hat{u}$.

Neyman and Scott example continued Transform the data $\{y_{i1}, y_{i2}\} \to \{x_{i1}, x_{i2}\}$ using the orthogonal transformation $x_{i1} = (y_{i1} - y_{i2})/\sqrt{2}$ and $x_{i2} = (y_{i1} + y_{i2})/\sqrt{2}$ so that $\{x_{i1}\}$ are i.i.d. Normal $(0, \sigma^2)$ and independent of $\{x_{i2}\}$ for which x_{i2} is Normal $(\sqrt{2}\mu_i, \sigma^2)$. The conditional likelihood of σ^2 is the density of $t = m\tilde{\sigma}^2 = \sum_{i=1}^m x_{i1}^2 \sim \sigma^2 \chi_m^2$. The conditional likelihood gets the degrees of freedom right and, if used in place of the full likelihood, (9.3) and (8.4) offer saddlepoint accuracy as would be seen if Lugannani and Rice were applied to the Gamma $(m/2, 1/2)$ distribution. The input values are

$$
\begin{aligned}
{}_c\hat{w}_\xi &= \text{sgn}(\tilde{\sigma}^2/\sigma^2 - 1)\sqrt{m\{\tilde{\sigma}^2/\sigma^2 - 1 - \ln(\tilde{\sigma}^2/\sigma^2)\}} \\
{}_c\hat{u} &= \{-{}_cl_{;tt}(\tilde{\sigma})\}^{-1/2}\{{}_cl_{;t}(\tilde{\sigma}) - {}_cl_{;t}(\sigma)\} = (\tilde{\sigma}^2/\sigma^2 - 1)m/\sqrt{2(m-2)},
\end{aligned}
$$

where $\tilde{\sigma}^2 = 2\hat{\sigma}^2$ is the maximum conditional likelihood estimate. Note that both ${}_c\hat{w}_\xi$ and ${}_c\hat{u}$ use the correct degrees of freedom m. Furthermore ${}_c\hat{u}$ converges to a Normal $(0, 1)$ since the asymptotic distribution of $\tilde{\sigma}^2$ is Normal $(\sigma^2, 2\sigma^4/m)$.

By contrast, the full likelihood, as used in (9.39) yields the values in (9.48) and (9.49) and fails to provide comparable accuracy. Both input terms incorrectly take the degrees of freedom as $2m$. The use of $\hat{\sigma}^2$ rather than $\tilde{\sigma}^2$ is most critical. Since t and complementary

u are independent, $J_{tu} = 0$ and the value on the right side of (9.50) is $\hat{j}_{tt}(\hat{\xi}_\psi) = -{}_c l_{;tt}(\sigma)$ which would be correct except for the fact that it is evaluated at σ rather than $\tilde{\sigma}$ as in ${}_c\hat{u}$. If $\hat{\xi}_\psi$ rather than $\hat{\xi}$ were used in the information determinants than the extra factor $(\hat{\sigma}^2/\sigma^2)^{m/2}$ would be 1 so that $\hat{u}$ would conform more closely with ${}_c\hat{u}$.

Sequential saddlepoint approximation In chapter 10, the sequential saddlepoint method is introduced. Its greatest asset is its ability to maintain accuracy in the presence of a large numbers of nuisance parameters. Indeed the sequential saddlepoint method may be thought of as a formal method for reducing the likelihood to an approximate conditional likelihood and then using it as the input into the saddlepoint approximation.

9.7 Examples

9.7.1 Generalized Behrens–Fisher problem

Suppose that data are samples from a treatment group and a control group and that the two groups may have different scale parameters. The Behrens–Fisher problem is concerned with testing that the two groups have the same location parameters and more generally in determining a confidence interval for the difference in location parameters with the different scaling factors.

Suppose the general location/scale model with independent responses

$$y_i = \mu + \sigma e_i \quad i = 1, \ldots, m$$
$$z_j = \mu + \psi + \tau f_j \quad j = 1, \ldots, n$$

for which $\{e_i\}$ and $\{f_j\}$ are i.i.d. with density g and let $h = \ln g$. If y denotes the vector of $\{y_i\}$, etc. then, under the full model with the interest parameter also unknown, the ancillaries are the m-vector $a = (y - \hat{\mu}1)/\hat{\sigma}$ and the n-vector $b = (z - \hat{\mu} - \hat{\psi})/\hat{\tau}$. The parameter is $\xi = (\chi, \psi)$ with $\chi = (\mu, \sigma, \tau)$ and the log-likelihood is

$$l(\xi) = -m \ln \sigma - n \ln \tau + \sum_{i=1}^{m} h(e_i) + \sum_{j=1}^{n} h(f_j).$$

Many of the details for this example are developed in Exercise 15. With ψ fixed, define the fitted residuals as

$$\hat{e}_{i\psi} = (y_i - \hat{\mu}_\psi)/\hat{\sigma}_\psi \quad \text{and} \quad \hat{f}_{j\psi} = (z_j - \hat{\mu}_\psi - \psi)/\hat{\tau}_\psi.$$

With ψ as a parameter free to vary, the information determinant is that for two independent samples and is therefore

$$|j_{\hat{\xi}}| = \hat{\sigma}^{-4}\hat{\tau}^{-4}|K_m(a)||K_n(b)| \tag{9.51}$$

where, for example, $K_m(a)$ is given in (9.81). With ψ fixed, then

$$|j_{\chi\chi}(\hat{\xi}_\psi)| = j_{\hat{\sigma}_\psi} j_{\hat{\tau}_\psi} \{j_{\hat{\mu}_\psi} - j^2_{\hat{\mu}_\psi \hat{\sigma}_\psi}/j_{\hat{\sigma}_\psi} - j^2_{\hat{\mu}_\psi \hat{\tau}_\psi}/j_{\hat{\tau}_\psi}\} \tag{9.52}$$

where expressions for each of the information components are given in (9.82). The variety of terms involved in these computations suggests that it is most efficient to use the following notation. Define 3×3 matrices

$$
{}_1^e H = \sum_{i=1}^m h'(\hat{e}_{i\psi}) \begin{pmatrix} 1 \\ \hat{e}_{i\psi} \\ a_i \end{pmatrix} (1 \ \hat{e}_{i\psi} \ a_i) = \begin{pmatrix} {}_1^e H_{11} & {}_1^e H_{1\hat{e}} & {}_1^e H_{1a} \\ - & {}_1^e H_{\hat{e}\hat{e}} & {}_1^e H_{\hat{e}a} \\ - & - & {}_1^e H_{aa} \end{pmatrix}
$$

in which left superscript e denotes the first group and left subscript 1 indicates that h', with one derivative, is the weight function. With weight function $h''(\hat{e}_{i\psi})$ the matrix would be ${}_2^e H$. Matrices ${}_1^f H$ and ${}_2^f H$ are also defined analogously using terms from the second sample. With quite a lot of effort the absolute determinant of the matrix of sample space derivatives may be computed as

$$
\hat{\sigma}_\psi^{-2} \hat{\tau}_\psi^{-2} \begin{Vmatrix} \frac{1}{\hat{\sigma}_\psi^2} {}_2^e H_{11} + \frac{1}{\hat{\tau}_\psi^2} {}_2^f H_{11} & \frac{1}{\hat{\sigma}_\psi^2} {}_2^e H_{1a} & \frac{1}{\hat{\tau}_\psi^2} {}_2^f H_{1b} & \frac{1}{\hat{\tau}_\psi^2} {}_2^f H_{11} \\ {}_1^e H_{11} + {}_2^e H_{1\hat{e}} & {}_1^e H_{1a} + {}_2^e H_{\hat{e}a} & 0 & 0 \\ {}_1^f H_{11} + {}_2^f H_{1\hat{f}} & 0 & {}_1^f H_{1b} + {}_2^f H_{\hat{f}b} & {}_1^f H_{11} + {}_2^f H_{1\hat{f}} \\ 0 & \frac{m}{\hat{\sigma}} + \frac{1}{\hat{\sigma}_\psi} {}_1^e H_{1a} & \frac{n}{\hat{\tau}} + \frac{1}{\hat{\tau}_\psi} {}_1^f H_{1b} & \frac{1}{\hat{\tau}_\psi} {}_1^f H_{11} \end{Vmatrix} \tag{9.53}
$$

where the row ordering is $\chi = (\mu, \sigma, \tau)^T$ followed by the change in $l_{;\hat{\xi}}$ and the columns are $\hat{\xi} = (\hat{\mu}, \hat{\sigma}, \hat{\tau}, \hat{\psi})^T$. In (9.53), negative signs have been removed from all entries for simplicity.

Normal errors

Let $\hat{\mu}, \hat{\sigma}, \hat{\psi}$, and $\hat{\tau}$ denote the ordinary unconstrained MLEs. Subscript with ψ for the constrained MLEs that satisfy (9.80) or, for normal errors,

$$
0 = \hat{\sigma}_\psi^{-2} \sum_{i=1}^m (y_i - \hat{\mu}_\psi) + \hat{\tau}_\psi^{-2} \sum_{j=1}^n (z_j - \hat{\mu}_\psi - \psi)
$$

$$
\hat{\sigma}_\psi^2 = m^{-1} \sum_{i=1}^m (y_i - \hat{\mu}_\psi)^2 \quad \hat{\tau}_\psi^2 = n^{-1} \sum_{j=1}^n (z_j - \hat{\mu}_\psi - \psi)^2. \tag{9.54}
$$

Using simple ANOVA identities it can be shown that

$$
\hat{w}_\psi = \mathrm{sgn}(\hat{\psi} - \psi) \sqrt{m \ln(\hat{\sigma}_\psi^2/\hat{\sigma}^2) + n \ln(\hat{\tau}_\psi^2/\hat{\tau}^2)}
$$
$$
= \mathrm{sgn}(\hat{\psi} - \psi) \sqrt{-m \ln(1 - \hat{\sigma}_\psi^2 P^2/m^2) - n \ln(1 - \hat{\tau}_\psi^2 P^2/n^2)}
$$

where

$$
P = -\frac{m}{\hat{\sigma}_\psi^2}(\bar{y} - \hat{\mu}_\psi) = \frac{n}{\hat{\tau}_\psi^2}(\bar{z} - \hat{\mu}_\psi - \psi)
$$
$$
= (\hat{\sigma}_\psi^2/m + \hat{\tau}_\psi^2/n)^{-1}(\bar{z} - \bar{y} - \psi) \tag{9.55}
$$

plays a role similar to a pivot even though it is not a standardized variable. Note that P is $\hat{\psi} - \psi = \bar{z} - \bar{y} - \psi$ divided by an estimate of its variance and not by its standard deviation.

The value of $\hat{u}$ is also expressed naturally in terms of P. The rather long and tedious derivation is outlined in Exercise 16 and leads to

$$
\hat{u} = P \frac{\hat{\sigma}\hat{\tau}}{\hat{\sigma}_\psi \hat{\tau}_\psi} \frac{\hat{\sigma}_\psi^2/m + \hat{\tau}_\psi^2/n - \hat{\sigma}_\psi^2 \hat{\tau}_\psi^2 P^2(1/m + 1/n)/(mn)}{\sqrt{\hat{\sigma}_\psi^2/m + \hat{\tau}_\psi^2/n - 2\hat{\sigma}_\psi^2 \hat{\tau}_\psi^2 P^2(1/m + 1/n)/(mn)}}. \tag{9.56}
$$

Table 9.3. *One-sided p-values for the test that $\psi = 0$ and 95% confidence intervals for the various methods listed in the columns.*

	Methods			
	(9.3)	Behrens–Fisher	Satterthwaite	Student-t
p-values	.04263	.05116	.04713	.08143
Conf. Ints.	$(-.309, 4.309)$	$(-.381, 4.381)$	$(-.412, 4.412)$	$(-.958, 4.958)$

The form of $\hat{u}$ may be understood by noting that the final ratio is, to high order of accuracy, simply $(\hat{\sigma}_\psi^2/m + \hat{\tau}_\psi^2/n)^{1/2}$. This is the term needed to standardize the leading factor P. The middle term $\hat{\sigma}\hat{\tau}/(\hat{\sigma}_\psi \hat{\tau}_\xi)$ affects the value to a lesser degree.

Jensen (1992) also considered this example with normal errors but rather approached it as a regular exponential family instead of as a linear model. Thus he computed $\hat{u}$ using (9.7) but did not reduce it to the explicit form in (9.56). Using the r^* form in (9.5), he showed that significance levels from r^* are closer to their nominal levels than for the Welsh–Aspin expansions to order $O(n^{-2})$ with $m = 10 = n$ and $m = 20 = n$. Furthermore, in his large deviation computations, he showed that r^* bounds relative error while the Welsh–Aspin approximation does not.

Numerical example Suppose the y-data are $\{1, 2, 3, 4\}$ with $\hat{\mu} = 5/2$ and $\hat{\sigma} = \sqrt{5/4}$, and the z-data are $\{1(1)8\}$ with $\bar{z} = 9/2$, $\hat{\psi} = 2$, and $\hat{\tau}^2 = 21/4$. A test that $\psi = 0$ leads to the following values in the computation of (9.3):

$$\hat{\mu}_0 = 3.1343 \qquad \hat{\sigma}_0 = 1.2854 \qquad \hat{\tau}_0 = 2.6674$$
$$P = 1.5355 \qquad \hat{w}_0 = 1.8836 \qquad \hat{u} = 1.3882.$$

Table 9.3 compares one-sided p-values of four methods for testing $\psi = 0$ as well as listing the associated 95% confidence intervals determined by inverting the tests. The Behrens–Fisher method approximates the pivotal quantity

$$d = \frac{\bar{z} - \bar{y} - \psi}{\sqrt{\hat{\sigma}^2/(m-1) + \hat{\tau}^2/(n-1)}}$$

as a weighted sum of two independent Student-t distributions; see Kendall and Stuart (1973, §21.26). In this instance, the small odd degrees of freedom allow for exact computation of the p-value and confidence intervals using the results in Fisher and Healy (1958) and are given in the table. The Satterthwaite method approximates the distribution of d as Student-t_f where f is estimated by the data and has been computed from the SPSS statistical package. The final method is the ordinary two sample t test assuming equal variances which gives quite different results from the other methods.

The two sample variances are not significantly different for this data. The F-test for equal variances has $F = 3.6$ and $p = 0.16$ while Levene's nonparametric test has $F = 2.42$ and $p = 0.15$.

9.7.2 Two samples with a common mean

In the context of the generalized Behrens–Fisher problem, assume $\psi = 0$ and suppose that inference concerns the common mean of two samples that have different scaling factors. A confidence interval for μ is determined by inverting the approximate distribution for the likelihood ratio statistic given in (9.3) when testing the various values of μ. The probability computation in (9.3) fixes ancillaries that are related to quantities from the Behrens–Fisher analysis. The constrained set of residuals evaluated with $\psi = 0$, or $\{a_i\} = \{\hat{e}_{i\psi}\}|_{\psi=0}$ and $\{b_j\} = \{\hat{f}_{j\psi}\}|_{\psi=0}$, now form the bulk of the ancillaries held fixed in this inference. The additional signed log-likelihood ratio statistic $\hat{w}_0$ for testing that $\psi = 0$ is the final ancillary included.

The three maximum likelihood equations for the parameters, ordered according to $\xi = (\sigma, \tau, \mu)^T$, are given in (9.80) with $\psi = 0$. The constrained MLEs that fix the value of μ are the bottom two equations only and determine the scale estimates $\hat{\sigma}_\mu$ and $\hat{\tau}_\mu$. In the most general setting with kernel density g and $h = \ln g$, the absolute determinant of the matrix of sample space derivatives is

$$\begin{Vmatrix} \frac{1}{\hat{\sigma}_\mu^2}\left({}_1^e H_{1a} + {}_2^e H_{\hat{e}a}\right) & 0 & \frac{1}{\hat{\sigma}_\mu^2}\left({}_1^e H_{11} + {}_2^e H_{1\hat{e}}\right) \\ 0 & \frac{1}{\hat{\tau}_\mu^2}\left({}_1^f H_{1b} + {}_2^f H_{\hat{f}b}\right) & \frac{1}{\hat{\tau}_\mu^2}\left({}_1^f H_{11} + {}_2^f H_{1\hat{f}}\right) \\ \frac{m}{\hat{\sigma}} + \frac{1}{\hat{\sigma}_\mu} {}_1^e H_{1a} & \frac{n}{\hat{\tau}} + \frac{1}{\hat{\tau}_\mu} {}_1^f H_{1b} & \frac{1}{\hat{\sigma}_\mu} {}_1^e H_{11} + \frac{1}{\hat{\tau}_\mu} {}_1^f H_{11} \end{Vmatrix}. \quad (9.57)$$

In (9.57), negative signs have been removed from all cell entries to simplify the expressions. The notation used has been explained in the last example and, for example ${}_2^f H_{\hat{f}b}$ defines $\sum_{j=1}^n h''(\hat{f}_{j\mu}) \hat{f}_{j\mu} b_j$. The information matrix $j_{\chi\chi}(\hat{\xi}_\mu)$ is the product of $j_\sigma(\hat{\xi}_\mu)$ and $j_\tau(\hat{\xi}_\mu)$ where

$$j_\sigma(\hat{\xi}_\mu) = \hat{\sigma}_\mu^{-2}\left(m - {}_2^e H_{\hat{e}\hat{e}}\right) \qquad j_\tau(\hat{\xi}_\mu) = \hat{\tau}_\mu^{-2}\left(n - {}_2^f H_{\hat{f}\hat{f}}\right).$$

The matrix j_ξ is given in (9.52) by taking $\psi = 0$.

Normal errors

Deep reductions occur in the expressions for this case. The MLEs $\hat{\sigma}$, $\hat{\tau}$, and $\hat{\mu}$ solve the equations in (9.54) with $\psi = 0$. The constrained MLEs $\hat{\sigma}_\mu^2$ and $\hat{\tau}_\mu^2$ are simply the average sums of squares about μ. The entries for (9.3) reduce to

$$\hat{w}_\mu = \text{sgn}(\hat{\mu} - \mu)\sqrt{m \ln\left(\hat{\sigma}_\mu^2/\hat{\sigma}^2\right) + n \ln\left(\hat{\tau}_\mu^2/\hat{\tau}^2\right)} \quad (9.58)$$

$$\hat{u} = \left\{\eta_z(\mu) m(\bar{y} - \mu)/\hat{\sigma}_\mu^2 + \eta_y(\mu) n(\bar{z} - \mu)/\hat{\tau}_\mu^2\right\} \hat{\vartheta}^{-1/2}$$

where

$$\eta_y(\mu) = 1 + (\bar{y} - \mu)(\mu - \hat{\mu})/\hat{\sigma}_\mu^2 \qquad \eta_z(\mu) = 1 + (\bar{z} - \mu)(\mu - \hat{\mu})/\hat{\tau}_\mu^2$$

$$\hat{\vartheta} = \frac{m}{\hat{\sigma}^2} + \frac{n}{\hat{\tau}^2} - 2P_0^2\left(\frac{1}{m} + \frac{1}{n}\right) \qquad P_0 = (\hat{\sigma}^2/m + \hat{\tau}^2/n)^{-1}(\bar{z} - \bar{y}).$$

At $\mu = \hat{\mu}$, $\eta_y(\hat{\mu}) = 1 = \eta_z(\hat{\mu})$ and $\hat{u} = 0$ since the term in the curly braces is the estimating equation for $\hat{\mu}$. The term in the curly braces has the same sign as $\hat{\mu} - \mu$.

Table 9.4. *Confidence intervals for μ using the various methods listed in the columns.*

	Methods		
95% Conf. Ints.	(9.3) (1.829, 5.378)	(9.5) (1.829, 5.366)	Normal (1.938, 4.963)

Numerical example continued The data from the Behrens–Fisher example are used to determine 95% confidence intervals for a common mean μ. Table 9.4 displays three intervals with nominal 95% coverage: intervals computed from (9.3) and its r^* version in (9.5) that are almost identical, and a normal interval that results from assuming that $\hat{w}_\mu$ has a Normal $(0, 1)$ distribution. The interpretations of the first two are different from the normal interval. In the first two, the 95% coverage is with respect to a reference set with a, b, and $\hat{w}_0$ fixed; the latter interval is generally not interpreted conditionally.

Also an interval was computed by using the approximate sample space derivatives suggested by Skovgaard in section 9.4.1. The curved exponential family structure of the model may be used to compute $\hat{u}$ by using (9.29). This leads to the interval (1.660, 5.582) and is more conservative than the exact sample space procedures. The normal approximation is too liberal, which is a policy best reserved for expressing political views.

9.7.3 Mixed normal linear models

The factorization of Patterson and Thompson (1971) that leads to *restricted maximum likelihood (REML) estimation* provides an example of conditional likelihood. Suppose that $y = X\beta + e$ with the fixed effects contained in the full rank $n \times p$ design matrix X and all random effects marginalized into the error $e \sim \text{Normal}_n(0, V)$. Covariance matrix $V = \sum_{i=1}^{m} \tau_i V_i$ with unknown variance components $\{\tau_i\}$ is typically a patterned matrix with the matrices $\{V_i\}$ expressing associated blocks of observations with ones and zeros according to the particular design. The framework for these models may be used to explain both balanced and unbalanced data sets.

The linear transformation

$$\begin{pmatrix} t \\ u \end{pmatrix} = \begin{pmatrix} I_n - P \\ X^T V^{-1} \end{pmatrix} y$$

with $P = X(X^T X)^{-1} X^T$ leads to t and u that are independent. The distribution of t is a singular normal with mean 0_n and covariance $(I_n - P)V(I_n - P)$ while $u \sim \text{Normal}_p (X^T V^{-1} X\beta, X^T V^{-1} X)$.

If the interest parameter depends only on the variance components $\{\tau_i\}$ in V, then the distribution of t is the conditional likelihood since it does not depend on the nuisance parameter β. According to the discussion above, it is perhaps better to compute $\hat{w}$ and $\hat{u}$ directly from the conditional likelihood. In this case the value of $\hat{w}^2$ is the conditional log-likelihood ratio based on REML estimates and $\hat{u}$ is accordingly computed directly from conditional likelihood.

The singularity in the distribution of t was dealt with in Patterson and Thompson (1971) by replacing the matrix $I_n - P$ with any set of $(n - p)$ orthonormal eigenvectors associated with the repeated eigenvalue 1. If these eigenvectors comprise the columns of $n \times (n - p)$ matrix O, then $z = O^T y \sim \text{Normal}_{n-p}(0, O^T V O)$ is the conditional likelihood. It is convenient to define

$$W = O^T V O = \sum_{i=1}^{m} \tau_i O^T V_i O = \sum_{i=1}^{m} \tau_i W_i$$

and denote $\tau = (\tau_1, \ldots, \tau_m)$.

Tests on individual variance components

Three methods are commonly used to perform such tests. Either approximate F-statistics are constructed by using method of moment estimators, or a full likelihood ratio test is used, or a conditional likelihood ratio test is used. We consider the latter two tests and attempt to improve upon χ_1^2 approximations for both by using (9.3). Often the test is for $H_0 : \tau_m = 0$ and confidence intervals for τ_m may be computed by inverting tests for the more general hypotheses $H_\tau : \tau_m = \tau$ versus $\tau_m \neq \tau$. For inference about ratios of variance components and other hypotheses, see Exercise 24. The development below assumes that $\tau_m = \tau \geq 0$.

Using conditional likelihood and REML

If the conditional MLE $\tilde{\tau}$ does not occur on the boundary of its parameter space, then it must solve the set of REML equations

$$\text{tr } \tilde{W}^{-1} W_i = z^T \tilde{W}^{-1} W_i \tilde{W}^{-1} z \qquad i = 1, \ldots, m \qquad (9.59)$$

where $\tilde{W} = \sum_i \tilde{\tau}_i W_i$. Additionally, let $\tilde{\tau}_\tau = \{\tilde{\tau}_{i\tau}\}$ denote the constrained MLE fixing $\tau_m = \tau$ that is determined by solving the first $m - 1$ equations in (9.59). For the moment, suppose the simple and more common situation in which all components of $\tilde{\tau}$ and $\tilde{\tau}_\tau$ are positive and solve their respective estimating equations. The signed conditional likelihood ratio statistic, used in (9.3), is $\text{sgn}(\tilde{\tau}_m - \tau)|_c \hat{w}_\tau|$ where

$$_c\hat{w}_\tau^2 = -2\{_cl(\tilde{\tau}_\tau) - {_c}l(\tilde{\tau})\} \qquad (9.60)$$

$$= \ln \frac{|\tilde{W}_\tau|}{|\tilde{W}|} - \tau \left\{ \text{tr}\left(\tilde{W}_\tau^{-1} W_m\right) - z^T \tilde{W}_\tau^{-1} W_m \tilde{W}_\tau^{-1} z \right\}$$

and

$$\tilde{W}_\tau = \sum_{i=1}^{m-1} \tilde{\tau}_{i\tau} W_i + \tau W_m \qquad \tilde{W} = \sum_{i=1}^{m} \tilde{\tau}_i W_i.$$

This value is derived in Exercise 20 from log-conditional likelihood

$$_cl(\tau) = -\tfrac{1}{2} \ln |W| - \tfrac{1}{2} \text{tr } W^{-1} z z^T \qquad (9.61)$$

that has the structure of a $\{(n - p)^2, m\}$ curved exponential family. The canonical parameter consists of the $(n - p)^2$ components of $-W^{-1}/2$ while the canonical sufficient statistic is the corresponding components of zz^T. Since there appears to be little hope of computing exact sample space derivatives, the approximate methods introduced by Skovgaard (1996) in section 9.4.1 are used with expression (9.29) for curved exponential families. Exercise 20 guides the reader through the details in deriving the information matrices and the approximate sample space derivative expressions. The $m \times m$ expected information matrix is

$$i_{\tilde{\tau}} = (i_{\tilde{\tau}_i;\tilde{\tau}_j}) = \tfrac{1}{2}\{\operatorname{tr}(\tilde{W}^{-1} W_i \tilde{W}^{-1} W_j)\} \tag{9.62}$$

with observed information as

$$j_{\tilde{\tau}} = (j_{\tilde{\tau}_i;\tilde{\tau}_j}) = (z^T \tilde{W}^{-1} W_i \tilde{W}^{-1} W_j \tilde{W}^{-1} z - i_{\tilde{\tau}_i \tilde{\tau}_j}). \tag{9.63}$$

The matrices of sample space derivative approximations for computing $_c\hat{u}$ are

$$_c l_{\tau_1,\ldots,\tau_{m-1};\tilde{\tau}}(\tilde{\tau}_\tau) \simeq i_{m-1,m}(\tilde{\tau}_\tau) i_{\tilde{\tau}}^{-1} j_{\tilde{\tau}} \tag{9.64}$$

$$_c l_{;\tilde{\tau}}(\tilde{\tau}) - {}_c l_{;\tilde{\tau}}(\tilde{\tau}_\tau) \simeq q(\hat{\tau}, \hat{\tau}_\tau)^T i_{\tilde{\tau}}^{-1} j_{\tilde{\tau}}$$

where

$$q(\tilde{\tau}, \tilde{\tau}_\tau)^T = -\tfrac{1}{2}\left[\operatorname{tr}\{(\tilde{W}^{-1} - \tilde{W}_\tau^{-1}) W_1\}, \ldots, \operatorname{tr}\{(\tilde{W}^{-1} - \tilde{W}_\tau^{-1}) W_m\}\right] \tag{9.65}$$

and $i_{m-1,m}(\tilde{\tau}_\tau)$ is the top $(m-1) \times m$ portion of the expected information matrix evaluated at $\tilde{\tau}_\tau$ or

$$i_{m-1,m}(\tilde{\tau}_\tau) = \tfrac{1}{2}\{\operatorname{tr}(\tilde{W}_\tau^{-1} W_i \tilde{W}_\tau^{-1} W_j)\} \qquad i = 1,\ldots,m-1; \quad j = 1,\ldots,m.$$

Putting it all together as in (9.29), then

$$_c\hat{u} = \operatorname{sgn}(\tilde{\tau}_m - \tau) \frac{|j_{\tilde{\tau}}|^{1/2}}{|i_{\tilde{\tau}}| |j_{\tau_1,\ldots,\tau_{m-1}}(\tilde{\tau}_\tau)|^{1/2}} \left\| \begin{array}{c} i_{m-1,m}(\tilde{\tau}_\tau) \\ q(\tilde{\tau}, \tilde{\tau}_\tau)^T \end{array} \right\| \tag{9.66}$$

where $j_{\tau_1,\ldots,\tau_{m-1}}(\tilde{\tau}_\tau)$ is the $(m-1) \times (m-1)$ nuisance parameter block of observed information j evaluated at $\tilde{\tau}_\tau$.

Confidence intervals from test inversion

In determining a confidence interval for τ_m, values of $\tau > 0$ are considered and this is the setting in which the asymptotic χ_1^2 distribution for $_c\hat{w}_\tau^2$ can be justified. A 95% two-sided confidence interval works with both tails of the distribution for $_c\hat{W}_\tau$ and selects the interval bounds (τ_L, τ_U) to satisfy

$$\widehat{\Pr}(_c\hat{W}_{\tau_L} < {}_c\hat{w}_{\tau_L}|a; \star\tau) = 0.025 = \widehat{\Pr}(_c\hat{W}_{\tau_U} > {}_c\hat{w}_{\tau_U}|a; \star\tau). \tag{9.67}$$

Often $\widehat{\Pr}(_c\hat{W}_0 < {}_c\hat{w}_0|a; \star\tau) \geq 0.025$ so that $\tau_L = 0$ is used instead.

The one-tailed χ_1^2 distribution can also be used by solving

$$\Pr\left(\chi_1^2 < {}_c\hat{w}_{\tau_L}^2\right) = 0.025 = \Pr\left(\chi_1^2 > {}_c\hat{w}_{\tau_U}^2\right).$$

If $\Pr(\chi_1^2 < {}_c\hat{w}_0^2) \geq 0.025$ then $\tau_L = 0$ is used.

One-sided tests of $H_0 : \tau_m = 0$ and estimates on the parameter boundary

When $\tau = 0$, the asymptotic distribution of $_c\hat{w}_0^2$ is no longer a χ_1^2 distribution. Furthermore $_c\hat{w}_0 \geq 0$ and cannot be negative. Saddlepoint expression (9.3) has the leading term $\Phi(_c\hat{w}_0) \geq$

1/2 and $_c\hat{W}_0$ has a one-tailed distribution on $[0, \infty)$. An appropriate p-value for testing $H_0 : \tau_m = 0$ versus $H_1 : \tau_m > 0$ is approximately $2\widehat{\Pr}(_c\hat{W}_0 > {}_c\hat{w}_0 | a; \star \tau)$ as discussed below.

The null hypothesis setting of $\tau = 0$ places the parameter τ_m on the boundary of the parameter space and the standard asymptotic argument leading to χ_1^2 for the likelihood ratio test does not apply. The asymptotic distribution of $_c\hat{W}_0$ is rather a mixture distribution that puts mass 1/2 at 0 and spreads mass 1/2 over $(0, \infty)$ as a χ_1^2; see Chernoff (1954). In this case Fisher's (1961) notion of the reference set becomes directly relevant. The sample space of y-values over which repeatability of the experiment should be considered and with respect to which p-values should be computed, is $\{y : \tilde{\tau}_m > 0\}$. When conditioning on the reference set, the point mass at zero is eliminated so that the conditional asymptotic distribution is χ_1^2 and not the mixture distribution. If $\{y : \tilde{\tau}_m > 0\} = \{y : {}_c\hat{w}_0 > 0\}$, then the ideal calculation with saddlepoint expression (9.3) would be

$$\widehat{\Pr}(_c\hat{W}_0 > {}_c\hat{w}_0 | a; \star \tau) / \widehat{\Pr}(_c\hat{W}_0 > 0 | a; \star \tau).$$

The denominator computation appears quite formidable since it needs to be computed as a limit. The leading term in its computation, however, is $\Phi(0) = 1/2$ and this term serves as the rationale for doubling the numerator in a one-sided p-value computation.

Of course a practical answer to all this is that we don't literally believe that τ_m is zero, just very small. In this instance the asymptotic χ_1^2 is still justified.

If a nuisance parameter, τ_1 say, is such that $\tilde{\tau}_1 = 0$ under the alternative hypothesis and $\tilde{\tau}_{10} = 0$ under the null model, then the conditional likelihood ratio statistic $_c\hat{w}_0$ is the same whether the random effect represented is included in the model or not. This supports the obvious solution to the problem of simply removing the effect from the model. See Exercise 23 for an example. Except for situations in which for example $\tilde{\tau}_1 = 0$ and $\tilde{\tau}_{10} > 0$, the common situation described above prevails in which all components of $\tilde{\tau}$ and $\tilde{\tau}_\tau$ are positive and solve their respective estimating equations.

Using the full likelihood

The degree to which the conditional likelihood analysis differs from the full likelihood analysis depends on (i) the number of fixed effect parameters p to be dealt with, and (ii) the informativeness of the data for the variance components. Large p and a lack of information for τ lead to a large difference in analyses. The full likelihood treatment of this testing problem, using Skovgaard's approximate sample space derivatives as outlined below, was first considered in Lyons and Peters (2000).

Suppose that the MLE $\hat{\tau}$ does not occur on the boundary of its parameter space. Then $(\hat{\beta}, \hat{\tau})$ solves the ML equations

$$\hat{\beta} = (X^T \hat{V}^{-1} X)^{-1} X^T \hat{V}^{-1} y \tag{9.68}$$
$$\text{tr}\, \hat{V}^{-1} V_i = (y - X\hat{\beta})^T \hat{V}^{-1} V_i \hat{V}^{-1} (y - X\hat{\beta}) \qquad i = 1, \ldots, m$$

where $\hat{V} = \sum_i \hat{\tau}_i V_i$. Additionally, let $\hat{\beta}_\tau$ and $\hat{\tau}_\tau = \{\hat{\tau}_{i\tau}\}$ denote the constrained MLEs fixing $\tau_m = \tau$ and assume the simple setting in which all components of $\hat{\tau}_\tau$ are positive. The likelihood ratio statistic is

$$\hat{w}_\tau^2 = -2\{l(\hat{\beta}_\tau, \hat{\tau}_\tau) - l(\hat{\beta}, \hat{\tau})\} \qquad (9.69)$$

$$= \ln \frac{|\hat{V}_\tau|}{|\hat{V}|} - \tau \left\{ \text{tr}\left(\hat{V}_\tau^{-1} V_m\right) - (y - X\hat{\beta}_\tau)^T \hat{V}_\tau^{-1} V_m \hat{V}_\tau^{-1} (y - X\hat{\beta}_\tau)\right\}.$$

Under H_τ, $\hat{w}_\tau^2$ is compared to a χ_1^2 distribution. Confidence intervals and p-values follow the same recommendations discussed for REML using the conditional likelihood.

Expressions for information matrices and sample space derivative approximations are given below with their derivations outlined in the steps of Exercise 21. The $(p+m) \times (p+m)$ expected information matrix $i_{(\beta,\tau)}$ is block diagonal with $i_{\beta\beta} = X^T V^{-1} X$, $i_{\beta\tau} = 0$, and

$$i_{\tau_i \tau_j} = \tfrac{1}{2} \text{tr}(V^{-1} V_i V^{-1} V_j). \qquad (9.70)$$

The observed information matrix $j_{(\beta,\tau)}$ is not zero on the off-diagonal and has values

$$j_{\beta\beta} = X^T V^{-1} X \qquad j_{\beta\tau_i} = X^T V^{-1} V_i V^{-1} (y - X\beta) \qquad (9.71)$$
$$j_{\tau_i \tau_j} = (y - X\beta)^T V^{-1} V_i V^{-1} V_j V^{-1} (y - X\beta) - i_{\tau_i \tau_j}.$$

Using the approximate sample space derivatives from (9.29), the value of $\hat{u}$ is

$$\hat{u} = \frac{|j_{(\hat{\beta},\hat{\tau})}|^{1/2}}{|i_{(\hat{\beta},\hat{\tau})}| \, |j_{\chi\chi}(\hat{\beta}_\tau, \hat{\tau}_\tau)|^{1/2}} |X^T \hat{V}_\tau^{-1} X| \left\| \begin{array}{c} i_{m-1,m}(\hat{\tau}_\tau) \\ q(\hat{\tau}, \hat{\tau}_\tau)^T \end{array} \right\|, \qquad (9.72)$$

where $\chi^T = (\beta^T, \tau_1, \ldots, \tau_{m-1})$, $i_{m-1,m}(\hat{\tau}_\tau)$ is the top $(m-1) \times m$ portion of the expected information matrix for τ evaluated at $\hat{\tau}_\tau$ or

$$i_{m-1,m}(\hat{\tau}_\tau) = \tfrac{1}{2}\left\{ \text{tr}\left(\hat{V}_\tau^{-1} V_i \hat{V}_\tau^{-1} V_j\right)\right\} \qquad i = 1, \ldots, m-1; \quad j = 1, \ldots, m$$

and

$$q(\hat{\tau}, \hat{\tau}_\tau)^T = -\tfrac{1}{2}\left[\text{tr}\left\{\left(\hat{V}^{-1} - \hat{V}_\tau^{-1}\right) V_1\right\}, \ldots, \text{tr}\left\{\left(\hat{V}^{-1} - \hat{V}_\tau^{-1}\right) V_m\right\}\right]. \qquad (9.73)$$

There are some striking similarities between (9.66) and (9.72) in terms of form. However, the latter expression has information matrices of dimension $p + m$ rather than m and depends on n-dimensional covariance matrix V rather than $(n - p)$-dimensional W.

Numerical example

The data below were taken from Milliken and Johnson (1984, p. 237) and record insect damage to four varieties of wheat.

	Variety		
A	B	C	D
3.90	3.60	4.15	3.35
4.05	4.20	4.60	
4.25	4.05	4.15	3.80
	3.85	4.40	

Table 9.5. *P-values for the test that $H_0 : \tau_\alpha = 0$ and confidence intervals at various levels for τ_α using the two likelihood approaches. The two p-values from (9.3) have been doubled.*

	Methods			
	Conditional Likelihood		Full Likelihood	
	(9.3)	χ_1^2	(9.3)	χ_1^2
P-values	0.0607	0.0808	0.0617	0.1406
97.5% Conf. Ints.	[0, 2.05)	[0, 1.26)	[0, 2.15)	[0, 0.814)
95% Conf. Ints.	[0, 1.20)	[0, 0.805)	[0, 1.33)	[0, 0.575)
90% Conf. Ints.	$(.0^2 388, 0.646)$	$(.0^2 306, 0.510)$	$(.0^3 813, 0.805)$	[0, 0.401)

The data are described by the unbalanced one-factor random model

$$y_{ij} = \mu + \alpha_i + e_{ij} \quad i = 1, \ldots, 4; \quad j = 1, \ldots, n_i$$

where $\{\alpha_i\}$ are i.i.d. Normal $(0, \tau_\alpha)$ independent of $\{e_{ij}\}$ i.i.d. Normal $(0, \tau_e)$. By equating mean squares to expected mean squares, Milliken and Johnson determine moment estimates of $\check{\tau}_\alpha = 0.06719$ and $\check{\tau}_e = 0.05644$.

Conditional likelihood (REML) analysis The REML estimates are $\tilde{\tau}_\alpha = 0.073155$ and $\tilde{\tau}_e = 0.057003$. These compare to the unrestricted MLEs reported by Milliken and Johnson as $\hat{\mu} = 3.991$, $\hat{\tau}_\alpha = 0.04855$ and $\hat{\tau}_e = 0.05749$. The REML estimate for τ_α is 51% larger.

The test of hypothesis $H_0 : \tau_\alpha = 0$ using conditional likelihood leads to

$$_c\hat{w}_0^2 = \ln|\hat{W}_0| - \ln|\hat{W}| = -23.9398 - (-26.9880) \simeq 3.048.$$

This value has a *p*-value computed from a χ_1^2 as $p = 0.0808$. Twice the right tail computation from (9.3), which uses the values $_c\hat{w}_0 = 1.74596$ and $_c\hat{u} = 2.1887$ as computed in (9.66), leads to $p = 0.0607$. Similar computation from $r^* = 1.87535$ gives $p = 0.0607$. Perhaps 6% is the best significance computation for the data.

Full likelihood analysis Milliken and Johnson compute $\hat{w}_0^2 = 2.171$ and determine MLEs under H_0 that are $\hat{\mu}_0 = 4.0269$ and $\hat{\tau}_{e0} = 0.10139$. When compared to a χ_1^2, then $p = 0.1406$ is a quite different significance level than occurs with REML. Twice the right tail probability from (9.3) gives $p = 0.0617$ which is almost the same answer as with the conditional likelihood analysis (0.0607). This computation was based on $\hat{u} = 2.5911$. The results are summarized in table 9.5. The F test for $H_0 : \tau_\alpha = 0$, that is equivalent to treating the random effects as fixed, leads to $F = 4.785$ with $p = 0.0293$ and is very comparable to the saddlepoint *p*-values without the doubling. Note the extremely short confidence intervals determined by using a χ_1^2 approximation to $\hat{w}_0^2$ and the much wider intervals that results from using the approximation in (9.3).

An asymptotic confidence interval based on assuming $\hat{\tau}_\alpha \sim$ Normal $(\tau_\alpha, \hat{\imath}_{\tau_\alpha}^{-1})$ was determined as $[0, 0.143)$ by Milliken and Johnson. The approximation $\ln \hat{\tau}_\alpha \sim$ Normal $(\ln \tau_\alpha, \hat{\jmath}_{\tau_\alpha}^{-1}/\hat{\tau}_\alpha^2)$ makes two improvements. Most importantly it attempts to remove the right skewness in the small sample distribution with the log transformation and it also replaces expected information with observed. The resulting 95% interval for τ_α is (0.00626, 0.377). Using conditional likelihood and the same procedure for $\ln \tilde{\tau}_\alpha$ produces the interval (0.00905, 0.592).

The distinction between conditional and full likelihood analysis should not be especially pronounced for this example since the two approaches differ only in terms of accounting for the location parameter. When inference uses (9.3) then the two likelihood analyses are similar whereas with a χ_1^2 approximation, the two approaches differ considerably. The χ_1^2 tails are clearly too short with both approaches.

For other examples, see Exercises 22 and 23. Other hypotheses are developed in Exercise 24.

9.8 Properties of (9.3)

9.8.1 Asymptotics

All the asymptotic properties described as holding for (8.1) in section 8.7.3 carry over to (9.3) in the presence of nuisance parameters. In a general (m, p) curved exponential family, a $(m - p)$-dimensional first-order likelihood ancillary a^0 determines a third-order ancillary a^* of the same dimension which, when fixed, determines a more refined conditionality resolution $\hat{W}_\psi | a^*; \star\psi$ by which accuracy may be judged. However, since a^* is virtually impossible to compute, the sample space derivatives required in (9.4) are computed by fixing the likelihood ancillary a^0 instead. Then (9.3) achieves saddlepoint accuracy and

$$\Pr(\hat{W}_\psi \leq \hat{w}_\psi | a^*; \star\psi) = \widehat{\Pr}(\hat{W}_\psi \leq \hat{w}_\psi | a^0; \star\psi)\{1 + O(n^{-1}\hat{w}_\psi) + O(n^{-3/2})\} \quad (9.74)$$

uniformly over $\hat{w}_\psi \in (w_0, \delta\sqrt{n})$ for fixed w_0 and $\delta > 0$ sufficiently small. The approximation on the right side of (9.74) is readily computed since it is based on fixed a^0, however it may be interpreted as being conditional on a^* since it reproduces this more refined conditionality resolution with the saddlepoint accuracy in (9.74).

If sample space derivatives are computed by fixing an affine ancillary a instead, then typically

$$\Pr(\hat{W}_\psi \leq \hat{w}_\psi | a^*; \star\psi) = \widehat{\Pr}(\hat{W}_\psi \leq \hat{w}_\psi | a; \star\psi)\{1 + O(n^{-1}\|a\|^2\,\hat{w}_\psi) + O(n^{-1}\hat{w}_\psi)\}$$

See Barndorff-Nielsen and Wood (1998) for more precise statements and details.

9.8.2 Symmetry

Some simple guidelines assure symmetry for the approximate distribution in (9.3) but it is difficult to foresee when these guidelines apply without having specific formulas for $\hat{w}_\psi$ and $\hat{u}$. For the true distribution of $\hat{W}_\psi$ and a particular value of ψ, symmetry is defined as $dF(\hat{w}_\psi; \psi) = dF(-\hat{w}_\psi; \psi)$ on a set of probability 1; thus it is the mapping $x \to \hat{w}_\psi$ which assures that the same mass occurs on either side of 0. A sufficient condition for symmetry of the approximate distribution in (9.3) requires consideration of the second mapping $x \to \hat{u}$. Consider any two data points with equal probability differentials that map to values of opposite sign, e.g. $x_1 \to \hat{w}_\psi$ and $x_2 \to -\hat{w}_\psi$. If in the second mapping, $x_1 \to \hat{u}$

and $x_2 \to -\hat{u}$ and this occurs for a set of x-values with probability 1, then (9.3) must be symmetric.

In practice this symmetry was seen because $\hat{w}_\psi$ and $\hat{u}$ were both functions of a certain pivotal. For example, in the normal linear regression and Fieller examples, a t pivot exists and this provides the structure for assuring that the oddness of $\hat{u}$ follows the same oddness pattern of $\hat{w}_\psi$.

9.9 Exercises

1. Consider the (curved) exponential family setting with nuisance parameters in which interest parameter ψ is not canonical.
   (a) Derive the expression for $\hat{u}$ in (9.7) for a regular exponential family.
   (b) Show that this expression does not depend on the particular complementary nuisance parameter χ chosen to complete the 1-1 transformation from canonical θ.
   (c) Derive (9.8) for a curve exponential family.
   (d) Show that (9.8) does not depend on the particular nuisance parameter χ used to complement ψ.
   (e) Prove more generally that the value of $\hat{u}$ in (9.4) is invariant under 1-1 transformations of the nuisance parameter.

2. Derive the expression for $\hat{u}$ in the independent linear regression model with general error described in section 9.2.3.
   (a) Derive the following intermediary expressions for the sample space derivatives:
   $$l_{\chi;\xi}(\hat{\xi}_\psi) = -\frac{1}{\hat{\sigma}_\psi^2} \sum_{i=1}^n \begin{pmatrix} h''(\hat{e}_{i\psi})x_{i\backslash} \\ h'(\hat{e}_{i\psi}) + \hat{e}_{i\psi}h''(\hat{e}_{i\psi}) \end{pmatrix} (x_i^T, a_i)$$
   $$l_{;\xi}(\hat{\xi}) - l_{;\xi}(\hat{\xi}_\psi) = \sum_{i=1}^n \{h'(a_i)/\hat{\sigma} - h'(\hat{e}_{i\psi})/\hat{\sigma}_\psi\}(x_i^T, a_i). \qquad (9.75)$$

   (b) Derive the following equations to determine the MLEs for ξ (left) and with ψ fixed (right):
   $$\begin{array}{ll} \sum_{i=1}^n h'(a_i)x_i = 0_p & \sum_{i=1}^n h'(\hat{e}_{i\psi})x_{i\backslash} = 0_{p-1} \\ \sum_{i=1}^n h'(a_i)a_i = -n & \sum_{i=1}^n h'(\hat{e}_{i\psi})\hat{e}_{i\psi} = -n. \end{array} \qquad (9.76)$$

   Symbol 0_p denotes a p-vector of zeros.

   (c) Use (9.76) to reduce the change of $l_{;\xi}$ in (9.75) to
   $$l_{;\xi}(\hat{\xi}) - l_{;\xi}(\hat{\xi}_\psi) = (0_{p-1}^T, -\hat{\rho}_{1\psi}/\hat{\sigma}_\psi, -n/\hat{\sigma} - \hat{\rho}_{2\psi}/\hat{\sigma}_\psi). \qquad (9.77)$$

   (d) Verify (9.10) using the simplification offered by (9.77) when computing the determinant.

3. Verify (9.14) in the normal linear model.
   (a) Show that $K_p(\hat{e}_\psi) = 2n|X_\backslash^T X_\backslash|$ and $K_{p+1}(a) = 2n|X^T X|$ where $X_\backslash$ denotes X without its last column x_p.
   (b) Show that
   $$\det \hat{D}_{1\psi} = (-1)^p (2n\hat{\sigma}/\hat{\sigma}_\psi)|X_\backslash^T X_\backslash|$$
   $$\det \hat{D}_{2\psi} = (-1)^p (2\hat{e}_\psi^T x_p)|X_\backslash^T X_\backslash|.$$

   (c) Show that $\hat{u}$ may be expressed as
   $$\mathrm{sgn}(\hat{\beta}_p - \beta_p)\frac{\hat{\sigma}^p}{\hat{\sigma}_\psi^p} \frac{|\hat{e}_\psi^T x_p|}{\sqrt{x_p^T(I_n - P_\backslash)x_p}} = \frac{\hat{\sigma}^p}{\hat{\sigma}_\psi^{p+1}}(\hat{\beta}_p - \beta_p)\sqrt{x_p^T(I_n - P_\backslash)x_p} \qquad (9.78)$$

and finally as given in (9.14). In (9.78), $P_\setminus = X_\setminus (X_\setminus^T X_\setminus)^{-1} X_\setminus$ projects onto the column space of $X_\setminus$.

4. The simplest case of the general linear model of section 9.2.3 is a location/scale model.
   (a) Simplify the expressions for $\hat{u}$ when the interest parameter is $\psi = \mu$ and there are no other location parameters.
   (b) Fit such a model to the ball bearing data of Exercise 8 in chapter 7. The log transformation of the data leads to a kernel g that is the left-skewed Gumbel (μ, σ).
   (c) Take $\psi = e^\mu$ and determine a 95% confidence interval. Compare it with the interval obtained in Exercise 8 and also to $(68.05, \infty)$ as determined by Lawless (1973, 1974).

5. Consider inference for the scale parameter in the independent linear regression model in section 9.2.3.
   (a) Take $\psi = \sigma$ and show that
   $$l_{;\hat{\beta}}(\hat{\xi}) - l_{;\hat{\beta}}(\hat{\xi}_\psi) = 0_p.$$
   (b) Using part (a), derive an explicit expression for $\hat{u}$.
   (c) Fit the special case of the location/scale model to the ball bearing data of Exercise 8 in chapter 7.
   (d) With $\psi = 1/\sigma$, determine a 95% confidence interval. Compare it with the interval obtained in Exercise 8 and also to $(1.52, 2.59)$ as determined by Lawless (1973, 1974).
   (e) Take kernel g as a standard normal distribution. Determine the form of $\hat{w}_\psi$ and show that $\hat{u}$ agrees with (9.40).

6. In the regular exponential family setting, show that the use of Skovgaard's approximate sample space derivatives in (9.24) and (9.25) lead to the exact expressions given in (9.6) and (9.7) for ψ as a canonical and general parameter respectively.

7. In the stationary setting of the $M/M/1$ queue considered in section 9.5.1, the total time for service, which includes both queueing and service time, has an Exponential $(\mu - \lambda)$ distribution.
   (a) Determine a pivot for $\psi = \exp\{-2(\mu - \lambda)\}$, the probability that this time exceeds 2.
   (b) Use the data in section 9.5.1 to determine 90% and 95% confidence intervals.
   (c) Assess the large deviation error of the approximation according to the methods used in figure 9.2.

8. Suppose that $y_1, \ldots, y_n$ are an i.i.d. sample from a Gamma (α, β) distribution.
   (a) If the interest parameter is the mean $\psi = \alpha/\beta$, determine (9.3) and (9.4) (Barndorff-Nielsen, 1986, Example 3.3).
   (b) Use the voltage data in table 9.2 as a sample. Determine a 95% confidence interval for the mean.
   (c) Now let the interest parameter be the variance $\psi = \alpha/\beta^2$. Repeat the tasks of parts (a) and (b).

9. Consider the general normal regression model in section 9.2.3.
   (a) Confirm the validity of the values in (9.14).
   (b) Use the Student-t_{n-p} density as the marginal likelihood and confirm the values in (9.16).

10. Consider the F pivot for the ratio of Poisson rates constructed from the two independent Exponential samples. Confirm that (9.31) and (9.32) may be derived directly by taking the F pivot as likelihood as has been justified using theorem 9.6.1.

11. Suppose $x_1, \ldots, x_n$ are the order statistics of an i.i.d. sample from density
$$f(x; \chi, \psi) = \psi^{-1} \exp\{-(x - \chi)/\psi\} \qquad x > \chi.$$

9.9 Exercises

Kalbfleisch and Sprott (1973) determine the conditional likelihood by transforming the data to $\{t_i = x_i - x_1\}$ and $u = x_1$. Then the likelihood separates as

$$\mathcal{L}(\chi, \psi) = \psi^{-n} \exp(-t/\psi) \times \exp\{-n(u - \chi)/\psi\} \qquad u > \chi$$

where $t = \sum_{i=2}^{n} t_i$. The first factor is the conditional likelihood of ψ with t and u independent in this instance.

(a) Explain why the various r^* methods that treat χ as a nuisance parameter cannot be used in this example.

(b) Starting directly from conditional likelihood without a nuisance parameter, show that r^* methods may be applied. Determine the values of the $\hat{w}$ and $\hat{u}$.

12. Suppose the marginal likelihood structure in (9.46).

    (a) Show the simplifications to the sample space determinant that result when it is assumed that $l_{\chi;\hat{\psi}}(\hat{\xi}_\psi; t|u) = 0$.

    (b) Consider a random sample of size n from a Normal (μ, σ^2) distribution. Show that if $\psi = \sigma^2$ (and $\chi = \mu$) then the condition in (a) holds. Show the simplification in (a) lead to expression (9.40).

    (c) For the normal sample in (b), show that if μ is the interest parameter, then the condition in (a) does not hold.

13. Consider the Neyman–Scott problem of section 9.6.4.

    (a) When dealing with the t pivot, show that the limits in (9.47) hold.

    (b) Using the fact that $1 - \Phi(z) \sim \phi(z)/z$ as $z \to \infty$, show that the limiting relative error as $z \to \infty$ is $100(1/\sqrt{2} - 1)\%$.

    (c) Graph $R(z)$ versus $z > 0$. For $z > 0$ show that $R'(z) < 0$ and $\lim_{z \to 0} R(z) = 0$. For appropriately defined $R(z)$ for $z < 0$, note that the graph is symmetric about $z = 0$.

    (d) Prove that the limiting relative error is $R(z)$ w.p. 1 and therefore that the limiting relative error is bounded between 0% and −29.3% w.p. 1.

    (e) When inferring σ^2, show that

    $$\hat{w}_\psi = O_p\{-\sqrt{2m(-1/2 + \ln 2)}\} \quad \text{and} \quad \hat{u} = O_p\{2^{-(m+1)/2}\sqrt{m}\}.$$

    (f) Use part (e) to conclude that expression (9.3) has limiting value 0 w.p. 1 for any σ^2.

14. Consider the general two sample model in which each sample has its own scale parameter. Comparison of the two scale parameters is of interest. Let treatment data $\{y_i\}$ and control data $\{z_j\}$ have the following the linear models

    $$y_i = \mu + \sigma e_i \qquad i = 1, \ldots, m$$
    $$z_j = \nu + \tau f_j \qquad j = 1, \ldots, n$$

    where $\{e_i\}$ and $\{f_j\}$ are i.i.d. with density g and $h = \ln g$.

    (a) If the interest parameter is $\psi = \sigma/\tau$, determine an explicit expression for $\hat{u}$.

    (b) Fit the audio analgesia data in table 4.8 using the logistic density kernel as given in section 2.4.6. Determine a 75% confidence interval for ψ.

    (c) Compute the values of $\hat{w}_\psi$ and $\hat{u}$ when g is the standard normal density. Write each explicitly in terms of the F statistic.

    (d) Reanalyze the data using the normal kernel and the expressions derived in part (c). Compute 75% confidence intervals and compare them to your answer in part (b).

    (e) For the sample sizes used in part (d), check the accuracy of the distribution for the signed likelihood ratio statistic in the normal case by comparing it to the F distribution. Plot the percentage relative error versus the value of F. Does the plot support the large deviation results described in section 9.8.1?

15. Derive the various expressions given for the generalized Behrens–Fisher problem.
    (a) Show that the set of four MLE equations in ξ are

    $$0 = \sum_{i=1}^{m} h'(a_i) \qquad\qquad 0 = \sum_{j=1}^{n} h'(b_j)$$
    $$1 = -m^{-1} \sum_{i=1}^{m} h'(a_i)a_i \qquad 1 = -n^{-1} \sum_{j=1}^{n} h'(b_j)b_j$$
    (9.79)

    and the MLE equations with ψ fixed are

    $$0 = \hat{\sigma}_\psi^{-1} \sum_{i=1}^{m} h'(\hat{e}_{i\psi}) + \hat{\tau}_\psi^{-1} \sum_{j=1}^{n} h'(\hat{f}_{j\psi})$$
    $$1 = -m^{-1} \sum_{i=1}^{m} h'(\hat{e}_{i\psi})\hat{e}_{i\psi} \qquad 1 = -n^{-1} \sum_{j=1}^{n} h'(\hat{f}_{j\psi})\hat{f}_{j\psi}.$$
    (9.80)

    (b) Show (9.51) where

    $$K_m(a) = \det\left\{ \sum_{i=1}^{m} \begin{pmatrix} h''(a_i) & h''(a_i)a_i \\ h''(a_i)a_i & -1 + h''(a_i)a_i^2 \end{pmatrix} \right\}. \tag{9.81}$$

    For this computation, it is helpful to order the parameters as $\xi^T = (\mu, \sigma, \psi, \tau)$. Then, when the third row is subtracted from the first row in j_ξ, the information matrix is transformed to block diagonal form.

    (c) For ψ fixed, show (9.52) where

    $$j_{\hat{\sigma}_\psi} = -\hat{\sigma}_\psi^{-2}\left(-m + \tfrac{e}{2}H_{\hat{e}\hat{e}}\right) \qquad j_{\hat{\tau}_\psi} = -\hat{\tau}_\psi^{-2}\left(-n + \tfrac{f}{2}H_{\hat{f}\hat{f}}\right)$$
    $$j_{\hat{\mu}_\psi \hat{\sigma}_\psi} = -\hat{\sigma}_\psi^{-2}\left({}_1^e H_{11} + \tfrac{e}{2}H_{1\hat{e}}\right) \qquad j_{\hat{\mu}_\psi \hat{\tau}_\psi} = -\hat{\tau}_\psi^{-2}\left({}_1^f H_{11} + \tfrac{f}{2}H_{1\hat{f}}\right)$$
    $$j_{\hat{\mu}_\psi} = -\hat{\sigma}_\psi^{-2} \tfrac{e}{2}H_{11} - \hat{\tau}_\psi^{-2} \tfrac{f}{2}H_{11}.$$
    (9.82)

16. Compute $\hat{w}_\psi$ and $\hat{u}$ for the normal case of the Behrens–Fisher problem.
    (a) Show that the matrix inside the determinant brackets in (9.53) is

    $$\begin{pmatrix} m/\hat{\sigma}_\psi^2 + n/\hat{\tau}_\psi^2 & 0 & 0 & n/\hat{\tau}_\psi^2 \\ -2P\hat{\sigma}_\psi & 2m\hat{\sigma}/\hat{\sigma}_\psi & 0 & 0 \\ 2P\hat{\tau}_\psi & 0 & 2n\hat{\tau}/\hat{\tau}_\psi & 2P\hat{\tau}_\psi \\ 0 & -m/\hat{\sigma} + m\hat{\sigma}/\hat{\sigma}_\psi^2 & -n/\hat{\tau} + n\hat{\tau}/\hat{\tau}_\psi^2 & P \end{pmatrix}$$

    when all entries have had their signs changed. Show that the absolute determinant in (9.53) with the leading factor $\hat{\sigma}_\psi^{-2}\hat{\tau}_\psi^{-2}$ is

    $$4mn|P| \frac{1}{\hat{\sigma}_\psi^2 \hat{\tau}_\psi^2} \left(m\frac{\hat{\sigma}\hat{\tau}_\psi}{\hat{\tau}\hat{\sigma}_\psi^3} + n\frac{\hat{\tau}\hat{\sigma}_\psi}{\hat{\sigma}\hat{\tau}_\psi^3} \right).$$

    (b) Show that

    $$|j_\xi| = \hat{\sigma}^{-4}\hat{\tau}^{-4}(4m^2n^2)$$

    and (9.52) reduces to

    $$|j_{\chi\chi}(\hat{\xi}_\psi)| = \frac{4mn}{\hat{\sigma}_\psi^2 \hat{\tau}_\psi^2} \left\{ m/\hat{\sigma}_\psi^2 + n/\hat{\tau}_\psi^2 - 2P^2(1/m + 1/n) \right\}. \tag{9.83}$$

    (c) Use the relationship of $\hat{\sigma}^2$ to $\hat{\sigma}_\psi^2$ and $\hat{\tau}^2$ to $\hat{\tau}_\psi^2$ to determine the final form for $\hat{w}_\psi$ and $\hat{u}$.

17. Two types of materials were used to make engine bearings (Rice, 1995). In order to compare their performances, ten bearings of each type were tested and their times to failure were recorded.

Type I failure times were recorded as

$$3.03 \quad 5.53 \quad 5.60 \quad 9.30 \quad 9.92 \quad 12.51 \quad 12.95 \quad 15.21 \quad 16.04 \quad 16.84$$

and type II are

$$3.19 \quad 4.26 \quad 4.47 \quad 4.53 \quad 4.67 \quad 4.69 \quad 6.79 \quad 9.37 \quad 12.75 \quad 12.78$$

(a) Compare the locations of the two groups by using the normal Behrens–Fisher model applied to the log transform of the data. This is equivalent to applying the generalized Behrens–Fisher model with kernel density g as log-normal. Use exact and approximate sample space derivatives.

(b) Compare the locations using the log transformation of the data but assuming that its kernel g is a left-skewed Gumbel $(0, 1)$. Use only exact sample space derivatives.

(c) Save the ancillaries from the two analyses in parts (a) and (b). Use them to assess whether the log-normal or the Weibull model fits the data better.

18. Derive the expressions for the example of two samples with a common mean.

    (a) Derive the matrix of sample space derivatives in (9.57).
    (b) In the normal setting, show that the expressions for $\hat{w}_\mu$ and $\hat{u}$ in (9.58) are correct. In determining $\hat{u}$, show that the matrix of sample space derivatives (without any sign changes) is

    $$\begin{pmatrix} \frac{2a}{\hat{\sigma}_\mu^3 \hat{\sigma}} & 0 & \frac{2m(\bar{y}-\mu)}{\hat{\sigma}_\mu^3} \\ 0 & \frac{2b}{\hat{\tau}_\mu^3 \hat{\tau}} & \frac{2n(\bar{z}-\mu)}{\hat{\tau}_\mu^3} \\ -\frac{m}{\hat{\sigma}} + \frac{a}{\hat{\sigma}_\mu^2 \hat{\sigma}} & -\frac{n}{\hat{\tau}} + \frac{b}{\hat{\tau}_\mu^2 \hat{\tau}} & \frac{m(\bar{y}-\mu)}{\hat{\sigma}_\mu^2} + \frac{n(\bar{z}-\mu)}{\hat{\tau}_\mu^2} \end{pmatrix}$$

    where

    $$a = \sum_{i=1}^m (y_i - \mu)(y_i - \hat{\mu}) \qquad b = \sum_{j=1}^n (z_j - \mu)(z_j - \hat{\mu}).$$

19. Two different materials, gold and glass, were used to determine the gravitational constant G (Lehmann and D'Abrera, 1975, p. 252). Using each material, the second and third digits in 5 measurement attempts were given as

    $$\begin{array}{lccccc} \text{Gold:} & 83 & 81 & 76 & 79 & 76 \\ \text{Glass:} & 78 & 71 & 75 & 72 & 74. \end{array}$$

    For example, the entry 83 corresponds to an observation of 6.683.

    (a) Using a normal kernel, find a 95% confidence interval for G assuming that the two samples have G as a common location parameter. Use exact and approximate sample space derivatives.

    (b) Suppose that a new third material, platinum, is introduced yielding the five measurements

    $$\text{Platinum:} \quad 61 \quad 61 \quad 67 \quad 67 \quad 64$$

    Assume that platinum also has the location parameter G. Extend the linear model to accommodate this third group and determine a new 95% confidence interval using data from all three materials. Consider exact and approximate sample space derivatives.

20. For the general mixed linear model in section 9.7.3, derive the values of $_c\hat{w}$ and $_c\hat{u}$ in (9.60) and (9.66) from the conditional likelihood in (9.61).

    (a) Use the matrix derivative identities $\partial |W|/\partial \tau_1 = \text{tr}(W^{-1}\dot{W})$ and $\partial W^{-1}/\partial \tau_1 = -W^{-1}\dot{W}W^{-1}$ with $\dot{W} = \partial W/\partial \tau_1$ to compute the REML equations in (9.59) and the observed and expected information matrices given in (9.63) and (9.62).

(b) Show that $_c\hat{w}$ is given as (9.60). Use the fact that

$$n - p = \operatorname{tr}\left(\tilde{W}^{-1}\sum_{i=1}^{m}\tilde{\tau}_i W_i\right)$$
$$= \operatorname{tr}\left(\tilde{W}_\tau^{-1}\sum_{i=1}^{m-1}\tilde{\tau}_{i\tau} W_i + \tau W_m\right)$$

along with the REML equations in (9.59).

(c) Expression (9.29) requires the computation of $\tilde{\theta}^T \dot{\mu}_{\tilde{\tau}}$ and $\tilde{\theta}_\tau^T \dot{\mu}_{\tilde{\tau}}$ where $\theta = \operatorname{vec}(-\frac{1}{2}W^{-1})$ and $\mu = \operatorname{vec} W$. Use the identity

$$\operatorname{vec}\left(-\tfrac{1}{2}\tilde{W}^{-1}\right)^T \operatorname{vec} \dot{W}_{\tilde{\tau}} = \operatorname{tr}\left(-\tfrac{1}{2}\tilde{W}^{-1}\dot{W}_{\tilde{\tau}}\right)$$

to validate the approximate sample space derivative in (9.65).

(d) Use the same methods as in part (c) to compute $\partial \theta^T/\partial \chi|_{\xi=\hat{\xi}_\psi} \dot{\mu}_{\tilde{\tau}}$ and show that it leads to the approximate sample space derivative (9.64).

21. For the general mixed linear model in section 9.7.3, derive the values of $\hat{w}_\tau^2$ and $\hat{u}$ in (9.69) and (9.72) from the full likelihood.

    (a) Compute the observed and expected information matrices making use of the methods noted in Exercise 20.

    (b) Show that

    $$n = (y - X\hat{\beta})^T \hat{V}^{-1}(y - X\hat{\beta})$$
    $$= (y - X\hat{\beta}_\tau)^T \hat{V}_\tau^{-1}(y - X\hat{\beta}_\tau)$$
    $$+ \tau \left\{\operatorname{tr}\left(\hat{V}_\tau^{-1} V_m\right) - (y - X\hat{\beta}_\tau)^T \hat{V}_\tau^{-1} V_m \hat{V}_\tau^{-1}(y - X\hat{\beta}_\tau)\right\}.$$

    Use these identities to derive (9.69).

    (c) Show that the likelihood is a $\{n + n^2, p + m\}$ curved exponential family with canonical parameter $\theta^T = \{\beta^T X^T V^{-1}, \operatorname{vec}(-V^{-1}/2)^T\}$ and canonical sufficient statistic $x^T = \{y^T, \operatorname{vec}(yy^T)^T\}$.

    (d) Show

    $$\dot{\mu}_{\beta\tau} = \begin{pmatrix} \partial\mu/\partial\beta^T & \partial\mu/\partial\tau^T \end{pmatrix}$$
    $$= \begin{pmatrix} x_1 & \cdots & x_p & 0_{n\times 1} & \cdots & 0_{n\times 1} \\ \operatorname{vec} B_1 & \cdots & \operatorname{vec} B_p & \operatorname{vec} V_1 & \cdots & \operatorname{vec} V_m \end{pmatrix}$$

    where, for example, $B_1 = X\{(\beta, 0_p, \ldots, 0_p) + (\beta, 0_p, \ldots, 0_p)^T\}X^T$ and B_i places the β in the ith column.

    (e) Show that

    $$(\hat{\theta} - \hat{\theta}_\tau)^T \dot{\mu}_{\hat{\beta}} = (\hat{\beta} - \hat{\beta}_\tau)^T X^T \hat{V}_\tau^{-1} X$$
    $$(\hat{\theta} - \hat{\theta}_\tau)^T \dot{\mu}_{\hat{\tau}} = q(\hat{\tau}, \hat{\tau}_\tau)^T$$

    as given in (9.73).

    (f) Show that

    $$\dot{\theta}_{\beta\tau}^T = \begin{pmatrix} \partial\theta^T/\partial\beta \\ \partial\theta^T/\partial\tau \end{pmatrix} = \begin{pmatrix} X^T V^{-1} & 0_{p\times n^2} \\ -\beta^T X^T V^{-1} V_1 V^{-1} & \tfrac{1}{2}\operatorname{vec}(V^{-1} V_1 V^{-1})^T \\ \vdots & \vdots \\ -\beta^T X^T V^{-1} V_m V^{-1} & \tfrac{1}{2}\operatorname{vec}(V^{-1} V_m V^{-1})^T \end{pmatrix}.$$

Hence conclude that

$$\dot{\theta}^T_{\hat{\beta}_\tau \hat{\tau}_\tau} \mu_{\hat{\beta}\hat{\tau}} = \begin{pmatrix} X^T \hat{V}^{-1}_\tau X & 0_{p \times m} \\ C & i(\hat{\tau}_\tau) \end{pmatrix}$$

where $i(\hat{\tau}_\tau)$ is the $m \times m$ expected information matrix evaluated at $\hat{\tau}_\tau$ and the ith row of C is $(\hat{\beta} - \hat{\beta}_\tau)^T X^T \hat{V}^{-1}_\tau V_i \hat{V}^{-1}_\tau X$. Verify that (9.72) is correct.

22. The data are given in Cunningham and Henderson (1968) and were first used to introduce the REML method in Patterson and Thompson (1971).

		Treatment	
		1	2
	1	2, 3	2, 3
Block	2	2, 3, 5, 6, 7	8, 8, 9
	3	3	2, 3, 4, 4, 5

The data conform to an unbalanced two-factor random model

$$y_{ijk} = \mu + \alpha_i + \beta_j + e_{ijk} \quad i = 1, 2; \quad j = 1, 2, 3 \quad k = 1, \ldots, n_{ij}$$

where $\alpha_1 + \alpha_2 = 0$ are fixed effects and block effects $\{\beta_j\}$ are i.i.d. Normal $(0, \tau_\beta)$ independent of $\{e_{ijk}\}$ i.i.d. Normal $(0, \tau_e)$. Patterson and Thompson determine unconditional MLEs as $\hat{\tau}_\beta = 1.0652$ and $\hat{\tau}_e = 2.3518$ and conditional REML estimates as $\tilde{\tau}_\beta = 1.5718$ and $\tilde{\tau}_e = 2.5185$.

(a) Confirm the validity of the unconditional and REML estimates determined by Patterson and Thompson.

(b) Test the hypothesis that $H_0 : \tau_\beta = 0$ by computing the signed unconditional likelihood ratio $\hat{w}_0$ and determining a p-value using $\hat{w}_0^2 \sim \chi_1^2$. Compute a p-value using (9.3) applied to the full likelihood.

(c) Using the conditional likelihood in place of the full likelihood, repeat part (b) by determining the conditional likelihood ratio test based on (i) a χ_1^2 approximation and (ii) using (9.3) applied to the conditional likelihood only.

(d) Compute $j_{\hat{\tau}_\beta}$, the observed Fisher information for τ_β. Determine asymptotic 95% confidence intervals of the form $\hat{\tau}_\beta \pm 1.96 j_{\hat{\tau}_\beta}^{-1/2}$ and, on the log-scale, $(\hat{\tau}_\beta/\hat{\varsigma}, \hat{\tau}_\beta \hat{\varsigma})$ where $\hat{\varsigma} = 1.96 j_{\hat{\tau}_\beta}^{-1/2}/\hat{\tau}_\beta$. Compare these to the intervals determined by inverting the test using (9.3) based on (i) the full likelihood and (ii) the conditional likelihood only.

23. Milliken and Johnson (1984, figure 19.2) provide the following data for an unbalanced two-factor ANOVA:

		Blocks		
		1	2	3
Treatments	1	10, 12, 11	13, 15	21, 19
	2	16, 18	13, 19, 14	11, 13

The random model is

$$y_{ijk} = \mu + \alpha_i + \beta_j + \gamma_{ij} + e_{ijk}$$

with treatment effects $\{\alpha_i\}$ i.i.d. Normal $(0, \tau_\alpha)$, block effects $\{\beta_j\}$ i.i.d. Normal $(0, \tau_\beta)$, interaction effects $\{\gamma_{ij}\}$ i.i.d. Normal $(0, \tau_\gamma)$ and errors $\{e_{ijk}\}$ i.i.d. Normal $(0, \tau_e)$. The method of moments estimates, determined from using Henderson's method III sums of squares, are

$$\check{\tau}_\alpha = -8.0329 \qquad \check{\tau}_\beta = -11.108$$
$$\check{\tau}_\gamma = 22.4620 \qquad \check{\tau}_e = 3.8333.$$

(a) Determine REML estimates for the four variance components and compare them to the unrestricted MLEs $\hat{\mu} = 18.846$, $\hat{\tau}_\alpha = 0.0$, $\hat{\tau}_\beta = 0.0$, $\hat{\tau}_\gamma = 7.407$, and $\hat{\tau}_e = 3.843$.

(b) For testing hypotheses, Milliken and Johnson compute $-2l(\hat{\xi}) = 68.726$. In testing $H_0 : \tau_\alpha = 0$ or $H_0 : \tau_\beta = 0$, the value of $-2l$ is unchanged due to $\hat{\tau}_\alpha = 0 = \hat{\tau}_\beta$ along with null estimates. Under $H_0 : \tau_\gamma = 0$, the estimates are $\mu_0 = 14.643$, $\hat{\tau}_{\alpha 0} = 0$, $\hat{\tau}_{\beta 0} = 0$ and $\hat{\tau}_{e0} = 11.087$ and lead to $-2l(\hat{\xi}_0) = 73.411$. This gives $\hat{w}_0^2 = 4.685$ with $p = 0.03043$ determined from a χ_1^2. Compute a p-value using (9.3) applied to the full likelihood.

(c) Use the conditional likelihood in place of the full likelihood. Repeat part (b) by determining the conditional likelihood ratio test for $H_0 : \tau_\gamma = 0$ based on (i) a χ_1^2 approximation and (ii) using (9.3) applied to the conditional likelihood only.

(d) Determine the asymptotic variance of $\hat{\tau}_\gamma$ as the inverse of its Fisher information. From this determine asymptotic 95% confidence intervals as $\hat{\tau}_\gamma \pm 1.96 j_{\hat{\tau}_\gamma}^{-1/2}$ and, on the log-scale, $(\hat{\tau}_\gamma / \hat{\varsigma}, \hat{\tau}_\gamma \hat{\varsigma})$ where $\hat{\varsigma} = 1.96 j_{\hat{\tau}_\gamma}^{-1/2} / \hat{\tau}_\gamma$. Compare this to the interval determined by inverting the test using (9.3) based on (i) the full likelihood and (ii) the conditional likelihood only.

24. Consider the mixed random model of section 9.7.1.

    (a) Develop the necessary formulas to find confidence intervals for a ratio of variance components, say τ_{m-1}/τ_m, by factoring $W = \tau_m \sum_{i=1}^{m} \tau_i / \tau_m W_i$. Consider both the conditional and full likelihood approaches.

    (b) Use these expressions to compute 95% confidence intervals for τ_α / τ_e in the insect damage data considered by Milliken and Johnson.

    (c) Describe how to derive formulas for linear combinations and more general functions of the variance components.

10

Sequential saddlepoint applications

When the joint MGF of (X, Y) is available, the conditional MGF of Y given $X = x$ can be approximated by using the sequential saddlepoint method. Apart from providing conditional moments, this approximate conditional MGF may also serve as a surrogate for the true conditional MGF with saddlepoint methods. In such a role, it can be the input into a single-saddlepoint method, such as the Lugananni and Rice approximation, to give an approximate conditional distribution. Thus, the resulting sequential saddlepoint approximation to $\Pr(Y \leq y | X = x)$ provides an alternative to the double-saddlepoint methods of sections 4.1–4.2.

Computation of a p-value for the Bartlett–Nandi–Pillai trace statistic in MANOVA provides a context in which the sequential saddlepoint approximation succeeds with high accuracy but the competing double-saddlepoint CDF approximation fails. Among the latter methods, only numerical integration of the double-saddlepoint density successfully replicates the accuracy of the sequential saddlepoint CDF approximation; see Butler *et al.* (1992b).

Another highly successful application of sequential saddlepoint methods occurs when approximating the power function of Wilks' likelihood ratio test in MANOVA. This example is deferred to section 11.3.1.

10.1 Sequential saddlepoint approximation

Suppose (X, Y) is a m-dimensional random vector with known joint CGF $K(s, t)$ where s and t are the respective components. The goal is to use the joint CGF to determine conditional probabilities and moments of Y given $X = x$. Theoretically K determines a unique conditional distribution with an associated conditional CGF but extracting that information is not at all obvious. Fraser *et al.* (1991) have shown how an approximate conditional CGF for Y given $X = x$ can be determined from K by way of saddlepoint density approximations. Their approach is the *sequential saddlepoint* method presented below.

Suppose the dimensions are dim $X = m_x$ and dim $Y = m_y$ which add to m. If $f(x, y; 0, 0)$ denotes the true density of (X, Y), then an (s, t)-tilt of this density is

$$f(x, y; s, t) = \exp\{s^T x + t^T y - K(s, t)\} f(x, y; 0, 0) \qquad (10.1)$$

as described in sections 5.3–5.4. The family of tilted distributions in (10.1) indexed in (s, t) form an exponential family of distributions. From the properties of such families in section 5.1.2, the t-tilt of the conditional density of Y given $X = x$ forms a separate exponential family of the form

$$f(y|x;t) = \exp\{t^T y - \mathcal{K}(t|x)\} f(y|x;0)$$
$$= f(y|x;0)\exp(t^T y)/\mathcal{M}(t|x) \qquad (10.2)$$

where $\mathcal{K}(t|x) = \ln \mathcal{M}(t|x)$ is the unknown conditional CGF of $Y|X = x$. While $\mathcal{K}(t|x)$ may be unknown, saddlepoint approximation to the remaining terms in (10.2), namely $f(y|x;0)$ and $f(y|x;t)$, follows directly from the double-saddlepoint density approximation derived from the joint exponential family in (10.1). Using (5.55) with $\mathcal{L}(s,t)$ as $f(x,y;s,t)$ in (10.1), then

$$\hat{f}(y|x;t) = (2\pi)^{-m_y/2} \left\{ \frac{|K''(\hat{s},\hat{t})|}{|K''_{ss}\hat{s}_t,t)|} \right\}^{-1/2} \frac{f(x,y;\hat{s}_t,t)}{f(x,y;\hat{s},\hat{t})} \qquad (10.3)$$

$$\stackrel{t}{\propto} f(x,y;\hat{s}_t,t)|K''_{ss}(\hat{s}_t,t)|^{1/2}. \qquad (10.4)$$

Expression (10.4) is often referred to as the approximate conditional likelihood since it is a saddlepoint approximation to the conditional likelihood of tilting parameter t using the conditional distribution of $Y|X = x$. Expression (10.3) at $t = 0$ also provides the saddlepoint approximation for $\hat{f}(y|x;0)$. The use of both these approximations as surrogates for their true counterparts in (10.2) provides an approximate MGF given as

$$\hat{\mathcal{M}}(t|x) = \exp(t^T y) \frac{f(x,y;\hat{s}_0,0)|K''_{ss}(\hat{s}_0,0)|^{1/2}}{f(x,y;\hat{s}_t,t)|K''_{ss}(\hat{s}_t,t)|^{1/2}} \qquad (10.5)$$

$$= \exp\{K(\hat{s}_t,t) - K(\hat{s}_0,0) - (\hat{s}_t - \hat{s}_0)^T x\} \qquad (10.6)$$

$$\times \left\{ \frac{|K''_{ss}(\hat{s}_0,0)|}{|K''_{ss}(\hat{s}_t,t)|} \right\}^{1/2}. \qquad (10.7)$$

The range of allowable t-values places $(\hat{s}_t, t)$ inside the convex hull of the convergence region for K.

From a practical point of view, sequential saddlepoint implementation is facilitated by the ability to solve for $\hat{s}_t$ as a function of t. In some examples $\hat{s}_t$ may be explicitly determined. For example, this was the case in the common odds testing of section 6.2.3 when considering the double-saddlepoint power approximation. This example is continued below from a sequential perspective.

10.2 Comparison to the double-saddlepoint approach

Suppose that $\hat{\mathcal{M}}(t|x)$ in (10.6) is used as a surrogate for the conditional MGF of Y given $X = x$ to determine either a density, mass function, or CDF using a single-saddlepoint procedure. Since the expression is not necessarily a proper MGF, then $\hat{\mathcal{K}}(t|x) = \ln \hat{\mathcal{M}}(t|x)$ is not necessarily convex however this is generally not a computational problem when working with applications.

The surrogate use of $\hat{\mathcal{M}}(t|x)$ is related to the double saddlepoint of sections 4.1–4.2 in the following way. Suppose that the Hessian contribution inside the curly brackets of (10.7) is left off and

$$\tilde{M}(t|x) = \exp\{K(\hat{s}_t, t) - K(\hat{s}_0, 0) - (\hat{s}_t - \hat{s}_0)^T x\} \tag{10.8}$$

is used instead as a surrogate for the true conditional CGF. Then this latter procedure is the double saddlepoint method as the reader is asked to show in Exercise 2. Note that the difference between the two methods has only to do with the inclusion/exclusion of the Hessian contribution. Sometimes its inclusion increases the accuracy of the single-saddlepoint approximation and sometimes it decreases accuracy. The benefit is particular to the application as seen in the examples.

10.3 Examples

The first example determines the MGF of a Binomial (n, θ) by conditioning a Poisson on the sum of Poissons. It shows that the sequential saddlepoint approach does not necessarily improve upon the double-saddlepoint approach.

The second example determines the MGF of a Beta (α, β) using gamma variables. The resulting approximation leads to a very accurate approximation to the beta MGF as a $_1F_1$ hypergeometric function.

10.3.1 Binomial (n, θ)

Section 4.1 determined a double saddlepoint approximation for this distribution by conditioning $Z_1 \sim$ Poisson (λ_1) on $Z_1 + Z_2 = n$ where $Z_2 \sim$ Poisson (λ_2) is independent of Z_1 and $\theta = \lambda_1/(\lambda_1 + \lambda_2)$. The sequential computation of (10.6) is left as Exercise 1 and leads to the value

$$\hat{M}(t|x) = (\theta e^t + 1 - \theta)^n \{(\theta e^t + 1 - \theta)e^{-t/2}\}. \tag{10.9}$$

The exact MGF is of course the leading term in (10.9) without the term in curly brackets. This extra term is contributed by the Hessian terms in (10.7). Were this term to be left out as occurs when taking the Skovgaard approach to the problem, then the double-saddlepoint replicates the true conditional MGF. This is coincident with the Skovgaard double-saddlepoint CDF approximation agreeing with the Lugannani and Rice single-saddlepoint approach for this example as discussed in section 4.1.

10.3.2 Beta (α, β) and $_1F_1(a; b; t)$

Approximation to this distribution was considered from a conditional point of view in section 4.2.3. Let $Z_1 \sim$ Gamma $(\alpha, 1)$ be independent of $Z_2 \sim$ Gamma $(\beta, 1)$. Then the conditional distribution of $Y = Z_1$ given $X = Z_1 + Z_2 = 1$ is Beta (α, β).

The importance of the sequential approach in this setting is that it provides a very accurate approximation for the MGF of the beta distribution, which is the special function

$$_1F_1(\alpha; \alpha + \beta; t) = \frac{\Gamma(\alpha + \beta)}{\Gamma(\alpha)\Gamma(\beta)} \int_0^1 e^{ty} y^{\alpha-1}(1-y)^{\beta-1} dy \tag{10.10}$$

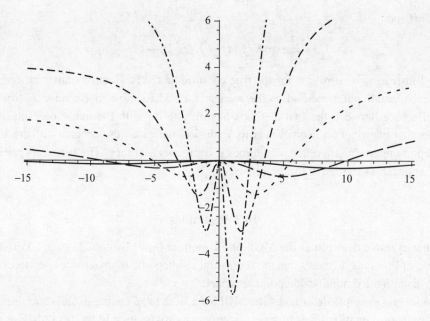

Figure 10.1. Percentage relative errors of ${}_1\hat{F}_1(\alpha;\alpha+\beta;t)$ versus t for values $(\alpha,\beta) = (0.3, 0.4), (0.6, 0.8), (1.2, 1.6), (2.4, 3.2)$, and $(4.8, 6.4)$. The respective lines show increasing accuracy and range from the least accurate (*dot-dot-dashed*) to the most accurate (*solid*).

known as the confluent hypergeometric function. This function may be readily computed in Maple however the computation can be slow.

An alternative is to use the explicit approximation

$${}_1\hat{F}_1(a;b;t) = \exp(t\hat{y}_t) \left(\frac{\hat{y}_t}{a}\right)^a \left(\frac{1-\hat{y}_t}{b-a}\right)^{b-a} b^{b-1/2} \frac{1}{\sqrt{\frac{\hat{y}_t^2}{a} + \frac{(1-\hat{y}_t)^2}{b-a}}} \quad (10.11)$$

that has been re-expressed instead in the arguments a and b with

$$\hat{y}_t = \frac{2a}{b-t+\sqrt{(t-b)^2+4at}}. \quad (10.12)$$

This expression was introduced in Butler and Wood (2002, §4.1) where two different approaches were used for its motivation. First, it results from the sequential saddlepoint approach and the somewhat tedious derivation has been outlined in Exercise 4. The second derivation is simpler and uses two Laplace approximations based on the integral definition given in (10.10). Exercise 5 provides the specifics.

Figure 10.1 plots the relative error of approximation ${}_1\hat{F}_1(\alpha;\alpha+\beta;t)$ versus t for five values of (α,β) pairs given as $(\alpha,\beta) = (0.3, 0.4), (0.6, 0.8), (1.2, 1.6), (2.4, 3.2)$, and $(4.8, 6.4)$.

The respective lines show increasing accuracy and range from the least accurate (*dot-dot-dashed*) to the most accurate (*solid*). For large values of t the relative errors also remain accurate with

$$\lim_{t\to\infty} \frac{{}_1\hat{F}_1(\alpha;\alpha+\beta;t)}{{}_1F_1(\alpha;\alpha+\beta;t)} - 1 = \frac{\hat{\Gamma}(\alpha+\beta)/\Gamma(\alpha+\beta)}{\hat{\Gamma}(\alpha)/\Gamma(\alpha)} - 1 \quad (10.13)$$

These values for the values of (α, β) shown in the plot are 13.07%, 7.44%, 3.92%, 1.99%, and 0.995%. The derivation of the finite limits in (10.13) is outlined in Exercise 5. It assures that relative error is uniform on the whole real line.

The difference between the usual double saddlepoint and sequential approaches has to do with the inclusion of the two Hessian components for conditional MGF approximation. The extent of this improvement may be seen with a simple numerical computation that approximates the MGF of a Uniform (0, 1) distribution with MGF $_1F_1(1; 2; t)$. For example $_1F_1(1; 2; 1) = e - 1 \simeq 1.718$ with $_1\hat{F}_1(1; 2; 1) = 1.705$ but when the Hessian components are excluded, the approximation is 1.752.

10.3.3 Pólya $(M, N; n)$ and $_2F_1(a, b; c; x)$

Let $Z_1 \sim$ Negative Binomial (M, θ) be independent of $Z_2 \sim$ Negative Binomial (N, θ). The conditional distribution of $Y = Z_1$ given $X = Z_1 + Z_2 = n$ is Pólya $(M, N; n)$ with mass function given in (4.25). Since the MGF of this mass function is expressed in terms of the Gauss hypergeometric function or $_2F_1$, application of the sequential saddlepoint approach should provide an accurate approximation to $_2F_1$ which is the main importance of this example. Johnson et al. (1992, p. 240) specify this MGF as

$$\mathcal{M}(t) = {}_2F_1(-n, M; M + N; 1 - e^t). \tag{10.14}$$

An explicit sequential saddlepoint approximation can be derived of the form

$$\hat{\mathcal{M}}(t) = {}_2\hat{F}_1(-n, M; M + N; 1 - e^t), \tag{10.15}$$

where, upon generalizing to arguments a, b, and c,

$$_2\hat{F}_1(a, b; c; x) = (1 - x\hat{y})^{-b} \left(\frac{\hat{y}}{a}\right)^a \left(\frac{1 - \hat{y}}{c - a}\right)^{c-a} c^{c-1/2} \hat{h}^{-1/2} \tag{10.16}$$

with

$$\hat{h} = \frac{\hat{y}^2}{a} + \frac{(1 - \hat{y})^2}{c - a} - \frac{bx^2}{(1 - x\hat{y})^2} \frac{\hat{y}^2}{a} \frac{(1 - \hat{y})^2}{c - a},$$

and

$$\hat{y} = \frac{2a}{\sqrt{\tau^2 - 4ax(c - b)} - \tau} \qquad \tau = x(b - a) - c. \tag{10.17}$$

The MGF approximation in (10.15) uses only integer values as arguments in $_2\hat{F}_1$, however the approximation has been stated as in (10.16) because of its applicability using a much wider range of argument values. In particular, for $x < 1$, the only restriction on the values of a, b, and c is that $0 < \min(|a|, |b|) < c$.

Approximation (10.16) was first derived in Butler and Wood (2002) as a Laplace approximation for the standard integral representation of $_2F_1$ as given in (10.34). For this reason, it is often referred to below as Laplace's approximation to $_2F_1$ and as a sequential Laplace method if used as a surrogate for the true MGF. Numerical computations with $_2\hat{F}_1$ are deferred to the next subsection.

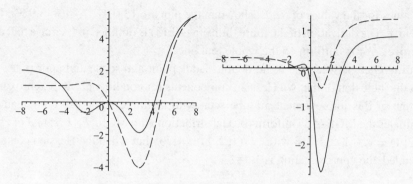

Figure 10.2. Plots of percentage relative error for the Laplace approximation $\hat{\mathcal{M}}(t)$ (*dashed*) and also for $\check{\mathcal{M}}(t)$ (*solid*). The left plot shows errors when considering the MGF of a Hypergeometric $(3, 6; 4)$ while the right plot shows the same for the MGF of a Pólya $(3, 6; 4)$.

10.3.4 Hypergeometric $(M, N; n)$ and $_2F_1(a, b; c; x)$

Let $Z_1 \sim$ Binomial (M, θ) be independent of $Z_2 \sim$ Binomial (N, θ). The conditional distribution of $Y = Z_1$ given $X = Z_1 + Z_2 = n$ is Hypergeometric $(M, N; n)$ as considered in section 4.2.3 in connection with the double-saddlepoint approximation. The MGF of this mass function is expressed in terms of $_2F_1$, and is given in Johnson *et al.* (1992, p. 238) as

$$\mathcal{M}(t) = \frac{_2F_1(-n, -M; N - n + 1; e^t)}{_2F_1(-n, -M; N - n + 1; 1)} \tag{10.18}$$

$$= C(M, N; n) \,_2F_1(-n, -M; N - n + 1; e^t), \tag{10.19}$$

where

$$C(M, N; n) = \frac{\Gamma(N + 1)\Gamma(M + N - n + 1)}{\Gamma(N - n + 1)\Gamma(M + N + 1)}. \tag{10.20}$$

In (10.19), the $_2F_1$ value at 1 in the denominator of (10.18) has been replaced with its equivalent gamma function expression as the inverse of (10.20) and also given in (15.1.20) of Abramowitz and Stegun (1972). An explicit sequential saddlepoint approximation for (10.18) can be derived as

$$\check{\mathcal{M}}(t) = C(M, N; n) \,_2\check{F}_1(-n, -M; N - n + 1; e^t) \tag{10.21}$$

where $_2\check{F}_1(a, b; c; x)$ is given in (10.36) of Exercise 11. Let $\hat{\mathcal{M}}(t)$ be the same expression but based on $_2\hat{F}_1$ instead.

Approximation $_2\check{F}_1(a, b; c; x)$ is not restricted to $a, b < 0$, $c > 0$, and $x > 0$. By using the tool of analytic continuation from complex variables, as discussed in section 6 of Butler and Wood (2002), the approximation can be "continued" to other argument values. The approximation is accurate for $x \in [0, 1]$ and $c - a - b - 1 > 0$, and also for $x < 0$ if continued using the relation in (10.37).

Figure 10.2 (left) plots the percentage relative error of $\hat{\mathcal{M}}(t)$ versus t (dashed line) and the relative error of $\check{\mathcal{M}}(t)$ vs. t (solid line) when approximating the MGF of a Hypergeometric $(3, 6; 4)$.

In this case, $\check{\mathcal{M}}(t)$ is slightly more accurate. Figure 10.2 (right) makes the same plots but when approximating the MGF of a Pólya $(3, 6; 4)$; now $\hat{\mathcal{M}}(t)$ is more accurate. Each

10.3 Examples

Table 10.1. *Values of $_2F_1(a, b; c; x)$ as compares to the Laplace $_2\hat{F}_1$ and sequential saddlepoint $_2\check{F}_1$ approximations.*

a	b	c	x	$_2F_1$	$_2\hat{F}_1$	$_2\check{F}_1$
−0.1	−0.5	1	0.6	1.03257	1.0337	1.0204
−1	−3	4	0.6	1.45	1.4496	1.4493
−3	−0.5	3	0.6	1.27885	1.27843	1.27812
−3	−5	12	0.6	1.89440	1.89437	1.89418
−10	−21	32	.06	21.5725	21.5713	21.5715
0.5	1.1	3	0.6	1.1489	1.1514	1.1997
1	2	5	0.6	1.3525	1.3550	1.3696
2.2	1.7	10	0.6	1.2966	1.2970	1.2963

Table 10.2. *P-value approximations in testing $\varphi = 1$ vs. $\varphi > 1$ for the penicillin data with the sequential saddlepoint (Seq. $_2\check{F}_1$) based on (10.21) and Seq. $_2\hat{F}_1$ that uses $_2\hat{F}_1$ instead.*

Type	Exact	Uncorrected	$\widehat{\Pr}_1$	$\widehat{\Pr}_2$
Single	0.0200	0.0^27735	0.02131	0.01992
Double	0.0200	0.0^27656	0.02214	0.02054
Seq. $_2\check{F}_1$	0.0200	0.0^27809	0.02152	0.01956
Seq. $_2\hat{F}_1$	0.0200	0.0^28829	0.02433	0.01850

approximation appears to work best when approximating the MGF from which it was derived.

More often than not, however, it is the Laplace approximation that is slightly better as may be seen in table 10.1. The entries are exact values for $_2F_1(a, b; c; x)$, the Laplace approximation $_2\hat{F}_1$ in (10.16) and $_2\check{F}_1$ in (10.36). Both approximations show remarkable accuracy. With values of a, b close to zero, the Laplace approximation maintains better accuracy. Also, in the sixth row, $c - a - b - 1 = 0.4$ which is close to the boundary on which $_2\check{F}_1(a, b; c; x)$ is defined. This affects its accuracy but not that of the Laplace approximation.

10.3.5 Common odds ratio

Reconsider the test for common odds ratio of section 6.2.3. The penicillin data in table 6.10 consist of three 2×2 tables and the *p*-value is computed as a probability for the sum of three independent Hypergeometric variables. Table 10.2 reproduces single- and double-saddlepoint approximations from table 6.11 along with two new approximations. The new rows include the sequential saddlepoint approximation "Seq. $_2\check{F}_1$" that uses (10.21) to approximate the MGF involved, and "Seq. $_2\hat{F}_1$" that uses the Laplace approximation of $_2F_1$ in (10.16).

The table suggests two points. First, the Seq. $_2\check{F}_1$ method is closer to the single saddlepoint that uses the exact $_2F_1$ than is the Seq. $_2\hat{F}_1$ method. This is consistent with figure 10.2.

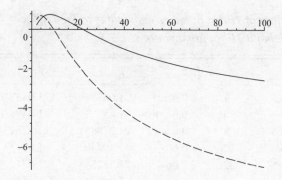

Figure 10.3. Percentage relative errors of the attained power function when using $\check{\mathcal{M}}(t)$ as a surrogate for $\mathcal{M}(t)$ (*solid*) and also when using the Laplace approximation $\hat{\mathcal{M}}(t)$ instead (*dashed*).

Secondly, the entries for Seq. $_2\check{F}_1$ are quite close to those for the single saddlepoint. This reinforces the notion that $\check{\mathcal{M}}(t)$ is a suitable surrogate for $\mathcal{M}(t)$ as concerns its use in a single-saddlepoint approximation.

The attained power curve of the test can also be based on using $\check{\mathcal{M}}(t)$ as a surrogate for $\mathcal{M}(t)$. The solid line in figure 10.3 is the percentage relative error in the power function computation when $\check{\mathcal{M}}(t)$ in (10.21) is used as a surrogate for $\mathcal{M}(t)$. The dashed line records relative error using $\hat{\mathcal{M}}(t)$ as a surrogate. The plot shows that either method of approximating the true MGF is completely adequate for practical applications as also was the double-saddlepoint method whose power curve was the simplest to compute.

10.4 P-values for the Bartlett–Nanda–Pillai trace statistic

This test statistic is one of the four main tests used in MANOVA and is perhaps the most versatile in term of maintaining power against a wide range of alternatives; see Pillai (1985). If $W_e \sim$ Wishart (n, Σ_e) and $W_h \sim$ Wishart (m, Σ_h) are the k-dimensional matrix sums of squares for error and hypothesis respectively, then the test statistic is $V = V_{k,m,n} = \text{tr}(W_e + W_h)^{-1} W_e$. Under the null hypothesis $\Sigma_e = \Sigma_h = \Sigma$, the distribution of V does not depend on Σ, and hence is ancillary. Since a complete sufficient statistic for Σ is $W_e + W_h$, then, by Basu's lemma, V is independent of $W_e + W_h$ so

$$\Pr(V \leq v) = \Pr\{V \leq v | W_e + W_h = I_k\} = \Pr(\text{tr } W_e \leq v | W_e + W_h = I_k).$$

This conditional characterization of the null distribution suggests pursuing a conditional and/or sequential approach for computing the p-value. The joint MGF of $Y = \text{tr } W_e$ and $k \times k$ matrix $X = W_e + W_h$ is computed from the MGF of the Wishart, as given in Muirhead (1982, §3.2.2), as

$$M(S, t) = |I_k - S - S_d|^{-m/2} |(1 - 2t)I_k - S - S_d|^{-n/2}, \tag{10.22}$$

where $S = (s_{ij})$ is symmetric, $S_d = \text{diag } S$ sets to zero the off-diagonal components of S, and the MGF is defined for all values of (S, t) over which the two matrices in (10.22) remain positive definite. The saddlepoint for fixed t is $(\hat{S} + \hat{S}_d)_t = 2\hat{c}_t I_k$ where scalar

$$\hat{c}_t = -\tfrac{1}{4}(2t + m + n - 2) - \tfrac{1}{4}\sqrt{(2t - m - n)^2 + 8nt}. \tag{10.23}$$

10.4 P-values for the Bartlett–Nanda–Pillai trace statistic

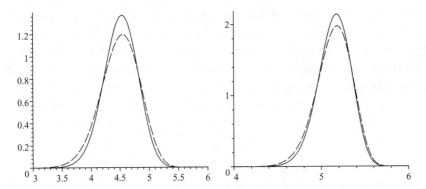

Figure 10.4. The normalized double-saddlepoint density (*solid*) and a $k \times$ Beta $(nk/2, mk/2)$ density (*dashed*) with $k = 6$, $m = 5$ and $n = 15$ (left) and $n = 30$ (right).

Substituting into (10.22) leads to an expression for (10.6) as

$$\hat{\mathcal{M}}(t|I_k) = \frac{(m+n)^{k(2m+2n-k-1)/4}}{(1-2\hat{c}_t)^{mk/2}(1-2\hat{c}_t-2t)^{nk/2}} e^{-k(\hat{c}_t-\hat{c}_0)} \quad (10.24)$$

$$\times \left\{ \frac{m}{(1-2\hat{c}_t)^2} + \frac{n}{(1-2\hat{c}_t-2t)^2} \right\}^{-k(k+1)/4}$$

and the steps and intermediate expressions are outlined in Exercise 13. Implementing the sequential saddlepoint method requires solving for a saddlepoint $\tilde{t}$ that is the root of

$$\hat{\mathcal{K}}'(\tilde{t}|I_k) = (\ln \hat{\mathcal{M}})'(\tilde{t}|I_k) = v$$

and has $\tilde{t} < 0$ when significance probabilities are in the left tail of the distribution of V.

Exercise 14 asks the reader to derive the double saddlepoint density as

$$\hat{f}(v|I_k) \propto (1-v/k)^{mk/2-1}(v/k)^{nk/2-1} \left\{ \frac{v^2}{n} + \frac{(k-v)^2}{m} \right\}^{-k(k+1)/4+1/2} \quad (10.25)$$

for $v \in (0, k)$ which appears as equation (3.8) in Butler *et al.* (1992b). Note that, apart from the term in curly braces, the density has V/k as a Beta $(nk/2, mk/2)$ density. When transformed to an $F_{nk,mk}$ statistic, it becomes the approximation of Pillai and Mijares (1959) that is commonly used in SAS.

When considered as a problem in conditional inference, the number of nuisance parameters is $k(k+1)/2$ and increases quadratically with k. Even the modest value of $k = 4$ has 10 nuisance parameters and it is informative to see which approximations maintain accuracy with larger values of k.

Figure 10.4 compares the normalized double-saddlepoint density (solid) in (10.25) with the $k \times$ Beta $(nk/2, mk/2)$ density (dashed) for $k = 6$ and $m = 5$. In the left plot $n = 15$ and there are pronounced differences. In the right plot $n = 30$ and the densities are coalescing.

10.4.1 Numerical comparisons in low dimensions

Tabulated percentage points at level 5% have been given for V in table 3 of Anderson (1984, p. 630) and in Kres (1983, p. 136). These values are used with various saddlepoint and *p*-value approximations in table 10.3 so that entries of 5% suggest high accuracy.

Table 10.3. *Percentage values of two saddlepoint approximations, the sequential saddlepoint (Seq. Sad.) and the integrated and normalized double-saddlepoint density (Double Den.), and the F approximation at approximate 5% significance levels. Entries have been partially drawn from table 6.2 in Butler et al. (1992b) with some numerical corrections made.*

k	m	n	Seq. Sad.	Double Den.	F
3	3	14	3.78	5.01	6.99
3	3	24	4.29	4.74	6.12
3	3	64	5.10	4.75	5.40
3	3	∞	5.00	5.00	5.00
3	6	14	4.68	5.21	6.70
3	6	24	4.77	4.96	6.06
3	6	64	4.99	4.87	5.41
5	3	16	2.62	5.30	8.28
5	3	26	3.81	4.85	6.92
5	3	66	5.13	4.74	5.71
5	6	16	4.45	5.42	7.77
5	6	26	4.70	5.06	6.76
5	6	66	4.98	4.88	5.71
10	3	27	0.081	5.69	9.34
10	3	71	5.35	4.84	6.41

These percentiles are only approximate but have been determined to be quite accurate from simulations reported in Butler *et al.* (1992b).

Noticeably absent from the table are entries for the double-saddlepoint CDF approximation in (4.12). All of these entries in the table would be 0.00% and are much too small.

Some conclusions may be drawn in the finite n settings. First, the most consistently accurate approximation is "Double Den." determined by integrating the normalized double-saddlepoint density in (10.25). It is uniformly more accurate in the table than the F approximation. The accuracy of Double Den. accompanied with the gross inaccuracy of the Skovgaard approximation suggests that the Temme argument, used to convert the conditional density to conditional CDF, has failed in this instance.

The sequential saddlepoint approximation "Seq. Sad." shows the greatest accuracy when n is large enough and there are sufficient error degrees of freedom to compensate for the fixed values of k and m. For smaller values of n relative to (k, m), it shows inaccuracy which becomes more marked as k increases.

For the setting $n = \infty$, the Double Den. computation is exact and Seq. Sad. has a quite small bias ($< 0.005\%$ in the table) as the reader is asked to prove in Exercise 16. The F approximation is of course exact in this case.

10.4.2 Numerical comparisons in high dimensions

The accuracy of the approximations when used in high dimensions is considered in table 10.4. Progressively larger values of k are used holding $m = 4$ fixed. As k increases,

Table 10.4. *Percentage values of two saddlepoint approximations, the sequential saddlepoint (Seq. Sad.) and the integrated and normalized double-saddlepoint density (Double Den.), and the F approximation at approximate 5% significance levels with* $m = 4$. *Partially drawn from table 6.3 in Butler et al. (1992b) with numerical corrections.*

k	n	Seq. Sad.	Double Den.	F	$\tilde{t}\{\hat{\mathcal{K}}''(0)\}^{1/2}$
15	56	0.01	5.30	7.94	−3.70
15	66	5.49	5.14	7.42	−1.71
15	96	5.04	4.95	6.58	−1.43
15	176	4.94	4.80	5.76	−1.40
35	66	0.00	7.55	12.60	−8.17
35	106	0.00	5.62	8.81	−9.21
35	156	5.05	5.12	7.45	−1.61
35	196	4.97	4.84	6.82	−1.55
45	66	0.00	10.3	16.68	−8.40
45	166	5.12	5.29	8.15	−1.68
75	196	5.82	5.85	9.90	−2.03
75	236	5.03	5.23	8.81	−1.73
125	286	5.78	5.09	11.00	−2.10

the dimensionality of the nuisance parameter $k(k+1)/2$ grows quadratically from 105 parameters at $k = 15$ up to 7875 at $k = 125$.

Similar conclusions to those of the previous section may be drawn. The integrated double-saddlepoint density Double Den. is consistently accurate for all values of (k, n). The sequential saddlepoint Seq. Sad. is quite accurate when n is above some threshold in magnitude. The F approximation has difficulty in high dimensions. Both saddlepoint methods are easily computed and Seq. Sad. can be trusted when it roughly agrees with the computation of Double Den.

An argument could be made that these examples are not high dimensional because fixing $m = 4$ can be used to restrict the dimension of the distribution theory to 4. This argument uses the distributional identity

$$V_{k,m,n} \sim V_{m,k,m+n-k} \tag{10.26}$$

given by Muirhead (1982, p. 454) so that $m = 4$ fixes the distribution to dimension to 4 on the right side. This argument, however, is not valid since the conditional saddlepoint methods do not show invariance to the distributional equivalence in (10.26) and depend on the conditioning event $W_e + W_h = I_k$. To illustrate this point, consider the third row of table 10.4 with $(k, m, n) = (15, 4, 96)$ and 5th percentile as 14.231 derived from Kres (1983, p. 144). Its distributional equivalent has $(k, m, n) = (4, 15, 85)$ and its 5th percentile 3.231 is derived from the same table entry in Kres (1983). In the latter case, the Seq. Sad. value is 4.99%, the Double Den. is 4.96%, and the F approximation is 5.40%, all different values from the third row. Note that the version with smaller k leads to more accurate approximations and suggests that the dimensionality of the application is affecting saddlepoint accuracy. Thus, for greatest accuracy, the saddlepoint approximation should work in the smallest dimension wherein the parameters (k, m, n) of $V_{k,m,n}$ are replaced by $\min(k, m)$, $\max(k, m)$, and $n - \max(k, m)$.

The success of the sequential saddlepoint CDF approximation sharply contrasts with the failure of the double-saddlepoint approximation and emphasizes their differences. The main difference occurs with the inclusion of the additional Hessian term $|K''_{ss}(\hat{s}_t, t)|^{-1/2}$ in (10.6). This is solely responsible for the difference between success and failure. Also the main difference between the double-saddlepoint density in (10.25) and the F approximation is the term with curly braces in (10.25). This term also results from the Hessian contribution $|K''(\hat{s}, \hat{t})|^{-1/2}$ as described in Exercise 14.

The success (failure) of the sequential saddlepoint when n is above (below) some threshold appears related to the large n and large k asymptotics of V. With fixed k, the limiting null distribution of $n(k - V)$ is χ^2_{km}; see Muirhead (1982, p. 479). If as n grows, k is also allowed to grow at a slightly slower rate, then the limiting distribution of standardized V should be the limit for a standardized χ^2_{km} or Normal $(0, 1)$. Evidence for this is provided in the last column of table 10.4. This column gives the approximate saddlepoint for the standardized value of V or $(V - \mu)/\sigma$. If $\tilde{t}$ is the saddlepoint for V, then, according to (2.1), $\tilde{t}\sigma$ is the saddlepoint for $(V - \mu)/\sigma$ and $\tilde{t}\sigma \simeq \tilde{t}\{\hat{\mathcal{K}}''(0)\}^{1/2}$. If indeed $(V - \mu)/\sigma$ is "close" to Normal $(0, 1)$, then its saddlepoint for the 5th percentile should be close to the saddlepoint for the 5th percentile of Normal $(0, 1)$ or -1.64. The last column shows that the standardized saddlepoints are tending to somewhere near -1.64 which is consistent with the conjectured central limit result.

10.5 Exercises

1. Derive the sequential saddlepoint approximation to the MGF of the Binomial (n, θ) distribution given in (10.9).
   (a) Show that
   $$\hat{s}_t = \ln \frac{n}{\lambda_1 e^t + \lambda_2}$$
   and that
   $$|K''_{ss}(\hat{s}_t, t)| = \lambda_1 \lambda_2 e^{2s+t}.$$
   (b) If $X \sim$ Binomial $(100, 0.03)$, show that the sequential saddlepoint approximation to $\Pr(X \geq 5) = 0.1821$ is $\widehat{\Pr}_2(X \geq 5) = 0.1279$. The extra Hessian contribution has diminished the accuracy as would be expected.
2. Prove that the double saddlepoint methods of sections 4.1–4.2 are equivalent to the two-step procedure of (i) using (10.8) as a surrogate for the true conditional MGF, and (ii) applying single-saddlepoint methods to the surrogate MGF as if it were the true conditional MGF. Show this equivalence for densities and CDFs and for all continuity-corrected approximations.
3. Suppose that Z_1 and Z_2 are i.i.d. with Laplace $(0, 1)$ density given in (1.10).
   (a) Transform to $Y = Z_1 + Z_2$ and $X = Z_1 - Z_2$, compute the conditional density of Y given $X = x$, and thus compute its MGF
   $$\mathcal{M}(t|x) = \frac{t \cosh(tx) + \sinh(tx)}{t(1 - t^2)(1 + x)} \qquad |t| < 1. \tag{10.27}$$
   as may be inferred from Bartlett (1938).

(b) Determine $\hat{\mathcal{M}}(t|x)$ as the sequential saddlepoint approximation to (10.27).
(c) Plot the relative error for saddlepoint accuracy that results from using $\hat{\mathcal{M}}(t|x)$ in a single-saddlepoint approximation. Compare this to the error of the single-saddlepoint approximation applied to (10.27) and also to the double-saddlepoint approximation.

4. Derive the sequential saddlepoint approximation to $_1F_1(\alpha; \alpha + \beta; t)$.
   (a) Show the saddlepoint is $\hat{s}_t = 1 - \hat{r}_t$ where
   $$\hat{r}_t = \tfrac{1}{2}(t + \alpha + \beta) + \tfrac{1}{2}\sqrt{(t + \alpha + \beta)^2 - 4\beta t}$$
   and the approximation is
   $$_1\hat{F}_1(\alpha; \alpha + \beta; t) = \frac{\exp\{\hat{r}_t - (\alpha + \beta)\}(\alpha + \beta)^{\alpha+\beta-1/2}}{(\hat{r}_t - t)^\alpha \hat{r}_t^\beta \sqrt{\alpha(\hat{r}_t - t)^{-2} + \beta \hat{r}_t^{-2}}}. \quad (10.28)$$
   (b) Show that (10.28) may be rewritten in terms of arguments a and b to give
   $$_1\hat{F}_1(a; b; t) = \frac{\exp\{\tilde{r}_t - b\}b^{b-1/2}}{(\tilde{r}_t - t)^a \tilde{r}_t^{b-a} \sqrt{a(\tilde{r}_t - t)^{-2} + (b - a)\tilde{r}_t^{-2}}} \quad (10.29)$$
   where $\tilde{r}_t$ solves
   $$\tilde{r}_t = \tfrac{1}{2}(t + b) + \tfrac{1}{2}\sqrt{(t - b)^2 + 4at}.$$
   (c) Now prove that (10.29) is equivalent to the expression in (10.11). In doing so show that the saddlepoints are related as $t\hat{y}_t = \tilde{r}_t - b$.

5. Derive (10.11) by using two Laplace approximations and prove the limiting relative error given in (10.13).
   (a) Consider a Laplace approximation to the integral
   $$_1F_1(a; b; t) = \frac{\Gamma(b)}{\Gamma(a)\Gamma(b-a)} \int_0^1 e^{ty} y^{a-1}(1-y)^{b-a-1} dy. \quad (10.30)$$
   Change to the variable of integration dx where
   $$x = \int_{1/2}^y t^{-1}(1-t)^{-1} dt = \ln\left(\frac{y}{1-y}\right)$$
   and thereby show that the Laplace approximation to (10.30) is
   $$_1\tilde{F}_1(a; b; t) = \exp(t\hat{y}_t)\hat{y}_t^a(1 - \hat{y}_t)^{b-a} \frac{(2\pi)^{1/2}\Gamma(b)}{\Gamma(a)\Gamma(b-a)} \hat{j}^{-1/2} \quad (10.31)$$
   with
   $$\hat{j} = a(1 - \hat{y}_t)^2 + (b-a)\hat{y}_t^2$$
   and $\hat{y}_t$ given in (10.12).
   (b) The second Laplace approximation required is (10.31) evaluated at $t = 0$. Since it is clear from (10.30) that $_1F_1(a; b; 0) = 1$, a calibrated Laplace approximation results from using
   $$_1\hat{F}_1(a; b; t) = {_1\tilde{F}_1(a; b; t)} / {_1\tilde{F}_1(a; b; 0)} \quad (10.32)$$
   instead of the numerator alone. Show that this calibration leads to the approximation in (10.11).
   (c) The limiting relative error as $t \to \infty$ requires asymptotic expressions for both $_1F_1$ and $_1\hat{F}_1$ that are comparable. From expression 13.5.1 of Abramowitz and Stegun(1972),
   $$_1F_1(a, b, t) = \frac{\Gamma(b)}{\Gamma(a)} e^t t^{a-b}\{1 + O(t^{-1})\}.$$

Show that

$$\sqrt{(t-b)^2 + 4at} = t + 2a - b + O(t^{-1})$$

and $\tilde{r}_t \sim t + a$. Hence show that

$$_1\hat{F}_1(a, b, t) = \frac{\hat{\Gamma}(b)}{\hat{\Gamma}(a)} e^t t^{a-b} \{1 + O(t^{-1})\}$$

where $\hat{\Gamma}(b)$ is Stirling's approximation.

6. Compute the sequential saddlepoint approximation of the beta distribution by using the conditional Mellin transform approach of Exercise 5 of chapter 4. In that context, the joint MGF of $Y = \ln Z_1$ and $X = Z_1 + Z_2$ is specified in (4.86) where Z_1 and Z_2 are the gamma variables of that exercise.

    (a) Determine an explicit expression for $\hat{s}_t$ and hence an explicit expression for the approximate conditional MGF of $\ln Z_1$ given that $Z_1 + Z_2 = 1$.

    (b) Determine the sequential saddlepoint approximation that results for the Beta (α, β) MGF. Compute the beta probabilities in table 1.6 of section 1.2.2 and compare its accuracy with that of the approximation in the table.

    (c) Consider the example in figure 2.1 of section 2.4.1. The figure shows the relative error that occurs in approximating the CDF of the Beta $(\alpha, .8)$ plotted versus α. Construct the same plot for the proposed sequential saddlepoint and compare it to the single-saddlepoint Mellin transform approach in section 2.4.1.

7. Compute the sequential saddlepoint approximation to a Uniform $(0, 1)$ by considering the conditional distribution of Z_1 given $Z_1 + Z_2 = 1$ with Z_1 and Z_2 as independent Exponential (1) variables.

    (a) Compare the relative errors of the sequential saddlepoint with the single-saddlepoint approximation computed from the exact uniform MGF. Plot both sets of relative errors on the same graph.

    (b) Plot the relative error of the double-saddlepoint approximation suggested in Exercise 5 of chapter 4.

    (c) When conditioning on $Z_1 + Z_2 = 1$, does a better approximation to the beta result if one considers $\ln Z_1$ rather than Z_1?

8. (a) Show that a second-order correction to the sequential saddlepoint approximation $\hat{\mathcal{M}}(t|x)$ in (10.5) is

$$\mathcal{M}(t|x) \simeq \hat{\mathcal{M}}(t|x) \frac{1 + O_{N;0} - O_{D;0}}{1 + O_{N;t} - O_{D;t}}, \qquad (10.33)$$

where $O_{N;t} - O_{D;t}$ is the second-order correction to $\hat{f}(y|x;t)$ in approximating $f(y|x;t)$, $O_{N;t}$ is given in (3.37), and $O_{D;t}$ is in (2.32).

    (b) Use (10.33) to compute a second-order correction term to be used in conjunction with $_1\hat{F}_1(\alpha, \alpha + \beta, t)$ in (10.28) and hence with $_1\hat{F}_1(a, b, t)$ in (10.29).

    (c) Derive a second-order correction term for the Laplace approximation. Note in (10.32) that both $_1\tilde{F}_1(a;b;t)$ and $_1\tilde{F}_1(a;b;0)$ require second-order corrections. Furthermore each of these corrections needs to be computed either in the x variable of integration, or from (2.20) which takes into account the portion of the integrand over which maximum y_t has been computed.

    (d) Compute the correction terms from the sequential and Laplace approaches to see whether or not they agree. Do they improve upon their first-order approximations or do they make them worse?

9. Derive the Laplace approximation to the Gauss hypergeometric function given in (10.16).
   (a) Abramowitz and Stegun (1972) give the integral representation for $_2F_1$ as

   $$_2F_1(a,b;c;x) = \frac{\Gamma(c)}{\Gamma(a)\Gamma(c-a)} \int_0^1 y^{a-1}(1-y)^{c-a-1}(1-xy)^{-b} dy. \quad (10.34)$$

   Use the change of variable in Exercise 5(a) to derive $_2\tilde{F}_1(a,b;c;x)$ as a Laplace approximation for (10.34). The approximation is

   $$_2\tilde{F}_1(a,b;c;x) = (1-x\hat{y})^{-b}\hat{y}^a(1-\hat{y})^{c-a}\hat{j}^{-1/2} \frac{\sqrt{2\pi}\,\Gamma(c)}{\Gamma(a)\Gamma(c-a)}$$

   where

   $$\hat{j} = a(1-\hat{y})^2 + (c-a)\hat{y}^2 - bx^2\hat{y}^2(1-\hat{y})^2/(1-x\hat{y})^2, \quad (10.35)$$

   and $\hat{y}$ is defined in (10.17).
   (b) Calibrate $_2\tilde{F}_1$ as describe in Exercise 5(b) by using the fact that $_2F_1(a,b;c;0) = 1$. Show that this leads to (10.16).

10. Consider the Pólya $(M,N;n)$ mass function given in (4.25). Derive the sequential saddlepoint approximation to its MGF as (10.15) using $\theta = 1/2$ as the parameter in the negative binomial MGFs.
    (a) Determine an explicit value for $\hat{u}_t = \exp(\hat{s}_t)$. Convert this expression into the notation of the Laplace expression by replacing n with $-a$, M with b, N with $c-b$ and

    $$\tau = n - N - e^t(M+n)$$
    $$x = 1 - e^t.$$

    After such replacement, show that

    $$\hat{u}_t = \frac{-4a}{\sqrt{\tau^2 - 4ax(c-b)} - \tau - 2a}$$
    $$= -2\hat{y}/(1-\hat{y}).$$

    Hence compute $K(\hat{s}_t, t)$ and show that

    $$\exp\{K(\hat{s}_t, t) - n\hat{s}_t\} \stackrel{t}{\propto} (1-x\hat{y})^{-b}\hat{y}^a(1-\hat{y})^{c-a}.$$

    (b) Show that

    $$K''_{ss}(\hat{s}_t, t) = \left[\frac{-Me^t}{\{1-(1-e^t)\hat{y}\}^2} - N\right]\hat{y}(1-\hat{y}).$$

    If $\hat{j}(t)$ is expression (10.35) from the Laplace approximation, show

    $$K''_{ss}(\hat{s}_t, t) = -\hat{j}(t).$$

    Hence complete the proof of equivalence.
    (c) Use (10.14) to compute the sequential probability entries for the examples in table 4.5. In particular, fill in entries for the continuity-corrected approximations. How does their accuracy compare to that of the double- and single-saddlepoint approximations listed in the table?

11. Derive the sequential saddlepoint approximation for the MGF of a Hypergeometric $(M, N; n)$ distribution given in (10.21).

(a) If $\hat{s}_t$ is the saddlepoint, solve for $\hat{u}_t = \exp(\hat{s}_t)$ as

$$\hat{u}_t = \frac{2n}{\tau + \sqrt{\tau^2 + 4ne^t(M + N - n)}}, \qquad \tau = N - n + e^t(M - n),$$

and $\hat{u}_0 = n/(M + N - n)$.

(b) Confirm (10.21) with $_2\check{F}_1$ given as

$$_2\check{F}_1(a, b; c; x) = (\check{v}_x x + 1)^{-b} \left(\frac{\check{v}_1}{\check{v}_x}\right)^{-a} (\check{v}_x + 1)^{c-a-1} \sqrt{\check{\eta}} \qquad (10.36)$$

$$\times \frac{\Gamma(c)\Gamma(c - a - b)}{\Gamma(c - a)\Gamma(c - b)} \left(1 + \frac{a}{c - a - b - 1}\right)^{c-a-b-1}$$

where

$$\check{\eta} = \frac{(c - a - b - 1)\check{v}_1/(\check{v}_1 + 1)^2}{(c - a - 1)\check{v}_x/(\check{v}_x + 1)^2 - b\check{v}_x x/(\check{v}_x x + 1)^2},$$

and

$$\check{v}_x = \frac{-2a}{\psi + \sqrt{\psi^2 - 4xa(c - b - 1)}} \qquad \psi = c - 1 + x(a - b),$$

and $\check{v}_1 = -a/(c - b - 1)$. Implicit in this confirmation is the validation of $_2\check{F}_1$ as an approximation to $_2F_1$. One might expect this approximation to only work with $a, b < 0$ since this is the constraint on the arguments in the sampling framework with M, N, and n. However, Butler and Wood (2002, §6) provide an analytic continuation argument that "continues" the approximation into argument values not allowed in its original context. Under this continuation, the method proves accurate for $x > 0$, and $c - a - b - 1 > 0$.

(c) Use the Maple function hypergeom($[a, b]$, $[c]$, x) to determine the numerical accuracy of (10.36) for an assortment of values for a, b, c, and $x \in (-1, 1)$. Compare accuracy to approximation (10.16). Continuation into negative values of x is made possible by letting $_2\check{F}_1$ conform to the Euler relation

$$_2F_1(a, b; c; x) = (1 - x)^{-b} {}_2F_1\{c - a, b; c; -x/(1 - x)\} \qquad (10.37)$$

given in (20) of Butler and Wood (2002).

12. Compute the sequential saddlepoint approximation for the common odds ratio test using the crying baby data of table 6.4.
    (a) Replicate the p-value computations of table 6.5 but for the sequential saddlepoint approximation using uncorrected and corrected methods.
    (b) Compute power function values in table 6.6 using the sequential methods and compare these to the corresponding single-saddlepoint computations.
    (c) In determining the power curve of the test, plot the percentage relative error incurred by the using the sequential saddlepoint method with $\check{\mathcal{M}}(t)$ instead of the true MGF $\mathcal{M}(t)$. Are the results comparable to those shown in figure 10.3?

13. In saddlepoint approximation for the Bartlett–Nanda–Pillai trace statistic, derive $\hat{\mathcal{M}}(t|I_k)$ given in (10.24)
    (a) Show that $(\hat{S} + \hat{S}_d)_t = 2\hat{c}_t I_k$ with $\hat{c}_t$ given in (10.23) solves the saddlepoint equation $K'_S(\hat{S}_t, t) = I_k$.
    (b) Show that $\hat{c}_0 = -(m + n - 1)/2$,

$$M(\hat{S}_t, t) = (1 - 2\hat{c}_t)^{-mk/2}(1 - 2\hat{c}_t - 2t)^{-nk/2}$$
$$M(\hat{S}_0, 0) = (m + n)^{-(m+n)k/2}.$$

Table 10.5. *First percentiles of the Bartlett–Nanda–Pillai trace statistic derived from table 7 of Kres (1983).*

	First percentiles of Pillai's trace						
k	3	5	10	15	15	45	15
m	6	6	3	4	4	4	4
n	64	66	71	96	176	166	66
1st percentile	2.528	4.327	9.35	14.153	14.521	43.715	13.812

(c) Show that the Hessian matrix $K''_{ss}(\hat{S}_t, t)$ is given in block form as

$$\left\{\frac{m}{(1-2\hat{c}_t)^2} + \frac{n}{(1-2\hat{c}_t - 2t)^2}\right\} \begin{pmatrix} 2I_k & 0 \\ 0 & I_{k(k-1)/2} \end{pmatrix}.$$

Hence complete the derivation of (10.24).

14. Derive the double-saddlepoint density in (10.25).

    (a) Compute

    $$\frac{\partial}{\partial t} K(S,t)|_{S=\hat{S}_t} = nk/(1 - 2\hat{c}_t - 2t).$$

    (b) Show that the double-saddlepoint solution to

    $$nk/(1 - 2\hat{c} - 2\hat{t}) = v$$
    $$m/(1 - 2\hat{c}) + n/(1 - 2\hat{c} - 2\hat{t}) = 1$$

    is

    $$\hat{c} = \hat{c}_{\hat{t}} = \frac{k - v - mk}{2(k-v)} \qquad \hat{t} = (1 - 2\hat{c} - nk/v)/2.$$

    (c) Compute the Hessian matrix $K''(\hat{S}, \hat{t})$ as

    $$\begin{pmatrix} kd_1 & d_1 1_k^T & 0 \\ d_1 1_k & 2d_2 I_k & 0 \\ 0 & 0 & d_2 I_{k(k-1)/2} \end{pmatrix}$$

    where

    $$d_1 = \frac{2n}{(1 - 2\hat{c} - 2\hat{t})^2} \qquad d_2 = \left\{\frac{m}{(1-2\hat{c})^2} + \frac{n}{(1-2\hat{c}-2\hat{t})^2}\right\}.$$

    (d) Finalize the determination of the density in (10.25).

15. (a) Compute the sequential and double-saddlepoint CDF values for the approximate first percentiles listed in table 10.5. What happens for the example in the last column?

    (b) Compute p-values for the table entries by numerically integrating the normalized double-saddlepoint density. Compared them to the p-values from the F approximation.

    (c) What can you conclude about the various methods for cases where the significance is extreme?

16. Derive the limiting values for the sequential and double-saddlepoint approximations for Pillai's trace statistic as $n \to \infty$.

(a) Transform from V to $Y = n(k - V)$ with $y = n(k - v)$. With y fixed, compute the following limits as $n \to \infty$.

$$\lim_{n \to \infty} e^{-\tilde{t}v} \hat{\mathcal{M}}(\tilde{t}|I_k) = \left(\frac{y}{mk}\right)^{mk/2} \exp(mk/2 - y/2)$$

$$\lim_{n \to \infty} \frac{1}{n} \hat{\mathcal{K}}''(\tilde{t}|I_k)^{-1/2} = \frac{1}{y}\left(\frac{mk}{2}\right)^{1/2}.$$

Here, $\tilde{t}$ is the sequential saddlepoint solving $\hat{\mathcal{K}}'(\tilde{t}|I_k) = v$.

(b) If

$$\hat{w} = \text{sgn}(\tilde{t})\sqrt{2\{\tilde{t}v - \mathcal{K}(\tilde{t}|I_k)\}}$$
$$\hat{u} = \tilde{t}\sqrt{\mathcal{K}''(\tilde{t}|I_k)},$$

use the results of (a) to show that

$$w_\infty = \lim_{n \to \infty} \hat{w} = -\text{sgn}(y - mk)\sqrt{y - mk - mk \ln\{y/(mk)\}}$$
$$u_\infty = \lim_{n \to \infty} \hat{u} = -(y - mk)/\sqrt{2mk}.$$

(c) Confirm that the limits in (b) are the inputs to the Lugannani and Rice formula for $Y \sim \chi^2_{mk}$. Thus the limiting value has the bias that occurs when approximating the χ^2_{mk} distribution.

(d) Again transform to Y and fix the value of y. Show that the limiting double-saddlepoint density is

$$\lim_{n \to \infty} \Pr\{Y \in (y, y + dy)\} = g(y)\Gamma(mk/2)/\hat{\Gamma}(mk/2)$$

where g is the density of a χ^2_{mk} and $\hat{\Gamma}$ is Stirling's approximation.

(e) Prove that the integrated normalized double-saddlepoint converges to the CDF of a χ^2_{mk}. (Butler *et al.* 1992b, §5).

11

Applications to multivariate testing

Saddlepoint methods are applied to many of the commonly used test statistics in MANOVA. The intent here is to highlight the usefulness of saddlepoint procedures in providing simple and accurate probability computations in classical multivariate normal theory models. Power curves and p-values are computed for tests in MANOVA and for tests of covariance. Convenient formulae are given which facilitate the practical implementation of the methods.

The null distributions of many multivariate tests are easily computed by using their Mellin transforms which admit saddlepoint approximations leading to very accurate p-values. The first section concentrates on the four important tests of MANOVA. Very accurate p-value computations are suggested for (i) the Wilks likelihood ratio for MANOVA, (ii) the Bartlett–Nanda–Pillai trace statistic, (iii) Roy's largest eigenvalue test, and (iv) the Lawley-Hotelling trace statistic. Saddlepoint methods for Wilks' likelihood ratio test were introduced in Srivastava and Yau (1989) and Butler *et al.* (1992a); approximations for (ii) were introduced in Butler *et al.* (1992b) and were also discussed in section 10.4; p-values for Roy's test do not use saddlepoint methods and are based on the development in Butler and Paige (2007) who extend the results of Gupta and Richards (1985); and p-values for Lawley–Hotelling trace are based on numerical inversion of its Laplace transform in work to be published by Butler and Paige.

The second section considers tests for covariance patterns. Saddlepoint approximations for p-values are suggested for the following tests: the likelihood ratio tests for (ii) block independence, (iii) sphericity, (iv) an intraclass correlation structure; and (v) the Bartlett–Box test for equal covariances. These applications were developed in Butler *et al.* (1993), and Booth *et al.* (1995).

Power curves for tests require noncentral distributions which are considerably more difficult to compute as may be seen in Muirhead (1982). However, the sequential saddlepoint methods developed in Butler and Wood (2002, 2004) provide simple power computations as outlined in section 11.3. Very accurate power curves of (a) Wilks likelihood ratio test in MANOVA, (b) the likelihood ratio for block independence, and (c) the Bartlett–Box test for equal covariances are presented along with simple computational formulas.

11.1 P-values in MANOVA

11.1.1 Wilks' likelihood ratio statistic Λ

This test was considered in section 2.4.1 where a single-saddlepoint approximation used the Mellin transform to provide accurate numerical probabilities as seen in table 2.2. The example is revisited in order to consider its computation from a conditional point of view. The leading conditional approximation and a new sequential method are slightly better than the single-saddlepoint approach although all three methods are exceedingly accurate. The conditional approximation is given in Butler *et al.* (1992a) where further conditional approaches were developed with other sorts of conditioning.

Suppose the error sum of squares matrix W_e is a $k \times k$ Wishart (n, Σ) and the hypothesis sum of squares matrix W_h is a $k \times k$ Wishart (m, Σ). Anderson (1958, chap. 8) gives two distributional characterizations for the null distribution of $\ln \Lambda_{k,m,n} \in (-\infty, 0)$ in terms of independent Beta variates $\{\beta_i\}$. They are

$$\ln \Lambda_{k,m,n} \sim \sum_{i=1}^{k} \ln \beta_i \{\tfrac{1}{2}a_i, \tfrac{1}{2}m\} \tag{11.1}$$

with $a_i = n - i + 1$, and

$$\ln \Lambda_{k,m,n} \sim 2 \sum_{i=1}^{r} \ln \beta_i(b_i, m) + w \ln \beta_{r+1}(c, \tfrac{1}{2}m), \tag{11.2}$$

where $r = \lfloor k/2 \rfloor$; $b_i = n - 2i + 1$ for $i = 1, \ldots, r$; $c = n/2 - r$; and w is the indicator that k is odd. Note the additional oddball term in (11.2) for k odd. It may be carried forward without duplication by using the indicator w.

The two characterizations lead to single-saddlepoint approximations that are the same analytically and which are therefore the approximation of section 2.4.1; see Exercise 1 for a proof. It is simple and expedient to use the CGF of (11.1) which is

$$K(t) = \sum_{i=1}^{k} \ln \frac{\Gamma\{\tfrac{1}{2}(a_i + m)\} \Gamma(t + \tfrac{1}{2}a_i)}{\Gamma\{t + \tfrac{1}{2}(a_i + m)\} \Gamma(\tfrac{1}{2}a_i)} \tag{11.3}$$

and defined for $t > -a_k/2 = -(n - k + 1)/2$.

Both characterizations also lead to double-saddlepoint approximations but these are not the same. To create this distribution theory, simply construct the distribution of each $\ln \beta_i$ by conditioning $\ln \chi_{1i}$ on $\chi_{i1} + \chi_{i2} = 1$ where χ_{i1} and χ_{i2} are the appropriate independent gamma variables. Of the two double-saddlepoint methods, the one based on (11.2) is more accurate so only it will be considered. In fact, this approximation turns out to be the most accurate overall. The second most accurate approximation is the sequential saddlepoint based on the conditional development of (11.2). It is hardly more difficult than the double-saddlepoint since the nuisance saddlepoint $\hat{s}_t$ is explicit in this case.

The distribution $\ln \beta_i \{b_i, m\}$ is created by conditioning $\ln \chi_{i1}$ on $\chi_{i1} + \chi_{i2} = 1$ where $\chi_{i1} \sim$ Gamma $(b_i, 1)$ and $\chi_{i2} \sim$ Gamma $(m, 1)$ for $i = 1, \ldots, r$. If k is odd, also include $\chi_{r+1,1} \sim$ Gamma $\{c, 1\}$ and $\chi_{r+1,2} \sim$ Gamma $(m/2, 1)$. The conditional characterization derived from (11.2) is that

$$\ln \Lambda_{k,m,n} \sim 2 \sum_{i=1}^{r} \ln \chi_{i1} + w \ln \chi_{r+1,1} \,|\, \{\chi_{j1} + \chi_{j2} = 1, \quad j = 1, \ldots, r+1\} \tag{11.4}$$

where each conditioning event affects only its corresponding term in the convolution. This structure leads to an explicit value for $\hat{s}_t$; the details of the derivation are outlined in Exercise 2.

Double-saddlepoint approximation

The test rejects for small values of $\ln \Lambda_{k,m,n}$ and $\Pr(\ln \Lambda_{k,m,n} \leq l)$ is computed from the Skovgaard approximation in (4.12) by using the following inputs:

$$\hat{w} = \text{sgn}(\hat{t})\sqrt{2\{\kappa(0) - \kappa(\hat{t}) - (r + w/2 - l/2)\hat{t}\}} \tag{11.5}$$

with

$$\kappa(t) = \sum_{i=1}^{r}\left[\ln\left\{\frac{\Gamma(t+b_i)}{\Gamma(b_i)}\right\} - (t+b_i+m)\ln(t+b_i+m)\right] \tag{11.6}$$

$$+ w\left\{\ln\left(\frac{\Gamma(t/2+c)}{\Gamma(c)}\right) - (t/2+c+m/2)\ln(t/2+c+m/2)\right\}.$$

Saddlepoint $\hat{t} \in (-n+k-1, \infty)$ is the root of

$$l/2 = \sum_{i=1}^{r}\{\psi(\hat{t}+b_i) - \ln(\hat{t}+b_i+m)\} + \tfrac{1}{2}w\{\psi(\hat{t}/2+c) - \ln(\hat{t}/2+c+m/2)\}, \tag{11.7}$$

where $\psi = (\ln \Gamma)'$ is the di-gamma function. The other input is

$$\hat{u} = \hat{t}\sqrt{|\hat{j}|/|\hat{j}_{ss}(\hat{s}_0, 0)|}$$

where

$$|\hat{j}_{ss}(\hat{s}_t, t)| = \prod_{i=1}^{r}(t+b_i+m)^{-1} \times (t/2+c+m/2)^{-w} \tag{11.8}$$

$$|\hat{j}| = |\hat{j}_{ss}(\hat{s}_{\hat{t}}, \hat{t})| \times \left[\sum_{i=1}^{r}\{\psi'(\hat{t}+b_i) - (\hat{t}+b_i+m)^{-1}\}\right.$$

$$\left. + \tfrac{1}{4}w\{\psi'(\hat{t}/2+c) - (\hat{t}/2+c+m/2)^{-1}\}\right]$$

and $\psi' = (\ln \Gamma)''$ is the tri-gamma function. Both the di- and tri-gamma functions are built-in functions in major computing environments so that the computation once programmed borders on trivial. Note, in particular, for any dimension k, the saddlepoint computation always entails finding only a one-dimensional root since $\hat{s}_t$ is explicit in t. All of these expressions appear in section 3.3 of Butler *et al.* (1992a) using quite different notation.

To understand why a double-saddlepoint approximation derived from (11.2) would be more accurate than one derived from (11.1), consider the distribution of $\ln \Lambda_{5,6,20}$. The characterizations are

$$\ln \Lambda_{5,6,20} \sim \ln \beta(10, 3) + \ln \beta(9.5, 3) + \cdots + \ln(8, 3) \tag{11.9}$$

$$\sim 2\{\ln \beta(19, 6) + \ln \beta(17, 6)\} + \ln \beta(8, 3). \tag{11.10}$$

The first characterization with five terms has roughly twice the number of terms as the second characterization with three, but these terms have half the degrees of freedom. The first reason (11.10) is more accurate is that it entails roughly half the amount of conditioning; dimensional reduction in conditioning generally leads to greater accuracy. The second reason is the higher degrees of freedom of the terms in (11.10) provide greater "balance" in the degrees of freedom for all of the beta terms.

Sequential saddlepoint approximation

This approximation uses the surrogate conditional CGF

$$\hat{\mathcal{K}}(t) = \kappa(t) - \kappa(0) + (r + w/2)t + \frac{1}{2}\ln\{|\hat{j}_{ss}(\hat{s}_0, 0)|/|\hat{j}_{ss}(\hat{s}_t, t)|\} \tag{11.11}$$

which is explicit in t. Numerical differentiation suffices for all its computations.

Numerical computations

Table 11.1 reconsiders the examples of table 2.2 for the double-, sequential-, and single-saddlepoint approximations. All three approximations were evaluated at the true percentiles given in the last column of table 2.2.

The last two examples deal with the distributions of $\ln \Lambda_{9,16,18}$ and $\ln \Lambda_{16,9,25}$ which are versions of the same distribution. More generally

$$\ln \Lambda_{k,m,n} \sim \ln \Lambda_{m,k,m+n-k} \tag{11.12}$$

and therefore one has the choice of computing the saddlepoint approximation either way. This choice makes no difference for the single-saddlepoint approximation since it is invariant to the version in (11.12) used. However, the double- and sequential saddlepoints do depend on the particular version. A comparison of the last two settings as well as other computations show that both approximations are better when the version with smallest dimensional parameter $\min(k, m)$ is used as the first index. If this strategy is followed then the double-saddlepoint approximation performs best overall, with sequential a slight second and single as a slight third. For practical purposes they are all incredibly accurate. Butler and Wood (2004, 2005) show that this accuracy improves with increasing dimension.

To better understand the need to reduce the dimension k, reconsider the example $\ln \Lambda_{5,6,20}$. The equivalent versions are

$$\ln \Lambda_{5,6,20} \sim 2\{\ln \beta(19, 6) + \ln \beta(17, 6)\} + \ln \beta(8, 3)$$
$$\ln \Lambda_{6,5,21} \sim 2\{\ln \beta(19, 5) + \ln \beta(17, 5)\} + \ln \beta(15, 5)\}.$$

The top characterization in which the second index is larger, provides better balance for the two degrees of freedom in the individual Betas leading to more accurate saddlepoint approximation.

11.1.2 Bartlett–Nanda–Pillai trace statistic

A complete discussion of this test was given in section 10.4. In practical computation, two approximations are recommended: (i) computation of the sequential CDF approximation

11.1 P-values in MANOVA

Table 11.1. *Percentage values of the double-, sequential- (Seq.), and single-saddlepoint approximations for various distributions of $\ln \Lambda_{k,m,n}$. The approximations were evaluated at the true percentiles provided in the last column of table 2.2.*

k	m	n	Exact	Double	Seq.	Single
3	5	7	10	10.04	10.06	9.957
			5	5.025	5.034	4.973
			2.5	2.509	2.513	2.479
			1	1.001	1.002	.9873
			0.5	.5021	.5028	.4947
5	5	9	10	10.02	10.09	9.976
			5	5.017	5.055	4.990
			2.5	2.506	2.527	2.491
			1	1.002	1.010	.9935
			0.5	.4983	.5022	.4935
9	16	18	10	10.06	10.09	10.06
			5	4.961	4.977	4.962
			2.5	2.520	2.528	2.520
			1	.9937	.9975	.9939
			0.5	.5037	.5058	.5038
16	9	25	10	9.945	10.10	10.06
			5	4.895	4.988	4.962
			2.5	2.482	2.535	2.520
			1	.9765	1.000	.9939
			0.5	.4943	.5073	.5038

using (10.24), and (ii) numerical integration of the normalized double-saddlepoint density in (10.25). With smaller n, the first approximation may not have the required accuracy in which case the double-saddlepoint density integration should be trusted instead.

The null distribution for this test statistic is the trace of a Matrix–Beta $(n/2, m/2)$ distribution as described in Muirhead (1982, p. 479). Using the multivariate Gamma function defined by

$$\Gamma_k(v) = \pi^{k(k-1)/4} \prod_{j=1}^{k} \Gamma\{v - \tfrac{1}{2}(j-1)\} \qquad v > (k-1)/2, \qquad (11.13)$$

the MGF is

$$E(e^{tV}) = \frac{\Gamma_k\{\tfrac{1}{2}(n+m)\}}{\Gamma_k(\tfrac{1}{2}n)\Gamma_k(\tfrac{1}{2}m)} \int_{0<U<I_k} e^{t \operatorname{tr} U} |U|^{\tfrac{1}{2}(n-k-1)} |I_k - U|^{\tfrac{1}{2}(m-k-1)} dU \qquad (11.14)$$

where $\{U : 0 < U < I_k\}$ is the set of Matrix Beta variates that are $k \times k$ symmetric with all eigenvalues in $(0, 1)$. In the univariate case with $k = 1$, the integral in (11.14) is the MGF for a univariate Beta $(n/2, m/2)$ which is known to be the $_1F_1\{n/2; (n+m)/2; t\}$ function. For $k \geq 1$ this function is again a special function that generalizes the univariate case and is known as $_1F_1\{n/2;(n+m)/2;tI_k\}$, the $_1F_1$ function again but with the $k \times k$ matrix argument tI_k.

In section 10.4, an approximation for this MGF has been derived using the sequential saddlepoint approximation; expression $\hat{\mathcal{M}}(t|I_k)$ in (10.24) approximates $_1F_1\{n/2;(n+m)/2;tI_k\}$. It turns out that this approximation is a special case of a very general Laplace approximation for matrix argument $_1F_1$ functions which has been described in Butler and Wood (2002). In particular

$$\hat{\mathcal{M}}(t|I_k) = {_1\hat{F}_1}\{n/2;(n+m)/2;tI_k\} \tag{11.15}$$

where $_1\hat{F}_1$ denotes a Laplace approximation for the integral in (11.14). A general expression for $_1\hat{F}_1$ with arbitrary $k \times k$ matrix argument is given in (11.33) of section 11.3.1.

11.1.3 Roy's maximum root test

In the MANOVA framework described in section 11.1.1, Roy's test statistic is the largest eigenvalue $\lambda_1 = \lambda_1(k,m,n)$ of the random Beta matrix $(W_e + W_h)^{-1/2} W_h (W_e + W_h)^{-1/2}$. Its null distribution depends only on dimension k and degrees of freedom parameters m and n. No saddlepoint approximations have been proposed to approximate this distribution. There is, in its place, an easily computed expression for the exact null CDF as stated in theorem 11.1.1 below. This expression is developed in Butler and Paige (2007) based on the theoretical derivation provided by Gupta and Richards (1985). An executable file for making the computation is posted at the author's website *http://www.smu.edu/statistics/faculty/butler.html*.

Suppose $n \geq k$ and, without loss in generality, $m \geq k$. The latter inequality can be assured to hold through the distributional equivalence

$$\lambda_1(k,m,n) \sim \lambda_1(m,k,m+n-k).$$

The following notation is used:

$$\alpha_i = \tfrac{1}{2}(m+k+1) - i \geq \tfrac{1}{2} \qquad i = 1,\ldots,k$$
$$\beta = \tfrac{1}{2}(n-k+1) \geq \tfrac{1}{2}.$$

Also define the multivariate beta function using the multivariate gamma function in (11.13) as

$$B_k(a,b) = \Gamma_k(a)\Gamma_k(b)/\Gamma_k(a+b).$$

Theorem 11.1.1 *The exact null CDF of $\lambda_1(k,m,n)$ is*

$$\Pr(\lambda_1 \leq r) = \frac{1}{B_k\left(\frac{m}{2},\frac{n}{2}\right)} \frac{\pi^{k^2/2}}{\Gamma_k(k/2)} r^{km/2} \sqrt{|A_r|} \qquad 0 < r < 1. \tag{11.16}$$

Here, $A_r = (a_{ij})$ is a skew-symmetric matrix ($a_{ij} = -a_{ji}$) whose structure is determined by whether k is even or odd. For k even, A is $k \times k$ with

$$a_{ij} = r^{-(\alpha_i+\alpha_j)} \left\{ 2\sum_{l=0}^{L_\beta}(-1)^l \binom{\beta-1}{l} \frac{C_r(\alpha_i+\alpha_j+l,\beta)}{\alpha_i+l} - C_r(\alpha_i,\beta)C_r(\alpha_j,\beta) \right\}, \tag{11.17}$$

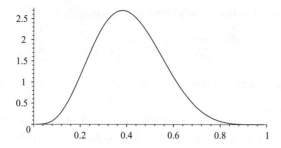

Figure 11.1. The density of λ_1 for $(k, m, n) = (2, 5, 13)$.

where

$$C_r(a, b) = B(a, b) I_r(a, b), \tag{11.18}$$

$B(a, b)$ is the beta function, and $I_r(\alpha_i, \beta)$ denotes the incomplete beta function as given in (26.5.1) of Abramowitz and Stegun (1972). The range of summation is

$$L_\beta = \begin{cases} \beta - 1 & \text{if } \beta \text{ is an integer} \\ \infty & \text{if } \text{otherwise.} \end{cases} \tag{11.19}$$

For the case in which k is odd, then A is $(k + 1) \times (k + 1)$. The upper left $k \times k$ block is the matrix $(a_{ij} : i, j = 1, \ldots, k)$ described above with a_{ij} given in (11.17). The $(k + 1)$st column (row) is determined as

$$a_{i,k+1} = -a_{k+1,i} = r^{-\alpha_i} C_r(\alpha_i, \beta) \qquad i = 1, \ldots, k. \tag{11.20}$$

Example

Kres (1983) tabulates 95th and 99th percentiles of the null distribution for a transformation of λ_1 for various combinations of (k, m, n) including the case $(2, 5, 13)$. When converted to percentiles of λ_1 these tabulated percentiles are 0.651 and 0.745. In two dimensions with $(m, n) = (5, 13)$, expression (11.17) is especially simple as

$$\Pr(\lambda_1 \leq r) = 21840 \left| 2 \sum_{l=0}^{5} (-1)^l \binom{5}{l} \frac{C_r(5 + l, 6)}{3 + l} - C_r(3, 6) C_r(2, 6) \right| \tag{11.21}$$

for $r \in (0, 1)$. This agrees with the $k = 2$ expression given in Johnson and Kotz (1972, §39.2, p. 185). Evaluation of (11.21) at $r = 0.651$ and 0.745 leads to probability values 0.9502 and 0.9901 respectively. The exact percentiles determined by inverting (11.21) are 0.65072 and 0.74461. A plot of the density taking numerical derivatives of (11.21) is given in figure 11.1.

11.1.4 Lawley–Hotelling trace statistic

The Lawley–Hotelling trace statistic $T_0^2 = \text{tr}(W_h W_e^{-1})$ has a null distribution whose MGF is specified in terms of a type II confluent hypergeometric function of matrix argument denoted by Ψ. Muirhead (1982, p. 474) derives the MGF as

$$E(e^{tT_0^2}) = \frac{\Gamma_k\left(\frac{n+m}{2}\right)}{\Gamma_k\left(\frac{n}{2}\right)} \Psi\left(\frac{m}{2}, \frac{k+1-n}{2}; -tI_k\right) \qquad \text{Re}(t) \le 0. \tag{11.22}$$

P-value computations are hampered by two difficulties: computation of Ψ and the inversion of (11.22).

Methods for exact and approximate computation of Ψ are provided in Butler and Wood (2003, 2005). Even with the computation of Ψ solved, inversion of (11.22) with saddlepoint methods is not possible since the MGF is not steep. Here, lack of steepness means that $\lim_{t\to 0} K'(t) = E(T_0^2) < \infty$ so the saddlepoint equation cannot be solved for T_0^2-values above the mean which is the rejection region for the test. Alternative numerical inversion methods are under development by Butler and Paige to determine approximate p-values for T_0^2. A technical report describing these procedures will be posted at *http://www.smu.edu/statistics/TechReports/tech-rpts.asp*

11.2 P-values in tests of covariance

11.2.1 Block independence

Suppose that W has a $k \times k$ Wishart (n, Σ) distribution. Let the k dimensions be partitioned into p blocks of sizes $k_1, \ldots, k_p$ so that $k = \sum_i k_i$. The likelihood ratio test for the null hypothesis that Σ is block diagonal and conformable to the partition rejects for small values of

$$\Lambda_I = |W| / \prod_{i=1}^{p} |W_i| \tag{11.23}$$

where W_i is the $k_i \times k_i$ submatrix of W associated with the ith block.

Srivastava and Khatri (1979, pp. 176 and 220) provide two distributional representations in terms of independent Wilks statistics denoted with indices $\Lambda_{k,m,n}$. If $\bar{k}_i = k_1 + \cdots + k_{i-1}$, then

$$\ln \Lambda_I \sim \sum_{i=2}^{p} \ln \Lambda_{k_i, \bar{k}_i, n-\bar{k}_i} \tag{11.24}$$

where it may be understood that the independent term $\Lambda_{k_i, \bar{k}_i, n-\bar{k}_i}$ tests that the ith block is independent of previous blocks $1, \ldots, i-1$ as described in Eaton (1983, p. 449). The second and equivalent representation is

$$\ln \Lambda_I \sim \sum_{i=2}^{p} \ln \Lambda_{m_i, M_i, n-M_i} \tag{11.25}$$

where $m_i = \min(k_i, \bar{k}_i)$ and $M_i = \max(k_i, \bar{k}_i)$. This version simply takes advantage of dimensional reduction using the equivalence in (11.12).

The single-saddlepoint works with the Mellin transform of $\ln \Lambda_I$ computed from the $p-1$ transforms of the independent Wilks statistics. Indeed, all of its properties derive from these independent components such as the fact that the approximation is invariant to which of the two representations is used. Its CGF is defined for $t > -(n-k+1)/2$ and determined from (11.24) as

$$K(t) = \sum_{i=2}^{p}\sum_{j=1}^{k_i} \ln \frac{\Gamma\{\frac{1}{2}(a_{ij}+\bar{k}_i)\}\Gamma(t+\frac{1}{2}a_{ij})}{\Gamma\{t+\frac{1}{2}(a_{ij}+\bar{k}_i)\}\Gamma(\frac{1}{2}a_{ij})}$$

where $a_{ij} = n - \bar{k}_i - j + 1$ and equivalently determined from (11.25) as

$$K(t) = \sum_{i=2}^{p}\sum_{j=1}^{m_i} \ln \frac{\Gamma\{\frac{1}{2}(\tilde{a}_{ij}+M_i)\}\Gamma(t+\frac{1}{2}\tilde{a}_{ij})}{\Gamma\{t+\frac{1}{2}(\tilde{a}_{ij}+M_i)\}\Gamma(\frac{1}{2}\tilde{a}_{ij})}$$

where $\tilde{a}_{ij} = n - M_i - j + 1$.

A double saddlepoint may be derived from (11.25) by specifying each independent Wilks statistics in terms of its underlying independent beta variables as

$$\ln \Lambda_I \sim \sum_{i=2}^{p}\sum_{j=1}^{r_i}\{2\ln\beta_{ij}(b_j - M_i, M_i) + w_i \ln \beta_{i,r_i+1}(c_i - M_i/2, M_i/2)\}. \quad (11.26)$$

Here, w_i is the indicator that m_i is odd, $m_i = 2r_i + w_i$, $\{\beta_{ij}\}$ are independent beta variables with the indicated parameters defined in terms of $b_j = n - 2j + 1$ and $c_i = n/2 - r_i$. If each beta variable is constructed using two independent gammas and the sum is conditioned to be 1, then the arguments used to determine the distribution of $\ln \Lambda_{k,m,n}$ would seem at least feasible and indeed are. The reader is asked to provide the details in Exercise 4.

The sequential saddlepoint method is even simpler. If $\hat{\mathcal{K}}_i(t)$ denotes the sequential approximation given in (11.11) adjusted for the MGF of $\ln \Lambda_{m_i, M_i, n-M_i}$, then the CGF approximation for $\ln \Lambda_I$ is $\sum_{i=2}^{p} \hat{\mathcal{K}}_i(t)$ to which the Lugannani and Rice formula may be applied.

Numerical computations

The test that Σ is diagonal is the special case in which $k_i \equiv 1$. Table 11.2 compares the single- and double-saddlepoint approximations with a normal approximation, after applying the best normalizing transformation, which is due to Mudholkar *et al.* (1982), and a $O(n^{-3})$ approximation due to Box (1949). Apart from the inclusion of the single saddlepoint, the table is a reduction of table 1 in Butler *et al.* (1993). The table entries are the approximations evaluated at the true percentiles listed in the last column which were taken from Mathai and Katiyar (1979) and converted from percentiles of $-\{n - (2p+5)/6\}\ln \Lambda$.

The single saddlepoint and the normalizing transformation were better than the double saddlepoint although the normalizing transformation has a peculiar behavior with small n.

Table 11.3 considers the general case with unequal block sizes and compares accuracy of the double-saddlepoint approximation with Box's $O(n^{-3})$ approximation. No exact percentiles have been tabulated in such unbalanced settings so the approximations have been inverted to provide approximate percentage points. The accuracy of these percentage points are judged by using simulations of 2 million repetitions and reporting the empirical coverages in table 11.3 which has been reprinted from table 2 in Butler *et al.* (1993). The saddlepoint approximation achieves high accuracy in all instances. The accuracy declines slightly with increasing n which is likely due to the increasing imbalance in the degrees of freedom that skews the beta variables in the conditional representation in (11.26).

Table 11.2. *Percentage values for the single- and double-saddlepoint approximations as compares to the best normalizing (NT) and Box approximations for the likelihood ratio test that the covariance is diagonal. Each approximation has been evaluated at the true percentiles (Perc.).*

n	$p = k$	Single	Double	NT	Box	Perc.
			Results for $\alpha = 5\%$			
3	3	4.97	5.11	4.92	4.05	−6.874
6	3	4.99	5.12	4.91	4.86	−1.875
5	5	4.92	4.95	4.98	3.62	−9.604
9	5	4.97	4.75	4.98	8.03	−2.892
7	7	4.91	4.89	5.01	1.50	−12.01
11	7	4.98	4.66	5.00	12.4	−4.376
9	9	4.93	4.90	95.0	0.46	−14.28
13	9	4.98	4.63	5.01	16.2	−5.941
11	10	4.89	4.85	95.0	0.23	−11.04
			Results for $\alpha = 1\%$			
6	3	.999	1.03	1.01	0.64	−2.722
5	5	.995	1.00	1.00	0.29	−12.86
9	5	.990	0.94	1.01	1.89	−3.685
9	9	.990	0.98	99.0	0.00	−17.59
11	10	.971	0.94	99.0	0.00	−17.59

11.2.2 Sphericity test

Suppose W has a $k \times k$ Wishart (n, Σ) distribution. The likelihood ratio test for the sphericity hypothesis $\Sigma = \sigma^2 I_k$ rejects for small values of

$$\Lambda_S = |W|/(\operatorname{tr} W/k)^k.$$

The test is routinely used in repeated measurement analysis to justify use of the standard F-test which requires that k orthogonal contrasts of the data have covariance of the form $\sigma^2 I_k$. See Morrison (1990) for discussion.

For single-saddlepoint approximation, Anderson (1984, p. 430) gives the CGF of $\ln \Lambda_S$ as

$$K(t) = kt \ln k + \ln \frac{\Gamma(\tfrac{1}{2}nk)}{\Gamma(tk + \tfrac{1}{2}nk)} + \sum_{i=1}^{k} \ln \frac{\Gamma(t + \tfrac{1}{2}a_i)}{\Gamma(\tfrac{1}{2}a_i)}$$

defined for $t > -(n - k + 1)/2$ where $a_i = n - i + 1$.

A double-saddlepoint approximation is given in Butler *et al.* (1993) and based on the null distributional characterization

$$\ln \Lambda_S \sim -2r \ln 2 + k \ln k + 2 \sum_{i=1}^{r} \ln D_i + w \ln D_{r+1} + k \ln \beta(\alpha, \gamma) \qquad (11.27)$$

11.2 P-values in tests of covariance

Table 11.3. *Empirical coverages for the double-saddlepoint approximation and the Box approximation for unbalanced block sizes.*

		$\alpha = 5\%$		$\alpha = 1\%$	
$k_1,\ldots,k_4$	n	Double	Box	Double	Box
4, 3, 2, 1	10	5.09	3.06	1.01	1.34
	20	5.02	0.45	1.01	0.06
	50	5.11	1.67	1.03	0.25
	75	5.13	2.38	1.03	0.38
1, 2, 3, 4	10	5.09	3.06	1.01	1.34
	20	5.15	0.45	1.04	0.06
	50	5.31	1.67	1.08	0.25
3, 3, 3, 3	12	5.10	5.21	1.02	2.62
	20	5.12	0.28	1.04	0.04
	50	5.26	1.06	1.08	0.16

where w is the indicator that k is odd, $k = 2r + w$, and $\{D_i\}$ form components of a Dirichlet vector independent of beta variable β with degrees of freedom

$$\alpha = r(n-r) + \tfrac{1}{2}w(n-2r) \qquad \gamma = r^2 + wr.$$

The dimension of $\{D_i\}$ depends on whether k is even or odd. If k is even, $D_{r+1} \equiv 1$, $D_r = 1 - \sum_{i=1}^{r-1} D_i$ and $D_1, \ldots, D_{r-1}$ has a Dirichlet $(n - 2i + 1 : i = 1, \ldots, r)$ distribution, whereas if k is odd, $D_{r+1} = 1 - \sum_{i=1}^{r} D_i$ and $D_1, \ldots, D_r$ has a Dirichlet $(n - 2i + 1 : i = 1, \ldots, r; (n - 2r)/2)$ distribution. The double-saddlepoint approach constructs the Dirichlet distribution by conditioning an independent vector of gammas so they sum to 1; see section 4.1.1.

Table 11.4 shows values of the single- and double-saddlepoint approximations along with Box's (1949) approximation to order $O(n^{-3})$, as given in Anderson (1983, p. 432) evaluated at the true fifth and first percentiles. The exact quantiles for $\ln \Lambda_S$ are listed in Exercise 5 and have been computed from the tables of Nagarsenker and Pillai (1973) for $k \geq 3$ and as $2\ln\alpha/(n-1)$ for $k = 2$; see Anderson (1984, p. 432).

The double-saddlepoint is more accurate than the single-saddlepoint for $k \geq 4$ at both levels however both are extremely accurate and the single-saddlepoint method is trivial to compute.

11.2.3 Equal variance, equal covariances test

This test is often referred to as a test for intraclass correlation structure. The structure for covariance $\Sigma = (\sigma_{ij})$ has $\sigma_{ii} \equiv \sigma$ for all i, and $\sigma_{ij} \equiv \rho\sigma$ for all $i \neq j$ where $|\rho| < 1$ is the common correlation. This covariance pattern is assumed, for example, in split-plot design models with randomization of treatments among subplots; see Cochran and Cox (1957, chap. 7). If $W = (w_{ij})$ has a $k \times k$ Wishart (n, Σ) distribution, the likelihood ratio test

Table 11.4. *Percentage values of the single- and double-saddlepoint values evaluated at the true percentiles of* $\ln \Lambda_S$ *listed in Exercise 5. Partly reproduced from Butler et al. (1993).*

		$\alpha = 5\%$			$\alpha = 1\%$		
n	k	Single	Double	Box	Single	Double	Box
5	2	5.03	4.96	5.00	1.01	1.06	1.00
7	3	4.98	4.96	4.98	.992	1.01	0.99
6	4	4.93	5.01	4.64	.980	0.99	0.85
6	5	4.90	4.97	3.32	.976	0.98	0.43
12	6	4.99	4.99	4.81	.996	1.00	0.93
8	7	4.89	4.94	1.92	.972	0.97	0.17
13	8	4.99	5.00	4.32	.995	1.00	0.76
11	9	4.94	4.97	2.20	.980	0.97	0.24
15	10	4.99	5.00	3.77	.996	1.00	0.61

rejects for small values of

$$\Lambda_{IC} = \frac{|W|}{w^{2k}(1-r)^{k-1}\{1+(k-1)r\}} \tag{11.28}$$

where $w^2 = \operatorname{tr} W/k$ and

$$w^2 r = \frac{1}{k(k-1)} \sum_{i \neq j} w_{ij};$$

see Morrison (1990, p. 294). The single-saddlepoint uses the CGF of $\ln \Lambda_{IC}$ which, with $l = k - 1$, was given by Wilks (1946) as

$$K(t) = tl \ln l + \ln \frac{\Gamma(\tfrac{1}{2}nl)}{\Gamma(tl + \tfrac{1}{2}nl)} + \sum_{i=1}^{l} \ln \frac{\Gamma\{t + \tfrac{1}{2}(n-i)\}}{\Gamma\{\tfrac{1}{2}(n-i)\}}$$

for $t > -(n - k + 1)/2$.

A double-saddlepoint approximation is given in Butler *et al.* (1993) and is based on the null distributional characterization

$$\ln \Lambda_{IC} \sim -2r \ln 2 + l \ln l + 2 \sum_{i=1}^{r} \ln D_i + w \ln D_{r+1} + l \ln \beta(\alpha, \gamma) \tag{11.29}$$

where w is the indicator that l is odd, $l = 2r + w$, and $\{D_i\}$ form components of a Dirichlet vector independent of beta variable β with degrees of freedom

$$\alpha = r(n - r - 1) + \tfrac{1}{2}w(n - l) \qquad \gamma = r^2 + r + \tfrac{1}{2}wl.$$

The dimension of $\{D_i\}$ depends on whether l is even or odd. If l is even, $D_{r+1} \equiv 1$, $D_r = 1 - \sum_{i=1}^{r-1} D_i$ and $D_1, \ldots, D_{r-1}$ has a Dirichlet $(n - 2i : i = 1, \ldots, r)$ distribution, whereas if l is odd, $D_{r+1} = 1 - \sum_{i=1}^{r} D_i$ and $D_1, \ldots, D_r$ has a Dirichlet $(n - 2i : i = 1, \ldots, r; (n-l)/2)$ distribution. The double-saddlepoint approach constructs the Dirichlet distribution by conditioning an independent vector of gammas so they sum to 1.

11.2 P-values in tests of covariance

Table 11.5. *Percentage values for the single- and double-saddlepoint approximations evaluated at the true percentiles (Perc.) of* $\ln \Lambda_{IC}$. *Partly reproduced from Butler et al. (1993).*

		$\alpha = 5\%$			$\alpha = 1\%$		
n	p	Single	Double	Perc.	Single	Double	Perc.
7	4	4.96	5.03	−3.0582	.986	1.01	−3.9908
7	5	4.93	5.02	−4.8711	.979	1.00	−6.1166
6	6	4.94	4.98	−11.594	1.03	1.03	−14.808
12	7	4.98	5.00	−4.3406	.994	1.00	−5.1225
9	8	4.89	4.95	−9.7475	.972	0.98	−11.569
11	9	4.94	4.98	−8.7729	.980	0.99	−10.142
11	10	4.90	4.93	−11.849	.973	0.98	−13.703

Table 11.5 shows values of the single- and double-saddlepoint approximations evaluated at the true fifth and first percentiles. The exact quantiles for $\ln \Lambda_{IC}$ are also listed and have been computed from the values of Λ_{IC} in Nagarsenker (1975). The double-saddlepoint is more accurate than the single-saddlepoint.

11.2.4 Tests for equal covariances

Saddlepoint approximation to p-values of the modified likelihood ratio test for homogeneity of covariance matrices across p normal populations is presented based on its development in Booth *et al.* (1995). For the ith population, suppose the within-error sample covariance matrix W_i has a $k \times k$ Wishart (n_i, Σ_i) distribution. Then the Bartlett–Box M-statistic that tests equality of the covariance matrices, or $H_0 : \Sigma_1 = \cdots = \Sigma_p$ rejects for small values of V or large values of M where

$$V = e^{-M/n} = |W/n|^{-1} \prod_{i=1}^{p} |W_i/n_i|^{n_i/n} \tag{11.30}$$

with $W = \sum_{i=1}^{p} W_i$ and $n = \sum_{i=1}^{p} n_i$. This test is generally preferred to the likelihood ratio test whose statistic would replace the degrees of freedom n_i with the sample size; thus n_i would be replaced by $n_i + 1$ when only a location parameter is fit.

The CGF of M under the null hypothesis is

$$\begin{aligned}
K(t) &= ktc + \ln \frac{\Gamma_k(\frac{1}{2}n)}{\Gamma_k(\frac{1}{2}n - nt)} + \sum_{i=1}^{p} \ln \frac{\Gamma_k(\frac{1}{2}n_i - n_i t)}{\Gamma_k(\frac{1}{2}n_i)} \\
&= ktc + \sum_{j=1}^{k} \left[\ln \frac{\Gamma\{\frac{1}{2}(n - j + 1)\}}{\Gamma\{\frac{1}{2}(n - j + 1) - nt\}} \right. \\
&\qquad \left. + \sum_{i=1}^{p} \ln \frac{\Gamma\{\frac{1}{2}(n_i - j + 1) - n_i t\}}{\Gamma\{\frac{1}{2}(n_i - j + 1)\}} \right]
\end{aligned}$$

Table 11.6. *Percentage tail values for the single- and double-saddlepoint approximations of the Bartlett–Box M test for equal covariances evaluated at exact 95th percentiles. Sample sizes are n_1 and balanced. Reproduced from table 1 in Booth et al. (1995).*

		Single Saddlepoint				Double Saddlepoint		
k	n_1	$p=3$	$p=7$	$p=10$		$p=3$	$p=7$	$p=10$
1	4	5.09	5.02	5.01		5.36	5.09	5.05
	7	5.07	5.01	5.01		5.30	5.08	5.05
	14	5.04	5.01	5.01		5.26	5.07	5.04
		$p=2$	$p=3$	$p=5$	SAS	$p=2$	$p=3$	$p=5$
3	5	4.99	5.00	5.00	3.72	5.08	5.03	5.02
	9	4.99	5.00	5.00	4.69	5.00	5.00	5.01
	15	5.00	4.99	5.00	4.48	4.94	5.00	4.98
6	10	4.99	5.01	5.01	2.63	4.97	5.01	5.00
	15	5.01	4.99	5.00	4.07	4.95	4.98	4.99
	20	5.00	5.02	4.98	4.50	4.92	4.99	4.96

and defined for $t < \{1 - (k-1)/\min(n_i)\}/2$, where Γ_k is the multivariate gamma function defined in (11.13) and $c = \sum_{i=1}^{p} n_i \ln(n_i/n)$.

A double-saddlepoint approximation is also given in Booth *et al.* (1995) which is based on the null distributional characterization

$$M \sim \sum_{i=1}^{p} \left\{ kn_i \ln \frac{n_i}{n} - 2n_i \sum_{j=1}^{r} \ln D_{ij} - wn_i \ln D_{i,r+1} \right\}. \quad (11.31)$$

Here, w is the indicator that k is odd, $k = 2r + w$, $\{(D_{1j}, \ldots, D_{pj}) : j = 1, \ldots r\}$ are independently distributed as Dirichlet $\{(n_i - 2j + 1) : i = 1, \ldots, p; (p-1)(2j-1)\}$, and $(D_{1,r+1}, \ldots, D_{p,r+1})$ is Dirichlet $\{\frac{1}{2}(n_i - 2r) : i = 1, \ldots, p; (p-1)r\}$ independent of the first r Dirichlet vectors. In the univariate setting, $k = 1$ so that $r = 0$ and $w = 1$; thus the middle summation in (11.31) is set to zero. A conditional saddlepoint approach constructs each of the $r + 1$ independent Dirichlet vectors by conditioning independent gamma variates to sum to one as explained in previous examples.

Tables 11.6 and 11.7 contain numerical comparisons for the accuracy of the suggested single- and double-saddlepoint approximations. Table 11.6 deals with the equal sample size case ($n_1 = \cdots = n_p$) in which the two approximations are evaluated at exact 95th percentiles listed in table 11.13 of Exercise 7. These exact values have been taken from tables given in Glaser (1976) for the univariate case, and Gupta and Tang (1984) for the multivariate case.

In the multivariate case with $k = 3$ and 6, the p-value from SAS is listed which uses the Box (1949) correction of a $\chi^2_{(p-1)k(k+1)/2}$ percentile as described in Morrison (1976, p. 153). Overall the single saddlepoint is most accurate. The double saddlepoint shows almost the same accuracy with some loss at the small parameter values $k = 1$ and $p = 3$.

Table 11.7. *Percentage tail values for the single- and double-saddlepoint approximations of the Bartlett–Box M test for equal covariances evaluated at simulated 95th percentiles. Sample sizes are given as n_i and are unbalanced. Reproduced from table 2 in Booth et al. (1995).*

k	p	n_i	Single	Double	SAS
1	3	2, 5, 8	5.07	4.69	5.42
		3, 4, 4	5.10	4.63	5.37
	6	2, 3, 4, 4, 5, 7	5.01	4.89	5.44
		3, 3, 3, 3, 5, 10	5.02	4.88	5.41
3	2	4, 9	4.92	5.02	3.69
	3	4, 7, 8	4.89	4.99	3.66
	6	4, 5, 5, 6, 7, 10	4.88	5.12	3.35
6	2	7, 11	4.90	5.07	1.70
	3	7, 9, 12	4.93	5.09	1.59
	5	6, 6, 8, 10, 10	4.90	5.17	0.09

Unequal sample sizes are considered in table 11.7. No exact percentage points are available so 95th percentiles were approximated by using empirical 95th percentiles of the Dirichlet representation in (11.31) based on 2,000,000 simulations. Both approximations show remarkable accuracy in settings where the SAS computation is inaccurate.

Further tests for homogeneity across populations are described in the Exercises.

11.3 Power functions for multivariate tests

The sequential saddlepoint method has been used in Butler and Wood (2002, 2004) to approximate the noncentral distributions for three multivariate tests. These approximations provide the power functions for: Wilks' likelihood ratio statistic Λ that tests the general linear hypothesis in MANOVA, the likelihood ratio statistic for testing block independence; and Bartlett's M-statistic that tests equality of covariance matrices. In each case, the two steps of the sequential approach include (i) Laplace approximation to the Mellin transform of the noncentral distribution followed by (ii) inversion of the approximate transform using a single-saddlepoint method. Since the Laplace approximations provide explicit MGF approximations for $\ln \Lambda$ and the other two statistics, the methods for determining the noncentral distributions require only slightly more computational effort than those for the null distributional setting of the previous section. Currently there are no other methods that are capable of delivering the virtually instantaneous two significant digit accuracy realized when using this simple approach.

The main complication for the non-null settings is the presence of intractable hypergeometric functions of matrix argument that are factors in the Mellin transforms. These functions have proved very difficult to compute and even to discuss; see Muirhead (1982, chap. 7). In lieu of any discussion of these functions, expressions representing extremely accurate Laplace approximations are presented that were derived in Butler and Wood (2002). The next three subsections consider the three multivariate tests and their power computations with numerical computations that demonstrate the remarkable accuracy of the Laplace approximations and their use with the sequential saddlepoint method.

11.3.1 Wilks' likelihood ratio test

Consider testing the general linear hypothesis in MANOVA using Wilks' Λ. Suppose the error sum of squares matrix W_e is a $k \times k$ Wishart (n, Σ) and the hypothesis sum of squares matrix W_h is a $k \times k$ Wishart (m, Σ) under the null hypothesis. Under model alternatives, the nonnull distribution of W_h is a noncentral Wishart (m, Σ, Ω) with noncentrality matrix Ω that is nonnegative definite, but not $\Omega = 0_{k \times k}$ as occurs under the null; see Muirhead (1982, §10.3).

The presentation to follow concentrates on the setting in which $n \geq k$ and $m \geq k$ and known as Case 1. The Case 2 setting, for which $n \geq k \geq m$, may be subsumed in Case 1 by using the substitution rule discussed in Muirhead (1982, pp. 454–5). Several facts about $\Lambda'_{k,m,n}(\Omega)$, the noncentral distribution of Λ, are required to understand the rule. First, if Ξ is the diagonal matrix consisting of the eigenvalues of Ω, then $\Lambda'_{k,m,n}(\Omega)$ has the same distribution as $\Lambda'_{k,m,n}(\Xi)$ and the dependence of the distribution on Ω is only through its eigenvalues. In Case 2 settings, Ω has between 1 and m non-zero eigenvalues whose $m \times m$ diagonal matrix we denote as Ξ_m. Then, the substitution rule for converting from Case 2 to Case 1 is

$$\Lambda'_{k,m,n}(\Omega) \sim \Lambda'_{m,k,m+n-k}(\Xi_m).$$

Approximate Mellin transform

For Case 1, the non-null Mellin transform or MGF for $\ln \Lambda$ is specified in theorem 10.5.1 of Muirhead (1982) as

$$M_{\ln \Lambda'}(t) = E(e^{t \ln \Lambda'})$$
$$= \frac{\Gamma_k\left(\frac{n}{2}+t\right) \Gamma_k\left\{\frac{1}{2}(n+m)\right\}}{\Gamma_k\left(\frac{n}{2}\right) \Gamma_k\left\{\frac{1}{2}(n+m)+t\right\}} \, _1F_1\left\{t; \frac{1}{2}(n+m)+t; -\frac{1}{2}\Xi\right\} \quad (11.32)$$

where $_1F_1$ is a hypergeometric function with the diagonal matrix argument $-\frac{1}{2}\Xi$.

The following Laplace approximation to $_1F_1$ has been given in Butler and Wood (2002). The approximation $_1\hat{F}_1$ has the explicit form given by

$$_1\hat{F}_1(a; b; X) = b^{kb - k(k+1)/4} R_{1,1}^{-1/2} \prod_{i=1}^{k} \left\{ \left(\frac{\hat{y}_i}{a}\right)^a \left(\frac{1-\hat{y}_i}{b-a}\right)^{b-a} e^{x_i \hat{y}_i} \right\} \quad (11.33)$$

where $X = \text{diag}(x_1, \ldots, x_k)$,

$$R_{1,1} = \prod_{i=1}^{k} \prod_{j=i}^{k} \left\{ \frac{\hat{y}_i \hat{y}_j}{a} + \frac{(1-\hat{y}_i)(1-\hat{y}_j)}{b-a} \right\},$$

and, for $i = 1, \ldots, k$,

$$\hat{y}_i = \frac{2a}{b - x_i + \sqrt{(x_i - b)^2 + 4ax_i}}.$$

It is difficult to appreciate much about this approximation except perhaps a demonstration of its numerical accuracy. For example, consider the mean of $\Lambda'_{7,8,39}(\Omega)$ or μ with $\Omega =$

11.3 Power functions for multivariate tests

diag$\{1/4, 1/2(1/2)3\}$ determined by evaluating (11.32) at $t = 1$. Then,

$$\hat{\mu} = \hat{E}(\Lambda'_{7,8,39}) = 0.19715 \qquad \tilde{\mu} = \tilde{E}(\Lambda'_{7,8,39}) = 0.19717$$

where $\hat{E}$ denotes the Laplace approximation that evaluates (11.32) at $t = 1$ with $_1\hat{F}_1$ replacing $_1F_1$, and $\tilde{E}$ is a simulated average of 1000, 000 values of Λ'. Even for those of us spoilt by saddlepoint performance, this degree of accuracy is unexpected and not atypical of the general performance of $_1\hat{F}_1$. A two-sided test of the null hypothesis $H_0 : \mu = \hat{E}(\Lambda'_{7,8,39})$, that the true mean of $\Lambda'_{7,8,39}$ is its Laplace approximation, can be performed by using the 1000, 000 simulated values of Λ' as a random sample. The t-test is not significant, $t = 0.682$ and $p = 0.495$, with 95% confidence interval (0.19715 ± 0.0^31146). While this is only a single simulation, such insignificance occurs quite often under repetition and suggests that the equivalent worth of the approximation, in terms of the simulation size required to replicate its accuracy, is quite high.

Sequential saddlepoint approximation

Let $\hat{K}_{\ln \Lambda'}(t) = \ln \hat{M}_{\ln \Lambda'}(t)$ where $\hat{M}_{\ln \Lambda'}$ is (11.32) with $_1\hat{F}_1$ replacing $_1F_1$. Then applying the single-saddlepoint approximation is straightforward when performed by using symbolic and/or numerical differentiation. In floating point computation, it is best to take the first derivative analytically so an expression for $\hat{K}'_{\ln \Lambda'}(t)$ is given in the appendix of section 11.5. Then, the second derivative required for $\hat{u}$ can be taken numerically from $\hat{K}'_{\ln \Lambda'}$ without any loss in accuracy.

The accuracy of the sequential saddlepoint approximation (Seq.) is revealed in table 11.8 using small to moderate eigenvalues for Ω. The table entries are the CDF approximations evaluated at empirical percentiles associated with the percentages listed in the top row from 1 to 99. These empirical percentiles were determined by simulating 10^6 independent values of $\log \Lambda'$. The entries display two and often three significant digit accuracy. The largest percentage absolute relative error for the whole table is $\leq 1\%$.

Also included are values for the $O(n^{-3})$ expansions of Sugiura and Fujikoshi (1969) (SF) in terms of noncentral chi-square distributions, and the Edgeworth-type expansions of Sugiura (1973a) (S) with error $O(n^{-3/2})$ for the case $p = 2$. Both expansions result in considerably less accurate percentages and are worse for every table entry. Moreover, these noncentral chi-square computations were quite difficult to perform as a result of register underflow problems in floating point computation. By contrast the sequential saddlepoint approximations were straightforward and instantaneous.

To evaluate the approximation with larger eigenvalues for Ω, Butler and Wood (2005) considered a balanced 1-way MANOVA design in k dimensions with I levels and J repetitions. If level i has mean $\mu + \alpha_i$, consider the noncentral distribution of $\ln \Lambda'$ when testing $H_0 : \alpha_1 = \cdots = \alpha_I = 0$. The MANOVA has $m = I - 1$ degrees of freedom for hypothesis, $n = I(J-1)$ degrees of freedom for error, and noncentrality matrix

$$\Omega = J \Sigma^{-1} \sum_{i=1}^{I} \alpha_i \alpha_i^T = J \Omega_1 \qquad (11.34)$$

which increases with J where Ω_1 is that portion which is fixed.

Table 11.8. *Percentage values for the sequential saddlepoint CDF approximation (Seq. sad.), the $O(n^{-3})$ expansion of Sugiura and Fujikoshi (1969) (SF), and the $O(n^{-3/2})$ expansion of Sugiura (1973a,b) (S) evaluated at simulated percentiles of noncentral* $\ln \Lambda$. *Reproduced from table 3 in Butler and Wood (2002).*

	1	5	10	30	50	70	90	95	99
$(k, m, n) = (2, 3, 10)$					$\Omega = \text{diag}\left\{\frac{1}{2}, 1\right\}$				
Seq.	1.011	5.021	9.987	30.07	50.06	70.00	89.99	94.99	98.99
SF	1.530	4.130	7.852	26.05	46.52	67.87	89.43	94.77	98.97
S	$.0^7 132$	$.0^2 193$	.1272	14.31	57.63	101.1	129.1	132.2	129.4
$(5, 5, 20)$					$\Omega = \text{diag}\left\{\frac{1}{4}(\frac{1}{4})1, 1\frac{1}{2}\right\}$				
Seq.	1.003	4.974	9.944	29.90	49.88	69.90	89.98	94.98	99.00
SF	1.512	4.703	8.502	25.52	44.86	65.86	88.32	94.13	98.83
$(8, 7, 40)$					$\Omega = \text{diag}\left\{0, \frac{1}{4}, \frac{1}{2}, 1\left(\frac{1}{2}\right)3\right\}$				
Seq.	.9919	4.996	9.992	29.96	49.93	70.00	90.04	95.05	99.01
SF	1.600	4.494	7.824	23.26	41.82	63.14	86.92	93.27	98.63
$(16, 14, 40)$				$\Omega_{16} = \text{diag}\left\{2(0), \frac{1}{4}, \frac{1}{2}\left(\frac{1}{2}\right)3, 3\frac{1}{4}\left(\frac{1}{4}\right)4\frac{3}{4}\right\}$					
Seq.	.9969	5.001	9.987	29.98	49.95	69.98	89.99	94.99	98.99
SF	16.50	29.02	30.64	20.51	18.13	31.94	67.54	81.56	95.56
$(32, 28, 60)$					$\Omega_{32} = I_2 \otimes \Omega_{16}$				
Seq.	1.006	5.034	10.00	29.98	49.97	69.98	90.04	95.03	99.01
SF	6.992	26.09	42.16	67.57	59.94	39.03	34.88	48.88	78.84
$(64, 56, 75)$					$\Omega_{64} = I_4 \otimes \Omega_{16}$				
Seq.	.9908	4.984	9.966	29.95	49.90	69.94	89.96	94.99	99.00
SF	$.0^{14} 68$	$.0^{11} 66$	$.0^9 19$	$.0^6 11$	$.0^5 56$	$.0^3 19$	.016	.102	1.884

Table 11.9 has been computed in the same manner as table 11.8. The entries are slightly less accurate using these larger entries for Ω but the absolute relative errors are still quite small. Also shown is the power approximation of Muller and Peterson (1984) and Muller et al. (1992) (F') which uses a noncentral F approximation. This approximation continues to decrease in accuracy for $k > 7$ and is therefore not listed. Also not given due to their inaccuracy are the expansions of Sugiura and Fujikoshi (1969), Sugiura (1973a) and also those of Kulp and Nagarsenkar (1984). The empirical quantiles at which these entries were computed are given in table 11.16.

Sample size determination

Consider the balanced one-way MANOVA design of the previous subsection in $k = 5$ dimensions with $I = 8$ levels and J repetitions per level where value J has yet to be determined. The equality of means across levels is tested with $m = 7$ degrees of freedom for hypothesis and $n = 8(J - 1)$ degrees of freedom for error. The noncentrality matrix in Wilks' test is given in (11.34) by $\Omega = J\Omega_1$ and increases with J where Ω_1 is the portion that is fixed.

11.3 Power functions for multivariate tests

Table 11.9. *One-way MANOVA. Sequential saddlepoint approximations (Seq.) and the noncentral F approximation of Muller and Peterson (1984) (F') for the noncentral CDF of Wilks'* $\ln \Lambda'$ *in one-way MANOVA hypothesis. Reproduced from table 1 of Butler and Wood (2005).*

	1	5	10	30	50	70	90	95	99
	(k, m, n)		(I, J)			$\Omega = J\Omega_1$			
	$(3, 7, 24)$		$(8, 4)$			$\Omega_1 = \text{diag}\{2, 4\frac{1}{2}, 7\}$			
Seq.	.9903	4.966	9.935	29.90	49.90	69.92	89.99	95.00	98.99
F'	.7662	4.000	8.190	25.91	44.84	64.95	87.06	93.12	98.40
	$(7, 7, 24)$		$(8, 4)$			$\Omega_1 = \text{diag}\{1(1)7\}$			
Seq.	.9633	4.883	9.762	29.62	49.67	69.78	89.89	94.95	99.00
F'	.1655	1.213	2.919	12.29	27.24	46.84	75.11	85.14	95.79
	$(7, 7, 24)$		$(8, 4)$			$\Omega_1 = \text{diag}\{2(5)32\}$			
Seq.	.8540	4.475	9.097	28.32	48.17	68.52	89.22	94.57	98.90
F'	.0220	.2384	.6950	4.573	12.43	26.82	56.02	70.06	89.00
	$(12, 7, 56)$		$(8, 8)$			$\Omega_1 = \{5(0), 1(1)7\}$			
Seq.	.9847	4.946	9.873	29.84	49.82	69.85	89.88	94.91	98.96
	$(25, 7, 56)$		$(8, 8)$			$\Omega_1 = \{18(0), 1(1)7\}$			
Seq.	1.007	4.979	9.995	29.92	49.93	69.94	89.99	95.00	98.99
	$(25, 7, 56)$		$(8, 8)$			$\Omega_1 = \{18(0), 2(5)32\}$			
Seq.	.9629	4.871	9.790	29.58	49.51	69.58	89.78	94.84	98.95
	$(50, 7, 88)$		$(8, 12)$			$\Omega_1 = \{43(0), 1(1)7\}$			
Seq.	.9983	5.015	10.02	29.93	49.92	69.95	89.98	95.01	99.01

For a given choice of $\Omega_1 \neq 0_{k \times k}$, determine the minimal value of J for which the 5% level test of the null hypothesis in a one-way MANOVA achieves 90% power at Ω_1? The power of the test depends only on the eigenvalues of Ω_1 and J. Therefore, consider five alternatives with the unequal complement of eigenvalues scaled by ω so that

$$\Omega_1 = \omega\{\left(\tfrac{5}{4}\right)^2, 1, \left(\tfrac{3}{4}\right)^2, \left(\tfrac{1}{2}\right)^2, \left(\tfrac{1}{4}\right)^2\} \quad \omega = 1, \tfrac{1}{2}, \tfrac{1}{4}, \tfrac{1}{8}, \tfrac{1}{16}. \tag{11.35}$$

Figure 11.2 plots the percentage power at Ω_1 vs. J for the 5% level test for each of the five scalings in ω. Reading from the scores on the horizontal axis, the design question is solved with replicate sizes of $J = 11, 20, 39$ for $\omega = 1, 1/2$, and $1/4$; also $J = 77$ and 152 for $\omega = 1/8$ and $1/16$ are not shown on the graph.

The determination of this plot requires two steps. First, a single-saddlepoint approximation for the null distribution of $\ln \Lambda_{5,7,8(J-1)}$ is used to determine the 5th percentile for the test. If

$$\hat{G}_0(\lambda; J) = \widehat{\text{Pr}}(\log \Lambda_{5,7,8(J-1)} \leq \lambda)$$

denotes the Lugannani and Rice approximation of the null distribution, then $\hat{\lambda}_{0J} = \hat{G}_0^{-1}(0.05; J)$ is the approximate cutoff. Secondly, the sequential saddlepoint provides the

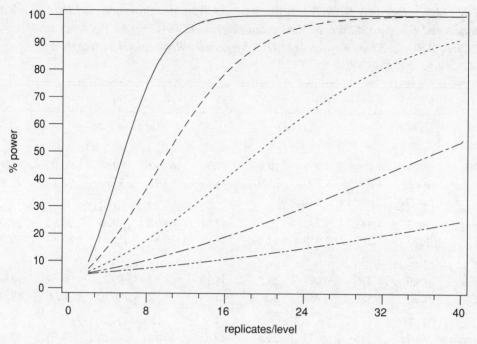

Figure 11.2. Power of Wilks' test for the one-way MANOVA versus replicates/level J with $k = 5$ and $I = 8$. The eigenvalues of Ω are given in (11.35) with $\omega = 1$ (*solid*), $1/2$ (*dashed*), $1/4$ (*dotted*), $1/8$ (*dashed one-dotted*), and $1/16$ (*dashed two-dotted*).

power function approximation as

$$\hat{G}_{\Omega_1}(\hat{\lambda}_{0J}, J) = \widehat{\Pr}\left\{\log \Lambda'_{5,7,8(J-1)}(J\Omega_1) \leq \hat{\lambda}_{0J}\right\}$$

when plotted versus J.

Behavior in high dimensions

The behavior of $\ln \Lambda'_{k,m,n}(\Omega)$ when the dimension $k \to \infty$ and $n \to \infty$ has been examined in Butler and Wood (2004). Under suitable growth conditions for Ω, the distribution is shown to have a central limit property. Evidence for this may be seen through the sequential saddlepoint procedures. Suppose the standardized value $Z = \{\ln \Lambda'_{k,m,n}(\Omega) - \mu\}/\sigma$ does approach a Normal $(0, 1)$ limit. Then the sequential saddlepoint approximation for the CGF of Z should begin looking like $t^2/2$ and any saddlepoints from it should approach the saddlepoints for Normal $(0, 1)$ or the appropriate z-scores. Table 11.10 shows this phenomenon quite clearly. The central limit property provides some explanation as to why these approximations are able to maintain such high accuracy with increasing dimensions.

11.3.2 Likelihood ratio test for block independence

Suppose that W has a $k \times k$ Wishart (n, Σ) distribution. Consider the case of $p = 2$ blocks of dimensions $k_1 \leq k_2$ and let $W = (W_{ij})$ and $\Sigma = (\Sigma_{ij})$ be the 2×2 block decompositions

Table 11.10. *Saddlepoints associated with the standardized value of* $\ln \Lambda'_{k,m,n}(\Omega)$ *and used in finding the saddlepoint approximations of table 11.9. Row z contains the various standard normal percentiles. Reproduced from table 1 of Butler and Wood (2004).*

k	1	5	10	30	50	70	90	95	99
3	−1.83	−1.37	−1.09	−.453	.050	.611	1.55	2.06	3.13
12	−2.06	−1.50	−1.19	−.491	.021	.561	1.39	1.81	2.65
25*	−2.06	−1.50	−1.18	−.490	.023	.562	1.39	1.81	2.64
50	−2.11	−1.52	−1.20	−.497	.018	.555	1.37	1.78	2.58
z	−2.33	−1.65	−1.28	−.524	0	.524	1.28	1.65	2.33

* Denotes the settings with smaller entries in Ω_1.

conformable with these sizes so Σ_{11} is $k_1 \times k_1$ etc. The nonnull Mellin transform for the likelihood ratio test statistic (11.23) of $H_0 : \Sigma_{12} = 0$ is given in Muirhead (1982, §11.2) as

$$M_{\ln \Lambda'_I}(t) = E(e^{t \log \Lambda'_I}) \qquad (11.36)$$

$$= \frac{\Gamma_{k_1}\left(\frac{n}{2}\right) \Gamma_{k_1}\left\{\frac{1}{2}(n - k_2) + t\right\}}{\Gamma_{k_1}\left(\frac{n}{2} + t\right) \Gamma_{k_1}\left\{\frac{1}{2}(n - k_2)\right\}}$$

$$\times \left|I_{k_1} - P^2\right|^{n/2} {}_2F_1\left(\tfrac{n}{2}, \tfrac{n}{2}; \tfrac{n}{2} + t; P^2\right),$$

where $P = \text{diag}\{\rho_1, \ldots, \rho_{k_1}\}$ and $\rho_1, \ldots, \rho_{k_1}$ are the eigenvalues of $\Sigma_{11}^{-1} \Sigma_{12} \Sigma_{22}^{-1} \Sigma_{21}$.

A Laplace approximation ${}_2\hat{F}_1$ for ${}_2F_1$ with $k \times k$ matrix argument is given in Butler and Wood (2002) as

$${}_2\hat{F}_1(a, b; c; X) = c^{kc - k(k+1)/4} R_{2,1}^{-1/2} \prod_{i=1}^{k} \left\{ \left(\frac{\hat{y}_i}{a}\right)^a \left(\frac{1 - \hat{y}_i}{c - a}\right)^{c-a} (1 - x_i \hat{y}_i)^{-b} \right\} \qquad (11.37)$$

where $X = \text{diag}(x_1, \ldots, x_k)$,

$$R_{2,1} = \prod_{i=1}^{k} \prod_{j=i}^{k} \left\{ \frac{\hat{y}_i \hat{y}_j}{a} + \frac{(1 - \hat{y}_i)(1 - \hat{y}_j)}{c - a} - \frac{b}{a(c-a)} S_i S_j \right\},$$

$$S_i = x_i \hat{y}_i (1 - \hat{y}_i) / (1 - x_i \hat{y}_i);$$

and, putting $\tau_i = x_i(b - a) - c$,

$$\hat{y}_i = \frac{2a}{\sqrt{\tau_i^2 - 4ax_i(c - b)} - \tau_i}.$$

The sequential saddlepoint leads to an approximation for the non-null distribution of the test statistic by applying a single-saddlepoint approximation to $\hat{M}_{\ln \Lambda'_I}(t)$ or (11.36) with ${}_2\hat{F}_1$ used as a surrogate for the intractable ${}_2F_1$. Table 11.11 shows the accuracy of this approximation and has similar structure and interpretation as table 11.8. The entries Seq. are the sequential saddlepoint approximation evaluated at empirical quantiles at the listed percentage levels. It shows remarkable accuracy with the largest percentage relative error as 2.1%. The empirical percentiles were determined by simulating 10^6 independent values of $\ln \Lambda'_I$ and are displayed in table 11.17.

Table 11.11. *Percentage values for the sequential saddlepoint CDF approximation (Seq. sad.) and two other approximations SF and LMS as described in the text. The approximations were evaluated at simulated percentiles of* $\ln \Lambda'_I$ *listed in table 11.17. Reproduced from table 2 in Butler and Wood (2004).*

	1	5	10	30	50	70	90	95	99
$(k_1, k_2, n) = (2, 3, 10)$					$P = \mathrm{diag}\{.3, .8\}$				
Seq.	.9883	4.965	9.936	29.90	49.89	69.88	89.95	94.98	99.00
SF	5.295	24.85	41.63	77.05	92.08	97.99	98.22	97.56	97.01
LMS	1.836	8.617	16.31	42.92	65.78	85.81	101.2	103.1	102.0
$(5, 7, 20)$					$P = \mathrm{diag}\{.4(.1).8\}$				
Seq.	.9915	4.977	9.968	30.02	50.03	70.04	90.02	95.00	99.00
SF	11.31	56.46	99.49	182.6	184.6	140.2	71.57	57.33	60.12
$(10, 13, 40)$					$P = \mathrm{diag}\{.1(.1).9, .95\}$				
Seq.	.9790	4.953	9.957	29.91	49.95	69.99	90.03	95.01	99.00
$(25, 28, 70)$		$P = \mathrm{diag}\{2(.1), 2(.2), 3(.3), \ldots, 3(.7), 2(.8), 2(.9), 2(.95)\}$							
Seq.	.9903	5.004	9.990	29.96	49.97	69.99	90.05	95.01	98.99

The two other approximations listed in the cases $k_1 = 2, 5$ are a $O(n^{-3/2})$ expansion based on the central limit tendency of $\ln \Lambda'_I$ by Sugiura and Fujikoshi (1969) (SF) and a $O(n^{-3})$ expansion under local alternatives in which $P \to 0$ proposed by Lee (1971), Muirhead (1972), and Sugiura (1973b) (LMS). These alternative approximations declined in accuracy for $k_1 \geq 5$ and do not appear capable of delivering the accurate probabilities seen with the sequential saddlepoint approximation.

11.3.3 Bartlett–Box M test for equality of covariances

The Bartlett–Box M test with $p = 2$ subpopulations tests $H_0 : \Sigma_1 = \Sigma_2$ and rejects for large values of M as described in section 11.2.4. Let $M'(\Delta)$ be the nonnull random variable where $\Delta = \mathrm{diag}\{\delta_1, \ldots, \delta_k\}$ and $\delta_1, \ldots, \delta_k$ are the eigenvalues of $\Sigma_1 \Sigma_2^{-1}$.

The MGF of $M'(\Delta)$ is given by theorem 8.2.11 of Muirhead (1982) as

$$M_{M'(\Delta)}(t) = E\left(e^{tM'(\Delta)}\right) \tag{11.38}$$
$$= e^{ktc} \frac{\Gamma_k\left(\frac{n}{2}\right) \Gamma_k\left(\frac{n_1}{2} - n_1 t\right) \Gamma_k\left(\frac{n_2}{2} - n_2 t\right)}{\Gamma_k\left(\frac{n}{2} - nt\right) \Gamma_k\left(\frac{n_1}{2}\right) \Gamma_k\left(\frac{n_2}{2}\right)}$$
$$\times |\Delta|^{-n_1 t} \, _2F_1\left(-nt, \frac{n_1}{2} - n_1 t; \frac{n}{2} - nt; I_k - \Delta\right)$$

where $c = \sum_{i=1}^{2} n_i \ln(n_i/n)$. The approximation $\hat{M}_{M'(\Delta)}(t)$ used in the sequential saddlepoint approximation replaces $_2F_1$ in (11.38) with $_2\hat{F}_1$.

Table 11.12 shows the accuracy attained using the sequential approximation. There is less accuracy in the center of the distribution than in the previous examples, but accuracy remains extremely high in the tails, and this should be acceptable for most practical purposes. The table also shows two expansions of order $O(n^{-3/2})$ for comparison in the cases $k = 3, 8$.

Table 11.12. *Percentage values for the sequential saddlepoint CDF approximation (Seq. sad.) and two other approximations S_F and S_1 as described in the text. The approximations were evaluated at simulated percentiles of $M'(\Delta)$ listed in table 11.18. Reproduced from table 3 in Butler and Wood (2005).*

	1	5	10	30	50	70	90	95	99
$(k, n_1, n_2) = (3, 5, 8)$					$\Delta = \text{diag}\{.2, .5, .7\}$				
Seq.	1.022	5.033	10.04	29.81	49.65	69.60	89.86	94.95	99.00
S_F	$.0^4 508$	.0800	.9811	19.92	53.08	86.37	112.0	114.7	111.1
S_1	.4049	3.113	6.921	23.47	41.75	61.90	85.41	92.26	98.31
(8, 15, 20)					$\Delta = \text{diag}\{.1(.1).8\}$				
Seq.	1.114	5.341	10.46	30.56	50.43	70.21	90.04	94.99	98.99
S_1	.1022	.6798	1.618	7.402	16.93	32.36	61.73	75.20	93.28
(16, 22, 27)					$\Delta_{16} = \text{diag}\{.05(.05).80\}$				
Seq.	1.257	5.866	11.43	32.39	52.43	71.93	90.75	95.40	99.09
(16, 50, 60)					$\Delta_{16} = \text{diag}\{.05(.05).80\}$				
Seq.	1.117	5.400	10.66	31.08	51.08	70.82	90.34	95.18	99.06
(32, 120, 125)					$\Delta_{32} = I_2 \otimes \Delta_{16}$				
Seq.	1.170	5.654	11.00	31.81	51.89	71.47	90.69	95.41	99.08

Approximation "S_F" is a $O(n^{-3/2})$ expansion for fixed Δ that was suggested by Sugiura (1974, Eq. 3.9) with a normal approximation as its leading term, and "S_1" is a local $O(n^{-3/2})$ expansion about the null hypothesis in which $\Delta - I = O(n^{-1/2})$ suggested by Sugiura (1974, Eq. 3.5) with a leading term that is noncentral χ^2. None of these expansions deliver the accuracy of the saddlepoint approximation.

11.4 Some multivariate saddlepoint densities

Multivariate analysis has densities such as the Wishart and Matrix Beta whose support is over some subset of symmetric positive definite matrices. Dealing with the true and saddlepoint densities as well as their MGFs can be a bookkeeping nightmare as a result of the repetition of components when using symmetric matrices. These two particular distributions are considered for saddlepoint approximation to illustrate that saddlepoint methods can be used to approximate such matrix distributions.

11.4.1 Wishart density

Suppose $W \sim \text{Wishart}(n, \Sigma)$ is a $k \times k$ sample covariance matrix with $n \geq k$ and density

$$f(W) = \frac{1}{2^{nk/2} \Gamma_k(n/2) |\Sigma|^{n/2}} |W|^{(n-k-1)/2} \exp\{-\text{tr}(\Sigma^{-1} W)/2\} \qquad W > 0$$

defined over W-values that are symmetric and positive definite.

Theorem 11.4.1 *The saddlepoint density for W is proportional to the exact Wishart density f and related as*

$$\hat{f}(W) = f(W) \frac{\Gamma_k(n/2)}{\hat{\Gamma}(n/2)^k} (8n)^{k(k-1)/4}, \tag{11.39}$$

where $\hat{\Gamma}(n/2)$ is Stirling's approximation to $\Gamma(n/2)$.

Note this agrees with the χ_n^2 approximation when $k = 1$ since $\Gamma_1(n/2) = \Gamma(n/2)$. The proportionality was first noted in Field (1982) and the details for the derivation of 11.39 are given in Butler and Wood (2000).

Proportionality of the saddlepoint density is shown below based on the following result taken from Anderson (1958, p. 162).

Lemma 11.4.2 *For $\Phi > 0$ and symmetric, the Jacobian for the transformation $W \to \Phi^{1/2} W \Phi^{1/2} = V$, is $|\Phi|^{(k+1)/2}$.*

Proof. If $\Sigma = I_k$ so that $W \sim$ Wishart (n, I_k), then $\Phi^{1/2} W \Phi^{1/2} \sim$ Wishart (n, Φ). The Jacobian of the transformation $|\partial(V)/\partial(W)|$ must be the ratio of densities or

$$f_W(W)/f_V(V) = |\Phi|^{(k+1)/2}. \tag{11.40}$$

Proof of (11.39). First note that it suffices to show the result when $\Sigma = I_k$. This is because, upon transforming $W \to \Sigma^{1/2} W \Sigma^{1/2}$, the Jacobian (11.40) transforms both f_W and $\hat{f}_W$ in the same manner.

The bookkeeping difficulties in Butler and Wood (2000) arose when trying to work with the MGF for the distinct elements of W. A much simpler approach, used by Field (1982) to show proportionality, makes use of the equivalent idea of exponential tilting. For symmetric $k \times k$ matrix $S < I_k/2$,

$$E\{e^{\mathrm{tr}(SW)}\} = |I_k - 2S|^{-n/2}$$

so that the S-tilt of $W \sim$ Wishart (n, I_k) is a Wishart $\{n, (I_k - 2S)^{-1}\}$ density. To determine the saddlepoint, the mean of the $\hat{S}$-tilted distribution of W is set to the generic value W

$$E(W; \hat{S}) = n(I_k - 2\hat{S})^{-1} = W$$

so that $\hat{S} = (I - nW^{-1})/2$ and the saddlepoint density is

$$\hat{f}(W) \propto \frac{1}{|\mathrm{Cov}(W; \hat{S})|^{1/2}} |I_k - 2\hat{S}|^{-n/2} \exp\{-\mathrm{tr}(\hat{S}W)\}$$

$$= \frac{1}{|\mathrm{Cov}(W; \hat{S})|^{1/2}} |W/n|^{n/2} \exp\left\{-\tfrac{1}{2} \mathrm{tr}(W - nI)\right\}. \tag{11.41}$$

The tilted covariance in (11.41) is $i_{\hat{S}}(0)$, the expected information about the distinct components of a Wishart $\{n, (I_k - 2\hat{S})^{-1}\}$ density. This is a difficult computation if attempted in the most straightforward manner. However first note that the covariance of the distinct elements of a Wishart (n, I_k) density, or $i(0)$, is proportional to an identity matrix. Upon using the relationship for the adjustment of expected information in (5.8) under the transformation $W \to (I_k - 2\hat{S})^{-1/2} W (I_k - 2\hat{S})^{-1/2}$, then

$$|\mathrm{Cov}(W; \hat{S})| = |i_{\hat{S}}(0)| = |i(0)| \, ||(I_k - 2\hat{S})||^{-(k+1)}, \tag{11.42}$$

where the latter term is the square of the determinant of the Jacobian in the lemma above. Substituting (11.42) into (11.41) yields the sought after proportionality.

11.4.2 Matrix Beta $(n/2, m/2)$

This distribution has density

$$f_B(B) = \frac{1}{B_k(n/2; m/2)} |B|^{(n-k-1)/2} |I_k - B|^{(m-k-1)/2} \qquad B > 0$$

for symmetric $k \times k$ matrix $B > 0$ where

$$B_k(n/2, m/2) = \frac{\Gamma_k(n/2)\Gamma_k(m/2)}{\Gamma_k\{(n+m)/2\}}.$$

Below it is shown that a double saddlepoint approximation for this density is exact up to a proportionality constant.

If $W_e \sim$ Wishart (n, I_k) and $W_h \sim$ Wishart (m, I_k) then $B = (W_e + W_h)^{-1/2} W_e (W_e + W_h)^{-1/2}$ has a Matrix Beta $(n/2, m/2)$ distribution when computed by using the Jacobian in the lemma above. The same computation also shows that B is independent of $W_e + W_h$ hence the marginal distribution of B is the conditional distribution of W_e given that $W_e + W_h = I_k$. A double-saddlepoint density approximation uses the ratio of the joint saddlepoint density of $(W_e, W_e + W_h)$ over the marginal density of $W_e + W_h$. Since the transformation $(W_e, W_h) \to (W_e, W_e + W_h)$ is linear with Jacobian factor 1, this is equivalent to

$$\hat{f}_B(B) = \frac{\hat{f}_{W_e}(B)\hat{f}_{W_h}(I - B)}{\hat{f}_{W_e + W_h}(I_k)} = f_B(B) \frac{B_k(n/2, m/2)}{\{\hat{B}(n/2, m/2)\}^k} \left(\frac{8nm}{n+m}\right)^{k(k-1)/4}$$

where $\hat{B}(n/2, m/2)$ is Stirling's approximation to $B_1(n/2, m/2)$.

11.5 Appendix

The expression needed to solve the saddlepoint equation in the case of Wilks' statistic is

$$\hat{K}'_{\ln \Lambda'}(t) = \sum_{i=1}^{k} \left\{ \psi\left[\tfrac{n}{2} + t - \tfrac{1}{2}(i-1)\right] - \psi\left[\tfrac{n+m}{2} + t - \tfrac{1}{2}(i-1)\right] \right\} + \nabla(t)$$

where $\psi(z) = d \ln \Gamma(z)/dz$ and

$$\nabla(t) = \frac{\partial}{\partial t} \ln {}_1\hat{F}_1\left(a, b; -\tfrac{1}{2}\Xi\right).$$

Here and also below we use $a = t$ and $b = (n+m)/2 + t$ to simplify the expressions. Derivative ∇ is computed using implicit differentiation and, after substantial simplification, yields

$$\nabla(t) = \left\{b - \tfrac{1}{4}(k+1)\right\} \frac{k}{b} + k \ln b$$

$$- \frac{1}{2} \sum_{i=1}^{k} \sum_{j=i}^{k} \left[\left\{ \frac{\hat{y}_i \hat{y}_j}{a} + \frac{(1-\hat{y}_i)(1-\hat{y}_j)}{b-a} \right\}^{-1} \left\{ \left(\frac{\hat{y}_i}{a} - \frac{1-\hat{y}_i}{b-a}\right) \frac{\partial \hat{y}_j}{\partial t} \right.\right.$$

$$\left.\left. + \left(\frac{\hat{y}_j}{a} - \frac{1-\hat{y}_j}{b-a}\right) \frac{\partial \hat{y}_i}{\partial t} - \frac{\hat{y}_i \hat{y}_j}{a^2} \right\} \right]$$

$$+ \sum_{i=1}^{k} \left\{ \ln\left(\frac{\hat{y}_i}{a}\right) - 1 + \left(\frac{a}{\hat{y}_i} - \frac{b-a}{1-\hat{y}_i} + x_i\right) \frac{\partial \hat{y}_i}{\partial t} \right\},$$

with
$$\frac{\partial \hat{y}_i}{\partial t} = 2 \frac{b - x_i + \sqrt{q_i} - a - a(x_i + b)/\sqrt{q_i}}{\left(b - x_i + \sqrt{q_i}\right)^2}$$

and
$$q_i = (x_i - b)^2 + 4ax_i.$$

11.6 Exercises

1. (a) Prove that the single-saddlepoint approximation from (11.2) agrees with the single-saddlepoint approximation using (11.1). Use the duplication formula for the gamma function.
   (b) Confirm the proof by computing a numerical entry from table 2.2.
   (c) Show with an example that the single-saddlepoint approximation does not preserve the distributional identity $\ln \Lambda_{k,m,n} \sim \ln \Lambda_{m,k,m+n-k}$ under the null hypothesis.

2. Derive the double and sequential saddlepoint approximations for the null distribution of $\frac{1}{2} \ln \Lambda_{k,m,n}$ using the conditional characterization in (11.4).
   (a) Let $X_i = \chi_{i1} + \chi_{i2}$ for $i = 1, \ldots, r+1$ with $s^T = (s_1, \ldots, ws_{r+1})$ its associated transform vector. Associate transform argument t with
   $$Y = \sum_{i=1}^{r} \ln \chi_{i1} + \tfrac{1}{2} w \ln \chi_{r+1,1}$$
   so that
   $$\Pr\left(\tfrac{1}{2} \ln \Lambda_{k,m,n} \leq l/2\right) = \Pr(Y \leq l/2 | X_i = 1, \quad i = 1, \ldots, r+1).$$
   Then show that the joint CGF of $X_1, \ldots, X_{r+1}, Y$ is
   $$K(s,t) = \sum_{i=1}^{r} \left\{ \ln \frac{\Gamma(t+b_i)}{\Gamma(b_i)} - (t + b_i + m) \ln(1 - s_i) \right\}$$
   $$+ w \left\{ \ln \frac{\Gamma(t/2+c)}{\Gamma(c)} - (t/2 + c + m/2) \ln(1 - s_{r+1}) \right\}$$
   and defined so that ln arguments are positive.
   (b) Hold t fixed and solve for the saddlepoint $\hat{s}_{it}$ as
   $$\hat{s}_{it} = 1 - (t + b_i + m) \quad i = 1, \ldots, r \qquad (11.43)$$
   $$\hat{s}_{r+1,t} = 1 - (t/2 + c + m/2).$$
   If $\hat{s}_t$ is the vector in (11.43) show that the solution to $\partial K(s,t)/\partial t|_{s=\hat{s}_t} = l/2$ is given by (11.7).
   (c) Show that $K''_{ss}(s,t)$ is a diagonal matrix. Use this to explicitly compute determinants of the Hessians in (11.8).
   (d) Use parts (a)–(c) to derive the final versions of the approximations in section 11.1.1 (Butler et al. 1992a).

3. The distribution of $\tau_1 = \tau_1(k, m)$, the largest eigenvalue of $W_h \sim \text{Wishart}_k(m, I_k)$ may be determined by taking the appropriate limit of the distribution of Roy's test statistic $\lambda_1(k, m, n)$ as $n \to \infty$. Using the weak convergence
   $$n\lambda_1(k, m, n) \xrightarrow{D} \tau_1(k, m),$$
   as $n \to \infty$, derive an expression for the CDF of τ_1 by taking the limits of the terms on the right side of (11.16) and (11.17) (Butler and Paige 2007).

11.6 Exercises

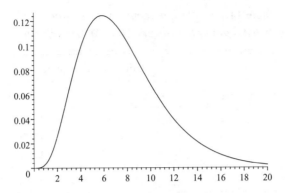

Figure 11.3. The density for τ_1, the largest eigenvalue of a Wishart$_2$ $(5, I_2)$.

(a) Prove that the exact CDF of $\tau_1(k, m)$ is

$$\Pr(\tau_1 \leq r) = \frac{1}{\Gamma_k(m/2)} \frac{\pi^{k^2/2}}{\Gamma_k(k/2)} (r/2)^{km/2} \sqrt{|B_r|} \qquad 0 < r < \infty. \tag{11.44}$$

Here, $B_r = (b_{ij})$ is a skew-symmetric matrix ($b_{ij} = -b_{ji}$) whose structure is determined by whether k is even or odd. For k even, B is $k \times k$ with

$$b_{ij} = (r/2)^{-(\alpha_i + \alpha_j)} \left\{ 2 \sum_{l=0}^{\infty} \frac{(-1)^l}{l!(\alpha_i + l)} \gamma(\alpha_i + \alpha_j + l, r/2) - \gamma(\alpha_i, r/2)\gamma(\alpha_j, r/2) \right\}, \tag{11.45}$$

where

$$\gamma(a, r/2) = \Gamma(a) P(a, r/2) = \int_0^{r/2} t^{a-1} e^{-t} dt \tag{11.46}$$

is the incomplete gamma function given in (6.5.2) of Abramowitz and Stegun (1972). For the case in which k is odd, then B is $(k+1) \times (k+1)$. The upper left $k \times k$ block is the matrix $(b_{ij} : i, j = 1, \ldots, k)$ described above with b_{ij} given in (11.45). The $(k+1)$st column (row) is determined as

$$b_{i,k+1} = -b_{k+1,i} = (r/2)^{-\alpha_i} \gamma(\alpha_i, r/2) \qquad i = 1, \ldots, k. \tag{11.47}$$

(b) Consider the case of $\tau_1(2, 5)$. Kres (1983, p. 278) provides the upper percentiles of this distribution as listed in the table below.

Probability	.95	.975	.99	.995
Percentile	14.49	16.36	18.73	20.48

Construct an algorithm to compute the CDF and confirm the accuracy of these percentiles. For example, by truncating the infinite summation in (11.45) at $l = 50$, show that

$$\Pr(\tau_1 \leq 14.49) = 0.950022.$$

(c) By taking numerical derivatives, confirm the density plot for τ_1 in figure 11.3.

4. Consider the likelihood ratio test for block independence.
   (a) Derive double and sequential saddlepoint approximations using the distributional characterization in (11.26).
   (b) Confirm some of the entries in table 11.2 for the double saddlepoint and complete the table for the sequential saddlepoint. Does the latter method improve upon the former?

Table 11.13. *95th percentiles for the null distribution of the Bartlett–Box M test statistic.*

		95th Percentiles of M		
k	n_1	$p=3$	$p=7$	$p=10$
1	4	6.616	13.72	18.39
	7	6.360	13.26	17.78
	14	6.179	12.93	17.36
		$p=2$	$p=3$	$p=5$
3	5	19.19	30.53	50.95
	9	15.46	25.23	42.90
	15	14.15	23.32	39.98
6	10	49.95	84.40	148.7
	15	42.01	72.60	129.9
	20	39.11	68.12	122.8

5. Consider the likelihood ratio test for sphericity.
   (a) Prove the distributional characterization in (11.27).
   (b) Derive double and sequential saddlepoint approximations using the distributional characterization in (11.27).
   (c) Confirm some of the entries in table 11.4 for the double saddlepoint and complete the table for the sequential saddlepoint. Does the latter method improve upon the former? The exact percentiles of $\ln \Lambda$ used to determine the table are

n	k	Fifth percentiles	First percentiles
5	2	-1.498	-2.303
7	3	-1.964	-2.686
6	4	-4.082	-5.292
6	5	-6.675	-8.429
12	6	-3.310	-3.974
8	7	-8.823	-10.628
13	8	-5.183	-6.001
11	9	-8.876	-10.248
15	10	-6.714	-7.588

6. Answer the questions of Exercise 5 for the equal variance, equal covariances test based on characterization (11.29).
7. Consider the Bartlett–Box M test for homogeneity of covariances across populations.
   (a) Use the uniqueness of the Mellin transform to prove that (11.31) is correct.
   (b) Derive a double-saddlepoint expression based on the characterization in (11.31).
   (c) Verify some of the entries in table 11.6 using the 95th percentiles of M given in table 11.13 and taken from Glaser (1976) for $k=1$ and Gupta and Tang (1984) for $k=3, 6$.

(d) Further percentage points are given in table 3 of section 8.2.4 in Muirhead (1982). Use these entries to investigate the accuracy of the single- and double-saddlepoint methods as they vary with k, p, and n_1. Describe your findings.

8. Extend the Bartlett–Box M test for testing homogeneity of covariances to a test for homogeneity of both covariances and means across normal populations. Thus, test $H_0 : (\mu_1, \Sigma_1) = \cdots = (\mu_p, \Sigma_p)$ where the ith population is Normal$_k$ (μ_i, Σ_i) (Booth et al. 1995).

   (a) Suppose that $Y_{i1}, \ldots, Y_{i,N_i}$ is a random sample from a Normal$_k$ (μ_i, Σ_i) population with $N_i = n_i + 1$ and $N = n + p$ where $n = \sum_i n_i$. Let $W_i \sim$ Wishart (n_i, Σ_i) denote the within matrix sums of squares of population i which agrees with the notation of section 11.2.4. Show that the modified likelihood ratio statistic for the testing situation here, is $V_1 = V \Lambda_1$ where V is given in (11.30) and Λ_1 is the likelihood ratio statistic for testing homogeneity of means given equal covariances with distribution $\Lambda \sim \Lambda_{k,p-1,n}$.

   (b) Prove that V and Λ_1 are independent. Show

   $$V_1 = e^{-M_1/n} = |T/n|^{-1} \prod_{i=1}^{p} |W_i/n_i|^{n_i/n}$$

   where $T = \sum_{i=1}^{p} \sum_{j=1}^{N_i} (Y_{ij} - \bar{Y}_{..})(Y_{ij} - \bar{Y}_{..})^T$ is the total matrix sum of squares. Derive the CGF of M_1 as

   $$K(t) = ktc + \ln \frac{\Gamma_k\{\tfrac{1}{2}(n+p-1)\}}{\Gamma_k\{\tfrac{1}{2}(n+p-1) - nt\}} + \sum_{i=1}^{p} \ln \frac{\Gamma_k(\tfrac{1}{2}n_i - n_i t)}{\Gamma_k(\tfrac{1}{2}n_i)}$$
   $$= ktc + \sum_{j=1}^{k} \left[\ln \frac{\Gamma\{\tfrac{1}{2}(n+p-j)\}}{\Gamma\{\tfrac{1}{2}(n+p-j) - nt\}} + \sum_{i=1}^{p} \ln \frac{\Gamma\{\tfrac{1}{2}(n_i - j + 1) - n_i t\}}{\Gamma\{\tfrac{1}{2}(n_i - j + 1)\}} \right]$$

   and defined for $t < \{1 - (k-1)/\min(n_i)\}/2$, where $c = \sum_{i=1}^{p} n_i \ln(n_i/n)$.

   (c) Booth et al. (1995) provide the following null distributional representation:

   $$M_1 \sim \sum_{i=1}^{p} \left\{ kn_i \ln \frac{n_i}{n} - 2n_i \sum_{j=1}^{r} \ln D_{ij} - wn_i \ln D_{i,r+1} \right\}. \tag{11.48}$$

   Here, w is the indicator that k is odd, $k = 2r + w$, $\{(D_{1j}, \ldots, D_{pj}) : j = 1, \ldots r\}$ are independently distributed as Dirichlet $\{(n_i - 2j + 1) : i = 1, \ldots, p; 2(p-1)j)\}$, and $(D_{1,r+1}, \ldots, D_{p,r+1})$ is Dirichlet $\{\tfrac{1}{2}(n_i - 2r) : i = 1, \ldots, p; \tfrac{1}{2}(p-1)k\}$ and independent of the first r Dirichlet vectors. In the univariate setting, $k = 1$ so that $r = 0$ and $w = 1$; thus the middle summation in (11.48) is set to zero. Prove that (11.48) agrees with the moment structure given by $K(t)$. Use (11.48) to determine a double-saddlepoint approximation for the CDF of M_1.

   (d) For each of the situations in table 11.14, determine "exact" 95th percentiles of M_1 by simulating 2000, 000 values of M_1 using the characterization in (11.48). Evaluate the single- and double-saddlepoint approximations to confirm the entries in table 11.14.

   (e) Compare the behaviors of the single- and double-saddlepoint approximations for various values of k, p, and $\{n_i\}$ with various amounts of balance and imbalance. What can be concluded?

9. Mendoza (1980) has suggested a test that extends the Bartlett–Box M statistic to test for simultaneous sphericity of population covariances, or $H_0 : \Sigma_1 = \cdots = \Sigma_p = \sigma^2 I_k$ for σ^2 unknown.

Table 11.14. *Percentage tail values for the single- and double-saddlepoint approximations of the Bartlett–Box M_1 test for equal means and covariances evaluated at simulated 95th percentiles. Reproduced from table 3 in Booth et al. (1995).*

k	p	n_i	Single-sad.	Double-sad.
1	3	4, 4, 4	4.92	4.84
	3	3, 5, 10	5.06	4.88
	6	5, 5, 5, 5, 5, 5	4.94	4.99
	6	3, 4, 4, 7, 8, 10	5.05	5.06
3	3	6, 6, 6	4.91	4.97
	3	4, 6, 10	4.93	4.98
	5	4, 4, 4, 4, 4	4.90	5.00
	5	4, 4, 5, 7, 10	5.05	5.03
4	2	5, 5	4.95	5.02
	2	5, 12	4.93	5.07
	6	5, 5, 5, 5, 5	4.91	5.06
	6	5, 6, 6, 6, 8, 10	4.87	5.20

Table 11.15. *Percentage tail values for the single- and double-saddlepoint approximations of the Bartlett–Box M_2 test for simultaneous sphericity evaluated at 95th percentiles based on 2 000, 000 simulations. Reproduced from table 4 in Booth et al. (1995).*

k	p	n_i	Single-sad.	Double-sad.
2	2	6, 6	4.96	4.91
	2	5, 10	4.92	4.92
	4	5, 5, 5, 5	5.01	5.01
	4	5, 7, 9, 11	4.94	5.04
3	3	6, 6, 6	5.00	5.06
	3	4, 6, 10	4.84	5.14
	5	4, 4, 4, 4, 4	4.84	5.11
	5	4, 4, 5, 7, 10	4.84	5.13
4	2	5, 5	5.02	5.04
	2	5, 12	4.94	5.16
	6	5, 5, 5, 5, 5	4.76	5.18
	6	5, 6, 6, 6, 8, 10	4.85	5.12

Testing this hypothesis is important in the context of repeated measurement designs with two or more repeated factors. Answer all of the questions of Exercise 8 in this new context. In particular, show that the hypothesis is tested by using the modified likelihood ratio criterion

$$V_2 = e^{-M_2/n} = \left(\frac{\operatorname{tr} W}{kn}\right)^{-k} \prod_{i=1}^{p} |W_i/n_i|^{n_i/n}$$

Table 11.16. *Empirical percentiles (top) and saddlepoint percentiles (bottom; not shown when identical to table accuracy) for noncentral* $\ln \Lambda'$. *The empirical percentiles were determined from 10^6 independent simulations of $\ln \Lambda'$. The approximate means and standard deviations $\hat{\mu}$ and $\hat{\sigma}$ were determined from the approximated CGF while those in parentheses were determined from simulation. The noncentrality matrices agree with those listed in table 11.9. Superscripts * and ** denote respectively the settings with the smaller and larger entries in Ω_1.*

1	5	10	30	50	70	90	95	99
	$(k,m,n) = (3,7,24)$			$\hat{\mu} = -2.184\, (-2.185)$		$\hat{\sigma} = .4260\, (.4264)$		
−3.273	−2.922	−2.744	−2.392	−2.163	−1.947	−1.654	−1.523	−1.294
−3.271	−2.920	−2.742	−2.391	−2.162	−1.946			−1.292
	$(7,7,24)^*$			$\hat{\mu} = -5.019\, (-5.025)$		$\hat{\sigma} = .6693\, (.6718)$		
−6.694	−6.168	−5.901	−5.359	−5.000	−4.655	−4.182	−3.964	−3.571
−6.683	−6.160	−5.891	−5.351	−4.994	−4.651	−4.178	−3.961	
	$(7,7,24)^{**}$			$\hat{\mu} = -9.910\, (-9.946)$		$\hat{\sigma} = .7757\, (.7814)$		
−11.85	−11.26	−10.96	−10.34	−9.926	−9.522	−8.961	−8.696	−8.214
−11.80	−11.22	−10.91	−10.30	−9.891	−9.490	−8.930	−8.669	−8.191
	$(12,7,56)$			$\hat{\mu} = -4.413\, (-4.415)$		$\hat{\sigma} = .4272\, (.4272)$		
−5.457	−5.136	−4.970	−4.631	−4.404	−4.183	−3.875	−3.732	−3.472
−5.454	−5.133	−4.967	−4.629	−4.402	−4.181	−3.873	−3.729	−3.467
	$(25,7,56)^*$			$\hat{\mu} = -6.684\, (-6.685)$		$\hat{\sigma} = .5452\, (.5548)$		
−8.009	−7.604	−7.391	−6.962	−6.672	−6.390	−5.995	−5.811	−5.477
−8.011	−7.603		−6.960	−6.671	−6.389			−5.475
	$(25,7,56)^{**}$			$\hat{\mu} = -11.42\, (-11.42)$		$\hat{\sigma} = .6061\, (.6063)$		
−12.88	−12.44	−12.21	−11.73	−11.41	−11.10	−10.65	−10.45	−10.06
−12.87	−12.43	−12.20	−11.73	−11.40	−11.09	−10.65	−10.44	−10.05
	$(50,7,88)$			$\hat{\mu} = -8.379\, (-8.379)$		$\hat{\sigma} = .5171\, (.5264)$		
−9.627	−9.245	−9.047	−8.643	−8.369	−8.101	−7.724	−7.546	−7.220
	−9.246		−8.642	−8.368	−8.100	−7.723		−7.221

where $W = \sum_{i=1}^{p} W_i$. Show this leads to a CGF for M_2 as

$$K(t) = tk \left(\sum_{i=1}^{p} n_i \ln \frac{n_i}{n} - n \ln k \right) + \ln \frac{\Gamma\left(\frac{1}{2}nk\right)}{\Gamma\left(\frac{1}{2}nk - nkt\right)}$$
$$+ \sum_{j=1}^{k} \sum_{i=1}^{p} \ln \frac{\Gamma\{\frac{1}{2}(n_i - j + 1) - n_i t\}}{\Gamma\{\frac{1}{2}(n_i - j + 1)\}}.$$

Prove this CGF is consistent with the distributional characterization

$$M_2 \sim k \sum_{i=1}^{p} n_i \ln \frac{n_i}{n} - kn \ln k - \sum_{i=1}^{p} \sum_{j=1}^{k} n_i \ln D_{ij} \qquad (11.49)$$

where $(D_{11}, \ldots, D_{p1}, \ldots, D_{1k}, \ldots, D_{pk})$ is distributed as Dirichlet $\{\frac{1}{2}(n_i - j + 1) : i = 1, \ldots, p,\ j = 1, \ldots, k;\ \frac{1}{4}pk(k-1)\}$. Note that this representation involves a single Dirichlet vector. Confirm the single- and double-saddlepoint entries in table 11.15.

Table 11.17. *Empirical percentiles (top) and saddlepoint percentiles (bottom; not shown when identical to table accuracy) for the noncentral distribution of* $\ln \Lambda'_I$. *The empirical percentiles were determined from* 10^6 *independent simulations of* $\ln \Lambda'_I$. *The case* $k_1 = 25$ *resulted in equal percentiles to table accuracy. The approximate means and standard deviations* $\hat{\mu}$ *and* $\hat{\sigma}$ *were determined from the approximated CGF while those in parentheses were determined with simulation.*

1	5	10	30	50	70	90	95	99
	$(k_1, k_2, n) = (2, 3, 10)$			$\hat{\mu} = -1.985\ (-1.985)$			$\hat{\sigma} = .7525\ (.7521)$	
−4.048	−3.328	−2.980	−2.322	−1.916	−1.547	−1.076	−.8775	−.5565
−4.043	−3.325	−2.976	−2.319	−1.913	−1.544	−1.074	−.8764	−.5570
	$(5, 7, 20)$			$\hat{\mu} = -5.284\ (-5.283)$			$\hat{\sigma} = .9098\ (.9104)$	
−7.594	−6.852	−6.475	−5.727	−5.240	−4.777	−4.149	−3.865	−3.359
−7.591	−6.850	−6.473		−5.241	−4.778	−4.150		−3.360
	$(10, 13, 40)$			$\hat{\mu} = -11.52\ (-11.52)$			$\hat{\sigma} = .8711\ (.8730)$	
−13.63	−12.99	−12.65	−11.97	−11.50	−11.05	−10.41	−10.12	−9.571
	−12.98		−11.96					
	$(25, 28, 70)$			$\hat{\mu} = -32.66\ (-32.66)$			$\hat{\sigma} = 1.221\ (1.223)$	
−35.57	−34.69	−34.23	−33.29	−32.64	−32.01	−31.10	−30.67	−29.89

Table 11.18. *Empirical percentiles (top) and saddlepoint percentiles (bottom; not shown when identical to table accuracy) for the noncentral distribution of* $M'(\Delta)$. *Computations of empirical percentiles and* $\hat{\mu}$ *and* $\hat{\sigma}$ *values were the same are described in tables 11.16 and 11.17.*

1	5	10	30	50	70	90	95	99
	$(k, n_1, n_2) = (3, 5, 8)$			$\hat{\mu} = 11.78\ (11.79)$			$\hat{\sigma} = 5.662\ (5.623)$	
2.407	4.136	5.319	8.374	10.97	14.03	19.28	22.22	28.47
2.394	4.123	5.293	8.322	10.92	13.99	19.29	22.25	28.55
	$(8, 15, 20)$			$\hat{\mu} = 82.98\ (82.77)$			$\hat{\sigma} = 14.74\ (14.57)$	
52.18	60.16	64.64	74.65	82.03	89.84	101.8	107.9	119.9
52.14	60.12	64.67	74.72	82.17	90.08	102.3	108.5	120.7
	$(16, 22, 27)$			$\hat{\mu} = 331.4\ (329.4)$			$\hat{\sigma} = 31.48\ (30.99)$	
261.8	280.4	290.7	312.3	328.5	345.1	369.6	381.9	405.9
262.8	281.4	291.7	314.2	330.4	347.1	369.9	384.8	408.8
	$(16, 50, 60)$			$\hat{\mu} = 448.3\ (447.2)$			$\hat{\sigma} = 32.80\ (32.58)$	
374.6	395.5	406.5	430.7	447.2	464.8	490.1	502.2	526.4
				448.3	465.9	491.2	503.3	527.5
	$(32, 120, 125)$			$\hat{\mu} = 1959.\ (1957.)$			$\hat{\sigma} = 69.73\ (69.26)$	
1797.	1844.	1868.	1920.	1957.	1991.	2045.	2072.	2121.
1800.	1846.	1871.	1922.	1959.	1996.	2050.	2074.	2123.

10. (a) Numerically confirm the equivalence in (11.15). In particular, compute $\hat{\mathcal{M}}(t|I_k)$ in (10.24), the sequential saddlepoint approximation for the null MGF of the Bartlett–Nanda–Pillai trace statistic, and show that it agrees with $_1\hat{F}_1\{n/2; (n+m)/2; tI_k\}$, where $_1\hat{F}_1$ is the Laplace approximation given in (11.33).
    (b) Prove the equivalence in (11.15).
11. Use the sequential saddlepoint method to investigate the accuracy and some properties of the noncentral distribution of Wilks' likelihood ratio test.
    (a) Verify the entries in table 11.9 for a particular choice of (k, m, n). Table 11.16 provides empirical estimates for the percentiles for the levels listed at the top.
    (b) Table 11.16 also provides saddlepoint approximations to the percentiles determined by inverting the sequential CDF approximation. Verify these values for the same choice of (k, m, n).
    (c) Once you have convinced yourself that the approximation is accurate, explore some aspect of the behavior of the power function for Wilks' Λ. For example, (i) how does the power of the 5% level test vary with the configuration of eigenvalues for noncentrality matrix Ω; (ii) with all other parameters fixed, how does power change with increasing dimension?
12. Answer the same questions as in Exercise 11 but consider instead the sequential saddlepoint approximation for the likelihood ratio test for block independence. Table 11.17 provides the necessary percentiles for verifying the entries of table 11.11.
13. Answer the same questions as in Exercise 11 but consider instead the sequential saddlepoint approximation for the Bartlett–Box test for equal covariances. Table 11.18 provides the necessary percentiles for verifying the entries of table 11.12.

12

Ratios and roots of estimating equations

The ratio $R = U/V$ of two random variables U and V, perhaps dependent, admits to saddlepoint approximation through the joint MGF of (U, V). If $V > 0$ with probability one, then the Lugannani and Rice approximation may be easily applied to approximate the associated CDF. Saddlepoint density approximation based on the joint MGF uses the Geary (1944) representation for its density. This approach was first noted in Daniels (1954, §9) and is discussed in section 12.1 below.

The ratio R is the root of the estimating equation $U - RV = 0$ and the distribution theory for ratios can be generalized to consider distributions for roots of general estimating equations. The results of section 12.1 are subsumed into the more general discussion of section 12.2 that provides approximate distributions for roots of general estimating equations. Saddlepoint approximations for these roots began in the robustness literature where M-estimates are the roots of certain estimating equations and the interest was in determining their distributions when sample sizes are small. Hampel (1973), Field and Hampel (1982), and Field (1982) were instrumental in developing this general approach.

Saddlepoint approximation for a vector of ratios, such as for example $(R_1, R_2, R_3) = \{U_1/V, U_2/V, U_3/V\}$, is presented in section 12.3 and generalizes the results of Geary (1944). An important class of such examples to be considered includes vector ratios of quadratic forms in normal variables. A particularly prominent example in times series which is treated in detail concerns approximation to the joint distribution for the sequence of lag correlations comprising the serial autocorrelation function. The conditional CDF for an autocorrelation of lag m given autocorrelations of lag $1, \ldots m - 1$ is also developed in section 12.5 and used for determining the autoregressive order.

Section 12.4 considers multivariate roots $R = (R_1, \ldots, R_m)^T$ to systems of estimating equations and section 12.5 includes a development for the conditional CDF of one root R_m given the others $R_1, \ldots, R_{m-1}$. The discussion in section 12.4 takes a somewhat new approach in the presentation of this material by basing it on the theoretical development of mixed Edgeworth saddlepoint expansions in Barndorff-Nielsen and Cox (1979, §3.3). The discussion still arrives at the most general expression for the multivariate density of a vector root as given in Field (1982, Eq. 6) but does so using an alternate route that is much simpler to motivate. Since the special cases that deal with vector ratios have already been worked out in section 12.3, they help to provide some guidance and understanding about the general methods as they are unfolded in section 12.4.

The connection between saddlepoint methods used for ratios and those applicable for roots of general estimating equations was clarified in Daniels (1983). Field and Ronchetti

12.1 Ratios

(1990) provide the most complete discussion about saddlepoint approximations for roots of estimating equations and their use in robust estimation.

12.1 Ratios

If ratio R has the form $R = U/V$ with $V > 0$ with probability one, then the CDF of R at r may be expressed as

$$\Pr(R \leq r) = \Pr(\Psi_r \leq 0) \tag{12.1}$$

where $\Psi_r = U - rV$ is a "constructed" random variable based on the particular cutoff r. For specific r, the MGF of Ψ_r is computed from the joint MGF of (U, V) and the probability in (12.1) follows directly by applying the Lugannani and Rice approximation at cutoff 0. If the joint MGF and CGF of (U, V) are denoted by $M(s, t)$ and $K(s, t)$ respectively, then

$$\widehat{\Pr}(R \leq r) = \Phi(\hat{w}) + \phi(\hat{w}) \left(\frac{1}{\hat{w}} - \frac{1}{\hat{u}} \right) \qquad \hat{s} \neq 0 \tag{12.2}$$

where

$$\hat{w} = \operatorname{sgn}(\hat{s}) \sqrt{-2 K_{\Psi_r}(\hat{s})} \qquad \hat{u} = \hat{s} \sqrt{K''_{\Psi_r}(\hat{s})}, \tag{12.3}$$

$\hat{s}$ is the root of $K'_{\Psi_r}(\hat{s}) = 0$, and $K_{\Psi_r}(s) = K(s, -rs)$.

The Beta (α, β) distribution in Example 5 of section 1.2.2 provides a simple example in which the CDF is constructed by using the ratio $Y_\alpha/(Y_\alpha + Y_\beta)$ where Y_α and Y_β are independent Gamma $(\alpha, 1)$ and Gamma $(\beta, 1)$ variables.

12.1.1 Density for a ratio

Less obvious is the manner in which the density of R at r is approximated. One approach uses a "constructed" random variable $\Psi_r^\star$ that was developed by Geary (1944) and which arises naturally when deriving $f_R(r)$ with a Jacobian transformation. If $V > 0$ with probability one, then a simple computation gives the marginal density of R as

$$f_R(r) = \int_0^\infty v f_{U,V}(rv, v) dv.$$

Define $(U^\star, V^\star)$ as the random vector whose density is proportional to $v f_{U,V}(u, v)$ so that

$$f_{U^\star, V^\star}(u, v) = v f_{U,V}(u, v) / E(V).$$

The "constructed" random variable in this instance is $\Psi_r^\star = U^\star - rV^\star$. Geary (1944) showed that

$$f_R(r) = E(V) f_{\Psi_r^\star}(0), \tag{12.4}$$

so that a saddlepoint density approximation $\hat{f}_R(r)$ can be computed as in (12.4) by finding the saddlepoint approximation $\hat{f}_{\Psi_r^\star}(0)$ instead. The basis for this latter computation is his derivation for the MGF for $\Psi_r^\star$ as

$$M_{\Psi_r^\star}(s) = \frac{1}{E(V)} \frac{\partial}{\partial t} M(s, t)\big|_{t=-rs}. \tag{12.5}$$

The proofs of both (12.4) and (12.5) are delayed so that a formulation of the saddlepoint density approximation for R and a simple example may be given first.

In determining (12.5), the computation of $\partial M / \partial t$ is expressed through $\partial K / \partial t$ according to

$$\frac{\partial}{\partial t} K(s,t) = \frac{1}{M(s,t)} \frac{\partial}{\partial t} M(s,t).$$

Thus, from (12.5),

$$M_{\Psi_r^\star}(s) = \frac{1}{E(V)} M(s, -rs) J_r(s) \qquad (12.6)$$

where

$$J_r(s) = \frac{\partial}{\partial t} K(s,t)|_{t=-rs}. \qquad (12.7)$$

However, note that $M(s, -rs)$ is $M_{\Psi_r}(s)$ and

$$E(V) = \left.\frac{\partial K(0,t)}{\partial t}\right|_{t=-rs=0} = J_r(0)$$

so (12.6) becomes

$$M_{\Psi_r^\star}(s) = M_{\Psi_r}(s) \{J_r(s)/J_r(0)\}. \qquad (12.8)$$

By performing the computation in this way, we can see that the leading term for the MGF of constructed variable $\Psi_r^\star$ is the MGF of the other constructed variable Ψ_r used to approximate the CDF. If, for purposes of saddlepoint determination, the latter term of (12.8) in curly braces is ignored and treated as constant in s, then the saddlepoint for the density approximation solves $K'_{\Psi_r}(\hat{s}) = 0$ and is the same saddlepoint as used in the CDF approximation in (12.2). Practical examples have shown that $\ln\{J_r(s)/J_r(0)\}$ is quite flat in s relative to $K_{\Psi_r}(s)$ so that the latter term is the dominant component of $K_{\Psi_r^\star}(s)$ when computing the saddlepoint. Proceeding in this manner, the saddlepoint density approximation for R at r is

$$\hat{f}_R(r) = E(V) \hat{f}_{\Psi_r^\star}(0) = \frac{M_{\Psi_r}(\hat{s}) J_r(\hat{s})}{\sqrt{2\pi K''_{\Psi_r}(\hat{s})}}. \qquad (12.9)$$

Example: Beta (α, β)

Here $V = Y_\alpha + Y_\beta$ and $E(V) = \alpha + \beta$ with

$$M(s,t) = E\{e^{sY_\alpha + t(Y_\alpha + Y_\beta)}\} = (1 - s - t)^{-\alpha}(1-t)^{-\beta}.$$

The MGF of $\Psi_r^\star$ as in (12.8) is

$$M_{\Psi_r^\star}(s) = M(s, -rs) \left\{ \left(\frac{\alpha}{1-s+rs} + \frac{\beta}{1+rs} \right) \Big/ (\alpha + \beta) \right\} \qquad (12.10)$$

and for α and β sufficiently large, the leading term $M(s, -rs)$ or M_{Ψ_r} is the dominant factor in (13.13). The saddlepoint $\hat{s}$ has been given in Example 5 of section 1.2.2 and leads to

$$\hat{f}_R(r) = \frac{\exp(-\hat{w}^2/2) J_r(\hat{s})}{\sqrt{2\pi K''_{\Psi_r}(\hat{s})}}$$

where $\hat{w}$ is given in (1.26) and $\sqrt{K''_{\Psi_r}(\hat{s})} = \hat{u}/\hat{s}$. Using the simplification $J_r(\hat{s}) = \alpha + \beta$ and combining terms, then

$$\hat{f}_R(r) = \frac{\hat{\Gamma}(\alpha+\beta)}{\hat{\Gamma}(\alpha)\hat{\Gamma}(\beta)} r^{\alpha-1}(1-r)^{\beta-1}$$
$$= \frac{B(\alpha,\beta)}{\hat{B}(\alpha,\beta)} f_R(r). \qquad (12.11)$$

The saddlepoint density is exact up to Stirling's approximation for the Beta function. At least for this example, the use of the dominant factor M_{Ψ_r} for determination of the saddlepoint is justified by the fact that it leads to the correct shape in (12.11). This would not have happened if the additional factor $J_r(s)$ were included to determine the saddlepoint.

Proof of (12.4) and (12.5)

The MGF of $(U^\star, V^\star)$ is

$$M_{U^\star,V^\star}(s,t) = E(e^{sU^\star+tV^\star}) = \frac{1}{E(V)} \int_0^\infty \int_{-\infty}^\infty e^{su+tv} v f_{U,V}(u,v) du dv$$
$$= \frac{1}{E(V)} \int_0^\infty \int_{-\infty}^\infty \frac{\partial}{\partial t}(e^{su+tv}) f_{U,V}(u,v) du dv$$
$$= \frac{1}{E(V)} \frac{\partial}{\partial t} M(s,t) \qquad (12.12)$$

if (s,t) is in the convergence strip of M. The last step in which the derivative operation is taken outside the integral follows from the dominated convergence theorem and the argument is left as Exercise 2. The MGF expression for $\Psi_r^\star$ in (12.5) follows from (12.12).

To show (12.4), note that the density of $\Psi_r^\star$ at zero is computed by marginalizing the joint density of $(U^\star - rV^\star, V^\star)$ and leads to

$$f_{\Psi_r^\star}(0) = \frac{1}{E(V)} \int_0^\infty v f_{U,V}(rv,v) dv = \frac{1}{E(V)} f_R(r).$$

12.1.2 Doubly and singly noncentral F

These two distributions determine the power of the F test for linear hypotheses in the analysis of variance. If the alternative model is true, then the singly noncentral F, denoted as $F^{(1)}$, determines the power. The doubly noncentral F, denoted as $F^{(2)}$, occurs when both the null and alternative models are special cases of the true model. This setting occurs when some relevant concomitant variables have been wrongly ignored in the modeling and testing. Both the null and alternative models treat these concomitant variables as if their non-zero regression coefficients are zero. In this context, both the power and level of the test are determined from the doubly noncentral F as discussed in Scheffé (1959, pp. 134–5) who gives a simple example.

A complete discussion of saddlepoint approximation for these two distributions has been given in Butler and Paolella (2002). They demonstrate that (i) the relative errors of tail probabilities using the second-order Lugannani and Rice approximation are typically around 0.03% and rarely much larger than 0.1% over a wide range of parameter values;

(ii) the saddlepoint approximation is an explicit computation that runs between 10 and 20 thousand times faster than the Imhof (1961) numerical integration approximation that has traditionally been used in this context; and (iii) the saddlepoint approximation is several orders of magnitude more accurate than Tiku's (1972) approximation, based on three moments, and the Tiku and Yip (1978) approximation based on four moments.

The doubly noncentral F distribution is determined as

$$F^{(2)} = \frac{U}{V} = \frac{\chi_1/n_1}{\chi_2/n_2} \sim F_{n_1,n_2}(\theta_1, \theta_2)$$

where U and V are independent with $\chi_1 \sim \chi^2(n_1, \theta_1)$ and $\chi_2 \sim \chi^2(n_2, \theta_2)$, n_i is the degrees of freedom, and θ_i is the noncentrality parameter of the noncentral χ^2 distribution. Taking $\theta_2 = 0$ results in the singly noncentral $F^{(1)}$ and the additional requirement $\theta_1 = 0$ leads to the central F.

The CDF of $F^{(2)}$ at r can be expressed as

$$\Pr\left(F^{(2)} \le r\right) = \Pr\left(\frac{n_2}{n_1}\chi_1 - r\chi_2 \le 0\right) = \Pr(\Psi_r < 0) \quad (12.13)$$

where Ψ_r is so defined. Saddlepoint CDF approximation is now based on the MGF of Ψ_r given in Exercise 3. The greatest accuracy is achieved by using the second-order approximation

$$\widehat{\Pr}_2(\Psi_r \le 0) = \widehat{\Pr}(\Psi_r \le 0) - \phi(\hat{w})\left\{\hat{u}^{-1}\left(\frac{\hat{\kappa}_4}{8} - \frac{5}{24}\hat{\kappa}_3^2\right) - \hat{u}^{-3} - \frac{\hat{\kappa}_3}{2\hat{u}^2} + \hat{w}^{-3}\right\} \quad (12.14)$$

taken from (2.48) for $0 \ne E(\Psi_r)$ where $\hat{\kappa}_i = K_{\Psi_r}^{(i)}(\hat{s})/K_{\Psi_r}''(\hat{s})^{i/2}$ is given in Exercise 3(b).

It is possible to determine an explicit saddlepoint $\hat{s}$ to facilitate this computation. The details are sufficiently technical to be deferred to the appendix in section 12.6.1.

12.1.3 Density approximation

Approximation to the density of $F^{(2)}$ follows directly from (12.9) and is outlined in Exercise 3(c). In the case of the central F, the saddlepoint density is

$$\hat{f}_R(r) = \left(\frac{n_2}{n_1}\right)^{n_2/2} \frac{1}{\hat{B}(n_1/2, n_2/2)} \frac{r^{n_1/2-1}}{(1 + n_1 r/n_2)^{(n_1+n_2)/2}}, \quad (12.15)$$

where $\hat{B}(n_1/2, n_2/2)$ is Stirling's approximation to the beta function $B(n_1/2, n_2/2)$. This is the exact F_{n_1,n_2} density apart from the use of Stirling's approximation. The result might have been anticipated once the exactness for the approximation to the central Beta in (12.11) was determined. The relationship of saddlepoint approximations for the doubly noncentral beta and the doubly noncentral F is considered in Exercises 5 and 6. There the reader is asked to prove that the CDFs are analytically equivalent. In addition, the densities are related by Jacobian transformation and it is this result which anticipates the exactness seen in (12.15).

Figure 12.1 plots the exact singly noncentral $F^{(1)}$ density $f_R(r)$ as the solid line and uses $n_1 = 1, n_2 = 12$, and $\theta_1 = 2316$ which corresponds to an application in Chow and Shao (1999). The normalized saddlepoint density $\bar{f}_R(r)$ (with constant of integration 1.00162) is

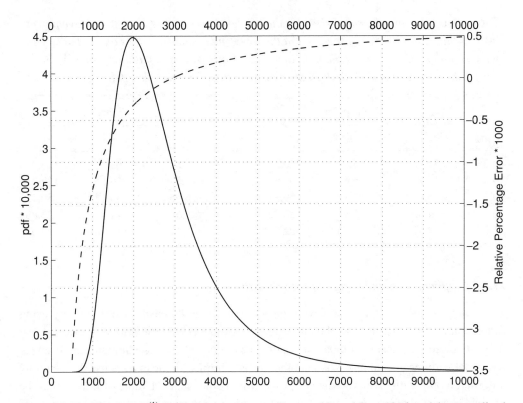

Figure 12.1. The exact $F^{(1)}$ density $f_R(r)$ with $n_1 = 1$, $n_2 = 12$, and $\theta_1 = 2316$, and the normalized saddlepoint density $\bar{f}_R(r)$ are graphically indistinguishable (*solid*). Also shown is the percentage relative error (*dashed*).

also plotted but is graphically indistinguishable. The dashed line plots the relative error of the approximation which is between $0.0^35\%$ and $-0.0^23\%$ as recorded using the scale on the far right.

Saddlepoint computation is attained by using a closed form expression whereas the exact evaluation has used an infinite series expansion from Johnson *et al.* (1995, Eq. 30.7). The point of truncation for this infinite expansion is often difficult to anticipate. This aspect, as well as the overall computational demand and the presence of a highly accurate saddlepoint approximation make exact computation somewhat impractical and unnecessary.

12.1.4 CDF approximation

Both the first- and second-order Lugannani and Rice approximations in (12.2) and (12.14) are very accurate for the singly and doubly noncentral F. Except for the case $n_1 = 1$ for which (12.2) is more accurate, it is generally the second-order approximation that has greater accuracy with relative error typically 32 times smaller than the first-order approximation. Butler and Paolella (2002) consider an example with $n_1 = 10$ and $n_2 = 20$ and $(\theta_1, \theta_2) \in (0, \infty)^2$ ranging over a grid of values. The absolute percentage relative error is mostly decreasing in both θ_1 and θ_2, has a maximum of 0.01% at $\theta_2 = 1 = \theta_1$, and has most of its values ranging between $0.0^21\%$ to $0.0^41\%$.

Table 12.1. *Required sample sizes for the 5% level two sample t-test to achieve power 90% at the values of (δ, σ^2) specified in the rows and columns. Values on the left (right) were computed using the saddlepoint approximation (exact computation) for $F^{(1)}$. To assess the accuracy achieved, the values are shown to four significant digits.*

	Saddlepoint			Exact		
$\delta \backslash \sigma^2$	0.1	0.5	1.0	0.1	0.5	1.0
0.2	53.50	263.6	526.2	53.52	263.6	526.3
0.4	14.16	66.63	132.3	14.17	66.65	132.3
0.6	6.952	30.17	59.34	6.955	30.18	59.35
0.8	4.500	17.43	33.82	4.501	17.44	33.83
1.0	3.413	11.55	22.02	3.415	11.56	22.02

12.1.5 Application to the design of experiments

The sample sizes required to achieve a preset power may be easily determined by using saddlepoint approximations for the noncentral F. Consider an agricultural experiment that uses a two sample t-test to compare a treatment to a control. Suppose n plants are used for treatment with mean response δ and another n for control with mean response 0. Assuming a normal model with error variance σ^2, then what is the smallest value of n such that the two-sided α-level test of $H_0 : \delta = 0$ attains power ρ at alternative value $\theta_1 = n\delta^2/(2\sigma^2)$? (See Lehmann, 1986, p. 373).

In this setting $T^2 \sim F_{1,2n-2}(\theta_1, 0)$ with power function $\Pr(T^2 > r)$, where r is the $100(1-\alpha)^{\text{th}}$ percentile of the central $F_{1,2n-2}$. Table 12.1 shows the required n that corresponds to $\alpha = 0.05$ and $\rho = 0.90$ for several δ and σ values. These sample sizes have been computed by using both the second-order saddlepoint approximation (left panel) and the exact singly-noncentral distribution function (right panel). In all cases, the saddlepoint approximation provides the correct value.

Suppose there are potential underlying block effects that are being ignored while using this t-test. In particular, suppose now that $n = 2m$ and that m plants of both treatment and control are tested in the morning and m in the afternoon. Let the block effects for morning and afternoon be 0 and η, respectively. The model is then a two way layout with two levels for the group factors and blocks and m observations per cell. Under these circumstances, straightforward calculation shows that $T^2 \sim F_{1,4m-2}(\theta_1, \theta_2)$ with $\theta_1 = m\delta^2/\sigma^2$ and $\theta_2 = m\eta^2/\sigma^2$. Larger values of η^2 tend to inflate the denominator of the t^2 statistic making it more difficult to reject regardless of the value of θ_1. This naturally decreases the power and diminishes the size of the F-test.

To determine the effect of nonzero values of η on the size and power of the test, figure 12.2 plots the size $\Pr(T^2 > r; 0, \theta_2)$ (solid) and power $\Pr(T^2 > r; \theta_1, \theta_2)$ (dashed) of the test versus η with r chosen as the appropriate α-level cutoff. The figure uses $\alpha = 0.05$, $\rho = 0.9$ and $\delta = \sigma = 1$. The power objective $\rho = 0.9$ is achieved by using $n = 22$ or $m = 11$ as read from table 12.1 and this sample size is also used in figure 12.2. The first-order saddlepoint approximation for size (dotted) closely follows its exact counterpart (solid) while for the power curve, its second-order saddlepoint (dotted) is almost indistinguishable from the exact computation (dashed).

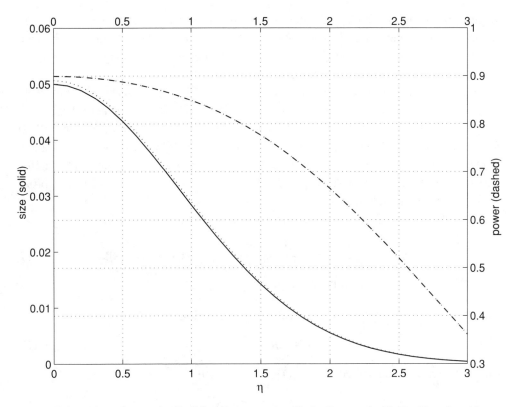

Figure 12.2. Plots of exact size (solid) and exact power (dashed) vs. η, the block effect size. Almost indistinguishable are the associated saddlepoint approximations (dotted).

The largest absolute relative percentage error in the saddlepoint size plot is 1.34% and occurs at $\eta = 0$. The saddlepoint power curve has a maximum absolute relative error of 0.0116%. The saddlepoint graphs took 5 seconds in Matlab while the exact plots required 14 minutes of CPU.

12.1.6 Ratio of quadratic forms in normal variables

Ratio of quadratic forms of the general form

$$R = \frac{\varepsilon^T A \varepsilon}{\varepsilon^T B \varepsilon} \tag{12.16}$$

occur throughout the study of linear models and time series analysis. In these subjects and elsewhere such ratios are the estimators of parameters and also are the statistics used when testing hypotheses. For example, in autoregressive time series models, such ratios include the Durbin–Watson test statistics to detect lag orders as well as the least squares, Yule–Walker, and Burg estimators of serial correlation with arbitrary lags. It therefore is an important practical consideration to determine the distribution of R under quite general distributional assumptions on ε.

Without loss in generality, suppose that A and B are $n \times n$ symmetric and $\varepsilon \sim N(\mu, I_n)$. Let B be positive semidefinite to assure that the denominator is positive with probability one. There is no loss in generality in taking the covariance of ε as the identity. This is because, if the distribution of ε were $N(\mu, \Sigma)$, then (12.16) describes the model with $\Sigma^{1/2} A \Sigma^{1/2}$ and $\Sigma^{1/2} B \Sigma^{1/2}$ replacing A and B respectively, and $\Sigma^{-1/2}\mu$ replacing μ in the distributional assumption on ε. Thus model (12.16) incorporates all dependence among the components of ε as well as any noncentrality that occurs when $\mu \neq 0$.

Various sorts of saddlepoint approximations for the CDF and density of R have been proposed beginning with the seminal work on serial correlations in Daniels (1956). Further marginal distributional approximations are given in McGregor (1960), Phillips (1978), Jensen (1988), Wang (1992), Lieberman (1994a,b), Marsh (1998), and Butler and Paolella (2002, 2004). The main contributions of the latter two papers are in establishing the uniformity of relative errors for the saddlepoint CDFs in (12.2) and (12.14) and density approximation (12.9) in the right tail of R.

The CDF for R in the most general setting with noncentrality is

$$\Pr(R \leq r) = \Pr\left(\frac{\varepsilon^T A \varepsilon}{\varepsilon^T B \varepsilon} \leq r\right) = \Pr\{\varepsilon^T (A - rB) \varepsilon \leq 0\} \qquad (12.17)$$
$$= \Pr(\Psi_r \leq 0)$$

where Ψ_r is so defined. Assume the spectral decomposition

$$A - rB = P_r^T \Lambda_r P_r \qquad (12.18)$$

where P_r is orthogonal and $\Lambda_r = \text{diag}(\lambda_1, \ldots, \lambda_n)$, with

$$\lambda_1 = \lambda_1(r) \leq \cdots \leq \lambda_n = \lambda_n(r),$$

consists of the ordered eigenvalues of (12.18). The distribution of Ψ_r is therefore

$$\Psi_r = \sum_{i=1}^n \lambda_i(r) \chi^2\left(1, v_i^2\right), \qquad (12.19)$$

where $\{v_i^2\}$ are determined as $(v_1, \ldots, v_n)^T = v = P_r \mu$ and represent the noncentrality parameters of the independent noncentral χ_1^2 variables specified in (12.19). The ordered values of $\{\lambda_i\}$ are in 1-1 correspondence with the components of v specified through the particular choice of P_r.

Before proceeding with the development of a saddlepoint approximation for the distribution of R, consideration is given to the support for R. The case in which R is degenerate at a single point is ruled out; this setting occurs if and only if $A = cB$ for some scalar constant c. If A (and B) have at least one nonzero (positive) eigenvalue, then the support of R is $(l, r) \subseteq (-\infty, \infty)$ with the particular values of l and r varying according to the properties of A and B; see Lemma 3 in Butler and Paolella (2004) for details. If cutoff $r \in (l, r)$ is an interior point, then the MGF of Ψ_r is

$$M_{\Psi_r}(s) = \left(\prod_{i=1}^n (1 - 2s\lambda_i)^{-1/2}\right) \exp\left\{s \sum_{i=1}^n \frac{\lambda_i v_i^2}{1 - 2s\lambda_i}\right\} \qquad (12.20)$$

and is convergent on the neighborhood of zero given as

$$\frac{1}{2\lambda_1(r)} < s < \frac{1}{2\lambda_n(r)}. \qquad (12.21)$$

The saddlepoint $\hat{s}$ is the unique root of

$$0 = K'_{\Psi_r}(\hat{s}) = \sum_{i=1}^{n} \left(\frac{\lambda_i}{1 - 2\hat{s}\lambda_i} + \frac{\lambda_i v_i^2}{(1 - 2\hat{s}\lambda_i)^2} \right) \quad (12.22)$$

in the range (12.21). The Lugannani and Rice approximation to first order (12.2) or second order (12.14) can be computed using the higher order derivatives of K_{Ψ_r} given as

$$K_{\Psi_r}^{(j)}(s) = 2^{j-1}(j-1)! \sum_{i=1}^{n} \lambda_i^j (1 - 2s\lambda_i)^{-j} \left(1 + \frac{j v_i^2}{1 - 2s\lambda_i} \right).$$

The saddlepoint density approximation for $f_R(r)$ is specified in (12.9) using $\hat{s}$ that solves (12.22) and factor $J_r(\hat{s})$ computed from

$$J_r(s) = \text{tr}(I - 2s\Lambda_r)^{-1} H_r + v_r^T (I - 2s\Lambda_r)^{-1} H_r (I - 2s\Lambda_r)^{-1} v_r \quad (12.23)$$

with $H_r = P_r B P_r^T$. The second-order saddlepoint density in this context is

$$\hat{f}_{R2}(r) = \hat{f}_R(r)(1 + O) \quad (12.24)$$

where

$$O = \left(\frac{\hat{\kappa}_4}{8} - \frac{5}{24} \hat{\kappa}_3^2 \right) + \frac{J'_r(\hat{s}) \hat{\kappa}_3}{2 J_r(\hat{s}) \sqrt{K''_{\Psi_r}(\hat{s})}} - \frac{J''_r(\hat{s})}{2 J_r(\hat{s}) K''_{\Psi_r}(\hat{s})}. \quad (12.25)$$

Exercise 7 outlines the derivation of the first-order expressions for the saddlepoint CDF and density. The second-order term in (12.25) is stated without derivation since it requires an argument that uses complex variables. Note however that it is simply the negative of the second-order expansion term in (2.20) for a comparable Laplace approximation. Indeed the complex variable argument leading to (12.25) is simply the complex analogue to the real variable argument that results in (2.20).

Numerical example

If $\varepsilon_1, \ldots, \varepsilon_n$ is a time series, the least square estimate for lag 1 correlation is

$$R = \sum_{i=1}^{n-1} \varepsilon_{i+1} \varepsilon_i \bigg/ \sum_{i=1}^{n-1} \varepsilon_i^2.$$

Consider the simplest possible case with $n = 2$ so that $R = \varepsilon_2/\varepsilon_1$. Suppose also that ε_1 and ε_2 are independent with $\varepsilon_i \sim N(\mu_i, 1)$ for $i = 1, 2$ so there are location effects for each term. The rationale for making the example so simple is that comparison with the exact density becomes possible since it can be expressed as

$$f_R(r) = \frac{1}{2\pi} \int_{-\infty}^{\infty} |x| \exp\left\{ -\frac{1}{2}(x - \mu_1)^2 \right\} \exp\left\{ -\frac{1}{2}(rx - \mu_2)^2 \right\} dx$$

$$= \frac{1}{\pi \delta_r} \exp\left\{ -\frac{\mu_1^2 + \mu_2^2}{2} \right\} + \frac{(\mu_1 + r\mu_2)}{\delta_r \sqrt{2\pi \delta_r}} \exp\left(-\frac{(\mu_1 r - \mu_2)^2}{2\delta_r} \right) \text{erf}\left(\frac{\mu_1 + r\mu_2}{\sqrt{2\delta_r}} \right)$$

$$(12.26)$$

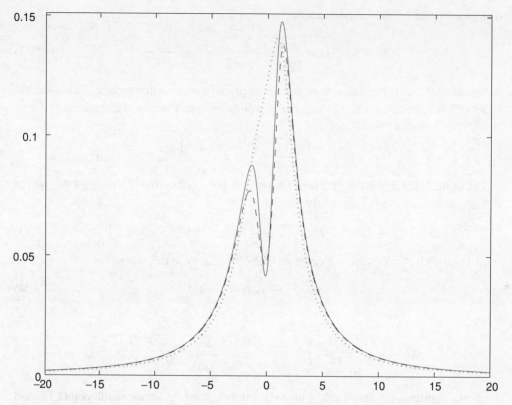

Figure 12.3. Exact density f_R (*solid*), second-order $\hat{f}_{R2}$ (*dashed*) and normalized $\bar{f}_R$ (*dotted*) saddlepoint densities.

where $\delta_r = 1 + r^2$. This density is both heavy tailed and bimodal for $\mu_1 = 0.2$ and $\mu_2 = 2$. Figure 12.3 plots the exact density (solid), the normalized version of $\hat{f}_R$ in (12.9) denoted as $\bar{f}_R$ (dotted), and the second order saddlepoint $\hat{f}_{R2}$ in (12.24) (dashed) for this case. While both appear highly accurate in the tails, only the latter captures the bimodality.

The true CDF of R, or $F_R(r)$, must be computed from (12.26) using numerical integration. Figure 12.4 plots

$$\frac{F_R(r)}{\hat{F}_R(r)} 1_{\{\hat{s} < 0\}} + \frac{1 - F_R(r)}{1 - \hat{F}_R(r)} 1_{\{\hat{s} > 0\}} \quad \text{vs.} \quad r \tag{12.27}$$

with $\hat{F}_R(r)$ as $\widehat{\Pr}_2$ in (12.14) (dashed) and as $\widehat{\Pr}$ in (12.2) (dot-dashed). For these values of μ_i, $\widehat{\Pr}$ is more accurate than $\widehat{\Pr}_2$ only in the range $-1.8 < r < 1.2$.

If $\mu_1 = \mu_2 = 0$, R is Cauchy and the exactness of this case is considered in Exercise 8.

12.2 Univariate roots of estimating equations

Suppose the random score Ψ_r is a general estimating equation in which random variable R is defined as its root that solves $\Psi_R = 0$. To make the problem well defined, suppose there is a unique root with probability 1, a result that is assured by also assuming that $\partial \Psi_r / \partial r < 0$ with probability 1. These probabilistic statements are made with reference to a continuous

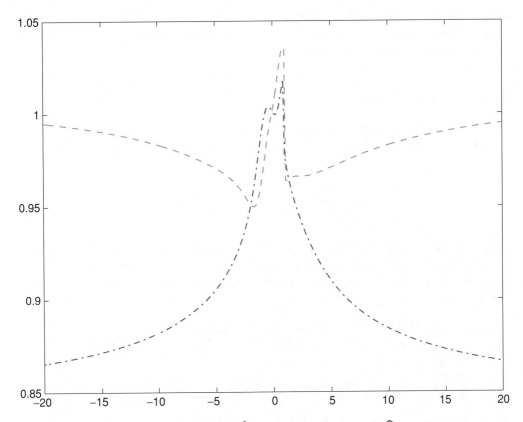

Figure 12.4. Plots of (12.27) with $\hat{F}_R(r)$ as $\widehat{\Pr}_2$ in (12.14) (*dashed*) and as $\widehat{\Pr}$ in (12.2) (*dot-dashed*).

random vector X that consists of more fundamental random variables used to construct Ψ_r. For example, in the ratio example, $\Psi_r = U - rV$ and $X^T = (U, V)$.

As a second example, consider an M-estimator for a location parameter using i.i.d. responses $X^T = (X_1, \ldots, X_n)$. The value R that solves

$$\Psi_R = \sum_{i=1}^{n} \psi(X_i - R) = 0$$

is the M-estimator based on the influence function ψ. When the influence function is strictly increasing, as occurs for least squares with $\psi(u) = u$, the conditions above are satisfied. With other choices of ψ, such as for example $\psi(u) = \text{sgn}(u)$ or Huber's M-estimator with

$$\psi(u) = u\, 1_{\{|u|<k\}} + k\, \text{sgn}(u)\, 1_{\{|u|\geq k\}}, \tag{12.28}$$

the strict monotonicity assumption is not satisfied. In these cases $\partial \Psi_r / \partial r \leq 0$ with probability 1 and the theory can still be used; see Field and Hampel (1982).

An approximation to the CDF of root R uses the fact that Ψ_R is decreasing in R to equate

$$\Pr(R \leq r) = \Pr(\Psi_r \leq 0). \tag{12.29}$$

The CGF of Ψ_r or $K_{\Psi_r}(s)$ allows approximation to (12.29) using the Lugannani and Rice approximation as given in (12.2) and (12.3). The saddlepoint equation $K'_{\Psi_r}(\hat{s}) = 0$ is solved so that $\hat{s}$ depends on the cutoff value r.

12.2.1 Saddlepoint density for R

A saddlepoint density approximation for R at r can be determined through the 1-1 relationship of $R = r$ and $\Psi_r = 0$ and using the CGF $K_{\Psi_r}(s)$. If saddlepoint $\hat{s}$ solves $K'_{\Psi_r}(\hat{s}) = 0$, then

$$\hat{f}_R(r) = \frac{M_{\Psi_r}(\hat{s})}{\sqrt{2\pi K''_{\Psi_r}(\hat{s})}} \left\{ -\frac{1}{\hat{s}} \frac{\partial K_{\Psi_r}(\hat{s})}{\partial r} \right\} \qquad \hat{s} \neq 0. \qquad (12.30)$$

This expression was given in Field and Hampel (1982, Eq. 4.3), Field (1982, Eq. 6), and Daniels (1983, Eq. 4.2) and is derived below. For the ratio setting in which $\Psi_r = U - rV$ this is the density for the ratio given in (12.9) if the last factor in (12.30) is $J_r(\hat{s})$. This follows since

$$-\frac{1}{\hat{s}} \frac{\partial K_{\Psi_r}(\hat{s})}{\partial r} = -\frac{1}{\hat{s}} \frac{\partial K_{U,V}(\hat{s}, -r\hat{s})}{\partial r} = \left. \frac{\partial K_{U,V}(\hat{s}, t)}{\partial t} \right|_{t=-r\hat{s}} = J_r(\hat{s}).$$

Derivation of the saddlepoint density

The simplest derivation is given in Daniels (1983) and exploits the CDF relationship of R and Ψ_r given in (12.29). First, however, some preliminary results on the transforms of CDFs and survival functions are needed. If random variable Y has MGF $M_Y(s)$, then Exercise 10 shows that the inversion of $M_Y(s)/s$ for $\text{Re}(s) > 0$ leads to the survival function $\Pr(Y > y)$.

The survival function $\Pr(R > r) = \Pr(\Psi_r > 0)$ is computed as the inversion of $M_{\Psi_r}(s)/s$ which gives

$$\Pr(R > r) = \frac{1}{2\pi i} \int_{c-i\infty}^{c+i\infty} \frac{M_{\Psi_r}(s)}{s} ds \qquad c > 0. \qquad (12.31)$$

Then

$$\begin{aligned} f_R(r) &= -\frac{\partial}{\partial r} \Pr(R > r) = \frac{1}{2\pi i} \int_{c-i\infty}^{c+i\infty} -\frac{1}{s} \frac{\partial M_{\Psi_r}(s)}{\partial r} ds \\ &= \frac{1}{2\pi i} \int_{c-i\infty}^{c+i\infty} M_{\Psi_r}(s) \left\{ -\frac{1}{s} \frac{\partial K_{\Psi_r}(s)}{\partial r} \right\} ds \end{aligned} \qquad (12.32)$$

where the derivative has been passed though the inversion integral in the top line. The dominant term of the integrand in (12.32) is $M_{\Psi_r}(s)$ which is used to determine the saddlepoint as the root to $K'_{\Psi_r}(\hat{s}) = 0$. The term in curly braces is of lesser importance and is shown to have a removable discontinuity at $s = 0$ below. Since the integrand has no poles, the ordinary saddlepoint density formula (1.4) may be used which leads to expression (12.30) as the approximate value of $f_R(r)$.

The apparent singularity in (12.32) at $s = 0$ is shown to be removable by passing to the limit as $s \to 0$. Then,

$$\begin{aligned}
\lim_{s \to 0} \frac{1}{s} \frac{\partial K_{\Psi_r}(s)}{\partial r} &= \lim_{s \to 0} \frac{1}{s M_{\Psi_r}(s)} \frac{\partial M_{\Psi_r}(s)}{\partial r} \\
&= \lim_{s \to 0} \frac{1}{s M_{\Psi_r}(s)} \frac{\partial}{\partial r} \int_{-\infty}^{\infty} e^{s \psi_r(x)} f(x) dx \\
&= \lim_{s \to 0} \frac{1}{s M_{\Psi_r}(s)} \int_{-\infty}^{\infty} s \frac{\partial \psi_r(x)}{\partial r} e^{s \psi_r(x)} f(x) dx \\
&= E \left\{ \frac{\partial \Psi_r}{\partial r}; s \right\}
\end{aligned} \qquad (12.33)$$

where $\Psi_r = \psi_r(X)$ makes explicit the dependence of the random score on X. The notation indicates that the expectation in (12.33) is with respect to the s-tilted distribution of $\Psi_r(X)$.

Interpretation of the saddlepoint density

The saddlepoint density in (12.30) has an interpretation in terms of the more standard method for deriving densities: using the Jacobian transformation of densities from Ψ_r to R. First, note that the leading term (not in curly braces) is the saddlepoint approximation for the density of Ψ_r at 0 or $\hat{f}_{\Psi_r}(0)$. The left-hand side $\hat{f}_R(r)$ should be related to $\hat{f}_{\Psi_r}(0)$ through the Jacobian $\partial \Psi_r(X)/\partial r$ since r and 0 are 1-1. However, the Jacobian is itself a random variable whose value is not completely specified with the constraint $\Psi_r(X) = 0$. Thus, in place of the random Jacobian, approximation (12.30) uses the expression in curly braces. According to the argument leading to (12.33),

$$-\frac{1}{\hat{s}} \frac{\partial K_{\Psi_r}(\hat{s})}{\partial r} = E\left(-\frac{\partial \Psi_r}{\partial r}; \hat{s}\right) = \int_{-\infty}^{\infty} -\frac{\partial \psi_r(x)}{\partial r} g_r(x; \hat{s}) dx \qquad (12.34)$$

where

$$g_r(x; \hat{s}) = \frac{e^{\hat{s} \psi_r(x)} f(x)}{M_{\Psi_r}(\hat{s})}, \qquad (12.35)$$

f is the density of X, and $\hat{s}$ is taken as fixed. The new density for X in (12.35) is referred to as its $\hat{s} \psi_r(X)$-tilted density. Because the Jacobian is random, its average value has been used instead and averaged with respect to the $\hat{s} \psi_r(X)$-tilted density of X. The lingering question however, is why should the averaging be with respect to this particular tilted distribution?

A more complete answer to this is deferred to the section 12.4.2 where the multivariate version of the problem is addressed. For now however, a partial answer is obtained by considering directly the Jacobian transformation required as it depends on differentials of probability in X. Skovgaard (1990, Eq. 2.3) and Spady (1991, Eq. 2.3) give a direct natural formula for the density of R as

$$f_R(r) = f_{\Psi_r}(0) E\left\{-\frac{\partial \Psi_r(X)}{\partial r} | \Psi_r = 0\right\}, \qquad (12.36)$$

where the assumption that $\partial \Psi_r(X)/\partial r < 0$ with probability 1 applies. Here, the average Jacobian value is used which has been averaged over $\{x : \psi_r(x) = 0\}$ or all the values of x which contribute to a score value of zero. Spady (1991, p. 882) speculates that (12.34)

Table 12.2. *Tail probability* Pr($R > r$) *of the M-estimators for the contaminated normal and Cauchy data respectively. The Lugannani and Rice (Lug-Rice) and numerically integrated saddlepoint density (NID) approximations are shown. Replicated from tables 1 and 2 of Daniels (1983).*

	Contaminated normal				Cauchy		
r	Exact	Lug-Rice	NID	r	Exact	Lug-Rice	NID
0.1	.43766	.43767	.43740	1	.11285	.11400	.11458
0.5	.21876	.21886	.21787	3	.00825	.00881	.00883
1.0	.06759	.06787	.06716	5	.00210	.00244	.00244
1.5	.01888	.01959	.01939	7	.00082	.00104	.00105
2.0	.00629	.00764	.00763	9	.00040	.00055	.00055
2.5	.00290	.00456	.00460				
3.0	.00205	.00368	.00373				

provides an accurate approximation for the average Jacobian in (12.36) and backs his claim by showing the accuracy achieved in density approximation for estimates in least absolute value regression. Taking this speculation as correct (see §12.4.2 for its derivation), so that

$$E\left\{-\frac{\partial \Psi_r(X)}{\partial r}|\Psi_r = 0\right\} \simeq E\left(-\frac{\partial \Psi_r}{\partial r}; \hat{s}\right),$$

then the approximation in (12.30) may be rewritten to mimic (12.36) as

$$\hat{f}_R(r) = \hat{f}_{\Psi_r}(0) E\left(-\frac{\partial \Psi_r}{\partial r}; \hat{s}\right). \quad (12.37)$$

This is the form in which the approximation appears in Field (1982, Eq. 6) and Field and Ronchetti (1990, theorem 4.3).

12.2.2 M-estimation of location

Field and Hampel (1982, tables 1 and 2) and Daniels (1983, tables 1 and 2) considered the Huber M-estimate in (12.28) when data are generated from a contaminated normal and also from a Cauchy. Two portions of their numerical work are reproduced in table 12.2.

In the left portion of the table, $n = 3$ observations were generated from density

$$f(x) = 0.95\phi(x) + 0.025\phi(x+12) + 0.025\phi(x-12).$$

Various approximations for the survival function of R, the M-estimator with $k = 1.4$, are listed and compared with the exact survival probabilities in column 2. Column 3 is the Lugannani and Rice (Lug-Rice) approximation in (12.2) and (12.3) and column 4 is the numerically integrated and normalized saddlepoint density (NID) in (12.30).

The right portion of the table shows the same approximations using $k = 1.5$ in the influence function and generating $n = 5$ observations from a Cauchy density. In both examples the number of observations represents the smallest sample sizes that display reasonable saddlepoint accuracy.

Table 12.3. *Bivariate data from Hinkley and Wei (1984).*

u	155	68	28	25	190	82	92	196	164
v	1546	505	280	410	1450	517	738	2225	1660
u	68	82	36	195	92	185	61	62	
v	505	680	145	224	733	1957	287	473	
u	71	207	185	475	155	29	96	699	
v	473	1260	1958	4375	1499	245	828	6361	

In these applications, the lack of strict monotonicity of the ψ function leads to a discrete component in the distribution for Ψ_r. As a result, the NID entries are not strictly justified since the saddlepoint density derivation supposed that Ψ_r was continuous. However, the NID values consistently track the Lug-Rice values and both are quite accurate despite any formal objections to their applicability. Discussion of this issue continues at the end of section 12.2.4.

12.2.3 Bootstrapped ratios of means

Suppose an estimator R is obtained as the root of an estimating equation

$$0 = \Psi_R = \sum_{i=1}^{n} \psi(X_i; R)$$

where $\{X_i\}$ may be i.i.d. vector responses from unknown CDF F. The proposed saddlepoint theory provides approximation to the distribution of R *if the CDF F is known*. The saddlepoint expressions use components such as M_{Ψ_R} which are given in terms of F.

Substitution of the empirical CDF $\hat{F}$ for F in the true distributional expressions for R results in the bootstrap distribution for F. The same substitutions into the corresponding saddlepoint expressions result in a saddlepoint approximation for that bootstrap distribution.

Davison and Hinkley (1988, §4) developed this idea in the context of estimating a ratio of means. Suppose the bivariate data $\{x_i = (u_i, v_i); i = 1, \ldots, 25\}$ in table 12.3 are used in the estimation of $E(V)/E(U)$ from the roots of estimating equations point of view.

Suppose that $X_1^*, \ldots, X_{25}^*$ are an i.i.d. sample from the bivariate data or from the bivariate empirical CDF $\hat{F}$ with $X_i^* = (U_i^*, V_i^*)$. A bootstrap estimator of the ratio of means is $R^* = \bar{V}^*/\bar{U}^*$ and the corresponding estimating equation is $\Psi_r^* = \sum_i (V_i^* - rU_i^*)$ with MGF

$$M_{\Psi_r^*}(s) = \left\{ \iint e^{s(v-ru)} d\hat{F}(u,v) \right\}^{25}. \tag{12.38}$$

Computation of a saddlepoint approximation for the true bootstrap CDF $\Pr(R^* \leq r)$ is now based upon working with (12.38) and using the univariate roots of estimating approach with $K_{\Psi_r}(s)$ in (12.3) taken as $\ln M_{\Psi_r^*}(s)$.

Table 12.4 has been extracted from Davison and Hinkley (1988, §4) and shows the accuracy obtained by using the saddlepoint CDF (Lug-Rice) to approximate the true bootstrap

Table 12.4. *Approximation for the bootstrap distribution* $\Pr(R^* \leq r)$.

r	Exact[†]	Lug-Rice	Normal	r	Exact[†]	Lug-Rice	Normal
7.8	.0016	.0013	$.0^3 1$	9.2	.7181	.7179	.7295
8.0	.0056	.0046	$.0^3 7$	9.4	.8920	.8893	.8929
8.2	.0149	.0146	.0055	9.6	.9705	.9682	.9695
8.4	.0408	.0428	.0279	9.8	.9922	.9932	.9939
8.6	.1080	.1121	.0999	10.0	.9988	.9989	.9991
8.8	.2480	.2540	.2575	10.2	$.9^3$	$.9^3$	$.9^3$
9.0	.4698	.4729	.4921				

[†]Computed using 10,000 bootstrap samples.

distribution (Exact) determined from 10,000 bootstrap samples. The "Normal" approximation uses the estimated variance $\sum_i (v_i - ru_i)/(25\bar{u})^2$.

12.2.4 Lattice-valued score functions

If Ψ_r takes its values on a lattice instead of a continuum, then a density for R can still be determined by modifying the previous arguments as suggested in Daniels (1983, §5). For some constant a, suppose Ψ_r takes on the values $\{k - a : k \in \mathcal{I}\}$ over a continuous range of r values.

As an example, consider the estimating equation required in order to estimate the $100p$ percentile of a distribution from an i.i.d. sample $\{x_1, \ldots, x_n\}$. Let

$$\Psi_r = \sum_{i=1}^{n} \psi(x_i - r) \qquad \psi(u) = \begin{cases} -(1-p) & \text{if } u \leq 0 \\ p & \text{if } u > 0 \end{cases} \qquad (12.39)$$

so that Ψ_r assumes the values $\{k - n(1-p) : k = 0, \ldots, n\}$. It is not generally possible to solve the equation $\Psi_r = 0$ in r since the steps of Ψ_r jump over the value zero unless np is an integer. However, by denoting $\hat{F}(x)$ as the empirical CDF of the data, the relationship

$$\{\Psi_r \leq 0\} = \{\hat{F}(r) \geq p\},$$

may be used to define a solution. The $100p$ empirical percentile is defined as

$$R_p = \inf\{r : \hat{F}(r) \geq p\} = \inf\{r : \Psi_r \leq 0\}$$

and this value can now represent the root for any p.

If events concerning R can be equated with events for Ψ_r, then a CDF approximation for R is provided by using a continuity-corrected version of the Lugannani and Rice approximation for the lattice variable Ψ_r. This holds for the percentile example through the equivalence

$$\{R_p \leq r\} = \{\hat{F}(r) \geq p\} = \{\Psi_r \leq 0\}.$$

Indeed the exact CDF is

$$\Pr(R_p \leq r) = \Pr(\Psi_r \leq 0) = \Pr(B \geq np)$$

where $B \sim$ Binomial $\{n, F(r)\}$ and $F(r) = \Pr(X_1 \le r)$. Taking $\bar{F}(r) = 1 - F(r)$, then the MGF of Ψ_r is

$$M_{\Psi_r}(s) = \left[E\{e^{s\psi(X_1-r)}\}\right]^n = \{F(r)e^{-(1-p)s} + \bar{F}(r)e^{ps}\}^n$$
$$= e^{nps}\{F(r)e^{-s} + \bar{F}(r)\}^n \qquad (12.40)$$

which shows that $\Psi_r \sim np - B$. A computation of $\Pr(-\Psi_r \ge 0)$ using the continuity-corrected saddlepoint approximations for the Binomial tail probability in Example 2 of section 1.2.4 gives the approximate CDF for R_p.

Saddlepoint density function

A saddlepoint density for root R can be determined when the score function Ψ_r has lattice support. The argument is given below using the same arguments as used with a continuous score but making the necessary adjustments for the discreteness. This leads to the saddlepoint density

$$\hat{f}_R(r) = \frac{M_{\Psi_r}(\hat{s})}{\sqrt{2\pi K''_{\Psi_r}(\hat{s})}} \left\{ -\frac{\exp(-\hat{s}l_a)}{1 - \exp(-\hat{s})} \frac{\partial K_{\Psi_r}(\hat{s})}{\partial r} \right\} \qquad \hat{s} \ne 0, \qquad (12.41)$$

where $l_a = \lfloor a \rfloor - a + 1$ and $\hat{s}$ solves $K'_{\Psi_r}(\hat{s}) = 0$. The expansion

$$\frac{\exp\{-l_a \hat{s}\}}{1 - \exp(-\hat{s})} = \frac{1}{\hat{s}} + \left(\frac{1}{2} - l_a\right) + O(\hat{s})$$

shows that the leading term in the continuity correction returns the expression associated with a continuous-valued Ψ_r.

If R_p is the $100p$ percentile, then a direct computation from (12.41) using the MGF in (12.40) gives

$$\hat{f}_{R_p}(r) = \left\{ \sqrt{\frac{n}{2\pi}} p^{l_a-np-1/2}(1-p)^{np-n-l_a+1/2} \right\} F(r)^{np-l_a} \bar{F}(r)^{n-np+l_a-1} f(r) \qquad (12.42)$$

where $f(r) = F'(r)$. If $np = m$ is an integer, then $l_a = 1$ and R_p is the mth largest order statistic whose saddlepoint density is

$$\hat{f}_{R_p}(r) = \frac{\widehat{n!}}{\hat{\Gamma}(m)\widehat{(n-m)!}} F(r)^{m-1} \bar{F}(r)^{n-m} f(r) \qquad (12.43)$$

where $\widehat{n!}$ and $\hat{\Gamma}(m)$ are Stirling approximations to $n!$ and $\Gamma(m)$ as given in (1.16) and (1.8) respectively. The density in (12.43) is exact up to Stirling's approximation, as noted by Daniels (1983, §5) and Field and Hampel (1982, §6).

Derivation of the saddlepoint density If constant $c > 0$, then the exact survival function of R is

$$\Pr(R > r) = \Pr(\Psi_r > 0) = \sum_{k=\lfloor a \rfloor+1}^{\infty} \Pr(\Psi_r = k - a)$$
$$= \sum_{k=\lfloor a \rfloor+1}^{\infty} \frac{1}{2\pi i} \int_{c-i\pi}^{c+i\pi} M_{\Psi_r}(s) e^{-s(k-a)} ds.$$

The necessity of taking $c > 0$ assures that $\text{Re}(s) > 0$ along the contour of integration so that the infinite summation is convergent. Therefore

$$\Pr(R > r) = \frac{1}{2\pi i} \int_{c-i\pi}^{c+i\pi} M_{\Psi_r}(s) \frac{e^{-s(\lfloor a \rfloor + 1 - a)}}{1 - e^{-s}} ds. \qquad (12.44)$$

Taking the derivative of both sides and passing the derivative through the contour integral gives

$$f_R(r) = \frac{1}{2\pi i} \int_{c-i\pi}^{c+i\pi} M_{\Psi_r}(s) \left\{ -\frac{e^{-sl_a}}{1 - e^{-s}} \frac{\partial K_{\Psi_r}(s)}{\partial r} \right\} ds. \qquad (12.45)$$

A slight modification to the argument of (12.33) shows that the integrand does not have a pole at $s = 0$. Hence a saddlepoint approximation to the integral in (12.45) leads to the expression (12.41).

Further comments Whether the score Ψ_r is continuous or lattice-valued, a saddlepoint approximation for the density of its root can be determined by using the methods discussed above. The existence of a true density for the root is a matter of whether the CDF, as expressed through its inversion integrals (12.31) and (12.44), is differentiable in r.

A third setting that may occur, one in which Ψ_r has both continuous and lattice support, is not nearly as straightforward as pointed out in Jensen (1995, p. 106). His discussion in this situation concerns two issues: (i) whether the root actually has a density and (ii) whether or not the saddlepoint density can be formally established.

Consider, for example, the Huber M-estimator using (12.28). The score Ψ_r assumes an integer value when all the data are at least distance k from r; otherwise the score is a continuous value. Formal justification of the saddlepoint density does not follow from either of the two derivations above since the true distribution of Ψ_r is a mixture of continuous and lattice-valued distributions. However, the lattice support has quite small probability, so that the density approximation (12.30) can be used without too much concern for its formal meaning. The numerical studies in table 12.2 with small n suggest that this approach accurately replicates the distribution of R when the saddlepoint density is integrated.

The issue of whether or not the Huber M-estimator actually has a true density appears to be connected to the lack of a precise definition for the M-estimator; it needs to be uniquely defined with probability 1. The setting $n = 2$ is considered in detail in Exercise 12 where a density is confirmed once the M-estimator has been uniquely defined. The delicacy of this issue mentioned in Jensen (1995, p. 106) appears to be due to the fact that the M-estimator has not been uniquely defined with probability 1.

12.3 Distributions for vector ratios

The saddlepoint theory for vector ratios with positive denominators may be developed by using the same arguments that worked in the case of a univariate ratio. An important example of such vector ratios is the serial correlogram up to lag m whose saddlepoint distribution theory was considered in Butler and Paolella (1998b). This particular setting provides the main example in the discussion below.

12.3 Distributions for vector ratios

A second objective in discussing vector ratios in detail is to use the resulting theory to help motivate and anticipate the more general theory for multivariate roots of estimating equations. Indeed the vector ratio expressions are special cases of the more general expressions.

12.3.1 Two simple illustrative examples

Two simple examples are used below to suggest the patterns that arise when considering ratios. The first example uses a common denominator in each ratio so $R_i = U_i/V$ for $i = 1, \ldots, m$ which is the vector ratio pattern for the serial correlogram. The second example allows two different denominators with

$$R_1 = U_1/V_1 \qquad R_2 = U_2/V_1 \qquad R_3 = U_3/V_2. \tag{12.46}$$

The formulas suggested by these patterns provide for an account of the theory for vector ratios that is much more comprehensible than general complicated formulas that would be hard to decipher.

Ratios with a common denominator

Suppose that $V > 0$ with probability 1 and collectively denote $U^T = (U_1, \ldots, U_m)$. If $r^T = (r_1, \ldots, r_m)$ is a particular value of $R^T = (R_1, \ldots, R_m)$, then the vector score equations are

$$\Psi_r^T = (\Psi_1 \ldots, \Psi_m) = (U_1 - r_1 V, \ldots, U_m - r_m V).$$

In the case of vector ratios, the multivariate survival of R is expressible directly in terms of the multivariate survival of Ψ_r or

$$\Pr(R > r) = \Pr(\Psi_r > 0). \tag{12.47}$$

This correspondence of regions along with the fact that

$$f_R(r) = (-1)^m \frac{\partial^m}{\partial r_1 \ldots \partial r_m} \Pr(R > r) \tag{12.48}$$

allows the arguments used in univariate roots of equations to be extended directly to these vector ratios.

If $M_{U,V}(s, t)$ and $K_{U,V} = \ln M_{U,V}$ denote the joint MGF and CGF for U and V, then the MGF of Ψ_r is

$$M_{\Psi_r}(s) = M_{U,V}(s, -r^T s).$$

The argument given below shows that the saddlepoint density approximation for R is

$$\hat{f}_R(r) = \frac{M_{\Psi_r}(\hat{s})}{(2\pi)^{m/2} \left| K''_{\Psi_r}(\hat{s}) \right|^{1/2}} J_r(\hat{s}) \tag{12.49}$$

where saddlepoint $\hat{s}$ solves the m equations $K'_{\Psi_r}(\hat{s}) = 0$ and

$$J_r(\hat{s}) = \left\{ \left. \frac{\partial K_{U,V}(\hat{s}, t)}{\partial t} \right|_{t=-r^T \hat{s}} \right\}^m.$$

The ratio in (12.46)

Suppose the denominators are positive with probability 1 and denote $U^T = (U_1, U_2, U_3)$ and $V^T = (V_1, V_2)^T$. Now the vector score equations for the three ratios are

$$\Psi_r^T = (\Psi_{1r}, \Psi_{2r}, \Psi_{3r}) = (U_1 - r_1 V_1, U_2 - r_2 V_1, U_3 - r_3 V_2).$$

with MGF

$$M_{\Psi_r}(s) = M_{U;V}(s; -r_1 s_1 - r_2 s_2, -r_3 s_3).$$

Based on the same arguments used for the first example, the saddlepoint density approximation for R is

$$\hat{f}_R(r) = \frac{M_{\Psi_r}(\hat{s})}{(2\pi)^{3/2} \left| K''_{\Psi_r}(\hat{s}) \right|^{1/2}} J_r(\hat{s}), \tag{12.50}$$

where saddlepoint $\hat{s}^T = (\hat{s}_1, \hat{s}_2, \hat{s}_3)$ solves $K'_{\Psi_r}(\hat{s}) = 0$, and

$$J_r(\hat{s}) = \left\{ \frac{\partial K_{U,V}(\hat{s}; \hat{t})}{\partial \hat{t}_1} \right\}^2 \frac{\partial K_{U,V}(\hat{s}, \hat{t})}{\partial \hat{t}_2} \tag{12.51}$$

where

$$\hat{t}^T = (\hat{t}_1, \hat{t}_2) = (-r_1 \hat{s}_1 - r_2 \hat{s}_2, -r_3 \hat{s}_3). \tag{12.52}$$

This is a direct generalization of the univariate expression (12.9) in which the power of $\partial K_{U,V}/\partial t_k$ in $J_r(\hat{s})$ is the number of denominators V_k appears in. Also the form of $\hat{t}$ in (12.52) reflects which of the denominators that V_1 and V_2 appear in.

Expression (12.50) serves as a template for writing down the saddlepoint density for any vector ratio in which the components of U and V appear respectively in the numerator and denominator. Exercises 13 and 14 consider examples in which a random variable appears in both the numerator of one ratio and the denominator of another.

12.3.2 Derivations of the saddlepoint density for vector ratios

The derivations are given for ratios with a common positive denominator although the arguments apply to all vector ratios with positive denominators. The first derivation provides a characterization of the density of R as developed for a univariate ratio in section 12.1.1. The second derivation uses multivariate inversions.

Characterization of the density of R

The density of R at r can be specified in terms of the density of a vector of constructed scores at 0. Suppose $U^\star, V^\star$ is a constructed vector of dimension $m+1$ with density

$$f_{U^\star, V^\star}(u, v) = v^m f_{U,V}(u, v) / E(V^m). \tag{12.53}$$

Let

$$\Psi_r^\star = (U_1^\star - r_1 V^\star, \ldots, U_m^\star - r_m V^\star)$$

be the constructed score vector. Then a Jacobian transformation and marginalization working from $U^\star, V^\star$ to $\Psi_r^\star$ shows that the density of R at r is related to the density of $\Psi_r^\star$ at 0 according to

$$f_R(r) = E(V^m) f_{\Psi_r^\star}(0). \tag{12.54}$$

Additionally, it can be shown that the MGF of $\Psi_r^\star$ is

$$M_{\Psi_r^\star}(s) = \frac{1}{E(V^m)} \frac{\partial^m}{\partial t^m} M_{U,V}(s,t) \Big|_{t=-r^T s} \tag{12.55}$$

so that

$$\hat{f}_R(r) = E(V^m) \hat{f}_{\Psi_r^\star}(0). \tag{12.56}$$

What remains is the inversion of the MGF in (12.55). The mth derivative in this expression is cumbersome and an expansion can be used so that only the dominant term in the expansion needs to be retained. The argument does not depend on the value of m so take $m = 3$, write $M = M_{U,V}(s,t) = \exp\{K\}$ and differentiate to get

$$\frac{\partial^3 M}{\partial t^3} = M \left\{ \left(\frac{\partial K}{\partial t} \right)^3 + 3 \frac{\partial K}{\partial t} \frac{\partial^2 K}{\partial t^2} + \frac{\partial^3 K}{\partial t^3} \right\}. \tag{12.57}$$

Only the leading term within the curly braces has been found to be important in the numerical work in Butler and Paolella (1998b) and so it alone is retained. Expression (12.49) results when a standard saddlepoint density approximation for $\hat{f}_{\Psi_r^\star}(0)$, as in (3.2), is computed and substituted into (12.56).

Multivariate inversion approach

Since $\{R > r\}$ and $\{\Psi_r > 0\}$ are the same events for vector ratios (see (12.47), their probability can be differentiated as in (12.48) to determine the density of R. An expression for $\Pr\{\Psi_r > 0\}$ is determined by inverting its multivariate transform as developed in Exercise 10. If $s_k > 0$ for $k = 1, \ldots, m$, then this multivariate transform is

$$\int \cdots \int_{\Re^m} e^{s^T \psi} \Pr(\Psi_r > \psi) d\psi = \frac{M_{\Psi_r}(s)}{s_1 \cdots s_m} = \frac{M_{U,V}(s, -r^T s)}{s_1 \cdots s_m}.$$

The inverse transform is

$$\Pr(R > r) = \Pr(\Psi_r > 0) = \frac{1}{(2\pi i)^m} \int_{c_1 - i\infty}^{c_1 + i\infty} \cdots \int_{c_m - i\infty}^{c_m + i\infty} \frac{M_{U,V}(s, -r^T s)}{s_1 \cdots s_m} ds$$

where $c_k > 0$ for each k so that the contour paths remain in the convergence strip of the transform. Taking the derivative in r and passing it through the contour integral gives

$$\begin{aligned} f_R(r) &= \frac{1}{(2\pi i)^m} \int_{c_1 - i\infty}^{c_1 + i\infty} \cdots \int_{c_m - i\infty}^{c_m + i\infty} \frac{\partial^m M_{U;V}(s;t)}{\partial t^m} \Big|_{t=-r^T s} ds \\ &\simeq \frac{1}{(2\pi i)^m} \int_{c_1 - i\infty}^{c_1 + i\infty} \cdots \int_{c_m - i\infty}^{c_m + i\infty} M_{\Psi_r}(s) \left\{ \frac{\partial K_{U;V}(s;t)}{\partial t} \Big|_{t=-r^T s} \right\}^m ds, \end{aligned} \tag{12.58}$$

where only the dominate term has been retained as described in (12.57). A standard saddlepoint approximation for (12.58), that determines the saddlepoint $\hat{s}$ by using only the dominant term $M_{\Psi_r}(s)$ in its integrand, leads to (12.49).

Domains of definition

Some caution is necessary when considering the domain on which $\hat{f}_R$ is defined since it may be larger than $\mathcal{R}$, the actual support of R. The density $\hat{f}_R$ is well defined for those values of r that admit unique solutions to $K'_{\Psi_r}(\hat{s}) = 0$ for $\hat{s}$ in the convergence strip of K_{Ψ_r}, e.g.

$$\{r : K'_{\Psi_r}(\hat{s}) = 0 \quad \exists_{\hat{s}}\} = \mathcal{D}. \tag{12.59}$$

The domain of $\hat{f}_R$ always includes the interior of $\mathcal{R}$ or

$$\mathcal{D} \supset \text{int}(\mathcal{R}) \tag{12.60}$$

but it is sometimes too large and includes additional regions of $\Re^m$ with nontrivial volume. To show (12.60), note that if $\rho \in \text{int}(\mathcal{R})$ then $f_R(\rho)dr > 0$ which is assumed mapped onto $f_{\Psi_\rho}(0)d\psi > 0$ so that $0 \in \text{int}\{\text{supp}(\Psi_\rho)\}$. Thus, $K'_{\Psi_\rho}(\hat{s}_\rho) = 0$ for some $\hat{s}_\rho$ and $\rho \in \mathcal{D}$.

As with all saddlepoint densities, excess support can result when $\mathcal{R}$ is not a convex set. Consider, for example, vector ratios with a common denominator V such that V has a degenerate distribution at 1. Suppose the support of U is a nonconvex region in $\Re^m$ centered about $(0, 0)$. Taking $\Pr(V = 1) = 1$, then $\Psi_r = U - r(1, 1)^T$ with probability 1 and has the nonconvex support $\mathcal{R} - r(1, 1)^T$ that is the translation of $\mathcal{R}$ and centered about $r(1, 1)^T$. Denoting $\mathcal{I}_\mathcal{R}$ as the interior of the convex hull of $\mathcal{R}$, then

$$\mathcal{D} = \left\{r : 0 \in \mathcal{I}_{\mathcal{R}-r(1,1)^T}\right\} = \mathcal{I}_\mathcal{R}$$

for this simple example. The saddlepoint density $\hat{f}_R$ is defined and positively valued beyond its true support $\mathcal{R}$ and on the additional nontrivial set $\mathcal{I}_\mathcal{R} \setminus \text{int}(\mathcal{R})$ where $f_R = 0$.

This is not a problem for normalization if $\mathcal{R}$ is known. Then, the normalization constant can be computed numerically as

$$\int \hat{f}_R(r) 1_{\{r \in \mathcal{R}\}} dr = \int \hat{f}_R(r) 1_{\{r \in \mathcal{R}\}} \left\| \frac{\partial r}{\partial \hat{s}^T} \right\| d\hat{s}$$

$$= \int \hat{f}_R(r) 1_{\{r \in \mathcal{R}\}} \left| K''_{\Psi_r}(\hat{s}) \right| \left\| \frac{\partial K'_{\Psi_r}(\hat{s})}{\partial r^T} \right\|^{-1} d\hat{s}.$$

The latter integral has used the change of variable with Jacobian determined by implicit differentiation of the saddlepoint equation $K'_{\Psi_r}(\hat{s}) = 0$. Integration $d\hat{s}$ is much simpler for settings in which r is an explicit function of $\hat{s}$ through the saddlepoint equation.

12.3.3 Vector ratios of quadratic forms

Consider the joint saddlepoint density for R with components

$$R_k = \frac{\varepsilon^T A_k \varepsilon}{\varepsilon^T \varepsilon} \qquad k = 1, \ldots, m, \tag{12.61}$$

where $\varepsilon \sim \text{Normal}_n(0, \Sigma)$, the $n \times n$ matrices $\{A_k\}$ are symmetric with nonzero rank, and are assumed to result in a full rank distribution for R. The serial correlogram is a particular setting of such vector ratios that is considered in detail below whose development has been given in Butler and Paolella (1998b). Consideration of other correlogram approximations by Daniels (1956) and Durbin (1980b) as well as saddlepoint CDF approximations for choosing the order of an autoregressive time series are deferred to section 12.5.6. See also Kay et al. (2001).

The vector ratios in (12.61) also include the case in which denominator $\varepsilon^T \varepsilon$ is replaced by $\varepsilon^T B \varepsilon$ for $B > 0$ and symmetric. The substitution $f = B^{1/2}\varepsilon$ allows R_k to be rewritten in terms of random vector $f \sim \text{Normal}_n(0, B^{1/2}\Sigma B^{1/2})$ with $B^{-1/2}A_k B^{-1/2}$ replacing A_k in the numerator.

The joint saddlepoint density is given in (12.49) and is based on the joint CGF of vector scores $\{\varepsilon^T(A_k - r_k I_n)\varepsilon : k = 1, \ldots, m\}$ given as

$$K_{\Psi_r}(s) = -\tfrac{1}{2}\ln\{|P(s)\Sigma|\} \qquad P(s) = \Sigma^{-1} + 2r^T s I_n - 2\sum_{k=1}^{m} s_k A_k.$$

The saddlepoint for the density at r solves

$$0 = \frac{\partial}{\partial \hat{s}_k} K_{\Psi_r}(\hat{s}) = \text{tr}\{P(\hat{s})^{-1}(A_k - r_k I_n)\} \qquad k = 1, \ldots, m$$

where $\hat{s}$ is constrained to $\{s : P(s) > 0\}$, the region over which the joint CGF is convergent. The Hessian matrix $K''_{\Psi_r}(\hat{s})$ has (i, j)th component

$$\frac{\partial^2}{\partial \hat{s}_i \partial \hat{s}_j} K_{\Psi_r}(\hat{s}) = 2\,\text{tr}\{P(\hat{s})^{-1}(A_i - r_i I_n)P(\hat{s})^{-1}(A_j - r_j I_n)\}$$

and $J_r(\hat{s}) = [\text{tr}\{P(\hat{s})^{-1}\}]^m$.

Dirichlet distribution

This the simplest vector ratio of independent gamma variables and two approaches are available for deriving its saddlepoint density. First, and simplest, the distribution may be built directly as a vector ratio of independent gammas input into (12.56). This approach supposes that $R_i = U_i/V$ for $i = 1, \ldots, m$ where $\{U_i : i = 1, \ldots, m+1\}$ are independent with $U_i \sim \text{Gamma}(\alpha_i, 1)$ and $V = \sum_{i=1}^{m+1} U_i$. The true density of $R = (R_1, \ldots, R_m)$ is Dirichlet $(\alpha_1, \ldots, \alpha_m; \alpha_{m+1})$ with density

$$f_R(r) = \Gamma(\alpha.)\prod_{i=1}^{m+1} \frac{1}{\Gamma(\alpha_i)} r_i^{\alpha_i - 1} \qquad r_{m+1} = 1 - \sum_{i=1}^{m} r_i \qquad (12.62)$$

where $\alpha. = \sum_{i=1}^{m+1} \alpha_i$. The saddlepoint density turns out to be exact up to Stirling's approximation for the gamma functions involved or

$$\hat{f}_R(r) = \frac{\hat{\Gamma}(\alpha.)}{\Gamma(\alpha.)}\left(\prod_{i=1}^{m+1} \frac{\Gamma(\alpha_i)}{\hat{\Gamma}(\alpha_i)}\right) f_R(r). \qquad (12.63)$$

12.3.4 Serial correlations and the serial correlogram

The lag k serial correlation computed from least square residuals $z^T = (z_1, \ldots, z_n)$ is

$$R_k = \frac{\sum_{i=k+1}^n z_i z_{i-k}}{\sum_{i=1}^n z_i^2} = \frac{z^T D_k z}{z^T z} \qquad k = 1, \ldots, m, \tag{12.64}$$

where $D_k = (d_{ijk})$ is the band matrix with $1/2$ in the kth off-diagonals or $d_{ijk} = 1_{\{|i-j|=k\}}/2$. If the regression is $y = X\beta + \zeta$ with $\zeta \sim \text{Normal}_n(0, \Psi)$ and X of rank p, then the residuals are $z = My = \{I_n - X(X^T X)^{-1} X\}y$. The ratios in (12.64) may be written in the form (12.61) by canonically reducing M using orthogonal matrix $L = (L_1, L_2)$ as

$$M = L \begin{pmatrix} I_{n-p} & 0 \\ 0 & 0 \end{pmatrix} L^T = L_1 L_1^T.$$

Then

$$R_k = \frac{y^T M D_k M y}{y^T M y} = \frac{\varepsilon^T L_1^T D_k L_1 \varepsilon}{\varepsilon^T \varepsilon}$$

where $\varepsilon = L_1^T y \sim \text{Normal}_{n-p}(0, L_1^T \Psi L_1)$. In conforming to (12.61), the value n is replaced by $n - p$, $\Sigma = L_1^T \Psi L_1$, and $A_k = L_1^T D_k L_1$. Note that the matrix L_1 is not uniquely determined when $p > 1$. This means that, unlike the true density of the correlogram, the saddlepoint density depends on the particular choice of L_1.

The support of R on $\mathcal{R} \subset \Re^m$ is the open convex set of values r identified as follows. If

$$Q_k = \begin{pmatrix} 1 & r_1 & \cdots & r_k \\ r_1 & 1 & \cdots & r_{k-1} \\ \vdots & & \ddots & \vdots \\ r_k & \cdots & r_1 & 1 \end{pmatrix} \qquad k = 1, \ldots, m, \tag{12.65}$$

then $\mathcal{R} = \{r : |Q_k| > 0 \quad k = 1, \ldots, m\}$. Butler and Paolella (1998b) prove that the mapping $\hat{s} \leftrightarrow r$ through the saddlepoint equation is a bijection from the open convex domain of $K_{\Psi_r}(s)$, the contiguous neighborhood of 0 over which $|P(s)| > 0$, onto the convex set $\mathcal{R}$. The importance of this is that the domain of definition for $\hat{f}_R$ is exactly the support of R. If, for example, the components of R are the various lagged orders of the Durbin–Watson statistics, then $\mathcal{R}$ is not convex and difficult to identify. In this instance, the density $\hat{f}_R$ is defined on a set larger than the true support and it is difficult to normalize due to the problem of identifying $\mathcal{R}$.

Univariate examples

Various univariate approximations for the lag 1 serial correlation are compared in table 12.5 using four AR(1) models. Suppose the true model is $y_i = \mu + \zeta_i$ with $\zeta_i = \alpha_1 \zeta_{i-1} + \xi_i$

12.3 Distributions for vector ratios

Table 12.5. *Univariate approximations for 95% probabilities of* $\Pr(R_1 \leq r_{10} : \alpha_1)$ *assuming the various values of n and* α_1 *listed in columns 1 and 2. Replicated from table 1 of Butler and Paolella (1998b).*

n	α_1	r_{10}	Lug-Rice	NID	Beta	Normal
10	0	.3471	95.01	94.96	95.01	86.38
	0.5	.6015	94.91	94.42	95.09	64.45
	1.0	.7368	94.75	94.48	Fail[a]	—
	1.1	.7460	94.62	94.64	Fail[a]	—
25	0	.2711	95.01	95.01	95.00	91.24
	0.5	.6467	95.09	95.01	95.01	80.29
	1.0	.8961	94.84	94.42	Fail[a]	—
	1.1	.8848	93.77	92.39	100.0	—

Note: [a]The beta approximation sometimes fails as a result of a negative degree of freedom estimate.

where $\{\xi_i\}$ are i.i.d. Normal (0, 1) errors. Approximations to $\Pr(R_1 \leq r_{10} : \alpha_1)$ are displayed for $n = 10$ and 25 in four settings: null ($\alpha_1 = 0$), stationary ($\alpha_1 = 0.5$), unit root ($\alpha_1 = 1$), and nonstationary ($\alpha_1 = 1.1$). The cutoffs r_{10} have been chosen as the "true" 95th percentiles determined by using the numerical inversion algorithm of Imhof (1961); Lug-Rice is the Lugannani and Rice approximation of (12.2); NID is the numerically integrated and normalized density in (12.9); Beta is Henshaw's (1966) beta distribution approximation that matches third and fourth moments; and Normal is the approximation $R_1 \sim$ Normal $(\alpha_1, 1/n)$ applicable only when $|\alpha_1| < 1$. The Lugannani and Rice approximation is quite accurate even in the unit root and non-stationary settings.

Bivariate examples

Bivariate probability approximations for $\Pr(R_1 \leq r_{10}, R_2 \leq r_{20}; \alpha_1, \alpha_2)$ are given in table 12.6. The cases considered include AR(0)-AR(2) models that require mean correction of the form $y_i = \mu + \zeta_i$ with $\zeta_i = \alpha_1 \zeta_{i-1} + \alpha_2 \zeta_{i-1} + \xi_i$ where $\{\xi_i\}$ are i.i.d. Normal (0, 1) errors. Included are the three stationary settings AR(0) with $\alpha_1 = 0 = \alpha_2$, AR(1) with $\alpha_1 = 0.5$, and AR(2) with $\alpha_1 = 1.2$ and $\alpha_2 = -0.8$. The entries in column NID are computed by numerically integrating and normalizing (12.49) and the Monte Carlo simulations use 2 million replications for each computation. The Normal entries are based on a Normal approximation that uses the theoretical autocorrelation function and Bartlett's formula (Priestley, 1981, §5.3.4).

Bivariate density approximations

Contour plots for the bivariate density of the first two serial correlations (R_1, R_2) are plotted in figures 12.5 and 12.6 in two different null settings, e.g. for i.i.d. sequences which are AR(0). The set of four plots was kindly provided by Marc Paolella.

Table 12.6. *Bivariate approximations for* $\Pr(R_1 \leq r_{10}, R_2 \leq r_{10}; \alpha_1, \alpha_2)$ *using AR(0)-AR(2) models. Replicated from table 2 of Butler and Paolella (1998b).*

n	α_1	α_2	r_{10}	r_{20}	NID	Simulation	Normal
10	0	0	0	0	38.35	38.54	25.00
	0.5	0	0	0	14.76	15.15	3.13
	1.2	-0.8	0.4	0.2	16.63	17.85	0.0344
25	0	0	0	0	33.26	33.25	25.00
	0.5	0	0	0	1.78	1.76	0.189
	1.2	-0.8	0.4	0.2	0.80	0.77	$0.0^5 401$
70	0	0	0	0	29.83	29.87	25.00
	0.5	0	0	0	$0.0^2 65$	$0.0^2 68$	$0.0^4 68$

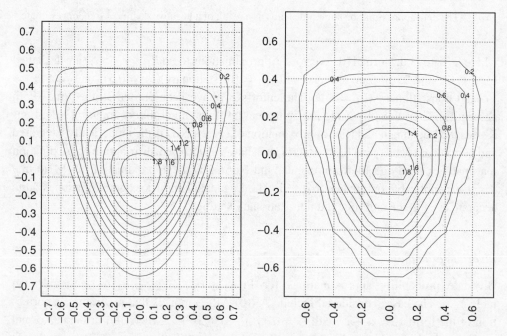

Figure 12.5. Saddlepoint density contours (left) for R_1 (horizontal axis) and R_2 (vertical axis) for i.i.d. data with mean zero and $n = 10$. The corresponding simulated density (right) based on 300,000 replications.

The first setting in figure 12.5 takes $n = 10$ and a mean of zero so there is no X matrix, hence no reduction in degrees of freedom for the sample size. The saddlepoint density on the left accurately replicates the simulated density on the right.

The second setting in figure 12.6 shows the same plots but with a location parameter included so that $X = 1$ is a column of ones. The 10 least squares residuals are correlated and the effective sample size is reduced to nine degrees of freedom. The inclusion of a location parameter along with the determination of serial correlation from residuals has lead to a dramatic shift in the shape of the bivariate density. Only locally about their modes do the plots have any resemblance to the shape of their asymptotic bivariate normal density.

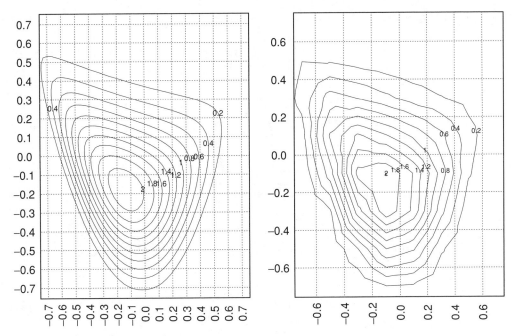

Figure 12.6. The same plots as in figure 12.5 but with a location parameter included in the model.

12.4 Multivariate roots of estimating equations

Let R be the m-dimensional root that solves the set of m-dimensional estimating equations $\Psi_R = 0$. The random vector score

$$\Psi_r^T = \Psi_r^T(X) = (\Psi_1, \ldots \Psi_m) = \{\Psi_1(X; r), \ldots, \Psi_m(X; r)\}$$

is dependent on a random vector X whose dimension is typically much larger than m. To assure that R is well defined, suppose there is a unique root to the equations with probability 1 and also that $|\partial \Psi_r / \partial r^T| < 0$ with probability 1. Weaker conditions were considered in Skovgaard (1990, Eq. 2.3), however under the above conditions, he has shown that the density of R at r is related to the density of Ψ_r at zero according to

$$f_R(r) = f_{\Psi_r}(0) E \left\{ \left\| \frac{\partial \Psi_r}{\partial r^T} \right\| \mid \Psi_r = 0 \right\}. \tag{12.66}$$

Expression (12.66) may be thought of as the average Jacobian transformation when making the transformation from Ψ_r at 0 to R at r. As x varies over $\{x : \Psi_r(x) = 0\}$, the density $f_{\Psi_r}(0)$ remains fixed however the Jacobian is random and (12.66) uses the average Jacobian value that contributes to the event $\Psi_r = 0$.

12.4.1 Multivariate saddlepoint density

A saddlepoint approximation to $f_R(r)$ is obtained by using $\hat{f}_{\Psi_r}(0)$ in place of $f_{\Psi_r}(0)$ and specifying an approximation for the averaged Jacobian. For the latter, Field (1982, Eq. 6) gave the approximation as

$$E\left\{\left\|\frac{\partial \Psi_r}{\partial r}\right\| \mid \Psi_r = 0\right\} \simeq \|J_r(\hat{s})\| \tag{12.67}$$

where

$$J_r(\hat{s}) = \int \cdots \int \frac{\partial \Psi_r(x)}{\partial r^T} g_r(x; \hat{s}) dx \tag{12.68}$$

is an $m \times m$ Jacobian matrix averaged with respect to

$$g_r(x; \hat{s}) = \frac{\exp\{\hat{s}^T \psi_r(x)\} f(x)}{M_{\Psi_r}(\hat{s})}, \tag{12.69}$$

the $\hat{s}^T \Psi_r(X)$-tilted density of X, and $\hat{s}$ is the saddlepoint that solves $K'_{\Psi_r}(\hat{s}) = 0$ in determining $\hat{f}_{\Psi_r}(0)$. With this approximation, Field (1982, Eq. 6) and Field and Ronchetti (1990, Eq. 4.25) provide the saddlepoint density as

$$\hat{f}_R(r) = \hat{f}_{\Psi_r}(0) \|J_r(\hat{s})\| = \frac{M_{\Psi_r}(\hat{s})}{(2\pi)^{m/2} |K''_{\Psi_r}(\hat{s})|^{1/2}} \|J_r(\hat{s})\|. \tag{12.70}$$

Examples

Suppose that R is an estimator of vector parameter θ, perhaps an MLE. Tingley and Field (1990) provide confidence intervals for a smooth scalar functional of θ in this roots of estimating equations context. Strawderman *et al.* (1996) consider examples in which R is the MLE for a generalized linear model with noncanonical link function.

Vector ratios This saddlepoint density generalizes the expressions for vector ratios in (12.49) and (12.50). In both cases, it is only the term $J_r(\hat{s})$ term that needs to be reconciled. The value of (12.68) in the case of the vector ratios in (12.46) is

$$J_r(\hat{s}) = \|\text{diag}\{-E(V_1; \hat{s}), -E(V_1; \hat{s}), -E(V_2; \hat{s})\}\|$$

where $E(V_1; \hat{s})$ denotes the mean of V_1 under the $\hat{s}$-tilted distribution in (12.69). In the notation used for this vector ratio example in (12.52), this is

$$E(V_1; \hat{s}) = \int\int v_1 e^{\hat{s}^T u + \hat{t}^T v} \frac{1}{M_{U,V}(\hat{s}, \hat{t})} f_{U,V}(u, v) du dv$$

$$= \frac{\partial}{\partial \hat{t}_1} K_{U,V}(\hat{s}, \hat{t})$$

so that $J_r(\hat{s})$ agrees with (12.51).

Standard multivariate normal distribution Suppose $X \sim \text{Normal}_m(0, I_m)$ and R is the root of the trivial multivariate estimating equation $\Psi_r = X - r$. Of course, the theory of estimating equations is hardly necessary and the expression in (12.70) leads to the exact normal density with $J_r(\hat{s}) = |I_m| = 1$. If $X_1, \ldots, X_n$ are i.i.d. Normal$_m(0, I_m)$, then the root of $\sum_{i=1}^n (X_i - r)$ again leads to $R \sim \text{Normal}_m(0, I_m/n)$.

Regular exponential family Suppose that a regular exponential family as in (5.1) with $m \times 1$ canonical parameter θ, canonical sufficient X, and score function

$$\Psi_{\hat{\theta}} = l'(\hat{\theta}) = X - c'(\hat{\theta}) = 0.$$

The MLE $\hat{\theta}$ is the root of the estimating equation. Using $\mu_{\hat{\theta}} = c'(\hat{\theta})$, then for fixed $\hat{\theta}$,

$$M_{\Psi_{\hat{\theta}}}(s) = E[\exp\{s^T(X - \mu_{\hat{\theta}})\}; \theta] = \exp\{-s^T \mu_{\hat{\theta}} + c(s + \theta) - c(\theta)\}.$$

The saddlepoint solves

$$0 = K'_{\Psi_{\hat{\theta}}}(\hat{s}) = -\mu_{\hat{\theta}} + c'(\hat{s} + \theta) = -X + c'(\hat{s} + \theta)$$

so that $\hat{s} = \hat{\theta} - \theta$. With $J_r(\hat{s}) = |c''(\hat{\theta})|$, then the saddlepoint density of $\hat{\theta}$ given in (12.70) is

$$\hat{f}(\hat{\theta}; \theta) = \frac{\exp\{-(\hat{\theta} - \theta)^T x + c(\hat{\theta}) - c(\theta)\}}{(2\pi)^{m/2}|j(\hat{\theta})|^{1/2}}|j(\hat{\theta})|$$

$$= (2\pi)^{-m/2}|j(\hat{\theta})|^{1/2}\frac{\mathcal{L}(\theta)}{\mathcal{L}(\hat{\theta})} = p^{\dagger}(\hat{\theta}).$$

12.4.2 Approximating the averaged Jacobian

The approximation made in (12.67) is the only really mysterious portion of (12.70). Explanations for it in the literature are quite difficult to understand and so an alternative motivation has been provided below that uses ideas from the mixed Edgeworth saddlepoint expansions in Barndorff-Nielsen and Cox (1979, §3.3).

Begin by considering what the random $m \times m$ Jacobian matrix in (12.66) means in terms of a reduction of X by sufficiency. Only the two mappings $X \to \Psi_r(X)$ and $X \to \partial \Psi_r(X)/\partial r^T = D_r$ are involved when determining $f_R(r)$. Typically X has a higher dimension than (Ψ_r, D_r) so that these two mappings can be thought of as a reduction in dimension of the sort that occurs with sufficiency. Indeed it is helpful to consider Ψ_r and D_r as the canonical sufficient statistics of an exponential family generated from the data X. In that regard, let T be $m \times m$ so that

$$h_r(x; s, T) = \frac{\exp[s^T \psi_r(x) + \text{tr}\{T D_r(x)\}]}{M_{\Psi_r, D_r}(s, T)} f(x) \tag{12.71}$$

is the $\{s^T \Psi_r, \text{tr}(T D_r)\}$-tilted exponential family assumed to converge in a contiguous neighborhood of $(0, 0)$. If the joint distribution of (Ψ_r, D_r) is determined by a mixed Edgeworth saddlepoint expansion that tilts by $\hat{s}$ in Ψ_r and Edgeworth expands at $T = 0$ in D_r, then the conditional distribution of D_r given Ψ_r is approximately

$$D_r | \Psi_r = 0 \sim \text{Normal}_{m^2}(\hat{\mu}_{D_r}; \hat{\Sigma}_{D_r|\Psi_r=0})$$

where $\hat{\mu}_{D_r} = J_r(\hat{s})$ as explained in the next subsection. Thus the conditional mean of D_r is $J_r(\hat{s})$ and is a small-deviation property of the expansion that allows it to retain excellent asymptotic accuracy in the theoretical development outlined in Field (1982) and Field and Ronchetti (1990).

Mixed Edgeworth saddlepoint expansions

Partial tilting may be used to determine the joint density of (Ψ_r, D_r) from the tilted exponential family in (12.71). Consider tilting as in (5.41) but using the parameter $(s, T) = (\hat{s}, 0)$ in which there is only a partial $\hat{s}$-tilt in the Ψ_r component such that

$$0 = E(\Psi_r; \hat{s}) = \partial K_{\Psi_r, D_r}(\hat{s}, 0)/\partial \hat{s}^T = K'_{\Psi_r}(\hat{s}).$$

Then the ratio of densities for sufficient statistic (Ψ_r, D_r) is the ratio of tilted likelihoods or

$$g_{\Psi_r, D_r}(\psi, d) = \frac{h_r(x; 0, 0)}{h_r(x; \hat{s}, 0)} g_{\Psi_r, D_r}(\psi, d; \hat{s}, 0)$$
$$= \exp\{K_{\Psi_r, D_r}(\hat{s}, 0) - \hat{s}^T \psi\} g_{\Psi_r, D_r}(\psi, d; \hat{s}, 0). \quad (12.72)$$

A general Edgeworth expression, as used in (7.53) and given in Barndorff-Nielsen and Cox (1979, Eq. 4.4), is applied to the tilted density for Ψ_r, D_r given on the right. The leading term is a Normal$_{m+m^2}\{(0, \hat{\mu}_{D_r})^T, \hat{\Sigma}\}$ density in which the $m \times m$ mean of D_r is

$$\hat{\mu}_{D_r} = \left.\frac{\partial K_{\Psi_r, D_r}(\hat{s}, T)}{\partial T}\right|_{T=0} = \int \cdots \int \frac{\partial \Psi_r(x)}{\partial r^T} \frac{\exp\{\hat{s}^T \psi_r(x)\} f(x)}{M_{\Psi_r}(\hat{s})} dx = J_r(\hat{s}).$$

Using vec(T) to stack the columns of T, the covariance $\hat{\Sigma}$ is

$$\left.\begin{pmatrix} K''_{\Psi_r}(\hat{s}) & \partial^2 K_{\Psi_r, D_r}(\hat{s}, T)\partial \hat{s}\partial(\text{vec } T)^T \\ \partial^2 K_{\Psi_r, D_r}(\hat{s}, T)/\partial(\text{vec } T)\partial \hat{s}^T & \partial^2 K_{\Psi_r, D_r}(\hat{s}, T)/\partial(\text{vec } T)\partial(\text{vec } T)^T \end{pmatrix}\right|_{T=0}$$

and $|\hat{\Sigma}| = |K''_{\Psi_r}(\hat{s})||\hat{\Sigma}_{D_r|\Psi_r=0}|$. Substituting into (12.72), the density for Ψ_r, D_r at $(0, d)$ is

$$g_{\Psi_r, D_r}(0, d) = \frac{M_{\Psi_r}(\hat{s}) \exp\left\{-\frac{1}{2} \text{vec}(d - \hat{\mu}_{D_r})^T \hat{\Sigma}^{-1}_{D_r|\Psi_r=0} \text{vec}(d - \hat{\mu}_{D_r})\right\}}{(2\pi)^{(m+m^2)/2} |K''_{\Psi_r}(\hat{s})|^{1/2} |\hat{\Sigma}_{D_r|\Psi_r=0}|^{1/2}}$$
$$= \hat{f}_{\Psi_r}(0) \times \text{Normal}_{m^2}\{d; \hat{\mu}_{D_r}, \hat{\Sigma}_{D_r|\Psi_r=0}\}. \quad (12.73)$$

The last factor in (12.73) is the multivariate normal density approximation for the conditional distribution of D_r given $\Psi_r = 0$.

The final portion of the argument is to approximate the conditional expectation of the determinant of $\partial \Psi_r(X)/\partial r^T$ by the determinant of the conditional expectation. At first this might seem to be a crude approximation. However, the δ-method allows for the determination of the asymptotic 1-dimensional conditional distribution for the random determinant; the mean of this asymptotic distribution is $|\hat{\mu}_{D_r}| = |J_r(\hat{s})|$.

12.4.3 Equivariance of the multivariate saddlepoint density

Exercises 5 and 6 and the discussion in section 12.1.3 determine that the saddlepoint density for the doubly noncentral F is a Jacobian transformation of the saddlepoint density of the equivalent doubly noncentral beta. Both of these random variables can be represented as the roots of estimating equations however the estimating equations are not the same.

The proposition below shows that the equivariance established for the F and beta is a much more general property of the saddlepoint density given in (12.70).

12.4 Multivariate roots of estimating equations

Proposition 12.4.1 *The multivariate saddlepoint density is equivariant under nonsingular transformation of the root. Suppose that m-vector R solves $\Psi_R = 0$ and m-vector Q solves $\Omega_Q = 0$ with $Q = g(R)$ a bijection. With $q = g(r)$, suppose also that the two sets of estimating equations are related according to*

$$\Omega_q = \Omega_{g(r)} = H_r \Psi_r \qquad (12.74)$$

where H_r is a fixed $m \times m$ nonsingular transformation for each r. Then the saddlepoint densities are related in the same manner as their true densities, e.g.

$$\hat{f}_R(r) = \hat{f}_Q(q) \left\| \partial q / \partial r^T \right\|.$$

Proof. The linearity in (12.74) assures that

$$\hat{f}_{\Psi_r}(0) = \hat{f}_{\Omega_q}(0) \|H_r\| \qquad (12.75)$$

as in (3.39). To compute $J_r(\hat{s})$, note that

$$\frac{\partial \Psi_r}{\partial r^T} = \frac{\partial}{\partial r^T}(H_r^{-1}\Omega_q) = H_r^{-1}\frac{\partial \Omega_q}{\partial q^T}\frac{\partial q}{\partial r^T} + H_r^{-1}\frac{\partial H_r}{\partial r^T}H_r^{-1}\Omega_q$$

$$= H_r^{-1}\frac{\partial \Omega_q}{\partial q^T}\frac{\partial q}{\partial r^T} + H_r^{-1}\frac{\partial H_r}{\partial r^T}\Psi_r.$$

The expectation of this needs to be computed under the $\hat{s}^T \Psi_r$-tilted distribution for X. Under this distribution

$$E(\Psi_r; \hat{s}) = K'_{\Psi_r}(\hat{s}) = 0$$

so that

$$E\left(\frac{\partial \Psi_r}{\partial r^T}; \hat{s}\right) = H_r^{-1} E\left(\frac{\partial \Omega_q}{\partial q^T}; \hat{s}\right) \frac{\partial q}{\partial r^T}. \qquad (12.76)$$

For the saddlepoint density of Q, the expectation in (12.76) needs to be with respect to the $\hat{t}^T \Omega_q$-tilt of the distribution of X. Since

$$K_{\Omega_q}(t) = E(e^{t^T \Omega_q}) = E(e^{t^T H_r \Psi_r}) = K_{\Psi_r}(H_r^T t),$$

$\hat{s} = H_r^T \hat{t}$ and the $\hat{t}^T \Omega_q$-tilted distribution of X is the same as the $\hat{s}^T \Psi_r$-tilted distribution of X. Thus the expectation is the required factor. Putting (12.75) and (12.76) together gives

$$\hat{f}_R(r) = \hat{f}_{\Psi_r}(0) \left\| E\left(\frac{\partial \Psi_r}{\partial r^T}; \hat{s}\right) \right\| = \hat{f}_{\Omega_q}(0)\|H_r\| \times \left\| H_r^{-1} E\left(\frac{\partial \Omega_q}{\partial q^T}; \hat{t}\right) \frac{\partial q}{\partial r^T} \right\|$$

$$= \hat{f}_{\Omega_q}(0) \left\| E\left(\frac{\partial \Omega_q}{\partial q^T}; \hat{t}\right) \right\| \left\| \frac{\partial q}{\partial r^T} \right\| = \hat{f}_Q(q) \left\| \frac{\partial q}{\partial r^T} \right\|.$$

The doubly noncentral F and beta densities

Suppose $R = (U/n_1)/(V/n_2)$ is a noncentral F with U and V as noncentral χ^2 variables having the degrees of freedom indicated. Ratio R is also the root of

$$\Psi_R = U/n_1 - RV/n_2 = 0.$$

Alternatively, $Q = U/(U + V)$ is the equivalent noncentral beta that solves $\Omega_Q = 0$ with components

$$\Omega_Q = U - Q(U + V) = 0.$$

Then $g(r) = r/(r + n_2/n_1)$ and

$$\Omega_q = \Psi_r/(r/n_2 + 1/n_1)$$

of the form (12.74).

The Dirichlet distribution and multivariate F

The multivariate generalization of the last example considers the components of $R = (R_1, \ldots, R_m)$, with the Dirichlet $(\alpha_1, \ldots, \alpha_m; \alpha_{m+1})$ distribution. Vector R comprises the roots of the m estimating equations $\Psi_R = 0$ whose components are

$$\Psi_i = U_i - r_i V = 0 \qquad i = 1, \ldots, m. \tag{12.77}$$

Here, $\{U_i\}$ are independent with $U_i \sim \text{Gamma}(\alpha_i, 1)$ and $V = \sum_{i=1}^{m+1} U_i$. Alternatively, the components of $Q = (Q_1, \ldots, Q_m)$ with a multivariate F distribution are the roots of the m estimating equations $\Omega_Q = 0$ with components

$$\Omega_i = U_i/\alpha_i - q_i U_{m+1}/\alpha_{m+1} = 0 \qquad i = 1, \ldots, m. \tag{12.78}$$

Exercise 16 asks the reader to show that the conditions of the proposition are met in this context.

Two immediate conclusions are (i) the saddlepoint density for the multivariate F density is a proper Jacobian transformation of the saddlepoint density for the Dirichlet density, and (ii) the former is exact up to Stirling's approximations for the gamma functions involved as a consequence of (i) and (12.63).

The ANOVA table

The Dirichlet density occurs in the ANOVA table when the vector for the various sums of squares is divided by the total sums of squares. In the "completely null" setting in which the ANOVA data are i.i.d., then this vector has a "central" Dirichlet density. If a particular set of F ratios is under consideration, then only those sum of squares that are used in the F ratios should be divided by their sum to get the central Dirichlet. With this in mind, the following conclusion may be drawn.

Proposition 12.4.2 *For any set of F ratios constructed from the ANOVA table, the saddlepoint density is exact upon normalization in the "completely null" setting. For a setting in which at least one F ratio is noncentral, then the multivariate F density is no longer exact, but it is the proper Jacobian transformation of the noncentral Dirichlet vector of the ANOVA table.*

The Proposition also allows for F ratios with different denominators although the argument above, derived from the multivariate F distribution using a common denominator, does not suffice. See Exercise 16.

12.4.4 Curved exponential families

In such settings the MLE solves an estimating equation and the roots of estimating approach provides a marginal density for the MLE. The resulting density approximation in (12.80) was first derived in Skovgaard (1985) and has also been discussed by Jensen (1995) both from an estimating equation (pp. 110–112) and a marginalization (pp. 98–100) points of view. This density is slightly different from the p^* density previously considered since the latter approximation is conditioned on a likelihood or affine ancillary determined by the nature of the curved exponential family.

In the notation of section 7.3, the $p \times 1$ MLE $\hat{\xi}$ solves the $p < m$ equations

$$0 = \dot{l}_\xi = \dot{\theta}_\xi^T (x - \mu_\xi) \tag{12.79}$$

where sufficient statistic x has dimension m and $\dot{\theta}_\xi^T$ is $p \times m$. The approximate marginal density of $\hat{\xi}$ is

$$\hat{f}(\hat{\xi}; \xi) = (2\pi)^{-p/2} |\tilde{\imath}|^{-1/2} |\tilde{\jmath}| \frac{\mathcal{L}(\theta_\xi)}{\mathcal{L}(\tilde{\theta})}. \tag{12.80}$$

Here the information matrices, $\tilde{\imath}$ in (12.83) and $\tilde{\jmath}$ in (12.84), and the canonical parameter estimate $\tilde{\theta}$ in (12.82), that solves (12.81), are based on the MLE $\hat{\xi}$, but are not the MLEs of their respective parameters. Were they to be based on the MLE, then (12.80) would be approximately $p^\dagger(\hat{\xi}|a; \xi)$ in theorem 7.3.2 for affine ancillary a, or $p^\dagger(\hat{\xi}|a_\chi^0; \xi)$ in theorem 7.3.3 for likelihood ancillary a_χ^0.

Derivation of (12.80)

For fixed $\hat{\xi}$, the MGF of the estimating equation in (12.79) is

$$M(s) = E(e^{s^T \dot{l}_\xi}) = E\{e^{s^T \dot{\theta}_\xi^T (X - \mu_\xi)}\} = \exp\{-s^T \dot{\theta}_\xi^T \mu_\xi + c(\dot{\theta}_\xi s + \theta_\xi) - c(\theta_\xi)\}.$$

The p equations for saddlepoint $\hat{s}$ are

$$0 = K'(\hat{s}) = -\dot{\theta}_\xi^T \mu_\xi + \dot{\theta}_\xi^T c'(\dot{\theta}_\xi \hat{s} + \theta_\xi)$$

or

$$\dot{\theta}_\xi^T \{x - c'(\dot{\theta}_\xi \hat{s} + \theta_\xi)\} = 0. \tag{12.81}$$

The MLE equation for $\hat{\xi}$ is the same equation but with the argument of c' as $\theta_{\hat{\xi}}$ so the domain of c' is the whole canonical parameter manifold and its image is the mean parameter manifold. However, in the case of (12.81), the domain of c' is restricted and is the θ_ξ-translation of the p-dimensional column space of $\dot{\theta}_\xi$. Thus if

$$\tilde{\theta} = \dot{\theta}_\xi \hat{s} + \theta_\xi, \tag{12.82}$$

then in the mean parameter space

$$\tilde{\mu} = c'(\tilde{\theta}) = \mu_\xi + Bv$$

where the $m - p$ columns of B are a basis for the null space of $\dot{\theta}_\xi^T$, e.g. $\dot{\theta}_\xi^T B = 0$, as would be needed to obtain the solution in (12.81). Jensen (1995, p. 98) provides further discussion of these issues.

The saddlepoint density depends on

$$K(\hat{s}) = -\hat{s}^T \dot{\theta}_\xi^T \mu_{\hat{\xi}} + c(\tilde{\theta}) - c(\theta_\xi) = -\hat{s}^T \dot{\theta}_\xi^T x + c(\tilde{\theta}) - c(\theta_\xi)$$
$$= -(\tilde{\theta} - \theta_\xi)^T x + c(\tilde{\theta}) - c(\theta_\xi) = \ln\{\mathcal{L}(\theta_\xi)/\mathcal{L}(\tilde{\theta})\}.$$

The Hessian is

$$\tilde{\imath} = K''(\hat{s}) = \dot{\theta}_\xi^T c''(\tilde{\theta}) \dot{\theta}_\xi. \tag{12.83}$$

The averaged Jacobian $J_{\hat{\xi}}(\hat{s})$ is $\|E(\ddot{l}_{\hat{\xi}})\|$ taken with respect to the $\hat{s}^T \dot{l}_{\hat{\xi}}$-tilted density of X. Using (7.30), then

$$J_{\hat{\xi}}(\hat{s}) = \left\| E\{-i_{\hat{\xi}} + (X - \mu_{\hat{\xi}}) \odot \ddot{\theta}_{\hat{\xi}}\} \right\|$$
$$= \left\| i_{\hat{\xi}} - (\tilde{\mu} - \mu_{\hat{\xi}}) \odot \ddot{\theta}_{\hat{\xi}} \right\| = |\tilde{j}|. \tag{12.84}$$

To see that $\tilde{\mu}$ is the correct mean for X, note that the tilted mean is

$$\frac{1}{M(\hat{s})} \int x \exp\left\{\hat{s}^T \dot{\theta}_\xi^T (x - \mu_{\hat{\xi}}) + \theta_\xi^T x - c(\theta_\xi) - d(x)\right\} dx.$$

This is the mean of the curved exponential family with canonical parameter $\tilde{\theta}$ so that $\tilde{\mu} = c'(\tilde{\theta})$.

Univariate CDF for $\hat{\xi}$

In the case $\dim(\hat{\xi}) = 1$, the Temme approximation may be used to convert density (12.80) into the approximate CDF

$$F(\hat{\xi}; \xi) = \Phi(\hat{w}) + \phi(\hat{w})(1/\hat{w} - 1/\hat{u}) \qquad \hat{s} \neq 0$$

where

$$\hat{w} = \text{sgn}(\hat{\xi} - \xi)\sqrt{-2\ln\{\mathcal{L}(\theta_\xi)/\mathcal{L}(\tilde{\theta})\}} \tag{12.85}$$
$$\hat{u} = \hat{s}\sqrt{\tilde{\imath}}.$$

The steps to this argument are outlined in Exercise 18.

Gamma hyperbola

Suppose a random sample of size n is drawn from this $(2, 1)$ curved exponential family as described in section 7.3.1. The regular exponential family within which it is constrained has canonical parameter $\theta = (\theta_1, \theta_2)^T$ and likelihood

$$\mathcal{L}(\theta) = (-\theta_1)^n (-\theta_2)^n \exp(\theta_1 x_1 + \theta_2 x_2)$$

where $\theta_1 = -\xi$ and $\theta_2 = -1/\xi$ for $\xi > 0$. The canonical sufficient statistics are $X_1 \sim$ Gamma (n, ξ) independently of $X_2 \sim$ Gamma $(n, 1/\xi)$. The MLE is $\hat{\xi} = \sqrt{x_2/x_1}$ whose distribution is $\xi \times \sqrt{F_{2n,2n}}$. Since the saddlepoint procedures are scale equivariant, the full generality of the example is maintained by taking $\xi = 1$ as the true parameter value.

The true density is

$$f(\hat{\xi}) = \frac{2\Gamma(2n)}{\Gamma(n)^2} \frac{\hat{\xi}^{2n-1}}{(1+\hat{\xi}^2)^{2n}} \qquad \hat{\xi} > 0 \tag{12.86}$$

and Exercise 19 provides the intermediate steps for showing that the saddlepoint density in (12.80) is (12.86) with the Gamma function terms replaced by their Stirling approximations. The saddlepoint CDF in (12.85) has

$$\hat{w} = \mathrm{sgn}(\hat{\xi} - 1)\sqrt{-4n \ln\{2\hat{\xi}/(1+\hat{\xi}^2)\}} \tag{12.87}$$
$$\hat{u} = \sqrt{2n}(\hat{\xi}^2 - 1)/(1+\hat{\xi}^2).$$

The exactness of the saddlepoint density upon normalization should not be surprising because the derivations used for the curved exponential theory are exactly those that were used to determine the density of a ratio or of any other root of estimating equation. Since the saddlepoint densities for the central beta in (12.11) and the central F in (12.15) are exact upon normalization and $\hat{\xi} = F_{2n,2n}^{1/2}$ is a bijection, then exactness for the density of $\hat{\xi}$ is assured by Proposition 12.4.1. The same argument applies to the CDF approximation in (12.87) which must also be the result from approximating $\Pr(X_2 - \hat{\xi}^2 X_1 \leq 0)$ by using the Lugannani and Rice formula.

12.4.5 Matrix T

Suppose $W \sim$ Wishart (n, I_k) with $n \geq k$ and is partitioned as

$$W = \begin{pmatrix} W_{11} & W_{12} \\ W_{21} & W_{22} \end{pmatrix}$$

where W_{11} is $l \times l$ and W_{12} is $l \times m$ with $l + m = k$. The $l \times m$ matrix $R = W_{11}^{-1} W_{12}$ is the least squares estimator when regressing normal data from the m-dimensional block 2 onto data from l-dimensional block 1. Saddlepoint approximation to the density of R has not been previously addressed and provides a more technically demanding example. Phillips (1985), however, has considered the exact inversion theory required to determine its density. Essentially it might be considered as a "matrix ratio" for which he extended the inversion expression of Geary (1944) in (12.4) and (12.5). The density of R is easily derived as

$$f(R) = \frac{\Gamma_l\{(n+m)/2\}}{\pi^{lm/2} \Gamma_l(n/2)} |R^T R + I_m|^{-(n+m)/2} \qquad R \in \Re^{lm}.$$

A saddlepoint approximation uses the root of the ml estimating equations

$$\Psi_R = W_{12} - W_{11} R = 0. \tag{12.88}$$

For $m \times l$ matrix S, in a suitably small neighborhood of $0 \in \Re^{lm}$, the MGF of Ψ_R is derived for the MGF of a Wishart by noting that

$$M_{\Psi_R}(S) = E\{e^{\mathrm{tr}(S\Psi_R)}\} = E\{e^{\mathrm{tr}(TW)}\}$$
$$= |I_k - 2T|^{-n/2} = |I_k + RS + S^T R^T - S^T S|^{-n/2} = |A(S)|^{-n/2}$$

where
$$T = \frac{1}{2}\begin{pmatrix} -(RS + S^T R^T) & S^T \\ S & 0 \end{pmatrix}.$$

Note that T is symmetric and the use of a nonsymmetric matrix leads to an incorrect expression.

The saddlepoint and Hessian determinant can be very difficult to derive without the use of differential elements as outlined in Magnus and Neudecker (1988, chap. 5). The saddlepoint $\hat{S}$ satisfies

$$\begin{aligned} 0 = dK_{\Psi_R}(\hat{S}) &= -n/2 \operatorname{tr}\{A(\hat{S})^{-1} dA(\hat{S})\} \\ &= -n/2 \operatorname{tr}[A(\hat{S})^{-1}\{Rd\hat{S} + (d\hat{S})^T R^T - (d\hat{S})^T \hat{S} - \hat{S}^T d\hat{S}\}] \\ &= -n \operatorname{tr}\{A(\hat{S})^{-1}(R - \hat{S}^T)d\hat{S}\} \end{aligned} \quad (12.89)$$

which is satisfied by $\hat{S} = R^T$. This gives

$$M_{\Psi_R}(\hat{S}) = |A(\hat{S})|^{-n/2} = |R^T R + I_m|^{-n/2}. \quad (12.90)$$

The Hessian is derived by computing the second differential or yet another differential of (12.89) which is

$$d^2 K_{\Psi_R}(S) = n \operatorname{tr}\{A(S)^{-1} dA(S) A(S)^{-1}(R - S^T) dS\} + n \operatorname{tr}\{A(S)^{-1}(dS)^T dS\}$$

so that

$$d^2 K_{\Psi_R}(\hat{S}) = n \operatorname{tr}\{A(\hat{S})^{-1}(d\hat{S})^T d\hat{S}\} = n \operatorname{tr}\{d\hat{S} A(\hat{S})^{-1}(d\hat{S})^T\}.$$

Using the identity

$$\operatorname{tr}(ABC) = \{\operatorname{vec}(A^T)\}^T (I \otimes B) \operatorname{vec}(C)$$

from Muirhead (1982, lemma 2.2.3), then

$$d^2 K_{\Psi_R}(\hat{S}) = n \{\operatorname{vec}(d\hat{S})^T\}^T \{I_m \otimes A(\hat{S})^{-1}\} \operatorname{vec}(d\hat{S})^T$$

and

$$|K''_{\Psi_R}(\hat{S})| = n^{lm}|I_m \otimes A(\hat{S})^{-1}| = n^{lm}|R^T R + I_m|^{-m}. \quad (12.91)$$

Finally the averaged Jacobian is computed by taking the differential of Ψ_R or $d\Psi_R = -W_{11} dR$. Then

$$\operatorname{vec}(d\Psi_R) = -(I_m \otimes W_{11}) \operatorname{vec}(dR)$$

as determined from Muirhead (1982, lemma 2.2.3), so that

$$|J_R(\hat{S})| = \left\| E\left\{ \frac{\partial \operatorname{vec}(\Psi_R)}{\partial \operatorname{vec}(dR)^T}; \hat{S} \right\} \right\| = |I_m \otimes E(W_{11}; \hat{S})|.$$

Here the mean of W_{11} is computed using the $\hat{T}$-tilt of a Wishart (n, I_k) which makes $W \sim$ Wishart $\{n, (I_k - 2\hat{T})^{-1}\}$. The $\hat{T}$-tilted mean of W_{11} is the upper left $l \times l$ block of the mean of W or

$$n(I_k - 2\hat{T})^{-1} = n \begin{pmatrix} I_l + 2RR^T & -R \\ -R^T & I_m \end{pmatrix}^{-1}$$

which, from block inversion, works out to be $n(I_l + RR^T)^{-1}$. Then
$$|J_R(\hat{S})| = |I_m \otimes n(I_l + RR^T)^{-1}| = n^{lm}|I_m + R^T R|^{-m}. \tag{12.92}$$

Combining (12.90), (12.91), and (12.92) gives
$$\hat{f}(R) = \left(\frac{n}{2\pi}\right)^{lm/2} |I_m + R^T R|^{-(n+m)/2} = f(R)\frac{\Gamma_l(n/2)(n/2)^{lm/2}}{\Gamma_l\{(n+m)/2\}}$$

so the density is exact upon normalization.

12.5 The conditional CDF of R_m given $R_1, \ldots, R_{m-1}$

A double-saddlepoint approximation for this CDF has been derived in Butler and Paolella (1998b) for the lag m serial correlation given lags 1 to $m-1$ from the correlogram. This setting is an example of a more general context for roots of estimation equations in which the root $R_- = (R_1, \ldots, R_{m-1})^T$, that solves the first $m-1$ equations, remains the same value when an mth equation is included and $R = (R_-, R_m)^T$ is its root.

The theory for the more general context is presented and derived below. Numerical applications include a sequence of forward selection significance tests for serial correlations of the correlogram intended to determine the lag order of an autoregressive time series. Gatto and Jammalamadaka (1999) have also derived a conditional CDF approximation in this context, however their expressions have the limited applicability explained in section 12.5.2 below.

12.5.1 The double-saddlepoint CDF approximation

Suppose R_- solves the $m - 1$ equations
$$0 = \Psi_{R_-}^- = \Psi_{R_-}^-(X; R_-) = \{\Psi_1(X; R_-), \ldots, \Psi_{m-1}(X; R_-)\}^T \tag{12.93}$$

and $R = (R_-^T, R_m)^T$ solves the m equations
$$0 = \Psi_R = \{\Psi_{R_-}^-(X; R_-)^T, \Psi_m(X; R_-, R_m)\}^T. \tag{12.94}$$

Most notably, (12.93) does not involve R_m, (12.94) involves R_m only in the mth equation Ψ_m, and R_- is the same root with probability 1 in both sets of equations.

Further notational simplification is needed in order to assure clarity of presentation. It is necessary to distinguish a generic value of $r = (r_-, r_m)$, in which only r_m is free to vary, from $r_0 = (r_-, r_{m0})$, the point of evaluation for
$$F(r_{m0}|r_-) = \Pr(R_m \leq r_{m0}|R_- = r_-).$$

Denote the MGF of $\Psi_r = \{(\Psi_{r_-}^-)^T, \Psi_{m;r}\}^T$ as
$$M_r(s, t) = E\{\exp(s^T \Psi_{r_-}^- + t \Psi_{m;r})\}$$

with CGF $K_r(s, t) = \ln M_r$. The CGF of $\Psi_{r_-}^-$ is now the marginal CGF $K_r(s, 0)$. At r_0, the "numerator" and "denominator" saddlepoints are $(\hat{s}, \hat{t})$ and $\hat{s}_0$ respectively and solve
$$\partial K_{r_0}(\hat{s}, \hat{t})/\partial(\hat{s}, \hat{t}) = 0 \tag{12.95}$$
$$\partial K_{r_0}(\hat{s}_0, 0)/\partial \hat{s}_0 = 0.$$

The averaged Jacobians used for the saddlepoint densities involved are the $m \times m$ and $(m-1) \times (m-1)$ matrices

$$J(s,t) = J_r(s,t) = \int \cdots \int \frac{\partial \Psi_r}{\partial r^T} g_r(x; s, t) dx$$

$$J^-(s,t) = J_{r_-}^-(s,t) = \int \cdots \int \frac{\partial \Psi_{r_-}^-}{\partial r_-^T} g_r(x; s, t) dx$$

where $g_r(x; s, t)$ is the $\{s^T \Psi_{r_-}^- + t \Psi_{m;r}\}$-tilted exponential family

$$g_r(x; s, t) = \exp\{s^T \Psi_{r_-}^-(x; r_-) + t \Psi_m(x; r)\} f(x) / M_r(s, t).$$

Note that $J^-(s,t)$ is the upper left $(m-1) \times (m-1)$ sub-matrix of $J(s,t)$.

The double-saddlepoint approximation has the familiar-looking form

$$\hat{F}(r_{m0}|r_-) = \Phi(\hat{w}) + \phi(\hat{w})(1/\hat{w} - 1/\hat{u}) \qquad \hat{t} \neq 0 \tag{12.96}$$

$$\hat{w} = \mathrm{sgn}(\hat{t})\sqrt{2\{K_{r_0}(\hat{s}_0, 0) - K_{r_0}(\hat{s}, \hat{t})\}} \tag{12.97}$$

$$\hat{u} = \hat{t} \left\{ \frac{|K_{r_0}''(\hat{s}, \hat{t})|}{|K_{r_0;ss}''(\hat{s}_0, 0)|} \right\}^{1/2} \frac{||J^-(\hat{s}_0, 0)||}{||J^-(\hat{s}, \hat{t})||} \tag{12.98}$$

where $r_0 = (r_-^T, r_{m0})^T$, K_{r_0}'' denotes the full $m \times m$ Hessian in (s, t), and $K_{r_0;ss}''(\hat{s}_0, 0) = \partial^2 K_{r_0}(\hat{s}_0, 0)/\partial \hat{s}_0 \partial \hat{s}_0^T$.

12.5.2 Nonrandom Jacobians

Consider the setting in which the two Jacobians involved when transforming $r_- \leftrightarrow \Psi_{r_-}^- = \Psi^-$ and $r \leftrightarrow \Psi_r$ have degenerate conditional distributions. Specifically, suppose that $\partial \Psi_{r_-}^-/\partial r_-^T$ has a degenerate distribution given $\{\Psi_{r_-}^- = 0\}$, and $\partial \Psi_r/\partial r^T$ has a degenerate distribution given $\{\Psi_r = 0\}$. Then it is shown below that

$$\Pr(R_m \leq r_{m0}|R_- = r_-) = \Pr(\Psi_{m;r} \leq 0|\Psi_{r_-}^- = 0). \tag{12.99}$$

The importance in having such an identity is that, when equipped with $K_r(s, t)$, the joint CGF of Ψ_r, then Skovgaard's approximation (4.12) is directly applicable to the right side of (12.99) without the need to first consider the complication of the densities that occur. A direct application of (4.12) leads to the value of $\hat{w}$ in (12.97) but the value of $\hat{u}$ in (12.98) no longer includes the ratio $||J^-(\hat{s}_0, 0)||/||J^-(\hat{s}, \hat{t})||$ which would attain equal values under these degeneracy conditions. Thus, if the equivalence in (12.99) holds, all the previous conditional theory may be applied and the derivation in section 12.5.5 below is unnecessary.

The error in Gatto and Jammalamadaka (1999) occurs when assuming that (12.99) holds even when the Jacobians do not necessarily have degenerate distributions. It should be intuitively clear that when the Jacobians are random, then some aspect of them needs to be taken into account for settings in which X is absolutely continuous. Interestingly, the degeneracy conditions stated above hold for all the examples considered by Gatto and Jammalamadaka (1999) in which X is absolutely continuous. This is, of course, why there was no indication of any particular problem with their approximations.

Wilks' likelihood ratio test Λ in MANOVA

The null distribution of $\ln \Lambda$ from (11.1) is $\ln \Lambda \sim \sum_{i=1}^{k} \ln \beta_i$ where $\{\beta_i\}$ represent independent beta variables. A double saddlepoint approximation for the CDF of $\ln \Lambda$ was developed in section 11.1.1 and that development is essentially a prototype for a large class of settings in which the Jacobians are degenerate.

Recall that the distribution of each $\ln \beta_i$ was constructed from two independent chi squares, χ_{i1} and χ_{i2}, by conditioning $\ln \chi_{1i}$ on $\chi_{i1} + \chi_{i2} = 1$. This conditional distribution setting can also be constructed from the $m = k + 1$ estimating equations

$$\Psi_i = \chi_{i1} + \chi_{i2} - r_i \qquad i = 1, \ldots k$$

$$\Psi_{k+1} = \sum_{i=1}^{k} \ln \chi_{i1} - r_{k+1}.$$

The p-value is $\Pr(R_{k+1} > r_{k+1} | r_-)$ where $r_- = (r_1, \ldots, r_k)^T = (1, \ldots, 1)^T$ and r_{k+1} is the observed value for $\ln \Lambda$. Note that the Jacobians are

$$\partial \Psi^- / \partial r_-^T = -I_k \qquad \partial \Psi_{m;r_{k+1}} / \partial r_{k+1} = -1$$

and not random. This particular context leads to an expression for $K_{\Psi_r}(s, t)$ that is linear in (s, t).

Goodness of fit tests

Gatto and Jammalamadaka (1999) use the spacings between order statistics to determine saddlepoint approximations for various goodness of fit tests. If $U_1, \ldots, U_{n-1}$ is an i.i.d. sample from a Uniform $(0, 1)$, then the order statistics from $\{U_i\}$ and the values 0, and 1 determine the spacings $D = (D_1, \ldots, D_n)$ with $\sum_{i=1}^{n} D_i = 1$. The distribution of D is an $(n-1)$-dimensional Dirichlet $(1, \ldots, 1)$ distribution. They consider a range of test statistics of the form $R_2 = \sum_{i=1}^{n} g(D_i)/n$ for various choices of g. The distribution of R_2 is constructed from $\{E_i : i = 1, \ldots, n\}$, a set of i.i.d. Exponential (1), as the distribution of $\sum_{i=1}^{n} g(E_i)/n$ given $\sum_{i=1}^{n} E_i = 1$. The two estimating equations required for constructing this distribution are

$$\Psi_1 = \sum_{i=1}^{n} E_i - r_1 \qquad \Psi_2 = \sum_{i=1}^{n} g(E_i)/n - r_2.$$

and $\Pr(R_2 > r_2 | R_1 = 1)$ gives the p-value of the test.

Proof of (12.99) in the degenerate case

With $\Psi^- = \Psi^-_{r_-}$, and degenerate values for $\partial \Psi^- / \partial r_-^T$ with $\Psi^-_{r_-} = 0$ and also for $\partial \Psi_r / \partial r^T$ with $\Psi_r = 0$ then

$$f_{R_-}(r_-) = f_{\Psi^-}(0) \left\| \partial \Psi^- / \partial r_-^T \right\| \qquad (12.100)$$

$$f_R(r) = f_{\Psi_r}(0) \left\| \partial \Psi_r / \partial r^T \right\|$$

$$= f_{\Psi_r}(0) \left\| \partial \Psi^-_r / \partial r_-^T \right\| \left\| \partial \Psi_{m;r} / \partial r_m \right\|.$$

Under the degeneracy, the determinant $\|\partial \Psi_r/\partial r^T\|$ cancels out in the ratio below and integration gives

$$\Pr(R_m \leq r_{m0}|r_-) = \int_{-\infty}^{r_{m0}} \frac{f_R(r)}{f_{R_-}(r_-)} dr_m = \int_{-\infty}^{r_{m0}} \frac{f_{\Psi_r}(0)}{f_{\Psi_-}(0)} \|\partial \Psi_{m;r}/\partial r_m\| dr_m$$

$$= \int_{-\infty}^{0} \frac{f_{\Psi_r}(0)}{f_{\Psi_-}(0)} d\Psi_{m;r} = \Pr(\Psi_{m;r} \leq 0|\Psi_{r_-}^- = 0).$$

12.5.3 Discrete X

Such settings involve mass functions rather than densities and so the continuous theory with its Jacobian factors is not formally applicable. Gatto and Jammalamadaka (1999) considered several such examples involving goodness of fit tests. In their settings and indeed in many such discrete settings, the conditional probability equality of (12.99) holds because Jacobians are not used when transforming discrete probability. The direct use of the Skovgaard formula for the right hand side of (12.99) without Jacobian correction is preferred.

Jackknifed ratios of means

The bootstrap distribution for a root of estimating equations was presented in section 12.2.3 based on the developments in Davison and Hinkley (1988). An alternate jackknifing approach developed in Booth and Butler (1990, §4) is now considered for the same context. Consider the true permutation distribution for the delete-k jackknife computation of the ratio $E(V)/E(U)$. The discussion is in terms of the bivariate sample of size $n = 25$ given in section 12.2.3.

Randomly sample $n - k$ values from $\{x_1, \ldots, x_n\}$ using *without replacement* sampling and denote the random indicators for the indices selected as $H = (H_1, \ldots, H_n)^T$. These indicators now consist of k zeros and $n - k$ ones. The delete-k jackknife distribution is the distribution for the ratio $R^J = \sum_i v_i H_i / \sum_i u_i H_i$. The mean of this distribution is the delete-k jackknife estimator. The aim in using saddlepoint approxiamtion here is to avoid the computation of the true permutation distribution and its mean by summing over simulated outcomes from the random sampling. Thus complicated simulations are replaced with simple analytical computation.

From the perspective of conditional distributions,

$$R^J \sim \sum_i v_i B_i / \sum_i u_i B_i \quad \text{given} \quad \sum_i B_i = n - k \qquad (12.101)$$

where $B = (B_1, \ldots, B_n)^T$ are i.i.d. Bernoulli $(1/2)$ and (12.101) has used the distributional identity $H \sim B \mid B^T 1 = n - k$. The two estimating equations

$$\Psi_1 = \sum_i B_i - r_1$$

$$\Psi_2 = \sum_i (v_i - r_2 u_i) B_i$$

suffice in computing the saddlepoint approximation by taking $r_1 = n - k$ and computing the conditional saddlepoint approximation for

$$\Pr(R^J \leq r_2) = \Pr(R_2 \leq r_2|R_1 = n - k) = \Pr(\Psi_2 \leq 0|\Psi_1 = 0). \qquad (12.102)$$

Table 12.7. *Approximations for the permutation distribution* $\Pr(R^J \le r_2)$.

r_2	Exact[†]	Skov.	Normal	r_2	Exact[†]	Skov.	Normal
6.6	$.0^3 1$	$.0^3 1$	$.0^3 8$	8.2	.2979	.2844	.2635
6.8	$.0^3 3$	$.0^3 4$	.0015	8.4	.4661	.4476	.4313
7.0	.0013	.0015	.0031	8.6	.6157	.6173	.6205
7.2	.0041	.0043	.0066	8.8	.7453	.7680	.7885
7.4	.0112	.0118	.0147	9.0	.8717	.8833	.9036
7.6	.0310	.0303	.0329	9.2	.9610	.9610	.9644
7.8	.0726	.0725	.0713	9.4	.9953	.9947	.9889
8.0	.1578	.1550	.1440	9.6	$.9^4$	$.9^3 8$	.9968

[†]Computed using 100,000 bootstrap samples.

Note that because of the discreteness in the problem, Jacobian factors do not enter into the saddlepoint expressions and Skovgaard's approximation is readily applied directly to the right side of (12.102).

Booth and Butler (1990) use the data from table 12.3 and saddlepoint approximations for the delete-10 jackknife distribution are replicated in table 12.7. The jackknife estimate as the mean of the true jackknife distribution was simulated as 8.454 whereas the value determined by using the entries of table 12.7 in a Riemann sum approximation is 8.451. The Normal approximation used

$$\Pr(R^J \le r_2) \simeq \Phi\left\{-(n-k)^{1/2}(\bar{v} - r_2\bar{u})/\hat{\sigma}(r_2)\right\}$$

with $\hat{\sigma}^2(r_2) = n^{-1}\sum_i\{v_i - \bar{v} - r_2(u_i - \bar{u})\}^2$. The saddlepoint approximation is almost uniformly better than the Normal approximation.

12.5.4 Independence of R_m and R^-

The identity (12.99) also holds when R_m is independent of R_- and correspondingly $\Psi_{m;r}$ is independent of Ψ^-. Examples of this are not especially common but see below. In this setting $(\hat{s}, \hat{t}) = (\hat{s}_0, 0)$ and the ratio $||J^-(\hat{s}_0, 0)||/||J^-(\hat{s}, \hat{t})|| = 1$ so (12.98) reduces to the Lugannani and Rice approximation applied to $\Pr(\Psi_{m;r} \le 0)$.

Dirichlet decomposition

Suppose that $D = (D_1, \ldots, D_m) \sim$ Dirichlet $(\alpha_1, \ldots, \alpha_{m+1})$ and transform D to

$$\begin{aligned} R_1 &= D_1 \\ R_2 &= D_2/(1 - D_1) \\ &\vdots \\ R_m &= D_m/(1 - D_1 - \cdots - D_{m-1}). \end{aligned} \quad (12.103)$$

The components of R are easily shown to be mutually independent.

Not all forms of equivalent estimating equations have $\Psi_{m;r}$ is independent of $\Psi^- = (\Psi_1^-, \ldots, \Psi_{m-1}^-)^T$. For example, simply undoing the ratios in (12.103) with

$$\Psi_j = D_j - r_j(1 - D_1 - \cdots - D_{j-1})$$

and similarly for $\Psi_{m;r}$ does not lead to such independence. To get independence, deconstruct D into the independent Gammas variates $\{G_i : i = 1, \ldots, m+1\}$ upon which it is built with $D_i = G_i/(G_1 + \cdots + G_{m+1})$. The estimating equation construction

$$\begin{aligned}\Psi_j &= G_j - r_j(G_j + \cdots + G_{m+1}) \qquad j = 1, \ldots, m-1 \\ \Psi_{m;r} &= G_m/(G_m + G_{m+1}) - r_m\end{aligned} \qquad (12.104)$$

has $\Psi^- = g(G_1, \ldots, G_{m-1}, G_m + G_{m+1})$ independent of $\Psi_{m;r}$ since the only common variable they share is $G_m + G_{m+1}$ and $\Psi_{m;r}$ is independent of its denominator. The form of (12.104) shows that $R_m \sim \text{Beta}(\alpha_m, \alpha_{m+1})$.

12.5.5 Derivation of the double-saddlepoint approximation

The expression in (12.96) is a result of three steps: approximation to the conditional saddlepoint density, integration to get the conditional CDF, and then approximation for the integral using the Temme approximation. The double saddlepoint density is the ratio of the two densities

$$\hat{f}_{R_m}(r_m|r_-) = \frac{\hat{f}_R(r)}{\hat{f}_{R_-}(r_-)} = (2\pi)^{-1/2} \left\{ \frac{|K_{r;ss}''(\hat{s}_0, 0)|}{|K_r''(\hat{s}, \hat{t})|} \right\}^{1/2} \frac{M_r(\hat{s}, \hat{t})}{M_r(\hat{s}_0, 0)} \frac{\|J_r(\hat{s}, \hat{t})\|}{\|J^-(\hat{s}_0, 0)\|} \qquad (12.105)$$

determined from (12.70). The conditional density in (12.105) has abused the notation by changing the meaning of the saddlepoints $\hat{s}$ and $\hat{t}$. On the right-hand side, the values of $\hat{s}$ and $\hat{t}$ represent the saddlepoints computed at the generic value r not at r_0, as indicated in (12.95). Thus, as r_m the argument of $\hat{f}_{R_m}$ changes, so do the values of $\hat{s}$ and $\hat{t}$. Integration yields an approximate conditional CDF

$$\hat{F}(r_{m0}|r_-) \simeq \int_{\mathcal{R}_m \cap (-\infty, r_{m0}]} \hat{f}_{R_m}(r_m|r_-) dr_m \qquad (12.106)$$

where $\mathcal{R}_m$ denotes the conditional support of R_m given r_-. The substitution

$$r_m \leftrightarrow w = \text{sgn}(\hat{t})\sqrt{2\{K_r(\hat{s}_0, 0) - K_r(\hat{s}, \hat{t})\}} \qquad (12.107)$$

is now used to change integration dr_m to dw. With this change, the numerator integral in (12.106) becomes $\int_{-\infty}^{\hat{w}} h(w)\phi(w)dw$ with

$$h(w) = \left\{ \frac{|K_{r;ss}''(\hat{s}_0, 0)|}{|K_r''(\hat{s}, \hat{t})|} \right\}^{1/2} \frac{\|J_r(\hat{s}, \hat{t})\|}{\|J^-(\hat{s}_0, 0)\|} \frac{\partial r_m}{\partial w} \qquad (12.108)$$

as needed for the Temme approximation in (2.34). Note in the substitution that the values of $\hat{s}$ and $\hat{t}$ within w continue to depend on r_m.

The determination of $\partial r_m/\partial w$ in (12.108) begins by expressing

$$w^2/2 = K_r(\hat{s}_0, 0) - K_r(\hat{s}, \hat{t})$$

12.5 The conditional CDF of R_m given $R_1, \ldots, R_{m-1}$

and differentiating $\partial/\partial r_m$ so that

$$w \frac{\partial w}{\partial r_m} = -\frac{\partial K_r(\hat{s}, \hat{t})}{\partial r_m} - K'_{r;s}(\hat{s}, \hat{t}) \frac{\partial \hat{s}}{\partial r_m} - K'_{r;t}(\hat{s}, \hat{t}) \frac{\partial \hat{t}}{\partial r_m}$$
$$= -\frac{\partial K_r(\hat{s}, \hat{t})}{\partial r_m} \tag{12.109}$$

since $K'_{r;s}(\hat{s}, \hat{t}) = 0 = K'_{r;t}(\hat{s}, \hat{t})$ are the saddlepoint equations. The partial derivative in (12.109) refers to differentiation with respect to the direct dependence of $K_r(\hat{s}, \hat{t})$ on r_m through its subscript r and holding $(\hat{s}, \hat{t})$ fixed. We defer to the appendix in section 12.6.2 the details of the arguments used to show that $h(w)$ has a removable discontinuity at $w = 0$ or $\lim_{w \to 0} h(w) = 1$.

The argument used in the Temme formula now applies with $\hat{w}$ as the value of w computed with $r = r_0$ and $\hat{u} = \hat{w}/h(\hat{w})$. At this point in the argument, $(\hat{s}, \hat{t})$ is evaluated at r_0 so it now refers to the saddlepoints used in the CDF approximation and given in (12.95). The value

$$w \frac{\partial w}{\partial r_m}\bigg|_{w=\hat{w}} = -\frac{\partial K_r(\hat{s}, \hat{t})}{\partial r_m}\bigg|_{r=r_0} = \frac{-1}{M_{r_0}(\hat{s}, \hat{t})} \frac{\partial M_r(\hat{s}, \hat{t})}{\partial r_m}\bigg|_{r=r_0}$$
$$= -\hat{t} \int \cdots \int \frac{\partial \Psi_m(x; r_0)}{\partial r_{m0}} g_r(x; \hat{s}, \hat{t}) dx$$
$$= -\hat{t} E \left\{ \frac{\partial \Psi_m(X; r_0)}{\partial r_{m0}}; \hat{s}, \hat{t} \right\}. \tag{12.110}$$

This mean, computed under the $\{\hat{s}^T \Psi_{r_-}^- + \hat{t}\Psi_{m;r_0}\}$-tilted distribution is the (m, m) entry of $J_{r_0}(\hat{s}, \hat{t})$ so that

$$\|J_{r_0}(\hat{s}, \hat{t})\| = \left\| \begin{matrix} J_{r_-}^-(\hat{s}, \hat{t}) & - \\ 0^T & E\left\{\frac{\partial \Psi_m(x; r_0)}{\partial r_{m0}}; \hat{s}, \hat{t}\right\} \end{matrix} \right\|$$
$$= -E\left\{\frac{\partial \Psi_m(x; r_0)}{\partial r_{m0}}; \hat{s}, \hat{t}\right\} \|J_{r_-}^-(\hat{s}, \hat{t})\|. \tag{12.111}$$

The value of $\hat{u} = \hat{w}/h(\hat{w})$ given in (12.98) follows from substituting (12.110) and (12.111) into (12.108).

12.5.6 Determination of autoregressive order by using optimal tests

The first m autocorrelations $R = (R_1, \ldots, R_m)^T$ of the correlogram have a joint saddlepoint density which has been discussed in section 12.3.4. Now the conditional CDF of R_m at r_{m0} given $R_- = (R_1, \ldots, R_{m-1})^T = r_-^T$ can be computed directly from (12.96). Such a conditional distribution is needed to determine the p-value for an approximately optimal test of the hypotheses AR $(m-1)$ versus AR (m). Using the standard exponential family theory for optimal unbiased tests, Anderson (1971, §6.3.2) has shown that the UMP unbiased test of AR $(m-1)$ versus AR (m) in the circular AR (m) model rejects for r_{m0} sufficiently deviant in either tail as determined by the conditional distribution of R_m given r_-. One-sided

p-values may be computed as $\min(\tau, 1 - \tau)$ where

$$\tau = \Pr(R_m > r_{m0} | r_-; \mathrm{H}_{m-1})$$

with the notation H_{m-1} used to indicate that the series is AR $(m - 1)$.

Two comments are needed before the application of Anderson's result can be meaningful. First, he assumes a circular AR (m) model whereas the theory of section 12.3.4 and the ordinary correlogram are based on a noncircular AR (m) model. For quite short times series such as $n = 11$ this is a concern and will be a source of inaccuracy. However, with moderately long series, such as $n = 41$, it is not especially important as will be seen in the examples below.

The second comment is that with circularity, the computation of τ under H_{m-1} is the same as it is under H_0, e.g. the setting in which the data are i.i.d. white noise. This occurs as a result of the unbiasedness of the test induced by conditioning on R_- so that the value of τ no longer depends on the autoregressive coeffficients of the AR $(m - 1)$ model. Thus, for moderately long series in which the circularity/noncircularity assumption is unimportant, the computations can be made simpler by taking $\Sigma = I_n$ without suffering a severe effect on the p-value computation. See Kay et al. (2001) for an alternative approach to model order selection in the circular setting.

The expressions for $\hat{w}$ and $\hat{u}$ given in section 12.3.4 make use of the following quantities whose notation has been changed to accommodate the isolation of the mth component:

$$K_{\Psi_r}(s, t) = -\tfrac{1}{2} \ln |P(s, t)\Sigma|$$
$$K_{\Psi_{r_-}^-}(s) = -\tfrac{1}{2} \ln |P(s, 0)\Sigma|$$

where

$$P(s, t) = \Sigma^{-1} + 2(s^T r_- + t r_{m0}) I_n - 2 \sum_{k=1}^{m-1} s_k A_k - 2 t A_m,$$

and

$$J^-(s, t) = [\mathrm{tr}\{P(s, t)^{-1}\}]^{m-1}.$$

The conditional support of R_m given r_- is easily determined for the correlogram based on the fact that R has support $\mathcal{R} = \{r : |Q_k| > 0 \; k = 1, \ldots, m\}$ with Q_k given in (12.65). If Q_{m-1}^- is Q_{m-1} evaluated at r_-, then the conditional support of $R_m | r_-$ is comprised of the values of r_m for which

$$0 < |Q_m| = |Q_{m-1}^-|\{1 - (r_-^T, r_m)(Q_{m-1}^-)^{-1}(r_-^T, r_m)^T\}.$$

The roots of the quadratic in r_m given in curly braces determine the endpoints of the conditional support. Finding the conditional support in other settings, such as for the various lagged orders of the Durbin–Watson statistics, appears to be quite difficult.

A short series for pear data

The annual pear data of Henshaw (1966) covers $n = 16$ years with $p = 5$ independent regressor variables inclusive of a location parameter. The net series length is 11 years once the five regressors are removed as described in section 12.3.4. The data set was analyzed

Table 12.8. *Progressive testing of AR $(m-1)$ vs. AR (m) using the distribution of $R_m|r_-$ for the pears data. Listed are the mth serial correlation r_{m0}, the mth serial partial correlation $r_{m\cdot}$, the support (Low, High) for the distribution of $R_m|r_-$, and various approximate p-values as determined using the conditional saddlepoint CDF (12.96), by numerically integrating the conditional saddlepoint density (NID), and using the Daniels–Durbin beta approximation.*

m	r_{m0}	$r_{m\cdot}$	Low	High	(12.96)	NID	D-D
1	.26335	.26335	-1	1	.0317	.0494	.0096
2	$-.20321$	$-.29288$	$-.86130$	1	.5665	.5711	.6249
3	$-.42921$	$-.33250$	$-.99714$	.70450	.1239	.1225	.4545
4	$-.24708$	$-.12116$	$-.91215$	.60137	.4339	.4045	.8469
5	.19903	.17838	.67962	.81167	.1151	.1178	.0238

in Butler and Paolella (1998b) and the results from the progressive testing of AR $(m-1)$ versus AR (m) for $m = 1, \ldots, 5$ are reproduced and displayed in table 12.8.

In each row m, values are given for the mth serial correlation r_{m0}, the mth serial partial correlation $r_{m\cdot}$, the support (Low, High) for the distribution of $R_m|r_-$, and various approximate p-values as determined three ways: (i) by using the conditional saddlepoint CDF in (12.96), (ii) by numerically integrating the conditional saddlepoint density (NID) in (12.105), and (iii) by using the Daniels–Durbin beta approximation (D-D) that is explained below.

The Daniels–Durbin beta approximation

The p-values in column D-D of table 12.8 make use of the Daniels (1956) and Durbin (1980b) circular distribution theory with $n = 16$ and $p = 5$. Their procedures use the set of partial serial correlations $\{r_{1\cdot}, \ldots, r_{5\cdot}\}$ displayed in the $r_{m\cdot}$ column. Their methods using partial correlations are applicable since the conditional test using $R_m|r_-$ is analytically equivalent to the partial correlation test based on $R_{m\cdot}|r_{1\cdot}, \ldots, r_{m-1\cdot}$. In the circular null setting, $R_{m\cdot}$ is approximately independent of $R_{1\cdot}, \ldots, R_{m-1\cdot}$ so that the p-value with $r_{m\cdot} < 0$ is determined from Durbin (1980b, Eq. 20) as

$$\hat{p} = \Pr(R_{m0} < r_{m0}) \simeq \begin{cases} \text{IB}\left(\frac{r_{m\cdot}+1}{2}; \frac{n-p+1}{2}, \frac{n+p+1}{2}\right) & \text{if } m \text{ is odd} \\ \text{IB}\left(\frac{r_{m\cdot}+1}{2}; \frac{n-p}{2}, \frac{n+p}{2}+1\right) & \text{if } m \text{ is even} \end{cases}$$

where IB $(x; \alpha, \beta)$ is the Beta (α, β) CDF at x. If $r_{m\cdot} > 0$ then the same distributions apply but using the right tails instead.

A moderately long series for the Vinod data

The same sequence of conditional tests to determine the order of serial dependence are applied to a moderately long sequence of quarterly measurements for investment data in Vinod (1973, table 1). The sequence spans $n = 44$ quarters with $p = 3$ regression variables

Table 12.9. *Progressive testing of AR* $(m-1)$ *vs. AR* (m) *using the distribution of* $R_m | r_-$ *for the Vinod data. See table 12.8 for a description.*

m	r_{m0}	$r.$	Low	High	(12.96)	NID	D-D
1	.01765	.01765	−1	1	.2784	.2797	.2846
2	.29141	.29119	−.99938	1	.00355	.00352	.00478
3	−.25872	−.29238	−.90614	.92371	.05924	.05914	.06195
4	.44199	.46192	−.78122	.89221	$.0^4 112$	$.0^4 120$	$.0^4 521$
5	−.42366	−.60055	−.68656	.62979	$.0^4 152$	$.0^4 200$	$.0^4 327$

including a mean location. Table 12.9 provide the same analysis as table 12.8 and has been extracted from Butler and Paolella (1998b).

The Daniels–Durbin Beta approximation provides p-values that are now quite similar to those from the noncircular saddlepoint methods. This is in constrast with the previous example where the data had a net length of 11. The moderately longer net length of 41 appears to be the determining factor in assuring similar p-values.

What emerges from these considerations are the following recommendations for judging the significance of progressive terms in the correlogram. First, with at least moderately long series, the Daniels–Durbin approximation is considerably more accurate and no more difficult to compute than the standard asymptotic normal bands $\pm 2/\sqrt{n}$ currently used. For short series however, the conditional saddlepoint CDF approximation in (12.96) provides the most meaningful analysis.

12.6 Appendix

12.6.1 Explicit value of $\hat{s}$ for the doubly noncentral F distribution

An explicit saddlepoint may be computed as the root to $K'_{\Psi_r}(\hat{s}) = 0$ in this instance. The solution to this equation reduces to finding a root of the cubic polynomial

$$0 = \hat{s}^3 + a_2 \hat{s}^2 + a_1 \hat{s} + a_0$$

in which the terms are

$$a \cdot a_2 = 8r(1-r)n_1 n_2^2 + 4r\left(n_2^3 + \theta_2 n_2^2 - n_1^2 n_2 r - n_1 n_2 \theta_1 r\right),$$
$$a \cdot a_1 = 2\left(n_2^2 n_1 + n_1^2 n_2 r^2\right) - 4r n_1 n_2 (n_1 + n_2 + \theta_1 + \theta_2),$$
$$a \cdot a_0 = r\theta_2 n_1^2 - (1-r)n_1^2 n_2 - n_1 n_2 \theta_1,$$
$$a = 8r^2 n_2^2 (n_1 + n_2).$$

Butler and Paolella (2002) show that this polynomial always leads to three real roots ordered as $z_2 < z_3 \le z_1$ and that $\hat{s}$ is always the middle root z_3. Upon further defining

$$q = \tfrac{1}{3} a_1 - \tfrac{1}{9} a_2^2, \quad p = \tfrac{1}{6}(a_1 a_2 - 3a_0) - \tfrac{1}{27} a_2^3, \quad m = q^3 + p^2 \qquad (12.112)$$

and the two values $s_{1,2} = \sqrt[3]{p \pm m^{1/2}}$, they give the saddlepoint as

$$\hat{s} = -\tfrac{1}{2}(s_1 + s_2) - \tfrac{1}{3}a_2 - \tfrac{1}{2}i\sqrt{3}(s_1 - s_2). \tag{12.113}$$

An alternative expression for the saddlepoint that is useful in an environment not supporting complex arithmetic is

$$\hat{s} = \sqrt{-q}\{\sqrt{3}\sin(\phi) - \cos(\phi)\} - \tfrac{1}{3}a_2 \tag{12.114}$$

where

$$\phi = \tfrac{1}{3}\arg\left(p + i\sqrt{-m}\right) = \begin{cases} \tan^{-1}(-m/p) & \text{if } p \geq 0, \\ \pi + \tan^{-1}(-m/p) & \text{if } p < 0, \end{cases}$$

and m and q are always < 0.

The saddlepoint for $F^{(1)}$ with $\theta_2 = 0$ is considerably simpler. The singly noncentral root is

$$\hat{s} = \frac{rn_1(n_1 + 2n_2 + \theta_1) - n_1 n_2 - \sqrt{n_1 y}}{4n_2 r (n_1 + n_2)}, \tag{12.115}$$

where

$$y = r^2 n_1^3 + 2r^2 n_1^2 \theta_1 + 2n_1^2 r n_2 + 4r^2 n_1 n_2 \theta_1 + n_1 \theta_1^2 r^2 + 2n_1 \theta_1 r n_2 + n_2^2 n_1 + 4r n_2^2 \theta_1.$$

Even simpler is the central F case with $\theta_1 = 0 = \theta_2$, in which

$$\hat{s} = \frac{n_1(r-1)}{2r(n_1 + n_2)}. \tag{12.116}$$

The saddlepoint $\hat{s} = 0$ occurs when $r = (1 + \theta_1/n_1)/(1 + \theta_2/n_2)$ for all parameter values. This occurs at the mean of Ψ_r which is not the mean of $F^{(2)}$. Of course for this value of r, the limiting approximation in (1.21) should be used. In the singly noncentral case, $\hat{s} = 0$ for $r = 1 + \theta_1/n_1$, which can be compared to the mean $E\{F^{(1)}\} = (1 + \theta_1/n_1)n_2/(n_2 - 2)$.

12.6.2 The removable singularity for $h(w)$ in (12.108)

As $w \to 0$ (or as $\hat{t} \to 0$), the terms in (12.108) pass to nonzero terms and only $\lim_{w \to 0} \partial r_m/\partial w$ needs to be determined. Expression (12.109) can be used to determine this limit by taking another derivative $\partial/\partial r_m$ to get

$$\left(\frac{\partial w}{\partial r_m}\right)^2 + w\frac{\partial^2 w}{\partial r_m^2} = -\frac{\partial^2 K_r(\hat{s},\hat{t})}{\partial r_m^2} - \frac{\partial}{\partial \hat{s}^T}\left\{\frac{\partial K_r(\hat{s},\hat{t})}{\partial r_m}\right\}\frac{\partial \hat{s}}{\partial r_m} - \frac{\partial}{\partial \hat{t}}\left\{\frac{\partial K_r(\hat{s},\hat{t})}{\partial r_m}\right\}\frac{\partial \hat{t}}{\partial r_m}. \tag{12.117}$$

The limit of each term on the right side is determined by expressing it as the derivative of the integral used for computing $M_r(\hat{s}, \hat{t})$. For the first term,

$$\frac{\partial^2 K_r(\hat{s},\hat{t})}{\partial r_m^2} = \frac{1}{M_r(\hat{s},\hat{t})}\frac{\partial^2 M_r(\hat{s},\hat{t})}{\partial r_m^2} - \frac{1}{M_r(\hat{s},\hat{t})^2}\left\{\frac{\partial M_r(\hat{s},\hat{t})}{\partial r_m}\right\}^2 \to 0$$

since the first and second derivatives $\partial/\partial r_m$ of M_r have a factor of $\hat{t}$. The next term in (12.117) has the factor $\hat{t}$ in each of its terms so it too vanishes. Only the last term contributes to the limit. Since

$$\frac{\partial}{\partial \hat{t}}\left\{\frac{\partial K_r(\hat{s},\hat{t})}{\partial r_m}\right\} = \frac{\partial}{\partial \hat{t}}\left\{\frac{1}{M_r(\hat{s},\hat{t})}\frac{\partial M_r(\hat{s},\hat{t})}{\partial r_m}\right\} \sim \frac{1}{M_r(\hat{s},\hat{t})}\frac{\partial}{\partial \hat{t}}\left\{\frac{\partial M_r(\hat{s},\hat{t})}{\partial r_m}\right\}$$
$$= \frac{1}{M_r(\hat{s},\hat{t})}\int\cdots\int\left\{\frac{\partial \Psi_m(x;r)}{\partial r_m}+O(\hat{t})\right\}g_r(x;\hat{s},\hat{t})dx$$
$$\to E\left\{\frac{\partial \Psi_m(X;r)}{\partial r_m};\hat{s}_0,0\right\} \qquad (12.118)$$

thus

$$\lim_{\hat{t}\to 0}\left(\frac{\partial w}{\partial r_m}\right)^2 = -E\left\{\frac{\partial \Psi_m(X;r)}{\partial r_m};\hat{s}_0,0\right\}\lim_{\hat{t}\to 0}\left(\frac{\partial \hat{t}}{\partial r_m}\right). \qquad (12.119)$$

The latter limit is computed by taking the derivative $\partial/\partial r_m$ of the m saddlepoint equations $K'_r(\hat{s},\hat{t})=0$ to get

$$0 = K''_r(\hat{s},\hat{t})\frac{\partial}{\partial r_m}\begin{pmatrix}\hat{s}\\\hat{t}\end{pmatrix}+\frac{\partial}{\partial r_m}\begin{pmatrix}K'_{r;s}(\hat{s},\hat{t})\\K'_{r;t}(\hat{s},\hat{t})\end{pmatrix}$$

or

$$\frac{\partial}{\partial r_m}\begin{pmatrix}\hat{s}\\\hat{t}\end{pmatrix} = -K''_r(\hat{s},\hat{t})^{-1}\frac{\partial}{\partial r_m}\begin{pmatrix}K'_{r;s}(\hat{s},\hat{t})\\K'_{r;t}(\hat{s},\hat{t})\end{pmatrix}$$
$$\sim -K''_r(\hat{s}_0,0)^{-1}\begin{pmatrix}0_{m-1}\\\partial K'_{r;t}(\hat{s},\hat{t})/\partial r_m\end{pmatrix}.$$

If $\{\cdot\}_{m,m}$ denotes the (m,m) component, then

$$\lim_{\hat{t}\to 0}\frac{\partial \hat{t}}{\partial r_m} = -\left\{K''_r(\hat{s}_0,0)^{-1}\right\}_{m,m}\lim_{\hat{t}\to 0}\frac{\partial K'_{r;t}(\hat{s},\hat{t})}{\partial r_m}$$
$$= -\frac{|K''_{r;ss}(\hat{s}_0,0)|}{|K''_r(\hat{s}_0,0)|}\times E\left\{\frac{\partial \Psi_m(X;r)}{\partial r_m};\hat{s}_0,0\right\} \qquad (12.120)$$

using (13.33).

Substituting (12.120) and (12.119) into (12.108) then $\lim_{w\to 0}h(w)$ is

$$\left\{\frac{|K''_{r;ss}(\hat{s}_0,0)|}{|K''_r(\hat{s}_0,0)|}\right\}^{1/2}\frac{\|J_r(\hat{s}_0,0)\|}{\|J^-(\hat{s}_0)\|}\times\sqrt{\frac{|K''_r(\hat{s}_0,0)|}{|K''_{r;ss}(\hat{s}_0,0)|}}\frac{1}{\left|E\left\{\frac{\partial \Psi_m(X;r)}{\partial r_m};\hat{s}_0,0\right\}\right|}$$

which is 1 upon using the identity in (12.111) that relates $\|J_r(\hat{s}_0,0)\|$ to $\|J^-(\hat{s}_0,0)\|$.

12.7 Exercises

1. For ratio $R=U/V$ with $V>0$, show that the Lugannani and Rice approximation for the CDF of R is given in (12.2) and (12.3).
2. Justify the step (12.12) in the derivation of the MGF for Ψ^*_r by using the dominated convergence theorem.

3. Derive the quantities required for the doubly and singly noncentral F approximation.
   (a) Use Exercise 4(a) of chapter 1 to show that the MGF of Ψ_r is
   $$M_{\Psi_r}(s) = \left(1 - 2s\frac{n_2}{n_1}\right)^{-n_1/2} (1 + 2sr)^{-n_2/2} \exp\left(\frac{s\theta_1 \frac{n_2}{n_1}}{1 - 2s\frac{n_2}{n_1}} - \frac{s\theta_2 r}{1 + 2sr}\right) \quad (12.121)$$
   and convergent for $-1/(2r) < s < n_1/(2n_2)$.
   (b) By defining $\ell_1 = n_2/n_1$ and $\ell_2 = -r$ and
   $$\vartheta_i = (1 - 2s\ell_i)^{-1},$$
   show that the dth order derivative of CGF $K_{\Psi_r}(s)$ is given by
   $$K_{\Psi_r}^{(d)}(s) = k_d \sum_{i=1}^{2} \ell_i^d \vartheta_i^d (n_i + d\theta_i \vartheta_i), \quad d \geq 1, \quad (12.122)$$
   where $k_1 = 1$ and $k_d = 2(d-1)k_{d-1}$ for $d > 1$. These expressions allow computation of
   $$\hat{\kappa}_i = K_{\Psi_r}^{(i)}(\hat{s})/K_{\Psi_r}''(\hat{s})^{i/2}$$
   as required for the second-order Lugannani and Rice formula given in (12.14).
   (c) Derive an expression for the saddlepoint density of $F^{(2)}$. Show that the constructed random variable Ψ_r^* has MGF $M_{\Psi_r}(s) J_r(s)/J_r(0)$ with M_{Ψ_r} given in (12.121) and
   $$J_r(s)/J_r(0) = \left\{\frac{\theta_2}{(1+2sr)^2} + \frac{n_2}{1+2sr}\right\}/(\theta_2 + n_2).$$
   (d) Use part (c) to derive the saddlepoint density for the central F. Show that it agrees with (12.15).
4. (a) Compute the exact singly noncentral F distribution at selected quantiles by using SAS. Do these computations for two values each of n_1, n_2, and θ_1 and therefore over the eight possible combinations.
   (b) Compare these values to the first- and second-order CDF approximations in (12.2) and (12.14).
5. If χ_1 and χ_2 are independent with $\chi_1 \sim \chi^2(n_1, \theta_1)$ and $\chi_2 \sim \chi^2(n_2, \theta_2)$, then define the doubly noncentral beta as the distribution of $B = \chi_1/(\chi_1 + \chi_2)$. Prove that saddlepoint CDF approximation for the doubly noncentral Beta is equivalent to CDF approximation of the doubly noncentral F. Hint: Let Ψ_b and Ψ_f be the constructed variables for the CDF of B and $F^{(2)}$ respectively. Show that
   $$\Pr(\Psi_b \leq 0) = \Pr\left(\frac{n_2}{n_1}\chi_1 - f\chi_2 \leq 0\right) = \Pr(\Psi_f \leq 0).$$
6. For the relationship in Exercise 5, show that the saddlepoint approximation for the doubly noncentral Beta density is a Jacobian transformation of the saddlepoint approximation for the doubly noncentral $F^{(2)}$ density.
   (a) Show that
   $$M_{\Psi_b}(s) = M_{\Psi_f}(s/c)$$
   where $c = f + n_2/n_1$. Hence conclude that $\hat{s}_b$, the saddlepoint for $B = b$, is related to $\hat{s}_f$, the saddlepoint for $F^{(2)} = f$, according to $\hat{s}_b/c = \hat{s}_f$.
   (b) Let $J_b(s)$ be (12.7) for the noncentral Beta density and $J_f(s)$ the same quantity for the noncentral $F^{(2)}$. Use the fact that $K'_{\Psi_b}(\hat{s}_b) = 0$ to reduce $J_b(\hat{s}_b)$ to
   $$J_b(\hat{s}_b) = \frac{n_1}{n_2}cK'_{\chi_2}(-f\hat{s}_f),$$
   where K_{χ_2} is the CGF of χ_2.

(c) Since $\partial f/\partial b = n_1 c^2/n_2$, show that $\hat{f}_{F^{(2)}}(f)$, the saddlepoint density for $F^{(2)}$ at f, is related to $\hat{f}_B(b)$ according to

$$\hat{f}_B(b) = \hat{f}_F(f) \frac{\partial f}{\partial b}.$$

(d) For Ψ_b, show that first- and second-order Lugannani and Rice approximations for B must be the same as those for $F^{(2)}$ when the cutoffs are appropriately converted. Hint: Explicit computation of the beta approximation is unnecessary.

7. Derive the saddlepoint approximations for the ratio of quadratic forms R in (12.16).

   (a) With $U = \varepsilon^T A \varepsilon$ and $V = \varepsilon^T B \varepsilon$, show that

   $$M(s,t) = E(e^{sU+tV})$$
   $$= |\Omega|^{-\frac{1}{2}} \exp\left\{-\tfrac{1}{2}\mu^T(I_n - \Omega^{-1})\mu\right\}$$

   where $\Omega = I_n - 2(sA + tB)$ and (s, t) is chosen in a neighborhood of $(0, 0)$ such that $\Omega > 0$.

   (b) Verify the correctness of the MGF for Ψ_r given in (12.20) along with its convergence region (12.21).

   (c) Using the rules of matrix differentiation, show that random variable Ψ_r^* has MGF

   $$\left.\frac{\partial M(s,t)}{\partial t}\right|_{t=-rs} = |\Xi_s|^{-\frac{1}{2}} \exp\left\{s\mu^T \Xi_s^{-1} D_r \mu\right\}\left\{\mu^T \Xi_s^{-1} B \Xi_s^{-1} \mu + \operatorname{tr} \Xi_s^{-1} B\right\}$$

   where $\Xi_s = I - 2sD_r$ and $D_r = A - rB$. Use the canonical reduction $P_r D_r P_r^T = \Lambda_r$ which also applies to Ξ_s, to rewrite this as

   $$= |I - 2s\Lambda_r|^{-\frac{1}{2}} \exp\left\{sv_r^T (I - 2s\Lambda_r)^{-1} \Lambda_r v_r\right\} J_r(s)$$
   $$= M_{\Psi_r}(s) J_r(s)$$

   where $J_r(s)$ is given in (12.23) and $M_{\Psi_r}(s)$ is given in (12.20).

8. In the ratio example of section 12.1.6, suppose that $\mu_1 = \mu_2 = 0$ so that R is Cauchy.

   (a) Show that the saddlepoint density reduces to

   $$\hat{f}_R(r) = \frac{1}{\sqrt{2\pi}\,(1+r^2)} = \sqrt{\frac{\pi}{2}} f_R(r), \tag{12.123}$$

   where $\sqrt{\pi/2}$ is $B(1/2, 1/2)/\hat{B}(1/2, 1/2)$ and $\hat{B}(1/2, 1/2)$ is Stirling's approximation. Thus $\hat{f}_R$ is exact and the saddlepoint solution to $0 = K'_{\Psi_r}(\hat{s})$ is given by $\hat{s} = r$.

   (b) For the CDF approximation, show that the inputs $\hat{w}$ and $\hat{u}$ for the Lugannani and Rice formula reduce to

   $$\hat{w} = \operatorname{sgn}(r)\sqrt{\ln(1+r^2)}, \quad \hat{u} = r(1+r^2)^{-1/2}.$$

   (c) Plot the relative error in part (b). Convince yourself that its limiting relative error as $r \to \infty$ is $\sqrt{\pi/2}$ and the same error as for (12.123).

9. Consider the linear model

   $$y = (y_1, \ldots, y_t)^T = X\beta + \varepsilon$$

   where X is a full rank $t \times p$ matrix and the components of ε are i.i.d. Normal $(0, \sigma^2)$. The Durbin–Watson test statistic for lag 1 autocorrelation was introduced in section 3.4.2 and rejects for large values of

   $$D = \frac{\sum_{i=2}^{t}(\hat{\varepsilon}_i - \hat{\varepsilon}_{i-1})^2}{\sum_{i=1}^{t}\hat{\varepsilon}_i^2}$$

   where $\hat{\varepsilon} = My = \{I_t - X(X^T X)^{-1} X^T\} y$ consists of the least squares residuals.

(a) Show that D may be expressed as a ratio of quadratic forms as in (12.16) with $B = M$ and $A = MCM$ where

$$C = \begin{pmatrix} 1 & -1 & 0 & \cdots & & 0 \\ -1 & 2 & -1 & \ddots & & \vdots \\ 0 & \ddots & \ddots & \ddots & & 0 \\ \vdots & & \ddots & -1 & 2 & -1 \\ 0 & \cdots & & 0 & -1 & 1 \end{pmatrix}$$

and is $t \times t$.

(b) Show that the distribution of D is given as

$$D \stackrel{\mathcal{D}}{=} \frac{\sum_{i=1}^{n} \omega_i \chi_i^2}{\sum_{i=1}^{n} \chi_i^2}$$

where $\{\omega_i\}$ are the $n = t - p$ positive eigenvalues of MC. (Hint: Decompose M into a diagonal matrix in it eigenvalues.)

(c) Note for this example that the eigenvalues $\{\lambda_i(r)\}$, given in the distributional characterization of (12.19), are

$$\lambda_i(r) = \omega_i - r.$$

Confirm that the resulting CDF approximation is that described in Exercise 16 of chapter 3.

(d) Part (c) describes a setting in which the complete collection of eigenvalues $\{\lambda_i(r)\}$ for all r is known as a result of computing a single determination of eigenvalues $\{\omega_i\}$ from MC. Specify general conditions on B and the distribution of ε that assure that this simplification is possible.

10. Suppose that continuous random variables X and Y have a joint density $f_{X,Y}(x, y)$ and a joint MGF $M_{XY}(s, t)$ that is convergent on an open neighborhood of $(0, 0)$.

(a) For $s > 0$ in the convergence strip of M_X, use integration by parts to show that

$$\int_{-\infty}^{\infty} e^{sx} \Pr(X > x) dx = M_X(s)/s.$$

(Hint: Use

$$e^{sx} \Pr(X > x) \leq \int_{x}^{\infty} e^{su} f_X(u) du$$

to show that the left hand side converges to 0 as $x \to \infty$.)

(b) For $s > 0 < t$ in the convergence strip of M_{XY}, indulge upon the argument used in part (a) to show that

$$\int_{-\infty}^{\infty} \int_{-\infty}^{\infty} e^{sx+ty} \Pr(X > x, Y > y) dy dx = M_{X,Y}(s, t)/(st).$$

(Hint: Apply integration by parts twice. Once to the inner integral for each x, and again for the outer integral.)

(c) Generalize the result to an m-dimensional random vector.

11. Derive the saddlepoint CDF and density for the $100p$ percentile of an i.i.d. sample with a continuous density.

(a) Use the MGF for the negative score $-\Psi_r$, e.g. the MGF of $B - np$ with $B \sim$ Binomial $\{n, F(r)\}$ to determine that the first continuity-corrected saddlepoint CDF for R_p, using

(1.27), is

$$\widehat{\Pr}_1(R_p \le r) = 1 - \Phi(\hat{w}) - \phi(\hat{w})\left(\frac{1}{\hat{w}} - \frac{1}{\tilde{u}_1}\right)$$

where

$$\hat{w} = \operatorname{sgn}(\hat{s})\sqrt{-2n\left\{p \ln \frac{F(r)}{p} + (1-p)\ln \frac{\bar{F}(r)}{1-p}\right\}}$$

$$\hat{s} = \ln \frac{p\bar{F}(r)}{(1-p)F(r)}$$

$$\tilde{u}_1 = \frac{p - F(r)}{p\bar{F}(r)}\sqrt{np(1-p)}.$$

(b) Derive the saddlepoint density of R_p using the MGF of Ψ_r given in (12.40). Confirm that it is (12.43) when $np = m$.

12. The Huber M-estimator using (12.28) with $n = 2$ is not well defined if $|X_1 - X_2| > 2k$. If $Y_1 < Y_2$ are the order statistics in this case, then any value in the range $(Y_1 + k, Y_2 - k)$ is an M-estimator.

    (a) Suppose that whenever a unique estimator is not defined, the mid-point of the range of roots is used as the estimator. Show that this more precise definition leads to $R = \bar{X}$ as the M-estimator.

    (b) Prove that the M-estimator has a density if X_1 has a density.

13. For the vector ratio with a common denominator, show that (12.54) is the relationship between the density for R at r and the density of the constructed score Ψ_r^* at 0.

    (a) Show that the marginal density of R is

    $$f_R(r) = \int_0^\infty v^m f_{U,V}(rv, v) dv.$$

    (b) Prove that the MGF of U^*, V^*, as defined by (12.53), is

    $$M_{U^*,V^*}(s, t) = \frac{1}{E(V^m)} \frac{\partial^m}{\partial t^m} M_{U,V}(s, t).$$

    (c) Using a Jacobian transformation, specify the marginal density of the constructed score in terms of the density of U, V, evaluate the density at 0, and hence confirm the relationship in (12.54).

    (d) Confirm that (12.55) is the MGF of Ψ_r^*.

14. Repeat the derivations of section 12.3.2 for the second vector ratio example in (12.46). Confirm that (12.50) is the appropriate saddlepoint density approximation.

15. Show that the saddlepoint approximation to the Dirichlet $(\alpha_1, \ldots, \alpha_m; \alpha_{m+1})$ is exact up to Stirling's approximation as given in (12.62).

    (a) Derive the result first by building it directly as a vector ratio of independent gammas using (12.56).

    (b) Repeat the derivation for half-integer values of $\{\alpha_i\}$, by using the formulae for the ratios of quadratic forms in section 12.3.3 with the $\{A_k\}$ partitioning subsets of the χ_1^2 variables involved.

16. Suppose $R = (R_1, \ldots, R_m)$ is Dirichlet $(\alpha_1, \ldots, \alpha_m; \alpha_{m+1})$ with R as the root of (12.77). Let the components of $Q = (Q_1, \ldots, Q_m)$ have the corresponding multivariate F distribution as the roots of (12.78).

    (a) For $m = 2$, show that the conditions of Proposition 12.4.1 are satisfied.

    (b) Generalize the argument for any m to concluded that the saddlepoint density for the multivariate F is exact up to Stirling's approximation for the Gamma functions involved.

(c) Consider an ANOVA, such as the one in Exercise 23 below, in which the F ratios of interest do not share the same denominator. Take a particular example and show that it satisfies the conditions of Proposition 12.4.1. Since the argument used does not depend on the particular example considered, Proposition 12.4.2 also applies to F ratios that lack common denominators.

17. Suppose that R and Q are roots of their respective estimating equations and $Q = g(R)$ is a smooth bijection as described in Proposition 12.4.1 of section 12.4.3. Of course the true density $f_Q(q)$ is related to $f_R(r)$ through a Jacobian transformation. However, suppose that $f_Q(q)$ and $f_R(r)$ are first expressed in terms of their respective estimating equations through (12.66). Show that their estimating equation representations also satisfy the same Jacobian relationship.

18. For a $(1, m)$ curved exponential family, use the Temme approximation to derive the CDF approximation for $\hat{\xi}$ in (12.85) by integrating density (12.80).

    (a) Express
    $$-\hat{w}^2/2 = -\hat{s}\dot{\theta}_{\hat{\xi}}^T \mu_{\hat{\xi}} + c(\tilde{\theta}) - c(\theta_{\hat{\xi}})$$
    and differentiate to show that
    $$\partial \hat{w}/\partial \hat{\xi} = \hat{s}\tilde{j}/\hat{w}.$$

    (b) Thus determine that $h(\hat{w}) = \hat{w}/\hat{u}$ to yield (12.85).

    (c) Show that
    $$\lim_{\hat{\xi} \to \xi} \left(\frac{\hat{w}}{\hat{s}}\right)^2 = \lim_{\hat{\xi} \to \xi} \left(\tilde{j}\, \partial \hat{\xi}/\partial \hat{s}\right).$$
    (Hint: See the arguments of section 2.3.2.)

    (d) To find $\partial \hat{s}/\partial \hat{\xi}$, differentiate the saddlepoint equation that determines $\hat{s}$. Show that
    $$\lim_{\hat{\xi} \to \xi} \partial \hat{s}/\partial \hat{\xi} = 1.$$

    (e) Show that
    $$\lim_{\hat{\xi} \to \xi} h(\hat{w}) = 1.$$

19. For the gamma hyperbola example in section 12.4.4 show that the saddlepoint density in (12.80) reduces to the exact density with Stirling's approximations replacing the Gamma functions in (12.86). Also derive the CDF approximation in (12.87).

    (a) Show that the saddlepoint is $\hat{s} = (\hat{\xi}^2 - 1)/2$ and
    $$\tilde{\theta}^T = -\tfrac{1}{2}(1 + \hat{\xi}^2, 1 + \hat{\xi}^{-2}) \qquad \theta_{\hat{\xi}} = -(\hat{\xi}, \hat{\xi}^{-1}).$$

    (b) Derive
    $$\tilde{i} = 8n(1 + \hat{\xi}^2)^{-2} \qquad \tilde{j} = 4n\hat{\xi}^{-1}(1 + \hat{\xi}^2)^{-1}$$
    and
    $$\mathcal{L}(\theta_{\hat{\xi}})/\mathcal{L}(\tilde{\theta}) = 4^n \hat{\xi}^{2n}(1 + \hat{\xi}^2)^{-2n}.$$

    (c) Combine parts (a) and (b) to determine the saddlepoint CDF in (12.87) and the saddlepoint density as
    $$\hat{f}(\hat{\xi}) = 4^n \sqrt{\frac{n}{\pi}} \frac{\hat{\xi}^{2n-1}}{(1 + \hat{\xi}^2)^{2n}} \qquad \hat{\xi} > 0$$

where
$$4^n\sqrt{\frac{n}{\pi}} = 2\hat{\Gamma}(2n)/\hat{\Gamma}(n)^2.$$

20. Consider the gamma exponential example of section 7.3.4 with true value $\xi = 1$. Plot the marginal density of $\hat{\xi}$ and also the CDF.
    (a) Determine that the MLE $\hat{\xi}$ solves
    $$e^{\hat{\xi}}x_2 + x_1 = 1/\hat{\xi} + 1. \tag{12.124}$$
    (b) Show that the saddlepoint $\hat{s}$ required in (12.81) is the solution to the equation
    $$e^{\hat{\xi}}x_2 + x_1 = (\hat{s}+1)^{-1} + (\hat{s}e^{\hat{\xi}} + e)^{-1}.$$
    (c) Combining (a) and (b), $\hat{s}$ is the root of a quadratic that has an explicit expression in terms $\hat{\xi}$. Determine this expression.
    (d) Complete the computation of the saddlepoint density and CDF.

21. Consider the marginal saddlepoint CDF for $\hat{\xi}$ in the bivariate normal correlation example of section 7.3. Take $n = 4$ and consider $\xi = 1/4$. In this example, $\hat{s}$ cannot be written explicitly in terms of $\hat{\xi}$. Equation (12.81) is quite complicated and must be solved numerically.
    (a) Compute a table of values for the marginal CDF of $\hat{\xi}$. (Hint: The $c(\theta)$ function may be difficult to determine from the structure of the exponential family. Note however that $c(s + \theta) - c(\theta)$ is the CGF of the canonical sufficient statistic which may be determined from the CGF of a 2×2 Wishart.)
    (b) Use Monte Carlo simulation to check the accuracy of the approximations computed in part (a).

22. A balanced two-factor ANOVA with repetitions has the model
    $$y_{ijk} = \mu + \alpha_i + \beta_j + \gamma_{ij} + e_{ijk}$$
    for $i = 1, \ldots, I$; $j = 1, \ldots, J$; and $k = 1, \ldots, K$. Suppose all factors are fixed effects with the usual constraints: $\alpha. = \sum_{i=1}^{I} \alpha_i = 0$, $\beta. = 0$, and $\gamma_{i.} = 0 = \gamma_{.j}$. The ANOVA table is

Source	d.f.	SS	E(MS)
Treatment	$I - 1$	SS_α	$\sigma^2 + \frac{KJ}{I-1}\sum_{i=1}^{I}\alpha_i^2$
Blocks	$J - 1$	SS_β	$\sigma^2 + \frac{KI}{J-1}\sum_{j=1}^{J}\beta_j^2$
Interaction	$(I-1)(J-1)$	SS_γ	$\sigma^2 + \frac{K}{(I-1)(J-1)}\sum_{i=1}^{I}\sum_{j=1}^{J}\gamma_{ij}^2$
Error	$IJ(K-1)$	SS_e	σ^2

    and all of the sums of squares are independent by Cochran's theorem. The F statistics for testing treatment and interaction effects are respectively $F_\alpha = MS_\alpha/MS_e$ and $F_\gamma = MS_\gamma/MS_e$ where $MS_\alpha = SS_\alpha/(I-1)$, etc. Assuming $\alpha_i \equiv 0 \equiv \gamma_{ij}$, then $SS_\alpha/\sigma^2 \sim \chi^2_{I-1}$, $SS_\gamma/\sigma^2 \sim \chi^2_{(I-1)(J-1)}$, and $SS_e/\sigma^2 \sim \chi^2_{IJ(K-1)}$.
    (a) For $I = J = K = 2$, compute the conditional saddlepoint CDF of F_α given $F_\gamma = f$ where f is the 95th percentile of an $F_{1,4}$. Display a table of values and plot the CDF.
    (b) Express the true conditional CDF of F_α given $F_\gamma = f$ as a double integral expression. Compute it for the values consider in part (a) and assess the accuracy of the conditional saddlepoint CDF approximation.

23. Consider the two-factor ANOVA model of Exercise 22 but now assume that treatment, block, and interaction effects are random. Suppose that $\{\alpha_i\}$, $\{\beta_j\}$, $\{\gamma_{ij}\}$, and $\{e_{ijk}\}$ are independent

normal with zero means and respective variances σ_α^2, σ_β^2, σ_γ^2, and σ^2. The important changes to the ANOVA table, as concerns the tests for treatment and interaction effects, are

$$E(MS_\alpha) = \sigma^2 + K\sigma_\gamma^2 + KJ\sigma_\alpha^2 \qquad (12.125)$$
$$E(MS_\gamma) = \sigma^2 + K\sigma_\gamma^2.$$

The new values in (12.125) motivate $F_\alpha = MS_\alpha/MS_\gamma$ as the test for treatment effects while the test for interaction remains as $F_\gamma = MS_\gamma/MS_e$. A balanced design assures that all mean squares are mutually independent. Repeat questions (a) and (b) in Exercise 22 assuming the null settings in which $\sigma_\alpha^2 = 0 = \sigma_\gamma^2$.

13

First passage and time to event distributions

In engineering reliability and multistate survival analysis, the machine or patient is viewed as a stochastic system or process which passes from one state to another over time. In many practical settings, the state space of this system is finite and the dynamics of the process are modelled as either a Markov process or alternatively as a semi-Markov process if aspects of the Markov assumption are unreasonable or too restrictive.

This chapter gives the CGFs connected with first passage times in general semi-Markov models with a finite number of states as developed in Butler (1997, 2000, 2001). Different formulae apply to different types of passage times, however all of these CGF formulae have one common feature: they are all represented as explicit matrix expressions that are ratios of matrix determinants. When inverting these CGFs using saddlepoint methods, the required first and second derivatives are also explicit so that the whole saddlepoint inversion becomes a simple explicit computation. These ingredients when used with the parametric plot routine in Maple, lead to explicit plots for the first passage density or CDF that completely avoid the burden of solving the saddlepoint equation.

Distributions for first passage or return times to a single absorbing state are considered in section 13.2 along with many examples derived from various stochastic models that arise in engineering reliability and queueing theory. Distributions for first passage to a subset of states require a different CGF formula which is developed in section 13.3. Passage times for birth and death processes can be approached from a slightly different perspective due to the movement restriction to neighboring states. Some recursions providing for their MGF and derivative computations are given in section 13.4 along with examples such as generalized random walks and Markov queues. The distinction between semi-Markov and Markov processes is clarified in section 13.1 and a detailed consideration of first passage in Markov processes, as a special case of semi-Markov processes, is discussed in section 13.5. The details for a more complicated example that deals with the reliability of a redundant and repairable system are consider in section 13.6.

13.1 Semi-Markov and Markov processes with finite state space

Suppose that a semi-Markov stochastic process has the state space $\mathcal{M} = \{1, \ldots, m\}$. The dynamics of the process can be characterized by specifying two $m \times m$ matrices. The first matrix is a probability transition matrix $P = (p_{ij})$ that oversees the probabilities for various state transitions. Thus P describes a discrete time Markov chain and is often referred to as the *jump chain* of the process; see Norris (1997). The second matrix is a matrix of

MGFs $M(s) = \{M_{ij}(s)\}$ whose entries correspond to the distributions for the various holding times before making the indexed state transition. For example, once destination j has been determined when exiting state i with probability p_{ij}, the random holding time in state i before exiting is characterized by the MGF $M_{ij}(s)$.

The dynamics can be made clear by considering the movement of a semi-Markov process that arrives in state 1 at time 0. Then, without any time delay, a Multinomial $(1; p_{11}, \ldots, p_{1m})$ trial is observed which determines state j as the next state with probability p_{1j}. With the destination thus determined, the holding time in state 1 until movement of state j is the random time associated with MGF $M_{1j}(s)$. Upon entering state j at time $X_1 \sim M_{1j}(s)$, the next state is determined (without any time delay) by a Multinomial $(1; p_{j1}, \ldots, p_{jm})$ trial that is independent of previous history, e.g. the previous multinomial trial and X_1. If state k is selected, then the system remains in state j for the random time $X_2 \sim M_{jk}(s)$ which is also independent of previous history. Thus the process continues onward with the jump chain determining state transitions and the matrix of MGFs specifying the random holding times in the various states before transition.

First passage properties of these processes always depend on these two matrices through their componentwise product

$$T(s) = \{T_{ij}(s)\} = P \odot M(s) = \{p_{ij} M_{ij}(s)\}$$

which is known as the *transmittance matrix* of the semi-Markov process. If P and $M(s)$ characterize the semi-Markov process, then so does the $m \times m$ transmittance matrix $T(s)$ since the two matrices can be recovered from $T(s)$ as $P = T(0)$ and $M(s) = T(s)/T(0)$. Those with a more traditional probability background would refer to $T(s)$ as the Laplace–Stieljes transform of the semi-Markov kernel matrix however the simpler transmittance terminology has been borrowed from electrical engineering and will be used here.

13.1.1 Examples

Negative binomial waiting time

A coin is tossed repeatedly with independent successive tosses and p the probability of heads. If Y is the waiting time for the third head, then $Y = 3 + X$ where X is the number of tails seen before the third head. The distribution of X is Negative Binomial $(3, p)$ as described in section 1.1.6 while Y is its alternative version with support translated to be over $\{3, 4, \ldots\}$.

A stochastic process for which Y is a first passage time has state space $\mathcal{M} = \{0, 1, 2, 3\}$. The state of the process at time k (after the kth toss) counts the cumulative number of heads seen up to time k. With each toss of the coin, the transmittance from current state i to state $i + 1$ is pe^s while the transmittance to remain in state i is qe^s where $q = 1 - p$. The 4×4 transmittance matrix of ordered states is

$$T(s) = \begin{pmatrix} qe^s & pe^s & 0 & 0 \\ 0 & qe^s & pe^s & 0 \\ 0 & 0 & qe^s & pe^s \\ 0 & 0 & 0 & 0 \end{pmatrix}. \tag{13.1}$$

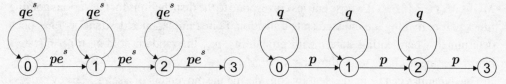

Figure 13.1. The waiting time Y is the first-passage distribution from $0 \to 3$ in the semi-Markov flowgraph on the left. Since the process is Markov, the unit step time MGF e^s can be removed to give the Markov graph on the right.

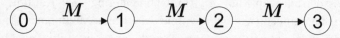

Figure 13.2. Semi-Markov flowgraph showing three successive Geometric (p) waiting times for the occurrance of a head.

The *flowgraph* in figure 13.1 (left) offers a pictorial approach for characterizing the same process. The nodes are the states and the labelled arrows show the possible transitions with the labels showing the transmittances. For this example, the holding times in states are the deterministic unit time 1 which makes the process a discrete time Markov chain. The usual Markov graph on the right-hand side is equivalent to the flowgraph with the MGF factor e^s removed.

A different process for which Y is also its passage time is shown in figure 13.2. The same state space $\mathcal{M}$ is used but now the common holding times in states are Geometric (p) waiting times that count the number of tosses until the first head with MGF

$$M(s) = pe^s(1 - qe^s)^{-1}. \tag{13.2}$$

An algebraic characterization of this process is the transmittance matrix T_1 given as

$$T_1(s) = \begin{pmatrix} 0 & M(s) & 0 & 0 \\ 0 & 0 & M(s) & 0 \\ 0 & 0 & 0 & M(s) \\ 0 & 0 & 0 & 0 \end{pmatrix}. \tag{13.3}$$

The process represented by $T_1(s)$ is a discrete time Markov process since the holding times in each states are Geometric distributions. The distinction between processes $T(s)$ and $T_1(s)$ in figures 13.1 and 13.2 is the presence of the self feedback loops attached to states of the former process. In $T(s)$, each unit tick of the clock leads to a potential state change whereas in $T_1(s)$ the self loops have been removed and states change only with a Geometric (p) number of ticks.

Waiting for three successive heads

Consider independent tosses of a coin as in the last example. Continue the tossing until the first run of three straight heads occurs. Also, for the purpose of discussion, suppose that it takes time a to toss a head and time b to toss a tail. The accumulated time until the first occurrence of three straight heads is the first passage time from state 0 to state 3 for the process shown in figure 13.3.

13.1 Semi-Markov and Markov processes with finite state space

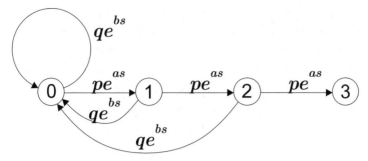

Figure 13.3. Flowgraph for a process that shows the first occurrance of three straight heads as a passage time from $0 \to 3$.

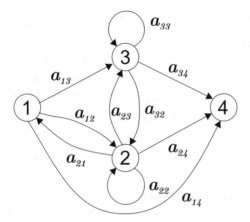

Figure 13.4. Markov graph for a continuous time Markov process.

The states are the current run length of heads. The transmittance of ordered states is

$$T(s) = \begin{pmatrix} qe^{bs} & pe^{as} & 0 & 0 \\ qe^{bs} & 0 & pe^{as} & 0 \\ qe^{bs} & 0 & 0 & pe^{as} \\ 0 & 0 & 0 & 0 \end{pmatrix}$$

and provides an algebraic characterization of the system in figure 13.3. When $a = b = 1$, the system is a Markov chain with state 3 as an absorbing end state. For more general values of a and b the process is not Markov but a semi-Markov process.

Continuous time Markov process

Figure 13.4 shows a Markov graph for a continuous time Markov process with state space $\mathcal{M} = \{1, 2, 3, 4\}$. The limitations of graphical display become obvious at this point due to the many different possible transitions. The transition arrows are labelled with the transition rates $\{a_{ij} > 0\}$ associated with independent Exponential (a_{ij}) transition times. The actual transition that occurs upon leaving state 1 say, is the result of the competition among the three independent Exponential times leaving state 1. With $X_{1i} \sim$ Exponential (a_{1i}) for $i = 2, 3, 4$ as these independent times, then the holding time in state 1 is $\min_i X_{1i} \sim$ Exponential $(a_{1\cdot})$

with $a_{1.} = a_{12} + a_{13} + a_{14}$. The probability the process goes to state i is proportional to the transition rate a_{1i} so that transition $1 \to 2$ occurs with probability $p_{12} = a_{12}/a_{1.}$.

The exit calculations out of state 1 apply to all state transitions in a Markov process and lead to the associated transmittance matrix

$$T(s) = \begin{pmatrix} 0 & p_{12}M_1 & p_{13}M_1 & p_{14}M_1 \\ p_{21}M_2 & p_{22}M_2 & p_{23}M_2 & p_{24}M_2 \\ 0 & p_{32}M_3 & p_{33}M_3 & p_{34}M_3 \\ 0 & 0 & 0 & 0 \end{pmatrix}$$

in which

$$p_{ij} = a_{ij}/a_{i.}. \qquad M_i(s) = (1 - s/a_{i.})^{-1}.$$

In the next section, the first passage time MGF from state $1 \to 4$ will be characterized in terms of the transmittance matrix $T(s)$.

Flowgraphs like figure 13.4 can get rather cluttered if all the transmittance labels are attached to the graph. A useful compromise is to use the flowgraph to help build the allowable transitions of states, but then summarize the quantitative aspects of these transitions by cataloging the transmittances in the matrix $T(s)$. Indeed only the transmittance matrix is needed to characterize the system however the flowgraph certainly brings the nature of the process to life.

13.1.2 Markov versus semi-Markov processes

The examples show the various forms of the transmittance matrix and also reveal when a semi-Markov process can be determined as Markov. For general semi-Markov transmittance $T(s) = \{T_{ij}(s)\}$ that is $m \times m$, there are three prototypes for Markov processes.

In the first case,

$$T_{ij}(s) = p_{ij}(1 - s/\lambda_i)^{-1} \qquad \lambda_i > 0 \qquad i, j = 1, \ldots, m$$

and is a continuous time Markov process in with the transition matrix $P = (p_{ij})$ determines the jump chain of state transitions and the holding time in state i is Exponential (λ_i).

The second case has

$$T_{ij}(s) = p_{ij}e^{as} \qquad i, j = 1, \ldots, m \qquad (13.4)$$

and is simply a Markov chain with time increment a instead of $a = 1$.

The third case has

$$T_{ij}(s) = \begin{cases} p_{ij}e^s\{1 - (1 - p_{i.})e^s\}^{-1} & \text{for } i \neq j = 1, \ldots, m \\ 0 & \text{for } i = j = 1, \ldots, m \end{cases} \qquad (13.5)$$

where $p_{i.} = \sum_{j \neq i} p_{ij}$. This is a discrete time Markov chain that does not allow self loops from $i \to i$, has transition probabilities $\{p_{ij}/p_{i.} : j \neq i\}$ out of state i, and has a holding time in state i that is Geometric $(p_{i.})$. Any deviations from these three prototypes assures that the Markov property cannot hold.

The discussion shows that there are two main qualitative differences between semi-Markov and Markov processes. First, apart from the chain in (13.4), Markov processes all have memoryless waiting times in the various states which are either exponential in continuous time or geometric in discrete time. Secondly, the holding time in each state is not allowed to depend on the destination state as it perhaps may for a semi-Markov process.

The remaining discussion of semi-Markov processes excludes the class of explosive processes. These are processes that allow for the possibility for an infinite number of state transitions in a finite amount of time. By and large, the majority of applications are not explosive.

13.2 Passage times with a single destination state

There is no loss in generality if the initial starting state is assumed to be 1 and the destination state is m in a general m-state semi-Markov process with transmittance $T(s)$.

The *first passage transmittance* from state 1 to state m is defined as the product of f_{1m}, the probability the process passes from $1 \to m$, times $\mathcal{F}_{1m}(s)$, the MGF for the first passage time given that arrival at state m is assured. In terms of expectation, this is

$$f_{1m}\mathcal{F}_{1m}(s) = E\left\{e^{sW_{1m}}1_{(W_{1m}<\infty)}\right\} \tag{13.6}$$

where W_{1m} is the first passage time from state 1 to state m. At $s = 0$, (13.6) is the identity $f_{1m} = \Pr(W_{1m} < \infty)$ giving the probability of passage, while $\mathcal{F}_{1m}(s)$ is the conditional MGF of W_{1m} given the event $\{W_{1m} < \infty\}$. When $f_{1m} < 1$, the distribution of W_{1m} is defective with $\Pr(W_{1m} = \infty) = 1 - f_{1m}$ the probability of never reaching state m.

13.2.1 First passage from 1 to $m \neq 1$

The following theorem specifies the first passage transmittance as a ratio of cofactors for the matrix $I_m - T(s)$. Various proofs of this result are provided in Butler (1997, 2000, 2001). The theorem supposes that the process state space $\mathcal{M} = \{1, \ldots, m\}$ is comprised of only states that are *relevant* to first passage from state 1 to state m. State i is considered a relevant state if it can be an intermediate state during such passage; thus some pathway $1 \to i \to m$ must have positive probability.

Theorem 13.2.1 *(Single destination cofactor rule). The first passage transmittance from state 1 to state $m \neq 1$ is*

$$f_{1m}\mathcal{F}_{1m}(s) = \frac{(m,1)\text{-cofactor of }\{I_m - T(s)\}}{(m,m)\text{-cofactor of }\{I_m - T(s)\}} = \frac{(-1)^{m+1}|\Psi_{m1}(s)|}{|\Psi_{mm}(s)|} \tag{13.7}$$

where $\Psi_{ij}(s)$ is the (i,j)th minor of $I_m - T(s)$, or the submatrix of $I_m - T(s)$ with the ith row and jth column removed. The ratio (13.7) is well defined over an maximal convergence neighborhood of 0 of the form $(-\infty, c)$ for some $c > 0$ under these conditions:

(1) *The system states $\mathcal{M} = \{1, \ldots, m\}$ are exactly those relevant to passage from $1 \to m$ with all relevant states and no nonrelevant states.*
(2) *The intersection of the maximal convergence neighborhoods for the MGFs in the first $m - 1$ rows of $T(s)$ is an open neighborhood of 0.*

As a simple example, consider the negative binomial transmittance in (13.1). According to theorem 13.2.1, the first passage transmittance from $0 \to 3$ has $f_{03} = 1$ and

$$\mathcal{F}_{03}(s) = (-1)^5 \begin{vmatrix} -pe^s & 0 & 0 \\ 1-qe^s & -pe^s & 0 \\ 0 & 1-qe^s & -pe^s \end{vmatrix} \begin{vmatrix} 1-qe^s & -pe^s & 0 \\ 0 & 1-qe^s & -pe^s \\ 0 & 0 & 1-qe^s \end{vmatrix}^{-1}$$
$$= \{pe^s/(1-qe^s)\}^3$$

which is the known MGF for Y. Alternatively, the process in figure 13.2 with transmittance (13.3) could have been considered which leads to

$$\mathcal{F}_{03}(s) = (-1)^5 \begin{vmatrix} -M(s) & 0 & 0 \\ 1 & -M(s) & 0 \\ 0 & 1 & -M(s) \end{vmatrix} \begin{vmatrix} 1 & -M(s) & 0 \\ 0 & 1 & -M(s) \\ 0 & 0 & 1 \end{vmatrix}^{-1}$$
$$= M(s)^3 \qquad (13.8)$$

which is the same result since $M(s)$ is a Geometric (p).

The flowgraph in figure 13.2 shows a series connection of the states $0 \to 1 \to 2 \to 3$. It is perhaps the simplest example of a *cascading* process defined as a process with no feedback loops and which therefore cannot repeat any of its states. For a cascading process, the theorem takes the following simpler form.

Corollary 13.2.2 *Suppose the semi-Markov process described in the theorem is also cascading. Then the first passage transmittance from $1 \to m$ in (13.7) further simplifies to*

$$f_{1m}\mathcal{F}_{1m}(s) = (-1)^{m+1} |\Psi_{m1}(s)| \qquad m \neq 1. \qquad (13.9)$$

The corollary follows by generalizing the computation made in (13.8).

13.2.2 Simple illustrative examples

Three successive heads

The first passage transmittance in this setting has $f_{03} = 1$ and

$$\mathcal{F}_{03}(s) = (-1)^5 \begin{vmatrix} -pe^{as} & 0 & 0 \\ 1 & -pe^{as} & 0 \\ 0 & 1 & -pe^{as} \end{vmatrix} \begin{vmatrix} 1-qe^{bs} & -pe^{as} & 0 \\ -qe^{bs} & 1 & -pe^{as} \\ -qe^{bs} & 0 & 1 \end{vmatrix}^{-1}$$
$$= p^3 e^{3as} \left\{ 1 - qe^{bs} - pqe^{(a+b)s} - p^2 qe^{(2a+b)s} \right\}^{-1}. \qquad (13.10)$$

With $a = 1 = b$, (13.10) is a well-known expression given in Feller (1968, XIII.7, (7.6)) in terms of its probability generating function $\mathcal{F}_{03}(\ln z)$. However it is worth comparing the different degrees of difficulty involved in deriving this result. Feller's development requires a good deal of understanding and thought within the context of his development of recurrent events. The derivation from theorem 13.2.1 is a simple elementary exercise in symbolic computation using Maple.

13.2 Passage times with a single destination state

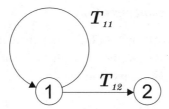

Figure 13.5. A simple self-loop.

Self-loop removal and compound distributions

Consider the simple self-loop in Figure 13.5 with $T_{1j} = p_{1j} M_{1j}(s)$ with $p_{12} = 1 - p_{11}$. First passage from $1 \to 2$ entails traversing the self-loop $1 \to 1$ for a Geometric $(p_{12}) - 1$ number of times with each loop requiring random time specified by $M_{11}(s)$. The final toss succeeds in reaching state 2 with holding time $M_{12}(s)$. Thus the first passage time has tranmittance

$$\mathcal{F}_{12}(s) = T_{12}(s)\{1 - T_{11}(s)\}^{-1} \qquad (13.11)$$

as may be confirmed from theorem 13.2.1.

In the special case in which $M_{11}(s) = M_{12}(s)$, then (13.11) is the MGF for a compound distribution. The Geometric (p_{12}) distribution has probability generating function $G(z) = p_{12}z(1 - p_{11}z)^{-1}$ and (13.11) is written as

$$\mathcal{F}_{12}(s) = G\{T_{11}(s)\}$$

and represents the distribution of $\sum_{i=1}^{N} Z_i$ where $N \sim$ Geometric (p_{12}) and $Z_1, Z_2, \ldots$ are i.i.d. with distribution represented by $M_{11}(s)$.

General self-loop removal

Consider a very general semi-Markov process with $m \times m$ transmittance matrix $T(s)$ in which any subset of the states $\mathcal{K} \subseteq \{1, \ldots, m-1\}$ have self loops $\{T_{kk}(s) : k \in \mathcal{K}\}$. These self loops are "removed" if another $m \times m$ semi-Markov transmittance $T^*(s)$ can be determined with the same states such that $T^*_{kk}(s) = 0$ for all $k \in \mathcal{K}$ and first passage from $1 \to m$ with $T(s)$ is the same distribution as with $T^*(s)$.

The previous example suggests the method for creating $T^*(s)$. Let $T^*(s)$ be $T(s)$ for all rows not indexed by $\mathcal{K}$. For each row $k \in \mathcal{K}$, set $T^*_{kk}(s) = 0$ and for each $l \neq k$ set $T^*_{kl}(s) = T_{kl}(s)/\{1 - T_{kk}(s)\}$. The idea is to modify each $T_{kl}(s)$ as in the simple loop removal above in order to account for time spent feeding back into state k. Let $\Psi_{m1}(s)$ and $\Psi^*_{m1}(s)$ represent the respective terms to be used in (13.7) for $T(s)$ and $T^*(s)$. Since the determinant function is linear in each row vector of its matrix argument, the new first passage transmittance from $T^*(s)$ is

$$(-1)^{m+1} \frac{|\Psi^*_{m1}(s)|}{|\Psi^*_{mm}(s)|} = (-1)^{m+1} \frac{\prod_{k \in \mathcal{K}}\{(1 - T_{kk})^{-1}\} |\Psi_{m1}(s)|}{\prod_{k \in \mathcal{K}}\{(1 - T_{kk})^{-1}\} |\Psi_{mm}(s)|} = f_{1m}\mathcal{F}_{1m}(s)$$

and the first passage transmittances are the same.

13.2.3 Explicitness of saddlepoint ingredients

To approximate the first passage density or CDF, only three ingredients are required and all are explicit matrix computations using the transmittance matrix $T(s)$. Suppose that $f_{1m} = 1$ and the passage time has a nondefective distribution with CGF

$$K_{1m}(s) = \ln \mathcal{F}_{1m}(s) = \ln \left\{ \frac{(-1)^{m+1} |\Psi_{m1}(s)|}{|\Psi_{mm}(s)|} \right\}. \tag{13.12}$$

Its explicit derivative for the saddlepoint equation, with the dependence on s suppressed, is

$$K'_{1m}(s) = \operatorname{tr}\left(\Psi_{m1}^{-1} \dot{\Psi}_{m1} - \Psi_{mm}^{-1} \dot{\Psi}_{mm}\right) \tag{13.13}$$

where $\dot{\Psi}_{m1} = d\Psi_{m1}(s)/ds$ is the $(m-1) \times (m-1)$ matrix of derivatives. One further derivative gives

$$K''_{1m}(s) = \operatorname{tr}\left\{\Psi_{m1}^{-1} \ddot{\Psi}_{m1} - \left(\Psi_{m1}^{-1} \dot{\Psi}_{m1}\right)^2 - \Psi_{mm}^{-1} \ddot{\Psi}_{mm} + \left(\Psi_{mm}^{-1} \dot{\Psi}_{mm}\right)^2\right\} \tag{13.14}$$

where $\ddot{\Psi}_{m1}(s) = d^2\Psi_{m1}(s)/ds^2$.

The explicitness of (13.12), (13.13), and (13.14) makes all saddlepoint computations for first passage distributions simple and easy if programmed by using a matrix manipulative language such as Maple. For most purposes, even the computation of K''_{1m} that uses (13.14) can be replaced by taking a numerical derivative of K'_{1m} in (13.13). Plots of the first passage density, CDF, and failure (hazard) rate are especially simple computations when the parametric plot routine in Maple is used. Suppose the convergence strip of $\mathcal{F}_{1m}(s)$ is $s \in (-\infty, c)$ and let

$$\hat{f}(s) = \frac{1}{\sqrt{2\pi K''_{1m}(s)}} \mathcal{F}_{1m}(s) \exp\left\{-s K'_{1m}(s)\right\}$$

be the saddlepoint density as a function of saddlepoint s. A plot of the coordinate pairs $\{K'_{1m}(s), \hat{f}(s)\}$ traced out for $s \in (-\infty, c)$ gives the density curve in 2-space without ever having to solve the saddlepoint equation. Likewise, if $\hat{F}(s)$ and $\hat{z}(s)$ are the CDF and failure rates in terms of s, then they too can be plotted the same way.

If $f_{1m} < 1$ and the passage time has a defective distribution, then saddlepoint methods are simply applied to the distribution of W_{1m} conditional on $\{W_{1m} < \infty\}$. The conditional CGF is

$$K_{1m}(s) = \ln \mathcal{F}_{1m}(s)$$

and the values of $K'_{1m}(s)$ and $K''_{1m}(s)$ above remain unchanged.

13.2.4 Numerical example: Reliability of systems

This example has been replicated from Butler (2001) and deals with the reliability of a system that consists of n independently operating components.

Determine the reliability for a system that consists of n components each of which can fail. With components labelled from 1 to n, suppose they fail independently and the failure time of component i is Exponential (λ_i). Suppose also that components in $C = \{1, \ldots, n-1\}$

are repairable while component n cannot be repaired. If component $i \in C$ fails, the system shuts down and is repaired in a time that has distribution H_i and is independent of the failure times. If component n fails however, the system cannot be repaired and there is system failure.

The system state space is $\mathcal{M} = \{0, 1, \ldots, n\}$ where 0 indicates that the system is up and $i \neq 0$ indicates that the system is down due to failure of component i. The lifetime of the system W including all repair times has a MGF as the first passage transmittance from $0 \to n$. Suppose that H_i has MGF $N_i(s)$ and denote the MGF of an Exponential $(\lambda.)$ with $\lambda. = \sum_i \lambda_i$ as $M(s)$. The system begins in state 0 and the first transition is $0 \to j$ if component j is the first to fail. This occurs with probability $p_j = \lambda_j/\lambda.$ and the waiting time for this failure is the holding time in state 0 which is Exponential $(\lambda.)$. Upon entering state j, the transmittance for return to state 0 is $N_j(s)$. Suppressing dependence on s, this leads to a transmittance with the components

$$T_{ij} = \begin{cases} 0 & \text{if } i = 0 \quad j = 0 \\ p_j M & \text{if } i = 0 \quad j \neq 0 \\ N_i & \text{if } i \in C \quad j = 0 \\ 0 & \text{if } i = n. \end{cases} \quad (13.15)$$

This is a semi-Markov process since some of the transmittances are not exponential.

Symbolic computation in Maple for this $(n+1)$-state system leads to the very simple form

$$\mathcal{F}_{0n}(s) = (-1)^{n+2} \frac{|\Psi_{n0}(s)|}{|\Psi_{nn}(s)|} = \frac{p_n M(s)}{1 - M(s) \sum_{i=1}^{n-1} p_i N_i(s)} \quad (13.16)$$

when computed using various values of n. Formal proof of (13.16) for all n follows from an inductive argument applied when computing the cofactors of the patterned matrix $T(s)$. The same expression also results from considering a simple two state system in which the states of $C \cup \{0\}$ are lumped into a "megastate" of a semi-Markov process having the 2×2 transmittance matrix

	$C \cup \{0\}$	n
$C \cup \{0\}$	$M(s) \sum_{i=1}^{n-1} p_i N_i(s)$	$p_n M(s)$
n	0	0

Self-loop removal from megastate $C \cup \{0\}$, as described in section 13.2.2, yields the same result as in (13.16). The essential idea of this equivalence is that the first passage transmittance from $0 \to n$ does not require that we distinguish the states in $C \cup \{0\}$ when the finite mixture transmittance $M \sum_{i=1}^{n-1} p_i N_i$ is the feedback transmittance of the megastate.

As a numerical example, suppose there are $n = 4$ components with $\lambda_1 = 0.22$, $\lambda_2 = 0.12$, $\lambda_3 = 0.04$ and $\lambda_4 = 0.02$. The system lifetime is the first passage from $0 \to 4$. Inverse Gaussian IG (μ, σ^2) repair times with mean μ and variance σ^2 are used so that $H_1 \sim$ IG $(1/2, 1/2)$, $H_2 \sim$ IG $(1, 1)$, and $H_3 \sim$ IG $(4, 4)$. These IG (μ, σ^2) distributions have the

440 First passage and time to event distributions

Table 13.1. *System reliabilities for W_{04}, the first passage time from $0 \to 4$. Saddlepoint reliabilities $\hat{R}(t)$ have been computed at the associated quantiles t determined by simulating 10^6 first passage times.*

Reliabilities	$R(t) = \Pr(W_{04} > t)$					
Tail probability	.30	.20	.10	.05	.01	.001
Simulated quantiles t	83.63	111.9	160.6	209.3	323.2	483.6
Sadpt. Rel. $\hat{R}(t)$	.3001	.2005	.1002	.05009	.009871	.001002

density

$$f(x;\mu,\sigma) = \frac{\mu^{3/2}}{\sqrt{2\pi}\sigma} \exp\left\{-\frac{\mu}{2\sigma^2 x}(x-\mu)^2\right\} \qquad x > 0 \qquad (13.17)$$

and all have CGFs that are convergent on $(-\infty, 1/2]$ with the form

$$\ln N(s) = \frac{\mu^2}{\sigma^2}\left\{1 - (1 - 2s\sigma^2/\mu)^{1/2}\right\} \qquad s \leq \mu/(2\sigma^2). \qquad (13.18)$$

The smallest positive root for the denominator of (13.16) is 0.01430 so $\mathcal{F}_{04}$ is convergent on $(-\infty, 0.01430)$.

Table 13.1 shows the accuracy of the Lugannani and Rice approximation $\hat{R}(t)$ for reliability $R(t)$ when inverting (13.16). The approximation $\hat{R}(t)$ has been evaluated at the distributional quantiles listed in row "Simulated quantiles t." The quantiles t are not exact but were determined by inverting the empirical CDF of 1 million simulations of the first passage time. A plot of the percentage relative error of $\hat{R}(t)$ as compares with $\tilde{R}(t)$, the empirical reliability function from the simulation, was reported in Butler (2001) and the graph never exceeds 3% in magnitude.

13.2.5 Numerical example: GI/M/1/4 queue

This queueing example has been reproduced from Butler (1997, 2000). In the interarrival renewal process, three different distributions are used for G which include Gamma (2, 2), compound Poisson, and inverse Gaussian.

Suppose that packets arrive at a server according to a renewal process in which the interarrival times are i.i.d. with distribution G. A single server provides service for the packets at rate μ with the service times of packets as i.i.d. Exponential (μ). Suppose the queue buffer can hold three packets and, with an additional packet under service, the maximum system capacity is four packets. If a new packet arrives when the system is at capacity 4, the queue overflows and the new packet is turned away. Let W be the waiting time to overflow for a system that starts at time 0 with an idle server, no packets in queue, and which uses G as the interarrival time for packets.

Take the state of the system as the queue length plus number in service so that $\mathcal{M} = \{0, \ldots, 5\}$ with state 5 representing overflow. The event $\{W \in (t, t + dt)\}$ occurs when a packet arrives during $(t, t + dt)$ which makes t an epoch for the renewal process of arrivals. It therefore suffices to devise a semi-Markov process in which state transitions occur only

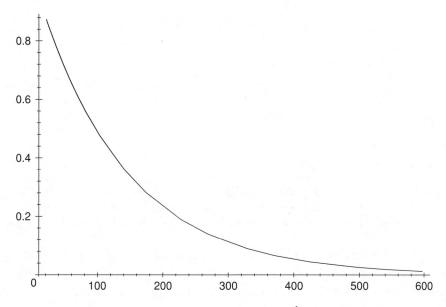

Figure 13.6. The almost indistinguishable reliability plots of $\hat{R}(t)$ (*dashed*) and $R(t)$ (*solid*) vs. t for the GI/M/1/4 queue with Gamma (2,2) interarrivals.

at the epochs of the arrival process. The 6×6 transmittance matrix for this system is

$$T(s) = \begin{pmatrix} 0 & U_0(s) & 0 & 0 & 0 & 0 \\ 0 & Q_{11}(s) & U_0(s-\mu) & 0 & 0 & 0 \\ 0 & Q_{21}(s) & U_1(s-\mu) & U_0(s-\mu) & 0 & 0 \\ 0 & Q_{31}(s) & U_2(s-\mu) & U_1(s-\mu) & U_0(s-\mu) & 0 \\ 0 & Q_{41}(s) & U_3(s-\mu) & U_2(s-\mu) & U_1(s-\mu) & U_0(s-\mu) \\ 0 & 0 & 0 & 0 & 0 & 0 \end{pmatrix} \quad (13.19)$$

where the individual entries are given as

$$\begin{aligned} U_k(s) &= \mu^k/k! \int_0^\infty w^k e^{sw} dG(w) & k &= 0, \ldots, 3 \\ Q_{k1}(s) &= U_0(s) - \sum_{j=0}^{k-1} U_j(s-\mu) & k &= 1, \ldots, 4. \end{aligned} \quad (13.20)$$

The reader is asked to verify this transmittance in Exercise 5. The time until queue overflow is $W = W_{05}$, the first passage time to state 5.

Gamma (α, β) *interarrivals* Suppose that $\mu = 2$ and G is a Gamma $(\alpha = 2, \beta = 2)$ distribution with mean 1 and variance $1/2$. The repair rate for the queue is 2 and the arrival rate is 1. The transmittances in (13.20) are based on the expression

$$U_k(s) = \frac{\mu^k \beta^\alpha}{(\beta-s)^{k+\alpha}} \binom{k+\alpha-1}{k} \quad k = 0, \ldots, 3. \quad (13.21)$$

For the integer values of α considered, these are rational expressions which, through the cofactor rule in theorem 13.2.1, determine the MGF of W as a rational function. Direct inversion of such rational MGFs is therefore possible by using a partial fraction expansion and leads to the exact calculation of reliability $R(t)$ and failure rate $z(t)$ which are shown in figures 13.6 and 13.7 as the solid lines. The saddlepoint approximation $\hat{R}(t)$ based on

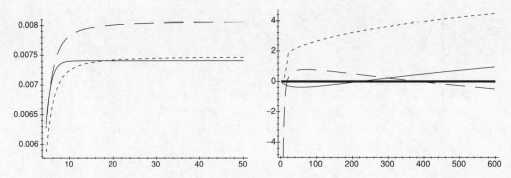

Figure 13.7. (Left) Plot of $z(t)$ (solid), $\hat{z}(t)$ (dashed), and $\bar{z}(t)$ (dotted) vs. t for the GI/M/1/4 queue with Gamma (2, 2) interarrivals. (Right) Percentage relative errors in saddlepoint approximation for $R(t)$ (solid = normal base, dotted = inverse Gaussian base (see section 16.2.2)) and for $z(t)$ (dashed).

inverting (13.7) is the almost indistinguishable dashed line in figure 13.6. Two saddlepoint failure rates are shown in figure 13.7 which include the unnormalized $\hat{z}(t)$ (dashed) and the normalized $\bar{z}(t)$ (dotted). For the range $t \in (0, 600)$, normalization is needed in order to accurately determine the failure rate.

The exact density, CDF, and failure rate are determined from theorem 13.2.1 as follows. $\mathcal{F}_{05}(s)$ is the (multiplicative) inverse of an order 10 polynomial and its ten roots are simple poles of $\mathcal{F}_{05}(s)$. These values are denoted as $v_1, \ldots, v_{10}$ and consist of four real roots and three complex conjugate pairs. Partial fraction expansion of $\mathcal{F}_{05}(s)$ yields $\mathcal{F}_{05}(s) = \sum_{i=1}^{10} c_i/(v_i - s)$, and the density of W from direct inversion is

$$f_W(t) = \sum_{i=1}^{10} c_i \exp(-v_i t) \tag{13.22}$$
$$= 0.00757 e^{-.00741 t} - 0.0441 e^{-.857 t} + 0.240 e^{-2.20 t}$$
$$- 0.624 e^{-3.39 t} + 0.468 e^{-4.05 t} \cos(1.04 t) + 0.110 e^{-4.05 t} \sin(1.04 t)$$
$$- 0.0544 e^{-5.05 t} \cos(2.16 t) - 0.0446 e^{-5.05 t} \sin(2.16 t)$$
$$+ 0.00682 e^{-5.67 t} \cos(2.98 t) + 0.00649 e^{-5.67 t} \sin(2.98 t).$$

The true $R(t)$ has been computed by using symbolic integration of (13.22).

The mean and standard deviation of W are determined from $\mathcal{F}_{05}$, as 138 and 135. An important characteristic of the system is its long term failure rate defined as the value of $z(t)$ as $t \to \infty$. From (13.22), this value is $z(\infty) = 0.00741$ and may also be seen as the asymptote for the plot of $z(t)$ in figure 13.7. It is the right edge of the convergence strip for $\mathcal{F}_{05}(s)$ which is a result that has been shown to hold quite generally in Butler and Bronson (2002) and Butler (2007).

Compound Poisson interarrivals

Suppose that G is the distribution of $Y = \sum_{i=1}^{N} X_i$ where $X_1, X_2, \ldots$ are i.i.d. Exponential (λ) and N as a Poisson (β) variable that is restricted to $N \geq 1$; the truncation prevents Y from having a point mass at 0. Lengthy calculations show in this setting that

$$U_k(s) = \frac{\mu^k e^{-\beta} \beta}{(1 - e^{-\beta}) \lambda^k} (1 - s/\lambda)^{-(k+1)} {}_1F_1\{k + 1, 2; \beta(1 - s/\lambda)^{-1}\} \quad s < \lambda$$

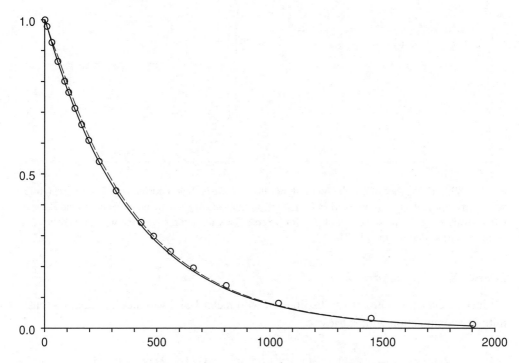

Figure 13.8. Comparison of saddlepoint reliability $\hat{R}(t)$ (*solid*) versus an empirical estimate of $R(t)$ (*circle centers*) for the $GI/M/1/4$ queue with compound Poisson interarrivals. The dashed line is an inverse Gaussian-based approximation for $R(t)$.

for $k = 0, \ldots, 3$, where $_1F_1$ is the confluent hypergeometric function. This expression reduces to simple forms for the integer values $k = 0, \ldots, 3$ assumed by k when the Kummer transform (13.1.27) and recurrence relation (13.4.4) of Abramowitz and Stegun (1970) are used.

Take $\mu = 2$, $\lambda = 1$, and $\beta = 1$ so that the repair rate is 2 and the arrival rate is $1/E(Y) = 1 - e^{-1} \simeq 0.632$. The resulting failure distribution has mean 392 and standard deviation 390. The mean and standard deviation are larger than for the queue with Gamma $(2, 2)$ interarrivals because the latter queue has the faster arrival rate of 1. The denominator of the MGF, as determined from cofactor rule (13.7), has its smallest positive real root as 0.00256 which is the asymptotic failure rate for the queue.

Figure 13.8 plots the saddlepoint reliability approximation $\hat{R}(t)$ (solid) against simulated approximations (circle centers) that used 10^6 failure runs through the system. The dashed curve $\check{R}(t)$ is an inverse Gaussian-based saddlepoint approximation introduced in Wood *et al.* (1993) and discussed in section 16.2.2.

Figure 13.9 compares the unnormalized $\hat{z}(t)$ (dashed) and normalized $\bar{z}(t)$ (solid) failure rate approximations with simulated approximations (circles) that use the 10^6 failure runs and implement kernel density approximation for the unknown density. The normalized saddlepoint approximation agrees to high accuracy with the simulation results. The saddlepoint computations required several minutes of computing time while the simulations took several hours. Also the saddlepoint plots do not suffer from the inherent roughness problems of kernel density estimation that exist even with samples of size 10^6. The normalization constant of the saddlepoint density is 1.083.

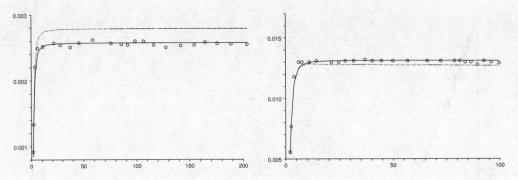

Figure 13.9. (Left) Plot of saddlepoint failure rate $\hat{z}(t)$ (*dashed*) and normalized failure $\bar{z}(t)$ (solid) versus an empirical approximation (*circle centers*) for the $GI/M/1/4$ queue with compound Poisson interarrivals. (Right) The same sort of plot for inverse Gaussian interarrivals but with $\bar{z}(t)$ (*solid*) and $\check{z}(t)$ (*dashed*) as defined in (13.23).

Inverse Gaussian interarrivals

An inverse Gaussian interarrival distribution with mean 1 and variance 1 leads to a transmittance computation with entries

$$U_k(s) = \frac{e\mu^k}{k!}\sqrt{\frac{2}{\pi}}(1-2s)^{-1/2(k-1/2)}K_{k-1/2}\bigl(\sqrt{1-2s}\bigr) \qquad s < 1/2$$

for $k = 0, \ldots, 3$ where K_ν denotes a Bessel K function. For half-integer values of ν, this function takes the simple form of the finite sum in (10.2.15) of Abramowitz and Stegun (1970). The convergence strip of the inverse Gaussian interarrival MGF is $(-\infty, 1/2]$ and not open; however the MGF of the failure distribution determined in (13.7) has an open convergence strip on $(-\infty, 0.0130)$ due to the zero in the denominator of the cofactor rule.

The mean failure time is 79.1 with standard deviation 76.7. Both moments are considerably smaller than the failure time with the other two interarrival distributions. This is also reflected by the much higher asymptotic failure rate of 0.013 that results from the inverse Gaussian interarrival. The higher failure rate tends to lead to a quicker failure than would occur with the other two systems.

Figure 13.10 plots $\hat{R}(t)$ (solid) and the inverse Gaussian-based approximation $\check{R}(t)$ (dashed) as previously described. Figure 13.9 (right) plots the normalized failure rate $\bar{z}(t)$ (solid) as well as the (dashed) normalized failure rate

$$\check{z}(t) = \bar{f}(t)/\check{R}(t) \tag{13.23}$$

based on $\check{R}(t)$ instead of $\hat{R}(t)$. Their empirical counterpart (circles) is based on 10^6 simulations that generate inverse Gaussian variates using the algorithm in Atkinson (1982). The normalization constant for the saddlepoint density is 1.080.

Final remarks

In each of the three different queueing models, the various saddlepoint approximations have been remarkable accurate in approximating the failure time distributions. The last model with inverse Gaussian interarrivals shows that the failure rate for the system can be quite

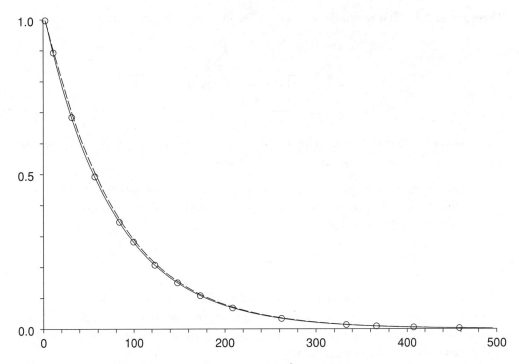

Figure 13.10. Comparison of saddlepoint reliability $\hat{R}(t)$ (*solid*) vs. an empirical estimate of $R(t)$ (*circle centers*) for the $GI/M/1/4$ queue with inverse Gaussian interarrivals. The dashed line is the inverse Gaussian-based approximation $\check{R}(t)$.

heavily dependent on the choice for the interarrival distribution and not just on the first two moments of that distribution.

Other characteristics of the $GI/M/1/4$ queue can be determined by using the cofactor rule in theorem 13.2.1. For example, the first passage transmittance from $4 \to 1$ from theorem 13.2.1 is $\mathcal{F}_{41}(s)$ and it is implicit in its computation that the system has avoided absorbing state 5. This is a setting for which $f_{41} < 1$ since $f_{45} > 0$ is the probability of absorption into 5 that preempts passage to state 0. For consideration of the second arrival time in state 1 after starting in state 0, see Exercise 8. The transmittance used with this semi-Markov process does not allow for computation of the busy period transmittance. This would be the first passage from 1 until the time that all service is complete but before arrival of the next packet. The first return time $1 \to 1$ records a time that is stochastically larger than the busy period and its computation is considered next.

13.2.6 First return to state 1

Consider the case in which W_{11} is the first return time to state 1, a situation that has been excluded from consideration in theorem 13.2.1. If the semi-Markov process enters state 1 at time 0, then f_{11} is $\Pr(W_{11} < \infty)$, the probability that the process returns to state 1 in finite time. The first return time MGF $\mathcal{F}_{11}(s)$ is the conditional MGF of W_{11} given that $W_{11} < \infty$. Various proofs of the following theorem are given in Butler (1997, 2000, 2001).

Theorem 13.2.3 *(First return cofactor rule). The first return transmittance for state 1 is*

$$f_{11}\mathcal{F}_{11}(s) = E\left\{e^{sW_{11}}1_{(W_{11}<\infty)}\right\}$$
$$= 1 - \frac{|I_m - T(s)|}{(1,1)\text{-cofactor of }\{I_m - T(s)\}}$$
$$= 1 - \frac{|I_m - T(s)|}{|\Psi_{11}(s)|}. \tag{13.24}$$

The ratio in (13.24) is well-defined over a maximal convergence neighborhood of 0 of the form $(-\infty, c)$ *for some* $c > 0$ *under these conditions:*

(1) *The system states* $\mathcal{M} = \{1, \ldots, m\}$ *are exactly those relevant to passage from* $1 \to 1$.
(2) *The intersection of the maximal convergence neighborhoods for the MGFs in* $T(s)$ *is an open neighborhood of 0.*

Simple examples

Cyclical chain The periodic chain in which the states in $\mathcal{M} = \{1, 2, 3\}$ are visited in the sequence $1 \to 2 \to 3 \to 1$ has a probability transition matrix that is a 3×3 circulant matrix. With general transmittances, the first return transmittance is clearly $T_{12}T_{23}T_{31}$. This is also the result of theorem 13.2.3 in which

$$f_{11}\mathcal{F}_{11}(s) = 1 - \begin{vmatrix} 1 & -T_{12} & 0 \\ 0 & 1 & -T_{23} \\ -T_{31} & 0 & 1 \end{vmatrix} \begin{vmatrix} 1 & -T_{23} \\ 0 & 1 \end{vmatrix}^{-1} = T_{12}T_{23}T_{31}.$$

Three successive heads Consider first passage from $0 \to 0$ with state 3, achieved with three straight heads, as an absorbing state. State 3 is not relevant to passage from $0 \to 0$ so it is excluded when determining $T(s)$ which now becomes 3×3. The first return transmittance is computed symbolically as

$$f_{00}\mathcal{F}_{00}(s) = 1 - \begin{vmatrix} 1-qe^{bs} & -pe^{as} & 0 \\ -qe^{bs} & 1 & -pe^{as} \\ -qe^{bs} & 0 & 1 \end{vmatrix} \begin{vmatrix} 1 & -pe^{as} \\ 0 & 1 \end{vmatrix}^{-1}$$
$$= qe^{bs} + qe^{bs}pe^{as} + qe^{bs}(pe^{as})^2. \tag{13.25}$$

The first return probability is obtained by setting $s = 0$ and gives $f_{00} = 1 - p^3$. This answer might have been expected because its complementary event is direct passage to state 1 with three straight heads which has probability p^3.

The transmittance computation in (13.25) removed state 3 since it was not relevant. If this state were left in, so $T(s)$ is now 4×4, then the same transmittance would result from expression (13.24). While state 3 is not needed in order to get the correct transmittance, ignoring the fact that it is not relevant "does no harm" except for increasing the dimensions of the matrices. Further comment on this is given in the next subsection.

Modified negative binomial Suppose the two flowgraphs in figures 13.1 and 13.2 are adapted in the following way. Upon arriving in state 3, two exit transmittances are added: a self-loop $T_{33}(s) = pe^s$ and $T_{30}(s) = qe^s$. These exit modifications out of state 3, assure return to state 0 for each of the flowgraphs, however the two systems have different first passage transmittances from $0 \to 0$.

For the flowgraph in figure 13.2, first passage from $0 \to 0$ is only achieved by first passing from $0 \to 3$ followed by a Geometric (q) waiting time to return to 0. The minimal number of tosses is 4 and the transmittance reflects the convolution of a Negative Binomial $(3, p)$ and a Geometric (q) as

$$\mathcal{F}_{00}(s) = \{pe^s(1 - qe^s)^{-1}\}^3 \{qe^s(1 - pe^s)^{-1}\}. \tag{13.26}$$

Symbolic computation from (13.24) confirms this result.

For the flowgraph in figure 13.1, first return to 0 can either occur directly in 1 step due to the self-loop on state 0, or, if this does not happen, the system can proceed in its first step to state 1 with transmittance pe^s. From state 1, the remaining portion of the return branch convolutes a Negative Binomial $(2, p)$ and a Geometric (q) so that

$$\mathcal{F}_{00}(s) = qe^s + pe^s\{pe^s(1 - qe^s)^{-1}\}^2\{qe^s(1 - pe^s)^{-1}\}. \tag{13.27}$$

Again this result can be confirmed directly from (13.24).

The difference between (13.26) and (13.27) raises a subtle and interesting point concerning the removal of self-loops. The removal of self-loops in the flowgraph of figure 13.1 leads to the flowgraph of figure 13.2 and both have the same first passage transmittance $\mathcal{F}_{03}(s)$. However, if first return to state 0 is considered for the process represented in Figure 13.1, the correct transmittance cannot be computed from (13.24) if the self-loop on state 0 has been removed beforehand. This idea is formalized in the next example.

Self-loop removals The first return transmittance from $1 \to 1$ in a general m-state system with transmittance matrix $T(s)$ is "not altered" if self-loops from states $2, \ldots, m$ are properly removed and compensated for in a revised m-state system $T^*(s) = \{T^*_{ij}(s)\}$ in which

$$T^*_{ij}(s) = \begin{cases} T_{ij}(s)\{1 - T_{ii}(s)\}^{-1} & \text{if } i \neq 1 \\ T_{ij}(s) & i = 1 \end{cases}$$

for all j. By "not altered", we mean that $f_{11}\mathcal{F}_{11}(s)$, as computed from theorem 13.2.3, is the same whether based on the system $T(s)$ or the revised system $T^*(s)$ as input into (13.24).

The proof of this result is exactly the same as that used for self loop removal in first passage from $1 \to m$ but now working with the expression for $f_{11}\mathcal{F}_{11}(s)$ in Theorem 13.2.3.

Reliability of systems Determine the first return transmittance $f_{11}\mathcal{F}_{11}(s)$ for the four component system of section 13.2.4 with state space $\mathcal{M} = \{0, 1, \ldots, 4\}$. When considering passage $1 \to 1$, state 4 is absorbing and not relevant so it can be deleted from the system. Theorem 13.2.3 applies to the transmittance (13.15) with $T(s)$ restricted to states $\{0, 1, 2, 3\}$ and gives

$$f_{11}\mathcal{F}_{11}(s) = 1 - \begin{vmatrix} 1 & -p_1 M & -p_2 M & -p_3 M \\ -N_1 & 1 & 0 & 0 \\ -N_2 & 0 & 1 & 0 \\ -N_3 & 0 & 0 & 1 \end{vmatrix} \begin{vmatrix} 1 & -p_2 M & -p_3 M \\ -N_2 & 1 & 0 \\ -N_3 & 0 & 1 \end{vmatrix}^{-1}$$

$$= p_1 M(s) N_1(s) [1 - M(s)\{p_2 N_2(s) + p_3 N_3(s)\}]^{-1}. \tag{13.28}$$

The form of the transmittance suggests a simpler derivation. States $\{2, 3\}$ can be removed from the system if a self loop is imposed on state 0 with transmittance $M(s)\{p_2 N_2(s) + p_3 N_3(s)\}$. Now consider the two state system $\{0, 1\}$ with transmittance

$$\begin{pmatrix} M(s)\{p_2 N_2(s) + p_3 N_3(s)\} & p_1 M \\ N_1 & 0 \end{pmatrix}.$$

For this two state system, the simpler first return transmittance computation of $1 \to 1$ from theorem 13.2.3 leads to the same expression in (13.28).

Explicit derivatives as saddlepoint inputs

The reader is asked to derive the first two derivatives of $K_{11}(s) = \ln \mathcal{F}_{11}(s)$ in Exercise 11. With argument s suppressed, the first derivative is

$$K'_{11} = \left\{ \frac{1}{f_{11} \mathcal{F}_{11}} - 1 \right\} \operatorname{tr} \left[(I - T)^{-1} \dot{T} + \Psi_{11}^{-1} \dot{\Psi}_{11} \right] \tag{13.29}$$

while the second derivative is

$$K''_{11} = \left\{ \frac{1}{f_{11} \mathcal{F}_{11}} - 1 \right\} \operatorname{tr} \left[(I - T)^{-1} \ddot{T} + \{(I - T)^{-1} \dot{T}\}^2 + \Psi_{11}^{-1} \ddot{\Psi}_{11} \right.$$
$$\left. - \{\Psi_{11}^{-1} \dot{\Psi}_{11}\}^2 \right] - K'_{11}(s)^2 / (1 - f_{11} \mathcal{F}_{11}). \tag{13.30}$$

At $s = 0$, these derivatives have removable singularities when $f_{11} = 1$. The first two cumulants of the return distribution are described below.

13.2.7 Further topics

Relationship of two cofactor rules

Expression (13.27) of the modified negative binomial example suggests a general relationship between the first return and first passage cofactor rules in theorems 13.2.1 and 13.2.3. Suppressing argument s, the expansion of $|I - T|$ using Cramer's rule in the first row gives

$$f_{11} \mathcal{F}_{11} = 1 - |I - T|/|\Psi_{11}|$$
$$= 1 - \frac{1}{|\Psi_{11}|} \left\{ (1 - T_{11}) |\Psi_{11}| - \sum_{j=2}^{m} T_{1j} (-1)^{j+1} |\Psi_{1j}| \right\}$$
$$= T_{11} + \sum_{j=2}^{m} T_{1j} f_{j1} \mathcal{F}_{j1}. \tag{13.31}$$

The relationship in (13.31) is an expression of the one-step analysis methods of Taylor and Karlin (1998, section 3.4.1). It says that the first return to state 1 proceeds directly in one step with T_{11} or, if the first step is to state $j \neq 1$, in two or more steps via first passage from state j. Reversing this one-step analysis actually proves that theorem 13.2.3 can be derived from theorem 13.2.1.

13.2 Passage times with a single destination state

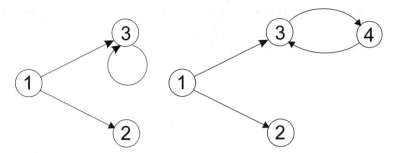

Figure 13.11. Systems with nonrelevant states for passage from $1 \to 2$.

First passage with an initial distribution

Suppose an m-state system starts in a random state $i \in \{1, \ldots, m-1\}$ with prior distribution $\pi^T = (\pi_1, \ldots, \pi_{m-1})$. Then a cofactor rule can be specified to deal with the first passage distribution from prior state $\pi \to m$ using the following adaptation of theorem 13.2.1.

Corollary 13.2.4 *The first passage transmittance from prior state π to state m is*

$$f_{\pi m} \mathcal{F}_{\pi m}(s) = \frac{1}{|\Psi_{mm}(s)|} \begin{vmatrix} \{I - T(s)\}_{-m} \\ \pi^T \end{vmatrix} \qquad (13.32)$$

where $\{I - T(s)\}_{-m}$ is the $(m-1) \times m$ matrix $I - T(s)$ with the mth row removed. The conditions required are that the system states include all states relevant to passage from $\{1, \ldots, m-1\} \to m$.

The proof follows from a superposition principle that averages over all starting points to give

$$f_{\pi m} \mathcal{F}_{\pi m} = \sum_{i=1}^{m-1} \pi_i f_{im} \mathcal{F}_{im} = \sum_{i=1}^{m-1} \pi_i (-1)^{m+i} \frac{|\Psi_{mi}(s)|}{|\Psi_{mm}(s)|}$$

which is (13.32) by Cramer's rule.

The main reason for formalizing the form of $f_{\pi m} \mathcal{F}_{\pi m}$ as a ratio of determinants is to make the derivative computations of $\ln \mathcal{F}_{\pi m}(s)$ easy through matrix computation as seen with previous cofactor rules. Its derivatives are essentially computed as in (13.13) and (13.14).

Need for relevant states

All of the cofactor rules provided above have made the following assumptions about relevant states: (i) that all relevant states are in the state space, and (ii) no nonrelevant states have been included. This common sense assumption keeps the size of the matrices involved to a minimum but assumption (ii) is certainly not necessary for the cofactor rules to hold.

The consequence of having non-relevant states in the cofactor rules is that there is the possibility that it will lead to a removable singularity at $s = 0$ which, from a practical point of view, can destabilize all numerical computations in a neighborhood of the mean passage time. To see this, consider the two examples that can be constructed from figure 13.11.

The first example considers the flowgraph in figure 13.11 (left) but without the self loop on state 3. Let T_{ij} be the transmittance from state $i \to j$. First passage from $1 \to 2$ according to the cofactor rule in Theorem 13.2.1 gives

$$f_{12}\mathcal{F}_{12} = (-1)^3 \begin{vmatrix} -T_{12} & -T_{13} \\ 0 & 1 \end{vmatrix} \begin{vmatrix} 1 & -T_{13} \\ 0 & 1 \end{vmatrix}^{-1} = T_{12}$$

as expected. However, when the self-loop $3 \to 3$ is included, the cofactor rule gives

$$f_{12}\mathcal{F}_{12} = (-1)^3 \begin{vmatrix} -T_{12} & -T_{13} \\ 0 & 1-T_{33} \end{vmatrix} \begin{vmatrix} 1 & -T_{13} \\ 0 & 1-T_{33} \end{vmatrix}^{-1} = T_{12}\frac{1-T_{33}}{1-T_{33}}$$

and produces a removable singularity at $s = 0$.

The second example uses the flowgraph in figure 13.11 (right) which, when used with theorem 13.2.1, yields

$$f_{12}\mathcal{F}_{12} = (-1)^3 \begin{vmatrix} -T_{12} & -T_{13} & 0 \\ 0 & 1 & -T_{34} \\ 0 & -T_{43} & 1 \end{vmatrix} \begin{vmatrix} 1 & -T_{13} & 0 \\ 0 & 1 & -T_{34} \\ 0 & -T_{43} & 1 \end{vmatrix}^{-1}$$

$$= T_{12}\frac{1-T_{34}T_{43}}{1-T_{34}T_{43}}. \tag{13.33}$$

If $T_{43} = 0$ then the removable singularity in (13.33) is removed.

The examples lead to the following general conclusions. The inclusion of nonrelevant states to the system does not affect the validity of the cofactor rules when $s \neq 0$. If feedback is not possible within the confines of the additional nonrelevant states, then no removable singularity results with their inclusion; with feedback however, a removable singularity occurs. Exercise 12 asks the reader to prove these results. The basis of the proofs may be seen in the examples above.

First passage moments

Often the detail of the entire first passage CDF and the accuracy obtained in its approximation by saddlepoint methods are unnecessary. In such cases, perhaps a simple description of this distribution can be summarized by computing the first few cumulants or moments obtainable through differentiation of the cofactor rules.

First passage moments from $1 \to m \neq 1$ In the case for which $f_{1m} = 1$, the mean first passage time is $K'_{1m}(0)$ as specified in (13.13). The variance of first passage is $K''_{1m}(0)$ and is computed as in (13.14). If $f_{1m} < 1$ then expression (13.13) gives $E\{W_{1m}|\, W_{1m} < \infty\}$, the first passage mean given that passage is assured, and (13.14) gives $Var\{W_{1m}|\, W_{1m} < \infty\}$.

If all nonrelevant states have been removed from the system, the matrices involved in computing $K'_{1m}(0)$ and $K''_{1m}(0)$ are all invertible with $s = 0$.

First return moments from $1 \to 1$ Determining the mean and variance of the first return distribution is more difficult and their computation depends on the status of state 1, e.g. whether it is a transient state or a member of an irreducible subchain.

13.2 Passage times with a single destination state

Suppose state 1 is transient and that $f_{11} < 1$. If the system contains irreducible subchains, then all states in these subchains are not relevant and should be removed from the system so that only transient states remain. In the system that results, the first two moments for the distribution of W_{11} given $W_{11} < \infty$ may be easily computed as $K'_{11}(0)$ and $K''_{11}(0)$ in (13.29) and (13.30) since both of these expressions are analytic at $s = 0$. The key to this is removing the nonrelevant states so that at least one row of the jump chain $T(0)$ does not add to 1. This assures that $I_m - T(0)$ is invertible.

If state 1 is a member of an irreducible chain and all nonrelevant states have been removed, then $I_m - T(0)$ is a singular matrix and the moments need to be computed by taking the limiting values of $K'_{11}(s)$ and $K''_{11}(s)$ as $s \to 0$. The moment derivations given below are derived in the appendix of section 13.7.1 for the setting in which the irreducible jump chain $T(0)$ has the discrete time stationary distribution $\vartheta^T = (\vartheta_1, \ldots, \vartheta_m)$ and $f_{ij} \equiv 1$. Then, the mean first return to state 1 is

$$E(W_{11}) = \frac{1}{\vartheta_1} \sum_{i=1}^{m} \vartheta_i T'_{i\cdot}(0) = \frac{1}{|\Psi_{11}(0)|} \sum_{i=1}^{m} |\Psi_{ii}(0)| \, T'_{i\cdot}(0) \qquad (13.34)$$

where

$$T'_{i\cdot}(0) = \sum_{i=1}^{m} p_{ij} M'_{ij}(0)$$

is the mean holding time in state i marginalized over the destination states. The middle expression in (13.34) is easily recognized as the mean for first return; see equation (2.8) of Barlow and Proschan (1965, p. 133). However the last expression is less well known and is related to the graphical approach used by Solberg (1975) to determine the stationary distribution. See section 13.7.1 for a proof of the rightmost expression in (13.34).

The second moment about zero is very difficult to derive through differentiation of $\mathcal{F}_{11}$ so the relationship between first return and first passages expressed in (13.31) is used instead. One derivative leads to the mean identity

$$\mathcal{F}'_{11}(0) = T'_{1\cdot}(0) + \sum_{j=2}^{m} p_{1j} \mathcal{F}'_{j1}(0)$$

whose intuitive meaning should be clear. A second derivative gives

$$E(W_{11}^2) = T''_{1\cdot}(0) + \sum_{j=2}^{m} p_{1j} \{\mathcal{F}''_{j1}(0) + 2 M'_{1j}(0) \mathcal{F}'_{j1}(0)\} \qquad (13.35)$$

where the terms $\mathcal{F}'_{j1}(0)$ and $\mathcal{F}''_{j1}(0)$ can be computed from (13.13) and (13.14) respectively. This expression however can be reduced to a much simpler expression involving ϑ, the stationary distribution of the jump chain, as

$$\begin{aligned} E(W_{11}^2) &= \frac{1}{\vartheta_1} \sum_{i=1}^{m} \vartheta_i \left\{ T''_{i\cdot}(0) + 2 \sum_{j=2}^{m} T'_{ij}(0) \mathcal{F}'_{j1}(0) \right\} \\ &= \frac{1}{|\Psi_{11}(0)|} \sum_{i=1}^{m} |\Psi_{ii}(0)| \left\{ T''_{i\cdot}(0) + 2 \sum_{j=2}^{m} T'_{ij}(0) \mathcal{F}'_{j1}(0) \right\}. \end{aligned} \qquad (13.36)$$

This agrees with expression (2.10) in Barlow and Proschan (1965, p. 134) and its derivation is provided in the appendix of section 13.7.1. Expression (13.36) is an elementary computation based on $T(s)$.

13.3 Passage times with many possible destination states

Suppose the destination of the semi-Markov process is the collections of system states $D = \{p+1, \ldots, m\} \subset \mathcal{M}$ rather than a single state. Theorem 13.3.1 gives the first passage transmittance to D assuming that the process starts by entering state 1 at time 0 with $C = \mathcal{M} \setminus D = \{1, \ldots, p\}$ as the non-destination states. Denote this transmittance as $\mathcal{F}_{1D}(s) = E\{e^{sW_{1D}}1_{(W_{1D} < \infty)}\}$ where W_{1D} is the first passage time from $1 \to D$.

In a reliability setting, D may be the distinct states for which the system is down and W_{1D} is the operating time for the system. In a multistate survival model, the states D could be the fatal states when considering some medical disorder so that W_{1D} is the lifetime of a patient.

Generally $\mathcal{F}_{1D}(s) \neq \sum_{j \in D} f_{1j} \mathcal{F}_{1j}(s)$ because the sum of the individual first passage transmittances has counted certain pathways that it should not. For example, suppose a pathway exists from $1 \to p+1$ that does not stop at state $p+2$ along the way, and also a pathway exists from $p+1$ to $p+2$ that avoids returning to state $p+1$. Then the pathway $1 \to p+1 \to p+2$ that combines the sequence of two paths is counted in $f_{1,p+2} \mathcal{F}_{1,p+2}(s)$ but it should not be counted for $\mathcal{F}_{1D}(s)$.

The system transmittance matrix $T(s)$ has block form

$$T(s) = \begin{pmatrix} T_{CC}(s) & T_{CD}(s) \\ T_{DC}(s) & T_{DD}(s) \end{pmatrix}$$

where T_{CC} is $p \times p$ taking C into C, T_{DD} is $(m-p) \times (m-p)$ taking D into D, etc. Denote the row sums of $T_{CD}(s)$ as

$$T_{CD}(s)\mathbf{1} = T_{C \cdot}(s) = (T_{1 \cdot} \cdots T_{p \cdot})^T,$$

and let $\{I_p - T_{CC}(s)\}_{\setminus 1}$ denote the $p \times (p-1)$ matrix $I_p - T_{CC}(s)$ with its first column removed.

Theorem 13.3.1 *The first passage transmittance from state $1 \in C = \{1, \ldots, p\}$ to subset $D = \{p+1, \ldots, m\}$ in an m-state system is*

$$f_{1D}\mathcal{F}_{1D}(s) = \frac{\left| T_{C \cdot}(s) \quad \{I_p - T_{CC}(s)\}_{\setminus 1} \right|}{\left| I_p - T_{CC}(s) \right|}, \tag{13.37}$$

under the conditions specified below. Expression (13.37) is well-defined over an maximal convergence neighborhood of 0 of the form $(-\infty, c)$ for some $c > 0$ under these conditions:

(1) *The system states $\{1, \ldots, m\}$ are exactly those relevant to passage from $1 \to D$.*
(2) *The intersection of the maximal convergence neighborhoods for the MGFs in the first p rows of $T(s)$ is an open neighborhood of 0.*

Theorem 13.3.1 takes a simpler form when the system is cascading in C.

13.3 Passage times with many possible destination states

Corollary 13.3.2 *Consider the m-state system of theorem 13.3.1 and suppose it is also cascading in subset C. Then the first passage transmittance from $1 \in C \to D$ in (13.37) further simplifies to*

$$f_{1D}\mathcal{F}_{1D}(s) = \left| T_{C \cdot}(s) \quad \{I_p - T_{CC}(s)\}_{\backslash 1} \right|.$$

The matrix is $I_p - T_{CC}(s)$ with its first column replaced with $T_{C \cdot}(s)$, the row sums of the $p \times (m - p)$ block $T_{CD}(s)$.

13.3.1 Examples

Three successive heads

Expression (13.37) is easily verified by taking $D = \{2, 3\}$ and computing the first passage transmittance from $0 \to D$ as

$$f_{0D}\mathcal{F}_{0D}(s) = \frac{\begin{vmatrix} 0 & -pe^{as} \\ pe^{as} & 1 \end{vmatrix}}{\begin{vmatrix} 1 - qe^{bs} & -pe^{as} \\ -qe^{bs} & 1 \end{vmatrix}} = \frac{p^2 e^{2as}}{1 - qe^{bs} - pqe^{(a+b)s}}. \qquad (13.38)$$

Since passage to state 3 is through 2, this transmittance must also be the first passage transmittance to state 2 and is easily shown to agree with $f_{02}\mathcal{F}_{02}(s)$ when computed from (13.7).

Reliability of systems

Previously, this n component system had only a single nonrepairable component. Now suppose that it has several nonrepairable components and labelled as the destination states $D = \{l+1, \ldots, n\}$. Suppose the repairable states are $C = \{1, \ldots, l\}$. First passage from $0 \to D$ is the total lifetime and fixing time of the system which has the $(n+1) \times (n+1)$ transmittance matrix

$$T_{ij} = \begin{cases} 0 & \text{if} \quad i = 0 \quad j = 0 \\ p_j M & \text{if} \quad i = 0 \quad j \neq 0 \\ N_i & \text{if} \quad i \in C \quad j = 0 \\ 0 & \text{if} \quad i \in D. \end{cases} \qquad (13.39)$$

The multistate destination formula (13.37) reduces to

$$\mathcal{F}_{0D}(s) = \frac{\sum_{i=l+1}^{n} p_i M(s)}{1 - M(s) \sum_{i=1}^{l} p_i N_i(s)}. \qquad (13.40)$$

A separate independent justification follows by considering a two state system that consists of two megastates $C \cup \{0\}$ and D as in the case when D was a single state. Formal proof for all l and n is again based on induction.

Consider the $n = 4$ state system from section 13.2.4 but now take $C = \{1, 2\}$ and $D = \{3, 4\}$. Use the same parameter values $\{\lambda_i : i = 1, \ldots, 4\}$ for the exponential failure times as well as the same inverse Gaussian repair times for components 1 and 2. Table 13.2 shows the accuracy achieved by the Lugannani and Rice approximation when used to approximate

Table 13.2. *System reliabilities for $W_{0,D}$, the first passage time from $0 \to D = \{3, 4\}$. Saddlepoint reliabilities $\hat{R}(t)$ have been computed at the associated quantiles t determined by simulating 10^6 first passage times.*

Reliabilities:			$R(t) = \Pr(W > t)$			
Tail probability	.30	.20	.10	.05	.01	.001
Simulated quantiles t	24.70	33.15	47.51	62.03	95.70	143.5
Saddlepoint Reliability $\hat{R}(t)$	.2997	.1966	.09986	.04959	$.0^2 9776$	$.0^3 9743$

the reliability function. One million first passage times from $0 \to \{3, 4\}$ were generated and the empirical quantiles for the indicated tail probabilities are listed as the values for t. The saddlepoint reliabilities are evaluated at these simulated quantiles and very accurately show the true tail probabilities.

The MGF $\mathcal{F}_{0D}$ involved is convergent on $(-\infty, 0.04837)$ and its distribution is stochastically smaller than the passage time from $0 \to 4$ since the additional risk of passage to 3 must now be considered. The asymptotic failure rate 0.04837 for $\mathcal{F}_{0D}$ is considerably larger than that for $\mathcal{F}_{04}$ which is 0.01430.

13.3.2 Further topics

Need for relevant states

Theorem 13.3.1 has the requirement that no nonrelevant states are present in the system and this requirement may be eliminated for $s \neq 0$. The inclusion of nonrelevant states to the system does not affect the validity of the cofactor rule in (13.37) if $s \neq 0$. With feedback possible within the confines of these additional nonrelevant states, then a removable singularity is entered into the cofactor expression (13.37) at $s = 0$, however without such feedback the expression is analytic at $s = 0$.

First passage moments

The mean and variance for W_{1D} given that $W_{1D} < \infty$ are determined from the CGF $K(s) = \ln \mathcal{F}_{1D}(s)$. They are computed as in (13.13) and (13.14) since $\mathcal{F}_{1D}(s)$ also takes the form of a ratio of determinants and the matrices involved are invertible at $s = 0$.

13.4 Birth and death processes and modular systems

Birth and death processes require an orderly transition in states in which the exit from state i leads only into states $i + 1$ and $i - 1$ with transmittances $T_{i,i+1}(s)$ and $T_{i,i+1}(s)$. As a consequence, the first passage and first return transmittances connected to these systems can be specified as recursions as shown in Butler (1997, 2000, 2001). The ingredients needed for saddlepoint approximation may also be computed recursively in the transform domain in much the way that Kalman recursions are used for time series computation in the time

13.4 Birth and death processes and modular systems

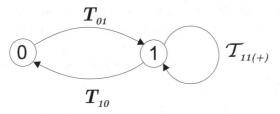

Figure 13.12. Modular system reflecting the recursions that describe the first return transmittance to state 0 in a birth and death process.

domain. General recursions for birth and death processes are given below that provide the first return transmittance to a state as well as the first passage transmittance to a new state.

These recursions have their graphical analogues in terms of flowgraphs that substitute first passage transmittances for single step branch transmittances in rather compact flowgraphs. The suppression of the detail for the first passage transmittance is an expression of the modular concept of subsystems that comprise a bigger system. The essential idea is that the single step traverse through a first passage transmittance branch represents the traverse through a subsystem with all the subsystem detail left out. Thus the flowgraph shows the overall general scheme by which the required passage transmittance is hierarchically achieved without all the subsystem details.

13.4.1 First return to state 0

Consider a semi-Markov process with state space $\mathcal{M}$ on the positive integers including the state 0. The first return transmittance to state 0 can be specified recursively using the flowgraph in figure 13.12. Suppose that $\mathcal{T}_{11(+)}(s) = f_{11(+)}\mathcal{F}_{11(+)}(s)$ is the first return transmittance to state 1 from the positively numbered states $\{2, 3, \ldots\}$ or the right-hand side.

Theorem 13.2.3 with cofactor rule (13.24) provides the first return transmittance to state 0 from the right side as

$$\mathcal{T}_{00(+)}(s) = \frac{\mathcal{T}_{01}(s)\,\mathcal{T}_{10}(s)}{1 - \mathcal{T}_{11(+)}(s)}. \tag{13.41}$$

This recursion for the first return $0 \to 0$ transmittance in terms of the $1 \to 1$ first return transmittance reflects the modular concept mentioned previously. The idea is to designate the first return $1 \to 1$ as a subsystem whose transmittance $\mathcal{T}_{11(+)}(s)$ summarizes all the activity that is required to bring the system back to state 1 so first return to state 0 is feasible in a single step. Of course $\mathcal{T}_{11(+)}(s)$ is itself a recursion in $\mathcal{T}_{22(+)}(s)$ according to a flowgraph just like that of figure 13.12, so in general

$$\mathcal{T}_{i,i(+)}(s) = \frac{\mathcal{T}_{i,i+1}(s)\,\mathcal{T}_{i+1,i}(s)}{1 - \mathcal{T}_{i+1,i+1(+)}(s)} \qquad i = 0, 1, \ldots \tag{13.42}$$

describes a recursive computation for $\mathcal{T}_{00(+)}(s)$. If the system state is finite, so $\mathcal{M} = \{0, 1, \ldots, n\}$, then the last recursion in (13.42) is for state $i = n - 1$ in terms of n. Otherwise the recursions continue indefinitely and their convergence depends on the situation. The recursion in (13.42) has an alternative derivation directly in terms of the two cofactor rules in theorems 13.2.1 and 13.2.3 which is given in the appendix of section 13.7.2.

The derivative functions $T'_{00(+)}(s)$ and $T''_{00(+)}(s)$, needed when using saddlepoint methods, are obtained by differentiating (13.42) twice and developing the recursions for $T'_{i,i(+)}$ and $T''_{i,i(+)}$ in terms of

$$\{T_{i+1,i+1(+)}, T'_{i+1,i+1(+)}, T''_{i+1,i+1(+)}\}$$

and derivatives of $T_{i,i+1}$ and $T_{i+1,i}$.

The two-sided state space $\mathcal{M} = \{-n, \ldots, -1, 0, 1, \ldots, n\}$ allows for first return from both sides. The first return transmittance is simply the sum of the first return transmittances from either side considered separately as disjoint subsystems. The first return from above is given by (13.41) while the first return from the negative integers has a flowgraph that is the mirror image of figure 13.12. Define $T_{i,i(-)}(s) = f_{i,i(-)} \mathcal{F}_{i,i(-)}(s)$ so $T_{0,0(-)}(s)$ may be computed recursively as

$$T_{i,i(-)}(s) = \frac{T_{i,i-1}(s) T_{i-1,i}(s)}{1 - T_{i-1,i-1(-)}(s)} \qquad i = 0, -1, \ldots \qquad (13.43)$$

with it derivatives also recursive. Hence, the overall first return transmittance is

$$T_{00}(s) = f_{00} \mathcal{F}_{00}(s) = T_{00(+)}(s) + T_{00(-)}(s).$$

General homogeneous random walks

For a classical random walk $\mathcal{M} = \{\ldots, -2, -1, 0, 1, 2, \ldots\}$ and each one-step transmittance is a coin toss with $T_{i,i+1}(s) = pe^s$ and $T_{i,i-1}(s) = qe^s$. With time homogeneity, the first return transmittances $T_{i,i(\pm)}(s)$ do not depend on the particular step i and also do not depend on the sign $\pm$ so the first return transmittance from either side is

$$T_{i,i(\pm)}(s) \equiv T_{0,0(+)}(s) \qquad \forall i.$$

The common value for $T_{0,0(+)}(s)$ can be resolved from recursions (13.42) and (13.43) by extracting the negative solution to the implied quadratic equations. This yields the well-known result

$$T_{0,0(+)}(s) = \frac{1}{2} - \frac{1}{2}\sqrt{1 - 4pqe^{2s}} \qquad (13.44)$$

found by Feller (1968, §XI.3, (3.13)) from a different point of view involving the renewal theory of generating functions.

The two-sided walk has

$$T_{0,0}(s) = 2T_{0,0(+)}(s) = 1 - \sqrt{1 - 4pqe^{2s}}$$

with convergence strip $(-\infty, c]$ that is not open where $c = -1/2 \ln(4pq) > 0$ if $p \neq 1/2$. Furthermore,

$$f_{00} = 1 - \sqrt{1 - 4pq} = 1 - |p - q|$$

which is less than 1 when $p \neq 1/2$. The case $p = 1/2$ and $c = 0$ results in a null persistent random walk for which $f_{00} = 1$ but with $\lim_{s \uparrow 0} \mathcal{F}'_{00}(s) = \infty$ so the saddlepoint equation can

13.4 Birth and death processes and modular systems

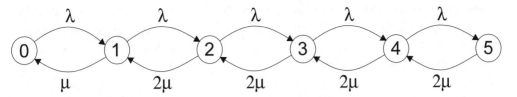

Figure 13.13. The Markov graph of a $M/M/2/5$ queue showing the exit rates out of the various states.

always be solved. However, for both the transient and null persistent settings in which p is in a neighborhood of $1/2$, the Lugannani and Rice approximation may exhibit uncharacteristic inaccuracy. These difficulties are overcome by switching to the inverse Gaussian-based saddlepoint CDF approximation as described in section 16.2.2.

A two-sided homogeneous random walk with general holding times in states occurs when $T_{i,i-1}(s) = p M_-(s)$ and $T_{i,i+1}(s) = q M_+(s)$. If the walk starts at 0, the first return transmittance from either side is

$$T_{0,0}(s) = 1 - \sqrt{1 - 4pq M_-(s) M_+(s)}. \tag{13.45}$$

This is twice the expression in (13.44) but with the fixed holding times MGFs e^{2s} replaced by $M_-(s) M_+(s)$. The same sort of saddlepoint inaccuracies occur when using the Lugannani and Rice approximation to invert (13.45) if p is anywhere near to $1/2$.

Busy period for a birth and death process

The *busy period* is defined as the first passage from $1 \to 0$ or the elapsed time from the moment state 1 is entered until state 0 is entered. The terminology is often used in the context of queueing systems as in the next example. The distribution for the busy period may be determined from its transmittance given by

$$T_{10}(s)\{1 - T_{1,1(+)}(s)\}^{-1} \tag{13.46}$$

where $T_{1,1(+)}(s)$ is the first return transmittance from the right as discussed above.

$M/M/q/n$ and $M/M/q/\infty$ queues

The $M/M/q$ queue is a Markov process that describes the dynamics of a stream of packets that arrive according to a Poisson (λ) process and are either distributed over the q independent servers or put into queue. All the servers complete tasks at rate μ with Exponential (μ) service times. The system state is the count of the number of jobs in queue and under service. The last symbol, n or ∞ in the queue designation, denotes the upper range for the system state so the $M/M/q/n$ queue has $\mathcal{M} = \{0, \ldots, n\}$ while the $M/M/q/\infty$ queue has $\mathcal{M}$ as the nonnegative integers. If the former queue is in state n and full up, any packet that arrives is lost. A full system consists of the queue length $n - q$ and all q servers busy.

Both of these queues are Markov birth and death processes and their first return transmittances are simple recursive computations. As a simple example consider a $M/M/2/5$ queue with the Markov graph shown in figure 13.13.

State 2 occurs when both servers are busy and the queue length is 0. Exit from state 2 is shown by the rates 2μ and λ reflecting a competition between two Exponential (μ) fixing times and the Exponential (λ) arrival time of an additional packet. With $r_2 = 2\mu + \lambda$, the corresponding transmittances are

$$T_{21}(s) = 2\mu/r_2(1 - s/r_2)^{-1} \qquad T_{23}(s) = \lambda/r_2(1 - s/r_2)^{-1}.$$

For the general $M/M/q/n$ queue,

$$T_{i,i-1}(s) = \begin{cases} i\mu/r_i(1 - s/r_i)^{-1} & \text{if } i \leq q \\ q\mu/r_q(1 - s/r_q)^{-1} & \text{if } i > q \end{cases} \qquad i = 1, \ldots, n \qquad (13.47)$$

and

$$T_{i,i+1}(s) = \begin{cases} \lambda/r_i(1 - s/r_i)^{-1} & \text{if } i \leq q \\ \lambda/r_q(1 - s/r_q)^{-1} & \text{if } i > q \end{cases} \qquad i = 0, \ldots, n-1 \qquad (13.48)$$

where $r_i = i\mu + \lambda$. The semi-Markov flowgraph would have the same structure as the Markov graph in figure 13.13 but would replace rates with transmittances.

For such a $M/M/q/n$ queue, the first return transmittance $0 \to 0$ is now a simple recursive computation.

A $M/M/2/\infty$ queue would have a Markov graph that continues the right side of figure 13.13 from state 6 onward using the same pattern of exit rates as used out of state 4. Thus, the portion of the process above state 2 is homogeneous and $T_{22(+)}(s)$, the first return transmittance to state 2 from above, is determined by using the homogeneous random walk example previously given. If first return $0 \to 0$ is the concern, then the infinite portion of the Markov graph from states 2 to ∞ can be replaced with a self-loop on state 2 with first return transmittance $T_{22(+)}(s)$. This modular idea leads to the shortened semi-Markov flowgraph in figure 13.14 as a replacement for the infinite Markov graph. The other transmittances in the figure that connect states 0, 1, and 2 are the same as those from the $M/M/2/5$ queue and are given in (13.47) and (13.48) with $i \leq q = 2$.

The first return transmittance $0 \to 0$ is now a simple computation based on theorem 13.2.3 that leads to

$$T_{00} = T_{01}T_{10}\frac{1 - T_{22(+)}}{1 - T_{22(+)} - T_{12}T_{21}}. \qquad (13.49)$$

The busy period from (13.46) is the simpler computation

$$\begin{aligned} T_{01} &= T_{10}\{1 - T_{11(+)}\}^{-1} \\ &= T_{10}\left\{1 - \frac{T_{12}T_{21}}{1 - T_{22(+)}}\right\}^{-1} \\ &= T_{10}\frac{1 - T_{22(+)}}{1 - T_{22(+)} - T_{12}T_{21}}. \end{aligned} \qquad (13.50)$$

The same expression in (13.50) results from using theorem 13.2.1 to compute the first passage transmittance from $1 \to 0$ in the finite system of figure 13.14.

13.4 Birth and death processes and modular systems

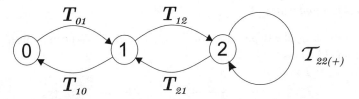

Figure 13.14. The flowgraph used to determine the first return transmittance $0 \to 0$ with module $\mathcal{T}_{22(+)}$ replacing the activity in states $\{3, 4, \ldots\}$ of the infinite Markov graph.

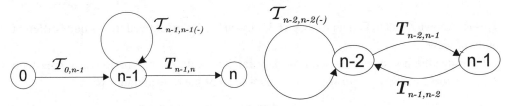

Figure 13.15. Modular systems showing a single step of the recursions needed to determine the first passage tranmittance from $0 \to m$ in a birth–death process.

13.4.2 First passage to state n

Similar recursions lead to the first passage transmittance from state 0 to n in a general birth and death process. Figure 13.15 shows the modular system flowgraphs needed to produce these recursions. The modular transmittances are shown in the leftmost flowgraph as $\mathcal{T}_{0,n-1}(s) = f_{0,n-1} \mathcal{F}_{0,n-1}(s)$, the first passage transmittance from state 0 to $n-1$, and $\mathcal{T}_{n-1,n-1(-)}$, the first return transmittance to state $n-1$ from the lower numbered states.

First passage $0 \to n$ is shown in the leftmost flowgraph and yields transmittance

$$\mathcal{T}_{0,n}(s) = \frac{\mathcal{T}_{0,n-1}(s) T_{n-1,n}(s)}{1 - \mathcal{T}_{n-1,n-1(-)}(s)} \tag{13.51}$$

as a recursion of $\mathcal{T}_{0,n}$ in n. Its computation also depends on a recursion for $\mathcal{T}_{n-1,n-1(-)}$ in $n-1$ specified in the rightmost flowgraph as

$$\mathcal{T}_{n-1,n-1(-)} = \frac{T_{n-1,n-2}(s) \, T_{n-2,n-1}(s)}{1 - \mathcal{T}_{n-2,n-2(-)}}. \tag{13.52}$$

Recursions (13.51) and (13.52) together along with their first two derivatives provide the inputs needed for saddlepoint inversion of $\mathcal{F}_{0,n}(s)$ even for quite large values of n such as $n = 1000$. The computation of $\mathcal{F}_{0,n}(s)$, $\mathcal{F}'_{0,n}(s)$, and $\mathcal{F}''_{0,n}(s)$ is implemented by carrying through the six-dimensional recursions for the terms

$$\mathcal{T}_{0,j} \; \mathcal{T}'_{0,j} \; \mathcal{T}''_{0,j} \quad \text{and} \quad \mathcal{T}_{j-1,j-1(-)} \; \mathcal{T}'_{j-1,j-1(-)} \; \mathcal{T}''_{j-1,j-1(-)} \tag{13.53}$$

for $j = 2, \ldots, n$. The only limitation in their use is the stability and accuracy of the computations that result from the accruement of roundoff error in the recursive computations. An alternative derivation for recursion (13.51) that uses the two cofactor rules in theorems 13.2.1 and 13.2.3 is given in the appendix.

Homogeneous random walks

In the coin toss setting, the expression for $\mathcal{T}_{n-1,n-1(-)}$ in (13.44) can be used to show that the recursion in (13.51) is

$$\mathcal{T}_{0,n}(s) = \frac{pe^s}{1 - \left(1/2 - 1/2\sqrt{1 - 4pqe^{2s}}\right)} \mathcal{T}_{0,n-1}(s)$$

$$= \left(\frac{1 - \sqrt{1 - 4pqe^{2s}}}{2qe^s}\right)^n. \tag{13.54}$$

This is a known result in Feller (1968, §XI.3, (3.6) and part (d)) derived from a quite different point of view.

Consider the homogeneous semi-Markov random walk of the previous example with $T_{i,i+1} = pM_+$ and $T_{i,i-1} = qM_-$. The general form of the transmittance is

$$\mathcal{T}_{0,n}(s) = \left(\frac{1 - \sqrt{1 - 4pq M_-(s) M_+(s)}}{2q M_-(s)}\right)^n.$$

$M/M/q/n$ and $M/M/q/\infty$ queues

For each of these queues, the first passage transmittance $\mathcal{T}_{0n}(s)$ is a straightforward recursive computation.

Repairman model

The classical repairman problem in Taylor and Karlin (1998, section 6.4) is a Markovian birth and death process. Suppose there are m operating machines, n spares, and q servers responsible for serving failed machines. Let spare machines be put into service immediately upon failure of a working machine. The state of the system may refer to the number broken machines so that $\mathcal{M} = \{0, \ldots, m+n\}$. If all machine failures are i.i.d. Exponential (λ) and independent of all server repair times, which are i.i.d. Exponential (μ), then the process is Markov.

In state $i \in \mathcal{M}$, the number of working machines is $m_i = \min(m, m+n-i)$, the number of spares is $n_i = \max(n-i, 0)$, and the number under repair is $q_i = \min(q, i)$. The transmittances from state i are

$$T_{i,i-1}(s) = \frac{q_i \mu}{r_i}(1 - s/r_i)^{-1} \qquad T_{i,i+1}(s) = \frac{m_i \lambda}{r_i}(1 - s/r_i)^{-1}$$

with $r_i = q_i \mu + m_i \lambda$. The rationale for these quantities is simply that while in state i, there is a competition between the repair of machines, for which q_i are under repair each at rate μ, and working machine failures, with m_i potentially failing each at rate λ. The holding time in state i is therefore Exponential (r_i) and the transition probabilities are as specified.

Some of the transmittances of practical interest include $\mathcal{T}_{0,m+n}(s)$, the time until complete system failure, $\mathcal{T}_{0,n+1}(s)$, the time until a diminished operating capacity in which only $m-1$ machines are working, and $\mathcal{T}_{1,0}(s)$ the busy time.

13.5 Markov processes

Markov processes may be identified within the class of semi-Markov processes by the holding time distributions in the various states as discussed in section 13.1.2. The special structure of these processes leads to methods for determining passage time distributions that are not applicable to the more general semi-Markov class. This section reviews these methods and connects them to the ($m \times m$) transmittance matrix $T(s)$ that summarizes the process.

In the discrete case for which $T_{ij}(s) = p_{ij} e^s$, the first passage distribution from $1 \to D = \{p+1, \ldots, m\}$ may be determined by creating a new process that is the same as the original process, with jump chain P, but has the states in D absorbing. If $C = \{1, \ldots, p\}$, then the block form of P and the jump chain $\mathcal{P}$ with D absorbing are

$$P = \begin{pmatrix} P_{CC} & P_{CD} \\ P_{DC} & P_{DD} \end{pmatrix} \quad \mathcal{P} = \begin{pmatrix} P_{CC} & P_{CD} \\ 0 & I_{m-p} \end{pmatrix}.$$

If the process starts in state 1, the probability of arriving in state $j \in C$ at step n and not entering any state in D during this time is $\pi_{1j}^{(n)}$ where $\{\pi_{ij}^{(n)}\} = \mathcal{P}^n$. If $P_{CD} \neq 0$, then simple computation gives

$$\mathcal{P}^n = \begin{pmatrix} P_{CC}^n & \{I_p - P_{CC}\}^{-1}\{I_p - P_{CC}^n\} P_{CD} \\ 0 & I_{m-p} \end{pmatrix}.$$

The distribution for W_{1D}, the first passage time from $1 \to D$, is

$$\Pr(W_{1D} \geq n) = \sum_{j=1}^{p} \{P_{CC}^n\}_{1j},$$

the first row sum of the matrix P_{CC}^n.

Suppose the Markov process is in discrete time with Geometric $(p_{i\cdot})$ holding times in states as in (13.5). Then the easiest way to determine exact passage time distributions is to revert back to the jump chain in which each tick of the clock is a potential state transition. The jump chain for this has the components

$$\begin{array}{ll} p_{ij} & \text{if } i \neq j \\ 1 - p_{i\cdot} & \text{if } i = j \end{array}$$

with $p_{i\cdot} = \sum_{j \neq i} p_{ij}$.

Continuous Markov chains are dealt with in the same manner as in discrete time but use matrix exponentials of the infinitesimal generator instead of powers of P. The semi-Markov transmittances $T_{ij}(s) = p_{ij}(1 - s/\lambda_i)^{-1}$ of the Markov chain are converted to the $m \times m$ infinitesimal generator $Q = (q_{ij})$ by setting

$$q_{ij} = \begin{cases} p_{ij} \lambda_i & \text{if } i \neq j \\ -\lambda_i \sum_{k \neq i} p_{ik} & \text{if } i = j. \end{cases} \quad (13.55)$$

This conversion is correct with or without any self loops that may exist within the transmittance matrix $T(s) = \{T_{ij}(s)\}$ as outlined in Exercise 21.

If a continuous time chain starts in state 1 at time zero, then the probability the chain is in state j at time t is $p_{1j}(t)$ where $\mathcal{P}(t) = \{p_{ij}(t)\} = \exp(Qt)$. To determine the first passage distribution from $1 \to D$, the Markov process is modified as in the discrete case to make the states in D absorbing. The infinitesimal generator $\mathcal{Q}$ for this new process is a modification of the block form of Q given by

$$\mathcal{Q} = \begin{pmatrix} Q_{CC} & Q_{CD} \\ 0 & 0 \end{pmatrix} \quad \text{with} \quad Q = \begin{pmatrix} Q_{CC} & Q_{CD} \\ Q_{DC} & Q_{DD} \end{pmatrix}.$$

Then, simple computations yield

$$\mathcal{P}(t) = \exp(\mathcal{Q}t) = \begin{pmatrix} \exp(Q_{CC}t) & Q_{CC}^{-1}\{\exp(Q_{CC}t) - I_p\}Q_{CD} \\ 0 & I_{m-p} \end{pmatrix}$$

as transition probability matrix for the chain with D absorbing. The distribution of W_{1D} is known as a *phase distribution* and is computed as

$$\Pr(W_{1D} > t) = \sum_{j=1}^{p} \{\exp(Q_{CC}t)\}_{1j} \tag{13.56}$$

or the sum of the first row of $\exp(Q_{CC}t)$.

The tidiness of these exact expressions belies the numerical difficulties that may be encountered when trying to make exact computations. Macros exist in most packages to compute matrix exponentials however these computations can be unstable when Q has high dimension. Also, in the discrete case, the computation of P^{200} can be quite unstable and inaccurate for large systems.

These difficulties do not occur when working with saddlepoint approximations that aim to invert first passage MGFs given by the various cofactor rules. Furthermore, saddlepoint approximations usually achieve such high accuracy in the Markov setting that it becomes difficult to decide whether their deviation from "exact" computation is due to error in the saddlepoint approximation or due to computational error when implementing the "exact" calculation. The next section provides a relatively complicated example.

13.6 A redundant and repairable system

Saddlepoint inversion of the first passage cofactor rules make it possible to easily compute the reliability for complex systems. An example is given below that was considered in Butler (1997, 2000) and was motivated by the practical examples found in Høyland and Rausand (1994, §§4.6, 4.7). The example considers a pumping station using repairable pumps but with imperfect switching that eventually leads to system failure.

A pumping station has four equivalent pumps available for use. Under ordinary operating conditions, there are two active pumps. One is designated as the primary pump and has an Exponential (λ) failure time. The second active pump is a partly loaded backup which means that it shares a reduced load with an Exponential (λ_1) failure time with $\lambda_1 < \lambda$. Spare pumps that have been repaired are as good as new. The pumps are repairable by four independent servers and we assume that all individual repair times are Exponential (μ). The system is subject to imperfect switching in the activation of replacement pumps. By this we mean that each switching attempt is an independent Bernoulli $(p = 1 - q)$ trial with p as the probability of a successful switching. The first switching mechanism failure triggers system failure after the currently active primary pump has failed. Finally, suppose that attempts to activate new pumps can only occur following the failure of the primary active pump.

13.6 A redundant and repairable system

The behavior of this pumping system is quite complicated and it is a major task to characterize the system as a semi-Markov process and to determine its transmittance matrix. For all of the details, the reader is referred to Butler (2000, Appendix C). In the present context, the overall structure and form of the process is determined. A numerical example is also examined to determine the accuracy achieved in approximating the system reliability function when saddlepoint methods are used to invert the cofactor rules.

Since system failure occurs with failure of the primary pump, primary pump failure can be used as the trigger for changes in system state. While waiting for the primary unit to fail in Exponential (λ) time, the backup pump has not failed in the interim with probability $r = \lambda/(\lambda_1 + \lambda) > 0.5$. Let the system states be $\mathcal{M} = \{0, \ldots, 5\}$ with $\{0, \ldots, 4\}$ designating the number of pumps that are under repair and state 5 as the state of system failure.

The destination states following primary pump failure can be states $j = 1, \ldots, 4$ under two conditions: (i) the activations of the new primary and/or backup pumps are successful, and (ii) subsequently there are j pumps undergoing repair. The destination state is failure state 5 when condition (i) fails, that is when the switching mechanism fails to successfully activate all the required pumps. The initial state is state 0. This specification of the states and the mechanisms by which state transitions occur characterizes the transmittance matrix of the process which is given in (13.57). The states for this matrix are ordered from $0, \ldots, 5$. Entry $M_\lambda = M_\lambda(s)$ refers to the MGF of an Exponential (λ) waiting time, $N_1 = N_1(s) = M_\lambda \{q + (1-r)pqM_\lambda\}$, and $N_2 = M_\lambda(a + bM_\lambda)$ for some probabilities a and b. Details for the determination of these forms, specification of the entries $\{p_{ij}\}$, and treatment of the n-unit generalization are given in Appendix C of Butler (2000). The large number of nonzero branch transmittances makes this system quite complicated. The system is semi-Markov and not Markov because, while in states 0, 1, 2, and 3, destination 5 has a different holding time than the other possible destinations.

$$T(s) = \begin{pmatrix} 0 & prM_\lambda & p^2(1-r)M_\lambda & 0 & 0 & N_1 \\ 0 & p_{10}M_\lambda & p_{11}M_\lambda & p_{12}M_\lambda & 0 & N_1 \\ 0 & p_{21}M_\lambda & p_{22}M_\lambda & p_{23}M_\lambda & p_{24}M_\lambda & N_1 \\ 0 & p_{31}M_\lambda & p_{32}M_\lambda & p_{33}M_\lambda & p_{34}M_\lambda & N_2 \\ 0 & 0 & 0 & pM_{4\mu} & 0 & qM_{4\mu} \\ 0 & 0 & 0 & 0 & 0 & 0 \end{pmatrix}. \quad (13.57)$$

As a numerical example, suppose the primary unit has failure rate $\lambda = 1$ and the single backup unit has the reduced rate $\lambda_1 = 1/2$. Let the fixing rate be $\mu = 2$ and suppose $p = 0.95$ is the probability for successful switching. The numerical value of $T(s)$ in this instance is

$$\begin{pmatrix} 0 & 0.6\overline{3}M_1 & 0.3008\overline{3}M_1 & 0 & 0 & N_1 \\ 0 & 0.36190M_1 & 0.52929M_1 & 0.042976M_1 & 0 & N_1 \\ 0 & 0.26320M_1 & 0.42869M_1 & 0.22586M_1 & 0.016\overline{409}M_1 & N_1 \\ 0 & 0.41257M_1 & 0.20629M_1 & 0.16286M_1 & 0.14286M_1 & N_2 \\ 0 & 0 & 0 & 0.95M_8 & 0 & 0.05M_8 \\ 0 & 0 & 0 & 0 & 0 & 0 \end{pmatrix} \quad (13.58)$$

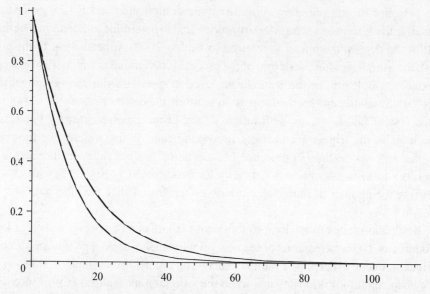

Figure 13.16. Plot of exact reliabilities $\Pr(W_{05} > t)$ (*upper solid*) and $\Pr(W_{0D} > t)$ (*lower solid*) versus their respective saddlepoint appproximations (*dashed*).

where

$$N_1(s) = 0.0158\bar{3}M_1^2(s) + 0.05M_1(s)$$
$$N_2(s) = 0.032571 M_1^2(s) + 0.042857 M_1(s).$$

Consider two different types of system failure: (i) passage to state 5 so failure is due to the switching mechanism, and (b) passage to either state 4 or 5 so that failure refers to the first stoppage of all pumps.

13.6.1 First passage to state 5

The cofactor rule in (13.7) gives a rational expression for $f_{05}\mathcal{F}_{05}(s)$ such that $f_{05} = 1$, so failure is ultimately assured. The mean time until failure is $\mathcal{F}'_{05}(0) = 14.8$, while the standard deviation may also be determined from the second derivative as 14.5. Approximate equality of the mean and standard deviation suggests an approximate exponential failure distribution, but the exact distribution is known to be a phase distribution; see Aalen (1995) and Neuts (1981). The convergence strip for $\mathcal{F}_{05}(s)$ is $s \in (-\infty, 0.0692)$.

An exact plot for $\Pr(W_{05} > t)$ (solid) and its saddlepoint approximation (dashed), as determined from transmittance (13.58), are shown in figure 13.16 as the upper pair of curves. The plot shows virtually no graphical difference. The exact reliability function has been determined as described in section 13.6.3.

In figure 13.17 (left), the lower triple of curves compares the failure rate approximations from section 1.3 for this example. Numerical integration yields 1.074 as the normalization constant for the saddlepoint density when computing $\bar{z}(t)$. The true asymptote is given by $z(t) = z(\infty) = 0.0692$. Portions of figure 13.17 (right) show various aspects of the percentage relative errors connected with the two previous graphs.

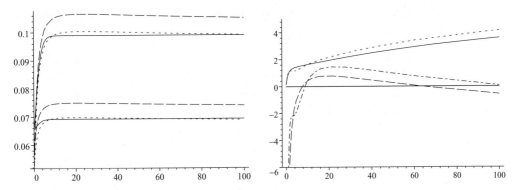

Figure 13.17. (Left) Plot of $z(t)$ (*solid*), $\hat{z}(t)$ (*dashed*), and $\bar{z}(t)$ (*dotted*) vs. t in the pumping system for passages $1 \to 5$ (lower) and $1 \to \{4, 5\}$ (upper). (Right) Percentage relative errors in saddlepoint approximation for system reliability vs. t for passage $1 \to 5$ (*solid*) and passage $1 \to \{4, 5\}$ (*dots*). Similarly, relative error plots in approximation for $z(t)$ with $\bar{z}(t)$ are the respective *dashed* and *dot-dashed* curves.

The solid and dashed lines give respectively the percentage relative errors for reliability approximation and failure rate approximation using $\bar{z}(t)$. The limiting relative error in failure rate approximation as $t \to 0$ is determined numerically as

$$\lim_{t \to 0} 100 \left(\frac{\bar{z}(t)}{z(t)} - 1 \right) = -23.3\%.$$

13.6.2 First passage to $D = \{4, 5\}$

The first passage transmittance expression in (13.37) yields $f_{1D}\mathcal{F}_{1D}(s)$. The first passage transmittance has $f_{1D} = 1$, mean time to failure of $\mathcal{F}'_{1D}(0) = 11.0$, standard deviation 10.2, and convergence region $(-\infty, 0.0990)$. The exact reliability function and failure rate have been determined as described in section 13.6.3 and used to check saddlepoint accuracy. This accuracy may be seen in the remaining graphs in figures 13.16, 13.17. Saddlepoint approximations show very close agreement. For figure 13.17 (left), the true asymptotic failure rate is $z(\infty) = 0.0990$. In figure 13.17 (right), the dotted and dotted-dashed lines plot the relative error in approximating the reliability and failure rate with

$$\lim_{t \to 0} 100 \left(\frac{\bar{z}(t)}{z(t)} - 1 \right) = -22.4\%$$

computed numerically.

13.6.3 Specifying the system as a Markov process

The system with transmittance (13.57) is semi-Markov but it can be made into a Markov process by including an additional state denoted as 6. This new state is entered when the switching mechanism succeeds in connecting up a new primary unit but fails in connecting a new backup unit. Transmittances of the former semi-Markov system into state 5, with $N_1(s)$ and $N_2(s)$, are now split into two separate pieces, one into 6 and the other into 5. With the states ordered as $\{0, \ldots, 4, 5, 6\}$, the infinitesimal generator for this Markov process in the example is

$$Q = \begin{pmatrix} -1.0 & 0.6\bar{3} & 0.300\,8\bar{3} & 0 & 0 & 0.05 & 0.015\bar{83} \\ 0 & -0.63810 & 0.52929 & 0.042976 & 0 & 0.05 & 0.015\bar{83} \\ 0 & 0.263\,20 & -0.57131 & 0.22586 & 0.0164\overline{09} & 0.05 & 0.015\bar{83} \\ 0 & 0.412\,57 & 0.20629 & -0.83714 & 0.14286 & 0.04286 & 0.03257 \\ 0 & 0 & 0 & 0.95(8) & -8 & 0.05(8) & 0 \\ 0 & 0 & 0 & 0 & 0 & 0 & 0 \\ 0 & 0 & 0 & 0 & 0 & 1.0 & -1.0 \end{pmatrix}.$$

(13.59)

Generator Q already has state 5 as absorbing and by removing the 5th row and 5th column and exponentiating the resulting matrix gives the exact passage time to state 5. The exact reliability as computed in (13.56) is

$$\Pr(W_{05} > t) = 1.01e^{-0.0692t} + 0.0133e^{-0.915t}\cos(0.171t) \qquad (13.60)$$
$$- 0.0338e^{-0.915t}\sin(0.171t) - 0.0220e^{-t} + .0^5 185e^{-8.15t}.$$

The leading term determines the asymptotic order in t and has an exponent which is the right edge of the convergence strip for the rational first passage MGF $\mathcal{F}_{05}(s)$. This calculation of reliability gives the true asymptotic failure rate as

$$z(\infty) = \lim_{t \to \infty} z(t) = \lim_{t \to \infty} \frac{-(d/dt)\Pr(W_{05} > t)}{\Pr(W_{05} > t)} = 0.0692.$$

The exact reliability and failure rate used in figures 13.16 and 13.17 were based upon (13.60).

For first passage to $D = \{5, 6\}$, the exact reliability function is

$$\Pr(W_{0D} > t) = 1.08e^{-0.0990t} - 0.213e^{-1.0t} +$$
$$+ 0.137e^{-0.974t}\cos(0.180t) - 0.220e^{-0.974t}\sin(0.180t).$$

The asymptote for the true failure rate is $z(t) = 0.0990$, the smallest real pole of $\mathcal{F}_{0D}(s)$.

13.7 Appendix

13.7.1 First return moments

Mean computations

Evaluation of the limiting derivatives of $K_{11}(s)$ as $s \to 0$ requires using the following key result from the limit theory connected to stationary semi-Markov processes:

$$\lim_{s \to 0} s\{I - T(s)\}^{-1} = \mathbf{1}\vartheta^T / \sum_{i=1}^{m} \vartheta_i T'_{i\cdot}(0) \qquad (13.61)$$

where $\mathbf{1}^T = (1, .., 1)$ is $1 \times m$, $\vartheta^T = (\vartheta_1, \ldots, \vartheta_m)^T$ is the stationary distribution for the jump chain $T(0)$, and $T'_{i\cdot}(0)$ is the mean waiting time in state i. The result is a standard theoretical result for which a simple proof is given in Medhi (1994, pp. 326–327).

To compute the limiting value of (13.29) as $s \to 0$, consider instead the limiting value of $\mathcal{F}'_{11}(s)$ expressed as

$$\mathcal{F}'_{11}(s) = \frac{|I - T|}{|\Psi_{11}|} \operatorname{tr}\{(I - T)^{-1}\dot{T} + \Psi_{11}^{-1}\dot{\Psi}_{11}\}$$

$$= \left\{\frac{|I - T|}{s|\Psi_{11}|}\right\} \operatorname{tr}\{s(I - T)^{-1}\dot{T} + s\Psi_{11}^{-1}\dot{\Psi}_{11}\} \quad (13.62)$$

where the dependence of T, Ψ_{11}, $\dot{T}$, etc. on s has been suppressed. Using (13.61) and Cramer's rule, the leading factor in curly braces is computed in terms of the limit of the (1, 1) entry of $s(I - T)^{-1}$ or

$$\frac{1}{\lim_{s\to 0} s(I - T)^{11}} = \frac{1}{\vartheta_1}\sum_{i=1}^{m} \vartheta_i T'_i(0). \quad (13.63)$$

Again substituting the limit in (13.61) for $s(I - T)^{-1}$ inside the trace operation gives the trace factor to $O(s)$ as

$$\frac{1}{\sum_{i=1}^{m}\vartheta_i T'_i(0)} \operatorname{tr}(\mathbf{1}\vartheta^T \dot{T}) = \frac{1}{\sum_{i=1}^{m}\vartheta_i T'_i(0)} \vartheta^T \dot{T}\mathbf{1} \to 1$$

as $s \to 0$. The first return mean is established as the right side of (13.63) which is the middle expression in (13.34).

The rightmost expression in (13.34) is proven by showing that the ith stationary probability is proportional to the (i, i) cofactor of $I - T(0)$ or

$$\vartheta_i \overset{i}{\propto} |\Psi_{ii}(0)|, \quad (13.64)$$

which is a result that is related to the graphical methods of Solberg (1975). This result is established by first proving the well-known result that

$$\vartheta \mathbf{1}^T = \lim_{n\to\infty} P^n = \lim_{z\to 1}(1 - z)(I - zP)^{-1}. \quad (13.65)$$

To derive this, first compute the z-transform of $\{P^n : n \geq 0\}$ as $(I - zP)^{-1}$. From this, the z-transform of the sequence of differences $\{I, P - I, P^2 - P, \ldots, P^n - P^{n-1}\}$ is $(1 - z)(I - zP)^{-1}$. The evaluation of a z-transform at $z = 1$ leads to the total sum of coefficients in the z-transform which is $\lim_{n\to\infty} P^n$ for the telescoping sum of differences and this establishes (13.65).

To establish (13.64), write

$$\lim_{z\to 1}(1 - z)(I - zP)^{-1} = \lim_{z\to 1} \frac{\operatorname{Adj}(I - zP)}{|I - zP|/(1 - z)}$$

$$= \operatorname{Adj}(I - P)/\lambda$$

where $\operatorname{Adj}(I - P)$ is the adjoint matrix for $I - P$, and λ is the product of the $m - 1$ nonzero eigenvalues of $I - P$. The result follows since the ith diagonal element of the adjoint is $|\Psi_{ii}(0)|$ as also for ith diagonal element of $\vartheta \mathbf{1}^T$ in (13.65).

Second moment computations

The derivation of $E(W_{11}^2)$ requires first developing expressions for $E(W_{ij}^2)$ for all i, j. Doing this in turn, requires a one step analysis of $f_{ij}\mathcal{F}_{ij}$ much as we did for $f_{11}\mathcal{F}_{11}$ in (13.31). In this analysis, the transition is either straight to j in one step with transmittance T_{ij} or through first step $k \neq j$ as indicated in

$$f_{ij}\mathcal{F}_{ij} = T_{ij} + \sum_{k \neq j} T_{ik} f_{kj}\mathcal{F}_{kj}.$$

Assuming that $f_{ij} \equiv 1$ for all i, j, two derivatives of this expression at 0 gives a generalization of (13.35) as

$$E(W_{ij}^2) = T_{i\cdot}''(0) + \sum_{k \neq j}^{m} p_{ik}\{\mathcal{F}_{kj}''(0) + 2M_{ik}'(0)\mathcal{F}_{kj}'(0)\}. \tag{13.66}$$

Take $\Sigma^{(1)} = \{E(W_{ij})\}$ and $\Sigma^{(2)} = \{E(W_{ij}^2)\}$ as the $m \times m$ matrices of first and second moments, $\delta^T = \{T_{1\cdot}''(0), \ldots, T_{1\cdot}''(0)\}$ as a $1 \times m$ vector, and let the d subscript notation $\Sigma_d^{(2)}$ denote the diagonal matrix that sets to zero all the off diagonal elements of $\Sigma^{(2)}$. In matrix notation, (13.66) may be rewritten as

$$\Sigma^{(2)} = \delta \mathbf{1}^T + P(\Sigma^{(2)} - \Sigma_d^{(2)}) + 2T'(0)\{\Sigma^{(1)} - \Sigma_d^{(1)}\}. \tag{13.67}$$

Left multiply (13.67) by ϑ^T to get

$$\vartheta^T \Sigma^{(2)} = \vartheta^T \delta \mathbf{1}^T + \vartheta^T (\Sigma^{(2)} - \Sigma_d^{(2)}) + 2\vartheta^T T'(0)\{\Sigma^{(1)} - \Sigma_d^{(1)}\}$$

where the second term on the right has P removed due to the fact that $\vartheta^T P = \vartheta^T$. Cancelling $\vartheta^T \Sigma^{(2)}$ on either side and solving for the ith column element of $\vartheta^T \Sigma_d^{(2)}$ leads to

$$E(W_{ii}^2) = \frac{1}{\vartheta_i} \sum_{j=1}^{m} \vartheta_j \left\{ T_{j\cdot}''(0) + 2\sum_{k \neq i}^{m} T_{jk}'(0)\mathcal{F}_{ki}'(0) \right\}$$

which is (13.36).

13.7.2 Derivation of birth and death recursions from theorems 13.2.1 and 13.2.3

All the recursions above may be derived directly from the two cofactor rules by using Cramer's rule. Consider the recursion for $\mathcal{T}_{i,i(+)}$ given in (13.42). Suppose the state space is $\mathcal{M} = \{0, \ldots, n\}$ and transmittance T is $(n+1) \times (n+1)$. From theorem 13.2.3,

$$\mathcal{T}_{0,0(+)} = 1 - |I - T|/|\Psi_{00}|.$$

Since the nonzero entries of T are only one step off the main diagonal, Cramer's rule applied to the first row of $I - T$ yields

$$\mathcal{T}_{0,0(+)} = 1 - \frac{|\Psi_{00}| + (-T_{01})(-1)^3|\Psi_{01}|}{|\Psi_{00}|} = T_{01}\frac{(-1)^3|\Psi_{01}|}{|\Psi_{00}|}$$
$$= T_{01}T_{10} = T_{01}T_{10}\{1 - \mathcal{T}_{11(+)}\}^{-1}$$

where the last two steps use theorem 13.2.1 and the busy period transmittance in (13.46). The busy period transmittance is a special case of the general first passage transmittance formula derived in the next paragraph.

To derive the first passage recursion for $\mathcal{T}_{0,n}$ in (13.51), let $\Psi_{ij}^{\backslash n}$ be the submatrix of $I - T$ that excludes state n (the last row and column) and also row i and column j. From theorem 13.2.1,

$$\mathcal{T}_{0,n} = (-1)^{n+2} \frac{|\Psi_{n0}|}{|\Psi_{nn}|} = (-1)^{n+2} \frac{-T_{n-1,n}(-1)^{2n} |\Psi_{n-1,0}^{\backslash n}|}{|\Psi_{nn}|}. \tag{13.68}$$

The last result is obtained by applying Cramer's rule to the last column of Ψ_{n0} since only the $-T_{n-1,n}$ entry is nonzero. Now substitute

$$|\Psi_{nn}| = \{1 - \mathcal{T}_{n-1,n-1(-)}\} |\Psi_{n-1,n-1}^{\backslash n}|,$$

which is a rearrangement of theorem 13.2.3 for first return to state $n - 1$ from below, into the denominator of (13.68) to get

$$\mathcal{T}_{0,n} = \frac{T_{n-1,n}}{1 - \mathcal{T}_{n-1,n-1(-)}} \frac{(-1)^{n+1} |\Psi_{n-1,0}^{\backslash n}|}{|\Psi_{n-1,n-1}^{\backslash n}|} = \frac{T_{n-1,n}}{1 - \mathcal{T}_{n-1,n-1(-)}} \mathcal{T}_{0,n-1}.$$

13.8 Exercises

1. (Pure death process). Suppose a system consists of n independently operating components each of which has an Exponential (λ) lifetime. If the system requires at least $k < n$ components to operate, show that the distribution for the system lifetime is Y with distribution

$$Y = \frac{1}{n} E_1 + \frac{1}{n-1} E_2 + \cdots + \frac{1}{k} E_{n-k+1}$$

where $E_1, E_2, \ldots, E_{n-k+1}$ are i.i.d. Exponential (λ). First derive the failure time distribution from basic principles. Then show that the same distribution can also be derived by using (13.7).

2. (Parallel system). Determine that the first passage transmittance from state 1 to 2 for the parallel system

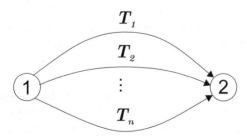

is $T_1(s) + \cdots + T_n(s)$. Of course this is a finite mixture distribution for the holding times expressed in the various transmittances. To derive this result from (13.7), relabel state 2 as state $n + 2$ and insert intermediate states in between states 1 and $n + 2$ along each of the n branches; label these new states as $2, \ldots, n + 1$. The dynamics of the process does not change if the transmittances from 1 to nodes $2, \ldots, n + 1$ are $T_1, \ldots, T_n$ respectively, and the transmittances from nodes $2, \ldots, n + 1$ to node $n + 2$ are all taken as $1 = 1e^0$ indicating certain passage with no time delay.

3. Consider the system failure time in the reliability example of section 13.2.4.
   (a) Use induction in n to prove (13.16) for a single nonrepairable component.
   (b) Consider a new representation for the system that begins in state $0 \in \mathcal{M} = \{0, 1, \ldots, n\}$ and, after transition out of 0, does not return to state 0. Rather than returning to state 0, the system

makes direct transitions among the transient states $C = \{1, \ldots, n-1\}$ until absorption into failure state n. Show this system has transmittances

$$T_{ij} = \begin{cases} 0 & \text{if} & j = 0 \\ p_j M N_j & \text{if} & i \in \{0\} \cup C \quad j \in C \\ p_n M & \text{if} & i \in \{0\} \cup C \quad j = n \\ 0 & \text{if} & i = n \end{cases} \quad (13.69)$$

where dependence on s has been suppressed.

(c) Use induction in n to prove that first passage transmittance from $0 \to n$ for the system in (13.69) agrees with (13.16) for a single nonrepairable component.

(d) Prove (13.40) when there are multiple nonrepairable components.

(e) In the context of multiple nonrepairable components, develop a new representation for the system that is similiar to that in part (b). Rather than returning to state 0, the new system should make direct transitions among the transient states C until absorption into the class of failure states in D. From the new system transmittance determine the first passage transmittance $0 \to D$ and show that it agrees with (13.40).

4. Consider the reliability of systems example of section 13.2.6.

   (a) Use the system transmittance in (13.69) to determine the first return transmittance for $1 \to 1$. Does this lead to the same transmittance as in (13.28)? Why or why not?

   (b) Suppose that states 2 and 3 are lumped together into a megastate. Show that the first return transmittance in part (a) can be derived by using a two state system with states 1 and $\{2, 3\}$.

5. Derive the transmittance entries for the $GI/M/1/4$ queue in section 13.2.5 given in (13.20).

   (a) If $X \sim G$, show that the transmittance from $i \in \{1, \ldots, 4\}$ to $j \in \{2, \ldots, i+1\}$ is

   $$T_{ij}(s) = E^X \left\{ \frac{(\mu X)^{i-j+1}}{(i-j+1)!} e^{-\mu X} e^{sX} \right\}$$
   $$= U_{i-j+1}(s - \mu).$$

   (b) Now derive the value of $T_{i1}(s)$ for $i \in \{1, \ldots, 4\}$.

   (c) Verify (13.21) in the case that G is Gamma (α, β).

6. Consider a first passage distribution for the Markov process shown in figure 13.4. Choose some rate parameters you think might be interesting and perform the following analyses.

   (a) For the first passage distribution from $1 \to 4$, compute the saddlepoint density and the normalized density by inverting $\mathcal{F}_{14}(s)$.

   (b) Compute the mean and standard deviation for this first passage distribution.

   (c) Since $\mathcal{F}_{14}(s)$ is rational, it may be expanded into partial fractions so that individual terms may be exactly inverted as was done in (13.22). Find the exact first passage density in this manner and plot it simultaneously with the saddlepoint densities from part (a).

   (d) Compute the Lugannani and Rice approximation to the CDF. Integrate the exact density from part (c) to get the exact CDF and plot it versus the saddlepoint CDF.

   (e) Compare the normalized $\bar{z}(t)$ and unnormalized $\hat{z}(t)$ saddlepoint failure approximations with the exact failure rate function by simultaneously plotting the three functions.

7. For the $GI/M/1/4$ queue of section 13.2.5 with Gamma $(2, 2)$ interarrivals, consider the first passage transmittance to the second occurrence of state 1, e.g. $0 \to 1 \to 1$.

   (a) Symbolically compute this transmittance from the transmittance matrix in (13.19). Note that state 5 is not relevant to the passage $1 \to 1$.

   (b) Compute the mean and standard deviation for this first passage distribution given that the passage $0 \to 1 \to 1$ is assured.

(c) Using the parameters for the example in section 13.2.5, invert the transmittance and plot the saddlepoint density and the normalized saddlepoint density.

(d) Simulate an empirical estimate for the reliability function by running through the system and making repeated attempts to pass from $0 \to 1 \to 1$. Since $100(1 - f_{11})\%$ fail to return to state 1, make $10^6/f_{11}$ attempts so that 10^6 passage times can be expected. Base the empirical CDF on the set of resulting passages times from $0 \to 1 \to 1$.

(e) Plot the empirical reliability from part (d) and compare it to the Lugannani and Rice approximation.

8. Determine the second passage transmittance for the $GI/M/1/4$ queue as in Exercise 7, but make the following modifications to the dynamics of the queue. Upon arrival in state 4, suppose the system waits in state 4 until a new packet arrives in queue. If during this wait no packets have yet been serviced, the system feeds back into state 4 and the new packet is removed from the queue. If $i \in \{1, \ldots, 4\}$ packets have been serviced, the system passes to state $5 - i \in \{1, \ldots, 4\}$.

    (a) Determine the necessary modifications to $T_{41}(s), \ldots, T_{44}(s)$ in the transmittance matrix (13.19) that reflect these exit dynamics out of state 4.

    (b) Do parts (a)–(e) of Exercise 7 with the new dynamics included in matrix $T(s)$.

9. Consider a first return distribution for the Markov process shown in figure 13.4. Choose some rate parameters you think might be interesting and perform all the same analyses as in Exercise 6.

10. Extend the first return analysis of the system reliability example of section 13.2.6 to n components as introduced in section 13.2.4.

    (a) If the first $n - 1$ components are repairable and the nth is not, use induction to prove that the first return transmittance for state 1 is
    $$f_{11}\mathcal{F}_{11}(s) = p_1 M(s) N_1(s) \left\{ 1 - M(s) \sum_{i=2}^{n} p_i N_i(s) \right\}^{-1}.$$

    (b) Show that the same result is derived from theorem 13.2.3 if the nonrelevant states 0 and n are mistakenly kept within the system transmittance $T(s)$.

    (c) For the specific example provided in section 13.2.4 with $n = 4$, compute f_{11} and determine the saddlepoint reliability function associated with $\mathcal{F}_{11}(s)$.

    (d) Starting in state 1, simulate $10^6/f_{11}$ attempts to return to state 1 and use the resulting passage times to determine an empirical estimate for the reliability function in part (c). Plot the saddlepoint and empirical reliability approximations and examine their accuracy.

11. Derive the derivatives for $\ln \mathcal{F}_{11}(s)$ given in (13.29) and (13.30).

12. Consider the influence that nonrelevant states have on the cofactor rules in theorems 13.2.1, 13.2.3, and 13.3.1.

    (a) Suppose that nonrelevant states are included into state space $\mathcal{M}$ one at a time and each additional state causes no internal feedback within the nonrelevant states. Use induction to prove that the cofactor rule in (13.7) is valid and that it is analytic at $s = 0$.

    (b) Suppose that at some point in the process of including nonrelevant states in part (a), a new nonrelevant state is added that leads to internal feedback within the set of nonrelevant states. With the inclusion of this nonrelevant state, prove that (13.7) is correct except at $s = 0$ where a removable singularity of order one occurs.

    (c) What circumstances would result in a removable singularity of order k in the computation of $\mathcal{F}_{1m}$ from (13.7)? Start by giving a simple example in the case $k = 2$.

    (d) Why do the remarks about the computation of $\mathcal{F}_{1m}$ in (13.7) automatically carry over and apply to $\mathcal{F}_{11}$ as computed from (13.24)?

    (e) Show that all the results of parts (a)–(c) also apply to the multiple destination cofactor rule in (13.37).

13. Consider an $M/M/1/6$ queue with an arrival rate of $\lambda = 1/4$ and a repair rate of 1. Consider first passage from state $0 \to 6$.
    (a) Compute $\mathcal{T}_{06}(s)$ and plot the saddlepoint density and saddlepoint CDF approximation. Tabulate selected values for the CDF that span the range of the distribution from the 0.5 to the 99.5 percentiles.
    (b) Determine the infinitessimal generator for this Markov chain and compute the exact density for first passage as well as the exact CDF. Overlay plots of these onto the two plots in part (a). Compare the exact CDF computations to the tabulated saddlepoint CDF computation in part (a).
14. Repeat all the components of Exercise 13 for the first return $0 \to 0$ distribution of the $M/M/1/6$ queue.
15. Repeat all the components of Exercise 13 for the busy period distribution of the $M/M/1/6$ queue.
16. Suppose an $M/M/1/\infty$ queue with an arrival rate of $\lambda = 1/4$ and a repair rate of 1. Consider first return to state 0.
    (a) Perform the computations of Exercise 14 (a) for this distribution.
    (b) Simulate 10^6 first return values for this queue. Compare the empirical CDF with the saddlepoint CDF graphically as well as in terms of the tabulated values.
17. Suppose an $M/M/6/10$ queue with an arrival rate of $\lambda = 1$ and a repair rate of $1/2$ for each of the six servers. Compare saddlepoint approximations versus exact computation as in Exercise 13 for the following distributions connected with the queue:
    (a) first passage from $0 \to 10$;
    (b) the busy period.
18. Make the comparisons of Exercise 16 for a $M/M/6/\infty$ queue with arrival rate $\lambda = 1$ and repair rate of $1/2$ for each of the 6 servers.
19. Consider the first passage transmittance $0 \to n$ for a $M/M/q/\infty$ queue with $n > q$. Is the following identity true of false?
$$\mathcal{T}_{0n}(s) = \mathcal{T}_{0q}(s)\{1 - \mathcal{T}_{qq(-)}(s)\}^{-1}\mathcal{T}_{qn}(s)$$
Why or why not?
20. Consider the Markov process in Exercise 6.
    (a) If you already did this exercise, then repeat the determination of the exact first passage CDF by basing the computation on e^{Qt} where Q is the infinitessimal generator of the process. Confirm that the computation based on e^{Qt} agrees with that from the partial fraction expansion.
    (b) If you did not do Exercise 6, then choose some transition rates and derive the exact CDF via both routes. Confirm that they are the same up to numerical accuracy.
21. Derive the infinitessimal generator Q in (13.55) from the transmittance matrix of a Markov chain given by $T_{ij}(s) = p_{ij}(1 - s/\lambda_i)^{-1}$.
    (a) If the transmittance has no self-loops, show that the rate matrix for state transitions is
$$\begin{cases} p_{ij}\lambda_i & \text{if } i \neq j \\ 0 & \text{if } i = j \end{cases}$$
    and that Q is given by (13.55).
    (b) Suppose that self-loops exist in the transmittance matrix. Show that the semi-Markov process with these self-loops removed has the transmittance
$$\frac{p_{ij}}{1 - p_{ii}}\left(1 - \frac{s}{\lambda_i(1 - p_{ii})}\right)^{-1} \qquad j \neq i \qquad (13.70)$$
    and transmittance 0 if $i = j$.

(c) The new transmittance in (13.70) characterizes a Markov process. Determine that this new process has the infinitessimal generator Q given in (13.55).

22. A discrete time Markov chain with states $\mathcal{M} = \{1, 2, 3, 4\}$ has the transition matrix

$$P = \begin{pmatrix} 0 & 0 & 1 & 0 \\ 0 & 0 & 2/3 & 1/3 \\ 1/4 & 3/4 & 0 & 0 \\ 0 & 1 & 0 & 0 \end{pmatrix}.$$

(a) Compute the probability generating function for the first passage time from $1 \to 4$. Expand it in a Taylor series to determine the exact mass function and compare it to the saddlepoint mass function. Compare the two continuity-corrected Lugannani and Rice approximations to the true CDF determined by summing mass points.

(b) Repeat part (a) for the first return distribution to state 0. Note the nature of the lattice support for this mass function.

23. (Discrete time queue). A queueing system consists of five servers that operate independently and identically. Suppose that the daily state of each server is either 0 (up) or 1 (down) and that daily changes in the status of each of the servers is a two state Markov chain with transition matrix

$$\begin{pmatrix} 0.90 & 0.10 \\ 0.80 & 0.20 \end{pmatrix}$$

over state space $\{0, 1\}$. Consider the process that counts the number of servers down at the beginning of each day. Also suppose the process starts at the beginning of day 1 with all five servers up.

(a) Determine its transition matrix in discrete time over the state space $\mathcal{M} = \{0, \ldots, 5\}$.

(b) Compute the saddlepoint mass function and the saddlepoint CDFs, $\widehat{\Pr}_1$ and $\widehat{\Pr}_2$ given in (1.27) and (1.32), for time to complete failure or first passage from $0 \to 5$. Compare these approximations to their exact counterparts in tabular form and graphically. Compute relative error plots.

(c) Repeat part (b) as concerns the duration of diminished capability as expressed in the first passage from $1 \to 0$.

(d) Repeat part (b) for the first return to state 0.

24. Suppose a population of N distinct flavors of ice cream is sampled with replacement. As more flavors are sampled, new flavors are less and less likely to be sampled over time. Suppose S_r is the number of ice cream cones it is necessary to sample in order to acquire the rth new flavor.

(a) Draw a flowgraph to describe the structure S_r and determine an explicit form for its MGF.

(b) Using $N = 10$ and $r = 4$, compute the saddlepoint CDFs, $\widehat{\Pr}_1$ and $\widehat{\Pr}_2$ given in (1.27) and (1.32), for S_4.

(c) Determine the exact CDF of S_4. Note that its MGF is a rational function of $t = e^s$ so that an exact inversion is possible using partial fractions. Compare the saddlepoint and exact CDF computations graphically and using a small table of values.

25. Consider a sequence of independent Bernoulli (p) trials with 1 denoting success and 0 failure.

(a) Use a flowgraph to determine the MGF for first occurrence of the pattern 101.

(b) Choose a value for p and compute the saddlepoint CDFs, $\widehat{\Pr}_1$ and $\widehat{\Pr}_2$ given in (1.27) and (1.32), for this waiting time distribution.

(c) Determine the exact waiting time distribution. Compare the saddlepoint and exact CDF computations graphically and using a small table of values.

14

Bootstrapping in the transform domain

Sometimes nonparametric bootstrap inference can be performed by using a saddlepoint inversion to replace the resampling that is normally required. This chapter considers this general concept and pursues the extent to which the general procedure can implemented.

The main set of examples have been taken from Butler and Bronson (2002) and Bronson (2001). They involve bootstrap inference for first passage characteristics and time to event distributions in the semi-Markov processes discussed in chapter 13. Saddlepoint approximations of nonparametric bootstrap estimates are given for the mean and standard deviation of first passage and as well as for the density, survival, and hazard functions associated with the passage events. Thus saddlepoint approximations become estimates for these moments and functions and represent approximations to the estimates that would normally be provided were a large amount of resampling to be implemented with "single" bootstrap resampling. In performing these computations however, no resampling has been used to compute the estimates, only saddlepoint inversions. The saddlepoint estimates utilize data from the process through an empirical estimate $\hat{T}(s)$ of the unknown system transmittance matrix $T(s)$. Empirical transmittance $\hat{T}(s)$ is used in place of $T(s)$ when implementing the saddlepoint inversions.

In addition to bootstrap estimates, confidence intervals for the moments and confidence bands for the functions may be computed by using the double bootstrap. Often the resampling demands inherent in the double bootstrap place its implementation beyond the range of practical computing (Booth and Presnell, 1998). The saddlepoint methods suggested here replace the majority of this resampling with analytical computation so the double bootstrap becomes a practical and relatively simple method to use. These saddlepoint methods provide approximations for double bootstrap confidence intervals and confidence bands of these semi-Markov first passage characteristics. The procedures and methods have been described in Butler and Bronson (2002) and Bronson (2001). Saddlepoint methods that facilitate implementation of the double bootstrap were first introduced by Hinkley and Shi (1989) and were further developed by DiCiccio *et al.* (1992a, b, 1994) in different contexts.

In many bootstrap applications, direct saddlepoint approximation for the bootstrap distribution of a statistic is not possible. As an example, Daniels and Young (1991) considered the studentized mean or t statistic. An indirect saddlepoint approximation has been developed in Daniels and Young (1991) by starting with a saddlepoint approximation for the bivariate bootstrap distribution of the sum and sums of squares and proceeding to transform and marginalize this to get an approximate CDF. Examples of such indirect saddlepoint approximations are considered below and include saddlepoint approximation to the bootstrap distribution of the t statistic as well as approximation to the CDF of the doubly noncentral t.

14.1 Saddlepoint approximation to the single bootstrap distribution

14.1.1 Direct and indirect saddlepoint approximation

Saddlepoint approximation, for a univariate bootstrap distribution or otherwise, may be arbitrarily divided into two cases: direct and indirect approximation. Direct approximation includes most all of the applications considered so far.

Direct approximation

Suppose that $X_1, \ldots, X_n$ is an i.i.d. sample from distribution F whose MGF $M(s)$ is available and known. Then saddlepoint approximation for the distribution of the mean $\bar{X} = \sum_{i=1}^{n} X_i/n$ is explicit and based on *direct* inversion of its MGF $M(s/n)^n$.

The first passage distributions for semi-Markov processes in chapter 13 were approximated by using direct saddlepoint methods. Suppose that an m-state process is characterized by its known $m \times m$ transmittance matrix $T(s)$. If the first passage time W_{1m} from state 1 to m has a nondefective distribution, then its MGF is given in (13.7) of Theorem 13.2.1 as

$$M_{W_{1m}}(s) = \frac{(-1)^{m+1} |\Psi_{m1}(s)|}{|\Psi_{mm}(s)|} \tag{14.1}$$

where $\Psi_{ij}(s)$ is the (i, j)th minor of $I_m - T(s)$. *Direct* saddlepoint inversion of (14.1) approximates the first passage distribution.

Indirect approximation

Direct saddlepoint approximation is not possible for the doubly noncentral t distribution defined by random variable $T = Z/\sqrt{Y/n}$, where $Z \sim N(\mu, 1)$ is independent of noncentral chi square variable $Y \sim \chi^2(n, \lambda)$. For the general context of this example, Daniels and Young (1991) have suggested an *indirect* approach that involves a sequence of steps. The details of the approach have been implemented for the noncentral t by Broda and Paolella (2004). These steps first recognize that $T = h(Y, Z)$ where the joint MGF of (Y, Z) is explicit. Jacobian transformation of the joint saddlepoint density of (Y, Z) under the transformation $(Y, Z) \to (T, Z)$ gives $\hat{f}(t, z)$, an approximation to the joint density of (T, Z). The two remaining steps consist of (i) computing $\int_0^\infty \hat{f}(t, z) dz$ by using Laplace approximation (2.19) to determine $\hat{f}(t)$, a marginal density approximation for T, and (ii) converting $\hat{f}(t)$ into a CDF using the Temme approximation in (2.34). Formulas for the final expressions in the general indirect case are given in section 14.4.1 while section 14.4.2 works out the specifics for the doubly noncentral t.

14.1.2 Bootstrap distribution for the sample mean and studentized mean

The bootstrap distribution of the sample mean admits direct saddlepoint approximation as was recognized in Davison and Hinkley (1988). Suppose data $x_1, \ldots, x_n$ are summarized in terms of the empirical CDF $\hat{F}(x)$ which estimates F. A resample $X_1^*, \ldots, X_n^*$ is an i.i.d. sample from $\hat{F}$ and the distribution of $\bar{X}^* = \sum_{i=1}^n X_i^*/n$ is the bootstrap distribution for the mean denoted by G. Each X_i^* has MGF given by the empirical MGF

$$\hat{M}(s) = \int e^{sx} d\hat{F}(x) = n^{-1} \sum_{i=1}^n e^{sx_i}, \qquad (14.2)$$

so the MGF of $\bar{X}^*$ is $\hat{M}(s/n)^n$. Thus, direct saddlepoint inversion of $\hat{M}(s/n)^n$, using the continuous Lugannani and Rice approximation, provides $\hat{G}(t)$ as an approximation for $G(t)$. This example was previously considered in section 3.4.1 in terms of approximating the bootstrap distribution of $\bar{X}^* - \bar{x}$ or $H(t) = G(t + \bar{x})$. The saddlepoint approximation is the translation equivariant value $\hat{H}(t) = \hat{G}(t + \bar{x})$.

A 90% confidence interval for the population mean μ of F can be computed from $\hat{H}$ or $\hat{G}$ by replacing the distribution of $\bar{X} - \mu$ with the saddlepoint approximation for the resampling distribution of $\bar{X}^* - \bar{x}$ assuming $\bar{x}$ fixed. In this case a 90% confidence interval is

$$[\bar{x} - \hat{H}^{-1}(0.95), \bar{x} - \hat{H}^{-1}(0.05)] = [2\bar{x} - \hat{G}^{-1}(0.95), 2\bar{x} - \hat{G}^{-1}(0.05)]. \qquad (14.3)$$

For the data in section 3.4.1, $\bar{x} = 17.87$ and H and $\hat{H}$ percentiles are given in table 3.4. The exact 90% interval and its saddlepoint approximation are respectively

$$[17.87 - 3.73, \; 17.87 + 3.34] = [14.14, \; 21.21]$$
$$[17.87 - 3.75, \; 17.87 + 3.33] = [14.12, \; 21.20].$$

Daniels and Young (1991) considered indirect saddlepoint approximation for the bootstrap distribution of the studentized mean or $t = \sqrt{n}(\bar{x} - \mu)/s$, with s^2 as the unbiased estimate of variance. The distribution of t is approximated by the distribution of $T^* = \sqrt{n}(\bar{X}^* - \bar{x})/S^*$ where $\bar{X}^*$ and S^* are resampled values of the sample mean and standard deviation. The steps in their approach involve recognizing first that if $U^* = \sum_{i=1}^n (X_i^* - \bar{x})$ and $V^* = \sum_{i=1}^n (X_i^* - \bar{x})^2$, then $T^* = h(U^*, V^*)$ with h a simple known function. Secondly, the joint MGF of (U^*, V^*) is $\hat{M}(s_1, s_2)^n$ where

$$\hat{M}(s_1, s_2) = n^{-1} \sum_{i=1}^n \exp\{s_1(x_i - \bar{x}) + s_2(x_i - \bar{x})^2\}. \qquad (14.4)$$

Finally, a marginal CDF approximation for the distribution of T^* can be determined from the joint MGF of (U^*, V^*) using the indirect method described above. In making this approximation, the methods are applied as if the bootstrap distributions are continuous. A detailed account is deferred to section 14.4.3.

14.1.3 *Bootstrap estimates of first passage characteristics in semi-Markov processes*

The data for constructing bootstrap estimates may come in many forms, however only the simplest case of complete data is considered here. Suppose the data consist of the full details of h sojourns through the semi-Markov process. Let $\mathcal{H}_1, \ldots, \mathcal{H}_h$ denote these sojourn histories that provide the sequence of states visited along with the holding times in each of these states. The sojourns need not have the same initial and ending states but they must all traverse the same system with a common transmittance matrix T. Any partial holding times in states are excluded from the data in this discussion.

14.1 Saddlepoint approximation to the single bootstrap distribution

Minimal sufficient statistic

Transmittance T characterizes the semi-Markov process and the data admit two sufficient statistics: the matrix of transition counts $N = (n_{ij})$, in which n_{ij} counts the total number of $i \to j$ transitions pooled across all h sojourns, and $\hat{M}(s) = \{\hat{M}_{ij}(s)\}$, in which $\hat{M}_{ij}$ is the empirical MGF for the n_{ij} holding times in state i before passing to state j. The empirical transition rate matrix $\hat{P} = (\hat{p}_{ij}) = (n_{ij}/n_{i\cdot})$ estimates $P = (p_{ij})$ while $\hat{M}$ is an estimate of M. Together they provide an estimate of the system transmittance given as

$$\hat{T}(s) = \hat{P} \odot \hat{M}(s) = \{\hat{p}_{ij}\hat{M}_{ij}(s)\} \tag{14.5}$$

and called the empirical transmittance. Since N and $\hat{M}(s)$ can be recovered from $\hat{T}(s)$, it is entirely appropriate to consider $\hat{T}(s)$ the minimal sufficient statistic for the data based on a nonparametric likelihood for the semi-Markov process. Note that $\hat{T}(s)$ is itself the transmittance matrix for a semi-Markov process with transition matrix $\hat{P}$ and holding time distributions associated with $\hat{M}(s)$.

Matching the relevant states for the $\hat{T}$- and T-processes

The transition $i \to j$ is not possible when $p_{ij} = 0$ and in such cases $\hat{p}_{ij} = 0$. When $p_{ij} > 0$, sufficient data is assumed to be available so that $\hat{p}_{ij} > 0$; should this not be the case then a nominal amount needs to be added to n_{ij} to make it so. The reason for making such assumptions and adjustments is to match the relevant states for the $\hat{T}$- and T-processes. The relevance of states to sojourns in a semi-Markov process has been defined in section 13.2.1 and the need for using only relevant states with the cofactor rules has been discussed in section 13.2.7. Indeed the relevant states of the $\hat{T}$-process are exactly those of the T-process when

$$\text{sgn}(\hat{p}_{ij}) = \text{sgn}(p_{ij}) \tag{14.6}$$

for all i and j, which confirms the assumptions and adjustments for matching the relevant states. See Seneta (1981, p. 3) for a proof that uses the theory of primitive matrices.

If, for example, (14.6) does not hold and a state of the $\hat{T}$-process is not relevant to the sojourn $1 \to m \neq 1$ but relevant for the T-process, then $\hat{f}_{1m}\hat{\mathcal{F}}_{1m}(s)$, the first passage transmittance for the $\hat{T}$-process, would have the possibility of admitting a removable singularity at $s = 0$ whereas $f_{im}\mathcal{F}_{1m}(s)$ for the T-process is analytic at $s = 0$. Here

$$\hat{f}_{1m}\hat{\mathcal{F}}_{1m}(s) = \frac{(-1)^{m+1}|\hat{\Psi}_{m1}(s)|}{|\hat{\Psi}_{mm}(s)|} \tag{14.7}$$

is the empirical transmittance with $\hat{\Psi}_{ij}(s)$ as the (i, j)th minor of $I_m - \hat{T}(s)$. The presence of a removable singularity at $s = 0$ for (14.7) renders its saddlepoint computation unstable within a neighborhood of the mean passage time.

Resampling first passage times

A resampled first passage time W^*_{1m} may be obtained by simulating the first passage $1 \to m$ through the $\hat{T}$-process. The total sojourn time is computed by using the semi-Markov

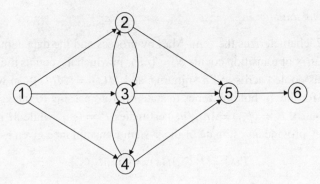

Figure 14.1. Flowgraph showing the degenerative states of dementia. Transmittance $T_{ij}(s)$ labels the state transitions $i \to j$ as indicated.

dynamics of the $\hat{T}$-process. At each step for which the process arrives in a new state i, the next state is chosen without time delay according to a Multinomial $(1; \hat{p}_{i1}, \ldots, \hat{p}_{im})$ trial with state j as the destination with probability $\hat{p}_{ij}$. Thereupon, the holding time in state i is randomly sampled from the mass function with MGF $\hat{M}_{ij}$. This procedure is repeated at each step to assure that the process traverses the system according to the probability laws given by transmittance $\hat{T}(s)$. Such dynamics make the following result immediate.

Theorem 14.1.1 *The first passage transmittance for resampled first passage time W_{1m}^* ($m \neq 1$) is $\hat{f}_{1m}\hat{\mathcal{F}}_{1m}(s)$ given in (14.7). The resampled times W_{11}^*, for first return $1 \to 1$, and W_{1D}^*, for first passage to a subset of states or $1 \to D = \{p+1, \ldots, m\}$, have first passage transmittances specified by using $\hat{T}(s)$ in place of $T(s)$ in their respective cofactor rules given in (13.24) of theorem 13.2.3 and (13.37) of theorem 13.3.1.*

The high accuracy achieved by saddlepoint methods when approximating time to event characteristics assures that saddlepoint inversion of $\hat{\mathcal{F}}_{1m}(s)$ is comparable to taking a very large number of bootstrap resamples. To see this, suppose L first passage times from $1 \to m$ are resampled with empirical CDF $\hat{F}_L^*$. As $L \to \infty$,

$$\hat{F}_L^*(t) \xrightarrow{p} \Pr(W_{1m}^* \leq t) \simeq \widehat{\Pr}(W_{1m}^* \leq t),$$

where Pr denotes the true bootstrap distribution assuming infinite resampling and $\widehat{\Pr}$ is its saddlepoint approximation from inverting $\hat{f}_{1m}\hat{\mathcal{F}}_{1m}(s)$. Thus, saddlepoint inversion approximates the true bootstrap limit. It is only the saddlepoint error of relation " $\simeq$ " that prevents saddlepoint inversion from replicating infinite resampling.

14.1.4 Example: Dementia

This example from Butler and Bronson (2002) considers the progressive states of dementia as shown in the flowgraph of figure 14.1. The model was considered by Commenges (1986) and follows patients from state 1, representing good health, to state 6, which is death. States 2–4 are increasingly severe states of dementia and state 5 is a terminal state. The data from this study were not available so simulated data were analyzed instead. This has the additional benefit of allowing for assessment of the accuracy achieved from saddlepoint and simulated bootstrap methods.

14.1 Saddlepoint approximation to the single bootstrap distribution

Table 14.1. *(Dementia survival distribution). Estimates and 90% double bootstrap confidence intervals for the various percentiles of W_{16} at the survival probabilities listed in the first column. For the double bootstrap, saddlepoint intervals are based on FRT and RRT outer layer resampling and are compared with double bootstrap intervals (Sim) using $10^3 \times 2 \times 10^5$ simulations of W_{16}^{**}.*

Survival Prob.	"Exact" Perc.	Estimate		Perc. Lower			Perc. Upper		
		SA	Sim	FRT	RRT	Sim	FRT	RRT	Sim
0.50	11.5	12.3	12.6	9.8	9.9	10.0	15.9	16.0	15.1
0.25	19.0	19.9	20.1	15.7	15.9	15.1	26.3	26.6	25.8
0.10	28.4	29.6	29.7	23.0	23.5	22.9	40.2	41.0	39.3
0.05	35.4	36.9	37.0	28.5	28.9	28.3	50.8	51.7	49.5
0.01	51.7	53.8	53.7	41.7	41.7	40.5	75.2	77.0	73.2

Suppose the true transmittance $T(s) = \{T_{ij}(s)\}$ for figure 14.1 is

$$\begin{pmatrix} 0 & 0.6\,\text{Ga}(2,2) & 0.25\,\text{Ga}(4,8) & 0.15\,\text{Ga}(9,27) & 0 & 0 \\ 0 & 0 & 0.85\,\text{Ra}(\tfrac{1}{2}\sqrt{\pi}) & 0 & 0.15\,\text{IG}(1,\tfrac{1}{2}) & 0 \\ 0 & 0.7\,\text{Ra}(\sqrt{\pi}) & 0 & 0.2\,\text{Ra}(\sqrt{\pi}) & 0.1\,\text{IG}(\tfrac{1}{2}\sqrt{2},\tfrac{1}{4}\sqrt{2}) & 0 \\ 0 & 0 & 0.7\,\text{Ra}(\tfrac{3}{2}\sqrt{\pi}) & 0 & 0.3\,\text{IG}(\sqrt{2},\sqrt{2}) & 0 \\ 0 & 0 & 0 & 0 & 0 & \text{IG}(1,1) \\ 0 & 0 & 0 & 0 & 0 & 0 \end{pmatrix} \quad (14.8)$$

where $\text{Ga}(\mu, \sigma^2)$ and $\text{IG}(\mu, \sigma^2)$ are gamma and inverse Gaussian MGFs with mean μ and variance σ^2, and $\text{Ra}(\mu)$ is the MGF of a Raleigh distribution given by

$$\text{Ra}(s;\mu) = 1 + \mu s \exp(\mu^2 s^2/\pi)\{1 + \text{erf}(\mu s/\sqrt{\pi})\} \qquad s \in (-\infty, \infty)$$

with mean $\mu > 0$ and density

$$f(t;\mu) = \pi t/(2\mu^2) \exp\{-\pi t^2/(4\mu^2)\} \qquad t > 0.$$

From this system, data are simulated that consist of the complete histories of 25 patients starting at state 1 and ending in absorbing state 6. The data that comprise these 25 histories determine the minimal sufficient statistic $\hat{T}(s)$. No adjustment to $\hat{P}$ for non-relevant states was necessary since the generated data satisfied (14.6). If W_{16} is the first passage time from $1 \to 6$, neither its density, survival, nor hazard function can be exactly determined from the transmittance in (14.8); however saddlepoint versions may be computed from the cofactor rules applied to $T(s)$. These saddlepoint versions are used as exact surrogates to assess the accuracy of the saddlepoint and bootstrap estimates.

Predictive inference

Table 14.1 shows the accuracy that is achieved when using saddlepoint inversion to determine an estimate for the survival percentiles of the distribution of W_{16}. The first column lists

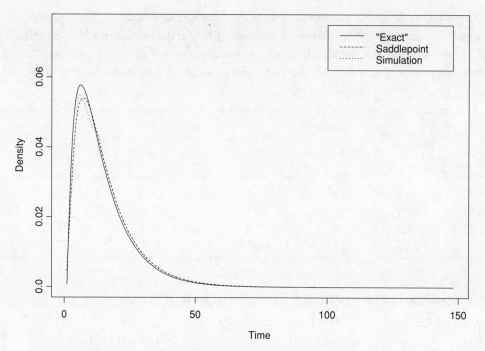

Figure 14.2. (Dementia). Density function estimates for W_{16}.

right tail probabilities with the "exact" percentiles in the second column as determined from saddlepoint inversion of $\mathcal{F}_{16}(s)$ based on the assumed $T(s)$. The third and fourth columns provide two estimates of the percentile. The first estimate is a saddlepoint approximation (SA) from inverting $\hat{\mathcal{F}}_{16}(s)$ and the second uses direct bootstrap resampling (Sim) of 2×10^5 values of W_{16}^* and taking the appropriate empirical percentile as an estimate. For example, the "true" 95th percentile of W_{16} is 35.4. The 95th percentile from saddlepoint inversion of $\hat{\mathcal{F}}_{16}(s)$ gives 36.9 which compares to 37.0, the 95th empirical percentile from resampling. The remaining portions of the table pertain to double bootstrap resampling as described in section 14.2.

Figure 14.2 compares the "exact" density of W_{16} (solid curve) with the saddlepoint approximation (short dashed) from inverting $\hat{\mathcal{F}}_{16}(s)$ and a bootstrap estimate based on simulating 2×10^5 values of W_{16}^* (dotted) and using kernel density estimation. Except near their modes, the dashed and dotted lines are virtually indistinguishable suggesting that the saddlepoint inversion has provided a very accurate substitute for the simulated bootstrap estimate that requires both resampling effort and density estimation methods.

Figure 14.3 shows the same three sets of survival curves with "exact" as the solid curve. The saddlepoint estimate (short dashed) and the empirical estimate using the 2×10^5 resampled values of W_{16}^* (dotted) were graphically indistinguishable and are shown overlain as a short dashed curve. The long dashes in the plot are 90% confidence saddlepoint bands from the double bootstrap described in section 14.2.

In figure 14.4, hazard rate estimates from saddlepoint inversion of $\hat{\mathcal{F}}_{16}(s)$ (short dashes) and a bootstrap estimate based on simulating 2×10^5 values of W_{16}^* (dotted) are compared to the exact hazard rate (solid). The stability and accuracy of saddlepoint estimates contrasts

14.1 Saddlepoint approximation to the single bootstrap distribution

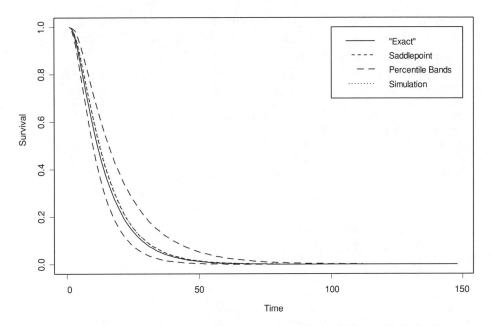

Figure 14.3. (Dementia). Survival function estimates with 90% confidence bands.

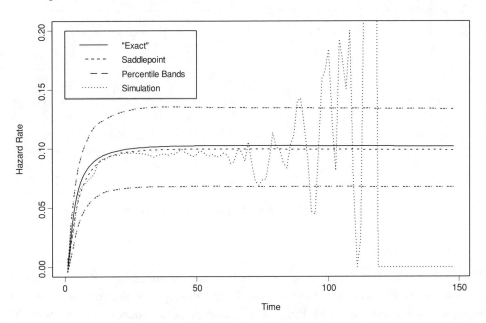

Figure 14.4. (Dementia). Hazard rate estimates and saddlepoint confidence bands.

dramatically with the simulation estimates particularly in the right tail. Saddlepoint confidence bands (long dashed) are described in section 14.2.

Relative errors in density and survival estimation, as compares with "exact" versions, are shown in figure 14.5 for saddlepoint and simulation estimates. The estimates constructed from saddlepoint inversions are smoother and more consistently accurate than the simulation estimates. The plots for saddlepoint estimates required 3.2% of the computer time used for the simulation estimates.

Table 14.2. *(Dementia parameter estimates). Estimates and confidence intervals for the mean and the standard deviation of W_{16}. The IID model assumes there are 25 i.i.d. survival times.*

Method	Mean			Standard Deviation		
	Exact	Est.	Conf. Int.	Exact	Est.	Conf. Int.
Using $\hat{K}_{16}(s)$	14.4	15.3	(12.3, 20.2)	10.7	11.1	(8.4, 15.7)
IID model		15.3	(11.9, 19.2)		10.6	(7.5, 13.1)

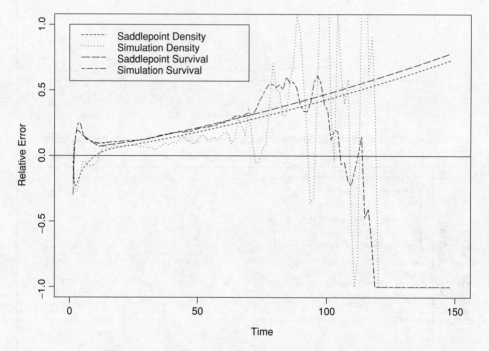

Figure 14.5. (Dementia). Relative errors in estimation.

Parametric inference

The mean and standard deviation for W_{16} can be easily estimated from $\hat{K}_{16}(s) = \ln \hat{\mathcal{F}}_{16}(s)$ using $\hat{K}'_{16}(0) = 15.3$ and $\{\hat{K}''_{16}(0)\}^{1/2} = 11.1$. The results appear in the first row of table 14.2 along with the exact values 14.4 and 10.7. Saddlepoint confidence intervals with 90% coverage are shown and constructed as described in section 14.2.

The second row in table 14.2 treats the survival data as an i.i.d. model in which their are 25 sojourn times. This approach effectively ignores the information in the data about the semi-Markov structure of the sojourns.

14.2 Saddlepoint approximations for double bootstrap confidence bands

The single bootstrap has provided (bootstrap) distribution estimates which, for settings that consider the sample mean and studentized mean, lead to nonparametric confidence intervals for the population mean. Double bootstrap resampling provides confidence bands for these bootstrap distributions that the single bootstrap has estimated.

Perhaps the benefit of using the double bootstrap is more apparent in the semi-Markov model setting. Here the single bootstrap provides survival and hazard function estimates for time to event phenomena in semi-Markov processes. However, in order to assess how accurately these functions may be estimated, the double bootstrap is used to provide confidence bands for these functions. The easiest way to understand the method is to simply apply it as described in the settings below.

14.2.1 The sample mean

In determining a 90% confidence interval for the population mean μ of F, the distribution of the pivot $\bar{X} - \mu$ is replaced with the resampling distribution of $\bar{X}^* - \bar{x}$ (given the data) as its surrogate. This is the single bootstrap concept in which this resampling distribution $H(t)$ is used to construct a confidence interval for μ. Since H has MGF $e^{-s\bar{x}} \hat{M}(s/n)^n$, with empirical MGF $\hat{M}(s)$ given in (14.2), H can be easily determined without simulation by using a saddlepoint inversion. Of course a second method for determining H is to simulate a large number of resamples $\bar{X}_1^* - \bar{x}, \ldots, \bar{X}_B^* - \bar{x}$ and use the empirical CDF $H_B^*(t)$ of the resamples as an estimate of $H(t)$.

The double bootstrap entails resampling each of the resamples used to determine $H_B^*(t)$. For example, for $i = 1, \ldots, B$, suppose that $\bar{X}_i^* = n^{-1} \sum_{k=1}^n X_{ik}^*$ and $X_{i1}^*, \ldots, X_{in}^*$ is the ith resample. Double bootstrap resampling consists of taking C resamples of $X_{i1}^*, \ldots, X_{in}^*$ for each i. Denote the sample means of these C resamples as $\bar{X}_{i1}^{**}, \ldots, \bar{X}_{iC}^{**}$ and let $H_i^{**}(t)$ denote the empirical CDF of the pivotal quantities $\bar{X}_{i1}^{**} - \bar{X}_i^*, \ldots, \bar{X}_{iC}^{**} - \bar{X}_i^*$. While the estimate of the pivotal distribution is $H_B^*(t)$, double bootstrap resampling has determined an ensemble of B additional CDFs $\{H_i^{**}(t) : i = 1, \ldots, B\}$ that fluctuate about $H_B^*(t)$ and can be used to determine confidence bands for H. For a particular value of t, a 90% confidence range for $H(t)$ is the probability range from the 5th to 95th empirical percentiles of the B values comprising $\{H_i^{**}(t) : i = 1, \ldots, B\}$. This is the percentile method for the double bootstrap which leads to a confidence band over t for the CDF $H(t)$.

To discuss this methodology further, it is convenient to refer to resample $X_{i1}^*, \ldots, X_{in}^*$ as the ith *outer* resample. The data yielding $\bar{X}_{ij}^{**}$, the jth mean sampled from the ith outer resample, is the jth *inner* resample nested inside the ith *outer* resample.

Saddlepoint approximation for inner resampling

Implementing the BC inner resamples of the double bootstrap in practical examples usually entails an enormous computational burden. Indeed B and C must often be chosen excessively large in order to achieve acceptable levels of Monte Carlo error (Booth and Presnell, 1998). For this reason, the double bootstrap method can often lie outside the realm of computability.

For this example, all of the inner resampling can be replaced with saddlepoint approximation as was first shown in Hinkley and Shi (1989). The ensemble $\{H_i^{**}(t) : i = 1, \ldots, B\}$ of inner resampled CDFs can be replaced with $\{\hat{H}_i^*(t) : i = 1, \ldots, B\}$, an ensemble of B saddlepoint approximations for the distributions that the former ensemble are meant to approximate. As $C \to \infty$, then

$$H_i^{**}(t) \xrightarrow{p} H_i^*(t) \simeq \hat{H}_i^*(t)$$

where distribution $H_i^*(t)$ has the associated MGF

$$E[\exp\{s(\bar{X}_i^{**} - \bar{X}_i^*)\}|\bar{X}_i^*] = e^{-s\bar{X}_i^*}\hat{M}_i^*(s/n)^n \tag{14.9}$$

and

$$\hat{M}_i^*(s) = n^{-1}\sum_{k=1}^{n} e^{sX_{ik}^*}.$$

Distribution $H_i^*(t)$ is the ith ensemble member with $C = \infty$ and infinite inner resampling, while its saddlepoint CDF $\hat{H}_i^*(t)$ requires no resampling. Expression (14.9), with the ith outer resample fixed, follows by analogy from the MGF of $\bar{X}^* - \bar{x}$, with the data fixed, which was determined as $e^{-s\bar{x}}\hat{M}(s/n)^n$.

The double bootstrap is often discussed in the context of prepivoting as considered in Davison and Hinkley (1997, §5.6). Indeed the original intent of Hinkley and Shi (1989), when introducing saddlepoint approximations to implement the double bootstrap, was to reduce the computational effort needed to facilitate prepivoting. Exercise 5 considers the connection of prepivoting to the double bootstrap and shows the saddlepoint implementation that is needed to facilitate prepivoting.

14.2.2 The studentized mean

If the studentized mean is used as a pivot, the bootstrap distribution of $t = \sqrt{n}(\bar{x} - \mu)/s$ becomes the pivotal surrogate in obtaining confidence intervals for μ. Thus the pivotal distribution is the distribution of $T^* = \sqrt{n}(\bar{X}^* - \bar{x})/S^*$, for fixed data. This distribution may be determined by using indirect saddlepoint methods based on the fact that $T^* = h(U^*, V^*)$ for $U^* = \sum_{i=1}^{n}(X_i^* - \bar{x})$ and $V^* = \sum_{i=1}^{n}(X_i^* - \bar{x})^2$, with the joint MGF of (U^*, V^*) determined by $\hat{M}(s, t)$ given in (14.4).

Double bootstrap resampling can be used to determine a confidence band for the pivotal CDF of t. If $T_i^* = \sqrt{n}(\bar{X}_i^* - \bar{x})/S_i^*$ is the ith outer resampled pivot for $i = 1, \ldots, B$, then all inner resampling may be avoided and replaced by B saddlepoint approximations. Inner resampling of the ith outer resample seeks to determine the distribution of $T_i^{**} = \sqrt{n}(\bar{X}_i^{**} - \bar{X}_i^*)/S_i^{**}$ with the ith outer resample fixed. However, an indirect saddlepoint approximation, of the sort used to determine the distribution of T^*, leads to $\hat{H}_i^*(t)$, the saddlepoint CDF approximation for $H_i^*(t)$, the true distribution of T_i^{**}. Now $\{\hat{H}_1^*(t), \ldots, \hat{H}_B^*(t)\}$ is an ensemble of CDFs that can be used to determine 90% confidence bands for the distribution of t.

14.2.3 Double bootstrap confidence bands for time to event distributions

A doubly resampled first passage time W_{1m}^{**} is the resampled passage time from $1 \to m$ through a resampled $\hat{T}^*$-process with $\hat{T}^*(s) = \{\hat{p}_{ij}^*\hat{M}_{ij}^*(s)\}$. The resampled $\hat{M}_{ij}^*(s)$ are the empirical MGFs based on sampling n_{ij} of the $i \to j$ transitions with replacement. The

14.2 Saddlepoint approximations for double bootstrap confidence bands

resampled transition probabilities are $\hat{p}_{ij}^* = n_{ij}^*/n_{i\cdot}$, where

$$\left(n_{i1}^*, \ldots, n_{im}^*\right) \mid \{n_{ij}\} \sim \text{Multinomial}\ (n_{i\cdot}; \hat{p}_{i1}, \ldots, \hat{p}_{im}). \tag{14.10}$$

The resampling in (14.10) can lead to zero values for n_{ij}^* when $n_{ij} > 0$. This can result in a resampled system $\hat{T}^*$ whose relevant states may not be the same as those for $\hat{T}$ which were assumed to agree with T. To avoid this difficulty, a nominal value $\varepsilon > 0$ is substituted for the zeros and the other multinomial values are proportionately diminished so that $n_{i\cdot}^* = n_{i\cdot}$. With this adjustment, all resampled versions of $\hat{T}^*$ maintain the same set of relevant states as $\hat{T}$.

Double bootstrap resampling can be used to determine a confidence band for the survival distribution of W_{1m}. Suppose the outer resampling consists of drawing B resampled transmittances $\hat{T}_1^*(s), \ldots, \hat{T}_B^*(s)$ all with the same set of relevant states as $\hat{T}(s)$. Inner resamples that take C passage times from $1 \to m$ over each system $\hat{T}_i^*$ result in passage times $W_{1m;i1}^{**}, \ldots, W_{1m;iC}^{**}$ whose empirical survival function is $S_i^{**}(t)$. The ensemble $\{S_1^{**}(t), \ldots, S_B^{**}(t)\}$ of doubly resampled survival functions are scattered about the true survival and for each value of t, the probability range of the 5th to 95th percentiles determines the confidence band at time t.

All of the BC inner resamples of W_{1m}^{**} may be avoided and replaced with an ensemble of B saddlepoint approximations $\{\hat{S}_1^*(t), \ldots, \hat{S}_B^*(t)\}$. Here $\hat{S}_i^*(t)$ is the saddlepoint inversion of $\hat{f}_{1m;i}^* \hat{\mathcal{F}}_{1m;i}^*(s)$, the first passage transmittance from $1 \to m$ for the resampled system with transmittance $\hat{T}_i^*(s)$. The following result is immediate.

Theorem 14.2.1 *Doubly resampled first passage time W_{1m}^{**} from $1 \to m \neq 1$ has transmittance*

$$\hat{f}_{1m}^* \hat{\mathcal{F}}_{1m}^*(s) = \frac{(-1)^{m+1} |\hat{\Psi}_{m1}^*(s)|}{|\hat{\Psi}_{mm}^*(s)|} \tag{14.11}$$

where $\hat{\Psi}_{ij}^*(s)$ is the (i, j)th minor of $I_m - \hat{T}^*(s)$. Doubly resampled W_{11}^{**} and W_{1D}^{**}, the first return time $1 \to 1$ and the first passage from 1 to subset D, have first passage transmittances specified by using resampled transmittance $\hat{T}^*(s)$ instead of $T(s)$ in their respective cofactor rules given in (13.24) of theorem 13.2.3 and (13.37) of theorem 13.3.1.

The resampling just described has not taken into account the random variation in the row totals $\{n_{i\cdot}\}$ and is therefore referred to as having *fixed row totals (FRT)*. A different sort of resampling might be used to help compensate for this. Consider first resampling the h patients, e.g. taking a random sample of size h from the indices $\{1, \ldots h\}$ with replacement. Using the selected patients, pool the transition rates out of the various states to determine the resampled values $\{n_{i\cdot}^*\}$ as a replacement for the fixed values of $\{n_{i\cdot}\}$ in (14.10). Now initiate each of the B resamples in (14.10) with its own resampled row totals $\{n_{i\cdot}^*\}$ to account for the random variation in $\{n_{i\cdot}\}$ when implementing the resampling scheme. This method is said to have *resampled row totals (RRT)*.

Predictive inference

An ensemble of B estimates for survival and hazard rate functions can be computed by resampling B determinations of the empirical transmittance $\{\hat{T}_1^*(s), \ldots, \hat{T}_B^*(s)\}$ using FRT and inverting the resulting first passage transmittances $\{\hat{f}_{1m;i}^* \hat{\mathcal{F}}_{1m,i}^*(s) : i = 1, \ldots, B\}$ using saddlepoint approximations. This leads to survival functions $\hat{S}^*_{16;1}(t), \ldots, \hat{S}^*_{16;B}(t)$ that are computed over a grid of time points $\{t_j\}$. The percentile method involves plotting the 5th and 95th percentiles over the grid of time points to determine lower and upper 90% confidence bands. For dementia passage distribution W_{16}, figures 14.3 and 14.4 show pointwise 90% confidence bands (long-dashed curves) of the survival function and hazard function respectively based on an ensemble of $B = 1000$ saddlepoint inversions.

The double bootstrap also provides 90% confidence intervals for the various survival percentiles of W_{16}. For example, the 25th survival percentile from table 14.1 is 19.0 and a 90% confidence interval has its range determined by selecting the 5th and 95th percentiles in the ensemble $\{(\hat{S}^*_{16;1})^{-1}(.25), \ldots, (\hat{S}^*_{16;B})^{-1}(.25)\}$. This results in the confidence interval (15.7, 26.3) listed under the FRT columns of table 14.1. As a check on saddlepoint accuracy, direct double bootstrap FRT resampling was implemented to determine the 90% percentile interval (15.1, 25.8) that is listed under the simulation (Sim) columns. Its computation consisted of simulating $B = 1000$ replications of $\hat{T}^*(s)$ at the outer level followed by the generation of 2×10^5 values of W^{**}_{16} at the inner level from each outer resample. The simulation bounds agree closely with the saddlepoint bounds for the 25th survival percentile and the other four survival percentiles listed in table 14.1.

The implementation of RRT instead of FRT resampling requires restarting the resampling from the very beginning. At the outer layer, the B resampled empirical transmittances $\{\hat{T}^*_1(s), \ldots, \hat{T}^*_B(s)\}$ need to be determined by using RRT resampling. The remaining computations are the same and lead to the saddlepoint confidence bands shown in table 14.1 under the RRT columns. The RRT intervals are quite close to the FRT intervals which suggests that the simpler FRT resampling scheme is adequate in this example, even with the rather small amount of data associated with 25 patient histories.

Butler and Bronson (2002, §3) presented the same dementia example but used the more complicated BC_a bootstrap confidence bands instead of the simple percentile method. Based on this example and others, the authors have found little difference between the two methods when used to determine confidence bands for time to event distributions. Thus, the suggested percentile method is preferred to the more complicated BC_a method based on its simplicity.

Parametric inference

Estimates for the mean and standard deviation of W_{16} have been provided in table 14.2 using the single bootstrap. Confidence intervals for these characteristics use the saddlepoint implementation of double bootstrap resampling. For example, if the outer resampling leads to the ensemble of CGFs $\hat{K}^*_{16;1}(s), \ldots, \hat{K}^*_{16;B}(s)$, then the 5th and 95th percentiles of $\hat{K}^{*\prime}_{16;1}(0), \ldots, \hat{K}^{*\prime}_{16;B}(0)$ determine the 90% confidence interval for the first passage mean. This range is (12.3, 20.2) as shown in table 14.2.

Table 14.2 also provides double bootstrap confidence intervals for the IID model that uses the 25 sojourn times as the data and which ignores the semi-Markov structure of the sojourn. These computations are performed in the same manner. The outer resample draws B sojourn distributions $\hat{F}^*_1(t), \ldots, \hat{F}^*_B(t)$ from the 25 sojourns. The saddlepoint method would then compute the B sample means as $\hat{K}^{*\prime}_1(0), \ldots, \hat{K}^{*\prime}_B(0)$ where $\hat{K}^*_i(s) = \ln\{\int e^{st} d\hat{F}^*_i(t)\}$ and extract the 5th and 95th percentiles.

The benefit of using the semi-Markov structure as opposed to the IID model when estimating the mean and standard deviation may be seen by considering the disparity in coverage accuracy for the two confidence methods. The original data of 25 complete histories were simulated 10,000 times and the 90% confidence intervals of table 14.2 were repeatedly computed. The saddlepoint confidence intervals for the mean using the semi-Markov structure attained empirical coverage 90.24% whereas the IID method had coverage 85.84%. The coverages for standard deviation were even further discrepant. Saddlepoint intervals using the semi-Markov structure exhibited coverage 90.09% whereas the IID intervals attained 70.39% coverage.

The same undercoverage when using the IID model was seen by Butler and Bronson (2002, §3.1.2) using BC_a intervals instead of the percentile method. The BC_a intervals from the IID model showed only slightly better attained coverage with 87.27% for the mean and 75.19% for the standard deviation.

Bootstrap intervals that used the semi-Markov structure for the sojourn provided intervals with very accurate coverage whereas the IID method showed undercoverage for the mean and substantial undercoverage for the standard deviation. This small sample undercoverage phenomenon seen with the IID method has also been noted recently in Polansky (1999). He explains that the undercoverage derives from the discrepancy between the bounded support of the bootstrap distribution as compares with the unbounded tail of the true (passage) distribution.

The undercoverage described by Polansky (1999) cannot occur when double bootstrap methods are implemented to determine sojourn characteristics of systems with feedback as in figure 14.1. This applies both to confidence bands for the survival function as well as for moments. The reason for this is that the presence of a feedback loop extends the support of W_{16}^* and W_{16}^{**} to ∞ so the right tail of support for W_{16}^* and W_{16}^{**} matches that of W_{16}. The situation differs however without such feedback, as would occur for a sojourn in a finite *cascading* semi-Markov system that prohibits repetition of states. Here, the number of states bounds the number of transitions, so that W^* and W^{**} have bounded support in the right tail. The extent to which the right tail support of W^* and W^{**} differ from W determines the level of susceptibility of the methods to undercoverage.

14.3 Semiparametric bootstrap

In some situations the transmittance $T(s)$ may also depend on parametric distributions along with the nonparametric distributions encountered in the previous section. Some examples include the $GI/M/1/4$ queue of section 13.2.5 as well as a $M/G/1$ queue. The Markov queues are entirely parametric. For such settings, the nonparametric bootstrap should be supplemented with the parametric bootstrap applied to the parametric portion of the stochastic system. Butler and Bronson (2002, §4) and Bronson (2001) have analyzed $GI/M/1/m$ queues for $m = 4$ and 9 and the results are reported below.

14.3.1 Example: $GI/M/1/m$ queue

This queue has states $\mathcal{M} = \{0, \ldots, m+1\}$ that describe the total number of packets in queue and under service with state $m + 1$ as the state of overflow. The process starts in state

0 at time 0 and overflow occurs when the queue arrives in state $m + 1$ at time $W_{0,m+1}$. The $(m + 2) \times (m + 2)$ transmittance matrix $T(s)$ is determined by two quantities: the interarrival distribution G for the renewal process representing the arrival of packets, and the service rate $\mu > 0$ for packets associated with the Exponential (μ) service time. Transmittance $T(s) = T(s; G, \mu)$ has the form given in (13.19) for the case $m = 4$ which is easily generalized to an $(m + 2) \times (m + 2)$ matrix with entries given in terms of

$$U_k(s) = \mu^k/k! \int_0^\infty w^k e^{sw} dG(w) \qquad k = 0, \ldots, m - 2$$
$$Q_{k1}(s) = U_0(s) - \sum_{j=0}^{k-1} U_j(s - \mu) \qquad k = 1, \ldots, m - 1. \qquad (14.12)$$

There are several data schemes that could be observed for this process. The simplest data takes $\{w_i, x_i : i = 1, \ldots, n\}$ as i.i.d. with w_i as the ith interarrival time and $x_i | w_i \sim$ Poisson (μw_i) counting the number of packets serviced during the interim. The data are summarized by two sufficient statistics: $\hat{G}$, the empirical distribution of $\{w_i\}$, and $\hat{\mu} = x./w.$, the MLE as a ratio of sums. The simplicity of this data structure serves the purpose of providing a simple data set that illustrates the bootstrap methods. However it does have the deficiency of not allowing for the possibility that the server sits idly during any of the interarrivals. The modifications to the data structure needed to accommodate this are considered in Exercise 3.

Transmittance matrix $T(s; \hat{G}, \hat{\mu})$ characterizes a semi-Markov process whose first passage transmittance from $0 \to m + 1$ is $\mathcal{F}_{0,m+1}(s; \hat{G}, \hat{\mu})$.

Theorem 14.3.1 *First passage transmittance $\mathcal{F}_{0,m+1}(s; \hat{G}, \hat{\mu})$ is the transmittance for the bootstrap distribution $W^*_{0,m+1}$.*

The distribution of $W^{**}_{0,m+1}$ in the double bootstrap, has the transmittance $\mathcal{F}_{0,m+1}(s; \hat{G}^*, \hat{\mu}^*)$, where the starred estimates are determined by resampling with replacement n pairs from $\{(w_i, x_i) : i = 1, \ldots, n\}$.

Predictive inference

Suppose G is a Gamma $(2, 2)$ distribution with mean 1 and variance $1/2$, the service rate is $\mu = 5/4$, and $n = 100$ values of $\{w_i, x_i\}$ have been generated as the data. Assume that the system capacity is $m = 4$ so the data provide a 6×6 empirical transmittance matrix $T(s; \hat{G}, \hat{\mu})$ whose first passage transmittance $\mathcal{F}_{05}(s; \hat{G}, \hat{\mu})$ can be inverted to determine the estimate of survival (short dashed line) shown in figure 14.6. The true survival function for W_{05} is the solid line and has been computed as outlined in Exercise 2. An empirical estimate (dotted line) is graphically indistinguishable from the saddlepoint estimate (short dashed) and was computed by using 2×10^5 simulated values for W^*_{05}. Double bootstrap 90% confidence bands are enclosed by the long dashes and were computed over a grid of 201 time points.

Figure 14.7 shows the same features for survival estimation but with the data set expanded to $n = 500$. The narrowing of the percentile bands is due to the greater informativeness of the larger data set. For the smaller $n = 100$ setting, figure 14.8 shows the saddlepoint density estimate (short dashed) for W_{05}, determined by inverting $\mathcal{F}_{05}(s; \hat{G}, \hat{\mu})$; the true density (solid); and a simulation estimate (dotted). Figure 14.9 shows the same sort of estimates for the hazard rate of W_{05} and also includes 90% percentile confidence bands from saddlepoint inversion (long dashed). Relative errors for the density and survival estimation are shown in

14.3 Semiparametric bootstrap

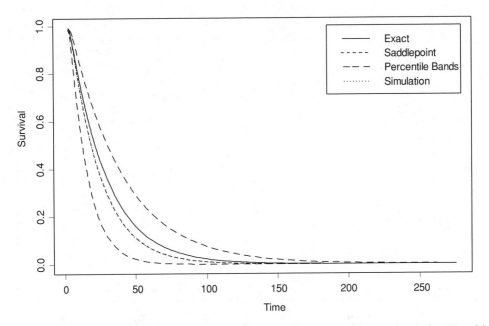

Figure 14.6. ($GI/M/1/4$ queue). Survival function estimates and confidence bands for W_{05} with $n = 100$.

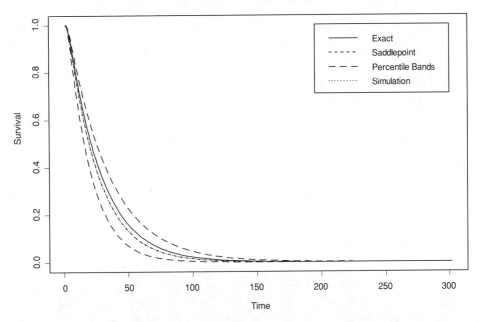

Figure 14.7. ($GI/M/1/4$ queue). Survival function estimates and confidence bands for W_{05} with $n = 500$.

figure 14.10. The saddlepoint estimates and confidence bands used only 2.9% of the CPU time required by the simulations.

The accuracy achieved by saddlepoint (SA) and simulation (Sim) estimates for various survival percentiles (Exact) of W_{05}^* is shown in table 14.3 using samples of 100 and 500. Just below the Exact percentile entries are saddlepoint approximations determined by inverting

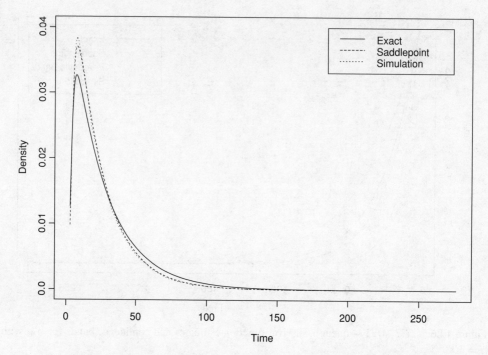

Figure 14.8. $(GI/M/1/4$ queue). Density function estimates for W_{05}.

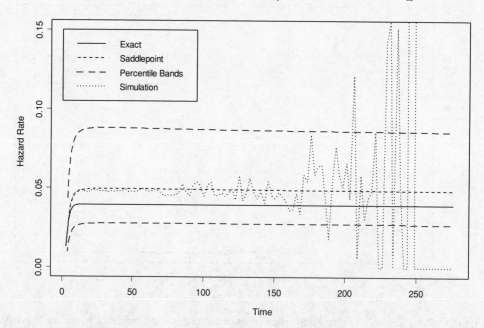

Figure 14.9. $(GI/M/1/4$ queue). Hazard rate estimates and saddlepoint confidence bands for W_{05}.

$\mathcal{F}_{05}(s; G, \mu)$ with G as Gamma $(2, 2)$ and $\mu = 5/4$. Single bootstrap point estimates and double bootstrap 90% percentile confidence intervals were computed using saddlepoint approximations (SA) and simulation (Sim). The close agreement between the SA and Sim entries suggests that the saddlepoint approximation may be used in place of simulations without much loss in accuracy and gaining substantially in computation speed.

Table 14.3. *($GI/M/1/4$ queue). Estimates and 90% percentile confidence intervals for various percentiles of W_{05} using the survival probabilities listed in the first column. The percentile intervals have been computed using saddlepoint methods (SA) and $10^4 \times 2 \times 10^5$ simulations (Sim) of W_{05}^{**}. The last column has cutoffs for 90% guaranteed coverage tolerance intervals of coverage $1-$ (survival prob.).*

Surv. Prob.	Exact SA	Sample Size	Estimate SA	Estimate Sim	Perc. Lower SA	Perc. Lower Sim	Perc. Upper SA	Perc. Upper Sim	Guar. Tol. SA	Guar. Tol. Sim
0.50	21.3	100	18.6	18.6	12.3	12.5	30.0	28.8	27.5	27.5
	21.2	500	19.6	19.6	15.8	15.8	25.4	25.4	24.1	24.1
0.25	38.8	100	33.1	33.0	20.7	20.9	55.6	53.8	50.1	50.1
	38.8	500	35.6	35.4	28.1	28.1	47.2	47.1	46.0	46.0
0.10	62.0	100	52.3	52.1	31.7	32.1	89.2	86.7	80.3	77.8
	62.0	500	56.6	56.3	44.0	44.2	75.8	75.5	70.7	70.7
0.05	79.5	100	66.8	66.4	40.0	40.5	114.6	111.8	105.4	100.4
	79.6	500	72.6	72.3	56.1	56.3	97.4	97.1	92.8	92.8
0.01	120.1	100	100.5	99.7	59.4	60.1	173.8	169.7	158.2	150.7
	120.4	500	109.5	110.0	84.1	84.5	147.6	147.7	139.6	139.6

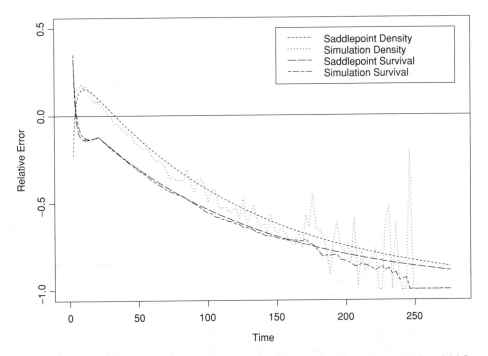

Figure 14.10. *($GI/M/1/4$ queue). Relative errors in estimation from figures 14.8 and 14.9.*

Table 14.4. *($GI/M/1/9$ queue). Estimates and 90% percentile confidence intervals for various percentiles of $W_{0,10}$ along with tolerance intervals as described in table 14.3.*

Right. Perc.	Exact SA	Sample Size	Estimate		Perc. Lower		Perc. Upper		Guar. Tol.	
			SA	Sim	SA	Sim	SA	Sim	SA	Sim
0.50	163.0	500	131.9	132.7	78.4	75.2	250.9	253.1	226.0	226.0
	162.3	1000	149.0	149.5	102.1	101.6	229.9	224.9	208.0	205.9
0.25	312.7	500	251.2	251.0	143.2	136.6	489.4	491.4	431.7	411.1
	312.6	1000	286.5	286.4	192.0	191.0	448.5	436.8	423.3	399.6
0.10	510.6	500	408.7	409.6	229.3	218.5	804.8	805.4	699.0	678.5
	511.4	1000	467.8	467.1	312.2	308.5	737.8	717.1	683.8	660.1
0.05	660.3	500	527.9	528.4	294.4	279.5	1043	1046.4	904.7	884.1
	661.7	1000	605.2	603.9	402.1	397.6	956.5	925.8	873.3	849.6
0.01	1007.9	500	804.2	810.9	445.0	422.0	1597	1601	1357	1337
	1010.2	1000	923.5	921.2	611.3	602.0	1464	1419	1323	1300.

Tolerance intervals with 90% guarantee of coverage $1-$ (survival prob.) are shown in the final column of table 14.3 based on saddlepoint and simulation methods. For example, with coverage $1-0.25$ and $n = 500$, 46.0 is found by both methods as the smallest $c > 0$ for which

$$\Pr\left\{\Pr\left(W_{05}^* < c\right) \geq 0.75\right\} \geq 0.9, \tag{14.13}$$

e.g. there is 90% guarantee of 75% coverage. Solution to (14.13) using saddlepoint approximations proceeds as follows: Take 1000 saddlepoint inversions of $\mathcal{F}_{05}(s; \hat{G}^*, \hat{\mu}^*)$ to determine 1000 survival approximations $\{\hat{S}_i^*\}$ from which we extract their 25th survival percentiles as $\{\hat{S}_i^{*-1}(.25)\}$. The 90th percentile from this sequence is the value of c that solves (14.13). Sim entries were determined by implementing the double bootstrap: For each of the 1000 resampled transmittances $\{T(s; \hat{G}_i^*, \hat{\mu}_i^*)\}$ with passage time W_{05}^*, 2×10^5 generations of W_{05}^{**} were simulated and the 25th percentile from these 2×10^5 generations was determined as $S_i^{**-1}(.25)$. The value c is the 90th percentile from the 1000 values in the set $\{S_i^{**-1}(.25)\}$. Note the striking agreement between the saddlepoint methods and the double bootstrap methods that they approximate.

Table 14.4 provides the same information as table 14.3 but pertains instead to the $GI/M/1/9$ queue. In this longer queue, first passage to breakdown occurs when the system size reaches 10. The table considers the estimation of the percentiles of $W_{0,10}$, the first passage to state 10. Since the same values of G and μ have been used, this passage requires considerably more time and state transitions for the queue to overflow. Also estimation of the percentiles of $W_{0,10}$ represents a considerably greater extrapolation from the estimated model, so any inaccuracies that occur in the estimation of G and μ are further accentuated and lead to substantially wider confidence bands for the percentiles. To compensate for this, sample sizes have been increased to $n = 500$ and 1000 in this example.

Table 14.5. *($GI/M/1/4$ and $GI/M/1/9$ queues). Percentile confidence intervals for the mean and the standard deviation of the breakdown times for the queues.*

Passage length	Sample size	Mean				Standard Deviation			
		Exact	Est.	Perc. lower	Perc. upper	Exact	Est.	Perc. lower	Perc. upper
W_{05}	100	29.0	25.0	16.0	41.6	25.3	21.0	12.0	37.1
	500		26.7	21.2	35.0		23.0	17.5	31.3
$W_{0,10}$	500	229.2	184.9	107.1	357.2	216.1	171.4	93.5	344.0
	1000		210.1	142.2	327.9		197.4	129.8	315.4

Table 14.6. *($GI/M/1/4$ and $GI/M/1/9$ queues). Coverage probabilities for 90% percentile confidence intervals using 10^4 repetitions of the data. The target coverage of 90% has been accurately achieved in all instances.*

Passage Length	Sample Size 100		Sample Size 500		Sample Size 1000	
	Mean	Std. Dev.	Mean	Std. Dev.	Mean	Std. Dev.
W_{05}	.8904	.8904	.9013	.9018		
$W_{0,10}$			.9024	.9028	.9031	.9026

Parametric inference

Table 14.5 provides saddlepoint estimates and confidence intervals for the means and standard deviations of the breakdown times for the two queues. The times to breakdown for the $GI/M/1/4$ and $GI/M/1/9$ queues are W_{05} and $W_{0,10}$ respectively.

To assess the coverage accuracy for the percentile intervals in table 14.5, 10^4 data sets were generated along with 10^4 determinations of the percentile intervals for mean and standard deviation of W_{05} and $W_{0,10}$. The coverage frequencies are listed in table 14.6.

The stationary distribution for a semi-Markov process has a mass function that can be estimated by using the bootstrap. As an example, consider adjusting the discipline of the nonstationary $GI/M/1/4$ queue so that it is stationary. Suppose upon arrival in state 5, the queue either feeds back to state 5 or is allowed to pass to states $1, \ldots, 4$. If the holding time in state 5 is G, then suppose the following discipline: it returns to state 5 when one or zero packets is serviced in the interim, and otherwise passes to states $1, \ldots, 4$ when two to five packets are serviced. This makes the queue a semi-Markov process whose stationary distribution depends only upon G and λ as specified in (13.64) or Ross (1983, §4.8). Table 14.7 displays 90% percentile confidence intervals on all of the queue's stationary probabilities.

The bootstrap can also be used to estimate the asymptotic hazard rate of W_{05}. The true asymptotic hazard rate is 0.0396 which is the smallest positive pole of $\mathcal{F}_{05}(s; G, \mu)$ or the smallest real root of $|\Psi_{55}(s)| = 0$. This rate is estimated by finding the smallest real pole of $\mathcal{F}_{05}(s; \hat{G}, \hat{\mu})$ or the smallest real root of $|\hat{\Psi}_{55}(s)| = 0$. With sample sizes 100 and

Table 14.7. *(Altered $GI/M/1/4$ queue). Ninety percent confidence intervals for the stationary distribution of the altered queue.*

State	Exact	Sample size	Estimate	Perc. Lower	Perc. Upper
0	0	100	0	0	0
		500	0	0	0
1	0.318	100	0.272	0.156	0.394
		500	0.296	0.237	0.358
2	0.235	100	0.222	0.166	0.251
		500	0.229	0.208	0.245
3	0.174	100	0.180	0.160	0.182
		500	0.177	0.167	0.180
4	0.127	100	0.145	0.102	0.182
		500	0.135	0.113	0.155
5	0.146	100	0.181	0.093	0.319
		500	0.163	0.116	0.219

500, the respective estimates are 0.0479 and 0.0437. Such roots, when computed for 1000 repetitions of the resampled $\mathcal{F}_{05}(s; \hat{G}^*, \hat{\mu}^*)$, allow computation of percentile confidence intervals. These intervals are (0.027, 0.084) and (0.032, 0.058) respectively for samples of $n = 100$ and 500. The coverage accuracy of these percentile intervals with sample size 100 was assessed by generating 2500 data sets and computing the corresponding percentile intervals. The resulting empirical coverage was 88.5% and reasonably close to the mark.

14.4 Indirect saddlepoint approximation

A very general indirect saddlepoint expression is given below for the marginal density and CDF of a nonlinear transformation of a joint saddlepoint density. The method follows the approach developed in Daniels and Young (1991) and further developed in Jing and Robinson (1994). Other approaches for the same problem have been considered by DiCiccio *et al.* (1990) and DiCiccio and Martin (1991). The general method is used in two applications: the doubly noncentral t distribution and the bootstrapped t pivotal.

Indirect conditional saddlepoint approximations, that approximate the conditional marginal density and CDF of a nonlinear transformation, have been considered in DiCiccio *et al.* (1993) and Jing and Robinson (1994).

14.4.1 General approach

Suppose random vector $X = (X_1, \ldots, X_m)^T$ has CGF $K(s)$ that is convergent on $s \in \mathcal{S} \subset \Re^m$. The general problem is to determine the marginal saddlepoint density and CDF for the scalar random variable $Y_1 = g_1(X)$ that is a nonlinear transformation of X. A smooth bijection $g(X) = Y = (Y_1, \ldots, Y_m)^T$ is assumed to exist such that Y_1 is the first component

14.4 Indirect saddlepoint approximation

of Y. Denote the nuisance variables of $Y = (Y_1, Y_b)^T$ as $Y_b = (Y_2, \ldots, Y_m)^T$ and suppose $E(X) = \mu = K'(0)$, $g(x) = y$, and $g(\mu) = \nu = (\nu_1, \ldots, \nu_m)^T$.

The saddlepoint density and CDF of Y_1 at y_0 are defined in terms of the saddlepoint values $\hat{s}_0 = (\hat{s}_{10}, \ldots, \hat{s}_{m0})^T$ and $(m-1) \times 1$ vector $\hat{y}_{b0}$ that solve

$$K'(\hat{s}_0) = \hat{x}_0 = g^{-1}(y_0, \hat{y}_{b0}) \tag{14.14}$$
$$0^T = \hat{s}_0^T \left(\partial x / \partial y_b^T\right)_0. \tag{14.15}$$

Equations (14.14) and (14.15) have m and $m-1$ dimensions respectively and the equations are solved with the constraints that $\hat{s}_0 \in \mathcal{S}$ and $(y_0, \hat{y}_{b0})$ is inside the joint support of Y. Term $(\partial x / \partial y_b^T)_0$ denotes the $m \times (m-1)$ matrix of derivatives $\partial x / \partial y_b^T$ evaluated at $\hat{y}_0 = (y_0, \hat{y}_{b0})^T$.

Theorem 14.4.1 *Expressions for the saddlepoint density and CDF of Y_1 at y_0 are given by*

$$\hat{f}_{Y_1}(y_0) = \phi(\hat{w}_0) \hat{s}_0^T (\partial x / \partial y_1)_0 / \hat{u}_0 \tag{14.16}$$

and

$$\widehat{\Pr}(Y_1 \leq y_0) = \Phi(\hat{w}_0) + \phi(\hat{w}_0) \{1/\hat{w}_0 - 1/\hat{u}_0\} \tag{14.17}$$

for $y_0 \neq \nu_1$. The various terms in (14.16) and (14.17) are

$$\hat{w}_0 = \operatorname{sgn}(y_0 - \nu_1) \sqrt{2\{\hat{s}_0^T \hat{x}_0 - K(\hat{s}_0)\}}$$
$$\hat{u}_0 = \hat{s}_0^T (\partial x / \partial y_1)_0 \sqrt{|K''(\hat{s}_0)| |\hat{H}_0|} \, \|(\partial y / \partial x^T)_0\| \tag{14.18}$$

where

$$\hat{H}_0 = \sum_{i=1}^m \hat{s}_{i0} \left(\partial^2 x_i / \partial y_b \partial y_b^T\right)_0 + \left(\frac{\partial x^T}{\partial y_b} K''(\hat{s}_0)^{-1} \frac{\partial x}{\partial y_b^T}\right)_0 \tag{14.19}$$

is a $(m-1) \times (m-1)$ Hessian matrix assumed to be positive definite.

Proof. Let $\hat{f}_X(x)$ denote the saddlepoint density of X at x determined by inverting K. Under Jacobian transformation, the joint saddlepoint density of Y at y is

$$\hat{f}_Y(y) = \hat{f}_X(x) \|\partial x / \partial y^T\| = (2\pi)^{-m/2} |K''(\hat{s})|^{-1/2} \exp\{K(\hat{s}) - \hat{s}^T x\} \|\partial x / \partial y^T\|.$$

Laplace's approximation (3.23) may be applied to the integral $\int \hat{f}_Y(y_1, y_b) dy_b$ to approximate the marginal density. Take $Q = \exp\{K(\hat{s}) - \hat{s}^T x\}$ as the dominant portion of the integrand over which the maximum is to be determined when using (3.23). Expression $\ln Q$ is convex in x (see Exercise 6), however it is not necessarily convex in y_b and this can lead to inaccuracy or failure of the approximation. Differentiating $\partial / \partial y_b$ and canceling terms leads to the expression

$$0^T = \partial (\ln Q) / \partial y_b^T = -\hat{s}^T \partial x / \partial y_b^T \tag{14.20}$$

that determines the critical value $\hat{y}_b$, as a function of y_1, at which the Laplace approximation is evaluated. Now set $y_1 = y_0$ so that $\hat{y}_b = \hat{y}_{b0}$ solves (14.20). Combine this with the need to have the saddlepoint density of X evaluated at $K'(\hat{s}_0) = \hat{x}_0 = g^{-1}(y_0, \hat{y}_{b0})$, so that $\hat{s}_0$ and $\hat{y}_{b0}$ necessarily solve both the equations specified in (14.14) and (14.15).

The Hessian computation for Laplace's approximation requires computing $-\partial/\partial y_b^T$ of the middle term in (14.20) which, upon using the chain rule, gives $\hat{H}_0$ in (14.19) when evaluated at $\hat{s}_0$ and $\hat{y}_{b0}$. The density in (14.16) follows directly from (3.23).

A Temme approximation (2.34) of the saddlepoint density produces the CDF approximation in (14.17). The approximate CDF

$$\widehat{\Pr}(Y_1 \leq y_0) = \int_{-\infty}^{y_0} \hat{f}_{Y_1}(y_1) dy_1 = \int_{-\infty}^{y_0} \phi(\hat{w}) \frac{\hat{s}^T \partial x / \partial y_1}{\hat{u}} dy_1$$

$$= \int_{-\infty}^{\hat{w}_0} \phi(\hat{w}) \frac{\hat{s}^T \partial x / \partial y_1}{\hat{u}} \frac{\partial y_1}{\partial \hat{w}} d\hat{w}$$

uses the Temme approximation by taking $h(\hat{w}) = (\hat{u}^{-1} \hat{s}^T \partial x / \partial y_1) \, \partial y_1 / \partial \hat{w}$. The latter derivative as a factor in $h(\hat{w})$ is computed by differentiating $\partial/\partial y_1$

$$-\hat{w}^2/2 = K(\hat{s}) - \hat{s}^T x \tag{14.21}$$

to get, after cancellation,

$$\frac{\partial y_1}{\partial \hat{w}} = \frac{\hat{w}}{\hat{s}^T \partial x / \partial y_1}. \tag{14.22}$$

Some longer computations given in the appendix of section 14.6.1 show that

$$\lim_{y_0 \to \nu_1} h(\hat{w}_0) = h(0) = 1 \tag{14.23}$$

so the Temme approximation (2.34) is

$$\widehat{\Pr}(Y_1 \leq y_0) = \Phi(\hat{w}_0) + \phi(\hat{w}_0)\{1/\hat{w}_0 - h(\hat{w}_0)/\hat{w}_0\}.$$

Substitution of $h(\hat{w}_0)$ using (14.22) gives (14.17).

The proof has made use of several assumptions concerning the relationship of X and Y. First, this relationship needs to be smooth, so $\partial x / \partial y^T$ is continuous, and also 1-1, so the Jacobian transformation gives the correct expression for the density of Y at y. Secondly, for given y_0, $\ln Q = K(\hat{s}) - \hat{s}^T x$ needs to be locally convex in y_b at $\hat{y}_0$ so that Laplace's approximation succeeds in providing a meaningful saddlepoint density at y_0. Finally, in deriving the CDF approximation, a meaningful density is required for $y_1 \in (-\infty, y_0)$ since it is integrated dy_1 to determine the Temme approximation. Even when such assumptions are met, there is no assurance that the density and CDF approximations will always achieve high accuracy. Daniels and Young (1991) suggest that both the Laplace and Temme approximations can result in inaccuracy and the success of the method necessarily depends on the particular context in which it is applied.

The computation of these approximations can lead to instability in root finding if a convenient starting point is not used for which the saddlepoint values are approximately known. Daniels and Young (1991) suggest starting with $y_0 = \nu_1$ wherein $\hat{s}_0 \simeq 0$ and $\hat{y}_{b0} \simeq (\nu_2, \ldots, \nu_m)^T$.

Ratios of random variables

Saddlepoint approximation for the distribution of the ratio $Y_1 = X_1/X_2$ was extensively discussed in section 12.1. This indirect method allows for a reconsideration of this problem by taking $Y_2 = X_2$ and considering the transformation from X to Y. Suppose $\Psi_{y_0} = X_1 - y_0 X_2$ has CGF $K_{y_0}(t) = K_X(t, -y_0 t)$ determined from the CGF of X.

Corollary 14.4.2 *The indirect method of theorem 14.4.1, when applied to the ratio $Y_1 = X_1/X_2$ leads to the same saddlepoint density and CDF approximations that are given in (12.9) and (12.3) of section 12.1. In particular,*

$$\hat{w}_0 = \operatorname{sgn}(\hat{t}_0)\sqrt{-2K_{y_0}(\hat{t}_0)} \qquad (14.24)$$

$$\hat{u}_0 = \hat{t}_0\sqrt{K''_{y_0}(\hat{t}_0)} \qquad (14.25)$$

where $\hat{t}_0$ solves the saddlepoint equation $K'_{y_0}(\hat{t}_0) = 0$.

Proof. Saddlepoint values $\hat{s}_0 = (\hat{s}_{10}, \hat{s}_{20})^T$ and $\hat{x}_{20}$ solve the equations

$$K'_{X;1}(\hat{s}_{10}, \hat{s}_{20}) = y_0 \hat{x}_{20}$$
$$K'_{X;2}(\hat{s}_{10}, \hat{s}_{20}) = \hat{x}_{20}$$
$$(\hat{s}_{10}, \hat{s}_{20})\{\partial(x_1, x_2)^T/\partial y_2\}_0 = \hat{s}_{10} y_0 + \hat{s}_{20} = 0 \qquad (14.26)$$

where $K'_{X;i}(s_1, s_2) = \partial K_X(s_1, s_2)/\partial s_i$ for $i = 1, 2$. Substituting (14.26) into the other two equations and combining the results gives

$$K'_{y_0}(\hat{s}_{10}) = K'_{X;1}(\hat{s}_{10}, -y_0 \hat{s}_{10}) - y_0 K'_{X;2}(\hat{s}_{10}, -y_0 \hat{s}_{10})$$
$$= y_0 \hat{x}_{20} - y_0 \hat{x}_{20} = 0.$$

Thus, by the uniqueness of the saddlepoint, $\hat{s}_{10} = \hat{t}_0$, $\hat{s}_{20} = -y_0 \hat{t}_0$, and $\hat{x}_{20} = K'_{X;2}(\hat{t}_0, -y_0 \hat{t}_0)$.

The remaining parts to the proof needed to show (14.24) and (14.25) have been outlined in Exercise 7.

Cauchy example Corollary 14.4.2 suggests that the methods for dealing with ratios X_1/X_2 can be extended to a denominator X_2 that is not a positive variable. The Cauchy ratio provides a cautionary note when trying to make this extension.

Suppose X_1 and X_2 are i.i.d. Normal $(0, 1)$ and $Y_1 = X_1/X_2$ is Cauchy. If the marginalization

$$f(y_1) = \frac{1}{2\pi} \int_{-\infty}^{\infty} |y_2| \exp\left\{-y_2^2(1 + y_1^2)\right\} dy_2 \qquad (14.27)$$

proceeds by using a Laplace approximation as indicated above, then $\hat{y}_{20}$ maximizes $\exp\{-y_2^2(1 + y_1^2)\}$ in y_2 so that $\hat{y}_{20} \equiv 0$ and $\hat{f}(y_1) \equiv 0$. The problem with this approach is that the integral in (14.27) needs to be partitioned into two integrals, over the positive and negative values of y_2, so that separate Laplace approximations can be used for the two parts.

The symmetry of the integrand in (14.27) allows both Laplace approximations to be computed as a single Laplace approximation to

$$f(y_1) = \frac{1}{\pi} \int_0^{\infty} y_2 \exp\left\{-y_2^2(1 + y_1^2)\right\} dy_2.$$

Applying (2.19) with $h(y_2) = 1/y_2$ and $g(y_2) = y_2^2(1 + y_1^2) - 2\ln y_2$, then

$$\hat{f}(y_1) = \hat{\Gamma}(1)f(y_1).$$

Thus two Laplace approximations were needed in order to derive an accurate marginal density approximation.

14.4.2 Doubly noncentral t distribution

As an application of this indirect method, Broda and Paolella (2004) have approximated the doubly noncentral t distribution and their results are replicated below. The distribution $t_n(\mu, \lambda)$ is constructed as $T = X_1/\sqrt{X_2/n}$ where $X_1 \sim$ Normal $(\mu, 1)$ and independent of $X_2 \sim \chi^2(n, \lambda)$, a noncentral chi-square variable with n degrees of freedom and noncentrality parameter $\lambda > 0$ as described in Exercise 4 of chapter 1.

Take $Y_1 = T$ and $Y_2 = \sqrt{X_2/n}$. The joint CGF of X is easily determined and the saddlepoint values solving (14.14) and (14.15) at $T = t_0$ are the values of $\hat{s}_0 = (\hat{s}_{10}, \hat{s}_{20})^T$ and $\hat{y}_{20}$ that satisfy the expressions given in section 14.6.2 of the appendix. If the approximation is restricted to the singly noncentral settings in which $\lambda = 0$, then the formulae reduce to the procedure suggested by DiCiccio and Martin (1991); see the appendix for details.

Broda and Paolella (2004) have evaluated the accuracy of the CDF approximation and find it typically provides a percentage relative error of less than 1% when $n > 4$ regardless of the values for μ and λ. They provide details for the particular case of a doubly noncentral $t_5(2, 5)$ that has been reproduced in figure 14.11. The figure uses the right scale when plotting the percentage relative errors for the density and CDF approximations over a range of ± 10 standard deviations about the mean. The error of the CDF approximation (14.17) is shown as the solid line and is considerably smaller than that of another approximation (dashed), due to Krishnan (1967), that was found to be the next most accurate amongst a larger set considered by Broda and Paolella (2004). The Krishnan procedure approximates $t_5(2, 5)$ by using a scale transformed singly noncentral $c \cdot t_f(2, 0)$ variable in which c and f are chosen to match the first two moments of each. The dotted line shows the percentage relative error of the normalized density estimate.

The left scale is used in plotting the exact density (solid), the saddlepoint density (dashed) in (14.16) and the normalized saddlepoint density (dotted), which is graphical indistinguishable from exact.

14.4.3 Bootstrapped t pivotal

The bootstrapped distribution of the studentized mean or $t = \sqrt{n}(\bar{x} - \mu)/s$ is the distribution of $T^* = \sqrt{n}(\bar{X}^* - \bar{x})/S^*$ where $\bar{X}^*$ and S^* are resampled values for the sample mean and standard deviation. The indirect saddlepoint method uses the nonlinear relationship

$$T^* = U^*\sqrt{1 - 1/n}/\sqrt{V^* - U^{*2}/n}$$

where $U^* = \sum_i (X_i^* - \bar{x})$ and $V^* = \sum_i (X_i^* - \bar{x})^2$ have the joint MGF $\hat{M}(s_1, s_2)^n$ with $\hat{M}(s_1, s_2)$ given in (14.4).

Daniels and Young (1991) apply the indirect method to the recentered values of the Davison and Hinkley data given in Exercise 4. They determine the marginal distribution of

14.4 Indirect saddlepoint approximation

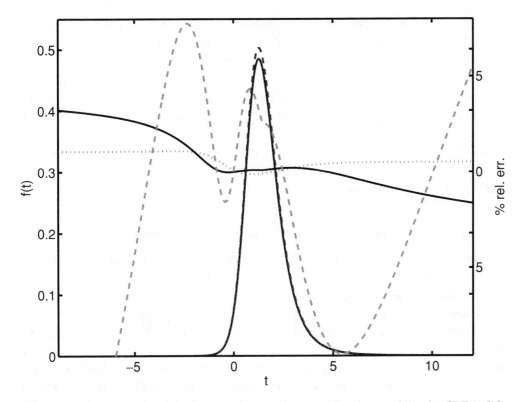

Figure 14.11. Percentage relative errors for the various approximations: saddlepoint CDF (*solid*), Krishnan (*dashed*), and the normalized saddlepoint density estimate (*dotted*) which are read from the right scale. The exact (*solid*), saddlepoint (*dashed*) and normalized saddlepoint densities (*dotted*, and not distinguishable from solid) are read from the left scale.

Table 14.8. *The survival function of $T^*/3$ using simulation (Sim.), saddlepoint CDF (14.17), and the integrated and normalized saddlepoint density (14.16).*

y_0	Sim.	Sad. CDF	Integrated Sad. Den.	y_0	Sim.	Sad. CDF	Integrated Sad. Den.
−1.8	.9978	.9978	.9983	−0.2	.6837	.6865	.6700
−1.6	.9953	.9954	.9961	0.0	.4769	.4715	.4624
−1.4	.9900	.9907	.9915	0.2	.2490	.2482	.2402
−1.2	.9807	.9820	.9830	0.4	.0910	.0896	.0884
−1.0	.9661	.9657	.9680	0.6	.0245	.0235	.0239
−0.8	.9390	.9383	.9417	0.8	.0053	.0050	.0053
−0.6	.8901	.8951	.8950	1.0	.0011	.0010	.0011
−0.4	.8136	.8203	.8118				

$Y_1 = T^*/\sqrt{n-1}$ using $Y_2 = S^*\sqrt{1-1/n}$ as the estimate of standard deviation with the denominator $\sqrt{n}$. Table 14.8 provides approximations for the survival function of Y_1 that have been reproduced from table 3 of Daniels and Young (1991).

The simulated estimates (Sim.) used 10^6 values of T^* and compare favorably with both the saddlepoint CDF approximation (14.17), shown in the middle columns, and an approximation from integrating and normalizing the saddlepoint density (14.16).

A linear interpolation of table 14.8 determines that the 5th and 95th percentiles for the distribution of T^* are -2.66 and 1.56 respectively. This leads to [14.3, 23.9] as a 90% confidence interval for μ computed from the bootstrapped t pivot. By comparison, the interval from §14.1.2 that lacks studentization is [14.1, 21.2].

The relative accuracy of these saddlepoint approximations for their bootstrap probabilities as well as the accuracy of the bootstrap itself in approximating the true signifcance of the t-pivotal have been discussed in Jing *et al.* (1994).

14.5 Empirical saddlepoint approximations

The computations for direct and indirect saddlepoint approximations require an underlying (multivariate) CGF that often may not be known. Substitution of an empirical CGF in place of the unknown true CGF has been utilized above for nonparametric inference in order to implement bootstrap computation. The resulting saddlepoint approximations have also been referred to as empirical saddlepoint approximations in the literature. Ronchetti and Welsh (1994) develop these ideas in the context of multivariate roots of estimating equations including M-estimators. Monti and Ronchetti (1993) show the relationship between these empirical saddlepoint approximations for multivariate M-estimators and empirical likelihood techniques.

14.6 Appendix

14.6.1 Proof that $h(0) = 1$ in (14.23)

In computing the limit of $h(\hat{w}_0) = \hat{u}_0^{-1} \hat{s}_0^T (\partial \hat{x}_0 / \partial y_0) \{ \partial y_0 / \partial \hat{w}_0 \}$ as $y_0 \to \nu_1$, note first that $\hat{s}_0 \to 0$, $\hat{w}_0 \to 0$, $\hat{x}_0 \to \mu$, and $\hat{y}_0 \to \nu$. The proof involves computing the limiting values for the components of the matrix

$$\hat{C}_0 = \left(\frac{\partial x^T}{\partial y} K''(\hat{s}_0)^{-1} \frac{\partial x}{\partial y^T} \right)_0 = \begin{pmatrix} \hat{c}_{11;0} & \hat{c}_{21;0}^T \\ \hat{c}_{21;0} & \hat{C}_{22;0} \end{pmatrix}$$

where $\hat{c}_{11;0}$ is a scalar and $\hat{C}_{22;0}$ is $(m-1) \times (m-1)$. Since $\hat{s}_0 \to 0$, note that

$$\hat{u}_0 / \hat{s}_0^T (\partial \hat{x}_0 / \partial y_0) = \{ \|(\partial y / \partial x^T)_0\| |K''(\hat{s}_0)|^{1/2} \} |\hat{H}_0|^{1/2}$$
$$\sim |\hat{C}_0|^{-1/2} |\hat{C}_{22;0}| \qquad (14.28)$$

as $y_0 \to \nu_1$. L'Hôpital's rule is used to find the limit of the second factor of $h(\hat{w}_0)$ in curly braces by differentiating $\partial \hat{w} / \partial y_1$, the inverse of (14.22), to get

$$\left(\frac{\partial \hat{w}}{\partial y_1} \right)^2 + \hat{w} \frac{\partial^2 \hat{w}}{\partial y_1^2} = \frac{\partial \hat{s}^T}{\partial y_1} \frac{\partial x}{\partial y_1} + \hat{s}^T \frac{\partial^2 x}{\partial y_1^2}.$$

As $y_0 \to \nu_1$,

$$\left(\frac{\partial \hat{w}_0}{\partial y_0} \right)^2 \sim \frac{\partial \hat{s}_0^T}{\partial y_0} \frac{\partial \hat{x}_0}{\partial y_0}.$$

To determine the former derivative, differentiate $K'(\hat{s}) = x$ so

$$\frac{\partial \hat{s}_0^T}{\partial y_0} = \frac{\partial \hat{x}_0^T}{\partial y_0} K''(\hat{s}_0)^{-1}. \tag{14.29}$$

Thus,

$$\frac{\partial \hat{w}_0}{\partial y_0} \sim \left\{ \frac{\partial \hat{x}_0^T}{\partial y_0} K''(\hat{s}_0)^{-1} \frac{\partial \hat{x}_0}{\partial y_0} \right\}^{1/2} = \sqrt{\hat{c}_{11;0}}. \tag{14.30}$$

Putting (14.28) and (14.30) together gives

$$h(\hat{w}_0) = \hat{u}_0^{-1} \hat{s}_0^T \frac{\partial \hat{x}_0}{\partial y_0} \frac{\partial y_0}{\partial \hat{w}_0} \sim \sqrt{|\hat{C}_0|/|\hat{C}_{22;0}|} \sqrt{1/\hat{c}_{11;0}}. \tag{14.31}$$

The right side of (14.31) converges to 1 if $\hat{c}_{21;0}^T$, the off-diagonal component of $\hat{C}_0$, converges to zero. To show this, differentiate (14.15) $\partial/\partial y_0$ to get

$$0^T = \frac{\partial \hat{s}_0^T}{\partial y_0} \left(\frac{\partial x}{\partial y_b^T} \right)_0 + \hat{s}_0^T \left(\frac{\partial^2 x}{\partial y_1 \partial y_b} \right)_0$$

so that

$$0^T \sim \frac{\partial \hat{s}_0^T}{\partial y_0} \left(\frac{\partial x}{\partial y_b^T} \right)_0. \tag{14.32}$$

Substituting (14.29) into (14.32) yields

$$0^T \sim \left(\frac{\partial x^T}{\partial y_1} \right)_0 K''(\hat{s}_0)^{-1} \left(\frac{\partial x}{\partial y_b^T} \right)_0 = \hat{c}_{21;0}^T.$$

Thus $|\hat{C}_0| \sim |\hat{C}_{22;0}| \hat{c}_{11;0}$ and $h(\hat{w}_0) \to 1$ as $y_0 \to v_1$.

14.6.2 Doubly noncentral t derivations

The saddlepoint values solving (14.14) and (14.15) are

$$\hat{s}_{10} = y_0 \hat{y}_{20} - \mu, \qquad \hat{s}_{20} = -y_0 \hat{s}_{10}/(2n \hat{y}_{20})$$

with $\hat{y}_{20}$ as the solution y_2 to the cubic equation

$$0 = y_2^3 + a_2 y_2^2 + a_1 y_2 + a_0 \tag{14.33}$$

in which the terms are

$$a_2 = -2 y_0 \mu / (y_0^2 + n)$$
$$a_1 = \{y_0^2(\mu^2 - n) - n(n + \lambda)\}/(y_0^2 + n)^2$$
$$a_0 = y_0 n \mu / (y_0^2 + n)^2.$$

Broda and Paolella (2004) follow the approach in Butler and Paolella (2002) and section 12.6.1 to show that the three roots of the cubic are real and ordered as $z_2 < z_3 < z_1$. However in this instance, z_1 is the correct choice for $\hat{y}_{20}$ whereas it was the middle root with the noncentral F.

Upon further defining

$$q = \frac{a_1}{3} - \frac{a_2^2}{9}, \qquad p = \frac{1}{6}(a_1 a_2 - 3a_0) - \frac{1}{27}a_2^3, \qquad m = q^3 + p^2$$

and $s_{1,2} = \sqrt[3]{p \pm m^{1/2}}$, the saddlepoint value is

$$\hat{y}_{20} = s_1 + s_2 - a_2/3.$$

Its value, without the complex arithmetic needed for computing the values $s_{1,2}$, is alternatively given by

$$\hat{y}_{20} = \sqrt{-4q} \cos \left\{ \cos^{-1}\left(p/\sqrt{-q^3}\right)/3 \right\} - \frac{a_2}{3}.$$

After some simplification, the following quantities result:

$$\hat{w}_0 = \text{sgn}\left(y_0 - \mu/\sqrt{1 + \lambda/n}\right)\sqrt{-\mu \hat{s}_{10} - n \ln \hat{\alpha} - 2\lambda \hat{\alpha} \hat{s}_{20}}$$
$$\hat{u}_0 = \hat{s}_{10} \hat{y}_{20} \left\{ \left(y_0^2 + 2n\hat{s}_{20}\right)\left(2n\hat{\alpha}^2 + 4\lambda \hat{\alpha}^3\right) + 4n^2 \hat{y}_{20}^2 \right\}^{1/2} / (2n \hat{y}_{20}^2)$$

with $\hat{s}_0^T (\partial x/\partial t_1)_0 = \hat{s}_{10} \hat{y}_{20}$ and $\hat{\alpha} = (1 - 2\hat{s}_{20})^{-1}$.

In the singly noncentral case for which $\lambda = 0$, these expressions simplify to

$$\hat{w}_0 = \text{sgn}(y_0 - \mu)\sqrt{-\mu \hat{s}_{10} - 2n \ln \hat{y}_{20}} \qquad (14.34)$$
$$\hat{u}_0 = \hat{s}_{10}\sqrt{(\mu y_0 \hat{y}_{20} + 2n)/2n}$$

where

$$\hat{y}_{20} = \frac{\mu y_0 + \sqrt{4n(y_0^2 + n) + \mu^2 y_0^2}}{2(y_0^2 + n)}.$$

This agrees with the approximation derived in DiCiccio and Martin (1991, Eq. 18) from the different point outlined in Exercise 10.

In the central case $\mu = 0$,

$$\hat{y}_{20} = \sqrt{n/(y_0^2 + n)}, \qquad \hat{w}_0 = \text{sgn}\, y_0 \sqrt{-2n \ln \hat{y}_{20}} \qquad (14.35)$$
$$\hat{u}_0 = y_0 \hat{y}_{20}, \qquad \hat{s}_0^T (\partial x/\partial t_1)_0 = y_0 \hat{y}_{20}^2$$

and the saddlepoint density is exact after normalization.

14.7 Exercises

1. Suppose a general semi-Markov process with m states and transmittance matrix $T(s)$. Let the data be first passage sojourns among states of the system as assumed in section 14.1.3.
   (a) Prove that the transition count matrix $N = (n_{ij})$ and the MGF matrix $\hat{M}(s) = \{\hat{M}_{ij}(s)\}$ are jointly sufficient in the nonparametric setting of the semi-Markov process.
   (b) Show that the empirical transmittance matrix (14.5) is minimal sufficient for $T(s)$.
2. The $GI/M/1/4$ queue in section 14.3.1 assumed a Gamma $(2, 2)$ interarrival time for G in the renewal arrival process. This particular choice of G leads to the computation of an exact density and survival function for W_{05} although generally such quantities are quite intractable.
   (a) The more general interarrival time $Y \sim \text{Gamma}\,(2, \beta)$ may be written as $Y = Z_1 + Z_2$ where Z_1 and Z_2 are i.i.d. Exponential (β). Use this decomposition to construct a Markov process

with state space $\mathcal{M}_1 = \{0, 0', 1, 1', \ldots, 4, 4', 5\}$ such that W_{05} is also the first passage time $0 \to 5$ for the new Markov process with $\beta = 2$.

(b) Prove that the process created in (a) is indeed Markov. In doing so, use the state ordering of $\mathcal{M}_1$ to show that the 11×11 transition rate matrix for the Markov process is the band matrix

$$\begin{pmatrix} 0 & \beta & 0 & & & & & & \\ 0 & 0 & \beta & 0 & & & & & \\ \mu & 0 & 0 & \beta & 0 & & & & \\ 0 & \mu & 0 & 0 & \beta & \ddots & & & \\ & \ddots & \ddots & \ddots & \ddots & \ddots & 0 & & \\ & & & 0 & \mu & 0 & 0 & \beta \\ & & & & 0 & 0 & 0 & 0 \end{pmatrix}.$$

The infinitessimal generator is constructed from the rate matrix by replacing its zero diagonal elements with the negative values of the row sums.

(c) Prove that the first passage time from $0 \to 5$ in the 11-state Markov process has the same distribution as W_{05} in the six-state semi-Markov process that describes the $GI/M/1/4$ queue.

(d) The first passage distribution in the Markov process is a phase distribution whose survival function is easily computed as described in section 13.5 (Bronson, 2001).

3. Suppose the $GI/M/1/4$ queue in section 14.3.1 starts in state 0 at time 0 and all details about the process are observed for the sojourn from $0 \to 5$.

(a) Describe the data that provide full likelihood information about μ and G.

(b) What are minimal sufficient statistics for μ and G based on (a)?

(c) With the data set in (a) and its summary in (b), how should the double bootstrap be implemented in order to infer characteristics concerning the semi-Markov process with transmittance T given in (13.19)?

4. The $n = 10$ data values of Davison and Hinkley (1988) given in section 3.4.1 are

$$-8.27, \; -7.47, \; -4.87, \; -2.87, \; -1.27, \; -0.67, \; -0.57, \; 3.93, \; 6.13, \; 15.93$$

when recentered to have mean zero. Denote these values as $z_1, \ldots, z_{10}$.

(a) Plot a saddlepoint approximation for the survival function of $\bar{X}^* - \bar{x} = \bar{Z}^*$.

(b) Take 20 resamples $\mathcal{D}_1^*, \ldots, \mathcal{D}_{20}^*$, each of size 10, of the centered data whose means can be denoted by $\bar{Z}_1^*, \ldots, \bar{Z}_{20}^*$. Superimpose on the graph in (a) an ensemble of 20 saddlepoint approximations for the survival functions of $\bar{Z}_1^{**} - \bar{Z}_1^*, \ldots, \bar{Z}_{20}^{**} - \bar{Z}_{20}^*$ where $\bar{Z}_i^{**}$ is the mean of a resample of 10 from $\mathcal{D}_i^*$.

(c) Repeat the same sort of computations in (b) but using 1000 resampled data sets. From these data sets, determine 90% percentile confidence bands for the survival function that underlies the original data.

5. For the data of Exercise 4, consider the use prepivoting, as described in Davison and Hinkley (1997, §5.6), A 90% confidence interval for μ, the mean of the original data set, has been given in (14.3) but prepivoting has not been used. When prepivoting a 90% confidence interval, the double bootstrap is applied to produce B doubly resampled confidence intervals of nominal coverage $100\alpha\%$ of the type given in (14.3) and denoted by

$$\hat{I}_i^{**}(\alpha) = [\bar{X}_i^* - H_i^{**-1}(1 - \alpha/2), \; \bar{X}_i^* - H_i^{**-1}(\alpha/2)] \qquad i = 1, \ldots, B.$$

Prepivoting consists of setting the nominal coverage level 90% to the value $\hat{\alpha}$ at which the attained empirical coverage frequency is 90%; e.g.

$$0.90 = B^{-1} \sum_{i=1}^{B} 1\{\bar{x} \in \hat{I}_i^{**}(\hat{\alpha})\}.$$

(a) Describe how saddlepoint approximations can be used to avoid the inner layer of resampling needed to implement prepivoting.

(b) Use prepivoting with $B = 1000$ and $B = 10,000$ outer resamples to determine 90% prepivoted confidence intervals for the mean μ.

6. Write out the details for the derivations of the indirect saddlepoint density and CDF in (14.16) and (14.17).

   (a) Prove that $\ln Q$ is convex in x but perhaps not so in y_b for fixed y_1. For fixed value y_0, what conditions would be needed in order to assure that Laplace's approximation is applicable?

   (b) Are these derivations invariant to the particular choice of the nuisance components in Y_b? Explain your answer.

   (c) Derive the saddlepoint expressions, and the values for $\hat{H}_0$, $\hat{u}_0$, and $\hat{w}_0$. Complete the details for showing that $h(0) = 1$ in section 14.6.1 of the Appendix.

7. Complete the proof of Corollary 14.4.2

   (a) Derive $\hat{w}_0$ in (14.24). Note, from the ratio approach in section 12.1, that $\hat{t} = 0$ if and only if $\mu_1 - y_0\mu_2 = 0$. In deriving $\hat{w}_0$, prove that $\hat{t} > 0$ if and only if $y_0 > \nu_1$ and therefore conclude that $\text{sgn}(\hat{t}) = \text{sgn}(y_0 - \nu_1)$.

   (b) Show that

   $$|\hat{H}_0| = K''_{y_0}(\hat{t}_0)/|K''_X(\hat{s}_0)|$$

   and hence derive $\hat{u}_0$ in (14.25).

   (c) Show that the saddlepoint density in (14.16) agrees with the ratio saddlepoint density given in (12.9).

8. Derive the doubly noncentral t approximations of section 14.4.2 at the t-value y_0.

   (a) Determine the joint CGF of X as

   $$K_X(s) = s_1\mu + s_1^2/2 - \frac{n}{2}\ln(1 - 2s_2) + s_2\lambda/(1 - 2s_2)$$

   for $s_2 < 1/2$.

   (b) For T and Y_2 as given in section 14.4.2, show that (14.14) and (14.15) are

   $$\mu + \hat{s}_{10} = y_0 \hat{y}_{20}$$
   $$n(1 - 2\hat{s}_{20})^{-1} + \lambda(1 - 2\hat{s}_{20})^{-2} = n\hat{y}_{20}^2$$
   $$\hat{s}_{10}y_0 + 2n\hat{s}_{20}\hat{y}_{20} = 0$$

   when solved with $\hat{s}_{20} < 1/2$ and $\hat{y}_{20} > 0$. Show this leads to the solution in (14.33).

9. Show that the central t approximation in (14.35) does not quite agree with the Skovgaard approximation developed in section 5.4.5. Convert to $n - 1$ degrees of freedom and show that the approximations to the density and CDF of a t_{n-1} at y_0 are related by

   $$\hat{w}_0 = \hat{w}\sqrt{1 - 1/n}, \qquad \hat{u}_0 = \hat{u}\sqrt{(y_0^2 + n - 1)/n}. \qquad (14.36)$$

   Here, $(\hat{w}, \hat{u})$ is the input to Skovgaard's approximation and the leftmost values in (14.36) refer to the entries used in (14.18) with the indirect saddlepoint methods.

10. Suppose the indirect method is used to approximate the density and CDF of Y_1 but the methods start from the exact density of X rather than its saddlepoint density.

(a) Derive expressions for the density and CDF approximations in terms of the various log-likelihoods involved.

(b) Consider using this indirect method for the singly noncentral t example with $X_1 \sim$ Normal $(\mu, 1)$ and independent of $X_2 \sim \chi_n^2$. The joint saddlepoint density for X is proportional to the exact density of X and this suggests that perhaps the true density approach can lead to the same approximations as given in (14.34). Show this is possible when $Y_2 = \ln(X_2/n)/2$, e.g. the integration for Laplace is performed on the log-scale (DiCiccio and Martin, 1991).

11. Derive an indirect saddlepoint approximation for the distribution of the two independent sample t pivotal with unequal variances.

(a) Suppose the data are the two samples $U_1, \ldots, U_n$ and $W_1, \ldots, W_m$ from Normal (μ, σ^2) and Normal (ξ, τ^2) distributions respectively. The two sample t pivotal is

$$T = \frac{\bar{U} - \bar{W} - (\mu - \xi)}{\sqrt{S_u^2/n + S_w^2/m}} \qquad (14.37)$$

where S_u^2 and S_w^2 are the unbiased sample variances. Derive an indirect saddlepoint approximation for the density and CDF of T.

(b) Compute the saddlepoint distribution of T for some parameter values and compare it to an empirical estimate using 10^6 simulated values of T.

12. Derive an indirect saddlepoint approximation for the bootstrap distribution of the T pivotal in (14.37).

(a) Suppose the data are the two samples $u_1, \ldots, u_n$ and $w_1, \ldots, w_m$. The bootstrap distribution of the t statistic is

$$T^* = \frac{\bar{U}^* - \bar{W}^* - (\bar{u} - \bar{w})}{\sqrt{S_u^{*2}/n + S_w^{*2}/m}}.$$

Derive an indirect saddlepoint approximation for the density and CDF of T^*.

(b) Compute the saddlepoint approximation for the distribution of T^* using the engine bearing data of chapter 9, Exercise 17.

13. The permutation distribution of T in (14.37) may be determined by dissociating the group membership labels of the data with their responses. Random assignment of such labels leads to $\binom{m+n}{n}$ possible assignments each with probability $\binom{m+n}{n}^{-1}$. Let $B_1, \ldots, B_{m+n}$ be i.i.d. Bernoulli $(1/2)$ variables. These random permutations are created by conditioning labels $B_1, \ldots, B_{m+n}$ on $\sum_{i=1}^{m+n} B_i = n$ to denote those assigned to the group of n with U-values. The conditioning is implemented by using the Skovgaard approximation.

(a) Using the approach of Daniels and Young (1991), develop conditional saddlepoint approximations for the permutation distribution of T in (14.37).

(b) Use the engine bearing data to compare your saddlepoint approximation with an empirical estimate based on 10^6 simulated values of the permuted value of T.

14. A marginal density approximation for the MLE in a curved exponential family was derived by Skovgaard (1985) and is given in (12.80).

(a) Use the indirect marginalization method to derive this saddlepoint density in a $(1, m)$ curved exponential family.

(b) Derive the CDF approximation given in (12.85).

15

Bayesian applications

Bayesian prediction of the predictand Z given data $X = x$ is considered for parametric settings in which the density and CDF of $Z|\theta$ are intractable. Here, intractable means that the distribution is theoretically specified but difficult at best to compute. The chapter mainly considers settings in which Z is a first passage time in a (semi-) Markov process. This is a setting for which the corresponding MGF

$$M_Z(s|\theta) = E\left(e^{s^T Z}|\theta\right) \tag{15.1}$$

has a computable and tractable expression that is available for saddlepoint inversion. In such instances, the saddlepoint density/mass function $\hat{g}(z|\theta)$ or CDF $\hat{G}(z|\theta)$ can be used as a surrogate in the predictive computation. Approximation to the predictive density $\hat{h}(z|x)$ or CDF $\hat{H}(z|x)$ relies on computing the posterior expectation of $\hat{g}(z|\theta)$ and $\hat{G}(z|\theta)$. When the posterior for $\theta|x$ is tractable, then Monte Carlo integration is implemented with respect to this posterior. Alternatively Laplace approximations, as in Davison (1986), can be used.

The posterior for $\theta|x$ may also be difficult to compute as a result of an intractable likelihood. If the intractable likelihood has MGF

$$M_X(s|\theta) = \exp\{K_X(s|\theta)\} = E\left(e^{s^T X}|\theta\right) \tag{15.2}$$

that is computable and tractable, then saddlepoint inversion is available to determine the approximate likelihood function $\hat{f}(x|\theta)$ for use as a surrogate. In such settings, x is fixed and $\hat{f}(x|\theta)$ is considered as a function of θ rather than x so the saddlepoint equation $K'_X(\hat{s}|\theta) = x$ is instead solved over a grid of θ-values. In the predictive context, computation of $\hat{h}(z|x)$ or CDF $\hat{H}(z|x)$ is now based on computing the prior expectation of $\hat{g}(z|\theta)\hat{f}(x|\theta)$ using Monte Carlo integration or Laplace approximation.

Many of the practical examples discussed in this chapter concern Markov processes in which Z and sometimes X are independent passage times. Also semi-Markov processes are considered that presume parametric transmittances although such models can quickly become over-parameterized due to the proliferation of parameters needed to model all of the one-step transmittance distributions. These passage times have MGFs that are given by the simple cofactor rules in chapter 13. Also chapter 13 gives simple matrix expressions for the first two derivatives $K'_X(s|\theta)$ and $K''_X(s|\theta)$ that are needed to implement the saddlepoint inversions. All of the examples below have appeared in Butler and Huzurbazar (1993, 1994, 1997, 2000) and the tables and density plots have been so reproduced.

15.1 Bayesian prediction with intractable predictand distribution

Let predictand Z have an intractable CDF $G(z|\theta)$ and a CGF $K_Z(s|\theta)$ that is available for saddlepoint inversion. In the most challenging case, the observed data $X = x$ would also have an intractable likelihood $f(x|\theta)$ and a readily available CGF $K_X(s|\theta)$. Let the prior distribution for θ be $\pi(\theta)$ on $\Re^m$. If posterior and prior expectation are denoted as $E_x^\theta\{\cdot\}$ and $E^\theta\{\cdot\}$, then the computation of the predictive distribution is

$$H(z|x) = E_x^\theta\{G(z|\theta)\} = \frac{\int G(z|\theta) f(x|\theta) \pi(\theta) d\theta}{\int f(x|\theta) \pi(\theta) d\theta} = \frac{E^\theta\{G(z|\theta) f(x|\theta)\}}{E^\theta\{f(x|\theta)\}}. \quad (15.3)$$

This computation is problematic with either an intractable predictand CDF $G(z|\theta)$ or likelihood $f(x|\theta)$. In such instances, saddlepoint surrogates can be used in (15.3) for the intractable terms along with various strategies for performing the posterior and prior expectations. Two settings are considered in turn: those in which $f(x|\theta)$ is available, and those in which it is intractable and must be saddlepoint approximated.

15.1.1 Tractable likelihood

Let $\hat{E}_x^\theta\{\cdot\}$ be an approximate strategy for computing the posterior expectation. Determine an appropriate grid of values $\mathfrak{z} = (z_1, \ldots, z_P)$ that adequately discretizes the predictive distribution. Then an approximate discretized predictive distribution for Z is

$$\hat{H}(\mathfrak{z}|x) = \{\hat{H}(z_1|x), \ldots, \hat{H}(z_P|x)\} = \hat{E}_x^\theta\{\hat{G}(\mathfrak{z}|\theta)\} = \hat{E}_x^\theta\{\hat{G}(z_1|\theta), \ldots, \hat{G}(z_P|\theta)\}.$$

The notation emphasizes the fact that the expectation can be performed in parallel over the entire grid $\mathfrak{z}$. In addition, the predictive density, denoted as $\hat{h}(\mathfrak{z}|x)$, can also be computed in parallel with the CDF.

The simplest expectation strategy involves simulating $\theta_1, \ldots, \theta_N$ as i.i.d. observations from posterior $\pi(\theta|x)$ and using

$$\hat{H}(\mathfrak{z}|x) = N^{-1} \sum_{i=1}^{N} \hat{G}(\mathfrak{z}|\theta_i). \quad (15.4)$$

The predictive density $\hat{h}(\mathfrak{z}|x)$ is best normalized to accommodate the discretization to give

$$\bar{h}(\mathfrak{z}|x) = \hat{h}(\mathfrak{z}|x) / \sum_{j=1}^{P} \hat{h}(z_j|x) \quad (15.5)$$

as the best estimate. A total of NP saddlepoints solutions are required in order to make these computations. A more complete discussion of the required algorithms is deferred to section 15.5.

Alternatively a Laplace approximation as in (2.5) can be used in place of the simulation from the posterior. The extra difficulty with this method is the need to maximize $\hat{G}(z_j|\theta) f(x|\theta)$ over θ with the saddlepoint computation of $\hat{G}(z_j|\theta)$ nested inside. For settings in which $f(x|\theta)$ is the dominate term in θ, the alternative Laplace (2.19) that maximizes $f(x|\theta)$ instead to determine the MLE $\hat{\theta}$ can be used as a simpler alternative although this method loses its accuracy when z is far into the tail of the distribution.

15.1.2 Intractable likelihood

For settings in which the likelihood must also be saddlepoint approximated using $\hat{f}(x|\theta)$, the predictive distributional approximation is achieved by performing the approximate prior expectations given by

$$\hat{H}(z|x) = \frac{\hat{E}^{\theta}\{\hat{G}(z|\theta)\hat{f}(x|\theta)\}}{\hat{E}^{\theta}\{\hat{f}(x|\theta)\}}.$$

The simplest strategy is to take $\theta_1, \ldots, \theta_N$ as i.i.d. observations from prior $\pi(\theta)$ and compute

$$\hat{H}(3|x) = \frac{\sum_{i=1}^{N} \hat{G}(3|\theta_i)\hat{f}(x|\theta_i)}{\sum_{i=1}^{N} \hat{f}(x|\theta_i)}. \tag{15.6}$$

The additional computational burden that results from not knowing the likelihood is $N \times \dim(x)$ additional saddlepoint approximations.

Monte Carlo integration based on simulating from the prior is efficient when the data are not highly informative about the parameter. In such instances, the simulated values from prior $\pi(\theta)$ have a reasonable chance of lying within the 99% credible region for the posterior $\pi(\theta|x)$. However, if the data are highly informative and lead to a rather concentrated posterior relative to the prior, then very few of the simulated prior values will fall within this concentrated portion of the posterior $\pi(\theta|x)$. The result is substantially lower efficiency than would occur if the simulation were taken from the posterior.

When the posterior is substantially more concentrated, a double Laplace approximation may serve as a more efficient method. First a Laplace approximation for the denominator integral is used by maximizing $\hat{f}(x|\theta)\pi(\theta)$ at $\hat{\theta}$. The numerator integral is also maximized as a function of both x and z. Extensive examples of this method are presented below. If the likelihood is highly informative and accuracy is not required for z in large deviations regions, then the alternative Laplace (2.19) can be effectively used by leaving the factor $\hat{G}(z|\theta)$ out of the maximization; thus the numerator maximum is the same as the denominator maximum $\hat{\theta}$ which greatly simplifies the double Laplace approximations. A second method, that has not been tried out, uses importance sampling for Monte Carlo integration, and is outlined in section 15.4.3.

15.1.3 Direct and indirect saddlepoint determination of predictive distributions

A simple conjugate Bayesian example clarifies the role of saddlepoint methods in performing Bayesian prediction.

A predictive F distribution

Suppose $Z = X_{n+1}$ is a future observable in the observed data sequence $x = (x_1, \ldots, x_n)$ in which the data are i.i.d. Exponential (θ). Suppose the prior on θ is Gamma (α, β) with mean α/β and variance α/β^2. If $\bar{x}$ is the mean of x, the posterior for θ is Gamma $(n + \alpha, n\bar{x} + \beta)$ and the predictive distribution of $(n + \alpha)Z/(n\bar{x} + \beta)$ is $F_{2, 2(n+\beta)}$.

For this particular example, the computation of the predictive density in (15.5) by simulating from the posterior leads to a numerically normalized F density. The saddlepoint

15.1 Bayesian prediction with intractable predictand distribution

density for the exponential density of $Z|\theta$ is $g(z|\theta)/\hat{\Gamma}(1)$ and

$$\bar{h}(\mathfrak{z}|x) \to h(\mathfrak{z}|x)/\sum_{j=1}^{P} h(z_j|x) \quad \text{as } N \to \infty$$

where $h(z|x)$ is the exact predictive density at z. The same convergence also applies when simulation is from the prior as shown in the density version of (15.6). The CDF version in (15.6) has convergence

$$\hat{H}(\mathfrak{z}|x) \to \frac{\int \hat{G}(\mathfrak{z}|\theta)\hat{f}(x|\theta)\pi(\theta)d\theta}{\int \hat{f}(x|\theta)\pi(\theta)d\theta} = \int \hat{G}(\mathfrak{z}|\theta)\pi(\theta|x)d\theta \quad \text{as } N \to \infty$$

which is not exact. The same limit is obtained by using simulation from the posterior as in (15.4).

A look at the inversion theory

The inversion theory of saddlepoint approximations provides some understanding about why heavy-tailed predictive distributions can be computed with such Monte Carlo integration but cannot be determined through direct inversion.

The saddlepoint density $\hat{g}(z|\theta)$ is an approximation to the inversion integral of MGF $M_Z(s|\theta)$ given as

$$g(z|\theta) = \frac{1}{2\pi i} \int_{c-i\infty}^{c+i\infty} M_Z(s|\theta) e^{-sz} ds. \tag{15.7}$$

The integral in (15.7) is well-defined if $c \in (a_\theta, b_\theta)$, the convergence strip of $M_Z(s|\theta)$, and assumes the true density value $g(z|\theta)$ if z is a continuity point for given θ. The predictive density is the posterior expectation of (15.7). If Fubini's theorem can be used to pass the expectation inside the contour integral, then

$$h(z|x) = E_x^\theta \{g(z|\theta)\} = \frac{1}{2\pi i} \int_{c-i\infty}^{c+i\infty} E_x^\theta \{M_Z(s|\theta)\} e^{-sz} ds$$

$$= \frac{1}{2\pi i} \int_{c-i\infty}^{c+i\infty} M_Z(s|x) e^{-sz} ds,$$

where $M_Z(s|x)$ is the MGF of the density $h(z|x)$. Fubini's theorem is only applicable if $M_Z(s|x)$ is well-defined along the contour path and, for the F distribution and other predictive passage time distributions, this requires that $c \leq 0$ and that the contour path remains on the imaginary axis or to its left.

With $c \leq 0$, direct saddlepoint approximation through the inversion of $M_Z(s|x)$ is only possible if $z \leq E(Z|x)$ but not otherwise. Furthermore $M_Z(s|x)$ is seldom a simple function. For the F distribution this MGF is a confluent Type II hypergeometric function (Phillips, 1982).

By contrast $E_x^\theta \{\hat{g}(z|\theta)\}$ is well-defined for all z including $x > E(Z|x)$ and is easily approximated with $\hat{E}_x^\theta \{\hat{g}(z|\theta)\}$ using Monte Carlo integration.

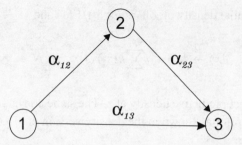

Figure 15.1. Markov graph for the cancer example showing transition rates. The states are presence of cancer (1), progressive state (2), and death (3).

15.2 Passage time examples for Markov processes

15.2.1 Example: Cancer

Lagakos (1976) and Johnson and Johnson (1980, chap. 5) describe a three state Markov process for the state of a cancer patient subjected to an experimental treatment. The Markov graph for a patient allocated to the treatment group is shown in figure 15.1.

The state of a cancer patient assigned to the control group follows the same Markov structure with transition rates $\{\beta_{ij}\}$ instead of the treatment rates $\{\alpha_{ij}\}$ shown in figure 15.1.

Let Z_T (Z_C) be the survival time for a patient from the treatment (control) group characterized as the passage from $1 \to 3$. The aim is to compare the posterior predictive distribution of Z_T with the posterior for Z_C as a way for judging the efficacy of the treatment in prolonging survival. The MGF for $Z_T|\theta$ is

$$M_{Z_T}(s|\theta) = \frac{\alpha_{13}}{\alpha_{1.}}(1 - s/\alpha_{1.})^{-1} + \frac{\alpha_{12}}{\alpha_{1.}}(1 - s/\alpha_{1.})^{-1}(1 - s/\alpha_{23})^{-1} \qquad (15.8)$$

with $\alpha_{1.} = \alpha_{12} + \alpha_{13}$. Control MGF $M_{Z_C}(s|\theta)$ has similar structure.

The system parameter is $\theta = (\alpha_{12}, \alpha_{13}, \alpha_{23}, \beta_{12}, \beta_{13}, \beta_{23})$ and consists of all the transition rates. Assuming a flat independent prior for θ on $(0, \infty)^6$, the posterior is proportional to the likelihood determined from the randomized clinical trial data x given in Lagakos (1976) that consists of 62 male lung cancer patients. Extrapolating from the sufficient statistics reported in Lagakos, the posterior has all six parameters independent with $\alpha_{12}|x \sim$ Gamma $(19, 300)$, $\alpha_{13}|x \sim$ Gamma $(5, 300)$, $\alpha_{23}|x \sim$ Gamma $(18, 254)$, $\beta_{12} \sim$ Gamma $(30, 680)$, $\beta_{13} \sim$ Gamma $(8, 680)$, and $\beta_{23}|x \sim$ Gamma $(26, 271)$.

The two predictive densities for $Z_T|x$ and $Z_C|x$ were computed as in (15.5) by taking the posterior expectation of the saddlepoint densities for $Z_T|\theta$ and $Z_C|\theta$ determined from (15.8). These graphs are shown as the dashed lines in figure 15.2 with treatment survival $Z_T|x$ as the taller graph.

The posterior expectation was based on simulating $N = 400$ values of θ from the independent component posterior. The closely matching solid lines are the corresponding "exact" predictive densities that were computed by using the same Monte Carlo integration but applied to the exact densities for $Z_T|\theta$ and $Z_C|\theta$. These exact densities are simple phase distributions that can easily be computed symbolically using partial fraction expansions. The two approximations are extremely close. The plots also show little indication of a treatment benefit. A similar conclusion was reached by Lagakos (1976, p. 558) who reported a log-likelihood ratio statistic of 2.6 whose p-value is 0.46 using a χ_3^2 distribution.

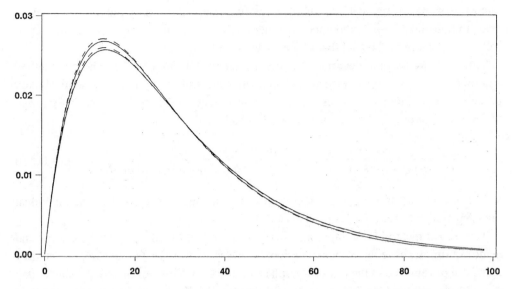

Figure 15.2. Posterior predictive distributions for treatment survival time Z_T (taller graphs) and control survival time Z_C. The saddlepoint densities (*dashed*) from (15.5) are compared to "exact" densities (*solid*).

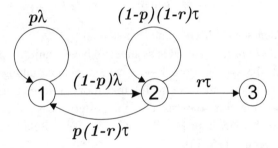

Figure 15.3. Markov graph for the reliability growth example. To convert to a semi-Markov graph, the branch transmittance from $1 \to 1$ is $p(1 - s/\lambda)^{-1}$ while the other branches are similarly converted.

An alternative to the normalization in (15.5) is the average over index i of the individually normalized saddlepoint approximations $\hat{h}(3|\theta_i)/\sum_{j=1}^{P} \hat{h}(z_j|\theta_i)$. This produced virtually identical results as (15.5). This suggests that the normalization constant is weakly dependent on θ and reinforces the theoretical conclusions in Barndorff-Nielsen and Cox (1984).

15.2.2 Reliability growth model

Lloyd and Lipow (1962) considered a Markov process that describes the time required to fix a device with a particular type of flaw while the device is maintained in operation. The device begins in state 1 of the Markov graph shown in figure 15.3 and is fixed once it reaches state 3. Feedback from $1 \to 1$ occurs when the device is successfully used with probability p in Exponential (λ) time; if unsuccessful, the device moves from $1 \to 2$ at rate $(1 - p)\lambda$. Repair activity commences in state 2, with time to repair plus subsequent usage requiring Exponential (τ) time. If a successful repair has probability r, then the transition $2 \to 3$ has rate $r\tau$ as shown in figure 15.3. An unsuccessful repair followed by failure of the device leads to transition $2 \to 2$ whereupon another repair is attempted. An unsuccessful repair followed by successful use of the device leads to the transition $2 \to 1$.

There are two ways to determine the first passage MGF from $1 \to 3$. The first method determines the 3×3 system transmittance matrix $T(s)$ of the system as a semi-Markov process with self-loops on states 1 and 2. Transmittance $T(s)$ specifies the first passage MGF $\mathcal{F}_{13}(s)$ through the cofactor rule (13.7) which gives $\mathcal{F}_{13}(s)$ as

$$\frac{\{\bar{p}M_\lambda(s)\}\{rM_\tau(s)\}}{1 - pM_\lambda(s) - \bar{p}\bar{r}M_\tau(s) - \{\bar{p}M_\lambda(s)\}\{p\bar{r}M_\tau(s)\} + \{pM_\lambda(s)\}\{\bar{p}\bar{r}M_\tau(s)\}} \tag{15.9}$$

where $\bar{p} = 1 - p$, $M_\lambda(s) = (1 - s/\lambda)^{-1}$ etc., and the curly braces enclose the individual one-step branch transmittances used in (13.7).

The second method for computing $\mathcal{F}_{13}(s)$ uses the fact that the process is Markov and leads to a simpler but equivalent expression. As a Markov process, the self-loops on states 1 and 2 can be removed in the Markov graph of Figure 15.3. This leaves only the single loop $1 \to 2 \to 1$. Without the self-loops, the first passage MGF has the simpler but equivalent expression

$$\mathcal{F}_{13}(s) = \frac{M_{\bar{p}\lambda}(s)\{r(r + p\bar{r})^{-1}M_{r\tau+p\bar{r}\tau}(s)\}}{1 - M_{\bar{p}\lambda}(s)\{p\bar{r}(r + p\bar{r})^{-1}M_{r\tau+p\bar{r}\tau}(s)\}}$$

where again $M_\psi(s)$ denotes the MGF of an Exponential (ψ).

The parameter is $\theta = (p, r, \lambda, \tau)$ and is assumed to have a uniform prior.

Nine runs through the system from $1 \to 3$ were simulated taking $p = 0.2$, $r = 0.7$, $\lambda = 1/3$, and $\tau = 2/3$. If the data x are the full transitional informational for the system in figure 15.3 concerning the nine sojourns, then the posterior has independent components which leads to marginal posteriors $p|x \sim$ Beta $(5, 15)$, $r|x \sim$ Beta $(10, 6)$, $\lambda|x \sim$ Gamma $(11, 21)$, and $\tau|x \sim$ Gamma $(15, 53)$.

Figure 15.4 compares the normalized saddlepoint approximation $\bar{h}(\mathfrak{z}|x)$ (dashed) in (15.5) using $N = 400$ with the "exact" density (solid) and a normalized Laplace approximation for $E_x^\theta\{\hat{g}(z|\theta)\}$ (dotted). The grid $\mathfrak{z} = \{0.2\,(0.1)\,50.2\}$ made use of $P = 500$ points. The "exact" density is a kernel density estimate computed by simulating 10^6 values from the distribution $Z|x$. The ith such observation entailed simulating θ_i from its posterior followed by recording Z_i as the time to traverse the system from $1 \to 3$ with parameter θ_i. The simulations $\{Z_i : i = 1, \ldots, 10^6\}$ are i.i.d. values from $h(z|x)$.

Table 15.1 shows quantiles computed by inverting the various approximations for $H(z|x)$. The "exact" percentiles are the appropriate sample percentiles from the simulation of 10^6 values of $\{Z_i\}$. The determination and inversion of $\hat{H}(\mathfrak{z}|x)$ used $N = 400$ simulations of $\{\theta_i\}$ from its posterior since this value of N was found sufficiently large to achieve convergence. The same simulations of $\{\theta_i\}$ were used to determine and invert the numerically integrated and normalized (NI&N) saddlepoint density $\bar{h}(\mathfrak{z}|x)$.

Of the remaining two approximations for $H(z|x)$, the first used Laplace's approximation (2.5) for the posterior integral $E_x^\theta\{\hat{G}(z|\theta)\}$ with Lugannani and Rice approximation $\hat{G}$ a factor in the maximization. The last approximation is a NI&N Laplace approximation (2.5) for $E_x^\theta\{\hat{g}(z|\theta)\}$ where $\hat{g}(z|\theta)$ has been used in the maximization.

Table 15.1. *Approximations for the predictive quantiles $H^{-1}(0.90|x)$, etc. in the reliability growth example determined by inverting the approximation for $H(z|x)$ described in the left-most column.*

$H(z|x)$ approximation	0.90	0.95	0.99	0.9^3	0.9^4	0.9^5
"Exact" from $\{Z_i\}$	20.0	25.9	41.2	67.3	101.8	146.0
$\hat{H}(3|x)$ in (15.4)	19.7	25.5	40.5	65.7	95.1	127.9
NI&N $\bar{h}(3|x)$ in (15.5)	19.9	25.7	40.8	66.0	95.5	128.2
Laplace for $E_x^\theta\{\hat{G}(z|\theta)\}$	19.8	25.5	37.9	56.4	73.2	88.5
NI&N Laplace for $E_x^\theta\{\hat{g}(z|\theta)\}$	20.5	26.5	42.3	69.6	104.3	152.0

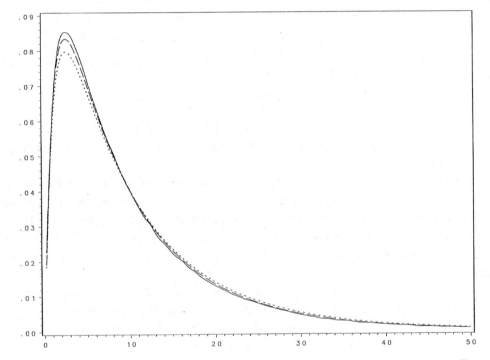

Figure 15.4. Approximate predictive densities for the Markov reliability growth example. Shown are the "exact" density (*solid*) from kernel density estimation, the normalized saddlepoint density (*dashed*) from (15.5), and the normalized Laplace approximation (*dotted*).

All the approximations show good accuracy except for Laplace's approximation of $E_x^\theta\{\hat{G}(z|\theta)\}$ above probability 0.99. This approximation resulted in values for $\hat{H}(z|x)$ that exceeded 1 and reached a height of 1.24 in the far right tail. In this particular case, the quantiles were determined by rescaling the approximation to attain the value 1 in the far right tail. This rescaling leads to the inaccuracy above probability 0.99.

15.2.3 Repairman model

This Markov queue was described in section 13.4.2 as a birth and death process. Suppose there are $m = 5$ operating machines, $n = 2$ spares, and $q = 2$ servers for fixing the failed

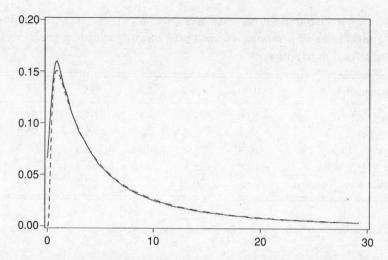

Figure 15.5. Saddlepoint predictive density approximation (15.5) (*dashed*) for first passage $0 \to 7$, the time to complete failure. The Laplace approximation (*dashed*) is indistinguishable and a kernel density estimate (*solid*) is based on 10^6 simulations from $h(z|x)$.

machines. The state of the system is the number of failed machines labelled $\{0, \ldots, 7\}$. The failure rate of each working machine is λ and the fixing rate is μ. The process parameter is $\theta = (\mu, \lambda)$.

Two predictive passage time distributions are considered with this example. First is the time to complete failure Z which is the first passage $0 \to 7$. This first passage MGF $\mathcal{F}_{0,7}(s|\theta)$ is determined through either the recursive computations described in section 13.4.2 or by the cofactor rule in (13.7). Second is the first return to zero or $0 \to 0$. Its MGF $\mathcal{F}_{0,0}(s|\theta)$ is determined through either the recursive computations in section 13.4.1 or by the cofactor rule in (13.24).

Both examples make use of simulated data x that consist of the complete sojourn information for a single traverse of $0 \to 7$. The data were generated by taking failure rate $\lambda = 3$ and service rate $\mu = 7$. Using a flat prior on θ, the posterior has independent components with $\lambda|x \sim$ Gamma (28, 9.1) and $\mu|x \sim$ Gamma (21, 3.2).

The final subsection considers the approximation of the likelihood for $\theta = (\mu, \lambda)$ when the data x have an intractable density $f(x|\theta)$. First passage data from $0 \to 7$ are used to make this construction using cofactor rule (13.7) to determine $M_X(s|\theta)$ over a grid of θ-values.

Time to complete failure

Monte Carlo simulation of $N = 1600$ θ-values from the posterior on θ was used to compute saddlepoint predictive density $\bar{h}(3|x)$ for first passage $0 \to 7$ as shown by the dashed curve in figure 15.5. The Laplace approximation (2.5) is graphically indistinguishable. This compares to a kernel density estimate (solid) constructed using 10^6 values of Z simulated from $h(z|x)$ as described in the previous example.

Quantiles for the predictive CDF $H(z|x)$ are shown in table 15.2.

15.2 Passage time examples for Markov processes

Table 15.2. *Approximate predictive quantiles $H^{-1}(0.90|x)$, etc. for the first passage distribution $0 \to 7$.*

$H(z|x)$ approximation	N	0.90	0.95	0.99	0.9^3	0.9^4
"Exact" from $\{Z_i\}$		29.5	50.3	143	495	1533
$\hat{H}(3|x)$ in (15.4)	1600	29.2	48.7	128	476	1354
$\hat{H}(3|x)$ in (15.4)	400	29.6	48.7	124	304	552

Table 15.3. *Approximate predictive quantiles $H^{-1}(0.90|x)$, etc. for the first return distribution $0 \to 0$.*

$H(z|x)$ approximation	N	0.90	0.95	0.99	0.9^3	0.9^4
"Exact" from $\{Z_i\}$		4.22	7.27	20.4	68.2	223
$\hat{H}(3|x)$ in (15.4)	1600	4.09	7.09	20.6	73.6	189

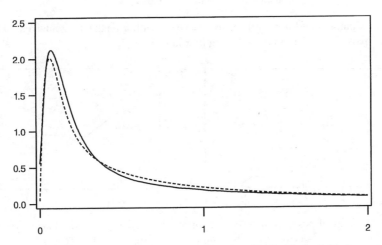

Figure 15.6. Saddlepoint predicitve density approximation (15.5) (*dashed*) for first return $0 \to 0$. The Laplace approximation (*dashed*) is indistinguisable. A kernel density estimate (*solid*) is based on 10^6 simulations from $h(z|x)$.

The empirical quantiles from the 10^6 simulated values are used as the "exact" quantiles. Quantiles are also given from the inversion of $\hat{H}(3|x)$ in (15.4) that was computed by using $N = 1600$ and $N = 400$ θ-values to perform the posterior expectation. Accuracy for the latter quantiles with $N = 400$ begins to break down at the 99.9 percentile.

First return to state 0

This example uses the same data and hence leads to the same posterior on θ. The same Monte Carlo integration methods as in the previous example were used to compute the first return $0 \to 0$ predictive density approximations as given in figure 15.6.

Quantiles extracted by inverting $H(z|x)$ approximations are given in table 15.3.

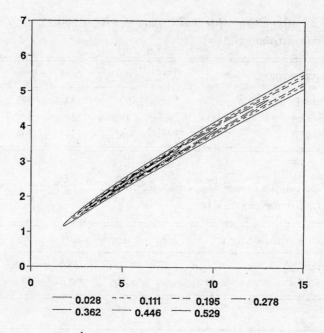

Figure 15.7. Contour plot for $\hat{f}(x|\theta)$ versus $\theta = (\mu, \lambda)$ with μ on the horizontal axis. The data are 100 i.i.d. first passage times from $0 \to 7$ using true value $\theta = (7, 3)$.

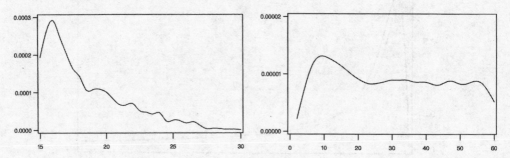

Figure 15.8. Marginal posterior distributions on the respective polar coordinates $\tan^{-1}(\lambda/\mu)$ (in degrees) and $(\mu^2 + \lambda^2)^{1/2}$ derived from the joint posterior proportional to $\hat{f}(x|\theta)$.

Saddlepoint approximation to the likelihood

A different data set is now used to approximate the likelihood function. Let the new data x consist of 100 i.i.d. first passage times from $0 \to 7$. Figure 15.7 provides a portion of the contour plot for the joint posterior on $\theta = (\mu, \lambda)$.

The posterior is a slightly bent streak extending about four times the length shown in the plot. The plot suggests a likelihood that is informative about the polar coordinate $\tan^{-1}(\lambda/\mu)$ but not about the radial coordinate. Taking the likelihood as the posterior, figure 15.8 plots the marginal densities for the polar coordinates $\tan^{-1}(\lambda/\mu)$ (in degrees) and $(\mu^2 + \lambda^2)^{1/2}$ respectively as the Jacobian transformation of the density seen in figure 15.7. These plots confirm that 100 i.i.d. first passage times carry information about the relative values of λ and μ as would be reflected in the force of movement from $0 \to 7$, however, this data is quite uninformative about the magnitude $(\mu^2 + \lambda^2)^{1/2}$. Note, for comparison, that the true values are $\tan^{-1}(3/7) = 23.2°$ and $(7^2 + 3^2)^{1/2} = 7.62$.

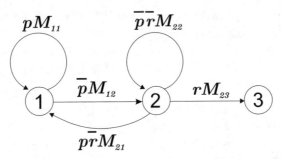

Figure 15.9. Semi-Markov flowgraph for the reliabilty growth example in which $M_{ij}(s) = (1 - s/\lambda_{ij})^{-1}$ and $\bar{p} = 1 - p$, etc.

15.2.4 General considerations

The examples show that if complete data are observed for sojourns in a Markov process, then a particular tractable posterior structure results for the system parameter θ. Suppose the process is parameterized so that θ consists of the $m \times m$ probability transition matrix P for the jump chain along with the exit rates $\Lambda = \text{diag}\{\lambda_1, \ldots, \lambda_m\}$ from the m states. The $m \times m$ matrix N that consists of the observed state transition counts has a Whittle distribution (Johnson and Kotz, 1972, §40.9) in which each row follows a Multinomial distribution with probabilities from the corresponding row of P. Independent of the Whittle distribution are the exponential holding times in states with their exit rates as scale parameters. Thus, the conjugate Bayesian analysis has the following posterior structure: mutually independent Dirichlet distributions for the rows of P that are independent of independent Gamma distributions for the components of Λ.

Incomplete data from Markov processes does not necessarily lead to such simple posterior structure and likelihood $f(x|\theta)$ may need to be saddlepoint approximated. An example of this occurs when the data are first passage times from $1 \to 7$ as considered for the repairman problem.

15.3 Passage time examples for semi-Markov processes

15.3.1 Reliability growth model

In this context the Markov model of §15.2.2 can be extended to a semi-Markov model by allowing the holding times in the states to depend on their destination states. A semi-Markov flowgraph that allows general exponential holding times before transition is shown in figure 15.9. Here, $M_{ij}(s)$ is the MGF for an Exponential (λ_{ij}) holding time.

The system parameter is $\theta = (p, r, \lambda_{11}, \lambda_{12}, \lambda_{21}, \lambda_{22}, \lambda_{23})$ and is assumed to have a uniform prior over $(0, 1)^2 \times (0, \infty)^4$. Suppose the data are the nine runs through the system from $1 \to 3$ previously used in §15.2.2 when the model was Markov. The resulting posterior on θ has mutually independent components with $p|x \sim$ Beta $(5, 15)$, $r|x \sim$ Beta $(10, 6)$, $\lambda_{11}|x \sim$ Gamma $(3, 8)$, $\lambda_{12}|x \sim$ Gamma $(9, 13)$, $\lambda_{21}|x \sim$ Gamma $(3, 7)$, $\lambda_{22}|x \sim$ Gamma $(4, 12)$, and $\lambda_{23}|x \sim$ Gamma $(9, 35)$.

Predictand Z, as the first passage time from $1 \to 3$, has MGF $M_Z(s|\theta)$ that is computed by using the cofactor rule in (13.7). This results in expression (15.9) with the appropriate adjustments in branch transmittances.

Table 15.4. *Approximate predictive quantiles $H^{-1}(0.90|x)$, etc. for first passage $1 \to 3$ in the semi-Markov reliability growth example.*

$H(z|x)$ approximation	N	0.90	0.95	0.99	0.9^3	0.9^4
"Exact" from $\{Z_i\}$		23.5	32.7	62.7	142.4	322.7
$\hat{H}(\mathfrak{z}|x)$ in (15.4)	400	24.0	33.4	62.3	128.8	236.0
NI&N $\bar{h}(\mathfrak{z}|x)$ in (15.5)	400	25.0	34.6	63.9	131.1	237.1
$\hat{H}(\mathfrak{z}|x)$ in (15.4)	1600	23.7	32.9	62.5	136.0	251.9
NI&N $\bar{h}(\mathfrak{z}|x)$ in (15.5)	1600	24.5	33.8	63.6	136.8	251.7

Table 15.5. *Approximate predictive quantiles $H^{-1}(0.90|x)$, etc. for first passage $1 \to 3$ in the semi-Markov reliability growth example. Approximations $\hat{H}(\mathfrak{z}|x)$ and NI&N $\bar{h}(\mathfrak{z}|x)$ were computed using $N = 400$.*

$H(z|x)$ approximation	0.90	0.95	0.99	0.9^3	0.9^4	0.9^5
"Exact" from $\{Z_i\}$	23.0	30.5	49.9	82.5	119.6	173.1
$\hat{H}(\mathfrak{z}|x)$ in (15.4)	23.5	31.2	50.8	84.3	126.9	181.1
NI&N $\bar{h}(\mathfrak{z}|x)$ in (15.5)	23.2	30.8	50.4	83.9	126.3	180.4
Laplace for $E_x^\theta\{\hat{G}(z|\theta)\}$	23.7	31.2	47.8	76.6	101.5	126.1
NI&N Laplace for $E_x^\theta\{\hat{g}(z|\theta)\}$	24.9	33.2	56.8	110.0	132.9	188.7

Approximate quantiles for $H(z|x)$, the posterior on Z, are shown in table 15.4 based on two sets of Monte Carlo simulations with $N = 400$ and $N = 1600$ observations drawn from the posterior of θ as explained for table 15.1. Note that the Monte Carlo approximations have deteriorating accuracy at 0.9^3 and beyond presumably because of the inadequate sample size N used in the Monte Carlo.

Alternative methods that use Laplace approximations were not computed in table 15.4 because the required iterative maximizations of $\hat{G}(z|\theta)f(x|\theta)$ and $\hat{g}(z|\theta)f(x|\theta)$ in θ failed to converge. This failure to locate maxima appears to be the result of likelihoods in θ that are not peaked enough in their seven dimensions.

The failed maximization suggests that more data should be simulated to give a more peaked likelihood. Doubling the number of runs through the system from 9 to 18 yielded a likelihood that could be maximized. Figure 15.10 shows the resulting predictive densities with the normalized Laplace approximation (dotted), the normalized saddlepoint density (dashed) from (15.5) using $N = 400$, and a kernel density estimate based on 10^6 simulated values.

Approximate quantiles for the CDF $H(z|x)$ are given in table 15.5 with x reflecting the 16 passages through the system. All the approximations are accurate except for Laplace for $E_x^\theta\{\hat{G}(z|\theta)\}$ which begins to show inaccuracy at 0.9^3. This approximation achieved the maximum height of 0.9971 in the right tail and was rescaled to 1 to determine the table entries. From this and previous examples, it is clear that the Laplace method is only successful in the right tail when used to approximate the predictive density.

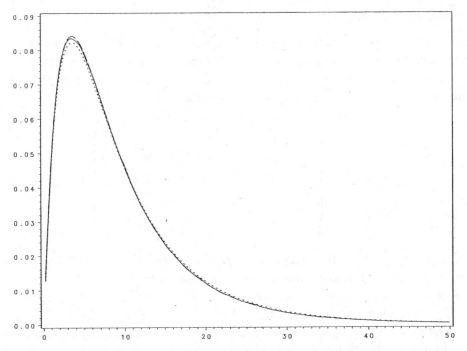

Figure 15.10. Approximate predictive densities for the semi-Markov reliability growth example. Shown are the "exact" density (*solid*) from kernel density estimation, the normalized saddlepoint density (*dashed*) from (15.5), and the normalized Laplace approximation (*dotted*).

15.3.2 M/G/1 queue

Suppose there are $m = 5$ operating machines and $n = 2$ spares with a single repairman. The machines operate independently with Exponential (λ) failure times and the repairman services machines in Inverse Gaussian (μ, σ^2) time. During periods of repair, failed machines do not enter the queue but do so only at the end of the service time in which they occur. Likewise spare machines are only brought on line at the end of a service time.

The queue begins with all machines operating (in state 0) and passes to state 1 when a machine fails after $E_{5\lambda} \sim$ Exponential (5λ) time. At this point, the server begins repair and completes this first service in Inverse Gaussian time G_1. During the repair time from $E_{5\lambda}$ to $E_{5\lambda} + G_1$, the system remains in state 1 despite the fact that anywhere from 0 to 5 additional machines may fail during the interim. At epoch $E_{5\lambda} + G_1$ all the failed machines enter the queue, the fixed machine is put on line, and spares are added to bring the number of operating machines up to 5 if possible. Under such dynamics, the system state is $\mathcal{M} = \{0, 1, \ldots, 6\}$ and counts the number of failed machines that the repairman has observed after last fixing a machine, or 0 if none have yet been fixed. The rationale for assuming these dynamics is that it leads to a process that is semi-Markov.

For the posterior Bayesian analysis, it is convenient to reparameterize the mean and variance of the Inverse Gaussian service time as $\mu = \tau_1/\tau_2$ and $\sigma^2 = \tau_1/\tau_2^3$ yielding the density

$$f(x|\tau_1, \tau_2) = \frac{\tau_1}{\sqrt{2\pi}} x^{-3/2} \exp\left\{ -\left(\tau_1^2/x + \tau_2^2 x\right)/2 + \tau_1\tau_2\right\}$$

with allowable parameter values $\tau_1 > 0$ and $\tau_2 \geq 0$. The CGF is specified in (13.18).

First passage from $0 \to 6$

Let Z be the first passage time from $0 \to 6$. The aim is to compute the posterior predictive distribution for Z. At time Z the repairman notes the system in state 6 for the first time which means that total breakdown occurred during the interim of the last repair. The system parameter is $\theta = (\lambda, \tau_1, \tau_2)$ and the system transmittance is

$$T(s) = \begin{pmatrix} 0 & M_{5\lambda} & 0 & 0 & 0 & 0 & 0 \\ T_{10} & T_{11} & T_{12} & T_{13} & T_{14} & T_{15} & 0 \\ 0 & T_{21} & T_{22} & T_{23} & T_{24} & T_{25} & T_{26} \\ 0 & 0 & T_{32} & T_{33} & T_{34} & T_{35} & T_{36} \\ 0 & 0 & 0 & T_{43} & T_{44} & T_{45} & T_{46} \\ 0 & 0 & 0 & 0 & T_{54} & T_{55} & T_{56} \\ 0 & 0 & 0 & 0 & 0 & T_{65} & T_{66} \end{pmatrix}$$

where $M_{5\lambda} = \{1 - s/(5\lambda)\}^{-1}$ and $T_{ij} = T_{ij}(s)$ is computed as described below. Cofactor rule (13.7) provides a simple computational expression for the MGF of $Z|\theta$.

To determine $T_{ij} = T_{ij}(s)$, fix the inverse Gaussian service time G so that the number of failed machines in time G is Binomial $(m_i; 1 - e^{-\lambda G})$, where $m_i = \min(m, m + n - i)$ is the number of machines operating while in state i. The transmittance is the expectation over G in the first line below and is computed by expanding the binomial term $(1 - e^{-\lambda G})^{j-i+1}$ to give the expression in the second line, or

$$T_{ij}(s) = E^G \left\{ e^{sG} \binom{m_i}{j-i+1} (1 - e^{-\lambda G})^{j-i+1} (e^{-\lambda G})^{m_i - (j-i+1)} \right\} \qquad (15.10)$$

$$= \binom{m_i}{j-i+1} \sum_{k=0}^{j-i+1} \binom{j-i+1}{k} (-1)^k \exp\left[\tau_1 \tau_2 - \tau_1 \{2(c - s + \lambda k)\}^{1/2}\right].$$

Expression $T_{ij}(s)$ is defined for $s \leq c = \tau_2^2/2 + \lambda(m_i - j + i - 1)$. The convergence of this transmittance at its endpoint c makes it a nonregular case for saddlepoints. However it can still be inverted by using saddlepoints; see Exercise 5.

Data x are simulated from this queue by recording the complete history of state transitions and holding times for a single passage from $0 \to 6$ using $\lambda = 0.4$, $\tau_1 = 3$, and $\tau_2 = 4$. The likelihood is comprised of three contributions. First, transition from $0 \to 1$ occurred $R_1 = 5$ times contributing five independent Exponential (5λ) variates totally 1.37 as an observed Gamma $(5, 5\lambda)$ term; secondly, $R_2 = 116$ machines were serviced contributing 116 i.i.d. inverse Gaussian variates $g_1, \ldots, g_{116}$ with sufficient statistics $u_1 = \sum_i 1/g_i = 166.0$ and $u_2 = \sum_i g_i = 86.6$; thirdly, there were R_2 binomial terms each counting the number of machine failures $\{n_i\}$ during service time i of the form

$$\prod_{i=1}^{R_2} \binom{o_i}{n_i} (1 - e^{-\lambda g_i})^{n_i} (e^{-\lambda g_i})^{o_i - n_i} \qquad (15.11)$$

where o_i is the number of machines operating at the beginning of the ith service time.

A flat prior on θ leads to a posterior in which $\lambda|x$ and $(\tau_1, \tau_2)|x$ are independent. The posterior on λ has the Gamma $(5, 5\lambda)$ term along with the binomial terms in (15.11) which essentially contribute the likelihood of a logistic regression of n_i on g_i using a canonical link function. The posterior on (τ_1, τ_2) can be shown to have the following structure: $\tau_1^2|x \sim$

Table 15.6. *Approximate predictive quantiles $H^{-1}(0.90|x)$, etc. for first passage $0 \to 6$ in the $M/G/1$ example. Approximation $\hat{H}(3|x)$ was computed using $N = 400$. The "Inverse Gaussian-based CDF" was computed like $\hat{H}(3|x)$ but with an inverse Gaussian-based CDF approximation (section 16.2.2) for $G(z|\theta)$ instead of Lugannani and Rice.*

$H(z|x)$ approximation	0.90	0.95	0.99	0.9^3	0.9^4
"Exact" from $\{Z_i\}$	140.1	192.7	340.9	642.7	1133
$\hat{H}(3|x)$ in (15.4)	143.0	196.5	350.7	687.2	1351
Inverse Gaussian-based CDF	140.7	193.8	347.0	680.7	1338

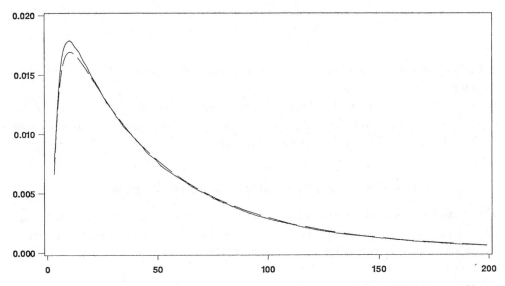

Figure 15.11. Approximate predictive densities for first passage $0 \to 6$ in the $M/G/1$ queue. Shown are $\bar{h}(3|x)$ (*dashed*) computed with $N = 400$ and a kernel density estimate (*solid*) using 10^6 simulated values. The normalized Laplace approximation is graphically indistinguisable from $\bar{h}(3|x)$.

Gamma $\{r_2/2 + 1, (u_1 - r_2^2/u_2)/2\}$ or Gamma $(59, 5.04)$ while $\tau_2|\tau_1, x \sim$ Truncated Normal $(r_2\tau_1/u_2, 1/u_2)$ or Normal $(1.34\tau_1, 0.11^2)$ constrained so that $\tau_2 > 0$.

Figure 15.11 shows the approximate predictive density $\bar{h}(3|x)$ (dashed) computed using $N = 400$ and compared with a kernel density estimate (solid) computed from 10^6 simulated values. The inverse Gaussian variates were generated by using the algorithm of Atkinson (1982). A normalized Laplace approximation was also computed and is graphically indistinguishable from $\bar{h}(3|x)$.

Table 15.6 shows approximate predictive quantiles for the first passage $0 \to 6$ in the $M/G/1$ example. Approximation $\hat{H}(3|x)$ was computed using $N = 400$. The "Inverse Gaussian-based CDF" was computed like $\hat{H}(3|x)$ but used an inverse Gaussian-based CDF approximation (section 16.2.2) for $G(z|\theta)$ instead of the Lugannani and Rice approximation. The inverse Gaussian-based approximation was slightly more accurate.

Table 15.7. *Predictive probabilities for the stationary distribution of the $M/G/1$ queue.*

State: $i =$	0	1	2	3	4	5	6
$\hat{E}_x^\theta\{\pi_i(\theta)\}$	.0136	.0976	.1813	.2887	.2703	.1268	.0216

Stationary distribution of $M/G/1$

The posterior stationary distribution can also be computed for this queue. Given system parameter θ, the stationary probability of state i is the mean waiting time in state i divided by the mean first return time to state i (Barlow and Proschan, 1965, p. 136). In terms of $T(s) = T(s|\theta)$, and $W_{ii}|\theta$, the first return time to state i, the stationary probability is

$$\pi_i(\theta) = T'_{i\cdot}(0|\theta)/E(W_{ii}|\theta) \qquad i = 0, \ldots, 6,$$
$$= T'_{i\cdot}(0|\theta)|\Psi_{ii}(0|\theta)|/\sum_{j=1}^{m} T'_{j\cdot}(0|\theta)|\Psi_{jj}(0|\theta)| \qquad (15.12)$$

if $E(W_{ii}|\theta)$ is replaced with (13.34). The determinant $|\Psi_{jj}(s|\theta)|$ in (15.12) is the (j, j) cofactor of $T(s|\theta)$.

The posterior stationary probabilities were computed by taking the approximate posterior expectation of (15.12). Simulating $N = 400$ θ-values from the posterior and averaging the resulting values in (15.12) leads to the entries in table 15.7.

15.4 Conclusions, applicability, and alternative methods

The examples considered in this chapter lead to several conclusions about Bayesian computations in stochastic processes. The limits of applicability for these methods are noted. Another Monte Carlo method for posterior expectation using importance sampling is briefly discussed.

15.4.1 Conclusions

All of the prediction examples considered had tractable posteriors on the system parameter θ in low dimensions. When posterior expectations were performed by taking i.i.d. samples from the posterior, the computations were stable, simple to perform, and accurate throughout when computing the predictive CDF $\hat{H}(\mathfrak{z}|x)$ in (15.4) and the predictive density $\bar{h}(\mathfrak{z}|x)$ in (15.5).

With tractable posteriors, the Laplace approximations were broadly successful and accurate when used to analytically compute $E_x^\theta\{\hat{g}(z|\theta)\}$. This resulted in accurate predictive densities as well as predictive CDFs when additional summation was used. By contrast, Laplace approximation for $E_x^\theta\{\hat{G}(z|\theta)\}$ was not widely successful. It was reasonably successful below the 99th percentile but its accuracy diminished further into the right tail of the predictive distribution.

It remains to be seen how well these predictive methods can perform with intractable likelihoods. An intractable likelihood was reconstructed in the repairman example of section 15.2.3 with posteriors shown in figures 15.7 and 15.8. Unfortunately there is no procedure for checking their accuracy. A predictive distribution was computed using (15.6) based on an intractable likelihood in Butler and Huzurbazar (1997, §8.2). The data were passage times in a semi-Markov process with parametric transmittances, however the accuracy of $\hat{H}(3|x)$ could not be determined since there are no other methods capable of providing similar or better accuracy to make comparisons.

15.4.2 Applicability

The emphasis has been on Markov and simple semi-Markov processes in which the processes have a system parameter of low dimension. This is the most applicable context for this methodology since complex systems with a high dimensional θ pose serious integration difficulties.

Perhaps what restricts the use of this methodology the most for general semi-Markov processes is the need to specify parametric branch transmittances for the system. It is generally difficult to determine what parametric forms to assume, and there is also the issue of the growing dimensionality of θ that occurs with increasing system complexity. This is a problem that was partially alleviated by adopting a Markov assumption, however such models are simply not realistic in many settings such as those dealing with multi-state survival analysis. The idea that hazard rates remain constant while holding in system states is unrealistic in most medical contexts and dubious at best in many reliability settings.

The bootstrap analysis of chapter 14 is the recommended inferential approach for more general semi-Markov models in which there is little information about the nature of the individual branch transmittances and the Markov assumption is deemed inappropriate. The general semi-Markov model is nonparametric and the bootstrap methodology provides an appropriate nonparametric analysis within the semi-Markov class of models.

15.4.3 Importance sampling integration with an intractable likelihood

Suppose the likelihood is intractable but its MGF is available for saddlepoint inversion. If the likelihood is not very informative, then sampling from the prior as specified in (15.6) leads to posterior predictive computations that are probably adequate if used with a sufficiently large sample size N.

However, if the likelihood is quite informative, so the posterior on θ is much more concentrated than the prior, then sampling from the prior can become quite inefficient as compares with sampling from the posterior. Without the ability to simulate from an exact posterior, a sample can still be drawn from an approximate posterior which can lead to the computation of posterior expectations through importance sampling.

A crude importance distribution can be determined by using the central limit tendencies of likelihoods and hence posteriors. Informative likelihoods in particular lead to posteriors that are approximately Normal as Bayesian central limit theorems (Berger, 1985, section 4.7.8). Suppose, by alternative means, it is possible to estimate θ by $\hat{\theta}_0$ and also estimate the asymptotic posterior variance $j(\hat{\theta}_0)^{-1}$. Then a m-dimensional Normal$_m\{\theta; \hat{\theta}_0, v^2 j(\hat{\theta}_0)^{-1}\}$ density, denoted as $\mathcal{N}(\theta)$, could be used where $v^2 \geq 1$ is an inflation factor used to assure good coverage and stability for the importance weights in the simulation. Alternatively stable importance weights may also result by taking v^{-2} to have a χ_5^2 distribution so that the marginal importance distribution on θ is a heavy-tailed multivariate t_5 distribution.

Simulating $\theta_1, \ldots, \theta_N$ as i.i.d. observations from the importance density $\mathcal{N}(\theta)$ leads to the importance sampling estimate

$$\check{H}_N(\mathfrak{Z}|x) = \frac{\sum_{i=1}^{N} \hat{G}(\mathfrak{Z}|\theta_i) w(\theta_i)}{\sum_{i=1}^{N} w(\theta_i)} \qquad w(\theta_i) = \frac{\hat{f}(x|\theta_i) \pi(\theta_i)}{\mathcal{N}(\theta_i)}.$$

where $\{w(\theta_i)\}$ are the importance weights. If $\theta_N \sim \mathcal{N}(\theta)$, the weak law of large numbers assures that

$$\check{H}_N(\mathfrak{Z}|x) \xrightarrow{p} \frac{E^{\theta_N}\{\hat{G}(\mathfrak{Z}|\theta_N) w(\theta_N)\}}{E^{\theta_N}\{w(\theta_N)\}} = \frac{\int \hat{G}(\mathfrak{Z}|\theta) \hat{f}(x|\theta) \pi(\theta) d\theta}{\int \hat{f}(x|\theta) \pi(\theta) d\theta}.$$

Such importance sampling methods were implemented in Butler *et al.* (2005) to perform Monte Carlo integration of saddlepoint densities. There is every reason to believe the same methods will lead to efficient computations of posterior expectations with intractable but informative likelihoods.

15.5 Computational algorithms

Implementation of these Bayesian methods can be quite unstable if not approached in the right way. The algorithm described below is the same as used for the bootstrap methods of chapter 14.

A predictand grid $\mathfrak{Z}$ may be determined by first computing the predictive mean $E(Z|x)$ and standard deviation $SD(Z|x)$ and letting $\mathfrak{Z}$ be equally spaced points ranging up to $E(Z|x) + 8SD(Z|x)$. With a tractable posterior, an initial computation of these moments would use

$$\hat{E}(Z|x) = N^{-1} \sum_{i=1}^{N} K'_Z(0|\theta_i) \qquad \widehat{SD}(Z|x)^2 = N^{-1} \sum_{i=1}^{N} K''_Z(0|\theta_i)$$

where $\theta_1, \ldots, \theta_N$ are sampled from the posterior.

15.5.1 Saddlepoint computations for $\hat{G}(\mathfrak{Z}|\theta)$ over grid $\mathfrak{Z}$

Computing $\hat{G}(\mathfrak{Z}|\theta_1), \ldots, \hat{G}(\mathfrak{Z}|\theta_N)$ for $\theta_1, \ldots, \theta_N$ simulated from the posterior can be an unstable computation due to the wide range of values for θ that could be generated. The algorithm below accommodates such a randomly selected θ and also avoids the main source of instability. The main difficulty is to determine the saddlepoints associated with grid points $z_1 < z_2 < \cdots < z_P$ using an efficient Newton iteration. With a randomly selected value of θ, there is a tendency for such iterative methods to jump outside of the convergence strip $(-\infty, c_\theta)$ of $K_Z(s|\theta)$ when finding saddlepoints in the right tail.

For randomly selected θ, compute the value of $K'_Z(0|\theta) = E(Z|\theta)$ and determine the closest grid point z_I to this mean value. By its determination, the saddlepoint $\hat{s}_I$ for z_I is very close to 0 and Newton's method to solve for $\hat{s}_I$ can be started at 0. Next the saddlepoint for z_{I+1} can be determined by starting Newton's algorithm at $\hat{s}_I$. Convergence is typically achieved in 2–4 iterations. The algorithm proceeds to compute the saddlepoint $\hat{s}_{j+1}$ for z_{j+1} by initiating the iterative solution at the previously determined $\hat{s}_j$ until the endpoint z_P of $\mathfrak{Z}$ is reached. If this careful algorithm still jumps outside of the convergence strip, then a finer grid 3 needs to be used. Coarse grids can be problematic. The same approach is useful in the left tail although the problem of leaving the convergence strip doesn't occur here.

15.5.2 Computation of intractable likelihoods

Computation of $\hat{f}(x|\theta)$ over a θ-grid $\{\theta_1, \theta_2, \ldots \theta_P\}$ for fixed data x has the same essential algorithm as section 15.5.1. The simplest method enumerates the grid $\{\theta_1, \theta_2, \ldots \theta_P\}$ so that neighboring θ-values are close together providing the linkage needed for making an iterative computation.

Suppose x is a single datum. An initial grid point θ_I along with its saddlepoint solution $\hat{s}_I$ to $K'_X(\hat{s}_I|\theta_I) = x$ is required to begin the set of computations. Often θ_I should be chosen as a parameter that "anticipates" datum x where "anticipates" means that the saddlepoint $\hat{s}_I$ is close to 0. This initial saddlepoint starts an iterative sweep through the grid of θ-values in both directions to determine the entire likelihood over the grid of θ-values.

15.6 Exercises

1. Consider the cancer example of section 15.2.1.
   (a) Derive the exact density for $Z_T|\theta$ using partial fraction expansions and check the computation using symbolic computation.
   (b) Determine the infinitessimal generator for the process and repeat the derivation for the density of $Z_T|\theta$ using matrix exponentials.
   (c) Determine the exact CDF for $Z_T|\theta$.
   (d) Use either exact integration or Monte Carlo simulation as in (15.4) to compute the saddlepoint predictive CDF for $Z_T|x$.
   (e) Use either exact integration or Monte Carlo simulation as in part (d) to compute the exact predictive CDF for $Z_T|x$. Plot the relative error for the approximation in part (d) and tabulate a range of interesting values out into the right tail of the distribution.
2. Consider the Markov reliability growth model of section 15.2.2.
   (a) Use the Markov graph to determine the exact survival function for $Z|\theta$ using matrix exponentials. Here, Z is the first passage time from $1 \to 3$ and $\theta = (p, r, \lambda, \tau)$. (Hint: The computation only requires exponentials of a 2×2 matrix so it can be easily computed symbolically.)
   (b) Differentiate the CDF in part (a) to determine the density for $Z|\theta$.
   (c) Using the posterior on θ, from section 15.2.2, simulate the predictive density and CDF of $Z|x$ and compare with figure 15.4 and table 15.1.
3. Consider the computation of the first return $1 \to 1$ predictive distribution $Z|x$ for the Markov reliability growth model of section 15.2.2.
   (a) Compute the first return transmittance for $Z|\theta$ using (13.24).
   (b) Let $F = \{1 \to 1\}$ be the event that the process starting in state 1 returns to state 1. Use simulations from the posterior on θ in section 15.2.2 to compute $\bar{h}(3|x, F)$ and $\hat{H}(3|x, F)$ the predictive first passage density and CDF given that return is assured.
   (c) Repeat the symbolic computations from Exercise 2 to determine $\Pr(F)$ and the exact CDF and density for $Z|\theta, F$. The hint also applies in this setting.
   (d) Simulate from the posterior on θ in section 15.2.2 to determine "exact" expressions for $h(3|x, F)$ and $H(3|x, F)$. Compare these with the expressions in part (b) using graphs for the densities and tables for the CDFs.

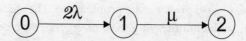

Figure 15.12. Markov graph for kidney example.

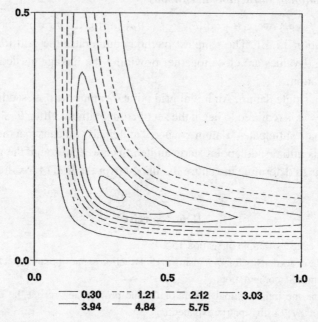

Figure 15.13. Likelihood contours in $\theta = (\mu, \lambda)$ that ignore the constraint $\lambda \leq \mu$ with μ plotted on the horizontal axis.

4. Derive the following results for the $M/G/1$ queue.
   (a) The values of $T_{ij}(s)$ given in (15.10).
   (b) The posterior distribution of $\theta = (\lambda, \tau_1, \tau_2)$.
5. In the $M/G/1$ queue, consider the CGF determined by $K(s) = \ln\{T_{ij}(s)/T_{ij}(0)\}$ in (15.10).
   (a) Show that
   $$K'(s) = O\{(c-s)^{-1/2}\} \qquad \text{as } s \to c.$$
   Hence show that K' maps $(-\infty, c)$ onto $(0, \infty)$.
   (b) Approximate the distribution using the Lugannani and Rice approximation and plot its saddlepoint density.
   (c) Simulate 10^5 observations from this distribution and compare the empirical distribution with Lugannani and Rice in part (b) and plot a kernel density estimate to compare with the saddlepoint density.
6. Gross et al. (1971) describe a three state Markov process for the condition of a patient suffering kidney failure. Let states $\mathcal{M} = \{0, 1, 2\}$ denote the number of nonfunctioning kidneys for a patient with Markov graph given in figure 15.12. Gross et al. report the values $\lambda = 1/15$ and $\mu = 1/3$ which are assumed to be the true values. Constrain the distributions on $\theta = (\mu, \lambda)$ so that $\lambda \leq \mu$.
   (a) Simulate 10 passage times from $0 \to 2$ and let these passage times denote the data as $x = (x_1, \ldots, x_{10})$. Compute the density for x_1 symbolically and use this to plot the true likelihood $f(x|\theta)$ for the data. Show that its contours resemble those in figure 15.13.

(b) Plot the likelihood surface obtained by using the saddlepoint density $\hat{f}(x|\theta)$ and compare to the plot from part (a).

(c) If Z is the passage time from $0 \to 2$, plot the exact posterior predictive density for $Z|x$ assuming a flat prior for θ on $(0, 3) \times (0, 1)$. Use either numerical integration or simulation from the prior.

(d) Simulate from the prior to determine an approximate predictive density using $\hat{g}(z|\theta)$ and $\hat{f}(x|\theta)$. Compare to part (c).

(e) Compute $\hat{H}(3|x)$ in (15.6) and the exact predictive CDF using numerical integration or simulation. Plot the relative error of $\hat{H}(3|x)$ over grid 3 (Butler and Huzurbazar, 1994, 1997).

7. A discrete time Markov chain with states $\mathcal{M} = \{1, 2, 3, 4\}$ has the transition matrix

$$P = \begin{pmatrix} 0 & 0 & 1 & 0 \\ 0 & 0 & \theta & 1-\theta \\ 1-\theta & \theta & 0 & 0 \\ 0 & 1 & 0 & 0 \end{pmatrix}$$

where $0 < \theta < 1$. Suppose the true value of θ is $2/3$.

(a) Simulate a first passage from $1 \to 4$ whose complete path represents the data x. Specify the resulting posterior on θ using a flat prior.

(b) Suppose Z is a first passage time from $1 \to 4$. Compute the saddlepoint mass function for $Z|x$ and its two continuity-corrected CDFs by taking the posterior expectation of the saddlepoint approximations for $Z|\theta$.

(c) Simulate 10^6 passage times from the distribution of $Z|x$ and compare these empirical probabilities with the mass function and CDFs of part (c).

8. Repeat all parts of Exercise 7 but instead let Z be the first return time to state 1. Note the periodic nature of the chain.

9. For the chain in Exercise 7, simulate five first passage times $x = (x_1, \ldots, x_5)$ from $1 \to 4$ as data.

(a) Plot the saddlepoint approximation $\hat{f}(x|\theta)$ for the likelihood. Attempt to plot the exact likelihood and make comparisons if this latter computation is feasible.

(b) If Z is the first passage time from $1 \to 4$, compute the continuity-corrected CDF approximations $\hat{H}(3|x)$ in (15.6) using saddlepoint approximations for both $Z|\theta$ and $\hat{f}(x|\theta)$.

(c) Simulate 10^6 values from the distribution $Z|x$ by using rejection sampling under the posterior density of θ. Use this simulation to assess the accuracy of $\hat{H}(3|x)$ in part (b).

10. Suppose a continuous time Markov chain with states $\mathcal{M} = \{1, 2, 3\}$ has the matrix of transition rates

$$\begin{pmatrix} 0 & \theta & 0 \\ 1 & 0 & \theta \\ 0 & 1 & 0 \end{pmatrix}$$

with $\theta > 0$ determining the transition rate for stepping up one state. Let the true value of θ be 2. Simulate a first passage from $1 \to 3$ and let the complete path denote data x. Answer all the questions in Exercise 7 as they pertain to first passage $1 \to 3$.

11. Repeat all parts of Exercise 10 but instead let Z be the first return time to state 1.

12. For the Markov process in Exercise 10, simulate 5 first passage times $x = (x_1, \ldots, x_5)$ from $1 \to 3$ and use this as data. Answer all the questions in Exercise 9 as they pertain to the first passage $1 \to 3$ in Exercise 10.

16

Nonnormal bases

Up to now, all of the saddlepoint formulas have involved the univariate normal density function $\phi(z)$ and its CDF $\Phi(z)$. These expressions are called normal-based saddlepoint approximations and, for the most part, they serve the majority of needs in a wide range of applications. In some specialized settings however, greater accuracy may be achieved by using saddlepoint approximations developed around the idea of using a different distributional base than the standard normal.

This chapter presents saddlepoint approximations that are based on distributions other than the standard normal distribution. Suppose this base distribution has density function $\lambda(z)$ and CDF $\Lambda(z)$ and define the saddlepoint approximations that use this distribution as (λ, Λ)-based saddlepoint approximations. Derivations and properties of such approximations are presented in section 16.1 along with some simple examples. Most of the development below is based on Wood *et al.* (1993).

Most prominent among the base distributions is the inverse Gaussian distribution. The importance of this base is that it provides very accurate probability approximations for certain heavy-tailed distributions in settings for which the usual normal-based saddlepoint approximations are not accurate. These distributions include various first passage times in random walks and queues in which the system is either unstable, so the first passage distribution may be defective, or stable and close to the border of stability. Examples include the first return distribution to state 0 in a random walk that is null persistent or close to being so in both discrete and continuous time. A second example considers a predictive Bayesian analysis for a Markov queue with infinite buffer capacity. Infinite buffer size results in a parameter space that may be partitioned into regions of stability and instability for the queue. Approximate Bayesian predictive distributions are computed by marginalizing the inverse-Gaussian-based saddlepoint approximations over the regions of stability and instability. This results in quite accurate predictive distributions for the queue.

16.1 Nonnormal-based saddlepoint expressions

The development of the (λ, Λ)-based saddlepoint density and CDF provided below uses real variable arguments rather than the original complex variable approach in Wood *et al.* (1993). Suppose random variable X has density/mass function $f(x)$, CDF $F(x)$, and CGF $K(s)$. Approximations are sought for f and F using K that make use of the continuous base distribution with density function $\lambda(z)$, CDF $\Lambda(z)$ and CGF $L(\mathfrak{s})$. Let Z denote a random variable with the base distribution.

16.1.1 λ-based saddlepoint density approximation

Consider first the normal-based approximations for $f(x)$ and $\lambda(z)$ given by $\hat{f}(x)$ and $\hat{\lambda}(z)$ respectively. Their relative values are set equal through the relation

$$\hat{f}(x)/f(x) \simeq 1 \simeq \hat{\lambda}(z)/\lambda(z)$$

for all "allowable" x, z meaning those values that lie within the interior of their respective support. This leads to

$$f(x) \simeq \lambda(z)\{\hat{f}(x)/\hat{\lambda}(z)\} \tag{16.1}$$

for all allowable z. The accuracy of the approximation in (16.1) can be optimized with a judicious choice for the value of z that indexes the approximation. Thus the right side of (16.1) is a z-tilted class of approximations, akin to the class of s-tilted densities used to motivate the saddlepoint density as the indirect Edgeworth expansion in (5.35) determined from the tilt in (5.34).

The approach of Wood et al. (1993) is equivalent to choosing z so that the dominate portions of $\hat{\lambda}(z)$ and $\hat{f}(x)$ agree. The dominate portion of

$$\hat{f}(x) = \exp\{K(\hat{s}) - \hat{s}x\}/\sqrt{2\pi K''(\hat{s})}$$

is due to the exponent $K(\hat{s}) - \hat{s}x$, where $\hat{s}$ solves $K'(\hat{s}) = x$. This is set equal to the dominant portion of $\hat{\lambda}(z)$ given as $L(\check{s}) - \check{s}z$ where $\check{s} = \check{s}(z)$ is the saddlepoint root of $L'(\check{s}) = z$. Thus, the optimal choice $\hat{z} = \hat{z}(x)$ is the root of

$$L\{\check{s}(z)\} - \check{s}(z)z = K(\hat{s}) - \hat{s}x. \tag{16.2}$$

The mapping $x \to \hat{z}$ can be made well defined and is discussed in the next paragraph. From (16.1), the optimally chosen λ-based saddlepoint density approximation is

$$f(x) \simeq \lambda(\hat{z})\{\hat{f}(x)/\hat{\lambda}(\hat{z})\} = \lambda(\hat{z})\sqrt{L''(\hat{\check{s}})/K''(\hat{s})} \tag{16.3}$$

where $\hat{\check{s}} = \check{s}(\hat{z})$.

For an understanding of the mapping $x \to \hat{z}$, consider $\hat{s}x - K(\hat{s})$ as a function of x and taking account of the implicit dependence through $\hat{s} = \hat{s}(x)$. This function is easily shown to be convex in x and is referred to as the Legendre–Fenchel transform of K (McCullagh, 1987, p. 174). Likewise $\check{s}(z)z - L\{\check{s}(z)\}$ is the Legendre-Fenchel transform for L. Therefore, both sides of (16.2) are concave and attain the common maximum of zero when $x = E(X)$, so that $\hat{s} = 0$, and $\hat{\check{s}} = \check{s}(\hat{z}) = 0$ with $\hat{z} = E(Z) = L'(0)$. For $x \neq E(X)$ there are two solutions $\hat{z}_-(x) < E(Z) < \hat{z}_+(x)$ on either side of the mean for the base distribution. The correct choice of root determines the mapping $x \to \hat{z}$ as

$$\hat{z}(x) = \begin{cases} \hat{z}_-(x) & \text{if } x < E(X) \\ L'(0) & \text{if } x = E(X) \\ \hat{z}_+(x) & \text{if } x > E(X). \end{cases} \tag{16.4}$$

When the base distribution is $-Z$ with CGF $L(-\mathfrak{s})$ rather than Z, then the roles of the two roots in the mapping $x \to \hat{z}$ get reversed.

Properties

If $f(x) \equiv \lambda(x)$, so the density to be approximated is also the base density, then the saddlepoint density on the right side of (16.3) is exact.

The density/mass function approximation in (16.3) is equivariant under linear transformation. This is a desirable property for continuous X however, in the lattice case, it forces the function to be considered on the integer lattice as described in section 2.1.1.

If density/mass function f is symmetric about its mean, then the density/mass function approximation also expresses this symmetry when a symmetric base distribution is used.

16.1.2 (λ, Λ)-based saddlepoint CDF approximation

A Temme-style derivation due to Wood (personal communication), which is a modification to the argument resulting in (2.34), leads to the CDF approximation in (16.5).

Theorem 16.1.1 *Suppose X has a continuous distribution $F(x)$ with CGF $K(s)$. The (λ, Λ)-based saddlepoint CDF approximation for $F(x)$ is*

$$\hat{F}(x) = \Lambda(\hat{z}) + \lambda(\hat{z}) \left\{ \frac{1}{\hat{s}} - \frac{\sqrt{L''(\hat{s})}}{\hat{u}} \right\} \quad x \neq E(X) \tag{16.5}$$

where $\hat{z}$ is given in (16.4), $\hat{s} = \check{s}(\hat{z})$ is the saddlepoint for $\hat{z}$ with respect to the base CGF, and $\hat{u} = \hat{s}\sqrt{K''(\hat{s})}$. The singularity at $x = E(X)$ is removable and the right side of (16.5) has the continuous limit

$$\hat{F}\{E(X)\} = \Lambda\{L'(0)\} + \tfrac{1}{6}\sqrt{L''(0)}\lambda\{L'(0)\} \left\{ \frac{K'''(0)}{\{K''(0)\}^{3/2}} - \frac{L'''(0)}{\{L''(0)\}^{3/2}} \right\}. \tag{16.6}$$

If Λ is the standard normal base distribution, then (16.5) and (16.6) are the ordinary Lugannani and Rice approximations given in (1.21). In this case $L(\mathsf{s}) = \mathsf{s}^2/2$, $\check{s}(z) = z$ and (16.2) is

$$z^2 - \hat{z}^2/2 = \hat{s}x - K(\hat{s})$$

so that $\hat{z} = \hat{w}$ as given in (1.22).

Proof of theorem 16.1.1. Integration of the λ-based density in (16.3) gives the approximation

$$F(y) \simeq \int_{-\infty}^{y} \lambda(\hat{z})\sqrt{L''(\hat{s})/K''(\hat{s})}\,dx. \tag{16.7}$$

The change of variable from x to $\hat{z} = \hat{z}(x)$ has differential $dx = \hat{s}d\hat{z}/\hat{s}$ obtained by differentiating (16.2). Changing variables with $\hat{z}_y = \hat{z}(y)$, gives

$$\begin{aligned}
F(y) &\simeq \int_{-\infty}^{\hat{z}_y} \lambda(\hat{z})\frac{\hat{s}}{\hat{s}}\left(\frac{L''(\hat{s})}{K''(\hat{s})}\right)^{1/2} d\hat{z} \\
&= \int_{-\infty}^{\hat{z}_y} \lambda(\hat{z})d\hat{z} + \int_{-\infty}^{\hat{z}_y} \lambda(\hat{z})\hat{s}\left\{\frac{1}{\hat{s}}\left(\frac{L''(\hat{s})}{K''(\hat{s})}\right)^{1/2} - \frac{1}{\hat{s}}\right\} d\hat{z} \\
&= \Lambda(\hat{z}_y) + \int_{-\infty}^{\hat{z}_y} \lambda(\hat{z})\hat{s}\left\{\frac{1}{\hat{s}}\left(\frac{L''(\hat{s})}{K''(\hat{s})}\right)^{1/2} - \frac{1}{\hat{s}}\right\} d\hat{z}.
\end{aligned} \tag{16.8}$$

16.1 Nonnormal-based saddlepoint expressions

In order to approximate the second term in (16.8), suppose the normal-based saddlepoint density is exact up to a normalization constant or

$$\lambda(\hat{z}) = c\hat{\lambda}(\hat{z}) = \frac{c}{\sqrt{2\pi L''(\hat{s})}} \exp\{L(\hat{s}) - \hat{s}\hat{z}\}.$$

A simple exercise shows that

$$\frac{\partial}{\partial \hat{z}}\{L(\hat{s}) - \hat{s}\hat{z}\} = -\hat{s} \tag{16.9}$$

so the second term in (16.8) is

$$-\int_{-\infty}^{\hat{z}_y} \frac{c}{\sqrt{2\pi L''(\hat{s})}} \left\{ \frac{1}{\hat{s}} \left(\frac{L''(\hat{s})}{K''(\hat{s})} \right)^{1/2} - \frac{1}{\hat{s}} \right\} d[\exp\{L(\hat{s}) - \hat{s}\hat{z}\}]. \tag{16.10}$$

Integration by parts, that retains the nonintegral term and ignores the integral term, yields the required correction term

$$-\lambda(\hat{z}) \left\{ \frac{1}{\hat{s}} \left(\frac{L''(\hat{s})}{K''(\hat{s})} \right)^{1/2} - \frac{1}{\hat{s}} \right\}. \tag{16.11}$$

If the normal-based saddlepoint density is not exact, then the same approximation in (16.5) can be derived but further approximations are needed. In this case the substitution $\lambda(\hat{z}) = c\hat{\lambda}(\hat{z})\{1 + e(\hat{z})\}$ is made with error $e(\hat{z})$. When the additional terms connected with $e(\hat{z})$ are ignored, the derivation leads to (16.5) with additional approximation. □

The discrete case

If X is a lattice random variable, then the same two continuity corrections that were used with the normal base can be made with nonnormal bases. The arguments for this are the same as those given by Daniels (1987, §6) when considering the normal base.

Analogous to the first continuity correction given in (1.27),

$$\widehat{\Pr}_1(X \geq x) = 1 - \Lambda(\hat{z}) - \lambda(\hat{z}) \left\{ \frac{1}{\hat{s}} - \frac{\sqrt{L''(\hat{s})}}{\tilde{u}_1} \right\} \quad x \neq E(X) \tag{16.12}$$

where $\tilde{u}_1$ is given in (1.28) and $x = k$ is the integer value used in the saddlepoint equation $K'(\hat{s}) = x$.

The second continuity correction uses $\tilde{u}_2$ in (1.31) instead of $\tilde{u}_1$ and also offsets $x = k$ to $k^- = k - 1/2$ in the saddlepoint equation $K'(\tilde{s}) = k^-$. This offset also affects the determination of $\hat{z}$ since the term x in (16.2) is offset to k^- when determining $\hat{z}$ and $\hat{s}$ is replaced by $\tilde{s}$ that solves $K'(\tilde{s}) = k^-$.

Properties

If $F(x) \equiv \Lambda(x)$ then approximation (16.5) is exact.

For continuous X with symmetric F, the approximation in (16.5) replicates this symmetry if a symmetric base distribution is used. In the lattice case, the same symmetry preservation holds for the second continuity correction $\widehat{\Pr}_2$ but not for the first correction $\widehat{\Pr}_1$.

It will be seen in the various examples below that non-normal-based CDF approximations can be quite accurate in settings for which the ordinary normal-based Lugannani and Rice approximation may be less than accurate or perhaps fail. This is especially true when using the inverse Gaussian base to approximate first passage distributions for systems that are either unstable or close to instability. The same cannot be said for non-normal based saddlepoint densities which are generally unnecessary in practice. This is due to the high accuracy achieved by the ordinary normal-based density/mass function approximation in most all practical settings. Such settings include those dealing with the approximation of first passage density/mass functions in (nearly) unstable systems.

16.2 Choice of base distribution

In the choice of base distribution, it is to helpful to first ask whether the use of a Normal (μ, σ^2) base leads to a different approximation than the Lugannani and Rice approximation using a Normal $(0, 1)$ base. A simple exercise shows that the two CDF approximations are identical. This motivates the following result which says that the CDF approximations in (16.5) are invariant to location and scale changes to the base distribution.

Proposition 16.2.1 *Suppose that* $Z \sim \Lambda$ *and* $\sigma Z + \mu \sim \Lambda(\mu, \sigma^2)$ *is a location and scale change with* $\sigma > 0$. *Then, the CDF approximations in (16.5) are the same whether* Λ *or* $\Lambda(\mu, \sigma^2)$ *is used as the base distribution.*

Proof. See Exercise 4. □

The proposition does not cover the case in which $\sigma < 0$. In this instance, base CGF $L(\mathfrak{s})$ is replaced by $L(-\mathfrak{s})$ and the roles of $\hat{z}_-(x)$ and $\hat{z}_+(x)$ get reversed in the mapping $x \to \hat{z}$. Thus the CDF approximations are different when the distribution of $L(\mathfrak{s})$ is asymmetric and the same when the base distribution is symmetric, as occurs in the normal case.

16.2.1 Chi-square base distribution

A χ_α^2 distribution has CGF $L(\mathfrak{s}; \alpha) = -(\alpha/2) \ln(1 - 2\mathfrak{s})$ and the question is how to choose the degrees of freedom parameter $\alpha > 0$. The following recommendation was given in Wood *et al.* (1993).

Match the tilted standardized skewness

The family of base distributions may be expanded to include the location and scale change $\sigma \chi_\alpha^2 + \mu$ with CGF $L(\mathfrak{s}; \alpha, \sigma, \mu)$. Then, the three parameters may be chosen to match the first three tilted moments

$$L^{(k)}(\hat{\mathfrak{s}}; \alpha, \sigma, \mu) = K^{(k)}(\hat{s}) \qquad k = 1, 2, 3$$

with the sign of σ chosen so the third derivatives have the same signs. In theory, the choice of base distribution is specific to the point x and different x-values may use different α-values, hence different base distributions.

16.2 Choice of base distribution

Table 16.1. *Values for the normal-based and $\chi^2_{\hat{\alpha}}$-based saddlepoint approximations evaluated at the exact 5th, 10th, 90th, and 95th percentiles of the $\chi^2(1, \lambda)$ distribution.*

Exact:	.05	.10	.90	.95
	$\lambda = 0.0$			
Normal base	.0557	.1094	.9003	.9498
$\chi^2_{\hat{\alpha}}$ Base	.0500	.1000	.9000	.9500
	$\lambda = 0.25$			
Normal base	.0557	.1093	.8969	.9478
$\chi^2_{\hat{\alpha}}$ Base	.0500	.1000	.8968	.9480
	$\lambda = 1.0$			
Normal base	.0555	.1083	.8969	.9484
$\chi^2_{\hat{\alpha}}$ Base	.0499	.0992	.8970	.9485

A simpler way to implement this is to match the third standardized cumulants since these quantities do not depend on σ and μ, only α. In this case

$$\frac{\{L''(\hat{s}; \alpha, \sigma, \mu)\}^3}{\{L'''(\hat{s}; \alpha, \sigma, \mu)\}^2} = \frac{\{L''(\hat{s}; \alpha)\}^3}{\{L'''(\hat{s}; \alpha)\}^2} = \frac{\alpha}{8}$$

irrespective of $\hat{s}$, σ, and μ. This motivates the choice

$$\hat{\alpha} = 8 \frac{\{K''(\hat{s})\}^3}{\{K'''(\hat{s})\}^2}. \tag{16.13}$$

By proposition 16.2.1, the choice of $\mu = \hat{\alpha}$ and $\sigma = 2\hat{\alpha}$ suffice and give the simplest expressions using based distribution $\chi^2_{\hat{\alpha}}$.

Three comments are in order.

(1) If $K'''(\hat{s}) = 0$, then $\hat{\alpha} = \infty$ and the chi square base is the normal base at this value of x due to the central limit theorem.
(2) If $K(s) = nK_0(s)$ is the CGF for a sum of i.i.d. $K_0(s)$ variates, then $\hat{\alpha} = O(n) \to \infty$ as $n \to \infty$ and the chi square-based approximation is likely to be very close to the normal-based approximation.
(3) If F is a location and positive scale change to a χ^2_{α} distribution for some $\alpha > 0$, then the choice in (16.13) leads to an approximation that is exact. See Exercise 5.

Noncentral chi-square example

Suppose that $F \sim \chi^2(1, \lambda)$ with the MGF for this distribution given in Exercise 4 of chapter 1. Wood et al. (1992) used a central χ^2_{α} base distribution with α chosen as in (16.13) to approximate this noncentral distribution. Table 16.1 has been reproduced from their technical report to compare the accuracy of $\chi^2_{\hat{\alpha}}$-base approximations with the ordinary Lugannani and Rice approximation using the normal base. In all but one entry, the approximations using the $\chi^2_{\hat\alpha}$-base are more accurate than their normal-base counterparts. The exactness of the $\chi^2_{\hat\alpha}$ base with $\lambda = 0$ was predicted in comment 3 above.

16.2.2 Inverse Gaussian base

The two parameter IG (μ, σ^2) distribution of section 13.2.4 has CGF given in (13.18). By scale invariance, this family may be reduced to a single parameter family with $\alpha > 0$. Setting $\mu = \alpha$ and $\sigma^2 = \alpha^3$ leads to the one parameter IG family in Wood *et al.* (1993) with CGF

$$L(\mathfrak{s}) = \alpha^{-1} - (\alpha^{-2} - 2\mathfrak{s})^{1/2} \quad \mathfrak{s} \le 1/(2\alpha^2). \tag{16.14}$$

Since the convergence strip for the CGF of this base distribution is $(-\infty, 1/(2\alpha^2)]$ and closed on the right edge, the distribution is not a "regular" setting for saddlepoint approximation. However, the saddlepoint equation always has a unique solution since $\lim_{\mathfrak{s} \to 1/(2\alpha^2)} L'(\mathfrak{s}) = \infty$ and L' is a bijection that maps the interior of the convergence strip onto $(0, \infty)$. This is referred to as the *steepness* property for the inverse Gaussian distribution in Barndorff-Nielsen (1978, p. 117).

Matching third standardized cumulants for the $\hat{\mathfrak{s}}$-tilted distribution of Λ with the $\hat{s}$-tilted distribution of F leads to the choice

$$\hat\alpha = \frac{\{K'''(\hat s)\}^2}{3\{K''(\hat s)\}^3}\left(3 + \hat w\sqrt{\frac{\{K'''(\hat s)\}^2}{\{K''(\hat s)\}^3}}\right)^{-1} \quad \text{if} \quad K'''(\hat s) > 0 \tag{16.15}$$

where $\hat w$ is given in (1.22). This agrees with expression (20) in Wood *et al.* (1993). The value $\hat z$, expressed in terms of $\hat\alpha$ in (16.15), is

$$\hat z = \hat\alpha + \tfrac{1}{2}\hat\alpha^2\bigl(\hat w^2 + \hat w\sqrt{\hat w^2 + 4/\hat\alpha}\bigr) \tag{16.16}$$

which agrees with expression (27) in Wood *et al.* (1993). Then

$$\hat{\mathfrak{s}} = \tfrac{1}{2}(\hat\alpha^{-2} - \hat z^{-2}). \tag{16.17}$$

Note that our use of the approximation has assumed that $K'''(\hat s) > 0$ as stated in (16.15). This has always been the case in the various first passage distribution applications for which this approximation has been used. The expression for $\hat\alpha$ in (16.15) must also be positive which also is generally true in these applications except when x lies in the extreme left tail of its first passage distribution. In such instances, taking $\hat\alpha$ as a small value like 0.1 generally suffices.

If F is any member of the two parameter IG (μ, σ^2) family, then the inverse Gaussian-based approximation is exact when $\hat\alpha$ is chosen as in (16.15).

Applications that use the inverse Gaussian base to approximate various first passage distributions in GI/M/1/4 queues have already been provided in section 13.2.5. These examples show good accuracy but not the extreme accuracy displayed by the normal-based approximations. The reason for this in this instance is that the systems of section 13.2.5 were specifically constructed to be quite stable. Thus, the superior accuracy of the normal-based approximations over those with an inverse Gaussian base could be expected when the systems do not border or approach instability.

16.2 Choice of base distribution

First passage times in a discrete time random walk

Wood *et al.* (1993) considered the first passage distribution for a classic random walk that is reproduced below. The random walk starts in state 0 and takes random incremental step $+1$ with probability p and step -1 with probability $q = 1 - p$. Let W_n be the first passage time to integer value $n > 0$ with distributional support $\{n, n+2, \ldots\}$. The true (perhaps defective) mass function has been given in Feller (1968, p. 275) as

$$\Pr(W_n = n + 2k) = \frac{n}{n+2k}\binom{n+2k}{k}p^{n+k}q^k. \tag{16.18}$$

Exact computations of $\Pr(W_n \geq n + 2k)$ can be made by summing (16.18) from $0, \ldots, k-1$ using exact symbolic computation.

The transmittance for W_n is given in (13.54) and, assuming $p \geq 1/2$, is a MGF. From (13.54), the CGF for $X_n = (W_n - n)/2$ is determined as

$$K(s) = n\bigl[\ln\bigl\{1 - \sqrt{1 - 4pqe^s}\bigr\} - s - \ln(2q)\bigr] \quad s < -\ln(4pq).$$

Solution to the saddlepoint equation leads to the explicit value

$$\hat{s} = \ln[(4pq)^{-1}\{1 - (y_k - 1)^{-2}\}]$$

for $k \geq 1$ where $y_k = 2(1 + k/n) > 2$. Further differentiation of K gives

$$K''(\hat{s}) = ny_k(y_k - 1)(y_k - 2)/4$$
$$K'''(\hat{s}) = K''(\hat{s})\{3(y_k - 1)^2 - 1\}/2 > 0.$$

These computations result in a normal-based lattice approximation $\widehat{\Pr}_1(X_n \geq k)$ in (1.27) that has explicit values for $\hat{w}$ and $\tilde{u}_1$.

An inverse Gaussian base is suggested by classical limit theory. Under certain limiting operations that are outlined in Wood *et al.* (1993, section 5.2), the random walk process becomes a Weiner process whose first passage to a fixed height has an inverse Gaussian distribution.

Table 16.2 displays right tail probabilities $\Pr(X_n \geq k)$ and their $\widehat{\Pr}_1$ approximations using both normal-based and inverse Gaussian-based approximations for three settings of the random walk. Note the exceptional accuracy of the inverse Gaussian-based approximation in all settings. The normal-based approximation achieves acceptable accuracy with larger values of p, which are the settings of quicker first passage transition, and larger n. However, the accuracy deteriorates with smaller n and p closer to $1/2$.

The central limit effect provides a simple explanation for the good performance of the normal-based approximation with larger values of n. Starting at 0, first passage W_n records the time needed for the recurrence of n record events in the random walk. As such, its distribution is $W_1^{(1)} + \cdots + W_1^{(n)}$ where $\{W_1^{(i)}\}$ are i.i.d. with distribution as the first passage from $0 \to 1$. The central limit effect with large values of n can be expected to improve the normal-based approximation.

The proximity of p to $1/2$, causes a total breakdown for the normal-based approximation. The case $p = 1/2$ forms the null persistent boundary, W_n has a defective distribution if $p < 1/2$, and W_n has all finite moments if $p > 1/2$. The inverse Gaussian approximation remains exceptionally stable even at $p = 0.51$.

Table 16.2. *Values for first passage probability* $\Pr(X_n \geq k)$ *computed using the* $\widehat{\Pr}_1$ *continuity correction in (16.12) with the inverse Gaussian-based and normal-based approximations.*

Approx.	p	n			Values of k			
	.9	1	1	2	3	4	5	6
Exact			.100	.0190	$.0^2442$	$.0^2114$	$.0^3313$	$.0^4896$
IG base			.100	.0191	$.0^2443$	$.0^2114$	$.0^3314$	$.0^4899$
Normal			.0871	.0157	$.0^2355$	$.0^3894$	$.0^3242$	$.0^4683$
	.9	10	1	3	6	9	12	
Exact			.651	.154	.0101	$.0^3524$	$.0^4247$	
IG base			.653	.154	.0101	$.0^3524$	$.0^4247$	
Normal			.647	.153	.0100	$.0^3517$	$.0^4244$	
	.51	1	1	23	70	515	942	2530
Exact			.490	.100	.0502	.0100	$.0^2500$	$.0^2100$
IG base			.513	.100	.0502	.0100	$.0^2500$	$.0^2100$
Normal			.209	$-.905$	$-.786$	$-.221$	$-.114$	$-.0235$

16.2.3 The mutual accommodation of base pairs

Consider two base distributions Λ_1 and Λ_2. If Λ_1 is successful as a base distribution in approximating Λ_2, the reverse approximation does not necessarily follow; e.g. Λ_2 does not necessarily succeed as a base distribution in approximating Λ_1. Examples of this for two pairs of base distributions are considered.

Normal and chi-square bases

This is a base pair in which the ordinary Lugannani and Rice approximation provides a very accurate approximation to the chi square distribution even down to a fractional degree of freedom; see section 1.2.2. Likewise, the chi square base provides an approximation for the normal distribution with arbitrarily small error due to the fact that $\hat{\alpha} = \infty$ for the normal distribution and the chi square base is essentially a normal base.

Normal and inverse Gaussian bases

These two bases are not mutually accommodating. First consider the normal base when used to approximate an IG $(4, 4^3)$ distribution as a member of the one parameter family in (16.14) with $\alpha = 4$. Figure 16.1 compares the true CDF (solid line) with its Lugannani and Rice approximation (dashed). The approximation fails and is quite inaccurate. Booth (1994) discusses this tail inaccuracy from an asymptotic point of view.

The inaccuracy in figure 16.1 seems even more perplexing when one notes that the normal-based saddlepoint density approximation is exact in this case. The reader is asked to show this in Exercise 7. The conclusion that may be drawn is that the Temme approximation

16.2 Choice of base distribution

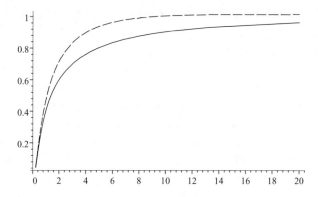

Figure 16.1. The CDF for an IG $(4, 4^3)$ (*solid*) and its Lugannani and Rice approximation (*dashed*).

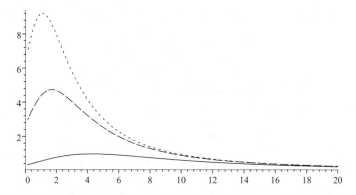

Figure 16.2. Percentage relative error versus x in approximating the normal distributuion using inverse Gaussian bases with $\hat{\alpha} = 1$ (*dotted*), 0.5 (*dashed*), and 0.1 (*solid*).

is failing in its accuracy when it is used to derive the Lugannani and Rice approximation through the integration of the saddlepoint density.

Now reverse the base distributions and use an inverse Gaussian base to approximate the standard normal distribution. The $\hat{\alpha}$ suggested in (16.15) is to take $\hat{\alpha} = 0$ which is not possible. However a sequence of α-values approaching 0 can be considered. Figure 16.2 shows the percentage relative error $100\{\hat{\Phi}(x)/\Phi(x) - 1\}$ over $x \in (0, 4)$ in approximating the normal distribution using inverse Gaussian bases with $\hat{\alpha} = 1$ (dotted line), 0.5 (dashed), and 0.1 (solid). The relative errors diminish with decreasing $\hat{\alpha}$.

The diminishing relative error is explained by the central limit effect as $\alpha \to 0$. If X_α has the CGF given in (16.14), then $(X_\alpha - \alpha)/\alpha^{3/2}$ converges in distribution to a Normal $(0, 1)$ as $\alpha \to 0$. Thus, at the extreme limit $\hat{\alpha} = 0$, the inverse Gaussian base achieves exact replication of the normal distribution.

Discussion

The mutual accommodation achieved by the normal (Λ_1) and chi square (Λ_2) bases suggests that they are essentially "equivalent" bases in the following sense. If a third distribution F were to be approximated by the Λ_1 base distribution, then it would be expected that the Λ_2

base would also be quite accurate. This is because the Λ_1 and λ_1 terms in its approximation to F can also be closely approximated by terms Λ_2 and λ_2 from the second base distribution.

The lack of mutual accommodation with the normal (Λ_1) and inverse Gaussian base (Λ_2) suggests that only the inverse Gaussian base can consistently replicate the accuracy of the normal base and not vice versa. Indeed the normal base fails to approximate the CDF of the inverse Gaussian. Also it fails to adequately approximate the first passage distribution for a random walk when the inverse Gaussian base is quite accurate.

We might conjecture the existence of "equivalence classes" of bases where the equivalence relation is the ability for bases to mutually accommodate accurate approximation for one another. In this sense the inverse Gaussian base seems to reside in a universe of its own.

16.3 Conditional distribution approximation

Let (X, Y) be a random vector in which $\dim(Y) = 1$ and suppose the joint CGF

$$K(s, t) = \ln \left\{ E\left(e^{s^T X + tY}\right) \right\}$$

converges in an open neighborhood of $(s, t) = (0, 0)$. If Y is either continuous or integer-valued, then the conditional density/mass function of $Y|X = x$ has a λ-based saddlepoint approximation given as

$$\begin{aligned} f(y|x) &\simeq \lambda(\hat{z}) \{\hat{f}(y|x)/\hat{\lambda}(\hat{z})\} \\ &= \lambda(\hat{z}) \sqrt{\frac{L''(\hat{z})|K''_{ss}(\hat{s}_0, 0)|}{|K''(\hat{s}, \hat{t})|}} \end{aligned} \qquad (16.19)$$

where $\hat{f}(y|x)$ is the double saddlepoint approximation in (4.7) and $\hat{\lambda}(z)$ is the single saddlepoint density approximation for the base density in (1.4). Value $\hat{z}$ is an optimally tilted z-value chosen so the dominate exponential portions of $\hat{f}(y|x)$ and $\hat{\lambda}(\hat{z})$ agree, e.g. $\hat{z}$ solves

$$L\{\check{s}(z)\} - \check{s}(z)z = K^*(\hat{t}) - \hat{t}y \qquad (16.20)$$

in z, with

$$K^*(t) = \left\{ K(\hat{s}_t, t) - \hat{s}_t^T x \right\} - \left\{ K(\hat{s}_0, 0) - \hat{s}_0^T x \right\}.$$

Here $(\hat{s}, \hat{t})$ is the numerator saddlepoint associated with (x, y), $\hat{s}_0$ is the denominator saddlepoint solving $K'_s(\hat{s}_0, 0) = x$, and $\hat{s}_t$ solves $K'_s(\hat{s}_t, t) = x$.

For fixed x, the mapping $y \to \hat{z}$ determined by the solution in (16.20) is a well-defined smooth bijection. The right side of (16.20) is a function of y that implicitly depends on y through $\hat{t}$ that solves $K'_t(\hat{s}, \hat{t}) = y$. The maximum of $K^*(\hat{t}) - \hat{t}y$ is 0 and occurs when $\hat{t} = 0$ and $y = K'_t(\hat{s}_0, 0) = v$. The mapped value for v is $\hat{z} = L'(0)$ that occurs when $0 = \check{s}(\hat{z}) = \hat{s}$. In all other cases, $K^*(\hat{t}) - \hat{t}y < 0$ and each negative value is associated with two values of y, y_- and y_+, for which $y_- < v < y_+$. The mapping associates $y_- \to \hat{z}_-$ and $y_+ \to z_+$ where $\hat{z}_- < L'(0) < \hat{z}_+$ and $\hat{z}_-$ and $\hat{z}_+$ are the two solutions to (16.20).

16.3.1 (λ, Λ)-based double saddlepoint CDF approximation

The same argument that proves theorem 16.1.1 can be used to derive the following double saddlepoint CDF approximation with a (λ, Λ) base.

Theorem 16.3.1 *For continuous Y and fixed $X = x$, the double saddlepoint CDF approximation for $F(y|x) = \Pr(Y \leq y | X = x)$ is*

$$\hat{F}(y|x) = \Lambda(\hat{z}) + \lambda(\hat{z}) \left\{ \frac{1}{\hat{s}} - \frac{\sqrt{L''(\hat{s})}}{\hat{u}} \right\} \quad \hat{t} \neq 0.$$

Here, $\hat{u}$ is the double saddlepoint term in (4.13), $\hat{z}$ is the value under the mapping $y \to \hat{z}$ that solves (16.20), and $\hat{s} = \check{s}(\hat{z})$ is the saddlepoint for $\hat{z}$.

The choice of a normal base leads to the Skovgaard approximation in (4.12).

To allow for integer-valued Y, $\hat{u}$ is replaced with $\tilde{u}_1$ in (4.16) to give the approximation

$$\widehat{\Pr}_1(Y \geq y | X = x) = 1 - \Lambda(\hat{z}) - \lambda(\hat{z}) \left\{ \frac{1}{\hat{s}} - \frac{\sqrt{L''(\hat{s})}}{\tilde{u}_1} \right\} \quad \hat{t} \neq 0.$$

The second continuity-corrected approximation $\widehat{\Pr}_2(Y \geq y|X = x)$ uses $\tilde{u}_2$ in (4.18) instead of $\tilde{u}_1$ and also offsets $y = k$ to $k^- = k - 1/2$ in the double saddlepoint equations to give the root $(\tilde{s}, \tilde{t})$. This offset also affects the determination of $\hat{z}$ since y in (16.20) is offset to k^- when determining $\hat{z}$ and the value $\hat{t}$ is also replaced by $\tilde{t}$ that solves $K'_t(\tilde{s}, \tilde{t}) = k^-$.

A sequential saddlepoint approximation entails using

$$K^{**}(t) = K^*(t) + \tfrac{1}{2} \ln\{|K''_{ss}(\hat{s}_0, 0)| / |K''_{ss}(\hat{s}_t, t)|\}$$

in place of $K^*(t)$ when matching the Legendre–Fenchel transforms in (16.20); see (10.6). In addition, $\hat{t}$ in (16.20) is the value that solves $K^{**\prime}(\hat{t}) = y$.

16.4 Examples

Approximations of first passage time distributions using normal-based saddlepoint methods can be inaccurate when the processes involved are either unstable, null persistent, or close to instability. This is shown in the next two examples that consider a null-persistent continuous time random walk given by Butler (2000) and a Markov queue with infinite buffer capacity evaluated in Butler and Huzurbazar (1994, 2000).

16.4.1 Null persistent random walk

Suppose the simple random walk of section 16.2.2 determines the state transitions of a continuous time Markov process that starts in state 0 at time zero and takes incremental steps ± 1 with probabilities $1/2$. In addition, suppose the holding time in each state is Exponential (1). If W is the first return time to state 0, then its first return transmittance is given in (13.45) as

$$M(s) = 1 - \sqrt{1 - (1 - s)^{-2}} \quad s \leq 0.$$

This transmittance is a MGF that is nonregular and steep so saddlepoint approximations are available.

Figure 16.3 compares the inverse Gaussian-based (dashed) and Lugannani and Rice (solid) approximations for the survival function of W with an empirical estimate (circles) based on simulating 10^6 returns.

540   Nonnormal bases

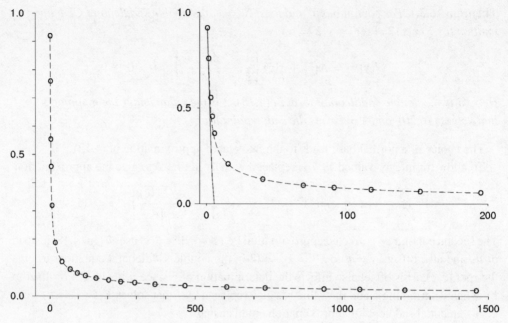

Figure 16.3. Saddlepoint approximations for Pr($W > t$) vs. t using the inverse Gaussian base (*dashed*) and the normal base (*solid*) along with an empirical estimate (*circles*).

The plot in the smaller inset shows the accuracy for $t \in (0, 200)$ and shows the failure of the Lugannani and Rice approximation (solid line) that turns negative at $t = 5$. Not shown is the fact that it continues to decrease in value to -16 at $t = 1516$. By contrast the inverse Gaussian-based approximation agrees quite closely with the simulation.

When fitting the inverse Gaussian approximation, the value for $\hat{\alpha}$ in (16.15) is 10^3 at $t = 8$ and continues to increase to 10^{10} at $t = 1516$. This suggests that the inverse Gaussian base is approaching the stable law with index $1/2$, which is a heavy-tailed distribution with mean ∞; see Barndorff-Nielsen (1978, p. 117).

Figure 16.4 compares the saddlepoint approximation $\check{z}(t)$ (solid) in (13.23) for the hazard rate of W with an empirical estimate (circles). Approximation $\check{z}(t)$ uses the normalized normal-based saddlepoint density in the numerator and the inverse Gaussian-based survival approximation in the denominator. The empirical estimate uses kernel density estimation and the empirical CDF based on 10^6 simulated returns. The plot indirectly shows the accuracy achieved by the normal-based saddlepoint density whose normalization constant in this instance was 1.000.

The best way to judge the accuracy of these approximations is to compare them directly with the true density and CDF for first return. This density is given in Feller (1971, §XIV.7, eqn. 6.16) as

$$f(t) = t^{-1} I_1(t) e^{-t} \quad t > 0$$

where $I_1(t)$ is a BessellI function of order 1. Integration, using Abramowitz and Stegun (1970, 11.3.14), gives the survival function as

$$R(t) = \{I_0(t) + I_1(t)\} e^{-t} \quad t > 0.$$

16.4 Examples

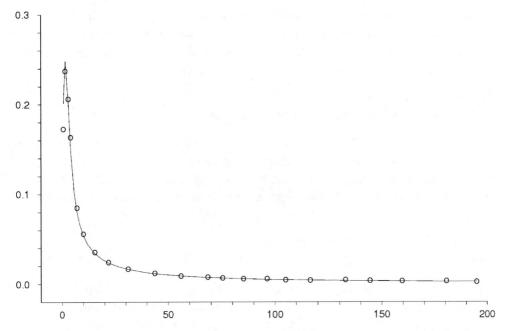

Figure 16.4. Plot of hazard approximation $\check{z}(t)$ vs. t (*solid*) as compares to empirical estimates (*circles*) based on 10^6 simulations.

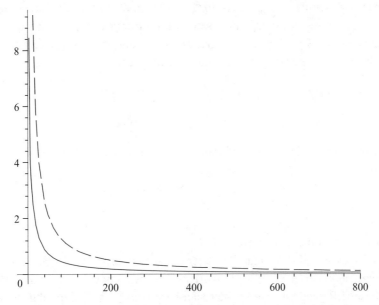

Figure 16.5. Percentage relative error for the survival approximation (solid) in figure 16.3 and the hazard approximation $\check{z}(t)$ (dashed) in figure 16.4.

These expressions allow exact computation of the percentage relative errors for the saddlepoint approximations. Figure 16.5 plots such error for the survival approximation (solid) in figure 16.3 and the hazard approximation $\check{z}(t)$ (dashed) in figure 16.4. Beyond $t = 800$ these errors continue to decrease until the respective errors of 0.05% and 0.03% are achieved at $t = 1516$.

16.4.2 Bayesian prediction in the M/M/2/∞ queue

This infinite state system has been described in section 13.4.1. The queue is stable if the arrival rate does not exceed the maximum fixing rate from the two servers or $\lambda \leq 2\mu$; otherwise the queue is unstable and increases in length with probability 1. This stability/instability is reflected in the first return transmittance value $\mathcal{T}_{00}(0|\lambda, \mu)$ in (13.49) that assumes value 1 if $\lambda \leq 2\mu$ and $\mu/(\lambda - \mu)$ otherwise.

Stable example

Assume the true model has $\lambda = 5$ and $\mu = 5.1$ so the queue is stable. Suppose data x are generated from this model and consist of the complete history of five first passage sojourns from $0 \to 0$. By choosing a flat diffuse prior on (μ, λ), the posterior has $\lambda|x$ and $\mu|x$ independent, and the data give $\lambda|x \sim$ Gamma (17, 3.08) and $\mu|x \sim$ Gamma (17, 2.58).

In conformance with the notation of chapter 15, let Z be the first return time to state zero so $Z|\mu, \lambda$ has the transmittance $\mathcal{T}_{00}(s|\mu, \lambda)$ in (13.49). Computation of the Bayesian predictive density for $Z|x$ requires marginalization over the posterior on (μ, λ) whose support includes both the stable region $S = \{\lambda \leq 2\mu\} \subset (0, \infty)^2$ and its unstable compliment S^C. If R is the event that the process returns to 0, its posterior probability is

$$\Pr(R|x) = \Pr(S|x) + \int_{S^C} \frac{\mu}{\lambda - \mu} f(\lambda|x) f(\mu|x) d\lambda d\mu$$
$$= 0.99345 + 0.00536 = 0.99881. \qquad (16.21)$$

The posterior probabilities of a stable and nonstable process are 0.99345 and 0.00655 respectively.

Figure 16.6 shows the conditional Bayesian predictive density $\bar{h}(\mathfrak{z}|x, R)$ in (15.5) (dashed line) that is conditional on R and plotted over a fine grid $\mathfrak{z} \subset (0, 2)$. Also shown is an empirical estimate (solid) from kernel density estimation that uses 10^6 values simulated from the true conditional density for $Z|x, R$.

Table 16.3 shows predictive quantiles determined from inverting the Bayesian predictive CDF approximations $\hat{H}(\mathfrak{z}|x, R)$ in (15.4). These quantiles have been determined by using a normal base (LR) and an inverse Gaussian base (IG) with N denoting the number of simulated (μ, λ)-points used in the posterior marginalization. The first set of rows deals with CDFs given R, thus they assume that the system returns to 0. The last two unconditional rows do not condition on R and approximate quantiles from the defective CDF approximation $\hat{H}(\mathfrak{z}|x)$. The simulation estimates are based on 10^6 values generated from the CDF $H(z|x)$. Among the 10^6 simulated returns, 99.8818% returned to the origin which agrees with (16.21).

The Lugannani and Rice approximation performs quite poorly due to (μ, λ)-values that were simulated near the boundary $\lambda = 2\mu$ and got averaged when determining $\hat{H}(\mathfrak{z}|x, R)$.

16.4 Examples

Table 16.3. *Conditional predictive quantiles for $Z|x, R$ determined from $\hat{H}^{-1}(3|x, R)$, and also unconditional quantiles from $\hat{H}^{-1}(3|x)$ using normal (LR) and inverse Gaussian-based (IG) saddlepoint approximations.*

	Percentage	N	90%	95%	99%
Conditional on R	Simulation		1.06	1.49	3.43
	LR[1]	400	0.93	1.22	1.81
	IG	400	1.03	1.43	3.01
	LR[1]	1600	0.95	1.26	1.64
	IG	1600	1.06	1.50	3.38
Unconditional	Simulation		1.07	1.51	3.68
	IG	1600	1.07	1.51	3.60

Note: [1] Indicates that the predictive CDFs eventually exceed 1 in value.

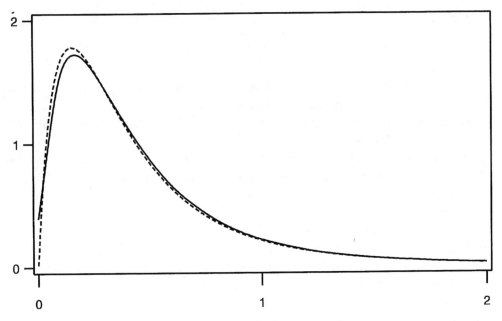

Figure 16.6. Conditional saddlepoint Bayesian predictive density $\bar{h}(3|x, R)$ given R (*dashed*) and an kernel-based empirical estimate (*solid*) using 10^6 simulated values.

Near this boundary, the Lugannani and Rice approximation often exceeded 1 and was not monotonic increasing. The inverse Gaussian approximation does not suffer such breakdown and is exceedingly accurate for $N = 1600$ and quite good even with $N = 400$.

Breakdown neighborhood for Lugannani and Rice

The range of inaccuracy for the normal-based approximation can be determined as a neighborhood in polar coordinates about the boundary $\lambda = 2\mu$. The distribution of λZ has transmittance $\mathcal{T}_{00}(\lambda s|\mu, \lambda)$ that depends only on the parameter ratio λ/μ. Thus, since saddlepoint

Table 16.4. *Quantiles for $Z|\mu, \lambda, R$ using simulation and normal (LR) and inverse Gaussian-based (IG) saddlepoint approximations.*

Angle θ	Method	90%	95%	99%
55°	Simulation	17.3	27.2	57.4
(stable)	IG	17.5	27.4	57.8
	LR	10.8	15.8	30.5
63°	Simulation	106	329	2461
(stable)	IG	106	327	2453
63.43°	Simulation	140	566	14013
(null persistent)	IG	143	570	12128
64°	Simulation	95.4	278	1836
(unstable)	IG	95.7	280	1832
75°	Simulation	8.96	13.1	25.7
(unstable)	IG	8.95	13.2	25.8
	LR	8.88	11.3	19.2

accuracy is invariant to scale changes, the saddlepoint accuracy only depends on the angular coordinate $\theta = \tan^{-1}(\lambda/\mu)$ of (μ, λ) and not on its radial coordinate.

This allows the inaccuracy of the Lugannani and Rice approximation to be tabulated according to angular deviation from the $\tan^{-1}(2) = 63.43°$ angle for the boundary of stability. Table 16.4 shows percentiles determined by inverting $\mathcal{T}_{00}(s|\mu, \lambda)/\mathcal{T}_{00}(0|\mu, \lambda)$ at various angles θ using saddlepoint approximations with both bases and also by simulating the percentiles. Division by $\mathcal{T}_{00}(0|\mu, \lambda)$ effectively conditions on R so that the time for first return is finite with probability 1.

The inverse Gaussian-based approximation maintains its accuracy over regions of stability, with angles 55° and 63°, at the null persistent boundary, and over regions of instability, with angles 64° and 75°. The Lugannani and Rice approximation is not accurate at 55° or 75° and for angles in between it exceeds 1 and is not monotonic which makes its inversion somewhat meaningless. The simulated 90th percentiles were determined as the $0.9 \times 10^6 \times \mathcal{T}_{00}(0|\mu, \lambda)$ order statistic in a simulation of 10^6 values for Z. Since $10^6 \times \mathcal{T}_{00}(0|\mu, \lambda)$ is the expected number to return to 0, the order statistic is the 90th percentile among the finite values of $Z|\mu, \lambda$.

Except at $\theta = 63.43°$, the convergence strip of $\mathcal{T}_{00}(s|\mu, \lambda)/\mathcal{T}_{00}(0|\mu, \lambda)$ is an open neighborhood of 0 so the saddlepoint inversion is regular. For $\theta > 63.43°$, the conditioning on R forces the distribution to have finite moments. The saddlepoint setting is regular although the Lugannani and Rice approximation is still not accurate. At $\theta = 63.43°$ the convergence strip is $(-\infty, 0]$ and nonregular and the inverse Gaussian base achieves high accuracy.

Unstable example

Consider a second example of this queue with a different posterior for which $\mu|x \sim$ Gamma (15, 12) independent of $\lambda|x \sim$ Gamma (15, 8). This posterior straddles the boundary of stability and places considerable posterior weight on regions where the Lugannani and Rice approximation fails. Now the posterior probability of R is

$$\Pr(R|x) = \Pr(S|x) + \Pr(R \cap S^c|x)$$
$$= 0.7823 + 0.1550 = 0.9373.$$

The probability of return to 0 when the queue is unstable is computed as

$$\Pr(R|x, S^c) = 0.1550/(1 - 0.7823) = 0.7120.$$

The posterior density $h(z|x, R)$ is approximated by mixing normal-based saddlepoint densities over S and S^c so that

$$\bar{h}(3|x, R) = \Pr(S|x)\bar{h}(3|x, R, S) + \Pr(S^c|x)\bar{h}(3|x, R, S^c). \qquad (16.22)$$

Figure 16.7 plots $\bar{h}(3|x, R, S)$, $\bar{h}(3|x, R, S^c)$, and $\bar{h}(3|x, R)$ respectively as the dashed lines. The weighted average of figures 16.7 (top and middle) according to (16.22) leads to the plot of $\bar{h}(3|x, R)$ in figure 16.7. Each approximation is compared with a kernel density approximation (solid line) based on simulating 10^6 values from the true densities $h(z|x, R, S)$ and $h(z|x, R, S^c)$. The kernel estimate in figure 16.7 (bottom) is the weighted average of the two kernel estimates used in figure 16.7 (top and middle). The normal-based saddlepoint density succeeds where the Lugannani and Rice approximation fails.

16.5 Exercises

1. Prove the properties for the non-normal based saddlepoint density approximation given in section 16.1.1.
   (a) If $f(x) \equiv \lambda(x)$, show that (16.3) is exact.
   (b) Show that the density/mass function approximation is equivariant under linear transformation.
   (c) Show that the density/mass function approximation expresses the underlying symmetry of f about its mean when a symmetric base distribution is used.
2. (a) Prove that the change of variable for the integration in (16.7) has $d\hat{z}/dx = \hat{s}/\hat{s}$.
   (b) Prove (16.9).
3. Prove the various exactness and symmetry properties of the approximation in (16.5). For parts (b)–(d) assume a symmetric base distribution.
   (a) If $F(x) \equiv \Lambda(x)$, show that (16.5) is exact.
   (b) Show that (16.5) replicates the symmetry of F in the continuous setting and in the lattice setting if the second continuity correction $\widehat{\Pr}_2$ is used.
   (c) Show that it does not preserve symmetry with the first correction $\widehat{\Pr}_1$.
   (d) In the lattice case, given a simple example that demonstrates the symmetry of $\widehat{\Pr}_2$ and the lack of symmetry for $\widehat{\Pr}_1$.
4. Show that a Normal (μ, σ^2) base distribution leads to the Lugannani and Rice approximation that assumed a Normal $(0, 1)$ base. Continue the argument and prove proposition 16.2.1 for arbitrary base distributions.
5. Suppose F is a location and positive scale change to a χ_α^2 distribution for some $\alpha > 0$. Prove that the choice of base distribution in (16.13) leads to a CDF approximation that is exact.
6. Consider the inverse Gaussian-based approximation.
   (a) Derive the expressions in (16.15), (16.16), and (16.17).

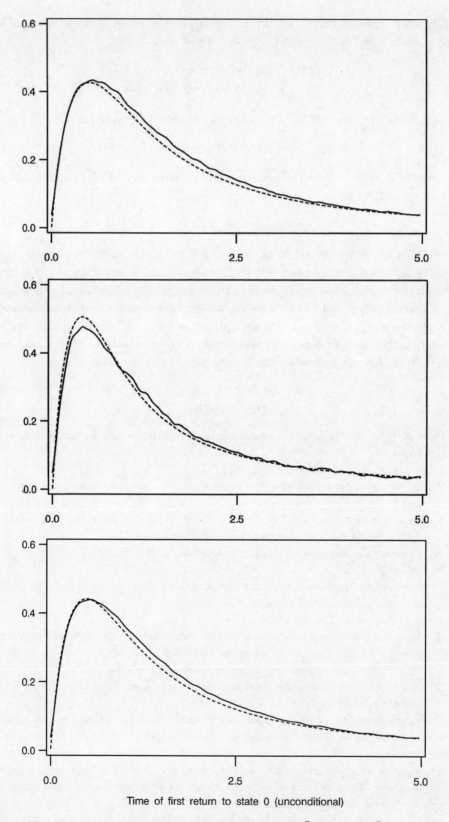

Figure 16.7. Plots of the normal-based saddlepoint densities $\bar{h}(3|x, R, S)$, $\bar{h}(3|x, R, S^c)$, and $\bar{h}(3|x, R)$ respectively (*dashed lines*). They are compared with kernel density approximations (*solid lines*) based on simulating 10^6 values from the true densities $h(z|x, R, S)$ and $h(z|x, R, S^c)$.

(b) If F is any member of the two parameter IG (μ, σ^2) family, show the inverse Gaussian-based approximation is exact when $\hat{\alpha}$ is chosen as in (16.15).
7. For the one parameter inverse Gaussian family in (16.14), show that the normal-based saddlepoint density is exact.
8. Suppose X_α has the CGF given in (16.14). Prove that $(X_\alpha - \alpha)/\alpha^{3/2}$ converges in distribution to a Normal $(0, 1)$ as $\alpha \to 0$.
9. Derive the conditional density and CDF approximations in section 16.3.
   (a) Prove the mapping defined by $y \to \hat{z}$ in (16.20) is well-defined. Also show that
   $$\frac{\partial}{\partial y}\{K^*(\hat{t}) - \hat{t}y\} = -\hat{t}.$$
   (b) Prove theorem 16.3.1 using the same argument that leads to theorem 16.1.1.
10. Compute a nonnormal-based double saddlepoint approximation for a simple conditional distribution.
    (a) Use a conditional distribution, such as a Beta (α, β) or Hypergeometric (M, N, n), so that exact computation can be used to assess accuracy. Tabulate enough values to determine the overall accuracy of the applications.
    (b) If a chi square or inverse Gaussian base is used, consider how the parameter α should be chosen.
    (c) Compare these results to the ordinary normal-based Skovgaard approximation.
11. The discussion of non-normal bases did not use a discrete distribution such as the Poisson distribution as a base. Explain why this might have difficulties.
12. Differentiate the (λ, Λ)-based CDF approximation in (16.5).
    (a) How does the derivative compare to the λ-based density approximation in (16.3)?
    (b) Does this relationship replicate the relationship found in Exercise 16 of chapter 2 for normal bases?
13. Develop a (λ, Λ)-based CDF approximation in which Λ is either the Gumbel or logistic distribution. Approximate various distributions using the generalized base procedure and assess its accuracy.

References

Aalen, O.O. (1995) Phase type distributions in survival analysis. *Scandinavian Journal of Statistics* **22**, 447–463.

Abramowitz, M. and Stegun, I.A. (1972) *Handbook of Mathematical Functions.* Ninth edition. New York: Dover.

Agresti, A. (1992) A survey of exact inference for contingency tables. *Statistical Science* **7**, 131–153.

Aitchison, J. (1986) *The Statistical Analysis of Compositional Data.* London: Chapman and Hall.

Anderson, T.W. (1958) *An Introduction to Multivariate Statistical Analysis.* New York: Wiley.

Anderson, T.W. (1971) *The Statistical Analysis of Time Series.* New York: Wiley.

Anderson, T.W. (1984) *An Introduction to Multivariate Statistical Analysis.* Second edition. New York: Wiley.

Anderson, T.W. (2003) *An Introduction to Multivariate Statistical Analysis.* Third edition. New York: Wiley.

Aroian, L.A. (1939) A study of Fisher's z distribution and the related F distribution, *Annals of Mathematical Statistics* **12**, 429–448.

Atkinson, A.C. (1982) The simulation of generalized inverse Gaussian and generalized hyperbolic random variables. *SIAM Journal on Scientific Computing* **3**, 502–515.

Barlow, R.E. and Proschan, F. (1965) *Mathematical Theory of Reliability.* New York: Wiley.

Barndorff-Nielsen, O.E. (1978) *Information and Exponential Families in Statistical Theory.* New York: Wiley.

Barndorff-Nielsen, O.E. (1980) Conditionality resolutions. *Biometrika* **67**, 293–310.

Barndorff-Nielsen, O.E. (1982) "Exponential Families" entry in the *Encyclopedia of Statistical Science*, ed. N. L. Johnson and S. Kotz. New York: Wiley.

Barndorff-Nielsen, O.E. (1983) On a formula for the conditional distribution of the maximum likelihood estimator. *Biometrika* **70**, 343–365.

Barndorff-Nielsen, O.E. (1986) Inference on full and partial parameters based on standardized signed log likelihood ratio. *Biometrika* **73**, 307–322.

Barndorff-Nielsen, O.E. (1988) Comment on Reid (1988). *Statistical Science* **3**, 228–229.

Barndorff-Nielsen, O.E. (1990) Approximate interval probabilities. *Journal of the Royal Statistical Society series B* **52**, 485–499.

Barndorff-Nielsen, O.E. (1991) Modified signed log likelihood ratio. *Biometrika* **78**, 557–563.

Barndorff-Nielsen, O.E. and Cox, D.R. (1979) Edgeworth and saddlepoint approximations with statistical applications (with Discussion) *Journal of the Royal Statistical Society series B* **41**, 279–312.

Barndorff-Nielsen, O.E. and Cox, D.R. (1984) Bartlett adjustments to the likelihood ratio statistic and the distribution of the maximum likelihood estimator. *Journal of the Royal Statistical Society series B* **46**, 483–495.

Barndorff-Nielsen, O.E. and Cox, D.R. (1989) *Asymptotic Techniques for Use in Statistics.* New York: Chapman and Hall.

Barndorff-Nielsen, O.E. and Cox, D.R. (1994) *Inference and Asymptotics.* New York: Chapman and Hall.

Barndorff-Nielsen, O.E. and Hall, P. (1988) On the level-error after Bartlett adjustment of the likelihood ratio statistic. *Biometrika* **75**, 374–378.

Barndorff-Nielsen, O.E. and Wood, A.T.A. (1998) On large deviations and choice of ancillary for p^* and r^*. *Bernoulli* **4**, 35–63.

Bartlett, M.S. (1937) Properties of sufficiency and statistical tests. *Proceedings of the Royal Society A* **160**, 268–282.

Bartlett, M.S. (1938) The characteristic function of a conditional statistic. *Journal of London Mathematical Society* **13**, 62–67.

Bedrick, E.J. and Hill, J.R. (1992) An empirical assessment of saddlepoint approximations for testing a logistic regression parameter. *Biometrics* **48**, 529–544.

Bellio, R. (2003) Likelihood methods for controlled calibration. *Scandinavian Journal of Statistics* **30**, 339–353.

Bellio, R. and Brazzale, A.R. (1999) Higher-order likelihood-based inference in non-linear regression. in *Proceedings of the 14th International Workshop on Statistical Modelling,* ed. H. Friedl, A. Berghold, and G. Kauermann, pp. 440–443. Technical University, Graz.

Bellio, R. and Brazzale, A.R. (2003) Higher-order asymptotics unleashed: software design for non-linear heteroscedastic models. *Journal of Computational and Graphical Statistics* **12**, 682–697.

Bellio, R., Jensen, J.E., and Seiden, P. (2000) Applications of likelihood asymptotics for nonlinear regressions in herbicide bioassays. *Biometrics* **56**, 1204–1212.

Berger, J.O. (1985) *Statistical Decision Theory and Bayesian Analysis.* New York: Springer-Verlag.

Berger, J.O. and Wolpert, R.L. (1988) *The Likelihood Principle,* Second edition. Hayward, CA: Institute of Mathematical Statistics.

Birnbaum, A. (1962) On the foundations of statistical inference (with discussion). *Journal of the American Statistical Association* **57**, 269–306.

Bjørnstad, J.F. (1996) On the generalization of the likelihood function and the likelihood principle. *Journal of the American Statistical Association* **91**, 791–806.

Blæsild, P. and Jensen, J.L. (1985) Saddlepoint formulas for reproductive exponential families. *Scandinavian Journal of Statistics* **12**, 193–202.

Booth, J.G. (1994) A note on the accuracy of two saddlepoint tail approximations. *Proceedings of the Section on Physical and Engineering Sciences* 56–58. Alexandria, VA: American Statistical Association.

Booth, J.G. and Butler, R.W. (1990) Randomization distributions and saddlepoint approximations in generalized linear models. *Biometrika* **77**, 787–796.

Booth, J.G. and Butler, R.W. (1998). Discussion of "Approximately exact inference for common odds ratio in several 2 tables." *Journal of the American Statistical Association* **93**, 1310–1313.

Booth, J.G., Butler, R.W., Huzurbazar, S. and Wood, A.T.A. (1995) Saddlepoint approximations for p-values of some tests of covariance matrices. *Journal of Statistical Computation and Simulation* **53**, 165–180.

Booth, J.G. and Presnell, B. (1998) Allocation of Monte Carlo resources for the iterated bootstrap. *Journal of Computational and Graphical Statistics* **7**, 92–112.

Box, G.E.P. (1949) A general distribution theory for a class of likelihood ratio criteria. *Biometrika* **36**, 317–346.

Broda, S. and Paolella, M.S. (2004) Saddlepoint approximations for the doubly noncentral t distribution. Technical Report, Swiss Banking Institute, University of Zurich, Switzerland.

Bronson, D.A. (2001) Bootstrapping stochastic systems in survival analysis. Ph.D. dissertation, Department of Statistics, Colorado State University.

Buck, R.C. (1965) *Advanced Calculus.* New York: McGraw-Hill.

Butler, R.W. (1989) Approximate predictive pivots and densities. *Biometrika* **76**, 489–501.

Butler, R.W. (1997) System reliability, flowgraphs, and saddlepoint approximation. Technical Report, Department of Statistics, Colorado State University.

Butler, R.W. (1998) Generalized inverse Gaussian distributions and their Wishart connections. *Scandinavian Journal of Statistics* **25**, 69–75.

Butler, R.W. (2000) Reliabilities for feedback systems and their saddlepoint approximation. *Statistical Science* **15**, 279–298.

Butler, R.W. (2001) First passage distributions in semi-Markov processes and their saddlepoint approximation. *Data Analysis from Statistical Foundations,* ed E. Saleh. Huntington, NY: Nova Science Publishers.

Butler, R.W. (2003) Comments on "Ancillaries and Conditional Inference" by D.A.S. Fraser. *Statistical Science* **19**, 351–355.

Butler, R.W. (2007) Asymptotic hazard rates and uniformity of saddlepoint approximations. Technical Report, Department of Statistical Science, Southern Methodist University.

Butler, R.W. and Bronson, D.A. (2002) Bootstrapping survival times in stochastic systems by using saddlepoint approximations. *Journal of the Royal Statistical Society series B* **64**, 31–49.

Butler, R.W. and Huzurbazar, A.V. (1993) Prediction in stochastic networks. In *Bulletin of the International Statistical Institute, Proceedings of the 49th Session, Firenze, Book 3, Topic 23.*

Butler, R.W. and Huzurbazar, A.V. (1994) Bayesian prediction in stochastic systems. Technical Report, Department of Statistics, Colorado State University.

Butler, R.W. and Huzurbazar, A.V. (1997) Stochastic network models for survival analysis. *Journal of the American Statistical Association* **92**, 246–257.

Butler. R.W. and Huzurbazar, A.V. (2000) Bayesian prediction of waiting times in stochastic models. *Canadian Journal of Statistics* **28**, 311–325.

Butler, R.W., Huzurbazar, S., and Booth, J.G. (1992a) Saddlepoint approximation to generalized variance and Wilks' statistic. *Journal of the Royal Statistical Society series B* **79**, 157–169.

Butler, R.W., Huzurbazar, S. and Booth, J.G. (1992b) Saddlepoint approximations for the Bartlett–Nanda–Pillai trace statistic in multivariate analysis. *Biometrika* **79**, 705–715.

Butler, R.W., Huzurbazar, S., and Booth, J.G. (1993) Saddlepoint approximations for tests of block independence, sphericity, and equal variances and covariances. *Journal of the Royal Statistical Society series B* **55**, 171–183.

Butler, R.W. and Paolella, M.S. (1998a) Saddlepoint approximations to the density and distribution of ratios of quadratic forms in normal variables with application to the sample autocorrelation function. Technical Report, Statistics Department, Colorado State University.

Butler, R.W. and Paolella, M.S. (1998b) Approximate distributions for the various serial correlograms. *Bernoulli* **4**, 497–518.

Butler, R.W. and Paolella, M.S. (2002) Calculating the density and distribution function for the singly and doubly noncentral F. *Statistics and Computing* **12**, 9–16.

Butler, R.W. and Paolella, M.S. (2004) Uniform saddlepoint approximations for ratios of quadratic forms. Technical Report, Colorado State University.

Butler, R.W. and Paige, R.L. (2007) Exact CDF computations for the largest eigenvalues of a matrix beta and Wishart. Technical Report, Department of Statistical Science, Southern Methodist University.

Butler, R.W. and Sutton, R.K. (1998) Saddlepoint approximation for multivariate CDF and probability computations in sampling theory and outlier testing. *Journal of the American Statistical Association* **93**, 596–604.

Butler, R.W., Sutton, R.K., Booth, J.G. and Ohman, P.A. (2005) Simulation-assisted saddlepoint approximation. Technical Report, Department of Statistics, Colorado State University.

Butler, R.W. and Wood, A.T.A. (2000) Laplace approximations for hypergeometric functions of matrix argument. Technical Report 00-05, Division of Statistics, University of Nottingham, available at: http://www.maths.nottingham.ac.uk/statsdiv/research/reports.html.

Butler, R.W. and Wood, A.T.A. (2002) Laplace approximation for hypergeometric functions of matrix argument. *Annals of Statistics* **30**, 1155–1177.

Butler, R.W. and Wood, A.T.A. (2003) Laplace approximations for Bessel functions of matrix argument. *Journal of Computational and Applied Mathematics* **155**, 359–382.

Butler, R.W. and Wood, A.T.A. (2004) A dimensional CLT for non-central Wilks' lambda in multivariate analysis. *Scandinavian Journal of Statistics* **31**, 585–601.

Butler, R.W. and Wood, A.T.A. (2005) Approximation of power in multivariate analysis. *Statistics and Computing* **15**, 281–287.

Chapman, P.L., Butler, R.W., and Paige, R.L. (2007) Confidence bands for LD-100α in logistic regression by inverting sadddlepoint mid-p-values. Technical Report, Colorado State University.

Chernoff, H. (1954) On the distribution of the likelihood ratio. *Annals of Mathematical Statistics* **25**, 573–578.

Chow, S.C. and Shao, J. (1999) On the difference between the classical and inverse methods of calibration. *Applied Statistics* **39**, 219–228.

Chung, K.L. (2001) *A Course in Probability Theory*. New York: Academic Press.

Cochran, W. and Cox, G. (1957) *Experimental Designs*, Second edition. New York: Wiley.

Commenges, D. (1986) Semi-Markov and non-homogeneous Markov models in medicine. In *Semi-Markov Models: Theory and Applications*, ed. J. Janssen, pp. 423-436. New York: Plenum Press.

Cox, D.R. (1970) *Analysis of Binary Data*. London: Chapman and Hall.

Cox, D.R. and Hinkley, D.V. (1974) *Theoretical Statistics*. London: Chapman and Hall.

Cox, D.R. and Lewis, P.A.W. (1966) *The Statistical Analysis of Series of Events*. London: Methuen.

Cunningham, E.P. and Henderson, C.R. (1968) An iterative procedure for estimating fixed effects and variance components in mixed model situations. *Biometrics* **24**, 13–25.

Daniels, H.E. (1954) Saddlepoint approximations in statistics. *Annals of Mathematical Statistics* **25**, 631–650.

Daniels, H.E. (1956) The approximate distribution of serial correlation coefficients. *Biometrika* **43**, 169–185.

Daniels, H.E. (1958) Discussion of paper by D.R. Cox, *Journal of the Royal Statistical Society series B* **20**, 236–238.

Daniels, H.E. (1980) Exact saddlepoint approximations. *Biometrika* **67**, 59–63.

Daniels, H.E. (1983) Saddlepoint approximations for estimating equations. *Biometrika* **70**, 89–96.

Daniels, H.E. (1987) Tail probability approximations. *International Statistical Review* **55**, 37–48.

Daniels, H.E. (1992) A class of nearly exact saddlepoint approximations. In *The Art of Statistical Science*, ed. K.V. Mardia. New York: Wiley.

Daniels, H.E. and Young, G.A. (1991) Saddlepoint approximation for the Studentized mean, with an application to the bootstrap. *Biometrika* **78**, 169–179.

Davison, A.C. (1986) Approximate predictive likelihood. *Biometrika* **73**, 323–332.

Davison, A.C. (1988) Approximate conditional inference in generalized linear models. *Journal of the Royal Statistical Society series B* **50**, 445–451.

Davison, A.C. and Hinkley, D.V. (1988) Saddlepoint approximations in resampling methods. *Biometrika* **75**, 417–431.

Davison, A.C. and Hinkley, D.V. (1997) *Bootstrap Methods and their Application*. Cambridge: Cambridge University Press.

Davison, A.C. and Wang, S. (2002) Saddlepoint approximations as smoothers. *Biometrika* **89**, 933–938.

DiCiccio, T.J., Field, C.A., and Fraser, D.A.S. (1990) Approximations for marginal tail probabilities and inference for scalar parameters. *Biometrika* **77**, 77–95.

DiCiccio, T.J. and Martin, M.A. (1991) Approximations of marginal tail probabilities for a class of smooth functions with applications to Bayesian and conditional inference. *Biometrika* **78**, 891–902.

DiCiccio, T.J., Martin, M.A., and Young, G.A. (1992a) Analytic approximations for iterated bootstrap confidence intervals. *Statistics and Computing* **2**, 161–171.

DiCiccio, T.J., Martin, M.A., and Young, G.A. (1992b) Fast and accurate approximate double bootstrap confidence intervals. *Biometrika* **79**, 285–295.

DiCiccio, T.J., Martin, M.A., and Young, G.A. (1993) Analytical approximations to conditional distribution functions. *Biometrika* **80**, 285–295.

DiCiccio, T.J., Martin, M.A., and Young, G.A. (1994) Analytic approximations to bootstrap distribution functions using saddlepoint approximations. *Statistica Sinica* **4**, 281–295.

Downton, F. (1970) Bivariate exponential distributions in reliability theory. *Journal of the Royal Statistical Society series B* **32**, 408–417.

Durbin, J. (1980a) Approximations for densities of sufficient statistics. *Biometrika* **67**, 311–333.

Durbin, J. (1980b) The approximate distribution of serial correlation coefficients calculated from residuals on Fourier series. *Biometrika* **67**, 335–350.

Durbin, J. and Watson, G.S. (1950) Testing for serial correlation in least squares regression I. *Biometrika* **37**, 409–428.

Durbin, J. and Watson, G.S. (1951) Testing for serial correlation in least squares regression II. *Biometrika* **38**, 159–178.

Eaton, M.L. (1983) *Multivariate Statistics*. Third edition. New York: McGraw-Hill.

Efron, B. (1975) Defining the curvature of a statistical problem (with applications to second-order efficiency) (with discussion). *Annals of Statistics* **3**, 1189–1242.

Efron, B. and Hinkley, D.V. (1978) Assessing the accuracy of the maximum likelihood estimator: Observed versus expected information (with discussion). *Biometrika* **65**, 457–487.

Esscher, F. (1932) On the probability function in the collective theory of risk. *Skand. Akt. Tidsskr.* **15**, 78–86.

Fahrmeir, L. and Tutz, G. (1994) *Multivariate Statistical Modeling Based on Generalized Linear Models*. New York: Springer-Verlag.

Feller, W. (1968) *An Introduction to Probability Theory and its Applications*, vol. I, Third edition. New York: Wiley.

Feller, W. (1971) *An Introduction to Probability Theory and its Applications*, vol. II, Second edition. New York: Wiley.

Feuerverger, A. (1989) On the empirical saddlepoint approximation, *Biometrika* **76**, 457–464.

Field, C. (1982) Small sample asymptotic expansions for the multivariate M-estimates. *Annals of Statistics* **10**, 672–689.

Field, C.A. and Hampel, F.R. (1982) Small-sample asymptotic distributions of M-estimates of location. *Biometrika* **69**, 29–46.

Field, C.A. and Ronchetti, E. (1990) *Small Sample Asymptotics*. Institute of Mathematical Statistics Lecture Notes **13**, Hayward, CA: Institute of Mathematical Statistics.

Fieller, E.C. (1954) Some problems on interval estimation. *Journal of the Royal Statistical Society series B* **16**, 175–185.

Fisher, R.A. (1934) Two new properties of mathematical likelihood. *Proceedings of the Royal Society A* **144**, 285–307.

Fisher, R.A. (1961) Sampling the reference set. *Sankhyā* **23**, 3–8.
Fisher, R.A. (1973) *Statistical Methods and Scientific Inference*, 3rd ed., Edinburgh: Oliver & Boyd.
Fisher, R.A. (1990) *Statistical Methods, Experimental Design, and Scientific Inference.* Oxford: Oxford University Press.
Fisher, R.A. and Healy, M.J.R. (1958) New tables of Behrens' test of significance. *Journal of the Royal Statistical Society series B* **18**, 212–216.
Fokianos, K. (2001) Truncated Poisson regression for time series of counts. *Scandinavian Journal of Statistics* **28**, 645–659.
Fraser, D.A.S. (1968) *The Structure of Inference.* Huntington, NY: Krieger.
Fraser, D.A.S. (1979) *Inference and Linear Models.* New York: McGraw-Hill.
Fraser, D.A.S. (1990) Tail probabilities from observe likelihoods. *Biometrika* **77**, 65–76.
Fraser, D.A.S. (2003) Likelihood for component parameters. *Biometrika* **90**, 327–339.
Fraser, D.A.S. (2004) Ancillaries and conditional inference. *Statistical Science* **19**, 333–369.
Fraser, D.A.S. and Reid, N. (1993) Third order asymptotic models: likelihood functions leading to accurate approximations for distribution functions. *Statistica Sinica* **3**, 67–82.
Fraser, D.A.S. and Reid, N. (1995) Ancillaries and third order inference. *Utilitas Mathematica* **47**, 33–53.
Fraser, D.A.S. and Reid, N. (2001) Ancillary information for statistical inference. *Empirical Bayes and Likelihood Inference. Lecture Notes in Statistics* **148**, 185–209. New York: Springer.
Fraser, D.A.S., Reid, N., and Wong, A. (1991) Exponential linear models: a two-pass procedure for saddlepoint approximation. *Journal of the Royal Statistical Society series B* **53**, 483–492.
Fraser, D.A.S., Reid, N., and Wu, J. (1999a) A simple general formula for tail probabilities for frequentist and Bayesian inference. *Biometrika* **86**, 249–264.
Fraser, D.A.S., Wong, A., and Wu, J. (1999b) Regression analysis, nonlinear or nonnormal: simple and accurate p values from likelihood analysis. *Journal of the American Statistical Association* **94**, 1286–1295.
Gatto, R. and Jammalamadaka, S.R. (1999) A conditional saddlepoint approximation for testing problems. *Journal of the American Statistical Association* **94**, 533–541.
Geary, R.C. (1944) Extension of a theorem by Harald Cramér on the frequency distribution of the quotient of two variables. *Journal of the Royal Statistical Society series B* **17**, 56–57.
Glaser, R.E. (1976) Exact critical values of Bartlett's test for homogeneity of variances. *Journal of the American Statistical Association* **71**, 488–490.
Gross, A.J., Clark, V.A., and Liu, V. (1970) Estimation of survival parameters where one or two organs must function for survival. *Biometrics* **27**, 369–377.
Gross, A.J., Clark, V.A., and Liu, V. (1971) Estimation of survival parameters where one or two organs must function for survival. *Biometrics* **27**, 369–377.
Gupta, A.K. and Tang, J. (1984) Distribution of likelihood ratio statistic for testing equality of covariance matrices of multivariate Gaussian models. *Biometrika* **71**, 555–559.
Gupta, R.D. and Richards, D.S.P. (1985) Hypergeometric functions of scalar matrix argument are expressible in terms of classical hypergeometric functions. *SIAM Journal on Mathematical Analysis* **16**, 852–858.
Haight, F.A. and Breuer, M.A. (1960) The Borel–Tanner distribution. *Biometrika* **47**, 143–150.
Hampel, F.R. (1973) Small sample asymptotics. In *Proceedings of the Prague Symposium on Asymptotic Statistics,* ed. J. Hajek. Prague: Charles University.
Hannig, J., Wang, C.M., and Iyer, H.K. (2003) Uncertainty calculation for the ratio of dependent measurements. *Metrologia* **40**, 177–183.
Henrici, P. (1977) *Applied and Computational Complex Analysis*, vol. II: *Special Functions–Integral Transforms–Asymptotics–Continued Fractions.* New York: Wiley.

Henshaw, R.C. (1966) Testing single equation least squares regression models for autocorrelated disturbances. *Econometrica* **34**, 646–660.

Hill, B.M. (1968) Posterior distribution of percentiles: Bayes' theorem for sampling from a population. *Journal of the American Statistical Association* **63**, 677–691.

Hinkley, D.V. (1980a) Likelihood. *Canadian Journal of Statistics* **2**, 151–163.

Hinkley D.V. (1980b) Likelihood as an approximate pivotal distribution. *Biometrika* **67**, 287–292.

Hinkley, D.V. and Shi, S. (1989) Importance sampling and the nested bootstrap. *Biometrika* **76**, 435–446.

Hinkley, D.V. and Wei, B.C. (1984) Improvement of jackknife confidence limit methods. *Biometrika* **71**, 331–339.

Holgate, P. (1964) Estimation for the bivariate Poisson distribution. *Biometrika* **51**, 241–245.

Hougaard, P. (1988) Comment on Reid (1988). *Statistical Science* **3**, 230–231.

Høyland, A. and Rausand, M. (1994) *System Reliability Theory, Models and Statistical Methods*. New York: Wiley.

Imhof, P. (1961) Computing the distribution of quadratic forms in normal variables. *Biometrika* **48**, 419–426.

Jeffreys, H. (1961) *Theory of Probability*. Oxford: Oxford University Press.

Jensen, J.L. (1988) Uniform saddlepoint approximations. *Advances in Applied Probability* **20**, 622–634.

Jensen, J.L. (1991a) Uniform saddlepoint approximations and log-concave densities. *Journal of the Royal Statistical Society series B* **53**, 157–172.

Jensen, J.L. (1991b) A large deviation type approximation for the 'Box class' of likelihood ratio criteria. *Journal of the American Statistical Association*, **86**, 437–440.

Jensen, J.L. (1992) The modified signed log likelihood statistic and saddlepoint approximations. *Biometrika* **79**, 693–703.

Jensen, J.L. (1995) *Saddlepoint Approximations*. Oxford: Clarendon Press.

Jing, B. and Robinson, J. (1994) Saddlepoint approximations for marginal and conditional probabilities of transformed variables. *Annals of Statistics* **22**, 1115–1132.

Jing, B., Feuerverger, A., and Robinson, J. (1994) On the bootstrap saddlepoint approximations. *Biometrika* **81**, 211–215.

Johnson, N.L. and Kotz, S. (1969) *Distributions in Statistics: Discrete Distributions*. New York: Wiley.

Johnson, N.L. and Kotz, S. (1970) *Distributions in Statistics: Continuous Univariate Distributions-2*. New York: Wiley.

Johnson, N.L. and Kotz, S. (1972) *Continuous Multivariate Distributions*. New York: Wiley.

Johnson, N.L. and Kotz, S. (1986) *Encyclopedia of Statistical Sciences*. New York: Wiley.

Johnson, N.L., Kotz, S., and Balakrishnan, N. (1995) *Continuous Univariate Distributions*, vol. II. Second edition. New York: Wiley.

Johnson, N.L., Kotz, S. and Kemp, A.W. (1992) *Univariate Discrete Distributions*. Second edition. New York: Wiley.

Johnson, R. and Johnson, N. (1980) *Survival Models and Data Analysis*. New York: Wiley.

Jørgensen, B. (1997) *The Theory of Dispersion Models*. London: Chapman and Hall.

Kalbfleisch, J.D. and Sprott, D.A. (1970) Application of likelihood methods to models involving large numbers of parameters (with Discussion). *Journal of the Royal Statistical Society series B* **32**, 175–208.

Kalbfleisch, J.D. and Sprott, D.A. (1973) Marginal and conditional likelihoods. *Sankhyā A* **35**, 311–328.

Karlin, S. and Taylor, H. (1981) *A Second Course in Stochastic Processes.* New York: Academic Press.

Kay, S.M., Nuttall, A.H., and Baggenstoss, P.M. (2001) Multidimensional probability density function approximations for detection, classification, and model order selection. *IEEE Transactions on Signal Processing* **49**, 2240–2252.

Kendall, M.G. and Stuart, A. (1969) *The Advanced Theory of Statistics*, vol. I. Third edition. New York: Hafner.

Kendall, M.G. and Stuart, A. (1973) *The Advanced Theory of Statistics*, vol. II, Third edition. New York: Hafner.

Kibble, W.F. (1941) A two-variate gamma type distribution. *Sankhyā* **5**, 137–150.

Kim, D. and Agresti, A. (1995) Improved exact inference about conditional association in three-way contingency tables. *Journal of the American Statistical Association* **90**, 632–639.

Kolassa, J.E. (1994) *Series Approximation Methods in Statistics.* New York: Springer-Verlag.

Kolassa, J.E. (1996) Higher-order approximations to conditional distribution functions. *Annals of Statistics,* **24**, 353–364.

Kolassa, J.E. (2003) Multivariate saddlepoint tail probability approximations. *Annals of Statistics*, **31**, 274–286.

Kolassa, J.E. (2004) Approximate multivariate conditional inference using the adjusted profile likelihood. *Canadian Journal of Statistics* **32**, 5–14.

Kolassa, J.E. and Robinson, J. (2006) Conditional saddlepoint approximations for non-continuous and non-lattice distributions. *Journal of Statistical Planning and Inference.* To appear.

Kres, H. (1983) *Statistical Tables for Multivariate Analysis.* New York: Springer-Verlag.

Krishnan, M. (1967) The moments of a doubly noncentral t-distribution. *Journal of the American Statistical Association* **62**, 278–287.

Kulp, R.W. and Nagarsenkar, B.N. (1984) An asymptotic expansion of the nonnull distribution of Wilks criterion for testing the multivariate linear hypothesis. *Annals of Statistics* **12**, 1576–1583.

Lagakos, S. (1976) A stochastic model for censored survival time in the presence of an auxiliary variable. *Biometrics* **32**, 551–559.

Lawless, J.F. (1973) Conditional versus unconditional confidence intervals for the parameters of the Weibull distribution. *Journal of the American Statistical Association* **68**, 665–669.

Lawless, J.F. (1974) Approximations to confidence intervals for parameters in the extreme value and Weibull distributions. *Biometrika* **61**, 123–129.

Lee, Y.-S. (1971) Distribution of the canonical correlations and asymptotic expansions for distributions of certain independence test statistics. *Annals of Mathematical Statistics* **42**, 526–537.

Lehmann, E.L. (1986) *Testing Statistical Hypotheses*, Second edition. New York: Wiley.

Lehmann, E.L. and D'Abrera, H.J.M. (1975) *Nonparametrics: Statistical Methods Based on Ranks.* San Francisco: Holden Day.

Lieberman, O. (1994a) Saddlepoint approximation for the least squares estimator in first-order autoregression. *Biometrika* **81**, 807-11. Corrigendum 1996 **83**, 246.

Lieberman, O. (1994b) Saddlepoint approximation for the distribution of a ratio of quadratic forms in normal variables. *Journal of the American Statistical Association* **89**, 924–928.

Lieblein, J. and Zelen, M. (1956) Statistical investigation of the fatigue life of deep groove ball bearings. *Journal of Research, National Bureau of Standards* **57**, 273–316.

Lloyd, D.K. and Lipow, M. (1962) *Reliability: Management, Methods, and Mathematics.* Englewood Cliffs: Prentice Hall.

Lugannani, R. and Rice, S.O. (1980) Saddlepoint approximations for the distribution of the sum of independent random variables. *Advances in Applied Probability* **12**, 475–490.

Lyons, B. and Peters, D. (2000). Applying Skovgaard's modified directed likelihood statistic to mixed linear models. *Journal of Statistical Computation and Simulation* **65**, 225–242.

Magnus, J.R. and Neudecker, H. (1988) *Matrix Differential Calculus with Applications in Statistics and Econometrics.* New York: Wiley.

Marsh, P.W.R. (1998) Saddlepoint approximations and non-central quadratic forms. *Econometric Theory* **14**, 539–559.

Martz, R.A. and Waller, H.F. (1982) *Bayesian Reliability Analysis.* New York: Wiley.

Mathai, A.M. and Katiyar, R.S. (1979) Exact percentage points for testing independence. *Biometrika* **66**, 353–356.

McCullagh, P. (1987) *Tensor Methods in Statistics.* New York: Chapman and Hall.

McCullagh, P. and Nelder, J.A. (1989) *Generalized Linear Models* 2nd ed. London: Chapman and Hall.

McGregor, J.R. (1960) An approximate test for serial correlation in polynomial regression. *Biometrika* **47**, 111–119.

Medhi, J. (1994) *Stochastic Processes,* Second edition. New York: Wiley.

Mendoza, J.L. (1980) A significance test for multisample sphericity. *Psychometrika* **45**, 495–498.

Milliken, G.A. and Johnson, D.E. (1984) *Analysis of Messy Data.* Belmont, CA: Wadsworth, Inc.

Monti, A.C. and Ronchetti, E. (1993) On the relationship between empirical likelihood and empirical saddlepoint approximation for multivariate M-estimators. *Biometrika* **80**, 329–338.

Morgan, B.J.T. (1992) *Analysis of Quantal Response Data.* London: Chapman and Hall.

Morrison, D.F. (1976). *Multivariate statistical methods*, 2nd ed. New York: McGraw-Hill.

Morrison, D.F. (1990) *Multivariate Statistical Methods,* 3rd ed. New York: McGraw-Hill.

Mudholkar, G.S., Trivedi, M.C., and Lin, C.T. (1982) On approximation to the distribution of the likelihood ratio statistic for testing complete independence. *Technometrics* **24**, 139–143.

Muirhead, R.J. (1972). On the test of independence between two sets of variates. *Annals of Mathematical Statistics* **43**, 1491–1497.

Muirhead, R.J. (1982) *Aspects of Multivariate Statistical Theory.* New York: Wiley.

Muller, K.E., LaVange, L.M., Ramey, S.L., and Ramey, C.T. (1992) Power calculations for general linear multivariate models including repeated measures applications. *Journal of the American Statistical Association* **87**, 1209–1226.

Muller, K.E. and Peterson, B.L. (1984) Practical methods for computing power in testing the multivariate general linear hypothesis. *Computational Statistics and Data Analysis* **2**, 143–158.

Nagarsenkar, B.N. (1975) Percentage points of Wilks' Λ_{vc} criterion. *Communications in Statistics* **4**, 629–641.

Nagarsenkar, B.N. and Pillai, K.C.S. (1973) The distribution of the sphericity test criterion. *Journal of Multivariate Analysis* **3**, 226–235.

Neuts, M.F. (1981) *Matrix-geometric Solutions in Stochastic Models.* Baltimore, MD: Johns Hopkins University Press.

Neyman, J. and Scott, E.L. (1948) Consistent estimates based on partially consistent observations. *Econometrica* **16**, 1–32.

Norris, J.R. (1997) *Markov Chains.* Cambridge: Cambridge University Press.

Pace, L. and Salvan, A. (1997) *Principles of Statistical Inference from a Neo-Fisherian Perspective.* Singapore: World Scientific Publishing.

Paolella, M.S. (1993) *Saddlepoint applications in regressions models.* M.S. Thesis, Department of Statistics, Colorado State University.

Paolella, M.S. and Butler, R.W. (1999) Approximating the density, distribution, and moments of ratios of quadratic forms in normal variables with application to the sample autocorrelation function. Technical Report, Department of Statistics, Colorado State University.

Patterson, H.D. and Thompson, R. (1971) Recovery of interblock information when block sizes are unequal. *Biometrika* **58**, 545–554.

Pedersen, B.V. (1981) A comparison of the Efron-Hinkley ancillary and the likelihood ratio ancillary in a particular example. *Annals of Statistics* **9**, 1328–1333.

Phillips, P.C.B. (1978) Edgeworth and saddlepoint approximations in the first-order non-circular autoregression. *Biometrika* **65**, 91–98.

Phillips, P.C.B. (1982) The true characteristic function of the F distribution. *Biometrika* **69**, 261–264.

Phillips, P.C.B. (1985) The distribution of matrix quotients. *Journal of Multivariate Analysis* **16**, 157–161.

Pierce, D.A. and Peters, D. (1992) Practical use of higher order asymptotics for multiparameter exponential families (with Discussion). *Journal of the Royal Statistical Society series B* **54**, 701–737.

Pillai, K.C.S. (1985) "Multivariate analysis of variance" entry in the *Encyclopedia of Statistical Science* ed. N.L. Johnson and S. Kotz. New York: Wiley.

Pillai, K.C.S. and Mijares, T.A. (1959) On the moments of the trace of a matrix and approximations to its distribution. *Annals of Mathematical Statistics* **30**, 1135–1140.

Polansky, A.M. (1999) Upper bounds on the true coverage of bootstrap percentile type confidence intervals. *American Statistician* **53**, 362–369.

Priestley, M.B. (1981) *Spectral Analysis and Time Series.* New York: Academic Press.

Press, S.J. (1972) *Applied Multivariate Analysis.* New York: Holt, Rinehart and Winston.

Quine, M.P. (1994) Probability approximations for divisible discrete distributions. *Australian Journal of Statistics* **36**, 339–349.

Reid, N. (1988) Saddlepoint methods and statistical inference. *Statistical Science* **3**, 213–238.

Reid, N. (1995) The roles of conditioning in inference. *Statistical Science* **10**, 138–157.

Reid, N. (1996) Likelihood and higher-order approximations to tail areas: A review and annotated bibliography. *Canadian Journal of Statistics* **24**, 141–166.

Reid, N. (2003) Asymptotics and the theory of inference. *Annals of Statistics* **31**, 1695–1731.

Rice, J.A. (1995) *Mathematical Statistics and Data Analysis.* Second edition. Belmont, CA: Wadsworth, Inc.

Robinson, J. (1982) Saddlepoint approximations for permutation tests and confidence intervals. *Journal of the Royal Statistical Society series B* **44**, 91–101.

Robinson, J., Höglund, T., Holst, L., and Quine, M.P. (1990) On approximating probabilities for small and large deviations in R^d. *Annals of Probability* **18**, 727–753.

Rockafellar, R.T. (1970) *Convex Analysis,* Princeton, NJ: Princeton University Press.

Ronchetti, E. and Welsh, A.H. (1994) Empirical saddlepoint approximations for multivariate M-estimators. *Journal of the Royal Statistical Society series B* **56**, 313–326.

Ross, S.L. (1974) *Differential Equations.* Second edition. Lexington, MA: Xerox College Publishing.

Ross, S.M. (1983) *Stochastic Processes.* New York: Wiley.

Routledge, R.D. (1994) Practicing safe statistics with the mid-p. *Canadian Journal of Statistics* **22**, 103–110.

Rubin, D.B. (1981) The Bayesian bootstrap. *Annals of Statistics* **9**, 130–134.

Sartori, N. (2003) Modified profile likelihoods in models with stratum nuisance parameters. *Biometrika* **90**, 533–549.

Sartori, N., Bellio, R., Pace, L., and Salvan, A. (1999) The directed modified profile likelihood in models with many nuisance parameters. *Biometrika* **86**, 735–742.

Scheffé, H. (1959) *The Analysis of Variance.* New York: Wiley.

Seneta, E. (1981) *Non-negative Matrices and Markov Chains.* New York: Springer-Verlag.

Severini, T.A. (2000a) *Likelihood Methods in Statistics.* Oxford: Oxford University Press.

Severini, T.A. (2000b) The likelihood ratio approximation to the conditional distribution of the maximum likelihood estimator in the discrete case. *Biometrika* **87**, 939–945.

Shun Z. and McCullagh, P. (1995) Laplace approximation of high dimensional integrals. *Journal of the Royal Statistical Society series B* **57**, 749–760.

Skovgaard, I.M. (1985) Large deviation approximations for maximum likelihood estimators. *Probability and Mathematical Statistics* **6**, 89–107.

Skovgaard, I.M. (1987) Saddlepoint expansions for conditional distributions. *Journal of Applied Probability* **24**, 875–887.

Skovgaard, I.M. (1990) On the density of minimum contrast estimators. *Annals of Statistics* **18**, 779–789.

Skovgaard, I.M. (1996) An explicit large-deviation approximation to one parameter tests. *Bernoulli* **2**, 145–165.

Skovgaard, I.M. (2001) Likelihood asymptotics. *Scandinavian Journal of Statistics* **28**, 3–32.

Solberg, J.J. (1975) A graph theoretic formula for the steady state distribution of finite Markov processes. *Management Science* **21**, 1040–1048.

Spady, R.H. (1991) Saddlepoint approximations for regression models. *Biometrika* **78**, 879–889.

Srivastava, M.S. and Khatri, C.G. (1979) *An Introduction to Multivariate Statistics.* New York: North-Holland.

Srivastava, M.S. and Yau, W.K. (1989) Saddlepoint method for obtaining tail probability of Wilks' likelihood ratio test. *Journal of Multivariate Analysis* **31**, 117–126.

Strawderman, R.L., Casella, G., and Wells, M.T. (1996) Practical small-sample asymptotics for regression problems. *Journal of the American Statistical Association* **91**, 643–654. Corrigendum 1997 **92**, 1657.

Strawderman, R.L. and Wells, M.T. (1998) Approximately exact inference for common odds ratio in several 2 × 2 tables. *Journal of the American Statistical Association* **93**, 1294–1307.

Sugiura, N. (1973a) Further asymptotic formulas for the non-null distributions of three statistics for multivariate linear hypothesis. *Annals of the Institute of Statistical Mathematics* **25**, 153–163.

Sugiura, N. (1973b) Asymptotic non-null distributions of the likelihood ratio criteria for covariance matrix under local alternatives. *Annals of Statistics* **1**, 718–728.

Sugiura, N. (1974) Asymptotic formulas for the hypergeometric function $_2F_1$ of matrix argument, useful in multivariate analysis. *Annals of the Institute of Statistical Mathematics* **26**, 117–125.

Sugiura, N. and Fujikoshi, Y. (1969) Asymptotic expansion of the non-null distributions of the likelihood ratio criteria for multivariate linear hypothesis and independence. *Annals of Mathematical Statistics* **40**, 942–952.

Sugiyama, T. (1967). Distribution of the largest latent root and the smallest latent root of the generalized B statistic and F statistic in multivariate analysis. *Annals of Mathematical Statistics* **38**, 1152–1159.

Taylor, H.M. and Karlin, S. (1998) *An Introduction to Stochastic Modeling.* Third edition. New York: Academic Press.

Temme, N.M. (1982) The uniform asymptotic expansion of a class of integrals related to cumulative distribution functions. *SIAM Journal on Mathematical Analysis* **13**, 239–253.

Thoman, D.R., Bain, L.J., and Antle, C.E. (1969) Inferences for the parameters of the Weibull distribution. *Technometrics* **11**, 445–460.

Tierney, L. and Kadane, J.B. (1986) Accurate approximations for posterior moments and marginal densities. *Journal of the American Statistical Association* **81**, 82–86.

Tiku, M.L. (1972) A note on the distribution of the doubly noncentral F distribution. *Australian Journal of Statistics* **14**, 37–40.

Tiku, M.L. and Yip, D.Y.N. (1978) A four moment approximation based on the F distribution. *Australian Journal of Statistics* **20**, 257–261.

Tingley, M. and Field, C. (1990) Small-sample confidence intervals. *Journal of the American Statistical Association* **85**, 427–434.

Vinod, H.D. (1973) Generalization of the Durbin-Watson statistic for higher order autoregressive processes. *Communications in Statistics A* **2**, 115–144.

Wang, S. (1990) Saddlepoint approximations or bivariate distributions. *Journal of Applied Probability* **27**, 586–597.

Wang, S. (1992) Tail probability approximations in the first-order non-circular autoregression. *Biometrika* **79**, 431–434.

Wang, S. and Carroll, R.J. (1999) High-order accurate methods for retrospective sampling problems. *Biometrika* **86**, 881–897.

Wicksell, S.D. (1933) On correlation surfaces of type III. *Biometrika* **25**, 121–136.

Wilks, S.S. (1946) Sample criteria for testing equality of means, equality of variances, and equality of covariances in a normal multivariate distribution. *Annals of Mathematical Statistics* **17**, 257–281.

Wood, A.T.A., Booth, J.G., and Butler, R.W. (1992) Saddlepoint approximations to the CDF for some statistics with non-normal limit distributions. Technical report, Colorado State University.

Wood, A.T.A., Booth, J.G., and Butler, R.W. (1993) Saddlepoint approximations to the CDF for some statistics with non-normal limit distributions. *Journal of the American Statistical Association* **88**, 680–686.

Zeger, S.L. (1988) A regression model for time series of counts. *Biometrika* **75**, 621–629.

Index

ancillary
  affine, 233, 234
  Efron–Hinkley, 234
  likelihood, 244
autocorrelation function, *see* serial
    correlogram
autoregression
  of counts, 208
  order determination, 417
    Daniels–Durbin approximation, 419
    optimal tests for, 417
    saddlepoint approximation for, 418
  serial correlogram, 398

Bartlett correction, 223
Bartlett–Box M test, 353
Bartlett–Nanda–Pillai trace, *see* distributions, multivariate
    analysis
Bayesian posterior, saddlepoint approximation of
    population mean
  finite population, 125
  infinite population, 124
Bayesian prediction, 506
  cancer example, 510
  intractable likelihood, 508
  intractable predictand density, 507
  kidney failure, 526
  $M/G/1$ queue, 519
  $M/M/2/\infty$ queue, 542
  reliability growth model, 511, 517
  repairman model, 513
  tractable likelihood, 507
  unstable queue, 544
  with importance sampling, 523
Behrens–Fisher problem, 234, 304
birth and death process, 454
  random walk, *see* random walk
  repairman model, 460
    Bayesian prediction, 513
  transmittance, busy period, 457
  transmittance, first passage
    recursive CGF, 459
    recursive saddlepoint computations, 459
  transmittance, first return
    recursive CGF, 455
    recursive saddlepoint computations, 456
block independence, test for, *see* distributions, multivariate
    analysis

bootstrap, double
  confidence bands for
    first passage bootstrap distribution, 485
    mean bootstrap distribution, 483
    studentized mean bootstrap distribution, 484
  confidence bands for single bootstrap, 483
  inner resampling, 483
  saddlepoint approximation for, 483, 485
bootstrap, saddlepoint approximation of
  Dirichlet bootstrap, 94, 124
  fist passage distribution
    single destination, 477
  mean, 93, 123, 475
  ratio of means, 389
  studentized mean, 476
  t pivotal, 498

CGF, *see* cumulant generating function
common odds ratio, 193, 329
compound distributions, 437
conditional density function
  saddlepoint approximation, 107, 108, 538
    asymptotics, 111
    canonical sufficient statistic, 162
    normalization, 112
    second-order correction, 112
conditional density functions, approximation to
  bivariate exponential, 110
  Dirichlet, 110
conditional distribution, continuous
  saddlepoint approximation, 113, 538
    asymptotics, 122
    canonical sufficient statistic, 170
    central limit connections, 126
    derivation, 119
    derivative, 123
    location of singularity, 122
    normalization, 130
    saddlepoint geometry, 129
    second-order correction, 122
  single-saddlepoint approximation, 131
conditional distribution, discrete
  saddlepoint approximation, 114, 539
    canonical sufficient statistic, 170
    first continuity correction, 114
    second continuity correction, 114
conditional distributions, approximation to
  continuous
    beta, 115

Index

Cauchy, 140
t distribution, 139, 163, 171
conditional distributions, approximation
    to discrete
  binomial, 115
  hypergeometric, 115
  Pólya, 118
conditional mass function
  saddlepoint approximation, 107, 108
    asymptotics, 111
    canonical sufficient statistic, 162
    normalization, 112
    second-order correction, 112
conditional mass functions, approximation to
  binomial, 109
  hypergeometric, 110
  multinomial, 110
  Pólya, 110
conditionality resolution, 230, 238
  from likelihood ancillaries, 244
confidence intervals
  mid-p-value inversion, 188
  p-value inversion, 187
covariance tests, *see* distributions, multivariate
    analysis
cumulant generating function, 1

densities, approximation to
  extreme value, 6
  gamma, 4
  Gumbel, 6
  logistic, 65
  noncentral beta, 73
  noncentral chi-square, 32
  normal, 4
  normal-Laplace convolution, 4
densities, approximation to multivariate
  bivariate exponential, 77
  Dirichlet, 397
  matrix beta, 365
  matrix T, 409
  multivariate F, 406
  multivariate gamma, 77
  multivariate logistic, 102
  multivariate normal, 76
  vector ratio of quadratic forms, 396
  Wishart, 363
density function
  saddlepoint approximation, 3
    canonical sufficient statistic, 159
    central limit connections, 60
    derivation, 47
    indirect Edgeworth expansion, as an, 157
    normalization, 64
    second-order correction, 49
density function, conditional, *see* conditional density function
density, multivariate
  saddlepoint approximation, 76
    derivation, 89
    second-order correction, 90
Dirichlet bootstrap, *see* bootstrap
distribution, continuous
  saddlepoint approximation, 12
    canonical suffcient statistic, 165
    central limit connections, 60
    derivation, 51
    derivative, 54
    second-order correction, 53
distribution, discrete
  δ-lattice saddlepoint approximations, 62
    consistency, 63
    first continuity-corrected, 63
    second continuity-corrected, 63
  saddlepoint approximation, 17
    canonical sufficient statistic, 166
    first continuity-corrected, 17
    first continuity-corrected, left tail, 24
    second continuity-corrected, 18
    second continuity-corrected, left tail, 25
distributions, approximation to continuous
  beta, 16
  extreme value, 15
  gamma, 13
  Gumbel, 15
  logarithmic series, 34
  noncentral beta, 73
  noncentral chi square, 32, 533
  noncentral F, doubly, 377
  noncentral t, doubly, 498
  normal, 13
  normal-Laplace convolution, 14
  ratios of normal quadratic forms, 381
distributions, approximation to discrete
  binomial, 20, 26
  Borel-Tanner, 36
  hypergeometric, 118
  negative binomial, 21, 26
  Neyman type A, 35
  Poisson, 19, 25
  Pólya, 118
  sum of binomials, 22
  uniform, 23
distributions, multivariate analysis
  covariance tests
    block independence, 348
    equal covariances, test for, 353
    equal variance, equal covariance, test for, 351
    intraclass correlation, test for, 351
    sphericity, 350
  homogeneity of covariances and means, 369
  matrix beta
    Bartlett-Nanda-Pillai trace, 330, 344
    Lawley-Hotelling trace test, 347
    likelihood ratio test, 56
    Roys' test, 346
    saddlepoint density, 365
    Wilks' test, 56, 342
  power functions, 355
    Bartlett-Box M test, 362
    block independence, 360
    Wilks' test, 356
  simultaneous sphericity, 369
  Wishart matrix
    generalized variance, 55
    largest eigenvalue, 366
    saddlepoint density, 363

double-saddlepoint approximation
  density approximation, 107, 108
    canonical sufficent statistic, 162
  distribution approximation, 113, 114
    canonical sufficent statistic, 170
  mass function approximation, 107, 108
    canonical sufficent statistic, 162
Durbin-Watson statistic, 94

Edgeworth expansion, 151
  Edgeworth series, 153
  Gram-Charlier series type A, 152
  indirect, 156
  Tchebycheff-Hermite polynomials, 151
empirical percentile, 390
empirical saddlepoint approximations, 500
empirical transmittance, 477
equivariance, see properties, saddlepoint approximation
Esscher tilting, see Edgeworth expansion, indirect
exponential family
  canonical sufficent statistic
    conditional saddlepoint density for, 162
    conditional saddlepoint distribution for, 170
    saddlepoint density for, 159
    saddlepoint distribution for, 165
  CGF, canonical suffcient statistic, 146
  conditional families, 150
  conditional power function, 206
  conditional power function, attained, 195
  curved families, 236
    Behrens-Fisher problem, see Behrens-Fisher problem
    bivariate normal correlation, 234, 240, 246, 272
    curvature, 235
    gamma exponential, 241, 246, 275
    gamma hyperbola, 232, 236, 239, 245, 262, 271
    maximum likelihood estimate, 232
    p^*, see p^*
    r^*, see r^*
  factorization theorem, 147
  Fisher information
    expected, 149
    observed, 149
  full, 145
  mean parameterization, 149
  p^*, see p^*
  power function, attained, 199
  regular, 64, 145
    canonical parameter, 145
    canonical sufficent statistic, 145
    maximum likelihood estimate, 148
  tilted family, 156
  uniformly most powerful (UMP) tests
    attained power function, 176
    power function approximation, 176
  uniformly most powerful (UMP) unbiased tests
    power function approximation, 176, 178

failure rate approximation, see hazard rate approximation
Fieller problem, 296
Finite population sampling
  distribution of mean, saddlepoint approximation, 125
first passage distribution
  CGF, single destination, 438
  cofactor rule

  CGF, first return, 446
  multiple destinations, 452
  single destination, 435
  single destination, random start, 449
  empirical cofactor rule
    single destination, 477
  transmittance, 435
flowgraph, 432

goodness of fit tests, 413

hazard rate approximation, 28
homogeneity of populations, tests for, see distributions, multivariate analysis
hypergeometric function
  matrix argument
    $_1F_1$, 356
    $_2F_1$, 361
  scalar argument
    $_1F_1$, 325
    $_2F_1$, 327, 328

importance sampling, 101, 523
incomplete beta function, 50
incomplete gamma function, 50
indirect saddlepoint approximation, 494
intraclass correlation, test for, 351
invariance, see properties, saddlepoint approximation

jackknife
  delete-k distribution, 414
  ratio of means
    saddlepoint approximation, 414

Laplace's approximation
  high dimensional, 97
    alternative second-order correction, 99
  multivariate, 83
    alternate expression, 86
    asymptotics, 87
    iterated usage, 86
    second-order correction, 88
    separability, 87
  univariate, 42
    alternative expression, 44
    alternative second-order correction, 44
    second-order correction, 44
large deviation, 53
lattice convention, 38
Lawley-Hotelling trace test, see distributions, multivariate analysis
likelihood, 146
  conditional, 300
    REML estimation, 309
  marginal, 168, 301
likelihood ratio statistic, signed, 64, 187, 285, 286
  r^* approximation for, 286
logistic regression, 183
  LD-50 dosage, 185
    confidence intervals, 190
Lugannani and Rice approximation, 12
  asymptotic order, 62
  removable singularity, 68

M-estimate, 388
marginal likelihood, *see* likelihood, marginal
Markov graph, 433
Markov process, 434, 461
mass function
   saddlepoint approximation, 8
      canonical sufficient statistic, 159
      high dimensional, 100
      high dimensional alternative, 100
      indirect Edgeworth expansion, as an, 157
      normalized, 64
mass function, conditional, *see* conditional mass function
mass functions, approximation to
   binomial, 10
   binomial sum, 11
   Borel-Tanner, 36
   logarithmic series, 34
   negative binomial, 10
   Neyman type A, 35
   Poisson, 9
mass functions, approximation to multivariate
   bivariate negative binomial, 102
   bivariate Poisson, 81
   multinomial, 80
maximum likelihood estimate
   curved exponential family, 232
      marginal saddlepoint CDF, 408
      marginal saddlepoint density, 407
   density for, *see* p^*
   regular exponential family, 148
Mellin transform, 54
MGF, *see* moment generating function
mid-p-value, 188
   confidence intervals from inversion, 188, 189
mixed normal linear models, 308
moment generating function, 1
multivariate tests, *see* distributions, multivariate analysis

Neyman and Scott problem, 302
nonnormal-base, 528
   chi square base, 532
   choice of, 532
   double saddlepoint approximation
      conditional density, 538
      conditional distribution, continuous, 538
   inverse Gaussian base, 534
   saddlepoint approximation
      density, 529
      distribution, continuous, 530
      distribution, discrete, 531
nonparametric tests
   Wilcoxon rank sum, 125
normalization, 64, 112
   higher-order asymptotics, 65

$p^\dagger$, 219
p^*, 219
   asymptotics, 248
   curved exponential families, 230
   group transformation models, 225
   hyperbolic secant, 221
   likelihood ratio statistic, 223
   Plank's radiation formula, 222
   regular exponential families, 219

Pascual sampling, 207
permutation tests
   two sample, 125
Pillai's trace, *see* distributions, multivariate analysis
Poisson process
   nonhomogeneous, 210
power functions, multivariate tests, *see* distributions, multivariate analysis
$\widehat{\Pr}_1$, 17, 114, 166, 170, 531, 539
$\widehat{\Pr}_2$, 18, 115, 166, 170, 531
properties, saddlepoint approximation
   equivariance, 38, 91, 111, 530
   independence, 91, 109
   invariance, 41, 120
   marginalization, 91
   symmetry, 39, 92, 111, 121, 530, 531
prospective sampling, 191

queue, *see* reliability, semi-Markov process

r^*, 259, 286
   Behrens Fisher problem, 304
   conditional likelihood, 300
   distribution approximation, 260, 286
   F distribution, 294
   Fieller problem, 296
   likelihood ratio, signed, 286
   marginal likelihood, 301
   mixed normal linear models, 308
   Neyman and Scott problem, 302
   properties
      asymptotics, 279, 314
      equivariance, 278
      symmetry, 278, 314
   regression model, 287
   REML, 308
   two samples, common mean, 307
   variation independence, 298
random walk
   first passage, discrete time
      inverse Gaussian base, 535
   first return, continuous time
      inverse Gaussian base, 539
   transmittance, first passage, 460
   transmittance, first return, 456
ratio, distributions of, *see* roots of estimating equations
regression model, 287
relevant state, 435, 449, 454
reliability, 2
   queue
      $GI/M/1/4$, 440, 487
      $M/M/q/n$, 457
      $M/G/1$, 519
      $M/M/2/\infty$, 542
      $M/M/q/\infty$, 458
   redundant system, 462
   reliability growth model
      Bayesian prediction, 511, 517
   repairable system, 462
   repairman model, 460
      Bayesian prediction, 513
   system, independent components, 438, 447, 453
   system, k-out-of-m, 2
REML estimation, 308

renewal process, 166, 174, 176
retrospective sampling, 191
roots of estimating equations
  MLE
    curved exponential family, 407
    regular exponential family, 403
  multivariate root, 401
    conditional saddlepoint CDF, 411
    density, 401
    equivariance, saddlepoint density, 404
    saddlepoint density, 402
  ratio
    density, 375
    distribution, 375, 496
    doubly noncentral F, 378
    quadratic normal forms, 381
    saddlepoint density, 376
  univariate, 384
    empirical percentiles, 390
    lattice-valued scores, 390
    M-estimate, 388
    ratio of means, boostrapped, 389
    saddlepoint density, 386
  vector ratio, 392
    density, 394
    normal quadratic forms, 396
    saddlepoint density, 395
Roy's maximum root test, *see* distributions, multivariate analysis

sample space derivatives, 259
  exact, 268
  Fraser, Reid, and Wu, 268, 293
  Skovgaard's approximate, 266, 292
semi-Markov process, 430
  birth and death process, *see* birth and death process
  bootstrap inference for, *see* bootstrap, saddlepoint approximation of
  flowgraph, 432
  Markov process as a, 434
  moments, first passage, 450, 454
  moments, first return, 450

queue
  $GI/M/1/4$, 440, 487
  $M/M/q/n$, 457
  $M/G/1$, 519
  $M/M/2/\infty$, 542
  $M/M/q/\infty$, 458
  transmittance matrix, 431
  transmittance, first passage, 435
    CGF, 438
    cofactor rule, 435, 449, 452
  transmittance, first return
    cofactor rule, 446
sequential saddlepoint, 323
  conditional CDF approximation, 324
  nonnormal base, 539
serial correlogram
  saddlepoint density, 398
Skovgaard approximation, *see* conditional distribution, continuous, saddlepoint approximation, *see* conditional distribution, discrete, saddlepoint approximation
sphericity, test for, *see* distributions, multivariate analysis
Stirling's approximation
  factorial, 9
  gamma function, 4
survival analysis, multistate
  dementia example, 478
symmetry, *see* properties, saddlepoint approximation
system, stochastic, 430

Tchebycheff-Hermite polynomials, *see* Edgeworth expansion
Temme approximation, 49
tilted density, 64
time series
  count data, 208
time to event, *see* first passage time
transmittance matrix, 431
truncation, 208, 212

Wilcoxon rank sum test, *see* nonparametric tests
Wilks' likelihood ratio statistic, *see* distributions, multivariate analysis
Wishart, largest eigenvalue distribution, 367